Chemical Vapour Deposition
Precursors, Processes and Applications

Chemical Vapour Deposition
Precursors, Processes and Applications

Edited by

Anthony C. Jones
Department of Chemistry, University of Liverpool, Liverpool, UK

Michael L. Hitchman
Thin Film Innovations Limited, Glasgow, UK

RSC Publishing

ISBN: 978-0-85404-465-8

A catalogue record for this book is available from the British Library

Published by The Royal Society of Chemistry,
Thomas Graham House, Science Park, Milton Road,
Cambridge CB4 0WF, UK

Registered Charity Number 207890

For further information see our web site at www.rsc.org

PREFACE

Chemical Vapour Deposition: Precursors, Processes and Applications

This book aims to give a comprehensive overview of Chemical Vapour Deposition (CVD). CVD involves the deposition of thin solid films from chemical precursors in the vapour phase, and encompasses various deposition techniques, including metal-organic chemical vapour deposition (MOCVD), plasma-enhanced CVD (PECVD), photo-initiated CVD and atomic layer deposition (ALD). The book has been written with the CVD practitioner in mind, such as the chemist who wishes to learn more about CVD process technology, or CVD technologists who wish to increase their knowledge of precursor chemistry. This book should prove useful to those who have recently entered the field, and certain aspects of the text may also be used in chemistry and materials science lecture courses at undergraduate and postgraduate level.

We have attempted to present a logical and progressive overview of the various aspects of CVD processes. Therefore, basic concepts, such as the various types of CVD processes, the design of CVD reactors, reaction modelling and CVD precursor chemistry, are covered in Chapters 1–5. This is followed in Chapters 6–12 by a detailed description of the use of various CVD techniques to deposit a wide range of materials, including semiconductors, metals, metal oxides and nitrides, protective coatings and functional coatings on glass. Finally, in Chapter 13, some commercial aspects of CVD are discussed. The development of CVD technology owes a great deal to collaboration between different scientific disciplines such as chemistry, physics, materials science, engineering and microelectronics, and it is hoped that this book will promote and stimulate continued dialogue between scientists from these different research areas.

We are greatly indebted to the chapter authors for their enormous effort in summarizing their extensive knowledge of many different aspects of CVD, especially in view of undoubted pressures from many directions. We are also grateful to the many members of our research staff, the unsung heroes of this volume, and it is hoped that the book will be a tribute to them. We must also acknowledge the hard work of the publishing staff at the Royal Society of Chemistry, in particular Mrs Annie Jacob and Mrs Janet Freshwater. Finally, our thanks go to our families and the families of our authors for their unstinting support and understanding.

Anthony C. Jones and Michael L. Hitchman
Liverpool and Glasgow

Chemical Vapour Deposition: Precursors, Processes and Applications
Edited by Anthony C. Jones and Michael L. Hitchman

Published by the Royal Society of Chemistry, www.rsc.org

Contents

Chemical Vapour Deposition: Precursors, Processes and Applications
Edited by Anthony C. Jones and Michael L. Hitchman

Published by the Royal Society of Chemistry, www.rsc.org

CHAPTER 1

Overview of Chemical Vapour Deposition

ANTHONY C. JONES[a] AND MICHAEL L. HITCHMAN[b]

[a] Department of Chemistry, University of Liverpool, Crown Street, Liverpool, L69 7ZD, UK; [b] Thin Film Innovations Ltd., Block 7, Kelvin Campus, West of Scotland Science Park, Glasgow, G20 0TH, UK

1.1 Basic Definitions

In the broadest sense chemical vapour deposition (CVD) involves the formation of a thin solid film on a substrate material by a *chemical reaction* of vapour-phase precursors. It can thus be distinguished from physical vapour deposition (PVD) processes, such as evaporation and reactive sputtering, which involve the adsorption of atomic or molecular species on the substrate. The chemical reactions of precursor species occur both in the gas phase and on the substrate. Reactions can be promoted or initiated by heat (thermal CVD), higher frequency radiation such as UV (photo-assisted CVD) or a plasma (plasma-enhanced CVD). There is a sometimes bewildering array of acronyms covered by the overall cachet of CVD and the interested reader is referred to several reviews.[1–4] Some of the more commonly used acronyms are defined below.

Metal-organic chemical vapour deposition (MOCVD) is a specific type of CVD that utilizes metal-organic precursors. In the strictest sense a metal-organic (or organometallic) compound contains a direct metal–carbon bond (σ or π) (*e.g.* metal alkyls, metal carbonyls). However, the definition of MOCVD has broadened to include precursors containing metal–oxygen bonds (*e.g.* metal-alkoxides, metal-β-diketonates) or metal–nitrogen bonds (*e.g.* metal alkylamides), and even metal hydrides (*e.g.* trimethylamine alane).

Metal-organic vapour phase epitaxy (MOVPE) or organometallic vapour phase epitaxy (OMVPE) is an MOCVD process that produces single crystal (*i.e.* epitaxial) films on single crystal substrates from metal-organic precursors. In MOCVD and MOVPE gas-phase reactions can sometimes play a significant role in the deposition chemistry.

Plasma-assisted or plasma-enhanced CVD (PECVD) is a technique in which electrical energy rather than thermal energy is used to initiate homogeneous reactions for the production of chemically active ions and radicals that can participate in heterogeneous reactions, which, in turn, lead to layer formation on the substrate. A major advantage of PECVD over thermal CVD processes is that

Chemical Vapour Deposition: Precursors, Processes and Applications
Edited by Anthony C. Jones and Michael L. Hitchman

Published by the Royal Society of Chemistry, www.rsc.org

deposition can occur at very low temperatures, even close to ambient, which allows temperature-sensitive substrates to be used.

Atomic layer deposition (ALD), sometimes called atomic layer epitaxy (ALE), alternatively-pulsed CVD, or atomic layer chemical vapour deposition (ALCVD), is a modification of the CVD process in which gaseous precursors are introduced sequentially to the substrate surface and the reactor is purged with an inert gas, or evacuated, between the precursor pulses. The chemical reactions leading to film deposition in ALD occur exclusively on the substrate at temperatures below the thermal decomposition temperature of the metal-containing precursor and gas-phase reactions are unimportant.

Chemical beam epitaxy (CBE) is high vacuum CVD technique that uses volatile metal-organic precursors and gaseous co-precursors. The closely related technique of metal-organic molecular beam epitaxy (MOMBE) uses volatile metal-organic precursors and co-precursor vapour derived from the solid element. In CBE and MOMBE the chemical reactions occur only on the substrate, leading to single crystal films and so gas-phase reactions play no significant role in film growth. Section 1.3 gives a more detailed description of these processes.

1.2 Historical Perspective

In common with many technologies, developments in CVD have largely arisen out of the requirements of society. These developments have been most rapid when other thin film deposition technologies have proved problematic or inadequate, for instance in the production of multiple thin films, as in modern semiconductor devices, or when the coating of large surface areas is required, as in large-scale functional coatings on glass. Several excellent reviews describe the historical development of CVD processes,[2,5,6] and the published literature from the earliest days to the mid-1960s is covered by a comprehensive review by Powell *et al.*[7] Therefore, this section gives only a brief description, highlighting some key advances.

Probably the earliest patent describing a CVD process was taken out by a certain John Howarth, for the production of "carbon black" for use as a pigment. Unfortunately, due to rather lax health and safety standards, the process only succeeded in burning down the wooden plant.[8] The early electric lamp industry provided another early impetus for CVD, and a patent issued in 1880 to Sawyer and Mann describes a process for the improvement of carbon fibre filaments.[9] However, these proved too fragile and later patents describe CVD processes for the deposition of various metals to produce more robust lamp filaments.[10,11]

One of the earliest examples of the CVD of metals is the deposition of tungsten, reported as early as 1855. Wöhler used WCl_6 with hydrogen carrier gas to deposit tungsten metal.[12] Later in the century (1890), the famous Mond Process was developed. This describes the deposition of pure nickel from nickel tetracarbonyl, $Ni(CO)_4$,[13,14] and was used for the refinement of nickel ore.[15]

The first reports of the deposition of silicon by CVD by the hydrogen reduction of $SiCl_4$ appear as early as 1909[16] and 1927,[17] and the widespread use of thin silicon films in the electronics industry is anticipated by the CVD of Si-based photo cells[18] and rectifiers[19] just after World War II.

During the late 1950s, triisobutylaluminium, $[Bu^t_3Al]$ began to be used extensively to catalyze the polymerization of olefins by the Ziegler–Natta process. At around the same time, it was found that the pyrolysis of $[Bu^t_3Al]$ gave high purity Al metal ($>$99 at.%). This led to its use in the early 1980s as a CVD precursor to Al metal for very large scale integration (VLSI) applications.[20,21] In patent literature of the late 1960s, aluminium trihydride (AlH_3, alane) was found to be useful for plating Al films from the vapour phase and by electroless deposition,[22–24] which led to the much later use of alane adducts such as $[AlH_3(NMe_3)]$ as CVD precursors for high purity Al thin films.[25] The reader is referred to Chapter 7 (Section 7.3) for recent developments in Al CVD.

Another important development in the history of CVD was the introduction of "on-line" CVD architectural coatings by Pilkington (now NSG Group). These coatings are deposited on a very large scale by atmospheric pressure CVD on a float glass production line.[26] By applying the coating directly to the float glass manufacturing line, economies of scale and production are achievable that are not possible with "off-line" deposition processes such as sputtering. Perhaps the most notable of these is fluorine-doped tin oxide, [SnO_2:F] developed by Pilkington in the mid-1980s ("Pilkington K-Glass"). This is a low thermal-emissivity (low–E) coating on windows, which prevents heat loss from the home and is essential to modern ecological energy saving efforts (Chapter 10, Section 10.1.1). It can be deposited using precursors such as [Me_4Sn], [$SnCl_4$] with halo-fluorocarbons or HF (Chapter 10, Section 10.2.2). A much more recent commercial product of Pilkington is "self-cleaning" glass. This has been coated on-line with a thin transparent film of TiO_2, and this chemically breaks down dirt by photocatalysis in sunlight (Chapter 10, Section 10.6).

Despite the various developments in CVD described above, the major impetus to the technology has undoubtedly been provided by the rapid development of the microelectronics industry since the mid-1970s. This has led to a requirement for very thin high purity films with precise control of uniformity, composition and doping.

Thin epitaxial films of n- or p-doped Si are the basic requirement for all Si integrated circuit technology. One of the earliest reports of silicon epitaxy was the closed tube transport of SiI_4 produced by heating solid Si in the presence of iodine.[27] Epitaxial Si films were later produced in the 1970s on a large commercial scale by the pyrolysis of monosilane (SiH_4) in H_2.[28]

Interest in the use of metal-organic compounds for CVD applications began in the early 1960s. The first reported preparation of a III-V material from a Group III metal-organic and a Group V hydride was by Didchenko *et al.* in 1960, who prepared InP in a closed tube by the thermal decomposition at 275–300 °C of a mixture of [Me_3In] and liquid [PH_3].[29] Next, in 1962, Harrison and Tomkins produced InSb in a closed tube by heating a mixture of [Me_3In] and [SbH_3] at 160 °C, and they also produced GaAs by heating a mixture of [Me_3Ga] and [AsH_3] at 200 °C.[30] In 1961 and 1965 patent applications by the Monsanto Co. claimed methods of depositing III-V compounds "suitable for use in semiconductor devices".[31,32] The processes involved the pyrolysis of volatile Group III and Group V compounds in an open tube system on a cubic crystal substrate to produce epitaxial films.

However, the Monsanto applications were of a rather general nature, listing a large range of volatile Group III compounds, and the few specific process examples given mainly involved Group III trihalides. In 1968, Manasevit and co-workers at the Rockwell Corporation gave the first clear description of the use of metal-organic compounds for the chemical vapour deposition of III-V materials. The first publication describes the deposition of GaAs by pyrolysis of a gas phase mixture of [Et_3Ga] and [AsH_3] in an open tube system using H_2 as the carrier gas.[33] Manasevit named the technique metal-organic chemical vapour deposition (MOCVD) and a patent was later filed for the MOCVD of a range of III-V materials and wide band-gap compound semiconductors.[34]

The emphasis in Manasevit's early work was on growth of non-epitaxial films on insulating substrates such as sapphire and spinel. However, in 1969 the growth of epitaxial GaAs on a GaAs substrate by metal-organic vapour phase epitaxy (MOVPE) was demonstrated.[35] Subsequently, a wide range of III-V compounds were deposited by MOCVD (or MOVPE), including AlGaAs,[36] InP, InAs, InGaAs,InAsP,[37,38] GaN and AlN,[39] although semiconductor device quality III-V materials still had not been produced. This was due largely to low purity precursors (often obtained from commercial suppliers of metal-organics for catalysis applications) and non-optimized MOCVD reactors and processes. In 1975, however, high-purity device quality GaAs films were grown[40] that had a low residual carrier concentration of $n = 7 \times 10^{13}\,cm^{-3}$ and high electron mobility ($\mu_{77K} = 120\,000\,cm^2\,V^{-1}\,s^{-1}$) (Section 1.7.2.4). Conventional techniques for the deposition of III-V materials such as liquid phase epitaxy (a combined melt of the components) proved incapable of producing the very thin multilayer structures required for efficient III-V devices and so MOVPE technology developed with ever increasing pace, and state-of-the-art GaAs photocathodes and field effect transistors (FETs) were soon produced,[41] as well as complex multilayer structures such as AlGaAs/GaAs/AlGaAs double heterojunction lasers.[42] A particularly significant advance in III-V technology was the discovery of how to p-dope GaN-based semiconductors grown by MOVPE[43] (Chapter 6, Section 6.6.4.1) as this spawned the growth of a large industry in full-colour high-brightness light emitting diodes for energy efficient displays (Chapter 13, Section 13.7).

The reader is referred to Chapter 6 for a detailed description of the MOCVD of III-V compound semiconductor materials, including (Section 6.3.1) further details on the historical development of the technology. In many ways this is now a mature technology, more the province of salesmen and chemical buyers than development scientists, and commercial developments in this technology are described in Chapter 13.

The discovery of high-T_c superconducting oxides in the mid-1980s led to intense efforts to prepare these materials as thin films. This led to the development, beginning in the late 1980s,[44] of MOCVD techniques for the deposition of oxides such as $YBa_2Cu_3O_{7-\delta}$, and other superconducting oxides.[45] The difficulty in transporting low volatility metal precursors was largely responsible for the introduction of liquid injection MOCVD (Section 1.5 Figure 1.15). There has also been a great deal of effort devoted to the MOCVD of ferroelectric oxides such as $Pb(Zr,Ti)O_3$ and $SrBi_2Ta_2O_9$, and early reports date back to the early 1990s.[46] More recent advances in the MOCVD of a range of ferroelectric oxide materials are given elsewhere[47] and in Chapter 8 (Section 8.4).

The rapid recent advances in Si integrated circuit technology have largely been achieved by aggressive shrinking or "scaling" of devices such as metal oxide semiconductor field effect transistors (MOSFETS) and dynamic random access memories (DRAMs) (Chapter 8, Section 8.3, and Chapter 9, Section 9.2.2 and Figure 9.2). This has led to a requirement for new high-permittivity (or high-κ) oxide insulating materials to replace the conventional SiO_2 insulating or capacitor layers. PVD techniques can not give the desired deposition control of the very thin films required, or the necessary step coverage in high aspect-ratio device structures such as trench- and stack-structured DRAMs. Therefore, over the last few years there has been an intense effort to develop CVD processes for the deposition of high-permittivity metal oxides, such as Al_2O_3, ZrO_2, HfO_2, Zr- and Hf-silicate and the lanthanide oxides, and many CVD developments in this area are also detailed in Chapter 8 (Section 8.3).

Shrinking device dimensions also make it necessary to modify existing multilevel metallization technologies. This has led to recent efforts to deposit Al and Cu by MOCVD (Chapter 7, Sections 7.3 and 7.4), as well as stimulating research into the MOCVD of diffusion barriers such as TiN and TaN (Chapter 9, Sections 9.3.1 and 9.3.2).

Atomic layer deposition (ALD) was first introduced by T. Suntola and co-workers in the early 1970s,[48,49] and was initially used for the manufacture of luminescent and dielectric films required in electroluminescent displays (Chapter 4, Section 4.5.1).[50,51] More recently, ALD has been used to deposit the very thin conformal oxide films required as gate insulators in CMOS technology and in DRAM capacitor layers; see Chapters 4 (Section 4.5.3) and 8 (Section 8.3).

It is impossible to do justice here to the huge volume of research and development carried out on CVD over the past 100 years or so, but hopefully this brief survey gives a flavour of the great advances made.

1.3 Chemical Vapour Deposition Processes

1.3.1 Conventional CVD Processes

CVD processes are extremely complex and involve a series of gas-phase and surface reactions. They are often summarized, though, by overall reaction schemes, as illustrated in Scheme 1.1.

An overall reaction scheme tells us little about the physicochemical processes and the gas-phase and surface reactions involved. A more informative illustration of a CVD process is illustrated by

$$[Me_3Ga]_{(g)} + [AsH_3]_{(g)} \xrightarrow{500\text{–}700^\circ C} GaAs_{(s)} + 3CH_{4(g)}\uparrow$$

$$[SiH_4]_{(g)} \xrightarrow{500\text{–}800^\circ C} Si_{(s)} + 2H_{2(g)}\uparrow$$

$$[SiH_4]_{(g)} + O_{2(g)} \xrightarrow{350\text{–}475^\circ C} SiO_{2(s)} + 2H_{2(g)}\uparrow$$

$$2[Ta(OEt)_5]_{(v)} + 5H_2O_{(v)} \xrightarrow{350\text{–}450^\circ C} Ta_2O_{5(s)} + 10EtOH_{(v)}\uparrow$$

$$[NbCl_5]_{(v)} + 5/2H_{2(g)} + \tfrac{1}{2}xN_{2(g)} \xrightarrow{960^\circ C} NbN_{x(s)} + 5HCl_{(v)}\uparrow$$

Scheme 1.1 Overall reaction schemes for a variety of CVD processes.

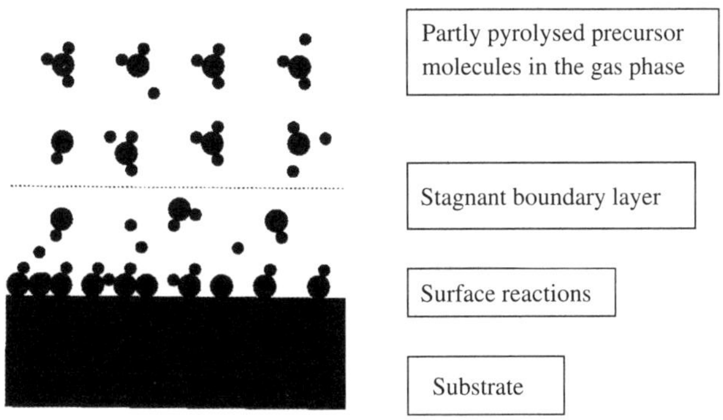

Figure 1.1 Simple schematic representation of the MOCVD process. (After ref. 52, Copyright John Wiley & Sons Limited, 1992. Reproduced with permission.)

the simple schematic[52] for an MOCVD reaction carried out at moderate pressures (*e.g.* 10–760 Torr) shown in Figure 1.1. A significant feature of the process is the presence of a hot layer of gas immediately above the substrate, termed the "boundary layer", and at these pressures gas-phase pyrolysis reactions occurring in the layer play a significant role in the MOCVD deposition process.

A more detailed picture of the basic physicochemical steps in an overall CVD reaction is illustrated in Figure 1.2, which indicates several key steps[4]:

1. Evaporation and transport of reagents (*i.e.* precursors) in the bulk gas flow region into the reactor;
2. Gas phase reactions of precursors in the reaction zone to produce reactive intermediates and gaseous by-products;
3. Mass transport of reactants to the substrate surface;
4. Adsorption of the reactants on the substrate surface;
5. Surface diffusion to growth sites, nucleation and surface chemical reactions leading to film formation;
6. Desorption and mass transport of remaining fragments of the decomposition away from the reaction zone.

In traditional thermal CVD, the film growth rate is determined by several parameters, the primary ones being the temperature of the substrate, the operating pressure of the reactor and the composition and chemistry of the gas-phase. The dependence of film growth rate on substrate temperature is typified by the growth of GaAs by MOCVD using $[Me_3Ga]$ and $[AsH_3]$[53,54] (Figure 1.3).

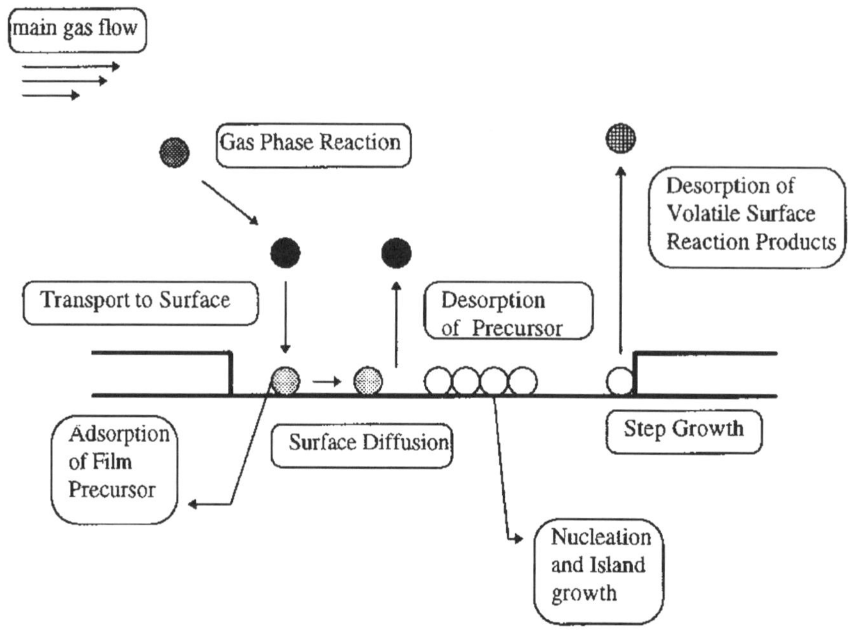

Figure 1.2 Precursor transport and reaction processes in CVD. (After ref. 4, p. 31, Copyright Wiley-VCH Verlag GmbH & Co. KGaA, 1997. Reproduced with permission.)

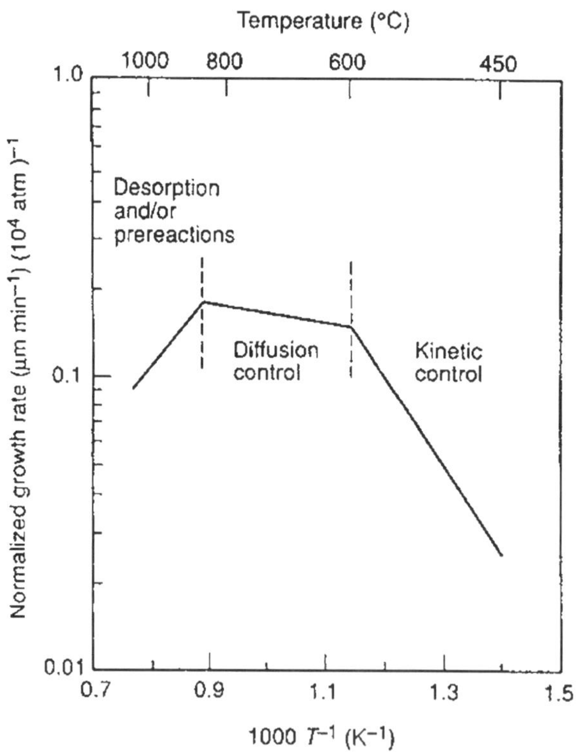

Figure 1.3 Plot of the normalized MOCVD growth rate of GaAs as a function of growth temperature. (After ref. 53, Copyright John Wiley & Sons Limited, 1987. Reproduced with permission.)

In this plot, of log of growth rate *vs.* reciprocal thermodynamic temperature, three regions are apparent. At lower growth temperatures the growth rate is controlled by the kinetics of chemical reactions occurring either in the gas-phase or on the substrate surface. This region is generally termed the region of *kinetic growth control* and the film growth rate increases exponentially with substrate temperature according to the Arrhenius equation:

$$\text{Growth rate} \propto \exp(E_A/RT) \tag{1.1}$$

where E_A is the apparent activation energy, R is the gas constant and T is the temperature. As the film growth rate is controlled by chemical kinetics, uniform film thickness can be achieved by minimizing temperature variations over the substrate surface, and this is the region utilized in hot-wall batch reactors used for the commercial production of Si epitaxial wafers by low pressure CVD.

As the temperature increases, the growth rate becomes nearly independent of temperature and is controlled by the mass transport of reagents through the boundary layer to the growth surface, and this is termed the region of *mass transport* or *diffusion-controlled* growth.

At even higher temperatures, the growth rate tends to decrease, due to an increased rate of desorption of film precursors or matrix elements from the growth surface and/or depletion of reagents on the reactor walls due to parasitic gas-phase side reactions. Gas-phase reactions become increasingly important with increasing temperature and higher partial pressures of the reactants.

Notably, Figure 1.3 is rather misleading in that it suggests there are sharp changeovers between the various regions. This is because of the nature of the plot where the log of the growth rate is used. If the growth rate itself were plotted then a much more gradual transition from one rate controlling step to the next would be seen.[2] An Arrhenius type plot as in Figure 1.3 has to be used with caution to obtain an activation energy for a kinetic process since there will be a contribution from mass transport. The slope of the "kinetic region" will not give a true value of the activation energy, but a lower value. Also, the ordinate contains the growth rate that is precursor concentration dependent and this may vary with temperature so, again, a true value of the activation energy will not be given from the slope of the "kinetic region".

A crucial factor that determines the relative importance of each regime is the pressure of the CVD reactor. From atmospheric pressure (760 Torr) to intermediate pressures (*e.g.* 10 Torr) gas phase reactions are important and, in addition, a significant boundary layer is present. Kinetics and mass transport can both play a significant role in deposition. As the pressure falls gas phase reactions tend to become less important, and particularly at pressures below 1 Torr layer growth is often controlled by surface reactions. At very low pressures (*e.g.* $<10^{-4}$ Torr) mass transport is completely absent and layer growth is primarily controlled by the gas and substrate temperature and by desorption of precursor fragments and matrix elements from the growth surface.

A much more detailed discussion of the modelling of CVD processes is given in Chapter 3, and in Chapter 6 there is a detailed description of the chemistry (Section 6.3.2), thermodynamics, kinetics and hydrodynamics (Section 6.3.3) involved in III-V MOCVD technology.

1.3.2 Variants of CVD

As the brief historical overview of CVD given in Section 1.2 shows, CVD has gone through wide-ranging developments over the years. Since most CVD reactions are thermodynamically endothermic and they also have a kinetic energy of activation then energy has to be supplied to the reacting system. Traditionally, CVD processes have been initiated by a thermal energy input, which can be inputted by several methods, *e.g.*:

- direct resistance heating of the substrate or substrate holder;
- rf induction of the substrate holder or susceptor;
- thermal radiation heating;
- photoradiation heating.

The use of thermal CVD can, however, be disadvantageous. For example, heat input can result in damage to temperature-sensitive substrates and so alternative forms of energy input have been developed which allow deposition at lower temperatures.

One way of reducing growth temperatures is to use plasma-assisted or plasma-enhanced CVD (PECVD).[55] With this technique deposition can occur at very low temperatures, even close to ambient, since electrical energy rather than thermal energy is used to initiate homogeneous reactions for the production of chemically active ions and radicals that can participate in heterogeneous reactions, which, in turn, lead to layer formation on the substrate. In non-thermal plasmas, which are typically generated by electrical discharges in the gas phase, the electron temperature is much higher than the gas temperature and inelastic collisions of the electrons with precursor molecules form the chemically active species. In addition, surfaces in the plasma can be bombarded with ions, electrons and photons, leading to changes in surface chemistry. Although PECVD usually allows lower temperature deposition than thermal CVD, the plasma bombardment of a surface often causes some substrate heating. PECVD processes have been widely used for the deposition of a large range of materials with standard and novel properties; both inorganic and organic materials, as well as polymers, have been prepared by the technique. There are complications, though, with PECVD, including plasma damage of the substrate and the growing film, and a strong process dependency on several parameters such as rf power and frequency, gas pressure, reagent flow rate, reactor geometry, *etc.* Chapter 12 reviews PECVD technology and its applications.

Another method of inputting energy to a CVD process is to use high energy photons. The process of photo-assisted CVD[56] involves interaction of light radiation with precursor molecules either in the gas phase or on the growth surface. Precursor molecules must absorb energy, and since traditionally simple inorganic precursors have been employed this necessitates the use of UV radiation. If more complex molecules are used as precursors, then photosensitizing agents may need to be added. The use of organometallic precursors (with π- and σ-bonded moieties) opens up the possibilities for a wider range of wavelengths, but this can lead to an increased potential for carbon incorporation. Photo-assisted CVD has similar potential advantages to those of PECVD; namely, low temperature deposition, modifications of properties of grown layers, *e.g.* dopant incorporation, and independent control of substrate temperature and dissociation of precursor. In addition, though, with masking or laser activation it is possible to achieve selected area growth. Chapter 11 considers photo-assisted CVD processes in detail.

A rather different variation of CVD is atomic layer deposition (ALD) and the specialist version atomic layer epitaxy (ALE).[57] In this modification of CVD, gaseous precursors are introduced alternately to the reaction chamber, where they reach a saturated adsorption level on the substrate surface. Introduction of the precursors is separated other by an inert gas purge, which removes any excess precursor molecules and volatile by-products from the reaction chamber, thus preventing unwanted gas phase reactions. In marked contrast to traditional thermal CVD, which involves pyrolysis of precursor molecules, ALD proceeds through surface exchange reactions, such as hydrolysis, between chemisorbed metal-containing precursor fragments and adsorbed nucleophilic reactant molecules. A typical growth cycle in the ALD of a lanthanide oxide from a precursor [LnL_3] (*e.g.* L = Cp, OR) occurs in the sequence H_2O pulse/[LnL_3] pulse/N_2 purge/H_2O pulse/N_2 purge (Figure 1.4).

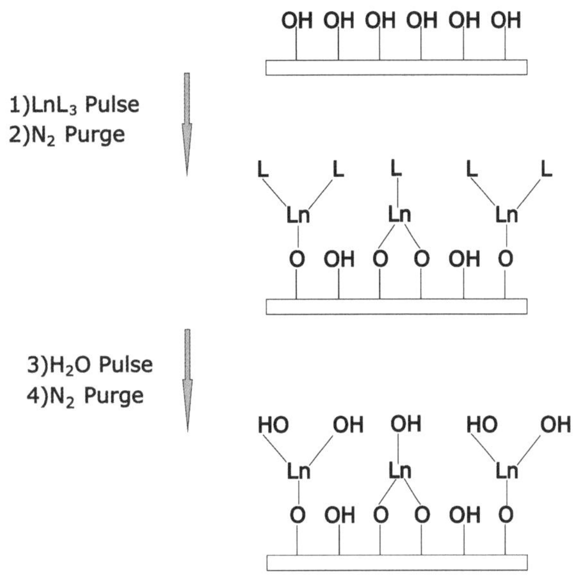

Figure 1.4 Schematic of an ALD growth cycle for deposition of a lanthanide oxide thin film from a lanthanide precursor $[LnL_3]$ (*e.g.* L = Cp, OR) and H_2O.

In the first step a pulse of H_2O gives a reactive [OH]-terminated surface. A pulse of the LnL_3 precursor then leads to a chemisorbed $[(L)_2Ln–O–]$ or $[(L)Ln–O_2–]$ surface species, and the liberated LH species are removed by a N_2 purge. The surface is then effectively terminated with unreactive L groups and growth self-limits. The next H_2O pulse removes the remaining L groups and regenerates a reactive [OH]-terminated surface.

Under optimum conditions, film growth proceeds through self-limiting surface reactions of a saturated adsorbent in which one ALD cycle produces one monolayer of material. However, due to steric hindrance or lack of reactive surface sites, the growth rate per cycle is often considerably less than one monolayer. This is illustrated in Figure 1.4 where, for steric reasons, the large LnL_3 molecules are unable to react with all the surface-bound OH groups. Nevertheless, the growth rate per cycle is constant, so the thickness of the thin film can be controlled simply and accurately by varying the number of deposition cycles. Because ALD reactions occur exclusively on the substrate surface, the process can give superior step-coverage (or improved conformality) to traditional CVD, and so ALD has become the technique of choice for film deposition on very high-aspect ratio substrates. In ALD it is important that surface reactions predominate and that thermal decomposition of the precursor is minimized or avoided altogether, otherwise self-limiting growth will break down. ALD processes are, therefore, generally carried out at substrate temperatures in the region 200–350 °C, which is below the thermal decomposition temperature of most precursors. Chapter 4 (Section 4.2.1) gives a much more detailed discussion about ALD processes.

Chemical beam epitaxy (CBE) is a rather specialized CVD technique. This is a high vacuum process that uses volatile metal-organic precursors (*e.g.* a Group III metal alkyl) and gaseous co-precursors (*e.g.* AsH_3 or PH_3). The closely related technique of metal-organic molecular beam epitaxy (MOMBE) uses volatile metal-organic precursors and a co-precursor vapour derived from the solid element (*e.g.* As, P).[58,59] In CBE and MOMBE the chemical reactions occur only on the substrate surface and so gas-phase reactions play no significant role in the growth process. The use of CBE has perhaps declined somewhat in recent years, but its aims are to combine the advantages of metal-organic vapour phase epitaxy (MOVPE) with those of molecular beam epitaxy (MBE). Since CBE is an ultrahigh vacuum (UHV) technique a potential advantage over conventional CVD is the ability to use vacuum *in situ* diagnostic techniques (*e.g.* RHEED, AES, MBMS) that provide real time analytical information on the growth process.

The growth kinetics of the CBE growth process are shown schematically in Figure 1.5, where it is compared to MOVPE and MBE processes.[60] Since CBE is an UHV technique, precursor desorption from the surface limits CBE growth processes to less than about 700 °C. It is therefore necessary to pre-pyrolyse thermally stable precursors. For example, in GaAs growth, thermally stable [AsH_3], has to be pre-decomposed to form the active surface species, As_2. A consequence of this is that there is a lack of active [AsH_x] species, which remove carbon-containing fragments in MOCVD,[61] and this leads to problems of carbon contamination in CBE.

There are still many other variants of CVD, but whatever the variant it is very apparent that, for a process to occur, starting materials, or precursors, are required, as is some form of reactor. The next two sections give a brief tour of these topics.

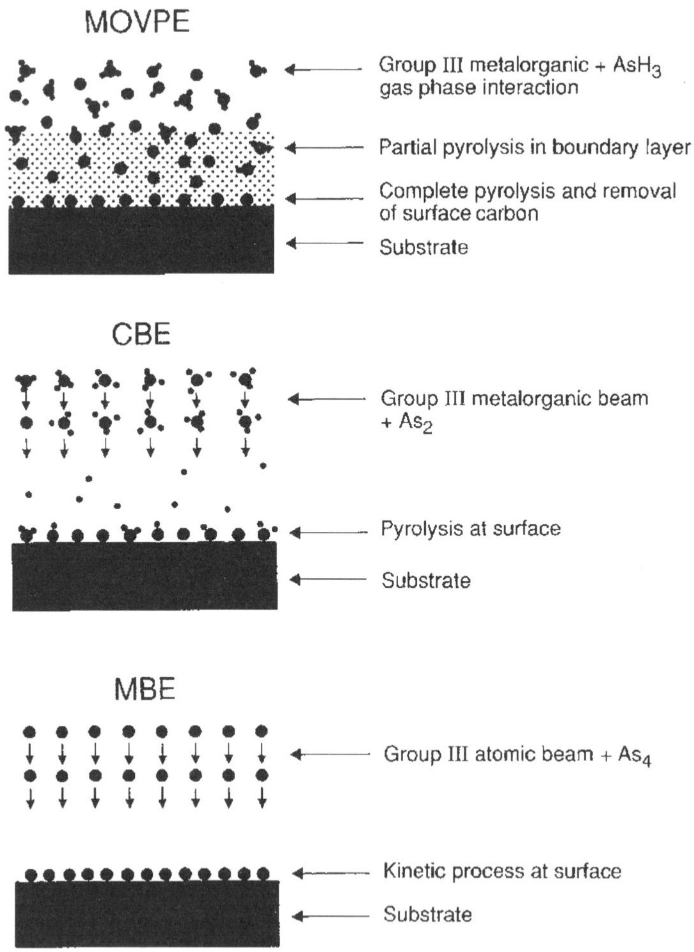

Figure 1.5 Schematic representation of the growth kinetics involved in MOVPE, CBE and MBE (molecular beam epitaxy, a PVD technique). (After ref. 60, Copyright John Wiley & Sons Limited, 1988. Reproduced with permission.)

1.4 CVD Precursors

1.4.1 Precursor Requirements

Whatever form a CVD process takes, the same precursor requirements generally apply. The characteristics of an "ideal" CVD precursor can be summarised as follows:

- Adequate volatility to achieve acceptable growth rates at moderate evaporation temperatures.
- Stability so that decomposition does not occur during evaporation.
- A sufficiently large temperature "window" between evaporation and decomposition for film deposition.
- High chemical purity.
- Clean decomposition without the incorporation of residual impurities.
- Good compatibility with co-precursors during the growth of complex materials.
- Long shelf-life with indefinite stability under ambient conditions, *i.e.* unaffected by air or moisture.
- Readily manufactured in high yield at low cost.
- Non-hazardous or with a low hazard risk.

Although these features are common for most CVD precursors, sometimes the precise precursor requirements can depend on the specific nature of the CVD process. For instance, in a traditional CVD processes it is a strong advantage if the precursor is relatively air-stable and not susceptible to reaction with water. This is a major disadvantage, though, in ALD, as it is essential that the surface exchange reaction with [OH] is facile, otherwise very low growth rates would result.

Another contrast between the requirements of a precursor for CVD in general and ALD is the size of the ligands attached to the central atom. In MOCVD, it is often advantageous to use very bulky ligands in the precursor, as this can often increase its vapour pressure, and also render it less air/moisture sensitive. In ALD, however, the presence of bulky or sterically demanding ligands around a precursor metal centre can lead to impractically low growth rates.

Also, a very high thermal stability (*e.g.* as in metal halides) can be a severe disadvantage in MOCVD, especially in microelectronics applications where low deposition temperatures ($< 500\ ^\circ C$) are often required. In contrast, for ALD high thermal stability can be an advantage, as long as the precursor is reactive to H_2O or surface [OH] species, as this will lead to exchange reactions predominating over thermal decomposition reactions, as required for self-limiting growth. For examples of the very different requirements for MOCVD and ALD precursors see ref. 62.

The high vacuum environment of CBE leads to the added precursor requirement that it should ideally decompose cleanly by a unimolecular process, without incorporating excessive amounts of carbon in the film. This tends to lead to the use of Group II and Group III metal alkyls with alkyl ligands containing an easily-eliminated β-hydride atom (*e.g.* ethyl, *iso*-propyl, *etc.*).

Throughout this volume the characteristics of precursors will be referred to in a wide range of contexts, and most of the requirements outlined above will become apparent. Here, though, some particular aspects of precursor chemistry mentioned above are considered in more detail.

1.4.2 Precursor Volatility

The main requirement of any precursor is that it has an adequate volatility. With precursors that occur naturally in a gaseous state (*e.g.* silane, diborane, ammonia, *etc.*) this is not a problem. However, for precursors that are liquids or solids, volatility often has to be enhanced by a reduction of the intermolecular forces that lead to dimer, oligomer or polymer formation. Good

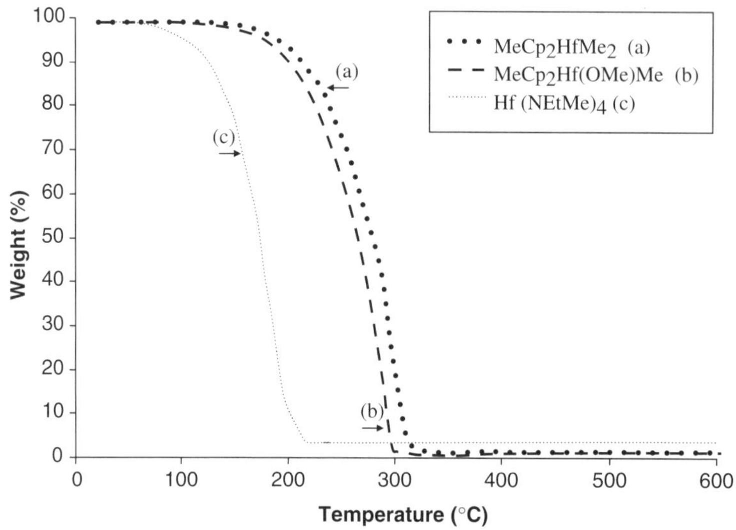

Figure 1.6 Thermogravimetric analytical (TGA) data for [Hf(NEtMe)$_4$], [(MeCp)$_2$HfMe$_2$] and [(MeCp)$_2$Hf(OMe)Me].

examples of how this is achieved are given in Chapter 8 (Section 8.2) and Chapter 5 (Section 5.6.1) for metal oxide precursor chemistry where neutral complexes [ML$_n$] are designed to have mono-anionic ligands that completely fill the coordination sphere of the metal.

The introduction of new chemicals into production processes can be an expensive, time consuming exercise. Therefore, a key parameter to ensure optimum process design is reliable precursor volatility data. A good idea of the temperature required to evaporate the precursor can be obtained by thermogravimetric analysis (TGA), in which the weight loss of a precursor is measured as a function of increasing temperature. Since many precursors are air sensitive they must be transferred to the TGA apparatus in an inert atmosphere, and hence the apparatus is generally enclosed in a nitrogen box. An example of the usefulness of TGA is illustrated by Figure 1.6, which shows the TGA data for three different precursors used in ALD, [Hf(NEtMe)$_4$], [(MeCp)$_2$HfMe$_2$] and [(MeCp)$_2$Hf(OMe)Me] (Chapter 8, Section 8.3.1.2).

The TGA data show that the [Hf(NEtMe)$_4$] complex evaporates in the temperature range ~100–225 °C, leaving a small amount of residue (~2.5%), possibly due to trace air ingress during sample handling. The organometallic complexes [(MeCp)$_2$Hf(OMe)Me] and [(MeCp)$_2$HfMe$_2$] are clearly less volatile, evaporating in the ranges ~150–300 °C and 150–320 °C, respectively. TGA data also provide information on precursor stability, with the presence of large amounts of residue indicating decomposition during evaporation.

However, although TGA is a useful indicator of precursor volatility and stability, it is no substitute for accurate vapour pressure determination. It is not a trivial excersise, though, to obtain accurate vapour pressure data, as shown by the large variation in literature values for even well-known precursors.[63–65] During vapour pressure measurements it is essential that the precursor is fully vapourized and that condensation of the precursor within the measurement equipment upstream of the pressure measurement device is avoided; this can be difficult, especially for low-volatility compounds. It is important too that the precursor is free from gaseous contaminants, deriving from precursor decomposition or from the inert gas blanket in the precursor storage container. It is also essential to calibrate the data for the precursor against a standard with a known accurately determined vapour pressure (*e.g.* naphthalene[66]). Figure 1.7 shows a schematic diagram

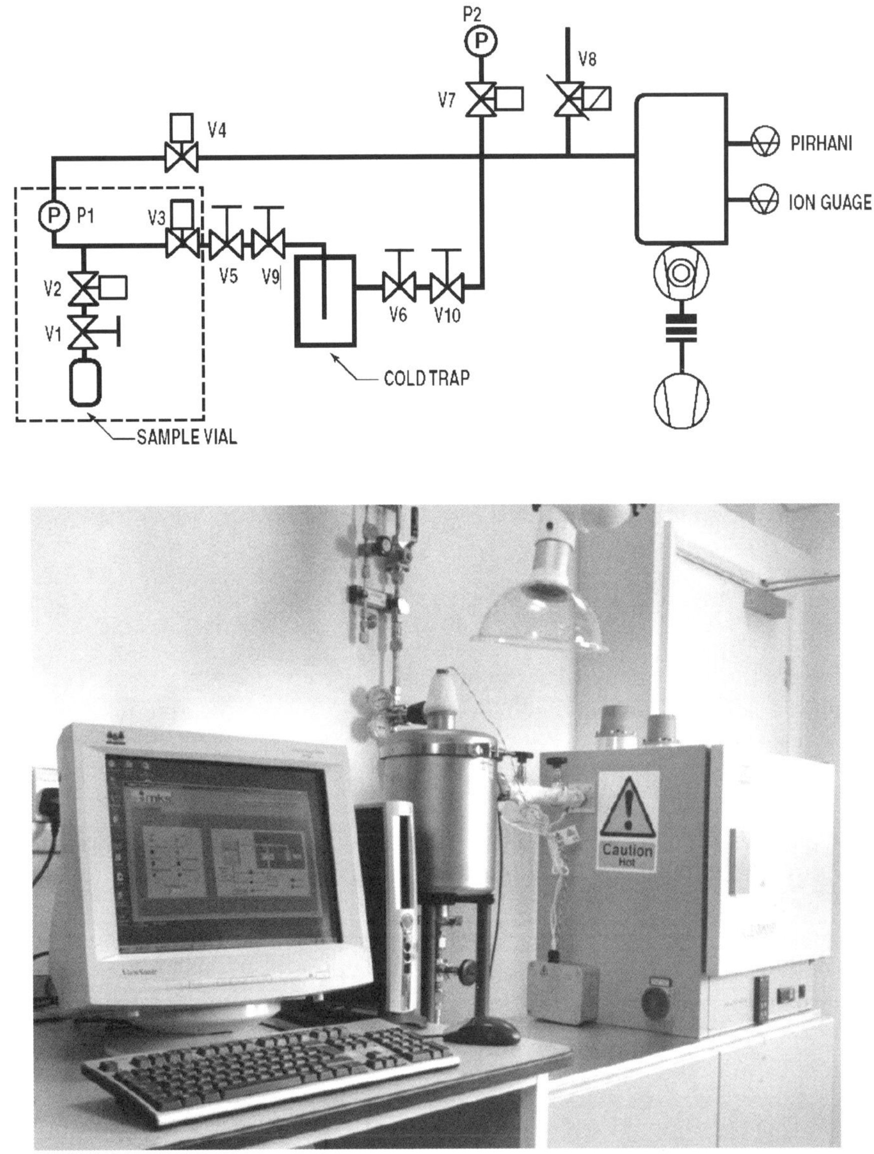

Figure 1.7 Apparatus used by a commercial manufacturer to accurately measure the vapour pressure of metalloorganic MOCVD and ALD precursors. (Reprinted from ref. 66, with permission from Elsevier.)

and photograph of a vapour pressure measurement system used by a commercial precursor manufacturer.[66] The instrument is constructed entirely of stainless steel with high integrity joints throughout. The key sections of the instrument are located inside an oven to prevent precursor condensation and vapour pressure is measured using a Baratron™ pressure gauge attached to a vacuum manifold.

The vapour pressure of the compound is measured at various temperatures (see Figure 1.8 for data for $[Ta(NMe_2)_5]$) and from a plot of $\log_{10}P$ against $1/T$ (in K) vapour pressures over a wide range of temperatures can be determined. The vapour pressure of $[Ta(NMe_2)_5]$ is given by Equation

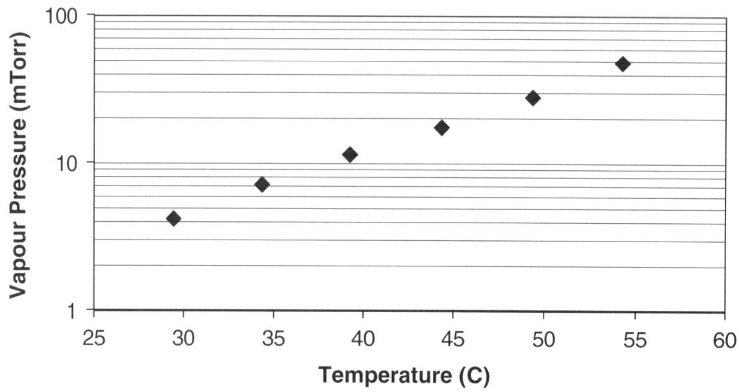

Figure 1.8 Vapour pressure data for $[Ta(NMe_2)_5]$. (Reprinted from ref. 66 with permission from Elsevier.)

(1.2),[66] which allows the calculation of vapour pressure at any specified temperature:

$$\mathrm{Log}_{10}P = -4124.9/T + 11.265 \tag{1.2}$$

Vapour pressure data for several metal oxide precursors are given in Chapter 8 (Table 8.5), while data for a selection of III-V semiconductor precursors are shown in Chapter 6 (Table 6.4).

1.4.3 Precursor Thermal Stability

As has been indicated earlier, thermal stability is also a crucial factor in determining the suitability of a precursor for CVD applications, and it is now routine for a commercial precursor manufacturer to carry out extensive thermal stability tests on their products. First, the precursor should be sufficiently stable for long-term storage at room temperature, and then it should not undergo any appreciable decomposition at the evaporation temperatures necessary to achieve adequate gas-phase transport. For safety reasons it is also essential to ensure that rapid thermal decomposition of a precursor does not occur by misuse or by accident, particularly when the precursor is manufactured and used in large multi-kilogram or ton quantities. In the case of some precursors, such as alane adducts (*e.g.* $[AlH_3(NMe)_3]$) – see Chapter 7 (Section 7.3.1.3) – poor storage practices can lead to decomposition, generating large volumes of gaseous products and pressures of hundreds of atmospheres in a sealed container.

Differential scanning calorimetry (DSC) is a common method of determining the thermal stability of precursors. DSC measures the difference in heat flux between a sample and a reference material as a function of temperature. Both the sample and reference material are subjected to a controlled temperature programme, and endothermal and exothermal changes in enthalpy can be observed. The heat changes reflect physical and/or chemical processes, such as phase transitions or chemical recations. Figure 1.9 shows a typical DSC curve for a CVD precursor, illustrating four common features commonly observed in DSC experiments.[67]

It can be seen that DSC provides information about the temperatures at which melting and decomposition begin (T_{onset},) and when the thermal event ends (T_{end}). T_{max} and T_{min} are the temperatures at the maximum of the exotherm and the minimum of the endotherm, respectively. The area under the exothermic and endothermic peaks are related to the total heat of reaction. These DSC parameters depend strongly on the rate of sample heating, with sample size being less critical, so that a low rate of heating ($\sim 2\,^{\circ}\mathrm{C\,min^{-1}}$) is generally used. Although it is difficult to

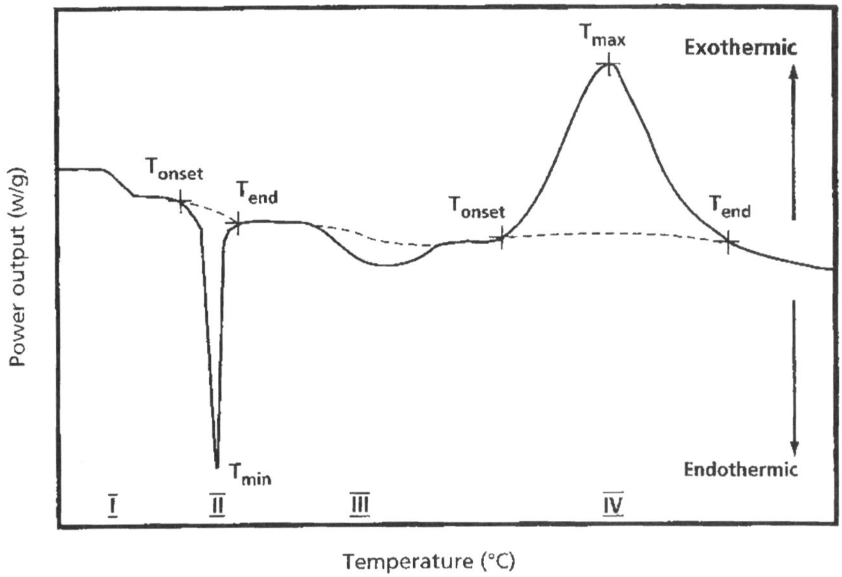

Figure 1.9 Typical DSC trace, illustrating four different types of transition: (I) a baseline shift caused by a glass transition; (II) an endothermal effect caused by melting; (III) an endothermal effect caused by dissociation of the compound; (IV) an exothermal effect caused by decomposition of the compound. (After ref. 67, Copyright John Wiley & Sons Limited, 1997. Reproduced with permission.)

derive absolute thermochemical values from DSC data, general trends in the thermal stability of CVD precursors can be established.[67]

1.4.4 Precursor Purity and Precursor Analysis

In any chemical process the purity of the reactants is of paramount importance. For thin film deposition by CVD this is especially true. The influence of precursor purity on the properties of the deposited film can be illustrated by considering the doping of III-V compound semiconductors (*e.g.* GaAs, InP), for which much work has been done on correlating impurity levels in the precursor with those in MOCVD-grown films, as described in an early review.[68] The presence of volatile trace metal impurities in the Group III metal-organic precursor (*e.g.* $[Me_3Ga]$, $[Me_3In]$) leads to the incorporation of shallow ionized impurities in the semiconductor layer. The energies of these impurities place them in the band gap of the III-V layer, and when introduced intentionally they lead to various sophisticated III-V devices, such as GaAs/AlGaAs/GaAs double heterojunction lasers and p/n doped GaN-based LEDs.

However, with unintentional doping the impurities can seriously degrade the properties of a III-V semiconductor by lowering the electron mobility through scattering,[69] and decreasing the photoluminescence by providing non-radiative recombination pathways.[70] GaAs and InP have intrinsic free charge carriers of $<10^{13}\,cm^{-3}$, and it has been found that as little as 1 ppm of an unintentional dopant impurity in the layer corresponds to an impurity concentration of $2.5\times10^{16}\,cm^{-3}$, which is unacceptable for most device applications.

Unintentional dopants are generally introduced during precursor synthesis, as is the case of volatile metal-organic impurities in $[R_3Ga]$ or $[R_3In]$ precursors.[71] For instance, $[Me_3Ga]$ can be synthesized from the reaction between a Grignard reagent, such as [MeMgI], and $[GaCl_3]$ (Chapter 6, Section 6.5.1). Magnesium is thus a common impurity in the $[Me_3Ga]$, along with Si and Zn,

Table 1.1 Common dopants in III-V semiconductors.

Impurity elements	*Group*	*Site occupied*	*Electrical behaviour*	*Typical dopant precursors*
Be, Mg	II A	III A (Al, Ga, In)	p-type Shallow ionized acceptor	[Et_2Be], [Cp_2Mg], [$(MeCp)_2Mg$]
Zn, Cd	II B	III A (Al, Ga, In)	p-type Shallow ionized acceptor	[Me_2Zn], [Et_2Zn] [Me_2Cd]
C	IV A	V A (As, P)	p-type Shallow ionized acceptor	[CCl_4], [CBr_4] ($CH_3^{\bullet}$)radicals from [Me_3M] precursor
Si, Sn, Ge	IV A	III A (Al, Ga, In)	n-type Shallow ionized donor	[SiR_4], [SiH_4], [R_4Sn], [R_4Ge]
S, Se, Te	VI A	V A (As, P)	n-type Shallow ionized donor	[H_2S], [H_2Se], [R_2S], [R_2Se], [R_2Te]

which originate in the Mg and/or the $GaCl_3$. Table 1.1 shows some common dopants in III-V materials.

As shown in Table 1.1, these impurities generally act substitutionally, occupying a lattice site normally belonging to a Group III or V element, which has the net effect of adding an electron (n-type charge carrier) or hole (p-type charge carrier). The concentration of these impurities in the III-V film is often a strong function of growth temperature, for instance the level of the n-type dopants Si and Sn tends to increase with increasing growth temperature, whilst the level of others, such a the p-type dopants Zn, Mg and Cd, tends to decrease.

To minimize unintentional doping, highly sophisticated analytical techniques have been developed for the determination of trace metal impurities in metal-organic precursors, which are often highly reactive or even pyrophoric (*e.g.* [Me_3Ga], [Me_3In], and [Me_3Al]). The techniques include flameless atomic absorption spectrophotometry,[72] direct injection inductively coupled plasma optical emission spectroscopy (ICPOES) and ICP-mass spectrometry (ICPMS).[73] These techniques can detect impurities at the ppb level (see examples for [Me_3Al], Chapter 6, Tables 6.5 and 6.6). Table 1.2 shows a correlation between impurities in the metal-organic precursor and the electrical properties of a III-V layer.[68]

As well as trace metal impurities affecting layer characteristics, carbon and oxygen contaminants also play a crucial role in determining layer properties. Carbon impurities are inherent to MOCVD and CBE growth processes, originating from the thermal decomposition of the organic groups in the precursor, and the level of carbon incorporation is strongly dependent on the molecular structure of the precursor. Oxygen may be introduced at various stages of an MOCVD process, through, for example, volatile oxygen-containing impurities in the precursor (*e.g.* [Me_2AlOMe] in [Me_3Al]), or from leaks in the source container or reactor inlet lines. The presence of oxygen is a particular problem in optical devices containing Al in the active layer, as this can seriously degrade their optical efficiency. Organic impurities are, perhaps, more difficult to detect and quantify than metallic impurities in precursors. Mass spectrometry is useful for identifying hydrocarbon impurities,[68] whilst Fourier- transform (FT) NMR has been used to detect oxygen-containing organic impurities at the low ppm level (Chapter 6, Section 6.5.2 and Table 6.7). Finally, hydrogen, present in the metal-organic precursor and in the MOCVD process gases, can also influence semiconductor layer properties, by passivating intentional dopants such as Mg in GaN.

Table 1.2 Effect of precursor purity on the electrical properties of a GaAs layer.[68]

Group III precursor	*ICP-ES analysis[a] (ppm)*	*III-V material [growth temp. (°C)]*	*Electron carrier concentration[b] [η_{77K} (cm^{-3})]*	*Electron mobility[b] [μ_{77K} ($cm^2 V^{-1} s^{-1}$)]*
[Me_3Ga] Batch 1	Si 1.2, Zn 2.0	GaAs (650)	1.1×10^{15}	42 600
[Me_3Ga] Batch 2	Si < 0.03 Zn < 0.2	GaAs (650)	1.4×10^{14}	137 000

[a]No other impurities detected.
[b]Low η values and high μ values indicate that the layer is of high purity with few extrinsic impurities.

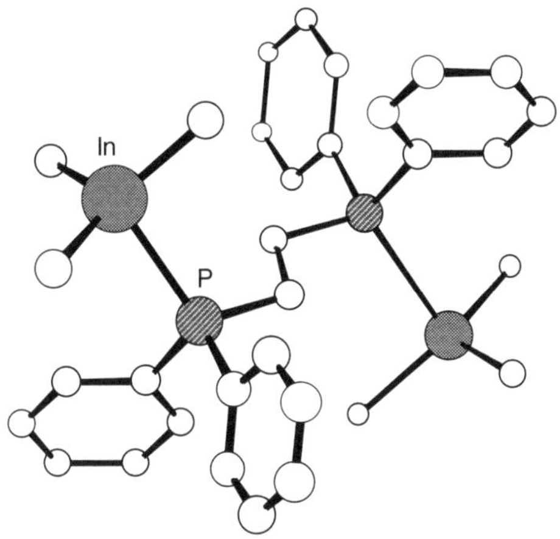

Figure 1.10 Crystal structure of the [$(Me_3In)_2 \cdot$ DIPHOS] adduct used to purify [Me_3In]. (Courtesy of the Cambridge Structural Database, see "The Cambridge Structural Database; a quarter of a million crystal structures and rising", F.H. Allen, *Acta Crystallogr., Sect. B: Struct. Sci.*, 2002, **58**, 380.)

1.4.5 Precursor Purification Techniques

The critical effect of precursor purity on layer properties, particularly in III-V MOCVD technology, has led to an intense effort to purify precursors to levels of up to 99.9999% purity (on a metal basis); the improvement of precursor purity, especially in the 1980s, played a critical role in the development of the compound semiconductor industry. Classical purification techniques[74] include sublimation, recrystallization, fractional distillation (or rectification),[75] preparative chromatography[76] and, for III-V precursors, distillation of the metal-organic compound from a reactive melt such as gallium[77] or sodium/potassium alloy.[78] The removal of trace metal and organic impurities from an organometallic compound can often be difficult using techniques such as distillation, especially when the contaminant has a similar boiling point to the organometallic compound (*e.g.* [R_4Si], [R_2Zn] *etc.*). This led to the development some years ago of "adduct-purification" techniques, which involve the formation of an involatile (Lewis acid–Lewis base) adduct as an intermediate in the purification process.[79,80] An example of this is the purification of trimethylindium by the formation of the [$(Me_3In)_2 \cdot$ DIPHOS] adduct (DIPHOS = 1,2-bis(diphenylphosphino)ethane) (Figure 1.10).[81] Volatile impurities such as diethyl ether solvent and trace metal contaminants such as [R_4Si] and [R_4Sn] which do not form adducts with DIPHOS can easily be removed under vacuum. Pure [Me_3In] is then obtained by thermal dissociation of the relatively weak [Me_3In]—[DIPHOS] bond and isolation by vacuum distillation[80] (Chapter 6, Section 6.5.1 and Figure 6.10).

The adduct purification process was also applied very successfully to $[Me_3Ga]$ using adducts with high boiling ethers.[71,79]

1.5 CVD Reactors

A very large range of reactors and several different precursor delivery systems have been used in the various areas of CVD technology. These are dealt with in detail in Chapter 2, but it may be useful to begin with a brief and general overview of CVD reactors. Irrespective of the variations in CVD processes, all thermal CVD reactors have common features:

- Precursor sources.
- Gas handling system to control input of precursor gases or vapours to the reaction zone.
- A reaction zone, usually within an enclosed cell, with a holder that can accommodate the substrate, and which is heated by a surrounding oven or furnace, or by external radiofrequency or infrared radiation.
- An exhaust system, which may include a vacuum pump for low pressure operation, to remove waste products and a waste treatment facility with any necessary waste monitoring devices.

A schematic of a typical MOCVD reactor used to deposit a III-V material from a Group III trialkyl compound and a Group V hydride is shown in Figure 1.11 (taken from Chapter 6, Figure 6.8).

The volatile precursors are contained in separate containers, commonly called "bubblers", usually made of stainless steel (although quartz or Pyrex glass can also used). The bubbler is equipped with inlet and outlet bellows-sealed or diaphragm-sealed valves and a dip-tube attached to the inlet valve (Figures 1.12 and Ch. 6, Figure 6.9). Precursor transport is controlled by passing an inert carrier gas (*e.g.* H_2, Ar, N_2) through the precursor in the bubbler. A mass flow controller upstream of the bubbler can accurately control the rate of precursor transport. Ideally, the carrier gas should become fully saturated with precursor vapour, but this is not absolutely essential as long as the pick-up by the carrier gas is reproducible. To facilitate this, the bubblers are held in temperature-controlled baths. For the most careful deposition control, reactors are generally operated in "vent-run" mode, in which a vent line allows the flow of reactants to be stabilized before entering the reactor. When the reactant is required the manifold very rapidly switches the reactant flow into the reactor, and this can allow the growth of complex multilayer materials.

The flow into a reactor can be in a vertical or horizontal manner and schematic illustrations of some common reactor types are shown in Figure 1.13.[53] As mentioned earlier, various modes of heating can be employed.

Precursors used for some growth processes, such as III-V and II-VI semiconductors generally have high vapour pressures (>1 Torr), at moderate source temperatures, typically $<50\,^{\circ}C$.[82,83] However, for other processes, such as metal oxide MOCVD, many precursors have only very low vapour pressures of $\ll 1$ Torr close to room temperature, so that high evaporation temperatures are needed for efficient deposition. Several strategies have been employed to deal with this situation. The chemistry-based approach, mentioned earlier, is to design precursor molecules with reduced intermolecular forces to overcome oligomerization. An equipment-based approach is to have not just the precursor container heated but also the pipe-work to the reactor heated to a temperature high enough to prevent precursor condensation; this would typically be 20–30 °C above the bubbler temperature. Pipe-work can be heated by using co-axial tubing with a heat transfer reagent (*e.g.* oil) passing through the annular region (Figure 1.14),[84] or simply by wrapping heating tapes around the pipes.

Both approaches have been successfully used for the growth of several complex ferroelectric oxides such as $PbTiO_3$, $Pb(Zr,Ti)O_3$ and high-Tc superconducting oxides.[84,85] A problem remains, though, in that some metal oxide precursors, such as metal alkoxides or β-diketonates, have

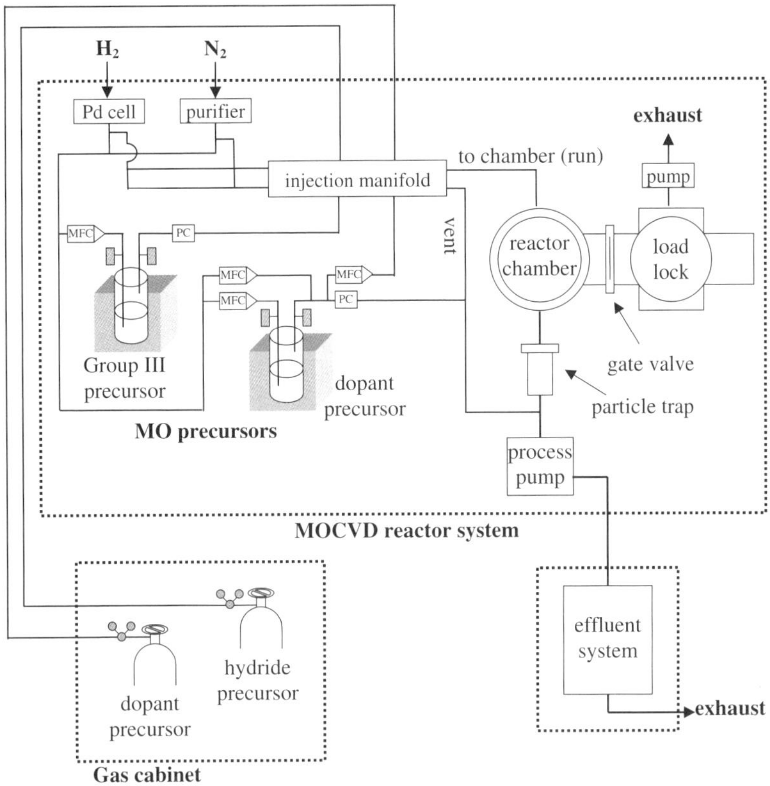

Figure 1.11 Schematic of a typical MOCVD reactor used for the deposition of III-V semiconductor films. Taken from Chapter 6 (Figure 6.8).

Figure 1.12 A commercial "bubbler" used to evaporate metal-organic compounds for MOCVD. (Courtesy of SAFC Hitech.)

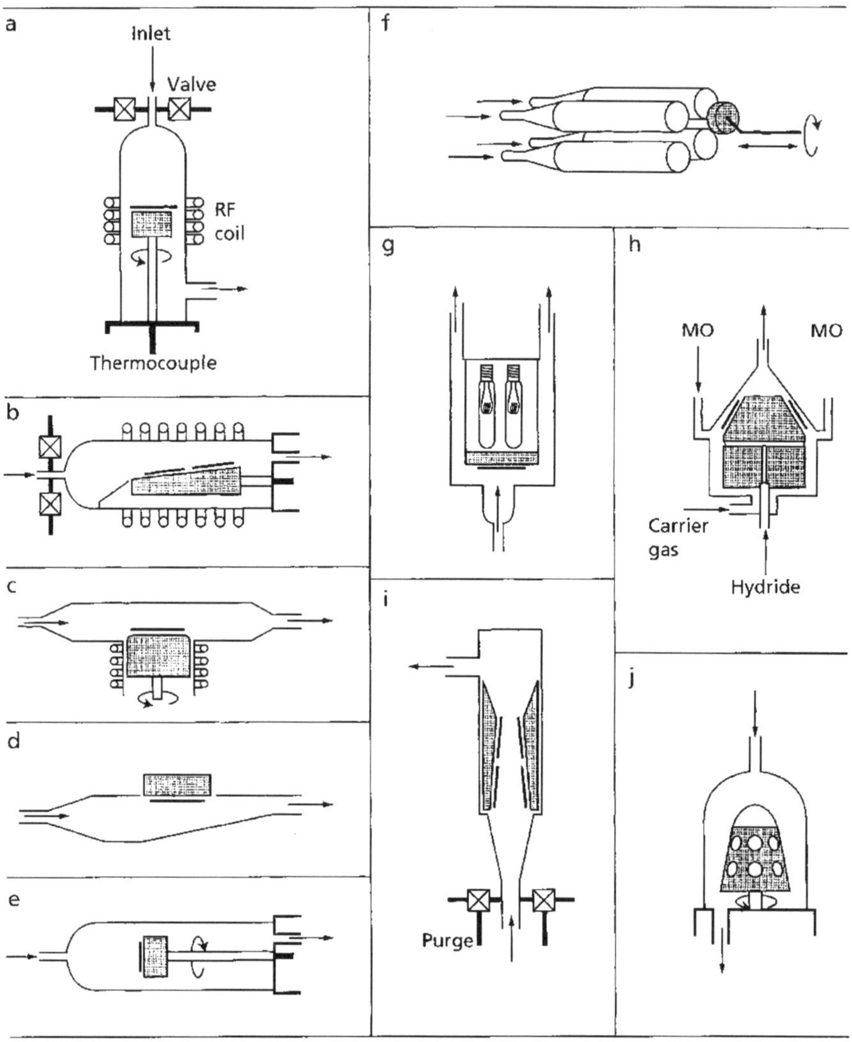

Figure 1.13 Some common reactor cell geometries used in MOCVD: (*a*) rotating disc, gas inlet at top, (*b*) horizontal reactor, (*c*) "T" reactor with rotating susceptor, (*d*) reactor with inverted wafer mounting, (*e*) horizontal reactor with substrate normal to gas flow, (*f*) multiple barrel reactor, (*g*) barrel-type reactor with separate inlets, (*h*) inverted stagnation point flow reactor, (*i*) chimney reactor, (*j*) large-scale barrel reactor. (After ref. 53, Copyright John Wiley & Sons Limited, 1987. Reproduced with permission.)

insufficient thermal stability to withstand heating for long periods. They can slowly decompose in the bubbler or in the inlet pipe-work, leading to poor oxide layer uniformity and reactor blockages, or undergo annealing and crystallization, so that carrier gas pick-up rates change with time. These problems associated with low volatility precursors can be largely overcome by the use of liquid injection MOCVD, in which the precursor is dissolved in an inert solvent, usually an ether (*e.g.* tetrahydrofuran) or a hydrocarbon (*e.g.* heptane or nonane).[47] The precursor solution is kept at room temperature and when required it is delivered at a precisely controlled rate and quantity into a heated evaporator and transported by a carrier gas into the reactor zone. For multi-component layers, separate precursor solutions can be used for growth, or several precursors can be dissolved

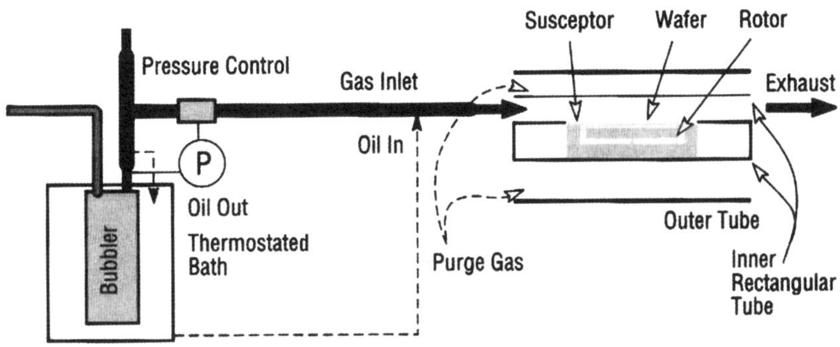

Figure 1.14 Schematic of a conventional MOCVD reactor with a heated delivery line used to deposit ferroelectric oxides [*e.g.* $Pb(Zr,Ti)O_3$] using low vapour pressure precursors. (After ref. 84. Reproduced by permission of the MRS Bulletin.)

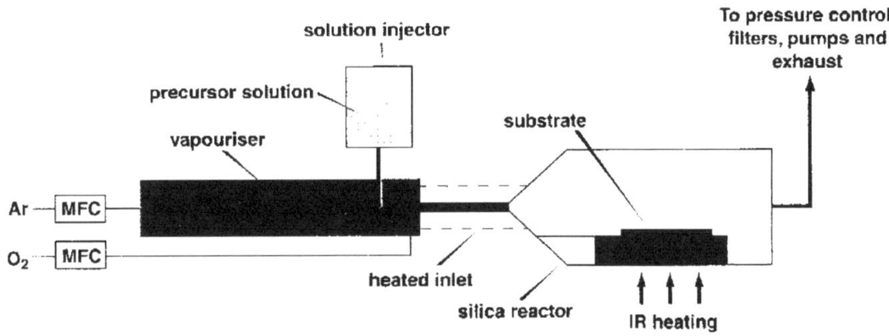

Figure 1.15 Schematic diagram of a simple liquid injection MOCVD reactor. (After ref. 47, reproduced by permission of The Royal Society of Chemistry.)

in an appropriate molar ratio in a single solution to form a "precursor cocktail". Figure 1.15 shows a schematic of a liquid injection MOCVD system.[47]

A potential disadvantage of liquid injection MOCVD is that the precursor solution is evaporated from a single heated evaporator, unlike in conventional MOCVD where each bubbler is separately heated. In liquid injection MOCVD it is therefore often necessary to select, or design, co-precursors that evaporate at similar temperatures.

In ALD a similar precursor delivery system to conventional CVD is generally used, but valves are added to shut off the precursor pulse or oxygen reactant pulse, and an inert gas purge line is added to remove materials desorbed during each ALD cycle. A full explanation of an inert gas valving system is given in Chapter 4 (Section 4.4).

With CBE, although film growth occurs in a UHV environment, the condensed phase metalorganic Group III precursors and gaseous Group V precursor are again held in containers external to the system. In this case, though, reagent flow into the growth chamber is in the molecular regime, allowing retention of the use of shutters, and *in situ* masks. An advantage of CBE is the use of *in situ* analytical equipment such as RHEED, mentioned above, and also of residual gas analysis for the detection of impurities. A typical CBE reactor is shown in Figure 1.16.[58,59]

Chapter 2 gives a highly detailed description of the large range of CVD reactors, injection systems and ancillary equipment, together with a discussion of gas dynamics, delivery systems and reactor cells.

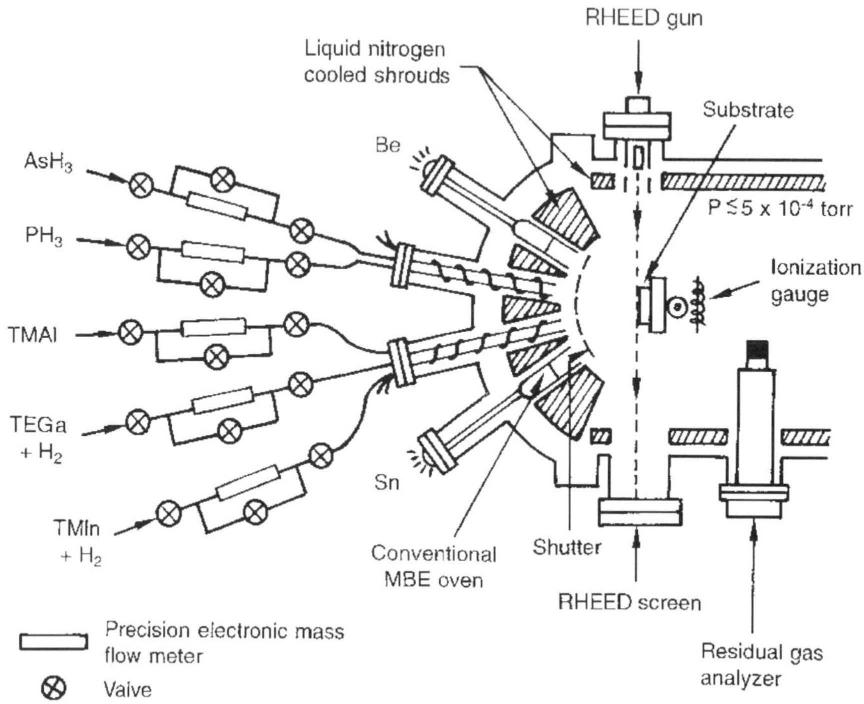

Figure 1.16 Schematic diagram of a typical CBE reactor system. (After ref. 60, Copyright John Wiley & Sons Limited, 1988. Reproduced with permission.)

1.6 Materials Deposited by CVD and Applications

A very extensive range of materials has been deposited using conventional CVD and its variants. These materials find applications in the following main areas of technology:

- microelectronics
- optoelectronics
- protective and decorative coatings
- optical coatings

Table 1.3 give examples of the classes of CVD-grown films, classified according to their chemical nature, together with applications and typical precursors used. The table is by no means exhaustive, but gives a good idea of the wide variety of precursors used in the various application areas of CVD. Although some precursors are more suited to a particular CVD technology (*e.g.* ALD rather than MOCVD), many of these precursors are utilized in all areas of the technology.

Detailed discussions of CVD grown materials are given in Chapters 4 and 5–12.

1.7 Materials Properties

1.7.1 Layer Morphology

Layer morphology is important in determining the physical characteristics and usefulness of a film. It is determined by surface diffusion and nucleation processes during deposition, which in turn are

Table 1.3 Selection of films grown by CVD, their applications and typical precursors used.

General class of material	*Specific material*	*Applications*	*Typical CVD precursors*
Dielectric oxides	TiO_2, ZrO_2, $ZrSi_xO_y$, HfO_2, $HfSi_xO_y$	High-κ gate dielectric layers in CMOS technology	Metal-β-diketonates (*e.g.* [M(β-dik)$_4$] (M = Zr, Hf), [Ln(OR)$_3$], [Ln(β-dik)$_3$] (Ln = lanthanide) + O_2/O_3
	Ln_2O_3, $LnSi_xO_y$, $LnAlO_3$	Dielectric capacitor layers in DRAMs	Metal alkoxides, (*e.g.* [M(OR)$_4$] (M = Si, Ti, Zr, Hf), [M(OR)$_5$] (M = Ta, Nb) + O_2/O_3
	SiO_2, Ta_2O_5, Nb_2O_5	Dielectric capacitor layers in DRAMs	
		Optical coatings	Metal alkylamides [M(NR$_2$)$_4$] (M = Ti, Zr, Hf) + O_2/O_3 Organometallics (*e.g.* [Cp$_2$MMe$_2$], [(RCp)$_2$MMe$_2$] (M = Zr, Hf) + O_2/O_3 [LaCp$_3$], [Ln(RCp)$_3$] + O_2/O_3
Ferroelectric oxides	$SrTiO_3$, $(Ba,Sr)TiO_3$, $Pb(Zr,Ti)O_3$, $SrBi_2(Ta_xNb_{1-x})_2O_9$, $Bi_4Ti_3O_{12}$, $Pb(Sc,Ta)O_3$, $Pb(Mg,Nb)O_3$	DRAMs, NVFERAM , computer memories, infrared detectors, microelectromechanical devices, transducers, ceramic capacitors,	Metal-β-diketonates ([Ba(thd)$_2$], [(Sr(thd)$_2$], [Ti(OPr^i)$_2$(thd)$_2$], [Bi(thd)$_3$] [Pb(thd)$_2$], [Sc(thd)$_3$], [Mg(thd)$_2$] + O_2 Metal alkoxides ([Bi(OR)$_3$], [Ta(OR)$_5$], [Nb(OR)$_5$]) + O_2
Ferrites	$(Ni,Zn)Fe_2O_4$, $(Mn,Zn)Fe_2O_4$	Recording media, high frequency read heads	Metal-β-diketonates ([Ni(thd)$_2$], [Zn(thd)$_2$], [Mn(thd)$_2$], [Fe(thd)$_3$]) + O_2
Superconducting oxides	$YBa_2Cu_2O_{7-x}$, Bi-Sr-Ca-Cu-O	Josephson junctions, bolometers, SQUIDS	Metal-β-diketonates ([Y(thd)$_3$], [Ba(thd)$_2$], [Sr(thd)$_2$], [Cu(thd)$_2$], [Ca(thd)$_2$]) + O_2
Conducting oxides	$(La,Sr)CoO_3$, $(La,Mn)O_3$ RuO_2, $SrRuO_3$	Ferroelectric capacitor electrodes	Metal-β-diketonates + O_2 [Ru$_3$(CO)$_{12}$], [Ru(β-dik)$_3$], [Ru(RCp)$_2$] (R = H, Et)
Low-emissivity and conducting oxides	F-doped SnO_2	Architectural coatings on flat glass (solar control, anti-reflective), display coatings on glass	[SnCl$_4$]/[NH$_4$F]/O_2
	Sn-doped In_2O_3		[InCl$_3$]/[SnCl$_4$]/O_2,[In(thd)$_3$]/[Bu$_2$Sn (acetate)]/O_2, [In(acac)$_3$]/[Sn(acac)$_2$]/O_2
Electrochromic and photochromic oxides	WO_3	Architectural coatings on flat glass (displays)	[WCl$_6$]/O_2, [W(CO)$_6$]/O_2
	MoO_3		[Mo(CO)$_6$]/O_2

(*Continued*)

Table 1.3 *(continued)*.

General class of material	*Specific material*	*Applications*	*Typical CVD precursors*
Thermochromic oxides	VO_2	Architectural coatings on flat glass	$[VCl_4]/O_2$, $[VOCl_3]/O_2$
Self-cleaning coatings	TiO_2	Architectural coatings on flat glass	$[TiCl_4]/[Ti(OPr^i)_4]/O_2$
Garnets	$Y_3Fe_5O_{12}$	Microwave elements, magneto-optic recording	Metal-β-diketonates + oxygen
Elemental semiconductors	Si	Microelectronic devices	$[SiH_4]$, $[Si_2H_6]$, $[SiH_xCl_{4-x}]$ ($x = 0$–3)
	Doped-Si		$[SiH_4]/[PH_3]/[AsH_3]/[B_2H_6]$
	Ge		$[GeH_4]$
	Diamond and diamond-like carbon	Semiconductor devices	$[C_nH_{2n+2}]$
III-V Compound Semiconductors	GaAs	Solar cells, LEDs	Gp. III trialkyls, $[R_3M]$ (M = Ga, In, Al; R = Me, Et) + Group V trihydrides ($[AsH_3]$, $[PH_3]$, $[NH_3]$ or organometallics ($[Bu^tAsH_2]$, $[Bu^tPH_2]$, $[R_3Sb]$)
	GaAs/AlGaAs	Heterostructure lasers, solar cells, HEMTS, HBTs, FETs	
	GaP	Red LED, photocathodes	
	GaAsP, InGaP	Red LEDs	
	AlGaInP	Yellow/green LED	
	GaN	Blue LED	
	InGaN	Green LED	
	InGaP/AlGaInP	Red laser pointers	
	InP	Gunn diodes, radar devices	
	InP/InGaAs	Detectors in optical fibre technology	
	InGaAsP	Emitters in optical fibre technology	
	GaSb/AlGaSb	Thermal imaging devices, environmental sensors	
II-VI compound semiconductors	ZnS	Blue phosphors, TFELs	Gp. II dialkyls, $[R_2M]$ (M = Zn, Cd; R = Me)
	ZnSe, ZnSSe, ZnMgSSe	Blue/green LEDs and lasers	+
	ZnCdS, CdS/CdTe	Solar cells	Gp VI dihydrides, $[H_2Se]$, $[H_2S]$ or
	CdTe, CdHgTe	Infrared detector, thermal imaging systems	Gp VI dialkyls, $[Et_2Se]$, $[Et_2S]$, $[Et_2Te]$, $[Pr^i_2Te]$
Metals	Al	Interconnects in microelectronic devices, metallized polymers, gas	$[Bu^i_3Al]$, $[AlH_3(NR_3)]$

		diffusion barriers, optical coatings, adhesion aid	
	W	Metallization in transistors, interconnects in ICs, wear and erosion protection	$[WF_6]$, $[WCl_6]$,/H_2; $[W(CO)_6]$; W-alkyls
	Cu	Interconnects in ICs	Cu(II) β-diketonates/H_2; [(hfac)Cu(I)L] (L = $[PMe_3]$, [1,5-COD], alkyne, VTMS)/H_2
	Au, Ag	Metallization in ULSI	Me_2Au(β-diketonate)]; [Ag(β-diketonate)]
	Pt, Pd, Ni	Metal contacts in microelectronic devices, catalysts, protective and decorative coatings	$[Pt(acac)_2]$, $[PtMe_3]$; [(1,5-COD)$PtMe_2$]; $[Pd(allyl)_2]$; [CpPd(allyl)]; $[Ni(CO)_4]$; $[NiCp_2]$; $[Ni(MeCp)_2]$
	Ti	Adhesion layers, metal foils, corrosion resistant coatings	$[TiI_4]$, $[TiBr_4]/H_2$
	Cr	Corrosion protection	$[CrF_2]$, $[CrCl_2]$, $[Cr(CO)_6]$, $[CrPh_2]/H_2$
	Mo	IC contact and gate metallization, wear resistant coatings, infrared reflector, laser mirror coating	$[MoF_5]$, $[MoCl_5]$, $[Mo(CO)_6]$
	Ru	IC contact and gate metallization, diffusion barriers	$[Ru_3(CO)_{12}]$, $[RuCp_2]$, $[Ru(acac)_2]/H_2$
	Ta	Capacitors, resistors, corrosion resistant coatings	$[TaCl_5]$, $[TaF_5]$, $[Ta(CO)_5]/H_2$
Metal nitrides	AlN	Surface acoustic wave devices, packaging material in microelectronic devices	$[AlBr_3]/[NH_3]$, $[AlMe_3/NH_3]$; $[Al(NMe_2)_3]/[NH_3]$; $[AlMe_3(NH_3)]$ adduct
	Si_3N_4	Chemical passivation, encapsulation of silicon bipolar and MOS devices	$[SiH_4]/[NH_3]$; $[Si_2Cl_6]/[NH_3]$; $[Si(NMe_2)_{4-n}H_n]$
	TiN	Wear-resistant and friction reducing coatings, transparent optical films, energy efficient windows, decorative coatings, low resistant contacts, diffusion barriers, and gate electrodes in microelectronic devices	$[TiCl_4]/[NH_3]$, $[TiI_4]/[NH_3]$; $[Ti(NR_2)_4]/[NH_3]$
	ZrN, HfN	Hard coatings for machine tools, diffusion barriers and gate electrodes in microelectronics	$[M(NR_2)_4]/[NH_3]$; $[M(NEt_2)_4]/[Me_2NNH_2]$; $[M(NR_2)_4/N_2]$(plasma-assisted)
	TaN, NbN	Diffusion barriers for Cu in ICs; gate electrodes	$[TaCl_5]/[NH_3]$, $[TaBr_5]/[NH_3]$; $[Ta(NMe_2)_5]/[NH_3]$, $[Ta(NEt_2)_5]/[NH_3]$; $[Ta(NEt_2)_3(N\text{-}Bu^t)]/[NH_3]$; $[Ta(NEt_2)_3(NCMe_2Et)]/[NH_3]$

(*Continued*)

Table 1.3 *(continued)*.

General class of material	*Specific material*	*Applications*	*Typical CVD precursors*
			$[NbCl_5]/[NH_3]$; $[NbCl_5]/[Me_2NNH_2]$; $[Nb(NR_2)_5]/[NH_3]$, $[Nb(NR_2)_4]/[NH_3]$
	WN	Diffusion barriers, gate electrodes in microelectronic devices	$[WCl_6]/[NH_3]$, $[WF_6]/[NH_3]$; $[W(CO)_6]/[NH_3]$; $[W(NBu^t)_2(NR_2)_2]/[NH_3]$; $[(RN)WCl_4(NCR)]/[NH_3]$; $[W_2(NMe_2)_6]/[NH_3]$
	MoN		$[MoF_6]/[NH_3]$, $[Mo(CO)_6]/[NH_3]$; $[Mo(NMe_2)_4]/[NH_3]$; $[Mo(NBu^t)_2(NMe_2)_2]/[NH_3]$
Metal carbides	TiC	Hard coating for cutting, milling, forming and stamping tools	$[TiCl_4]/[CH_4]$; $[Cp_2TiCl_2]/[H_2]$
	ZrC	Coating for nuclear fuel particles	$[ZrCl_4]$, $[ZrBr_4]/[CH_4]$ or $[C_3H_{11}]$
	HfC	Oxidation-resistant coating for composites, coating for superalloys	$[HfCl_4]/[CH_4]$
	Cr_7C_3, Cr_3C_2	Corrosion- and wear-resistant coatings	$[CrCl_3]$/n-butane; $[Cr(C_6H_5\text{-i-Pr})_2]$; $[CrBu^t_2]$
	WC, W_2C, W_3C	Tool coatings, catalysts	$[WF_6]$/cyclopropane; $[WCl_6]/[CH_4]$
	V-carbide	Wear- and corrosion-resistant coatings	$[VCl_4]/[CH_4]$; $[VCp_2]$
	Ta, Nb-carbide	Protective coatings	$[TaCl_5]/[MeCl]/[H_2]$; $[NbCl_5]/[CCl_4]/[H_2]$
	SiC	Blue LEDs, heat sinks, protective coatings	$[MeSiCl_3]/[H_2]$
	GeC	Optical coatings	$[C_nH_{2n+2}]/[GeH_4]$

Abbreviations: β-dik=β-diketonate, acac=acetylacetonate, thd = tetramethylheptanedionate, hfac=hexa fluoroacetylacetonate (Ch. 8, Table 8.2), $C_p=C_5H_5$, $RC_p=RC_5H_4$, 1,5-COD=1,5 cyclooctadiene.

significantly influenced by CVD process parameters such as growth temperature, partial pressures of gaseous species and the total pressure of the system. A detailed description of how such parameters affect the microstructure of CVD-grown layers is not yet available, but a few useful reviews describe the current state of understanding.[86–89] Three main types of morphology are briefly discussed below. Various analytical techniques are used to determine layer morphology. Scanning electron microscopy (SEM) and atomic force microscopy (AFM) give information about crystallinity to the sub-micrometer level, whilst crystalline orientation can be determined by X-ray diffraction (XRD), which also allows the observations of the transition when amorphous layers, with no features in XRD, crystallize to give clearly defined X-ray diffraction peaks.

1.7.1.1 Epitaxial Layers

Epitaxial layers are single crystal films grown by lattice matching the crystalline spacings of the film and underlying substrate. To achieve this it is important that the substrate is free from defects and surface contamination. Also for epitaxy, low growth rates are required so that surface diffusion is fast relative to the arrival of new growth initiating species on the surface. This allows adsorbed species to diffuse to step growth sites and form a layer that replicates the underlying substrate. Growth at high temperatures (typically $>700\ ^{\circ}C$) is generally required to promote epitaxy since this increases the desorption of impurities as well as the surface mobility of adsorbed precursor species. Growth on a substrate of the same material is called homoepitaxy and growth on a different material, but with very similar crystal structure, is known as heteroepitaxy. A good review on the epitaxial growth of silicon by CVD has been published,[28] and the importance of epitaxial growth of compound semiconductors is illustrated in Chapter 6 of this book.

1.7.1.2 Amorphous Layers

The formation of amorphous films is promoted by very high growth rates and low substrate temperatures, where the arrival of film precursors is much more rapid than the diffusion of surface species. However, care has to be taken not to leave films grown in the amorphous state in a hot reactor for any length of time since this can lead to annealing and crystallization.

1.7.1.3 Polycrystalline Layers

Polycrystalline layers are often deposited at intermediate temperatures and growth rates between those used for growth of single crystal and amorphous films. Polycrystalline growth is facilitated by polycrystalline surfaces where nucleation occurs at many different surface sites, leading to growth of islands that coalesce to form a polycrystalline layer. The control of the size and nature of the crystallites is important in determining the properties of CVD films.

In Figure 1.17, the SEM data for an amorphous data film of Hf-silicate are compared to the SEM data for a polycrystalline film of HfO_2. The columnar microstructure of the oxide film is in marked contrast to the featureless structure of the Hf-silicate film. The transition from an amorphous to a polycrystalline phase is illustrated by XRD data for a HfO_2 film deposited at various substrate temperatures (Figure 1.18), from which it can be seen that the crystallinity of the material is a strong function of growth temperature, with the polycrystalline monoclinic phase forming at $450\ ^{\circ}C$ and above.

1.7.2 Layer Properties

The aim of depositing a film by CVD is to obtain a functional material with specific mechanical, electrical, magnetic, optical or chemical properties, or a combination of several of these properties.

(a)

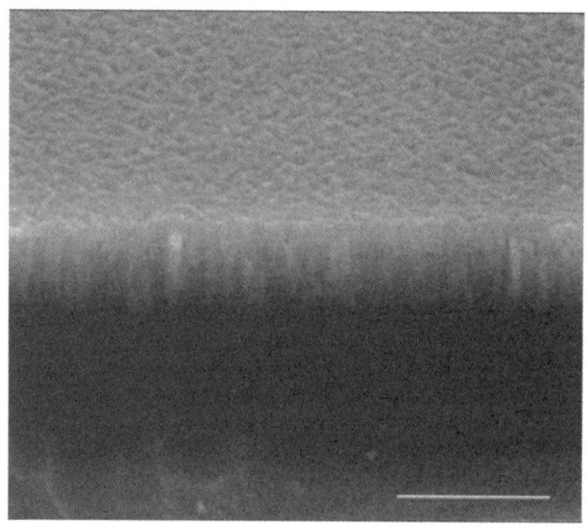

(b)

Figure 1.17 SEM data for (*a*) an amorphous HF-silicate film and (*b*) a polycrystalline HfO_2 film.

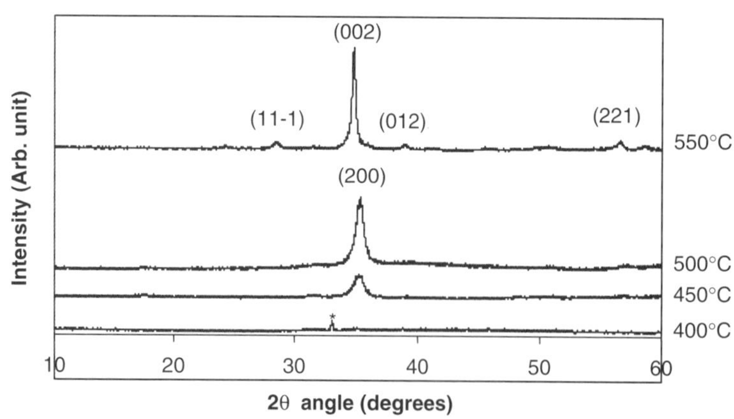

Figure 1.18 Variation of crystallinity of HfO_2 grown by MOCVD with growth temperature.

Layer properties can be strongly influenced by several film characteristics common to all CVD-grown layers.

1.7.2.1 Layer Thickness and Density

The thickness of layers deposited by CVD can vary over a wide range, from single atomic layers (*e.g.* in ALD), through a few nanometres in, for example, multi-quantum well structures, to hundreds of nanometres for optical and semiconductor applications, and even to greater than 100 μm for some applications such as wear resistant coatings. Layer thickness is dependent on both the CVD technique employed and the range of deposition parameters.

Thick films result from high growth rates often associated with high temperatures and pressures of many thermal CVD and PECVD processes, whilst thin films are obtained from the low growth rates and nano-scale control that are a feature of ALD and CBE techniques. Layer thickness is generally measured after deposition by optical techniques (*e.g.* interferometry and ellipsometry), scattering techniques (*e.g.* Rutherford scattering and β back scattering), electromechanical profilometry or microscopy (using an optical microscope or scanning electron microscope after layer sectioning). Simple weighing of the substrate using an accurate microbalance before and after film deposition can also be used to measure layer thickness, provided the density of the layer is known. The measure of weight variation as a function of time using a quartz crystal oscillator, or by deposition on a substrate suspended from the arm of a microbalance, allows the *in situ* monitoring of layer growth.

Film density is a measure of the quality and potential functionality of the film, with low density films indicative of high porosity and the incorporation of impurities in the crystal lattice, possibly due to incomplete decomposition of the precursor, as often observed in films grown at low temperature (*e.g.* by ALD). Film density generally increases as the growth temperature increases, or by high temperature annealing of the film, and the refractive index of the film can give a measure of film density, with dense films having high refractive indices.

1.7.2.2 Layer Adhesion

For any practical use, it is essential that a CVD film adheres well to the substrate. However, the factors that influence film adhesion are not totally understood, although it is generally accepted that to promote good film adhesion it is important that the substrate surface is thoroughly cleaned before film growth. The reason for this is that impurities on the surface can inhibit deposition of a coating and can also significantly affect the mode and amount of nucleation of depositing material. The measurement of adhesion is still very much based on empiricism, as it is difficult to quantitatively model the forces governing adhesion. Consequently, for example, qualitative assessment of how well a film is adhering to a substrate is made by the rather unsophisticated technique of the "scratch test"[90] and the even less sophisticated "Scotch tape test."[91] The real test, though, of any film adhesion comes when the layer is put to a practical application. A good review of the qualities necessary in effective protective coatings is available.[92]

1.7.2.3 Layer Composition and Purity

Layer composition and purity are vital factors in determining the functionality of a CVD film. Layer composition can be measured by several techniques such as Auger electron spectroscopy (AES), X-ray photoelectron spectroscopy (XPS), Rutherford back scattering (RBS), time-of-flight elastic recoil spectrometry (TOF-ERDA), energy or wavelength dispersive X-ray analysis (EDX or WDX) and secondary ion or secondary neutral mass spectrometry (SIMS or SNMS). The first five of these techniques can determine bulk stoichiometry of a layer and impurities, such as carbon and nitrogen, down to levels of about 1 at.%. SIMS is a much more sensitive technique, capable of detecting impurities in the film down to the ppm, or even ppb, levels, although quantitative analysis needs careful calibration samples to be made. SNMS, on the other hand, can measure from the high atomic % down to ppm much more readily.

The reason that it is important to be able to determine layer composition at very low levels is that even very small traces of an impurity species can dramatically affect layer characteristics. This is true for layers that are grown for optical, mechanical, chemical and magnetic applications, but it is particularly so for materials that are deposited with specific electrical properties in mind. This is illustrated in the next section, which considers the electrical characteristics of compound semiconductor layers.

A rather specialized technique used to investigate the thickness, composition and crystallinity of thin films is medium energy ion scattering (MEIS). This involves bombarding the film with energetic He^+ ions ($\sim$200 keV), and then analysing the energy of the ions scattered from the film. Owing to the inelastic scattering that the He^+ ions undergo as a function of depth below the sample surface, the energy distribution gives an effective depth profile of the target atoms. MEIS also provides valuable information on the extent of interaction between the film and the substrate, particularly valuable in assessing the stability of dielectric oxide films on Si for CMOS and DRAM applications, with film crystallization often leading to an increased film–substrate interaction.

The method used for determining the properties and functionality of a thin film clearly depends on the specific application. However, due to the increasing importance in microelectronics of thin films grown by CVD, some methods used for the characterization of two types of film from this area of technology are discussed below.

1.7.2.4 Electrical Characterization of Compound Semiconductor Films

The principal electrical parameters used to characterize III–V or II–VI compound semiconductor materials are the carrier concentration, n (*i.e.* the number of charge carriers per cm^3) and the carrier mobility, which is a measure of the ease with which the carriers can move under the influence of an electrical field. The mobility of the charge carriers in a semiconductor is a more complicated function of temperature than the conductivity of a metal because, besides the temperature dependent scattering processes found in metals the actual number of carriers (n) and their energy distribution are temperature dependent.

The carrier concentration and carrier mobility in a semiconductor can be determined from the conductivity of the film, which is derived from resistivity measurements on the layer using the four-point probe method[93] and Hall measurements. The measured value of n represents the *net* value of residual charge carriers and is thus a balance between the total concentration of ionized donor (N_D) and acceptor impurities (N_A). Once the carrier mobility has been determined, it is possible to determine the concentrations of ionized donor and acceptor impurities and the ratio N_A/N_D (the compensation ratio) is often used as an indication of layer purity. For n-type material the residual electron concentration is often expressed as n, although the term $(N_D - N_A)$ is more accurate, and residual hole density can be expressed as p or $(N_A - N_D)$.

High purity compound semiconductor material will have a low residual carrier concentration and a correspondingly high carrier mobility. For instance a good quality GaAs layer might have a residual carrier concentration (n) of $1 \times 10^{14}\,cm^{-3}$ and a mobility (μ) at 77 K of $130\,000\,cm^2\,V^{-1}\,s^{-1}$, with the material having a relatively low level of donor impurities ($N_D = 1.4 \times 10^{14}\,cm^{-3}$) and an even lower level of acceptor impurities ($N_A = 0.4 \times 10^{14}\,cm^{-3}$). The low compensation ratio (N_A/N_D) of 0.28 shows that the film is high purity. By contrast, a low purity sample of GaAs may still have a low *residual* carrier concentration (n) of, for example, $1 \times 10^{14}\,cm^{-3}$, but a low electron mobility (*e.g.* 26 500 $cm^2 V^{-1} s^{-1}$) at 77 K reflects the low purity of the sample. This is due to the presence of high levels of donor *and* acceptor impurities ($N_D = 2 \times 10^{15} cm^{-3}$; $N_A = 1.9 \times 10^{15} cm^{-3}$), which lead to the very low charge mobility because of impurity scattering. In this case, the compensation ratio of 0.96 is high.

The dramatic effect of impurities on layer properties can be seen by relating carrier concentrations to the level of impurity incorporated in the layer. For example, a GaAs layer with a carrier mobility of 130 000 $cm^2 V^{-1} s^{-1}$ and a carrier concentration of $1 \times 10^{14} cm^{-3}$ has one carrier atom per 2.167×10^8 gallium atoms, or 4×10^{-7} mol.% of carriers.

1.7.2.5 Electrical Characterization of Dielectric Thin Films

A detailed review covering the physics of metal-oxide semiconductors has been published.[94] However, in view of the great importance of dielectric films in microelectronics applications,[95] [see, for example, Chapters 4 (Section 4.5.3) and 8 (Section 8.3)] it may be useful to briefly describe how the films are frequently characterized.

For dielectric thin films, key parameters are the permittivity or dielectric constant (κ) and leakage current. For insulating gate oxides (*e.g.* SiO_2, HfO_2) deposited on a Si substrate, high frequency capacitance–voltage (*C-V*) data on a metal/oxide/semiconductor capacitor (MOSC) can be used to determine film quality.[96] Figure 1.19 shows the high-frequency *C-V* data for a [Al/HfO_2/SiO_2/p-Si] MOSC structure, before and after annealing. The HfO_2 dielectric layer was deposited by ALD at 360 °C using alternating pulses of [Hf(mmp)$_4$] (mmp = $OCMe_2CH_2OMe$) (Chapter 8, Section 8.3.1.2) and water.[97]

The *C-V* curves show features characteristic of a well-insulating dielectric material with a high capacitance *accumulation region* at negative bias voltages, a *depletion region* of decreasing capacitance, and a low capacitance *inversion region* at positive biases. The capacitance of the MOSC structure is a series combination of metal oxide capacitance and the Si semiconductor depletion layer capacitance. At negative bias voltages, the majority charge carriers (holes) in the p-type Si are attracted towards the gate oxide. The holes accumulate at the HfO_2/Si interface and

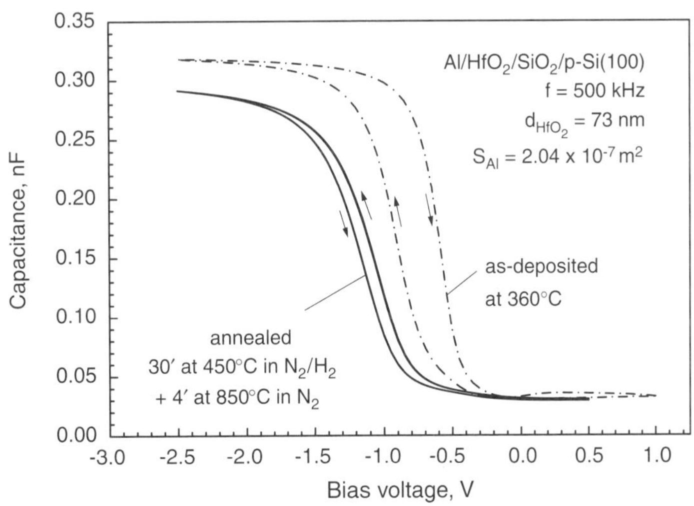

Figure 1.19 High-frequency (500 kHz) *C-V* data for a [Al/HfO_2/SiO_2/p-Si] MOS structure fabricated by ALD using [Hf(mmp)$_4$] and H_2O. (Reprinted with permission from ref. 97. Copyright 2003, American Chemical Society.)

there is no depletion region. As a result, the measured capacitance is close to the HfO_2 film capacitance and the effective dielectric constant, or permittivity, (κ) of the HfO_2 film is given by Equation (1.3):

$$\kappa = C_{ox} d / \varepsilon_0 A \tag{1.3}$$

where C_{ox} = capacitance value in accumulation region, d = oxide layer thickness, A = area of gate contact (in this case 1.96×10^{-7} m^2) and ε_0 = permittivity of free space (8.854×10^{-12} F m^{-1}).

As the gate voltage becomes more positive, the positive gate electrostatically repels the holes away from the HfO_2/Si interface. A depletion zone penetrates more deeply into the Si semiconductor and the depletion capacitance becomes smaller, reducing the overall capacitance of the MOSC stack. The depletion layer broadens, until there is an accumulation of minority charge carriers (electrons) at the Si/insulating oxide interface. The accumulated minority carrier layer is called the inversion layer, because the original polarity of the surface has now become inverted. In equilibrium conditions, the minority carriers in the inversion layer inhibit further penetration of the electric field in the semiconductor, leading to the minimum capacitance values in the C-V curve.

The amount of hysteresis in the C-V curve during a backwards and forwards voltage sweep across the capacitor provides information about the amount of mobile ionic charge (*e.g.* H^+ ions) and/or slow states in the dielectric oxide film. The flatband voltage shift (ΔV_{FB}) is the difference in bias voltage measured at the flat band capacitance in the depletion region in the C-V curve from the flatband voltage calculated for the ideal case (~–0.9 to –1.0 V), and is a measure of interface quality, indicating the amount and nature of fixed charge at the dielectric/Si interface. The C-V data for the [Al/HfO_2/SiO_2/p-Si] structure shown in Figure 1.19 clearly show that the quality of the HfO_2 film improves after annealing, with a marked decrease in hysteresis and a reduced flatband voltage shift, indicating a reduction in mobile ions and interface fixed charge.

The C-V data for an n-type MOSC such as [Al/HfO_2/SiO_2/n-Si], in which the majority charge carriers in the Si are electrons, are essentially a mirror image (in the y-axis) of the p-type MOSC curve shape, with the accumulation region occurring at positive polarities and the inversion region at negative polarities.

When comparing different dielectric materials, and especially high-κ dielectric oxide films used in gate dielectric and capacitor applications (Chapter 8, Section 8.3), the films are often characterized by quoting the "equivalent oxide thickness" (EOT), which is the theoretical thickness of SiO_2 that would provide the same capacitance density as the alternative high-κ dielectric (ignoring the effect of leakage current and reliability).

Current–voltage (I-V) data obtained on MOSC structures are also frequently used to assess the quality of an insulating dielectric film. I-V data give a measure of the leakage of current through the film at a particular applied voltage and they are critically dependent on film thickness, polarity of applied voltage, gate electrode material and the type and nature of defects in the film, with thin films generally giving higher leakage currents than thick films of the same material. As an approximate guide, leakage current densities for a thin film (*e.g.* ~5 nm) of a high quality dielectric oxide are generally below $\sim 5 \times 10^{-5}$ A cm^{-2}. For instance, Figure 1.20 shows a plot of leakage current density (J) against electric field (E_{ox}) for a series of HfO_2 films grown by liquid injection MOCVD and ALD using [(MeCp)$_2$HfMe(OR)] (OR = OPr^i, mmp).[98] The data show that all films showed good electrical integrity ($< 2 \times 10^{-5}$ A cm^{-2}) at 1 MV cm^{-1}, although it must be noted that the films grown by ALD are much thinner (1.7 and 9.6 nm) than those grown by MOCVD (76 and 102 nm), indicating that they are better insulators. At higher electric fields (~2.5 MV cm^{-1}) the leakage current density of three of the films rose considerably to $\sim 1 \times 10^{-4}$ to 5×10^{-3} A cm^{-2}, indicating some breakdown in the integrity of these HfO_2 films.

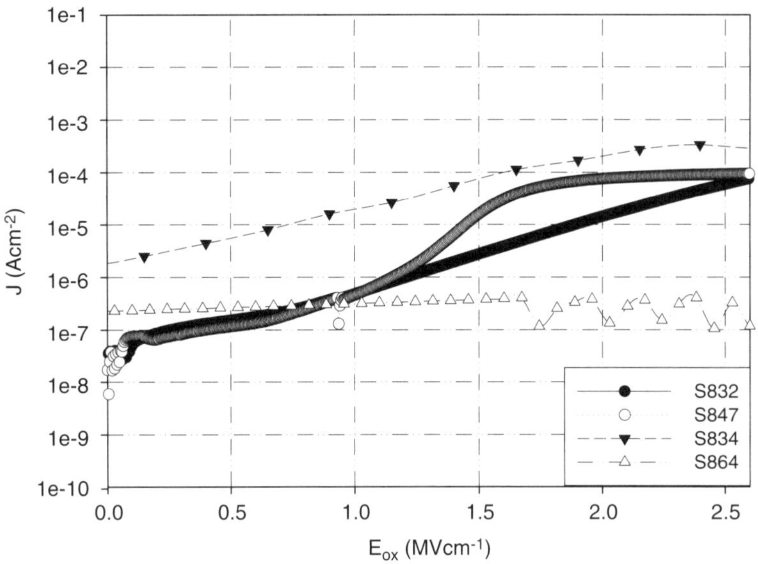

Figure 1.20 Leakage current density (J) versus electric field (E_{ox}) for HfO_2 films deposited by MOCVD (●,○) (film thicknesses 76 and 102 nm) and ALD (▼, △) (film thicknesses 1.7 and 9.6 nm). (After ref. 98, Copyright Wiley-VCH Verlag GmbH & Co. KGaA, 2007. Reproduced with permission.)

1.8 Postscript

Although it is impossible in a chapter of this length to give a comprehensive overview, it is hoped that this brief description of some of the main elements of CVD gives a flavour of the complex interweaving threads associated with the technology. For more detailed descriptions of the many various aspects of CVD processes read on!

References

1. G. B. Stringfellow, *Chapter 1 in Organometallic Vapor Phase Epitaxy*, Academic Press, New York, 1989.
2. M. L. Hitchman and K. F. Jensen, in *Chemical Vapor Deposition*, eds. M. L. Hitchman and K. F. Jensen, Academic Press, New York, 1989, Chapter 1.
3. T. T. Kodas and M. J. Hampden-Smith, *Chapter 1 in The Chemistry of Metal CVD*, VCH, Weinheim, 1994.
4. A. C. Jones and P. O'Brien, in *CVD of Compound Semiconductors*, VCH, Weinheim, 1997, and references therein.
5. R. L. Moon and Y.-M. Houng, in *Chemical Vapor Deposition*, Eds. M.L. Hitchman and K.F. Jensen, Academic Press, New York, 1989, Chapter 6.
6. E. M. Sherwood and J. M. Blocher, *J. Metals*, 1965, **17**, 594.
7. C. F. Powell, J. H. Oxley and J. M. Blocher, *Vapour Deposition*, The Electrochemical Society, Pennington NJ, 1966.
8. J. F. Gallie, *Petroleum Refiner*, 1944, **23**, 176.
9. W. E. Sawyer and A. Mann, *US Patent 229,335* (1880).
10. J. W. Aylsworth, *US Patent, 553,296* (1896).
11. A. de Lodyguine, *US Patent 575,002* (1897).

12. F. Wöhler and L. Uslar, *Lieb. Ann.*, 1855, **94**, 255.
13. L. Mond, C. Langer and F. Quincke, *J. Chem. Soc.*, 1890, **57**, 749.
14. L. Mond, *U.S. Patent, 455,230*, 1891.
15. R. Mond, *Chem. Ind.*, 1930, **49T**, 371.
16. J. N. Pring and W. Fielding, *J. Chem. Soc.*, 1909, **95**, 1497.
17. R. Hoelbling, *Z. Angew.Chem.*, 1927, **40**, 655.
18. G. K. Teal, J. R. Fisher and A. W. Treptow, *J. Appl. Phys.*, 1946, **17**, 879.
19. K. H. Storks and G. K. Teal, *US Patent 2,441,603* (1948).
20. M. J. Cook, R. A. Heinecke, R. C. Stern and J. W. Maes, *Solid State Technol.*, 1982, **25**(12), 62.
21. M. L. Green, R. A. Levy, R. G. Nuzzo and E. Coleman, *Thin Solid Films*, 1984, **114**, 367.
22. T. P. Whaley and V. Norman, *US Patent 3,209,326* (1965).
23. D. R. Carley and J. H. Dunn, *US Patent 3,375,129* (1968).
24. D. L. Schmidt and R. Hellmann, *US patent 3,462,288* (1969).
25. W. L. Gladfelter, D. C. Boyd and K. F. Jensen, *Chem. Mater.*, 1989, **1**, 339.
26. Pilkington Brothers Ltd., *Proceedings of The Royal Society of London Series A-Mathematical And Physical Sciences*, 1969, **314**, 1.
27. R. Glang and B. W. Kippenham, *IBM J. Res. Devel.*, 1960, 299.
28. B. S. Meyerson, in *Chemical Vapor Deposition*, eds. M. L. Hitchman and K. F. Jensen, Academic Press, New York, 1989, Chapter 5.
29. R. Didchenko, J. D. Alix and R. H. Toeniskoettler, *J. Inorg. Chem.*, 1960, **4**, 35.
30. B. Harrison and E. H. Tomkins, *Inorg. Chem.*, 1962, **1**, 951.
31. Monsanto Co. *US Patent Applications 113,108 and 129,919* (1961).
32. Monsanto Co. *UK Patent 1,011,979* (1965).
33. H. M. Manasevit, *Appl. Phys. Lett.*, 1968, **12**, 156.
34. H. M. Manasevit, *U.S. Patent, 4,368,098C*, 1983.
35. P. Rhai Choudry, *J. Electrochem. Soc.*, 1969, **116**, 1745.
36. H. M. Manasevit, *J. Electrochem. Soc.*, 1971, **118**, 647.
37. H. M. Manasevit and W. I. Simpson, *J. Electrochem. Soc.*, 1971, **118**, C291.
38. H. M. Manasevit and W. I. Simpson, *J. Electrochem. Soc.*, 1973, **120**, 135.
39. H. M. Manasevit and W. I. Simpson, *J. Electrochem. Soc.*, 1971, **118**, 1864.
40. Y. Seki, K. Tanno, K. Iida and E. Ichiki, *J. Electrochem. Soc.*, 1975, **122**, 1108.
41. S. J. Bass, *J. Crystal Growth*, 1975, **31**, 172.
42. R. D. Dupuis and P. D. Dapkus, *Appl. Phys. Lett.*, 1977, **31**, 466.
43. S. Nakamura, T. Mukai and M. Senoh, *Appl. Phys. Lett.*, 1994, **64**, 1687.
44. A. D. Berry, D. K. Gaskill, R. T. Holm, E. J. Cukauskas, R. Kaplan and R. L. Henry, *Appl. Phys. Lett.*, 1988, **52**, 1743.
45. *High Temperature Superconductors: Fundamental Properties and Novel Materials Processing*, eds. D. Christen, J. Narayan and L. Scneemeyer, Material Research Society Symposium Proceeding, 1990, Vol. **169**, and references therein.
46. G. J. M. Dormans, P. J. van Veldhoven and M. de Keijser, *J Crystal Growth*, 1992, **123**, 537.
47. A. C. Jones, *J. Mater. Chem.*, 2002, **12**, 2576.
48. T. Suntola and J. Antson, *US Patent 4,058,430* (1977).
49. T. Suntola, *Mater. Sci. Rep.*, 1989, **4**, 261.
50. T. Suntola, J. Antson, A. Pakkala and S. Lindfors, *SID 80 Digest*, 1980, **11**, 108.
51. M. Leskelä and L. Niinistö, Electroluminescence, *Proc. 6th Int. Workshop on Electroluminescence*, Cino Puntos Press, El-Paso, 1992, pp. 249–246.
52. P. O'Brien, in *Inorganic Materials*, Chapter 9, eds. D.W. Bruce and D. O'Hare, Wiley, New York, 1992, p. 500.
53. M. R. Leys, *Chemtronics*, 1987, **2**, 155.
54. R. H. Reep and S. K. Ghandi, *J. Crystal Growth*, 1983, **61**, 499.

55. D. W. Hess and D. B. Graves, in *Chemical Vapor Deposition*, Eds. M. L. Hitchman and K. F. Jensen, Academic Press, New York, 1989, Chapter 7, and refs. therein.
56. V. R. McCrary and V. M. Donnelly, in *Chemical Vapor Deposition*, eds. M. L. Hitchman and K. F. Jensen, Academic Press, New York, 1989, Chapter 8, and refs. therein.
57. M. Leskelä and M. Ritala, in *Handbook of Thin Film Materials*, ed. H.S. Nalwa, Academic Press, 2002, Vol. **1**, p. 103.
58. W. T. Tsang, *J. Crystal Growth*, 1990, **105**, 1.
59. W. T. Tsang, *J. Electron. Mater.*, 1986, **15**, 235.
60. G. J. Davies and D. A. Andrews, *Chemtronics*, 1988, **3**, 3.
61. T. F. Keuch and E. Veuhoff, *J. Crystal Growth*, 1984, **68**, 148.
62. A. C. Jones, H. C. Aspinall, P. R. Chalker, R. J. Potter, T. D. Manning, Y. F. Loo, R. O'Kane, J. M. Gaskell and L. M. Smith, *Chem. Vap. Deposition*, 2006, **12**, 83.
63. M. Fulem K. Růžiča, E. Hulicius, T. Šimeček, K. Melichar, J. Pangrác, S. A. Rushworth and L. M. Smith, *J. Crystal Growth*, 2003, **248**, 99.
64. C. L. Griffiths, A. Stafford, S. J. C. Irvine, N. Maung, A. C. Jones, L. M. Smith and S. A. Rushworth, *Appl. Phys. Lett.*, 1996, **68**, 194.
65. O. Kayser, *Chemtronics*, 1988, **3**, 90.
66. S. A. Rushworth, L. M. Smith, A. J. Kingsley, R. Odedra, R. Nickson and P. Hughes, *Microelectronics and Reliability*, 2005, **45**, 1000.
67. S. Duffy, P. F. Nolan, S. A. Rushworth, A. B. Leese and A. C. Jones, *Adv. Mater. Opt. and Electron.*, 1997, **7**, 233.
68. A. C. Jones, *Chemtronics*, 1989, **4**, 15.
69. S. Fujita, Y. Uemoto, S. Araki, M. Imaizumi, Y. Takeda and A. Sarasaki, *Jpn. J. Appl. Phys.*, 1988, **27**, 1151.
70. R. A. Smith, in *Semiconductors* (Cambridge University Press, Cambridge, 1978) Chapter 3.
71. A. C. Jones, G. Wales, P. J. Wright and P. E. Oliver, *Chemtronics*, 1987, **2**, 83.
72. T. Nakanisi, T. Udagawa, A. Tanaka and K. Kamei, *J. Crystal Growth*, 1981, **55**, 255.
73. A. C. Jones and P. O'Brien, *Chapter 2 in CVD of Compound Semiconductors*, VCH, Weinheim, 1997.
74. A. A. Efremov, V. A. Federov and E. E. Grinberg, translated from *Vysokchiste Veschestra*, No. 3, May-June 1998, pp 5-43, Plenum Publishing Corp. 1989.
75. E. E. Grinberg and A. A. Efremov, *Electron. Tekh. Ser. Mat.*, 1975, **7**, 62.
76. V. G. Makarenko, V. A. Federov and A. A. Efremov, *Electron. Tekh. Ser. Mat.*, 1974, **11**, 71.
77. A. A. Efremov, E. E. Grinberg, Yu M. Fetisov, USSR Pat. 546,617, *Byull. Izobrett.*, 1977, **6**.
78. H. Kodakura, K. Sawara and T. Yako, *U. K. Pat. Application. GB 2,183,651(A)*, (1987).
79. A. C. Jones, D. J. Cole-Hamilton, A. K. Holliday and M. Munir Ahmad, *J. Chem. Soc., Dalton Trans.*, 1983, 1047.
80. A. H. Moore, M. D. Scott, J. I. Davies, D. C. Bradley, M. M. Faktor and H. Chudzynska, *J. Crystal Growth*, 1986, **77**, 13.
81. D. C. Bradley, H. Chudzynska, M. M. Faktor, D. M. Frigo, M. B. Hursthouse, B. Hussein and L. M. Smith, *Polyhedron*, 1988, **7**, 1289.
82. A. C. Jones, *Chem. Soc. Rev.*, 1997, **26**, 101.
83. A. C. Jones, *J Crystal Growth*, 1993, **129**, 728.
84. M. de Keijser and G. J. M. Dormans, *Mater. Res. Soc. Bulletin*, 1996, **21**, 37.
85. I. M. Watson, *Chem. Vap. Deposition*, 1997, **3**, 9.
86. J. Bloem, *J. Crystal Growth*, 1980, **50**, 581.
87. G. H. Gilmer and J. Q. Broughton, *Ann. Rev. Mater. Sci.*, 1986, **16**, 487.
88. J. A. Venables, G. D. T. Spiller and M. Hanbrucka, *Rep. Progr. Phys.*, 1984, **47**, 399.
89. A. Madhukar and S. A. Chaisas, in *CRC Crit. Rev. Soild State Mater. Sci.*, 1988, **14**, 1.
90. H. E. Hintermann, *Ann. C I R P*, 1982, **31**, 405.

91. S. Wolf and R. N. Tauber, in *Silicon Processing for the VLSI era* (Lattice Press, Sunnat Beach, CA, 1986).
92. G. Wahl, in *Chemical Vapor Deposition*, eds. M. L. Hitchman and K. F. Jensen, Academic Press, New York, 1989, Chapter 10, and refs. therein.
93. L. J. van der Pauw, *Phys. Tech. Rev.*, 1958, **20**, 220.
94. E. R. Nicollian and J. Brews, in *MOS Physics and Technology* (Wiley, 1982).
95. G. D. Wilk, R. M. Wallace and J. M. Anthony, *J. Appl. Phys.*, 2001, **89**, 5243.
96. S. Taylor, P. A. Williams, J. L. Roberts, A. C. Jones and P. R. Chalker, *Electron. Lett.*, 2002, **38**, 1285.
97. K. Kukli, M. Ritala, M. Leskelä, T. Sajavaara, J. Keinonen, A. C. Jones and J. L. Roberts, *Chem. Mater.*, 2003, **15**, 1722.
98. R. O'Kane, J. Gaskell, A. C. Jones, P. R. Chalker, K. Black, M. Werner, P. Taechakumput, S. Taylor, P. N. Heys and R. Odedra, *Chem. Vap. Deposition*, 2007, **13**, 609.

CHAPTER 2
CVD Reactors and Delivery System Technology

SUSAN P. KRUMDIECK

Department of Mechanical Engineering, Private Bag 4800, University of Canterbury, Christchurch 8014, New Zealand

2.1 CVD System Overview

From the engineering perspective, the CVD system is essentially a "molecular scale" manufacturing facility. Just like any factory, the production rate is determined by the slowest step in the manufacturing process. The product quality, production economy and efficiency are assured by initial process design and equipment selection and by continuous monitoring and control of every aspect of the complex manufacturing system. However, unlike the typical manufacturing plant, we do not readily have access or much direct information about what is happening in the assembly area. It often seems that all the action in CVD occurs at the deposition surface: adsorption, reaction, incorporation, crystal growth. However, a successful deposition process is the outcome of well designed and controlled delivery, transport, thermal and exhaust management systems. This is the unique challenge of CVD on three levels – reactor design, system platform and process control – to achieve molecular-scale control at the deposition surface through macro-scale control in the systems outside the reaction chamber.

This chapter is set out as a guide for the majority of CVD practitioners and researchers who are more focused on the molecules and materials than on the machines. You do not want to take up mechanical or chemical engineering as a second career, you just want to cook up new precursors and materials, analyze their properties and develop exciting new thin film devices. Yet you could pose a danger to yourself and others, and you could easily waste a lot of time and money, if you do not have a basic understanding of what is going on in the thin film production machine, often called the *CVD tool*. If you are working in industry, you will surely be using one of the engineering masterpieces from CVD tool manufacturers like Aixtron, Applied Materials, Novellus, Genus, CVD Equipment Corporation, ASM International, or Sierra Therm. The basic approach of this chapter should help you understand how the CVD system works, make some sense of the technical documentation and communicate with the equipment engineers.

Chemical Vapour Deposition: Precursors, Processes and Applications
Edited by Anthony C. Jones and Michael L. Hitchman

Published by the Royal Society of Chemistry, www.rsc.org

A good deal of device R&D takes place in existing industrial facilities, using commercial systems that were specifically developed to deliver a particular step in the chip manufacture process. Most manufacturers make allowances for flexibility in processing. For example, Applied Materials' modular system platforms allow small batch runs of one or two wafers in machines that are capable of full-scale production in the range of 100 wafers per hour (wph). If you are one of these researchers, and you have ever thought about running the system outside of the specified ranges, you should definitely read this chapter.

If you are lucky enough to be carrying out research at one of the top-flight universities or industrial research centers, then you may be using a research or niche market-scale commercial system such as the Thomas Swan Scientific (TSS) GaN MOCVD system at Cambridge. Companies like Aixtron, Varian, Jipelec, Oxford Instruments, and Picosun also manufacture smaller CVD, MOCVD and ALD systems for R&D. However, a significant portion of basic research is carried out with systems built in-house, by post-graduate students and technicians. It would be highly informative for you "do-it-yourself" system designers to be able to see the details of the well-engineered commercial systems. Unfortunately, this chapter will only be able to describe the different CVD reactors in a generic sense according to their fundamental operating principles and process control strategies. An impressive research and development effort goes into each industrial CVD system, and it is understandable that companies in a highly competitive environment are not enthusiastic about publishing their trade secrets in a handbook.

The introductory section of this chapter sets out the performance considerations for CVD operations, Section 2.2 explains the basics of system design, and Section 2.3 explains component operation and selection. The rest of the chapter examines the operating principle and system configurations for conventional CVD reactors, liquid injection and aerosol spray delivery systems, atomic layer deposition (ALD) and chemical beam epitaxy (CBE) reactors.

2.1.1 CVD System Performance

Millions of dollars worth of electronic and photonic devices, solar photovoltaic cells and other products are manufactured using commercially mature CVD technology. It is important that researchers understand the performance requirements for commercial CVD processes. All new research developments will eventually have to meet technical and economic feasibility requirements to bridge the "valley of death" between innovation and commercial production. There are three main performance aspects of any CVD manufacturing process:

- quality
- uniformity
- throughput.

Table 2.1 outlines the details of these process performance requirements. CVD system engineering aims to optimize all three performance requirements, but each technology platform will involve compromise of some sort. For example, the Aixtron planetary MOCVD reactor has undergone extensive computer modeling and design development to optimize the cost of ownership (CoO), throughput and yield.[1] The precursor conversion efficiency and consumption of high purity gases are important cost factors. The reactor geometry, flow and temperature field are designed to minimize parasitic precursor losses on hot surfaces and to suppress pre-reactions in the gas phase by minimizing the mixing region. Stable operation is important for uniformity and quality, and a cooled ceramic gas injector nozzle is used in the Aixtron system to eliminate deposition on the nozzle, which would change the flow characteristics over time. The elimination of gas-phase reactions also improves quality by reducing particle contamination, but the low pressures that

Table 2.1 CVD performance considerations.

Quality	*Uniformity*	*Throughput*
Composition of thin film material, and interfaces	Yield: number of devices within specifications per wafer and per production cycle	Production rate: wafers per hour (wph) factor of deposition rate and cycle time
Crystal structure, orientation or epitaxy	Thickness uniformity across the wafer surface	Pump-down, bake-out, and heat-up time on start-up for each production cycle
Cleanliness-free from environmental particulates, powder formation or flakes from walls	Wafer-to-wafer thickness uniformity	Equipment capital and cost of operation (CoO) per production unit. Fab Lab Footprint
Lattice defects, contamination or doping	Uniformity of composition and phase	Precursor conversion efficiency
Material phase	Run-to-run: process reliability and repeatability	Waste production, treatment and handling
Adhesion and film service life	Conformality or step-coverage on features and shapes	Equipment cleaning and maintenance, drift in measurement or operation
Thermal or residual stress in film or substrate	Microstructure and grain size uniformity	Safety, clean-room, and power service failure risks

reduce particle formation also mean slower growth rate. The uniformity is greatly affected by the reactor gas velocity and pressure and the Aixtron approach is to tune the mechanical satellite manipulation of substrates to gas flow conditions to optimize uniformity and efficiency.

Commercial systems must also balance engineering for robust operation with the cost of the equipment. This is the usual cost *vs.* quality engineering problem, but for semiconductor manufacturing quality is paramount and the equipment is expensive. The industry measures of robust operation are the mean time between failure (MTBF) and the "utilization factor", which is reported as a percentage (actual production run time/scheduled production run time). All systems must have regular maintenance for cleaning, replacing heating and other components, and testing measurement systems. A well engineered machine will have a minimized number of components, will suffer few unexpected breakdowns and will have easy access and rapid replacement components so that down-time is minimized during scheduled maintenance.

The equipment footprint is another aspect of cost. All commercial equipment manufacturers dedicate engineering effort to reducing the equipment footprint in the semiconductor fabrication facility, often called the "fab lab." This is because the fab lab is usually a clean-room facility. Clean-rooms attract extraordinary construction and operating costs per square meter of floor space with maintenance, cooling and air handling energy consumption often more than ten times higher than ordinary office space. The throughput, measured in wafers per hour (wph), depends not only on the deposition rate but also on the time it takes to load wafers into the machine, heat up the reaction chamber and establish stable gas flows. Mean time between runs (MTBR) is the industry measure of this aspect of system design and operation. Most high volume production systems have features like clean-room interfaces where many wafers can be loaded then vacuum purged and, in some cases, even preheated. Another trend is robotic wafer loading, which allows numerous deposition runs to be carried out without opening or cooling down the reactor. Applied Materials has integrated rapid cleaning cycles, and there are industry-wide trends to combine several of the device manufacturing steps into the same reactor platform to maximize productivity and fab lab space utilization. In general, all commercial equipment suppliers are continuously pursuing optimization and system engineering advances to meet ever more stringent performance requirements, to increase wph, and to minimize footprint, CoO, MTBF and MTBR.

2.1.2 Historical Perspective on Reactor Diversity

Several decades ago, different approaches to film deposition were being explored that could be classified by pressure range and heating strategy.[2] Atmospheric pressure CVD gave high growth rates, but uniformity was difficult due to the diffusion limited growth regime. APCVD also tended to have quality issues due to thermal gradients and particle formation in the gas. Low pressure CVD offered better uniformity due to higher diffusivity in the gas. Lower temperature processing with surface reaction kinetics being the rate-controlling step also provided good uniformity. However, the challenges with these approaches were then increasing the throughput and reducing the cost per unit. The extremely complex and expensive ultrahigh vacuum approach of MBE provided superior film quality, but major innovations were required to achieve commercial production rates.

The hot-wall reactor, where the whole reactor is essentially in an oven, provided good uniformity of the deposition temperature, but presented challenges in quality and throughput. Deposition on hot surfaces other than the substrate could become a source of particle contamination and increased costs due to cleaning and lower precursor conversion efficiency. The cold-wall reactor was not as susceptible to particle contamination, and could often be constructed at lower cost as lower temperature reactor components could be used. However, the temperature field in the reactor presented engineering challenges for controlling the flow field and heating the substrate uniformly.[3] The plasma enhanced heating systems were more expensive and could cause substrate and chamber damage, but allowed lower substrate temperature.

Today, all of the CVD reactor configurations as well as physical vapor deposition are still in use, and different kinds of CVD tools are often integrated into the production and device fabrication process. Research continues into engineering and control improvements and modeling of heat and mass transport for new reactor designs. New ideas for system components such as precursor delivery, automation and reactor cleaning are active areas of industrial research. The CVD equipment suppliers tend to focus on related system platforms. For example, the Producer® platform by Applied Materials is an industry leader in electronic IC devices. The Producer® is a modular system for simultaneous deposition on two large wafers that are vacuum lock-loaded from the clean room, pre-heated and robot-manipulated in and out of the ceramic lined deposition chambers. The AM Producer® platform also includes integrated rapid cleaning using NF_3 gas, built in metrology, and control and factory interface software. The engineering development in this LPCVD platform has been adapted to specialized machines such as the SE for dielectric film deposition, and the HARP for transistor gap fill which has an optional integrated Chemical Mechanical Planarization (CMP) process. The Aixtron Planetary Reactor® platform is an MOCVD system for LED fabrication. Because LEDs are currently manufactured from the smaller wafers, Aixtron has worked to increase throughput by increasing the number of wafers in the reactor during each deposition run.

There is not really a competition between reactor types for the same devices, because each product, whether an LED or an IC, requires very specific processes. It is not likely that any one particular CVD configuration (*e.g.* MOCVD, ALD, APCVD, LPCVD) will become the dominant process in the future with the others fading into disuse. Thus, a CVD practitioner would be well advised to understand the different approaches, why each approach is appropriate for different applications, and how the processing issues for each technology platform are being addressed to achieve economical manufacturing.

2.2 CVD Process Fundamentals

All of the action is at the deposition surface. The thin film manufacturing process takes place at the molecular level, driven by the thermodynamics and kinetics of the precursor chemistry, the dynamics

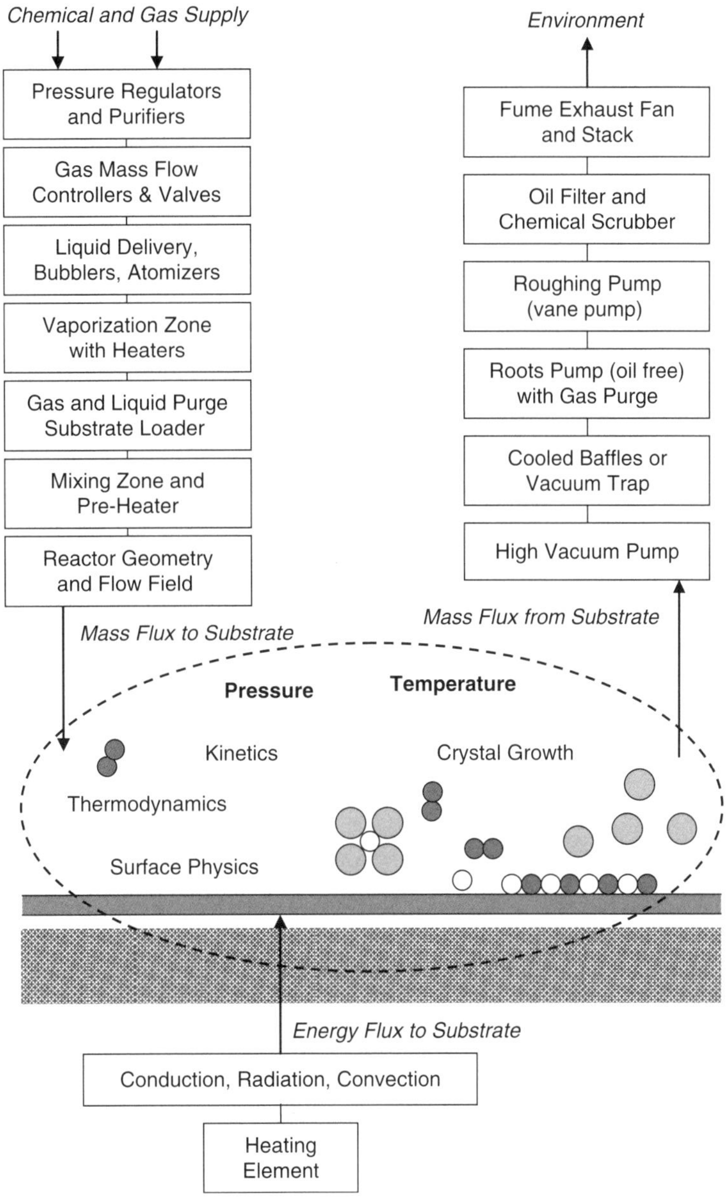

Figure 2.1 Schematic diagram of the CVD thin film molecular-scale manufacturing process (in the dashed oval) and the macro-scale equipment for precursor delivery, product exhaust and substrate heating.

of heat and mass transport, and the physics of surface adsorption and crystal growth. Figure 2.1 gives a schematic of the representative system components that ultimately control the deposition process. Unlike "macro-scale" manufacturing processes, like assembling a cell phone, it is difficult to measure, monitor, and control the deposition process directly. That is, we can not store, feed and manipulate individual molecules in the way we could the parts of a cell phone. Rather, the uniformity, quality and repeatability for a given thin film process are the result of previous experience with that particular system. The control strategies for CVD are developed through experimentation and modeling, with feedback from material analysis. This is why today's CVD tool manufacturers maintain research engineering and materials science as part of their core business.

2.2.1 Research and Development of CVD Technology

The CVD development project goal is to develop a machine that delivers the precursor vapor, supplies the activation and crystallization energy, and evacuates the products, all at just the right rates for a desired film deposition on a particular substrate. We understand that the temperature and precursor vapor pressure at the substrate determine the morphology and properties of the film deposit. However, the particular set of conditions at the substrate that produces the desired film can only be achieved indirectly through controlling the heater temperature, reactor pressure and precursor flow rate. These *process control parameters* of temperature, pressure and flow are directly related to all of the reactor variables that affect the conditions at the substrate. The *reactor variables* include flow velocity, flow regime, residence time, vapor temperature and flow patterns in a given reactor geometry. A successful deposition system for a particular material on a particular substrate is developed through modeling and empirical research. Once a process recipe in a given reactor system is developed, engineering research and modeling are again required to develop economically viable process equipment for different applications.

Metrology is the science of measurement. Metrology signals are used to continuously adjust the manufacturing processes to maintain product specifications. The metrology signals may be gained by any combination of process parameter measurements, or by taking measurements of some aspect of sample products, and relating these measurements back to processing steps. The main sources of information about what is happening in the deposition zone are heater temperature and the reactor total pressure measurements. Chapter 3 describes some of the methods that are used to analyze the depositing film, and which can be used to monitor the system for problems or processing drift. To understand the measurement and control equipment, we need a basic understanding of rarefied gas and fluid dynamics, which will be covered in the rest of this section. The next section will focus on the equipment that delivers the precursors and energy to the deposition zone, manages the waste stream, and monitors and controls the feed rate, reactor pressure and temperature.

If you do not know what the Knudsen (Kn) number is, you may want to consult a vacuum science resource for some background reading.[4,5] If you have never heard of the Reynolds number (Re) you should just skip the next few sections. We will look at the flow dynamics fundamentals and mass transport with the purpose of understanding how the control parameters of reactor pressure and gas supply rate relate to the system performance, in particular the uniformity and throughput. If you are going to skip ahead, then the one thing you should take away is this cautionary note: if the system you are using is producing good uniformity, do not change the feed rate or pressure, and if you are having uniformity problems, call the engineering support team – do not attempt a remedy yourself.

2.2.2 Gas Dynamics and Flow Basics

The Knudsen number, Kn, is the ratio of the mean free path, λ, to the characteristic reactor dimension perpendicular to the flow direction, D (*e.g.* the diameter of a tube). The "mean free path" is the average distance traveled by a vapor molecule between collisions with another molecule and is a well-known parameter in gas dynamics.[6] At room temperature, a nitrogen molecule has a thermal velocity around 470 m s^{-1}. To get a quick idea of the conditions at any point in a system you can use an estimate of the mean free path (m) for air at room temperature and pressure, P (in Pa):

$$\lambda_{\text{air}} = 6.6 \times 10^{-3}/P \tag{2.1}$$

At atmospheric pressure, λ is only about 100$\times$ the molecular diameter, while at pressures below 10^{-3} Pa, the mean free path is larger than most reactor dimensions. This has interesting implications for mass transfer of vapors from the reactor inlet to the substrate surface. At atmospheric pressure, we must move the precursor to the deposition zone by moving the carrier gas, because the precursor is effectively trapped in the bulk flow. At very low pressure, the precursor molecules can move across a one meter space in a few milliseconds. Of course, there are only a few molecules arriving very quickly at such a low pressure. Many electronic devices now require deposition on micro-and nano-scale 3D features that may not be evenly coated if internal features are smaller than the mean free path; λ is clearly a very important parameter in the deposition physics but it can only be determined indirectly by pressure measurement.

The rate that a gas impinges on a surface is an important factor in the deposition rate and thus the throughput. The well-known Knudsen equation gives the molecular flux per unit area, J (molecules $m^{-2}\,s^{-1}$), from a quiescent gas:

$$J = \frac{N_A P}{\sqrt{2\pi MRT}} \tag{2.2}$$

where N_A is Avogadro's number, M is the molar mass ($kg\,mol^{-1}$), R is the ideal gas constant and T is the temperature in Kelvin. The gas pressure is clearly the most important processing parameter in determining the arrival rate of precursor vapors at the surface. Equation (2.2) is expressed for a total gas pressure, but for a mixture the partial pressure of the reactant gas would be used. We can usually control the mixture concentration of the reactant gas in a diluting carrier gas by controlling the respective mass flow rates. However, as we will see in the later discussion of delivery systems, liquid bubbler and solid evaporation into a carrier gas stream is more difficult to control.

The temperature of the gas in a CVD system is often difficult to determine with precision. The ideal gas relation indicates that the temperature will drop substantially as a certain number of moles of gas, N_m, expand from the high pressure supply tank into the low pressure reactor system with volume, V.

$$P = \frac{N_m RT}{V} \tag{2.3}$$

We could calculate the post-expansion temperature, but then the gas flows into a thermal field where it will be heated by radiation, depending on its absorption characteristics, or by conduction through collision with other heated molecules and heated surfaces. It is difficult to measure the gas temperature directly at low pressures because the temperature probe (*e.g.* a thermocouple) must be in thermal equilibrium with the gas. When a thermocouple is placed in a vacuum chamber, it is actually in thermal equilibrium with the walls of the chamber through radiation heat exchange. At very low pressures, the gas molecules are moving so fast that it can be assumed they have contacted one or more of the walls and so are at the oven temperature. In a cold-wall reactor it may be a better approximation to use the room temperature. In any case, the molecular flux depends more on the pressure than on the vapor temperature.

We use $Kn = \lambda/D$ to determine the mass transport characteristics of the reactor and exhaust components. The physics of the gas flow changes with decreasing pressure from the viscous regime where intermolecular collisions dominate, through a transition regime to the molecular flow regime where intermolecular collisions are rare. Figure 2.2 shows the relationship between gas pressure, component dimension perpendicular to the flow direction and the flow regime. Kn was calculated using nitrogen at room temperature: $Kn < 0.01$ viscous flow; $0.01 < Kn < 1$ transition flow; $Kn > 1$ molecular flow.

Any of these flow conditions can occur in different components in a given CVD system. Vapor is delivered from a higher pressure source at the reactor inlet and removed by vacuum pumps from the reactor outlet. In the reactor chamber, the gas very rapidly eliminates pressure variations, establishing a uniform molecular density. The details of the flow dynamics that lead to local variations in arrival at the substrate depend almost entirely on the solid geometry, the pressure and the total mass flow rate.

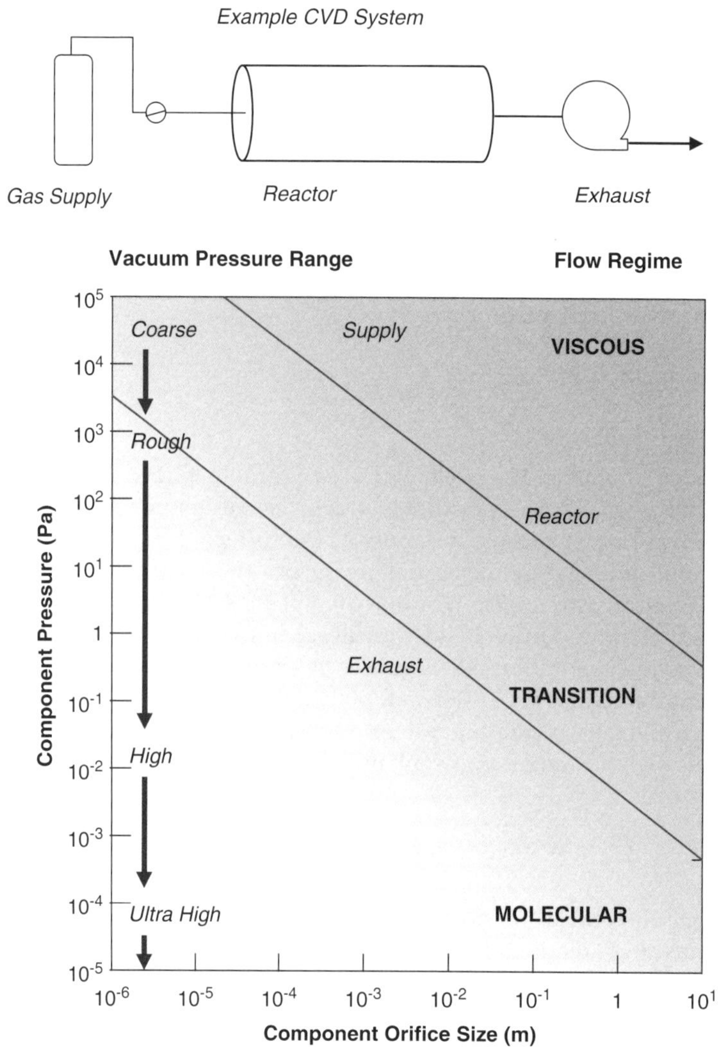

Figure 2.2 The three gas flow regimes as a function of system component dimensions and gas pressure. Also shown is an example CVD schematic and possible different flow conditions throughout the system.

Often, in equipment specifications, this flow rate is expressed in reference to the "standard" conditions of $T = 0\,^{\circ}\mathrm{C}$ and $P = 1$ atm, designated as stp. At stp the ideal gas law, Equation (2.3), gives a molecular density of 2.69×10^{29} molecules cm^{-3} and the molar volume is $22\,440\,\mathrm{cm}^3$. In industry, units of cm are often preferred over m, and pressure may be expressed in Torr or mmHg, where 1 Torr = 133 Pa = 1 mmHg. Scientists will note that the "reference" conditions in thermodynamic data are $T = 25\,^{\circ}\mathrm{C}$ and $P = 1$ bar, which can be a source of mysterious errors in communication between scientists using reference properties and engineers using stp.

2.2.3 Molecular and Transition Flow Regimes

When $\mathrm{Kn} > 1$, gas molecules are just as likely to strike a reactor surface as another gas molecule and we are in the *molecular flow regime*. The flow is a result of the random motion of molecules (Figure 2.3). If P_1 is greater than P_2 then there is a probability that molecules will move from the higher density area

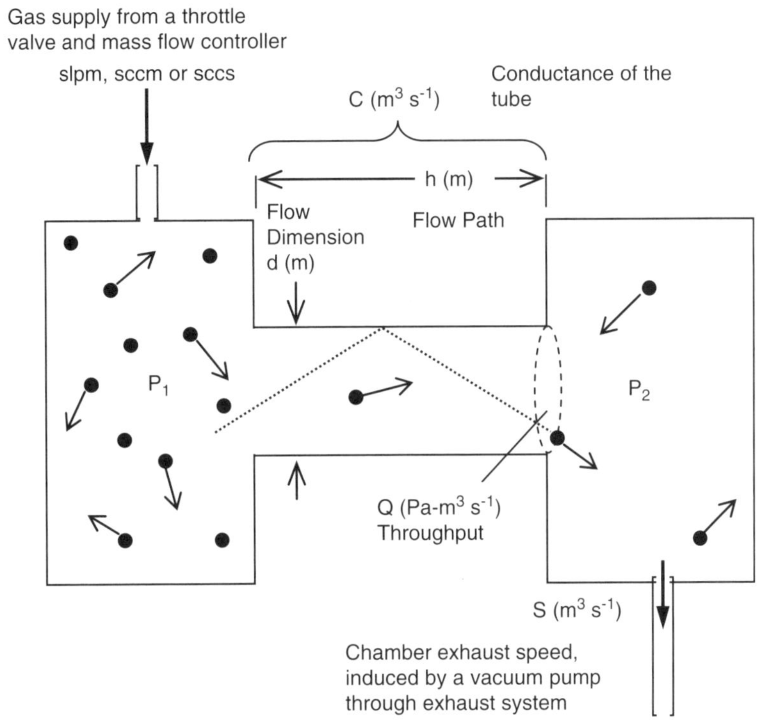

Figure 2.3 Schematic diagram of the molecular flow (Kn > 1) of molecules due to pressure gradient induced by a pumping/feed system with flow rate Q, through a system geometry with conductance C.

to the lower density area. The rate at which gas will flow due to these random motions is a function of the pressure difference, and the externally imposed steady state flow rate or *throughput*, Q (Pa m^3 s^{-1}). The free molecular motion is highly dependent on the "line of sight" geometry between the two pressures, which is characterized by the component conductance, C (m^3 s^{-1}):

$$Q = C(P_1 - P_2) \tag{2.4}$$

Relations for conductance can be found in any vacuum science handbook for simple geometries like an orifice, baffle, tube and annulus for free molecular flow. As an important example, let us consider the molecular flow conductance through a length of vacuum tube, h (m), connecting the reactor to the pump. The component dimension, d (m), is the tube diameter, and the standard formula from gas dynamics using air at room temperature is $C_{\text{tube}} = 12.2 \times 10^4 d^3/h$. The conductance (m^3 s^{-1}) represents a limit on the possible molecular flow through the tube. If we get a bigger vacuum pump to increase the throughput, Q, it won't change the conductance, but may change the reactor pressure, P_1. If we get a pump with a lower ultimate pressure, P_2, this won't change the conductance either, but would change the flow and the reactor pressure according to Equation (2.4). The conductance into the exhaust system entrance is directly proportional to the orifice area. Thus, once you hook up a certain length of vacuum tube, both the flow and rector pressure are thereafter functions of that exhaust system conductance. It is *much* cheaper to use a larger orifice exhaust system and to put the pump as close as possible to the reactor than to try to get the flow and pressure you want through buying a new pump. The units for conductance and pump speed are chosen for convenience and may be expressed as (L s^{-1}) or (cm^3 s^{-1}).

There are other components in the exhaust system besides just the tubing, including traps, valves, baffles, bends and filters. Most vacuum components are supplied with measured conductance values. Conductance is not a function of pressure as long as the flow is in the molecular regime (Kn > 1); however, this condition depends on the local pressure and the component geometry (Figure 2.2). Correction factors for conductance in the transition regime are usually provided by component manufacturers and are available in vacuum science references. The conductance of the entire exhaust system, C_{sys}, for all of the components connected in series is given by:

$$\frac{1}{C_{sys}} = \frac{1}{C_1} + \frac{1}{C_2} + \frac{1}{C_3} + \cdots \tag{2.5}$$

which clearly shows that one component with a small conductance will limit the conductance of the whole exhaust system. It may be possible to increase the flow by constructing parallel paths to the pump through duplicate sets of traps and filters. In this case the total exhaust system conductance for parallel exhaust lines can be increased as:

$$C_{sys} = C_1 + C_2 + C_3 + \cdots \tag{2.6}$$

where in this case the component conductances, C_1, C_2, *etc*., are actually system line conductances as given in Equation (2.5).

Figure 2.3 also gives the definition of some of the terms associated with the system flow control. There are three different perspectives on flow represented in a CVD system, and the units of measure are different for the injection, reaction chamber and exhaust system. Gas supply monitoring and control equipment use stp reference because the ideal gas law can be used to convert the equipment specifications into throughput at the specific reactor conditions. Equipment specifications of the "standard mass flow rate" use measurement units of "sccm" = standard cm^3 per minute, "sccs" = standard cm^3 per second, or "slpm" = standard liters per minute, where standard means stp (0 °C and 1 atm). The reader will be warned of this terminology discontinuity as clearly the units of measure for what engineers call the mass flow rate are actually volume, not mass. The reason for this is that the actual mass flow rate is easily calculated for the actual flow pressure from the stp measurement by applying the ideal gas law.

The throughput, Q, is the same at the inlet, the outlet and any other cross-flow plane in the system. Throughput is expressed in units of pressure times volume per unit time ($\mathrm{Pa\,m^3\,s^{-1}}$), ($\mathrm{Pa\,L\,s^{-1}}$) or ($\mathrm{Torr\,L\,s^{-1}}$) where L = liters. There is no measuring device for throughput, so it is determined from the input mass flow rate and the local pressure measurement. Using the number of molecules at stp from the ideal gas law, the standard mass flow rate from the precursor injection system can also be used to determine the throughput in terms of molecular or molar flow rate, Q_m, which is often the quantity of interest at the substrate: 1 sccm = 4.48×10^{17} molecules s^{-1}.

Finally, the exhaust system removes vapor from the reactor at the *exhaust speed*, S ($\mathrm{m^3\,s^{-1}}$ or *lpm*), and reactor pressure, P, where:

$$S = Q/P \tag{2.7}$$

The exhaust speed is produced by the evacuation system. The evacuation equipment consists of a pump with specified volumetric pumping speed, S_P ($\mathrm{m^3\,s^{-1}}$ or *lpm*), at the nominal operating pressure, P_o, and traps, tubes and valves with a total system conductance given by Equations (2.5) and (2.6).

The deposition process is controlled remotely through control of the injection and exhaust systems. The precursor mass flow rate and the pump speed balance to produce the reactor pressure. A pneumatically controlled leak valve on the gas inlet and a throttle valve in the vacuum exhaust line can be used to continuously adjust the throughput and the reactor outlet speed to achieve the desired reactor pressure as in Equation (2.7) by changing the exhaust line conductance as in Equation (2.8):

$$S = \frac{S_P}{1 + S_P/C_{sys}} \tag{2.8}$$

2.2.4 High Vacuum CVD Reactor Deposition Model

We now turn our attention to what is happening in the LPCVD reactor or "deposition chamber." In the molecular flow regime, and the lower pressure range of the transition regime, the volume of precursor and carrier gas vapor in the reactor can be modeled as a "well-stirred-reactor." This means that there are no pressure, velocity, density or, most importantly, concentration gradients in the reactor. Inside the reactor, the Knudsen equation (Equation 2.2) is essentially the model for mass transport to the substrate surface regardless of deposition chamber configuration as long as $Kn > 1$.

Figure 2.4 gives a simple schematic diagram of the reactor with definitions of the important parameters. The carrier gas flow enters the reactor chamber of volume, V, and exits at the same volume flow rate, S, while the precursor vapor enters at the injection molar concentration, n_{m-in}, and exits at the average chamber concentration, n_{m-ch}. The flat substrate surface of area, A_s, exposed to the vapor at pressure, P, and molar concentration, n_{m-ch}, will experience a molecular flux, J, proportional to the reactant partial pressure as in Equation (2.2). A proportion of these molecules will be adsorbed on the surface and react at a rate characterized by the thermodynamic

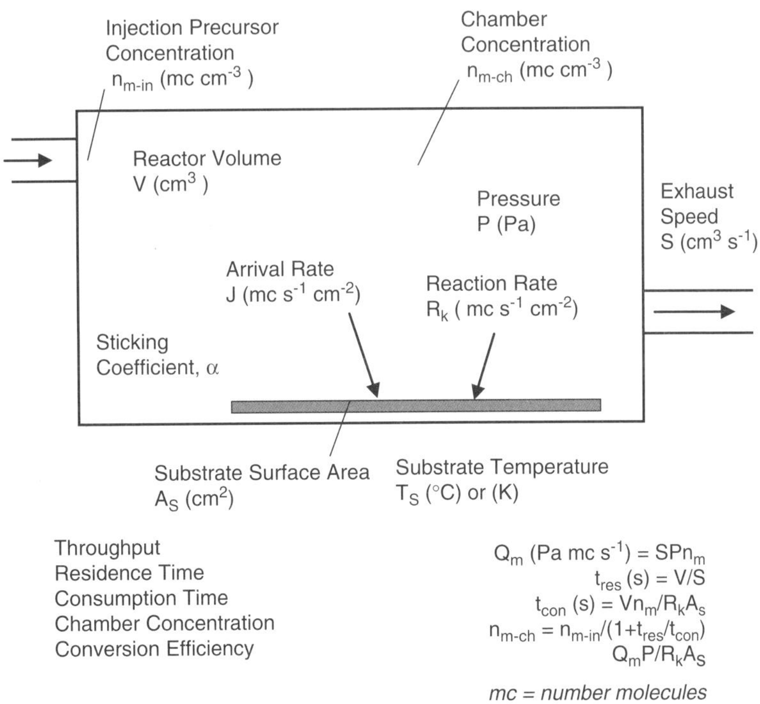

Figure 2.4 Reactor transport model for molecular and transition regime flow ($Kn > 0.01$) and definition of important terms. The model assumes the "well-mixed-reactor" condition with no density, velocity or concentration gradients.

chemical reaction rate, R_k (mc s^{-1} cm^{-2}). A steady-state mass balance on the reactor is straightforward as long as the substrate is the only heated surface consuming precursor:

$$Sn_{\text{m-in}} = R_k A_s n_{\text{m-ch}} + Sn_{\text{m-ch}} \tag{2.9}$$

The deposition kinetics can be understood in terms of the precursor reactor residence times. The nominal time that the vapor stays in the reactor is a simple function of the reactor volume and the throughput:

$$t_{\text{res}} = \frac{V}{S} = \frac{VPn_{\text{m}}}{Q_{\text{m}}} \tag{2.10}$$

The average time that precursor resides in the reactor before being adsorbed and incorporated into the depositing film can be estimated by:

$$t_{\text{con}} = \frac{Vn_{\text{m}}}{R_k A_{\text{s}}} \tag{2.11}$$

where A_s is the substrate surface area. Note that units of cm are common in CVD rather than m, as a matter of convenience.

At low temperatures the reaction rate will be slow and the consumption residence time will be large. If $t_{\text{res}} \gg t_{\text{con}}$, then we have surface limited growth and the conversion efficiency will be low as the exit precursor concentration is nearly equal to the injected concentration. This reactor condition is called a *differential reactor*.

At high temperatures, the reaction rate may be high enough that nearly all the precursor injected is consumed. When $t_{\text{res}} \ll t_{\text{con}}$, the film growth is mass-transport limited and this condition is called a "starved reactor." This process has high conversion efficiency, but may have an unnecessarily slow processing time if the gas throughput is too low so that the reactor is "starved" of reactants. The process engineering goal is to tune the residence time to the reaction rate so that the growth rate is as high as possible, while the wastage of precursor is minimized.

The heated substrate holder, and any other surfaces of the heater or walls that are at temperatures high enough to activate the deposition reaction, will essentially be sinks for reactants. This parasitic consumption of reactants has implications for system design and affects the conversion efficiency and growth rate. For a hot-wall reactor, the total reactor internal surface area consumes precursors, not just the substrate and heater stage.

The exact details of the motion of molecules in a high vacuum reactor cannot be known because of the inherent random behavior of the rarefied gases. The modeling technique commonly used for this flow regime is Monte Carlo simulation, which takes advantage of the fact that the governing flow characteristic is the random motion of individual molecules. Monte Carlo and Direct Simulation Monte Carlo (DSMC) modeling studies have been carried out for many LPCVD processes.[7] Commercial software is available that can be used for studying details of reactor and substrate configurations.

A note about surface adsorption is required at this point as it is a contributing factor to growth rate, wph throughput, and cleanliness, particularly in LPCVD and UHV-CVD. Surface science is a whole field and, again, you are encouraged to find a reference book if you need more information for your work.[4] The term "adsorption" refers to the accumulation of gas on a solid surface and is measured in units called the langmuir, L (molecules cm^{-2}). Any gas molecule, even inert gasses, when they come within a few molecular diameters of a solid material will begin to interact with that surface through the particular intermolecular forces of the respective materials, such as van der Waals forces. Owing to the attraction force, the vapor molecule will accelerate toward the surface, but will then encounter the repellant force of the solid nuclei. Depending on the trajectory and momentum of the gas molecule during this interaction, it may reflect away from the surface or it may dissipate some momentum as it interacts with the surface and oscillate on the surface under the influence of the attractive and repellant forces. The adsorbed gas molecule may then gain enough energy while hopping about on the surface (surface diffusion) to escape the surface, or "desorb," if the surface is hot. The sticking coefficient, usually designated by α, is the experimentally observed difference between the molecular arrival rate calculated from Equation (2.2) and the deposition rate. If the adsorbed precursor molecule resides long enough to undergo the deposition reaction, then it becomes permanently fixed at the surface as part of the new material deposit.

An adsorbed gas molecule is not a solid, but is in a new, reversible state of physisorption where it is oscillating on the surface and is quite mobile. Note that "absorption" refers to gas molecules that are taken up into the bulk material of the solid or dissolved in a liquid. Condensation is a special case of adsorption where the solid material is the same as the vapor. Chemisorption refers to vapor that has undergone an irreversible chemical reaction and become incorporated in a solid deposit.

Process control for quality, uniformity and throughput in the molecular flow regime is relatively straightforward. The film quality can be controlled by controlling contamination through good vacuum system design and practice, as we will see in the next section. The mass transport to the substrate is inherently uniform due to the random nature of the molecular movements, and as long as the geometry of the reactor has a relatively open shape above the substrate. The throughput is necessarily low as the mass flow rate must be kept low enough to maintain the low pressure. However, the throughput can be maximized by operating the reactor in the mass transport limited range. The main process control strategy is to set the injection mass flow rate with a mass flow controller, and to maintain the reactor pressure through a throttle valve on the exhaust entrance. Consideration in reactor design can be given to balancing the need for thermal uniformity and reducing the heated surface area to increase total conversion efficiency.

2.2.5 Viscous Flow Regime Convection Mass Transport

If $Kn < 0.01$ in the reactor or a component, then the reactant vapors are being *convected* through the reactor by other molecules in the carrier gas flow, and the mass transport is in the viscous flow regime. The precursor then diffuses from the bulk gas flow to the surface due to a concentration gradient. Convection mass transport is the general term for this area of study. Whereas the gas molecules in molecular flow act as if the other gas molecules do not exist, in viscous flow all of the gas molecules behave as a continuum. The Navier–Stokes model of continuum flow is amenable to numerical modeling, and commercial software packages, such as FLUENT©, can be used to investigate the flow patterns in a particular reactor configuration. There have been many treatments of the models for reactor flow in the literature for the range of reactor geometries.[8,9] In fact, detailed modeling is necessary to develop a reactor design for uniformity in the viscous flow regime.

The flow patterns in continuum flow CVD reactors are completely dependent on the bulk flow velocity, the fluid physical properties, the driving pressure gradient from inlet to outlet, and the geometry of the solid surfaces that contain and interact with the flow. These local details of the flow pattern determine the arrival rate of precursor vapor to the substrate at each point on the surface because of the solid–fluid interaction zone called the boundary layer.[10]

On the molecular scale, the gas molecules within one mean free path of the surface are subjected to the same physics as the molecules in our previous discussion of molecular flow. If these molecules impact on the solid and are adsorbed then their velocity becomes zero during the few microseconds that they are adsorbed. When they are desorbed back into the flow, they will have a random direction and a new thermal velocity due to any temperature difference between the surface and the gas. We should note that carrier gas is adsorbed and desorbed on the surface as well as precursor.

The molecules at the surface have zero velocity, but as soon as they desorb from the surface again they bump into molecules near the surface, resulting in momentum transfer, and the bulk flow "drags" on the solid surface. The way that a gas transmits this drag, or shear force, through the flow is characterized by the viscosity, $\mu\,(\mathrm{N\,m^{-2}})$, which is measured in units of shear force per unit area. The way in which this intermolecular momentum diffusion occurs also depends on the gas density, $\rho\,(\mathrm{kg\,m^{-2}})$. The kinematic viscosity, $v=\mu/\rho\,(\mathrm{cm^2\,s^{-1}}$ or Stokes), is essentially the momentum diffusivity, and is the property usually used to model the fluid dynamics. Viscosity, μ, is not sensitive to pressure, but of course density is very dependant on pressure, so the kinematic viscosity increases with decreasing pressure.

Figure 2.5 shows a basic flow model for a flat substrate in a horizontal tube reactor in the viscous regime. The Reynolds number, Re, is the non-dimensional parameter used to describe the flow characteristics:

$$\mathrm{Re}=\frac{U_{\infty}L}{\nu} \tag{2.12}$$

The bulk velocity, U_{∞}, is not measured, but is estimated from the injection mass flow rate, sccm, converted into the reactor volume flow rate, $Q\,(\mathrm{cm^3\,s^{-1}})$ at the reactor pressure through the ideal gas law, and the reactor flow cross section area, $A_c\,(\mathrm{cm^2})$ by $U_{\infty}=Q/A_c$. The characteristic length for flow inside a tube, L, is the tube diameter. For any other internal flow cross section shape, the characteristic length is called the "hydraulic diameter" and is estimated by the flow cross section area divided by the wetted perimeter length. The characteristic length for bulk flow over a flat surface is the length of the surface in the flow direction, x. The Reynolds number is always very low, below a few hundred, in CVD reactors, and the flow is laminar. The transition to turbulent flow is in the range of $\mathrm{Re}=2100$ for internal flows. The mixing and unpredictable swirling patterns in turbulent flow would not be desirable in CVD because of the need for growth rate control and uniformity. Eddies can also "trap" precursor and cause local concentrations high enough for gas-phase decomposition, and can transport particles around the reactor.

The bulk flow contains a carrier gas and one or more reactant species. The reactants are being consumed on the substrate surface, and so the gas concentration on the surface is assumed to be zero. Mass diffusion of reactant gasses from the high concentration in the bulk flow to the zero concentration at the substrate generates a concentration boundary layer. Diffusivity, $D\,(\mathrm{cm^2\,s^{-1}})$, is usually measured for one gas species diffusing through another by measuring the mixing rate of two gases which are initially separated. Diffusivity data are usually not readily available for the CVD reactants, so the approximation from kinetic gas theory is useful, $D\sim T^{3/2}/P$. Binary gas diffusivity is in the range 0.1–10 $\mathrm{cm^2\,s^{-1}}$ for viscous flow reactors. The momentum boundary layer is usually assumed to be the same thickness as the mass diffusion boundary layer because the physics that determine the momentum and mass transport are similar.

2.2.6 LPCVD Reactor Deposition Model

Precursor molecules in the bulk flow diffuse to the heated surfaces where they are adsorbed and consumed across a boundary layer. The most important thing to know about viscous flow without having to be a fluid dynamics specialist is that the uniformity of the deposition depends on the local diffusion flux to the surface, and that flux depends on the local boundary layer thickness and the local precursor concentration in the bulk flow as shown in Figure 2.5. The molecular flux at any point in a flow depends on the diffusivity and the concentration gradient, $J_i=h_m(n_{\infty}-n_s)$, where $h_m\approx D/\delta$ is the mass transfer coefficient and n_s is the surface concentration. In ALD or low temperature deposition processes, the reaction rate may be slower than the arrival rate, and the

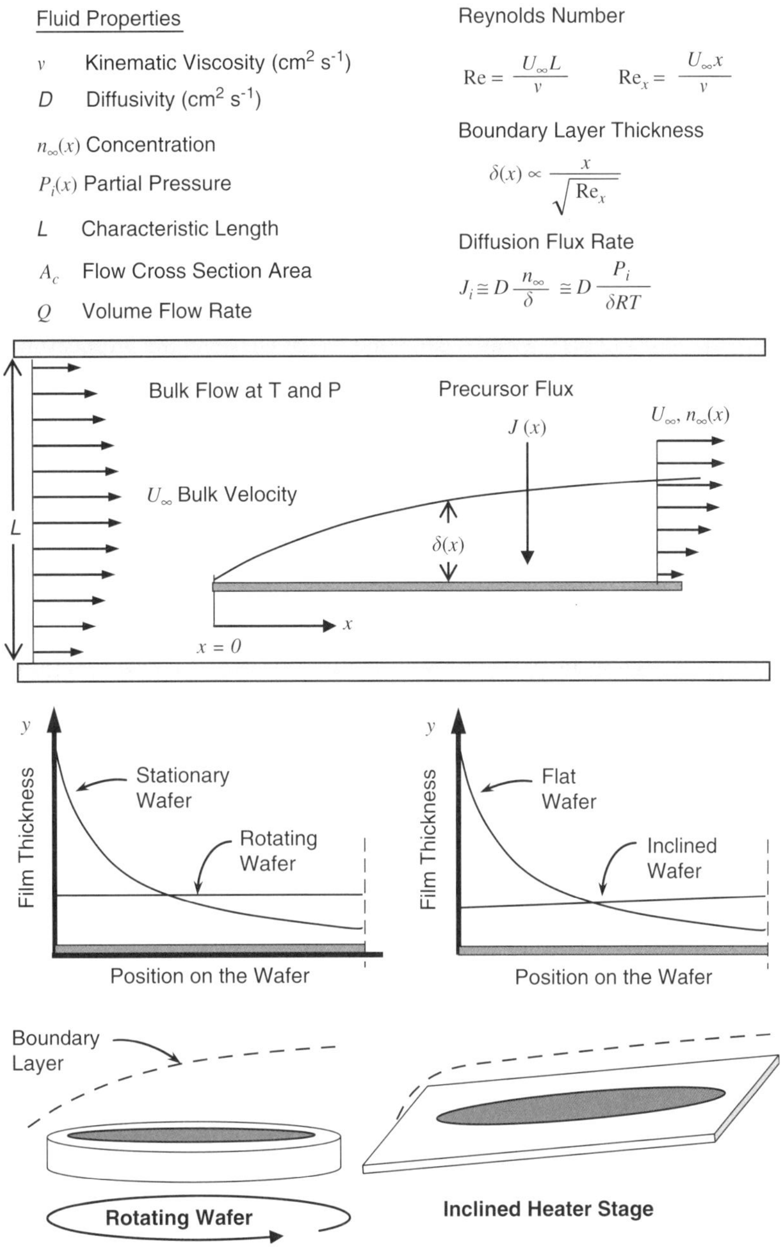

Figure 2.5 Schematic diagram of the laminar viscous flow ($Kn < 0.01$, $Re \ll 2000$) through a tube reactor with a horizontal substrate, illustrating the dominance of boundary layers in the deposition model. The lower figure shows two design mechanisms used to produce uniform deposition in the environment of non-uniform flow conditions.

model for deposition is the kinetic controlled growth, which depends only on surface temperature for uniformity and is not affected by concentration gradients in the reactor. However, the low temperature growth rate is exponentially slower than the mass transport, or diffusion limited growth, where all of the precursor reaching the surface is consumed. The model for local diffusion limited growth assumes that the surface precursor concentration is zero, that the diffusivity may vary locally with temperature ($D \approx T^2$), and that the local viscous boundary layer thickness will depend on details of the fluid dynamics. In terms of the local bulk partial pressure, P_i, and the local boundary layer thickness, $\delta(x)$, the local molecular flux of species i can be estimated by:

$$J_i(x) = \frac{DP_i(x)}{\delta(x)RT} \tag{2.13}$$

The typical viscous regime CVD reactor operates in the laminar flow regime where the bulk flow Reynolds number, $\mathrm{Re} = U_\infty L/\nu$, is less than 100. The thickness of the boundary layer, $\delta(x)$, at any position, x, on the substrate in the flow direction is estimated by $\delta(x) \sim x/\mathrm{Re}^{1/2}$. The dilute precursor gas flow typical of CVD processing means that we can assume the momentum and molecular diffusivity are of the same order, and thus the momentum boundary layer is about the same thickness as the diffusion boundary layer. In reactor configurations where the gas is flowing over the substrate, a faster flow velocity and narrower reactor dimension (*e.g.* L in Figure 2.5) result in a thinner boundary layer.

Higher velocity is not the only source of turbulence and recirculation. Reactors are carefully designed and tested with flow visualization techniques to ensure that there are no shapes inside the reactor that trip the flow into recirculation. In particular, the gas inlet needs to have a smoothly diverging shape rather than an abrupt sharp corner.

Depending on the precursor concentration, substrate loading and flow path length, the precursor concentration may vary significantly from the inlet to the outlet. This can be quite significant as seen in the stationary wafer deposition profile in Figure 2.5. The leading edge of the wafer faces the highest concentration flow and has the thinnest boundary layer, and so has the highest deposition rate. The highest efficiency CVD reactor would have nearly all of the reactants consumed by the time the carrier gas reaches the exit port. This would result in virtually no deposition on wafer surfaces near the exit and thus, a starved reactor condition.

Since diffusivity varies inversely as the pressure and increases with temperature, to some degree the effects of depleting reactants along the flow path and increasing boundary layer thickness can be compensated by increasing temperature and flow velocity in the flow direction. In practice, engineering the thermal field is much easier to achieve for uniform temperature than for a temperature gradient. The most common strategy for achieving uniformity, while optimizing the conversion efficiency and production throughput, is to manipulate the substrate during the deposition. Figure 2.5 shows the effect of rotating the wafer in cross flow. The rotation rate is tuned to the deposition rate so that each area of the substrate passes through the high deposition zone for the same length of time over the deposition run. There are other "tricks" of reactor design to work with the physics of boundary layer flow. As rotating wafer chucks are not an option for most research-scale tube reactors, the substrate holder is usually inclined so that the effective flow passage narrows along the flow direction. The substrate is positioned away from the leading edge of the inclined stage in an area with relatively uniform boundary layer thickness, as shown in Figure 2.5. The inclined stage increases the bulk flow velocity in the x direction and decreases the boundary layer thickness, with the effect that the boundary does not continue to grow along the substrate. Vertical tube reactors with flow that impinges on the substrate also use rotation of the wafer to accelerate the flow as it crosses the wafer with the effect of suppressing boundary layer growth.

Free convection, or natural convection, is a fluid motion induced by a density gradient in a gravitation field. For example, in a horizontal cold walled reactor, as in Figure 2.5, the substrate will locally heat the gas flowing over it. This hot gas is less dense than the cold gas above it, and so the buoyancy force causes the hot gas to rise. In a reactor configuration like Figure 2.5, this would cause the boundary layer to be thicker due to heating, and could even cause a disruption to the flow if the buoyancy force is greater than the pressure force driving the axial flow. The remedy for controlling free convection effects is the same as for reducing the boundary layer thickness, *i.e.*, to increase Re.

The viscous flow conditions in the reactor mean that the exhaust system will also be in the viscous flow regime. The conductance of long tubes (length, $h \gg$ diameter, d) is given by the fluid dynamics relation:

$$C_{\text{viscous}} = 1.41 d^4 P_{\text{avg}}/h \tag{2.14}$$

where P_{avg} is the average pressure (in Pa) from end to end of the tube. As with molecular flow conductance, the tube diameter should be as large as possible. The tube should be sized so that the conductance is at least as large as the pump speed.

2.3 System Platform and Equipment Considerations

The *system platform* is the basic configuration for the CVD equipment and it embodies all of the decisions and component choices that determine the nature of the process. Figure 2.6 gives an idea

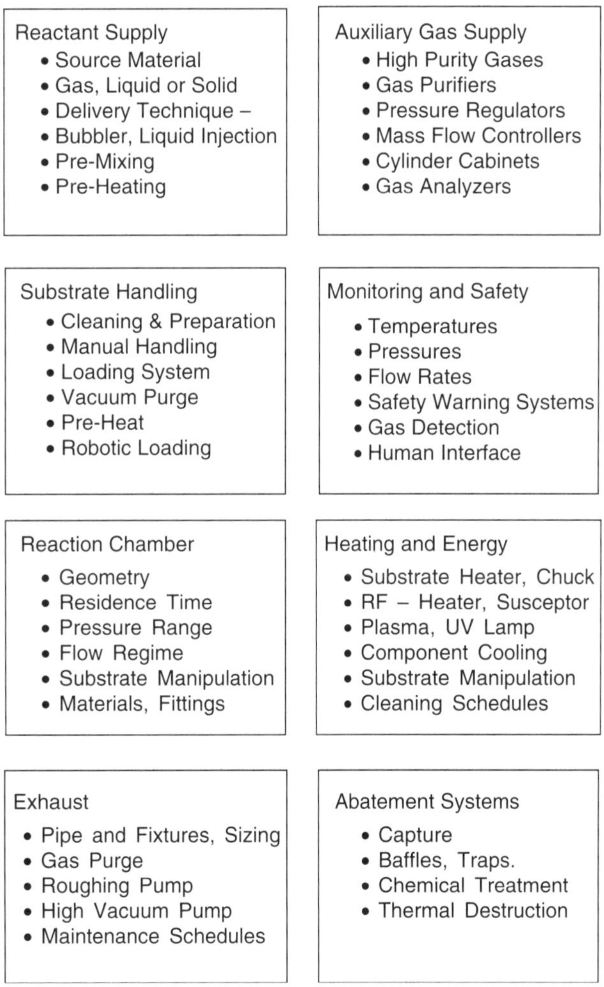

Figure 2.6 Survey of the typical CVD system platform engineering choices.

of the range of engineering choices that go into a system platform. As we have seen earlier, much of the system platform is determined by the particular CVD process. For example, a CBE platform will require high vacuum components, materials and pumps. In the next section, the platforms for the different CVD processes will be treated individually. The rest of this section highlights some of the essential system-level considerations for equipment, components and materials used in CVD.

2.3.1 Contamination: Material Selection and Design Considerations

The first CVD performance factor is quality, and the best approach to ensure quality is to choose high purity precursor materials and gasses, and to ensure proper handling and cleaning of substrate surfaces before introduction into the reactor. Sources of film contamination during deposition are vapors and solid dust particles. Contaminating vapors can arise from out-gassing of adsorbed species, particularly water vapor, from reactor surfaces, and from evaporation of materials in the reactor. A few considerations and good practices can address the most common vapor and dust contamination sources.

2.3.1.1 Outgassing

All surfaces in the reactor, in the precursor and gas delivery supply lines, and in the exhaust piping are sources of contaminating vapors through out-gassing of adsorbed gas and vapors. When exposed to the laboratory environment, a monolayer of oxygen, nitrogen, CO_2 and water vapor adsorb on the reactor and component surfaces. The number of atoms in the monolayer is determined by the equilibrium concentration at the atmospheric partial pressure of each constituent and the room temperature. As the reactor is brought down to pressure, the equilibrium adsorption concentration on the surface will decrease to the new partial pressure and temperature conditions. Out-gassing occurs as these adsorbed gasses are desorbed. Although the pressure in the reactor drops to the base pressure quickly as the vacuum pump is switched on, the out-gassing actually takes much longer. The desorption rate increases exponentially with temperature, so the pre-deposition procedure is to bake-out the reactor at low pressure, being careful not to exceed the temperature specifications of seals and other components. Epitaxial films will require deposition conditions free of out-gas contaminants, which can mean pumping down for several days at room temperature or for several hours at 130 °C. Smooth, shiny inert surfaces such as glass, fully dense or glazed ceramics and polished stainless steel make good reactor surfaces because they have minimal surface roughness and lower adsorption surface area that will need to be out-gassed.

2.3.1.2 Leaks

System leaks and virtual leaks are vapor contamination sources that are addressed through design and system construction practice. Vacuum system sealing is a field unto itself, and the do-it-yourself researcher is advised to read a good book on vacuum technology and find a knowledgeable technician to help with system design and construction. The special issue with CVD vacuum systems is the presence of dangerous and reactive gasses. Many CVD processes are carried out at moderate vacuum, but welded supply lines and fittings, and high vacuum seals on the reactor should be used. In the case that an elastomer seal must be used, as for connecting to glass or ceramic, an inert Viton® O-ring should be used. Care should be taken that the Viton® seal is not exposed to temperatures above 150 °C, which may require cooling around the seal. Swagelok® stainless compression fittings and stainless or Teflon® tubing are convenient, but should only be used for purge gas supply because with even a moderate pressure the sealing ferrule can come unseated from the tube and cause a leak. A leak can even be caused by over-tightening the fitting.

This would be a disaster in a silane supply line. The best rule of thumb for in-house built research equipment is to use only welded fittings on all lines that supply reactive gasses or vapors, and to secure all tubing and compression fittings to a panel or fixed frame where it is difficult for someone to grasp the tubing and put pressure on the fittings.

Measurement and manipulation vacuum feed-throughs are another potential leak hazard. Using high vacuum specification components is warranted for any pressure CVD system that involves dangerous gasses. For example, lower cost medium pressure thermocouple and electrical feed-throughs may use a PVC sealant that can absorb water and solvent and the wires may not be stripped bare. Higher cost high vacuum feed-throughs use a highly chemically and thermally stable sealing material like Kapton® or Tefzel® ETFE resin around stripped wires.

Virtual leaks often originate from sources of trapped air inside the reactor or from materials that evaporate readily at the temperature and pressure of deposition. Substrate holder and substrate heater design must allow sufficient openings for internal gasses to escape during pump-down and bake-out. Remember that as the reactor pressure transitions into the molecular flow range, the gas trapped in a cavity can only escape through openings at a rate proportional to the orifice size. Thus, a hole drilled through the housing will outgas much faster than a press-fit gap between parts. Mounting screws may even be a source of long-term outgassing that could effect epitaxial layer growth. Special vented screws should be specified in these more sensitive CVD systems.

Virtual leaks can also originate from water, solvents and gasses that are actually absorbed into the bulk materials in the reactor and seals. For example, Viton® and Teflon® PFA absorb solvents during cleaning, which will take much longer to diffuse to the surface and desorb. Cleaning can be done first with a dry microfiber cloth then with a quick wipe of methanol dampened lint-free cloth for de-greasing. This is opposed to the undergraduate intern's favored method of using the methanol squirt bottle to soak components then wipe the parts with a paper towel.

Alloyed metals may have trace elements that actually have quite high vapor pressures (p_v). Although the metal will appear to be stable, these alloying elements will gradually diffuse to the surface and desorb. A vacuum science book will list the vapor pressures of the elements, but a good rule of thumb is that brass fittings should be avoided in the reactor. Both 304(L) and 316(L) stainless steel are the right materials for CVD parts below 450 °C, and molybdenum for heater casings and sample holders. Electrical connectors and insulators need attention. Copper has a reasonable vapor pressure, but needs to be kept below 600 °C for $p_v = 10^{-8}$ Pa. Electrical connections should be made with screw pins not soft solder as Pb has high p_v (another undergraduate intern experience to be wary of). Ceramics like silica, alumina and boron nitride make good electrical insulators, as do Pyrex® and quartz glass.

2.3.1.3 Back-streaming

Back-streaming is a source of contamination that must be dealt with through the exhaust system design and careful shut-down procedures. Some pumps have features to reduce back-streaming, but waste products accumulated in lines, traps and filters can also back-stream into the reactor and become deposited on surfaces where they can be evaporated again during the next run. Figure 2.7 shows an exhaust system configuration used to control contamination and safely handle exhaust reactants in a simple research-scale CVD system in the viscous flow regime. A throttle valve in the exhaust line can be used in conjunction with a reactor pressure measurement and a simple differential controller to maintain a constant reactor pressure. The isolation valve is important to protect the reactor from back-streaming species in the case of coolant loss or pump failure, and during the shut down procedure. The isolation valve could be an electric solenoid valve that is tied into the control system as an automatic protection. The cryogenic trap is filled with liquid N_2 after the reactor has been pumped down and will serve to capture precursor and solvent vapors to protect

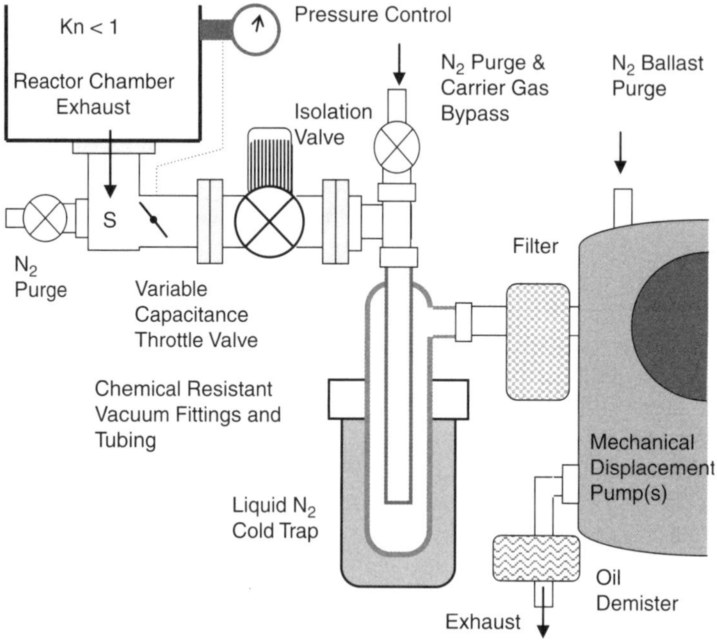

Figure 2.7 Schematic diagram of a typical reactor exhaust system for low-pressure processing where the flow is in the viscous flow regime and transition regime. The system features a capacitance control valve to control reactor pressure.

the pump. Note that if a N_2 carrier gas is used it will accumulate in the trap. The N_2 purge and carrier gas bypass is used to keep the mechanical pump operating in the viscous flow regime and to prevent oil back-streaming. N_2 is also used as a part of the carrier gas control system, *e.g.*, during start up and shut-down to purge out lines.

The Knudsen number should be calculated for the exhaust system tubing diameter (Kn will be different for the exhaust tube than for the reactor even though the pressure is the same). Low pressure CVD systems have two pumps and a more complex exhaust system to prevent contamination (Figure 2.8). The roughing line is used first in isolation to pump down the reactor system. The roughing pump is usually an oil-sealed pump or uses an oil lubricant, so must be isolated from the reactor once the low pressure is reached. After the roughing pressure is reached, the roughing line valve is closed, and the foreline and high vacuum line valves are opened. The high vacuum exhaust line is normally in the transition or molecular flow regime ($Kn \gg 1$). Therefore, vapors move by random motion, being adsorbed and re-evaporated from surfaces and not colliding with other molecules. Thus, hot oil vapors from the high vacuum pump (if it uses oil like an oil diffusion pump) will travel towards the reactor by diffusion. One remedy to this is to reduce the oil vapor molecule's chances of making it to the reactor by putting a cold trap and/or cooled chevron baffle in the line. A trap at pressures below about 10^{-3} Pa requires that there is no line-of-sight flight path, and that oil vapors that hit the trap surface and stay there because it is below the oil vapor pressure, p_v, at that temperature. This means we need to use a cooled baffle trap between a pump that uses oil and the reactor as well as any components that would be degraded by oil, including a turbo pump in high vacuum systems. The oil trap will become saturated with oil over time and will itself become a source of back-streaming, so part of the deposition process cycle must include cleaning of the exhaust line traps.

Back-streaming is not an issue when the exhaust line is in viscous flow as the oil vapors are swept back downstream through collisions with the exhaust gas. In high vacuum systems, a purge gas is

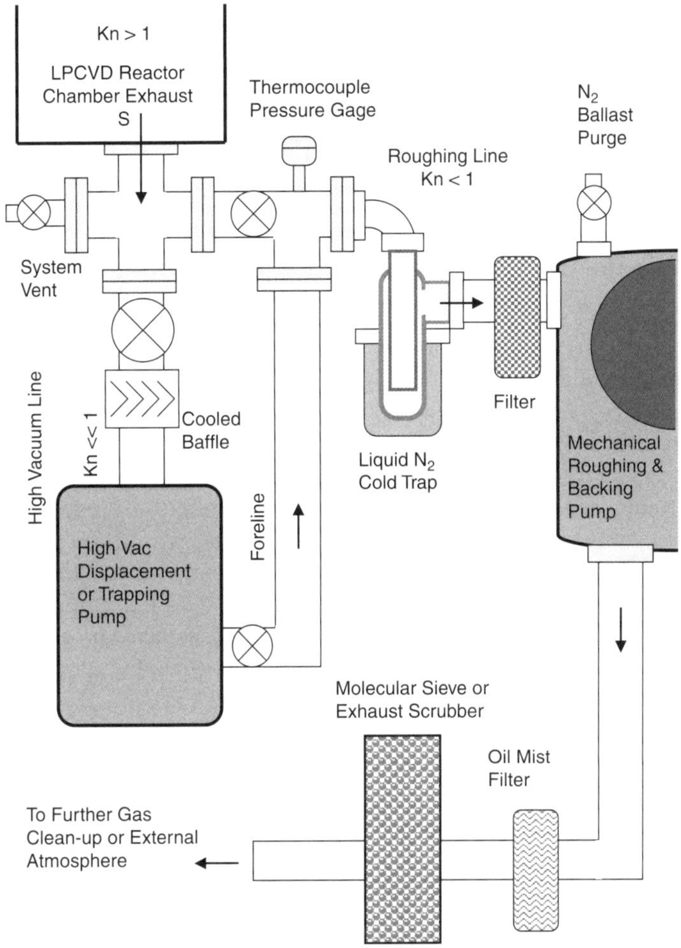

Figure 2.8 Schematic diagram of a typical high or ultrahigh vacuum reactor exhaust system. The system features a cooled baffle to protect against oil back-diffusion and a high vacuum gate valve on the high vacuum line as would be used with an oil diffusion pump.

injected into the line between the oil pump and oil-free components to effectively prevent back-streaming. The N_2 purging at the pump ballast will also sweep any corrosive or solvent vapors through the pump. The purge gas dilutes the vapors, lowering the partial pressure, so that when it is compressed it does not reach its saturation pressure and condense in the pump or dissolve in the pump oil. Supplying a purge gas will require one more pressure measurement and flow control valve in the system. The purge gas will also back-stream, so a gas must be chosen that will not interfere with the film deposition chemistry.

Whenever the reactor is shut down after a deposition run, a vacuum valve at the exit of the reactor should be closed to isolate the reactor from the hot oil vapors that will rapidly fill the exhaust line once the pump is shut off. The reactor should be brought up to unloading pressure by slowly filling with dry inert process gas through a choked orifice leading into a diffuser from the clean end of the system. This is a high risk time for film contamination by dust and particles. Although the desired film may be adherent, parasitic film deposition on other components will most likely have occurred at lower temperatures. These parasitic film depositions can spall off during cooling and be stirred up in the reactor by the filling gas. The gas velocity necessary to carry a particle is proportional to the particle size. There may also be powders that have formed in the vapor in sections of the reactor with the right conditions for homogeneous reactions. Some of these particles will be so small that they cannot be observed except under the SEM. These particles are also a health hazard, particularly if they consist of toxic materials like arsenic. Particle contamination and management is another area that commercial CVD systems deal with through investment in engineering research and development.

2.3.2 Establishing the Vacuum: Pumps and Components

All CVD systems, with the possible exception of some APCVD systems, use vacuum technology, pumps and components. Figures 2.7 and 2.8 illustrate two example exhaust equipment configurations for low pressure and high vacuum processes, respectively. All of the components shown are commercially available, and sales engineers at vacuum component supply companies such as MKS, Varian, Kurt J. Lesker, BOC and Shimadzu can help with selecting the appropriate components. We want to maximize the conductance of the exhaust system so that it is at least the same magnitude as the pump speed, with larger conductance than the pump speed being the ideal situation to handle the range of operating conditions. High exhaust line conductance is achieved by maximizing the tube and component orifice diameters, and by minimizing the tube length, the number of components and the number of bends. In the vacuum component catalogue you will notice that there is a significant price escalation with exhaust system diameter, but this is not the place to try to save money. Paying for larger orifice exhaust system components is necessary as the exhaust system conductance limits the available reactor exhaust speed for a given vacuum pump as in Equation (2.8). Money invested in an expensive vacuum pump with high pumping speed will not improve the system performance if lower cost, smaller diameter fittings are used in the exhaust system. In general, it is advisable to use high quality, high vacuum components that have not been previously used for some other purpose to reduce contamination risk.

2.3.2.1 Gas Pumping Load

Before calling the component supplier, you need to determine the working pressure and gas pumping load for the CVD system. As we have already seen, the reactor pressure is a critical processing parameter that is complexly related to the system performance. Figure 2.9 gives a highly simplified picture of the CVD system that is used to size the vacuum pump and to determine the exhaust system conductance and the out-gassing and leak rates. In some processes, the reaction pressure must be controlled precisely, and equipment is included to control the process gas supply, Q_s, and the exhaust system conductance, C_{sys}, during the deposition. The exhaust conductance is changed *via* a variable throttle valve, C_1, which adds to the exhaust conductance, C_2, according to Equation (2.5). The minimum pump pressure is called the base pressure, and the pumping speed is often called the gas load. As discussed above, Equations (2.4), (2.7) and (2.8) and the ideal gas law can be used to model the steady state gas throughput, but we will need a few more relations because there are too many things we do not know in the system, *e.g.*, the system conductance.

The mass balance on the entire CVD system, shown in Figure 2.9, can be expressed in a general form that fundamentally applies to any system:

$$\begin{gathered}\text{Input} + \text{Leaks} + \text{Outgassing} + \text{Generation} = \text{Output} + \text{Consumption}: \\ Q_s + Q_l + Q_d + Q_p = S_p P_o + Q_R\end{gathered} \tag{2.15}$$

When determining the deposition system pump speed and base pressure we tend to neglect the leak and outgassing loads, as we choose system components to minimize these sources, and then we deal

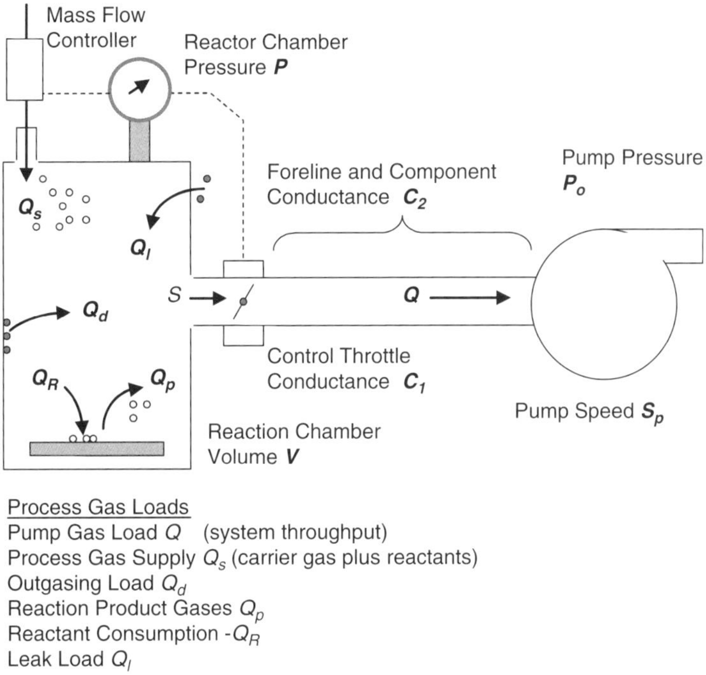

Figure 2.9 Schematic diagram for gas load calculation.

with them during the start-up procedures. We typically also neglect the reactant and product gas loads as they form minor volumetric constituents in dilute carrier gas. It may be very difficult to accurately determine the exhaust conductance. Exhaust component conductance is usually estimated from molecular flow relations in handbooks, corrected for transition pressures, or calculated from the viscous flow relation, Equation (2.14). We can get a reasonable idea about the system conductance by doing some simple experiments. The pump-down relation for the reactor volume, V, is derived from the mass balance with no input gas flow, and can be used to experimentally investigate leak and outgassing rates and the conductance:

$$P(t) = P(0)\exp\left(\frac{-t}{V/S}\right) \tag{2.16}$$

where $P(t)$ is the measured pressure during the pump-down time, t, from the initial pressure, $P(0)$. Equation (2.16) can be used to find the reactor exit speed, S, by fitting a curve to the pump-down measurement, then the conductance can be determined from Equation (2.8) by using the known pump speed, S_p, supplied by the manufacturer. The pump-down from atmosphere will be in the viscous flow regime until about 10 Pa is reached in the reactor. At this point, the outgassing and leak loads will start to become evident as the pressure approaches a stable value. The outgassing rate will reduce over time and the pressure will gradually reduce. But at the bottom of the pump-down curve the outgas and leak loads can be estimated by the mass balance on the closed reactor and exhaust system:

$$Q_d + Q_l = S_p P_o \tag{2.17}$$

Leaks and virtual leaks can be diagnosed after de-gassing and pump-down by shutting the reactor isolation valve and measuring the pressure rise with time, called the *leak-up* rate. If the leak-up rate is appreciable, then the fitting or connection that is leaking can be found by squirting ethanol around the individual connections in turn and watching for a pressure spike in the reactor. For UHV systems, helium gas is often used instead of ethanol as the vacuum measurement system is sensitive to helium, and thus can detect even very small leaks. When leaks have been eliminated and outgassing is completed, then the ultimate pressure of the pump, P_o, will be reached. The total time for pump-down will be a factor in wafer processing throughput, and can be modeled by considering the mass balance on the entire system:

$$\frac{P(t) - P_o}{P(0) - P_o} = \exp\left(\frac{-t}{V/S_p}\right) \tag{2.18}$$

The simple mathematics of the exponential decay pump-down equation shows that 98% of the pumping has been completed when $t = 4(V/S_p)$ or when $t = 4\tau_R$, where τ_R is the reactor system time constant.

At this point, we can see why the move in commercial system platforms has been toward reducing reactor volumes. In the previous section on viscous flow we saw how a smaller flow dimension increased Re and thus compressed the boundary layer and increased the flux to the surface. We have also discussed how the reactor surfaces contribute to out-gassing and how heated reactor surfaces reduce precursor efficiency through parasitic deposition. Here we see that the pump-down (and conversely the gas re-pressurization time for unloading) is impacted by the total reactor volume and pump speed. All of these factors point to minimizing reactor volume as a platform design consideration.

2.3.2.2 *Vacuum Pumps*

The vacuum pump is the prime-mover for the CVD system. The different types of pumps and considerations for pump operation will briefly be discussed here with the aim being general knowledge rather than expertise. Vacuum pumps are critically important, very expensive and easily damaged, so if you are new to the responsibility for selecting or operating the exhaust configuration for a CVD system, consult a vacuum technology text and get advice from an experienced vacuum engineer.

Vacuum pumps operate differently than fluid pumps or fans that "do work" on the fluid according to the first law of thermodynamics to induce a pressure rise to drive the flow forward. In fact, the gas in the vacuum system is not even aware of the work of the vacuum pump, and is moving according to the second law of thermodynamics to try to establish equilibrium by eliminating pressure, density and concentration gradients. The particular gradient of interest is the molecular density gradient created when the vacuum pump sweeps out or captures gas molecules. Thus, Maxwell–Boltzmann-type gas dynamics behavior is the dominant phenomenon in vacuum pumping of the reactor volume.

There are two general categories for vacuum pumps, according to whether they displace or trap gas molecules. Displacement pumps are also called gas-transfer pumps because they physically remove gas at the low pressure state, then transfer it to the ambient state after compression. Gas trapping pumps rely on condensation or chemisorption of gas molecules within the low pressure state, and do not actually exhaust the trapped molecules until after the deposition process. In general, displacement pumps have higher pump speed and gas trapping pumps can produce lower pressures. No one pump can cover the entire range of operation for vacuum systems from coarse and rough vacuum to high and ultrahigh vacuum.

Vacuum pumps are also classified by pressure operating range. Roughing and backing pumps are usually displacement pumps, while high vacuum pumps are capture pumps. Manufacturers' specifications for pumps focus on the ultimate pressure, P_o, and the nominal pumping speed, S_P ($L\,s^{-1}$ or $m^3\,h^{-1}$), but a pumping speed curve over the operating pressure range should also be available. The ultimate pressure is measured by connecting a vacuum gage to a blanked (dead end) fitting at the pump inlet and recording the pressure achieved after the pump is well warmed up. The pumping speed is measured under ideal conditions and the value reported by manufacturers is the highest value, which may only be achievable over a certain pressure range. For example, the roots pump has a definite pumping speed maxima in the mid range of operating pressures, while mechanical pumps have a rather constant pump speed that drops off to zero at the lower end of the operating pressure range. Pumps are also available as *wet or dry* versions, depending on whether they use oil for lubrication and sealing or not. *Corrosive* versions of displacement pumps are available that have various design and material strategies for dealing with aggressive chemicals.

Mechanical displacement pumps create a new volume of empty space within the sealed system with each stroke then move that volume to isolate it from the system. After the isolation, a mechanical compression process raises the gas pressure to above ambient so it can be exhausted to the atmospheric pressure gas clean-up system. Figure 2.10 shows the operating principles of mechanical pumps that require oil lubricants and sealants. Most vacuum pumps, for example the rotary vane pump, simultaneously create new vacuum space in the deposition system and compress and exhaust to the ambient, completing two intake and exhaust "strokes" with each pump shaft revolution. This produces a relatively smooth pumping action with moderate vibration. Isolation of the swept volume occurs at a point in the shaft rotation where the displacement volume is at a maximum. This displacement volume times the shaft rotation speed times two displacement strokes per rotation gives an indication of the pump speed. Higher pump speed can be gained by faster rotation or by a larger pump. Faster rotation produces more waste heat and larger pumps affect the system footprint.

The mechanical pump ballast purge gas flow is injected after the isolation, during the compression process. This raises the total pressure, preventing condensation of product gasses and solvents in the pump during the compression. The ballast gas also dilutes contaminants to keep the pump oil clean and reduce corrosion in pump and exhaust components. Some pumps have a variable gas ballast valve, because with the full ballast gas flow the ultimate pressure in a rotary pump is raised by as much as a factor of 10. Inert pump oil can be used instead of hydrocarbon oils when oxygen or other flammable gasses are being pumped. Perfluoropolyethers (PFPE) are completely inert, but if they are to replace an existing oil a full disassemble and solvent cleaning of the pump is called for, not simply a drain and refill.

Oil back-streaming is an issue for low pressure systems, so an oil-free or "dry" pump may be used. Figure 2.11 shows the operating principles of pump designs that are also available as *dry mechanical pumps*. The reason that these pumps can be run without oil on the contact surfaces is that the surfaces do not so much slide against each other as roll along each other, or they may not touch at all. The roots rotary pump can also be made with special tribological materials so that it does not have oil, but this increases the price substantially. The roots pump lobe surfaces are oil-free, but the shaft lubricating oil may still pose a contamination risk.

Non-mechanical gas transfer pumps include the molecular drag, turbo-molecular, oil diffusion and the newer turbo-drag hybrid. All non-mechanical transfer pumps use momentum transfer to gas molecules from the reactor to "knock" molecules from the chamber into the exhaust system where they are then captured and exhausted by a mechanical backing pump. Figure 2.12 shows the basic configuration for the high vacuum gas transfer pumps. All of these pumps should be located in line of sight of the chamber, and must use a line of sight isolation valve such as a butterfly or gate valve. Because they operate on the principle of momentum transfer, all of these gas transfer pumps have reduced performance with low molecular weight reactor gasses, particularly He and H_2. Gas transfer pumps require a mechanical backing pump, sometimes also called a foreline pump, because they cannot exhaust to atmospheric pressure.

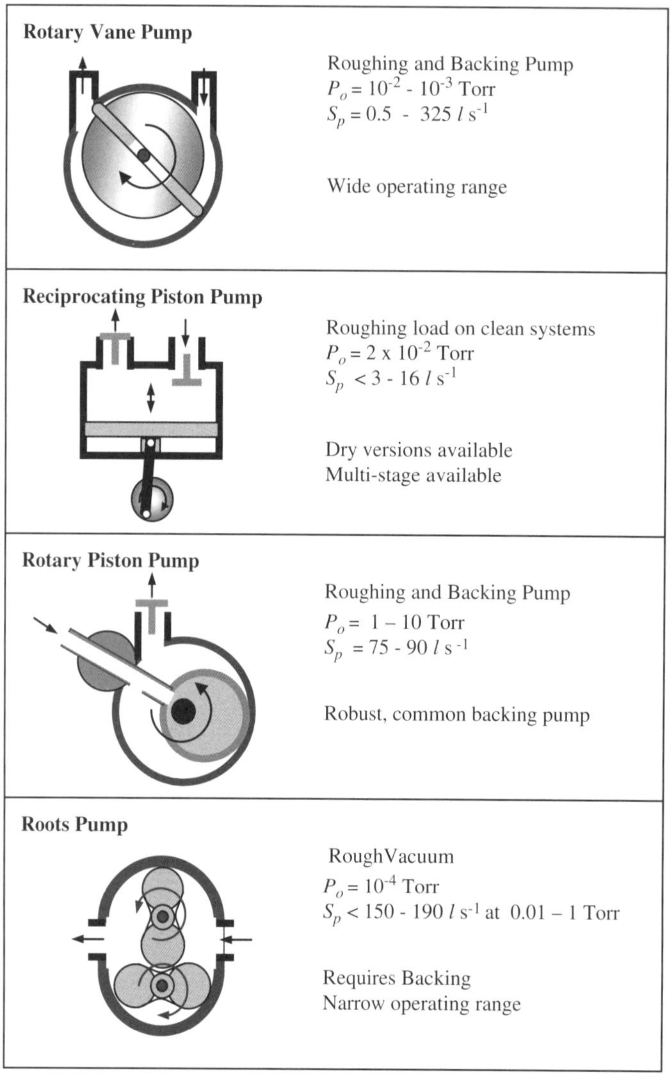

Figure 2.10 Operating principles and specifications for "wet" mechanical displacement pumps, which are generally sealed and lubricated with vacuum pump oil.

The oil diffusion pump is the oldest, lowest cost and most widely used high vacuum pump, but has a high risk of oil back-streaming. The diffusion pump operates with no moving parts by boiling the transfer oil in an enclosed space, and using the induced pressure to supply downward directed curtain nozzles. The high molecular weight oil vapor molecules entrain product gasses and direct them to the cooled condensation zone near the pump outlet. The turbomolecular pump resembles a jet engine with multiple angled blades on a very high speed shaft. Molecules randomly entering the swept area of the first blade will get hit by the underside of the blade with a resulting velocity direction through slots in stationary stator blades and toward sequential rotor blades. Multiple sets of rotor and stator blades produce very high ratio of inlet and exit pressure (called compression ratio), in the range of 100 for N_2. The shaft speed can be very high, up to 90 000 rpm, and very high pumping speeds can be achieved. Oil free versions of the pump use magnetically levitated bearings. The molecular drag pump can tolerate higher backing pressure, but also has lower pumping speed.

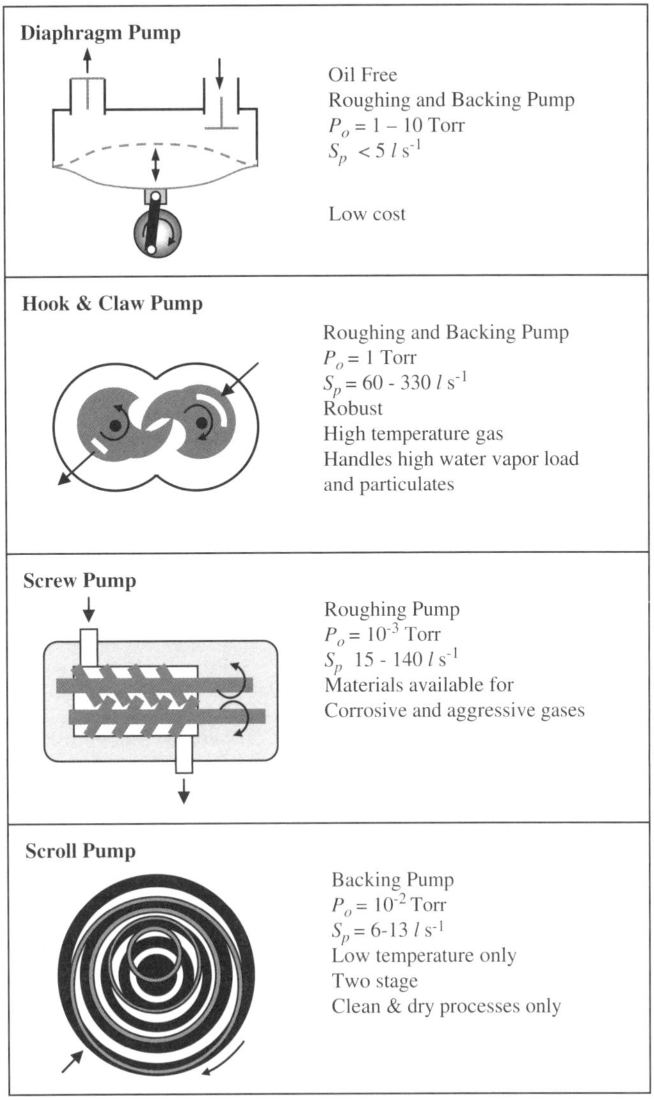

Figure 2.11 Operating principles and specifications for "dry" mechanical displacement or gas transfer pumps that have low oil back-streaming risk.

The drag pump has a high speed shaft with tangential speed approaching the molecular speed of the gas. The shaft rotates inside a stationary cylinder with helical groves that direct deflected molecules toward the pump exit. A very small clearance between the shaft and stator reduces molecular back-diffusion. However, as with the turbo pump, any failure of the pump due to bearing freezing or power failure will result in very rapid backstreaming.

Trapping pumps utilize the processes of molecular impingement, adsorption and chemisorption to trap molecules. The adsorbed molecules are chemically and energetically bound to the pumping surface in the pump. There is no outlet on trapping pumps, and so the pumping load is a one-time volume of gas at a certain pressure. The gas trapping pumps do not have oil but do pose a risk of contamination to the reactor with previously captured products in the event of loss of coolant or power. All gas trapping pumps pose a hazard of explosion of trapped flammables during the

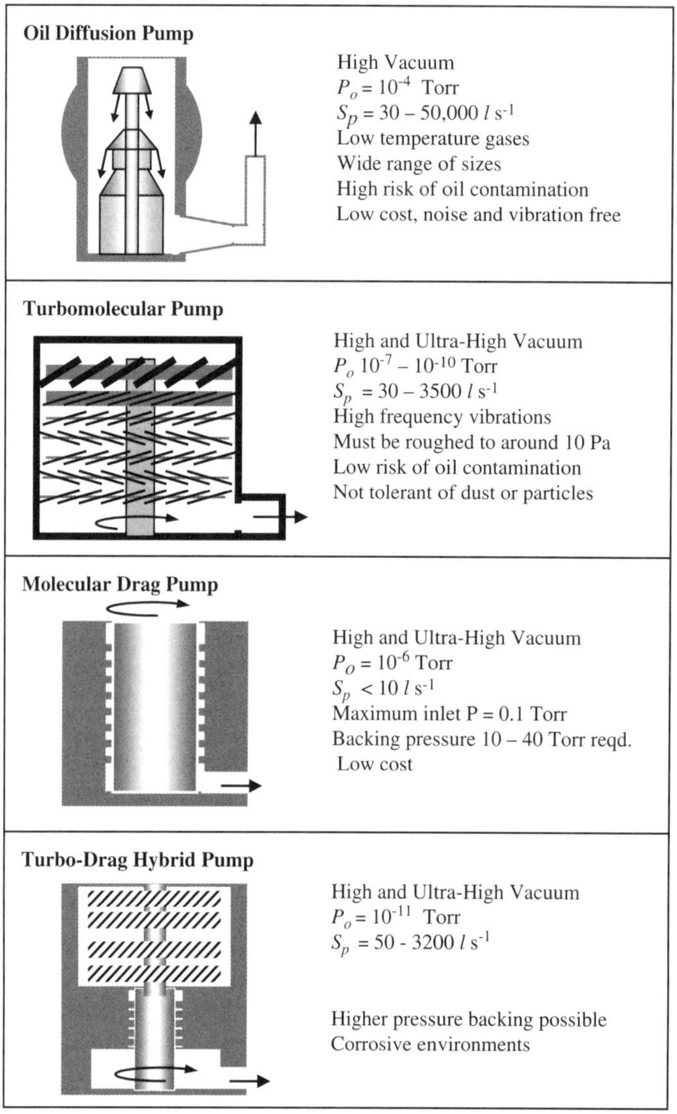

Figure 2.12 Operating principles and specifications for non-mechanical gas transfer vacuum pumps.

regeneration process at higher temperature and pressure, so adequate inert gas purge must be used and manufacturer's instructions followed.

The cryosorption pump is essentially a closed-end cylinder filled with molecular sieve pellets that are cooled with liquid N_2 in a surrounding Dewar sleeve. The pumping capability of the cryosorption pump is determined by the surface area of sieve material, which relates to cylinder size and sieve bead diameter. Cryosorption pumps are regenerable, low cost, easy to use and oil-free. The cryogenic pump, usually just called a "cryo-pump," is used for high and ultrahigh vacuum gas capture, and is also oil-free. A typical cryo-pump has entrance baffles cooled to 80 K to trap H_2O, then much cooler inner surfaces cooled around 20 K to trap other gasses including O_2, N_2 and Ar. One of these surfaces is activated carbon, which will trap H_2. The pump is kept cool by a He vapor-compression cycle with coolant supplied through insulated lines, and has a pumping range 10^{-6}–10^{-9} Torr.

The chilling cycle compressor vibrations may not be acceptable for some applications, and the pump is not suitable for aggressive gasses. The cryo-pump has found widespread application due to oil-free operation, high pumping speed and rapid recovery of pumping after high pressure operation.

Ion pumps are widely used to deliver UHV clean and vibration-free operation to 10^{-6}–10^{-11} Torr base pressure. Ion pumps are well known to be robust with long operating lives and low power input. A bank of stainless steel tubes are arrayed on the anode plate of a pair of Ti plates, and placed in a strong magnetic field parallel to the tube axes. Electrons from the cathode plate ionize any reactor gasses in the volume of the pump as they travel through the magnetic field to the tubes and on to the anode plate. The ionized gasses then are strongly attracted to the anode where they strike the tubes or plate with enough force to sputter the Ti, chemically react, or they can also be implanted (ion or neutron burial) in the cathode. The ion pump can pump inert gasses, and does not pose a de-gassing risk as the trapping is permanent. All trapping pumps require initial pump-down of the reactor and system lines with a mechanical pump before they are activated or they could be saturated with gas.

2.3.3 Pressure Measurement

Control of the CVD process depends entirely on the vapor supply system and the exhaust system. Pressure measurement is often the only feedback we have to the operation of each part of the system. There is no one vacuum pressure gage that can universally be used throughout the system. Figure 2.13 shows the basic operating principle and specifications for the most commonly used vacuum gages. There are three main classifications for gages, depending on how they detect the vacuum: mechanical gages, gas property gages and ionization gages. There are different sources of error for each gage, such as temperature and orientation, and a discussion with the technical specialist from the vacuum component supplier would be a good way to make sure the particular gage is installed properly, and that the reading is as accurate as possible.

Mechanical gages, also called direct pressure gages, undergo physical changes induced by exposure of the transducer to the CVD system. Mechanical gages are usually not accurate below 10^{-5} Torr, and are not sensitive to temperature. The Bourdon gage has a long curved oval cross section metal tube connected to the vacuum. The degree of vacuum causes deflection of the tube, and a fine mechanical linkage amplifies the deflection to a needle reading on a dial. The Bourdon gage is a low cost and very common gage with robust operation and little drift over a long device life. The Bourdon gage does not have electronic output, so is used for visual verification of the vacuum rather than for precise measurement or control. A Piezo sensor is a very small gage with a highly sensitive strain gage machined onto the surface of a silicon wafer in a way so that the deflection of the thin silicon crystal is the pressure transducer. The sensor can be used to supply an electronic signal and can be placed totally within the vacuum with only the wires fed through. In a capacitance manometer, a standard gas at known pressure is separated from the vacuum environment by a thin metal diaphragm. Deflection of the diaphragm is measured using the electrical capacitance field, which can be detected into the nanometer range of diaphragm deflection. Individual capacitance monometer gages have a measurement range of 10^3–10^4 Torr. The gages can be designed with maximum pressures over a huge range from 10^4 to 10^{-1} Torr. Thus, more than one capacitance manometer will be needed to measure pressures that range more than three orders of magnitude. They are the most accurate mechanical gage, and can be used for almost any gas or vapor. The diaphragm manometer operates the same way as the capacitance manometer, but the diaphragm deflection is measured by an attached strain gage. The pressure range is not as large, but these gages are robust and reliable. Mechanical gages are typically used for the roughing line, the reactor and in the delivery system.

Gas property gages, also called indirect pressure gages, measure heat transfer effects related to a bulk property in the vicinity of the gage such as viscosity or thermal conductivity. Each kind of

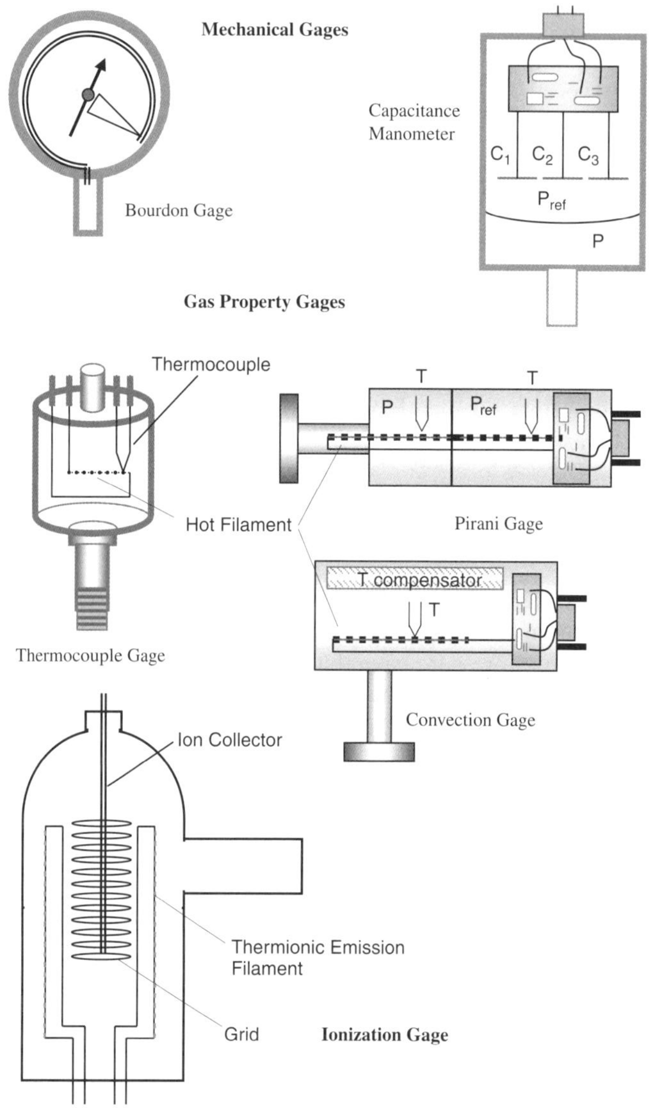

Figure 2.13 Mechanical gages are used for atmospheric, coarse, and rough to low vacuum. Gas property gages operate in the low pressure range, and the ionization gage can only be used in UH and UHV vacuum.

gage has a relatively narrow operating pressure range corresponding to the particular property variance with pressure. Property gages are inherently inaccurate for determining the actual pressure value because they are interacting with the reactor gasses. However, for a given system they can be used to measure a certain condition that can be used for system control with respectable repeatability. Thermocouple gages (T/C) have a constant current supplied to a high resistance heated filament, and the temperature of the filament is measured by an attached thermocouple. If there is no gas present, the filament will establish thermal equilibrium with the housing by radiation exchange, and the thermocouple reading will have no relation to pressure. In the molecular flow regime, the rate of heat transfer to the gas is a function of the molecular flux to the surface, and so the cooling effect is directly proportional to the pressure. The temperature will decrease with higher pressure until the viscous flow regime is reached and, because there is no bulk flow past the filament, the heat transfer rate is simply due to the conductivity of the gas. However, the property of thermal conductivity is essentially independent of pressure. The pressure range is 1–10^{-4} Torr and the gage is usually orientation dependent. Despite the disadvantages of nonlinearity of reading and inaccuracy, the thermocouple gage is low cost and commonly used, *e.g.*, to signal when the roughing process is complete and the high vacuum operation can be commenced. The Pirani gage uses the same principal as the T/C gage, but one heated filament is exposed to the system vapors and another reference filament is enclosed in a standard gas. Unlike the T/C gage, the filament is not set at a constant voltage or current but is varied through a feedback circuit to maintain a constant filament temperature. This increases the measurement range to 10–10^{-5} Torr and gives the Pirani a much faster response, if no greater accuracy. The convection gage uses a temperature compensator to remove heat from the filament by natural convection. The gage is strictly orientation dependent and works in the viscous range of 10 Torr to 1 atm.

Ionization gages include a family of gages with different high vacuum and UHV ranges, but with common operating principles. The reactor gasses are ionized by energetic electrons and the positive ions are collected at an electrode with the current generated being used to generate the pressure measurement. As a result, these gages are highly dependent on the gas composition. They cannot be used to give absolute pressure measurements except with a pure calibration gas. The gage has a linear response with pressure in the molecular range up to about 1 Pa. Hot filament gages use thermionic emission of electrons, and include Bayard–Alpert and Schulz–Phelps models. Cold cathode gages use a glow discharge to produce the electrons within a magnetic field, and models include Penning and inverted magnetron. Care must be taken not to expose an operating ion gage to high pressure. Ion gages are quite fragile, as the housing is made of glass, and they are quite expensive. Some ion gage designs have redundant filaments built into them as the lifetime and risk of failure of filaments is an operation issue.

System and process diagnostics can help to maintain quality production, and one important tool, the residual gas analyzer (RGA), is essentially an ionization gage integrated with a mass spectrometer. Water vapor, pump oil, air and leaking carrier gas and process gas valves can be detected at high vacuum operating conditions using a system-installed RGA. The instrument operates by examining the gas spectrum generated by separation of the ions by use of a quadrupole electrostatic rf field and impact on a sensor. Leaks from the atmosphere can be pinpointed in high vacuum systems by using a very fine tube supplying a tiny flow of He. The He tracer gas is passed over the exterior of each fixture in turn while the system is at vacuum, and the analyzer observed for traces of He. The number of ions of a given species detected is proportional to the partial pressure of the gas, so while the analysis of the mass spectrum signals is complex, RGAs can be used for a range of diagnostics. As with the gas property gages, ion gages can be used to control the CVD system through observation of the reading that correlates with the optimal deposition performance even if we do not know what the actual reactor pressure is with great precision.

Clearly, none of the vacuum gages can be used in the high temperature environment of the hot wall furnace. The gage is located outside the high temperature zone and is connected to the reactor *via* a tube. With a significant tube length, the temperature is reduced to room temperature. However, we know that the impingement rate in molecular flow is related to the local gas temperature by Equation (2.2). The pressure reading in the low temperature environment outside the reactor will not equal the pressure inside the hot reactor.

2.3.4 Flow Control

Mass flow controllers are used to maintain system operation in the processing window. If the reactants and precursors in the CVD process are gasses at room temperature, then a mass flow controller (MFC) can be calibrated to the gas or to the diluting carrier gas, and can be directly used to control the injection process. Precursors that are normally liquid or solid at room temperature and pressure have traditionally been heated and evaporated into a carrier gas stream. The flow cannot be controlled down-stream of the bubbler or sublimation source vessels, due to the risk of vapor condensation, so the MFC on the carrier gas supply line and the precursor reservoir temperature are the primary means of controlling the injection rate.

Figure 2.14(a) gives a simple schematic diagram of the measurement principle of a MFC and a sketch of the most common configuration of the device. The standard MFC uses a feedback control of a leak valve to achieve a temperature difference along a sample flow tube for a gas of known

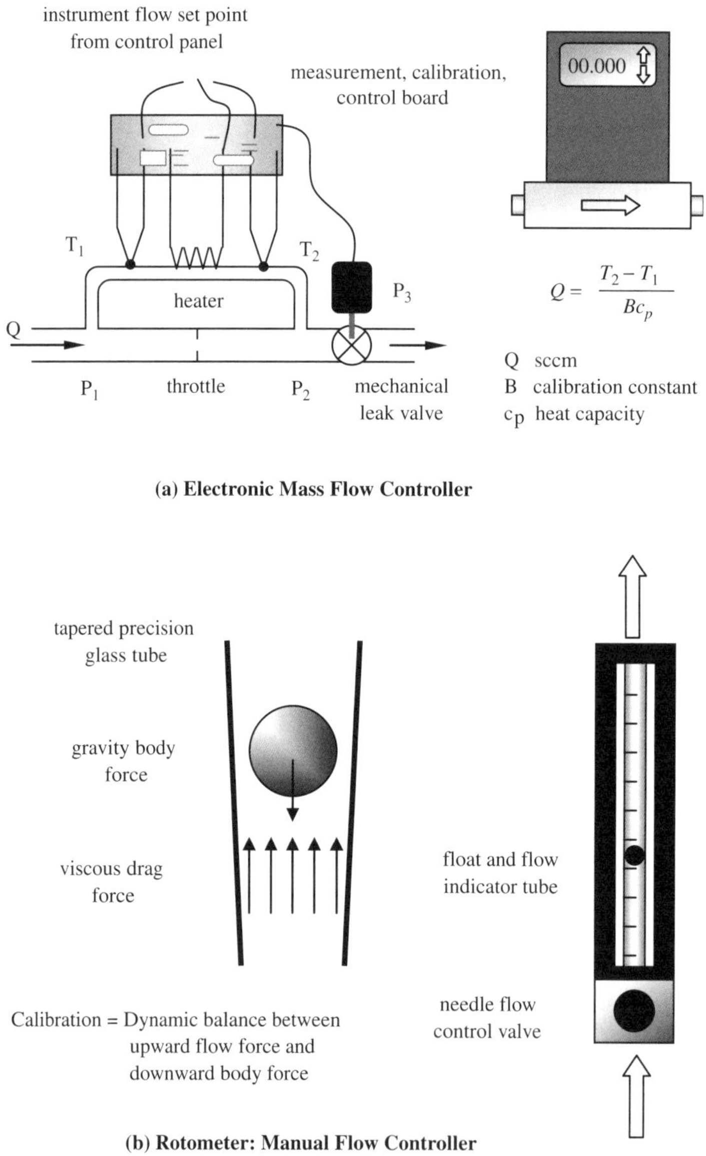

Figure 2.14 Schematic diagram of mass flow controllers: (*a*) differential temperature operating principle and sketch of a common MFC device; (*b*) manual tapered tube and float flow meter and sketch of a rotometer with needle valve.

properties. A small heating element with a controlled current and voltage will raise the temperature of a certain flow of gas by a predictable degree for laminar flow. Modern MFCs come with multiple pre-programmed calibration constants for standard clean gasses at room temperature. The MFC must be on the supply line, and protected from back diffusion of precursor chemicals and oils and from high temperatures. The MFC is not used as the gas safety or shut-off valve. Leak valves that control flow into the vacuum chamber may be down-stream of the MFC. If the CVD process uses a hazardous gas, a calibration of the MFC will be required. This is relatively easily done by pumping down the reactor, then using the pressure–time measurement and the leak-up rate, Equation (2.18), for different settings of the reactor leak valve. This simple method can also be used periodically to check the calibration of the MFC, which will drift over time and require re-calibration by the manufacturer.

Figure 2.14(b) shows the operating principle and instrument sketch of the manual mass flow meter/controller, or rotometer. The principle of operation is straightforward, but the device must be calibrated for non-standard gasses. The basic configuration is a tapered glass tube with a metal or PTFE float, which are available in a wide range of sizes for both liquids and gasses. The gas enters from the base, and upward pressure on the spherical or cylindrical float causes it to rise. The dynamic equilibrium between the viscous drag force of the flow around the float and the downward gravity force causes the float to assume a vertical position in the tapered tube that corresponds to a particular flow rate. The viscous force on the float is dependent on the density of the gas, so the calibration is pressure and temperature dependent. Most rotometers must be installed in a vertical position, but there are some models that use a spring to resist the drag on the float, and so are not orientation sensitive. The rotometer does not have electronic output or control, but is low cost and reliable for research laboratory applications and allows visual verification that the flow is present.

2.3.4.1 Valves, Traps and Filters

We have already mentioned valves and traps as important components in the vacuum system during a description of pumps and the exhaust system configuration. This short section is dedicated to describing the function of these components, given their importance in good system design as well as in economical and reliable operation. A general rule is that every component in the exhaust system represents an added conductance, and thus we work to minimize conductance penalties while optimizing system control, safety and protection for equipment and the environment.

A valve closes by pressing a gate against a seat, usually with an elastomer seal between the two parts. Vacuum valves are categorized according to their physical structure and application. In Figures 2.7 and 2.8, there are valves in different positions in the exhaust system and each of them would be a different type and have different characteristics. Valves commonly used in CVD systems are collated in Figure 2.15 with a simple diagram, function description and usage specifications. Electromagnetic solenoid actuators are available for most valves to allow automatic operation by the control system. The pump isolation valve in particular is a normally open solenoid valve that closes automatically in the event of a power failure to protect the system from oil vapor. Pressure-limited and flow-limited valves that automatically shut-off should pre-set conditions be exceeded are used in the supply line for safety.

Vacuum valves are factory tested before delivery and a nominal leak rate is supplied with the specifications. Contamination of valve seals over time can cause increased leak rates so that a maintenance and cleaning schedule should be part of CVD system operation.

Diaphragm valves are the most common type of valve in chemical processing applications. However, they do not generally have acceptable performance for use in vacuum exhaust and low pressure gas control. A mechanism presses an elastomer diaphragm against a seat to seal off a small flow hole or holes. The diaphragm valve has much too low conductance and too high leak rate for vacuum application.

Leak Valve

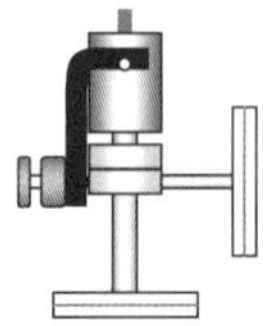

Also called a gas control valve.
Admits a small, adjustable flow of source gas into the vacuum system.
Soft metal pad is compressed against a hard metal seat with varying degrees of pressure.
High temperature and corrosive gas models.

Gate Valves

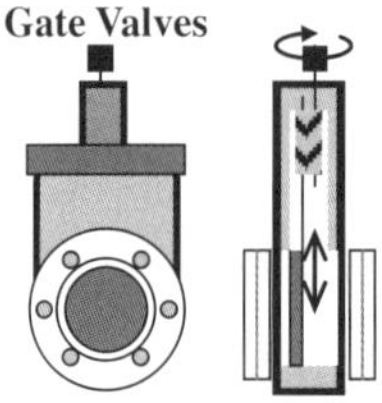

A *line of sight* valve.
Unobstructed access and high conductance.
Valve gate is raised out of the flow path.
Used above HV pumps, instrument access.
Common in CVD systems, expensive.
Rectangular gate used for wafer loading port.

Butterfly Valve

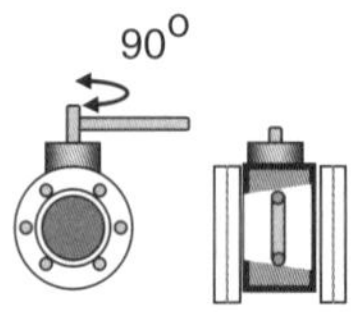

Also variable conductance throttle valve.
Valve gate rotates to open position.
Line of sight, low obstruction at full open.
Easy removal and cleaning.
Common HV and UHV valve.
Roughing and foreline.

Block Valves

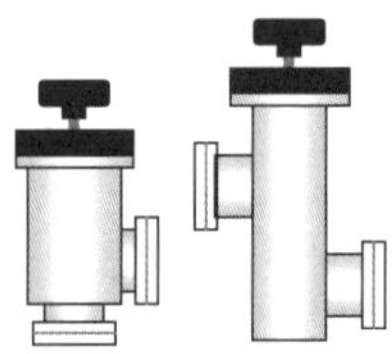

Includes right angle and inline valves.
Cylindrical gate seals against bore seat.
Low cost and robust, tolerant of particles.
Lower conductance due to bends inside valve.
UHV shutoff, bake-out, foreline valve.
Isolation valves for traps and filters.

Ball Valve

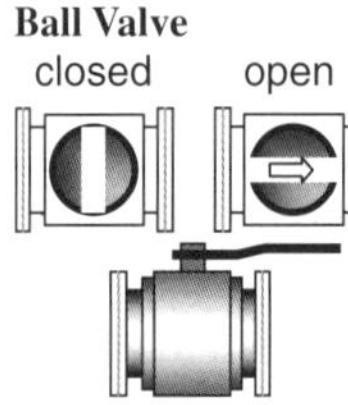

Reliable and robust operation, Low cost.
Gate ball with bore-through rotates in seat.
Foreline or roughing line valve *only*. Not used in high vacuum due to leaks and virtual leaks.
Line of sight, conductance determined by bore diameter.

Figure 2.15 Common valves used in CVD systems with functional sketch and application notes.

Traps collect droplets, vapors and gasses but the most important function is protection of the vacuum pumps which should not pump large quantities of condensable vapors. A system trap is located at the system outlet, and a foreline trap is placed between the high-vacuum pump and the mechanical backing pump, as close as possible to the mechanical pump. Traps normally work over all flow regimes, and necessarily have low conductance due to the change of direction of the flow. However, the direction changes and large surface area are what causes aerosols and vapors to impinge on a surface and become trapped. The rule of thumb is to get the highest trap conductance possible by using a large diameter trap. The cooled chevron baffle trap is a common system trap used to protect the reactor from back-diffusion of vacuum pump oil vapors. The foreline trap protects the high vacuum pump from oil vapors from the mechanical pump. To trap oil vapors, a foreline trap may be filled with high surface area fiber packing, porous materials, molecular sieves or some other absorbent material. These absorbent materials must be replaced or regenerated regularly under high temperature and purging gas as directed by the manufacturer.

Cryogenic traps capture aggressive vapors and solvents by condensation that could become dissolved in the pump oil, causing thinning and loss of lubrication. A liquid N_2 (LN_2) trap requires additional operation and safety procedures. During operation, the LN_2 level must be kept constant to avoid re-evaporation of frozen gasses above the LN_2 level. After the deposition run, the cryogenic trap is a potentially dangerous source of a significant amount of gas and toxic materials. The evolution of previously frozen trap gases can turn the CVD system into a pressure bomb if even part of the LN_2 coolant is lost. The shut-down procedure should include valve isolation of the trap, and heating and re-generation by low pressure, dry purge N_2 gas to an exhaust gas handling system. While the chemistry-style glass LN_2 trap is common laboratory equipment, commercial systems will employ more robust cryogenic traps like a Meissner trap that uses a LN_2 blow-off flow from a getter or even from a refrigeration cycle through multiple loops of coils with fins.

Filters capture particles and aerosols, and can be used in the foreline when powders are a major issue. An oil mist eliminator is a necessary component at the mechanical pump exit. A rather large load of oil droplets can be present in the pump exhaust, and the oil de-mister traps and returns most of the oil to the pump, rather than sending it into the gas clean-up and exhaust system. Particles and powders form inside the reactor and also in the exhaust lines. Most of the pumps in Figures 2.10 and 2.11 have high speed mechanical parts with very tight tolerances. Even very tiny particles and droplets can cause damage to pumps. Since particles and aerosols cannot be carried in molecular flow, filters are only used in transition or viscous flow lines. Like traps, filters also have many changes of flow direction and no direct line of sight flow path. This is provided by fiber materials and/or cyclone separator. Commercially available filters have multiple cyclone stages, cooled coils and filter elements to remove a large range of particulate sizes and condensable vapors from the foreline.

2.3.5 Temperature Measurement

Temperatures in the precursor supply system and, of course, in the reactor and at the substrate are critical to the CVD system performance. Temperature measurement techniques are well known, and this section is intended as an introduction and overview, not as an expert guide. The basic principle of temperature measurement is observation of a known and calibrated thermo-electrical property response of a transducer material to its surrounding temperature field. The tricky part is not the transducer or its response. The issue with temperature measurement is getting the transducer probe into thermal equilibrium with the solid or gas of interest. The probe will reliably report what ever temperature it is at. However, it is often not a simple matter to get the probe to be the same temperature as the substrate or low pressure vapor.

The most common temperature measurement device by far is the thermocouple. There is an electrical potential difference between dissimilar metals that varies with temperature of the point of contact (called the *junction*). Thermocouples are available as bare wires, insulated wires, spot welded junctions and a wide range of probes. The common standard *type K* thermocouple uses a junction between chemically resistant Chromel and Alumel alloys that produces an open circuit voltage of about $0.04\,mV\,K^{-1}$ with linear behavior over a wide operating range up to 1200 °C. A range of other thermocouple types are available, and they all have standard color-coded wire insulation and connecting plugs. As most laboratories will have several types of thermocouples and readers, a new experimentalist should acquaint themselves with good measurement practice by reading a measurement text and the supplier's technical notes.

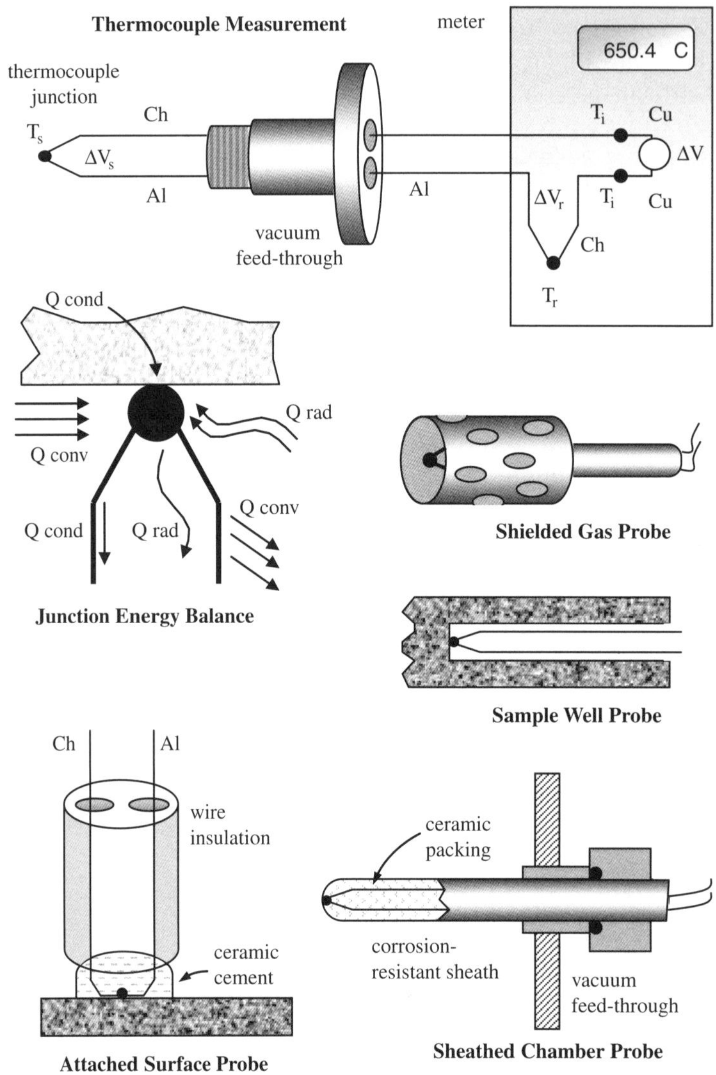

Figure 2.16 Measurement principle for a typical *type K* thermocouple, the energy balance on the junction, and four different probe types for different applications. Al = Alumel, Ch = Chromel.

Figure 2.16 shows the basic measurement principle and a range of thermocouple probe types. Although we know $d(\Delta V)/dT$ for a metal pair, we do not know what voltage the sample junction, T_s, will produce [*e.g.*, we do not know $V(T)$]. A second junction at T_r, a known reference temperature, allows calculation of the sample temperature because the voltage will be proportional to the temperature difference by a calibration constant. An ice bath is the common reference temperature, but modern meters have an electronic reference generator in the measurement circuit. The thermocouple wires must be connected to the meter, which has copper wiring. These junctions will also produce a voltage, so they are typically mounted on a heat sink block to ensure they are the same temperature and thus their voltages cancel each other out. It should be stressed that the meter voltage reading is the sum of all dissimilar metal junctions in the circuit. If the Ch-Al wires touch each other anywhere else, as in twisted wires, then a mysterious reading will occur that indicates a temperature value between the junction temperature and that at the crossed wires. Vacuum feed-throughs are available for all types of thermocouple wires and care must be taken to attach the wires correctly.

Obtaining a precise measurement of the deposition temperature is usually more difficult than obtaining a repeatable temperature measurement for process control. For example, the thermocouple embedded in a well in a wafer chuck will give a temperature reading that may be 50–150 °C different than the actual wafer temperature, depending on the pressure and flow conditions. However, CVD processes are usually developed experimentally, and a particular reading from the chuck thermocouple can be correlated with the conditions for a high quality film, and thus used as the process control measurement. The accuracy and repeatability of the temperature measurement depend on the thermal interaction between the probe and its environment. The junction energy balance sketch in Figure 2.16 illustrates all of the heat transfer modes affecting the equilibrium temperature of the junction. The strategy for a surface temperature measurement is to fix the junction to the surface with high refractory ceramic cement, which eliminates radiant exchange with the surroundings and has good thermal contact with the surface. Insulating the wires reduces the heat loss along the length of the wires. We usually also use the smallest gage wire possible to reduce heat loss from the junction, recognizing that smaller wires are more fragile so care will need to be taken. Temperature measurement on supply lines and precursor supply bottles are usually gained by adhesive attachment to the outside of supply lines. Gas temperature in the viscous regime uses a shielded thermocouple to reduce radiation exchange with surrounding solid surfaces. The substrate heater temperature is often monitored by drilling a small diameter hole as deep as possible into the housing material or into the susceptor on which the substrate is mounted. The junction is buried deep in the well, and may be sheathed with ceramic tubes with separate holes for the wires to prevent twisting and inadvertent touching of the wires. The low-pressure, hot-wall chamber temperature can be monitored using a probe packed in a refractory ceramic and protected by a corrosion resistant alloy sheath that also can be sealed into a vacuum feed-through.

An infrared non-contact optical pyrometer may be employed if contact with the substrate is not acceptable and optical access to the substrate surface is possible. The radiation emitted from a surface is directly related to its temperature and emissivity. Photo-detectors are readily available but few can be used in high temperature, corrosive or vacuum environments. A quartz window will transmit enough IR radiation to use the reading to control a process but, again, the actual temperature may be quite different from the reading. Interestingly, because the depositing film layer will nearly always have different emissivity than the substrate, the measured radiation will vary with deposition, and the pyrometer can be used to monitor the deposition process if not the actual temperature.

2.3.6 Heating Strategies

Heating can seem deceptively simple, after all, passing electric current through a resistive wire produces heat. But thermal engineering and control is one of the most troublesome aspects of CVD system design, often on a par with the flow dynamics. It should be clear to all CVD specialists, regardless of discipline, that the deposition temperature and the energy input to activate the deposition reaction are the critical requirements for deposition. We already know that uniformity, throughput and cleanliness are complexly related to the temperature and energy fields in the reactor. This brief discussion of heating strategies will cover a few of the important issues but will not give instruction on heat transfer, which is a critical discipline for those responsible for design or development of new reactors.[11]

The precursor and gas supply in many CVD processes are heated to aid evaporation of precursor into the carrier gas, and to reduce temperature gradients in the reactor. Temperature controlled heating jackets are standard chemistry equipment, and are commonly used for precursor evaporators and bubblers. Heating tape with several thermocouples along the tape length can be wrapped around inlet gas tubing to heat the process gas.

The outgassing procedure prior to deposition normally involves bakeout of the reactor and vacuum system. The higher bakeout temperature increases the desorption rate of adsorbed vapors and gases on reactor surfaces. As a rule of thumb, stainless steel reactor chambers must be baked out to at least 200 °C for one hour to reach pressures in the 10^{-7} Torr range, and 400 °C for many hours to reach 10^{-11} Torr. The reactor needs to be well insulted to reach higher bake-out temperatures, and the process of wrapping a research reactor for bake-out poses risk to delicate instruments and feed-throughs. Commercial systems would already be insulated and have the bake-out process programmed into the production cycle. Some commercial systems have gone to plasma or reactive gas clean-out to decrease the time needed for bake-out. Depending on the size and platform of the CVD reactor, the bake out heater can be either external or internal. Internal bake-out heaters are usually an infrared or quartz lamp called a *stab-in heater*. Sometimes the process heater is used as the bake-out heater. MBE, CBE and other HV and UHV systems tend to use external insulated blankets wrapped over heater tape or ceramic heaters on individual parts, essentially creating a virtual oven around the whole system.

Sample heaters, also called substrate heaters, elevate the temperature of the substrate by one of two methods, surface radiation by a lamp or conduction from the sample holder or susceptor. A range of HV and UHV compatible quartz lamps are available in different sizes and shapes that can be mounted inside the vacuum chamber and directed at the substrate. Lamps can heat wafers to 1000 °C with good uniformity over a specific area. Radiation heat transfer analysis and design are important for uniform heating by lamps.

Hot wall and plasma assisted processes often also employ a platform substrate heater, also called a heater stage. The substrate heater consists of either bare heating wires coiled or strung in a coverage pattern below the substrate or heating wires packed (or sintered) into refractory ceramic material. The substrate heater may heat either the substrate directly or a flat platform made of high temperature molybdenum or alloy onto which the substrate is placed. Larger commercial wafer heaters may employ boron-nitride or other high temperature ceramic heater stage, or a metal susceptor that is remotely heated by a radio frequency coil.

Contact thermal resistance between the stage and the substrate is a function of the mounting pressure and the interstitial substance. Even very flat clean surfaces that are not forced together actually have a very limited number of contact points where conduction heat transfer can occur between the solids directly. We cannot usually apply a lot of pressure onto the substrate or use a thermal contact grease in CVD, so we must accept that the substrate will be a different temperature than the heater stage, which is usually where the process thermocouple is located. A vacuum makes a reasonable insulating medium to convection heat transfer, and radiation is the only mode of heat transfer in UH and UHV applications.

In a cold-walled reactor with no other sources of heating from above, the substrate may be 50–200 °C cooler than the heater stage, depending on the temperature, reactor pressure and material properties. If the mean free path is on the order of the gap size, then heat transfer can be greatly enhanced, as trapped gas molecules can very rapidly move back and forth between the heater stage and substrate, transferring thermal energy with each impact. Of course, the heat transfer from the top surface will also be increased at higher pressure, but the temperature difference between the substrate and the susceptor will not be as large as at high vacuum. In the viscous regime, the heat transfer rate between the surfaces is limited by the conductivity of the trapped and stagnant gas. The substrate will again be at a lower temperature than the heater stage because the moving gas above the substrate can carry away heat more readily.

The hot-wall reactor platform places the whole reactor, or the deposition zone of a tube reactor, inside an oven. The uniform thermal field only exists where the radiation view factor of any openings or cold surfaces are small, *e.g.*, in the central section of a tube furnace. The induction coil is widely used to heat susceptors placed around the reactor wall or in the heater stage. Plasma is another means to provide the energy for activating the deposition reactions, and this is the subject of Chapter 12 of this book.

2.4 Conventional CVD Reactors

When you walk up to a commercial CVD system, you will typically see a nice cabinet with a user interface computer screen and a wafer loading port. If you open the cabinet, the maze of stainless tubing, fittings and electronics is truly stunning. Minimizing the system footprint and the maintenance down-time are design drivers for the compact yet accessible layout of the CVD system inside the cabinet, and these are hallmarks of industry-leading deposition tools. Here we will look at the design considerations of safety, process control and quality for the conventional CVD system that uses a toxic reactive gas and a liquid or solid precursor in the traditional bubbler and carrier gas delivery system. Figure 2.17 shows a very simple CVD system with only one reactive gas and one liquid precursor supply to one reaction chamber. These basic configurations would be replicated for more reactants and for multiple deposition chambers.

The first consideration is safety, and the cabinets themselves are part of the safety system. The cabinet structures are exhausted to the atmosphere outside the fab lab to keep them at constant negative pressure. This contains and vents any leaks during operation. It also draws gas, vapor or particulates out of the work area, which may result from line bleed during precursor, gas bottle or filter change and maintenance, and during rector unloading and cleaning. If the reactant gasses are very dangerous or aggressive, the gas supply cabinet may be located outside the fab lab to protect personnel and other sensitive equipment. Cabinets are fitted with gas detectors that trigger the control system to shut off the electronically activated safety valves on the gas sources and reactor supply and sound an alarm if hazardous levels are detected. High pressure gas bottles are fitted with mechanical pressure regulators that step down the very high bottle pressure to the system supply pressure. These gas regulators should have the feature of a safety lock that can be set at the maximum supply pressure to eliminate operator error. A mechanical gas-flow-limiting safety valve is an important safety feature on the gas supply line. This type of valve has a ball gate held open with a spring that is sealed against the seat when a pre-set high flow rate is reached and must be manually re-set. The flow limiting shut-off is a low cost, last line of defense in case of power or controller failure and line leak or supply valve failure. Flexible tubing or a helical loop of stainless tubing connects the gas bottles to a change-out valve fixed to a panel, which allows the bottle to be removed. All lines, valves and components in a CVD cabinet are clearly labeled, including on/off positions of manual valves. All hot surfaces, including the rotary vacuum pump, are also labeled and adhesive surface temperature indicators attached.

New, safer gas supply bottles are coming into the market, which use a high quality adsorbent-filled cylinder. More gas can actually be packed into the adsorbent, and it is supplied at a much lower pressure with fewer gas bottle change-overs. The selector valves on the inert gas line in the precursor cabinet are used to flush the lines after deposition and while any plumbing work is being done to minimize water vapor intrusion into the system. Pneumatic valves are operated by the control system to isolate the gas cabinet and the precursor cabinet during shut-down, or in the event of any other system irregularity without needing to open the cabinets. The pneumatic valves in the exhaust system are part of the exhaust system operation as described in the previous section, but also slam shut to isolate traps, filters and pumps in the event of equipment or power failure. The pneumatic valves on the precursor gas inlets to the CVD reactor are also safety valves to protect precursor supply equipment from back-flow of mixed gases in the event of coolant loss in the exhaust system, pump failure or any other event that causes a high reactor pressure. The reactor inlet valves also stop any detected supply system leaks from becoming atmospheric gas sources into the vacuum chamber once the source gas pressure is shut off. Clearly, the consequences of worker exposure to dangerous substances and equipment damage are more costly than the price of sensors

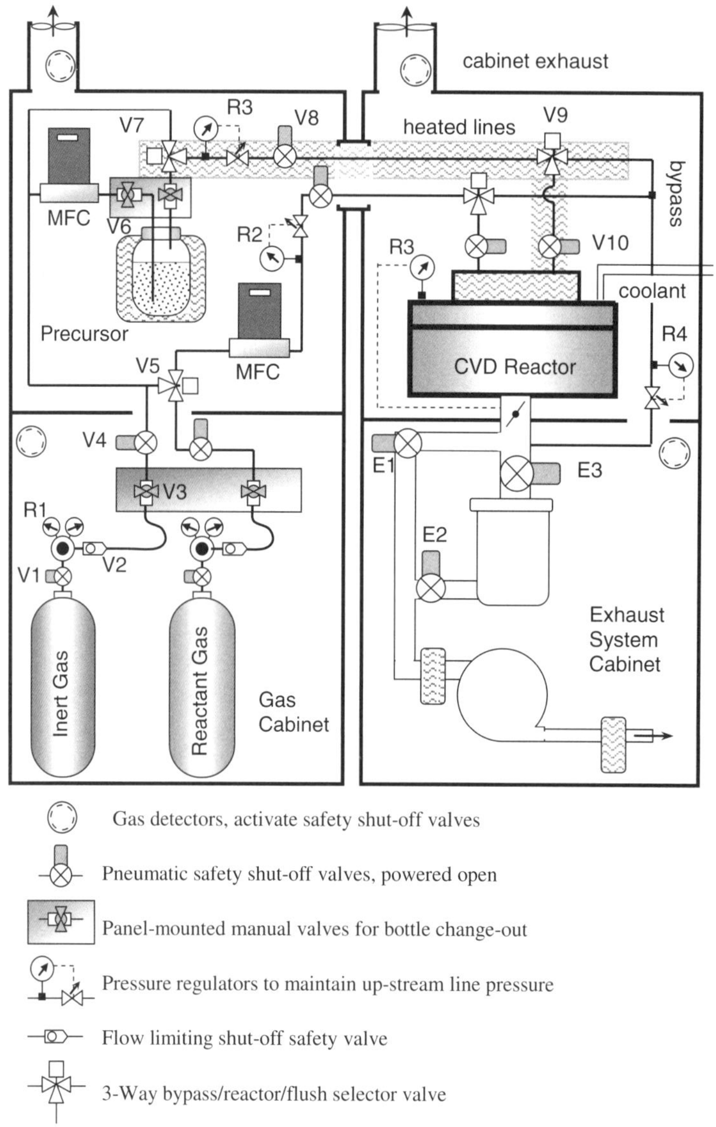

Figure 2.17 Schematic diagram of the typical elements and layout inside CVD cabinets.

and safety valves in commercial system design. The research laboratory should have the same focus on safety as the production fab lab.

Process control depends on control of reactor pressure, gas throughput and temperature. The mass flow controllers are essential for all gas supply, but they require a rather large pressure drop (1 kPa) across them to stay within calibration. Thus, we need pressure control and step-down to the reactor pressure provided by pressure regulators after the MFCs. Liquid and solid source precursors, when intimately contacted with the carrier gas flow, will evaporate to the saturation state at the evaporator temperature and pressure. The carrier gas evaporator pressure is controlled by the down-stream pressure regulator, and the source vessel temperature is controlled by a heating (or cooling) jacket. The temperature and pressure desired will depend entirely on the precursor vapor pressure, p_v, which is a strong function of temperature and the desired precursor concentration in the reactor. If the precursor needs to be heated, then the lines and fittings between the evaporator and the reactor will need to be insulated and maintained at the saturation temperature to prevent condensation. A leak-valve at the reactor inlet may be required to maintain pressure and prevent condensation. During the system start-up, the temperatures, flows and pressures must be stabilized prior to feeding into the reactor, so precursor flows are shunted to the bypass manifold and dealt with through the exhaust system. Also during the start-up, the degassing process is greatly speeded up by purging with either dry carrier gas or a reactive clean-up gas that reacts with water. During the shut-down, the precursor flow must be flushed out of the system and, again, the pneumatic 3-way selector valves at the reactor inlet will flush carrier gas through to the bypass and exhaust. As discussed in the previous section, the run-to-run and surface area uniformity are determined by flow details in the reactor, and contamination is eliminated through use of quality components and a well designed exhaust system. Now we can see how it all fits together into the CVD system platform. In the following sections we turn our focus onto the details of the reactor itself, first looking at small-scale research reactors, then the two most common reactors used in the semiconductor industry.

2.4.1 Research-scale Reactors

If you are a CVD technologist, you flip through the Experimental section of journal papers looking for new ideas and details of equipment design and components, and you are likely to be disappointed. Because most research CVD reactors are built in-house, there are a nearly inexhaustible range of variations on several basic themes. The substrate size is usually quite small in research settings, from a few millimetres to a few centimeters, and the reactor is also as small as possible to minimize costs and precursor usage. CVD reactors are cylindrical for flow symmetry and due to requirements for safe vacuum vessel shapes. High vacuum systems require a high vacuum vessel, usually machined and welded stainless steel. Inlet and outlet flanges can be welded into the end flanges or the sidewalls. LPCVD and APCVD research reactors are often fabricated from tempered low sodium glass tubes (Pyrex®) or from quartz tubes if thermal conditions require it. Elastomer O-rings (Viton®) must be used for vacuum sealing connections to glass, so the connection areas are kept out of high temperature zones. Care must be taken that the reactor does not experience steep thermal gradients in the glass, that metal parts do not contact the glass wall, and that safety measures are taken to prevent positive pressure build-up in the glass reactor. Glass wall reactors are not strong enough for high vacuum deposition. Hot-wall and cold-wall reactors are the two thermal reactor classifications. The most common approach for a hot-wall reactor is to have a rather long tube chamber with the deposition zone inside a tube furnace.

The main two reactor configurations are horizontal or vertical tube. The vertical tube reactor normally has a feed tube for the precursor, which generates a stagnation flow on the substrate. The substrate stage is sometimes on a rotating shaft vacuum feed-through to improve deposition uniformity. The horizontal tube reactor may have substrates mounted horizontally or vertically depending on reactor pressure. Temperature gradients, which drive natural convection flow, preclude use of cold-wall horizontal reactors in the viscous regime. Figure 2.18 shows the basic reactor configurations commonly found in research laboratories.

2.4.2 Tube Reactor

The horizontal hot-wall tube reactor has been an industry standard for 40 years. The tube reactor is operated at low pressure, below several hundred Pa, and operates in the laminar flow regime. Figure 2.19 shows the reactor detail, a schematic of the flow field and an example horizontal tube reactor platform for a commercial system.

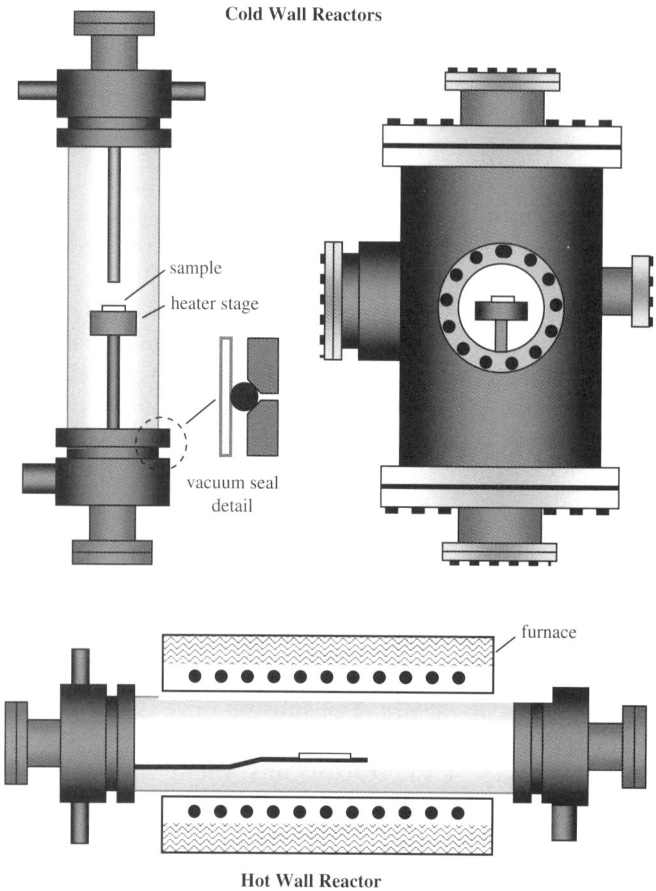

Figure 2.18 Schematic diagrams of the most common research-scale reactor configurations.

2.4.2.1 Uniformity

The uniformity problem is addressed by "not" having boundary layer flow over the wafer surface. Because there is no radial pressure gradient in axial flow in a symmetric geometry, then there will be no flow between the wafers.[12] At the normally low operating pressures, the diffusion of precursor from the annular flow down the tube walls into the space between the wafers is sufficiently fast to produce uniform thickness on most of the wafers. Sacrificial or "dummy" wafers are put in the first and last wafer positions where the flow conditions are not uniform. The wafers must be carefully positioned for radial symmetry to maintain the annular flow pattern. There will be a depletion of precursor in the axial direction but, again, due to the high diffusion rate at low pressure, axial diffusion helps to minimize the effect on wafer-to-wafer uniformity. However, the reactor cannot be operated near starvation conditions. Low conversion efficiency is inherent in the tube reactor both due to the excess precursor in the exhaust and the parasitic deposition on the hot walls and wafer boat.

2.4.2.2 Quality

Cleanliness is an issue as with all hot-wall reactors, with spalling of wall deposited film material a major issue during cool-down and manipulation of the wafer boat. Historically, break-down of the

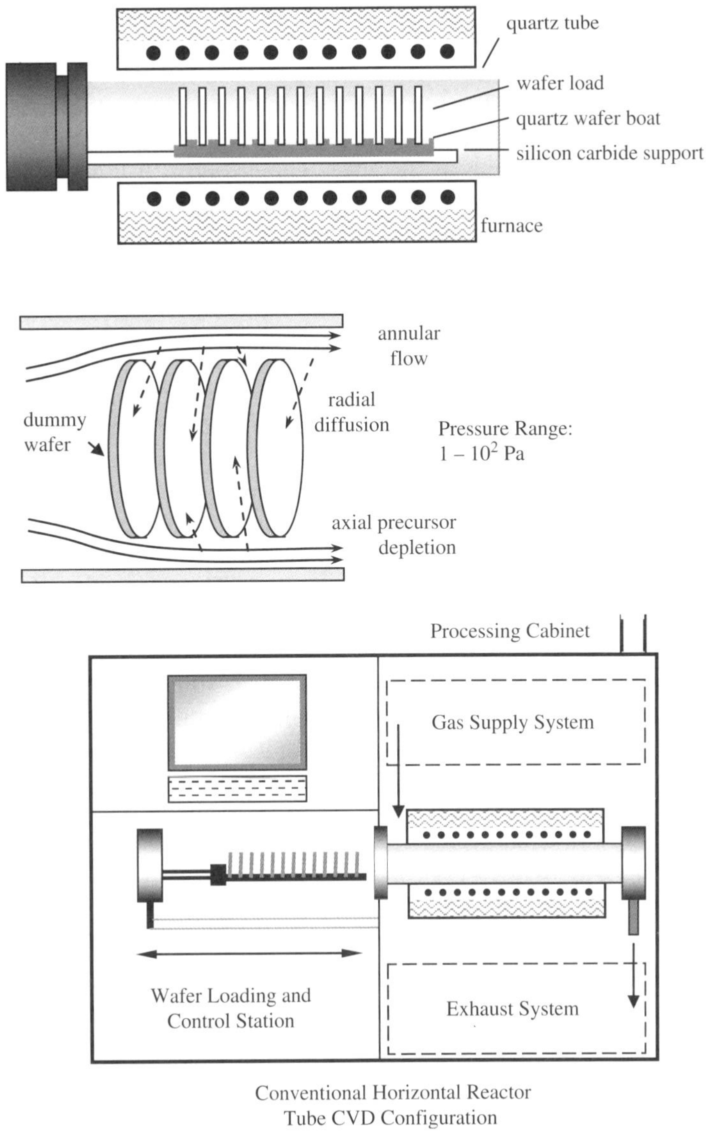

Figure 2.19 Schematic diagram of horizontal hot wall tube reactor with the basic flow dynamics and system platform configuration.

system for chemical cleaning has been necessary, but some commercial systems are now available with reactive gas and high temperature cleaning cycles. Of course, the configuration of the reactor precludes use of plasma for processing or cleaning. The vacuum seals to the quartz tube must be kept cool, so the tube extends out from the furnace and the end ports are cooled. Because elastomer seals are used, water and solvent out-gassing may need to be considered at the low end of the pressure operating range. The normal issues of contamination in a furnace system must be addressed.

2.4.2.3 *Thermal Management*

The thermal management of the tube reactor is straightforward, using the furnace and no substrate heater. Tube furnaces have a uniform axial temperature zone in the central section of the tube, and have very good radial temperature uniformity. The dummy wafers on the ends help to establish thermal uniformity in the product wafers, as well as flow uniformity, as they are exposed to the cooled tube supply and exhaust fittings at the ends. The gas flow, if not pre-heated, can cause a temperature gradient, but this is usually compensated for with different heating zones in the furnace which are tuned to the flow conditions and continuously controlled. The tube furnace has high temperature capability.

2.4.2.4 *Throughput*

The reactor tube is fabricated of high quality quartz, and quartz or silicon carbide are used for other holder and support components. The size of the wafer load that can be processed is effectively limited by the mechanical design of the cantilevered wafer boat and loader, and the fab lab footprint. The cantilever wafer holder presents a design problem because of the materials that can be used, and also because of dynamic stability during loading. That is, as the loader is being moved into position, even very small accelerations or vibrations can be amplified into large deflections of the boat holder. This poses a risk of hitting the tube wall. The properties of the cantilever will change with temperature, meaning it will sag, so the design must compensate so that at operating conditions the wafers are positioned in the center of the tube. Commercial tube reactor systems are quite long to accommodate the wafer loading, taking up several metres of space. Multiple tube furnaces can be stacked into the same processing platform, and this is how some commercial systems achieve high throughput even for low deposition rate processes. The horizontal configuration usually requires manual loading, but new designs may feature robotic loaders and ASM International is currently manufacturing a vertical furnace configuration to reduce footprint and with a reported throughput of 40–60 wph for LPCVD poly-SiGe.[13]

2.4.3 Showerhead and Planetary Reactors

The numerical modeling capabilities of the last few decades have been a key factor in the development and proliferation of showerhead and planetary CVD reactor platforms.[14] The combination of modeling and experimental research has led to development of a good number of high-performance commercial CVD and MOCVD systems built on the showerhead and planetary reactor configurations shown schematically in Figure 2.20. The low pressure reactor usually operates at pressures from a few to several hundred Pa with operation in the laminar viscous flow regime, as the cross-flow dimension is kept small. The reactor design is usually cold-wall to reduce parasitic deposition and to reduce heating of the precursor to a point to cause gas-phase reactions. As wafers have gotten bigger, the showerhead reactor was adapted to fit individual substrates. Showerhead reactors are popular for deposition of intermetal dielectric layers, tungsten and other metals, and the final passivation materials. Planetary reactors have been developed for multiple smaller wafers. For example, one of the planetary reactor systems developed by Aixtron is used for LED fabrication on 2 inch and 3 inch wafers.

The reactors are typically made from stainless steel with hinged lids for easy access. Some elegant engineering solutions for internal ducting of coolants and precursor feed channels have eliminated electrical and tubing connections from the reactor access area. This is highly important for planetary reactors which require manual loading. The shape and the use of metal lend this reactor to plasma assisted processing and cleaning. However, the use of stainless steel limits processing temperatures.

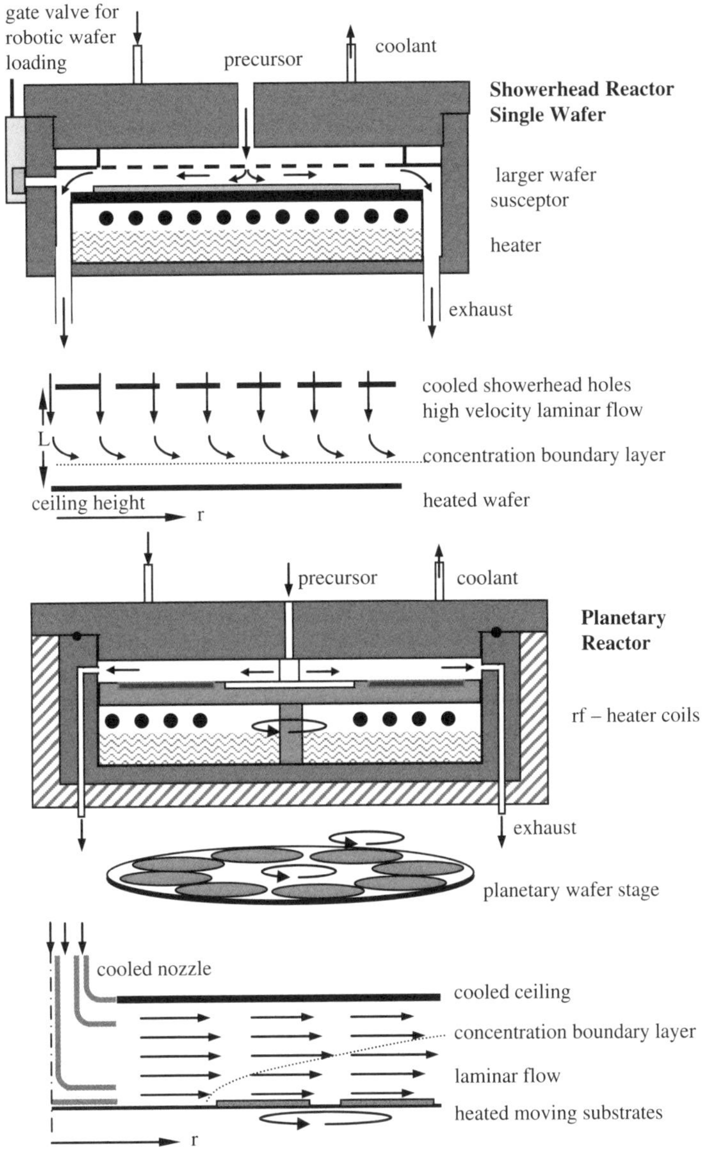

Figure 2.20 Schematic diagram of showerhead and planetary reactors with illustration of the basic flow dynamics.

2.4.3.1 Uniformity

In the showerhead reactor, the strategy for uniformity is running the reactor with excess precursor so that the residence time of vapor in the reactor is much less than the absorption and reaction time. The low pressure operation means high diffusivity, but the flow velocity is normally very high so that large concentration gradients exist in the reactor. The vertical flow direction at the showerhead holes compresses the boundary layer thickness as does the high axial flow velocity. New precursor is being added to the flow all across the wafer surface, keeping the concentration gradient normal to the substrate surface high throughout the reactor. This strategy is the starting point for achieving uniformity and represents a trade-off between low precursor conversion efficiency and high growth rate. In practice the uniformity will be developed through initial and continuous experimentation. The arrival rate at each point on the substrate is highly dependent on convection due to the local flow conditions which are, in turn, a complex result of flow rate, ceiling height, hole diameter, hole configuration and temperature. The hole shapes and diameter can change over time either from deposition or erosion, and even minute changes can affect uniformity of the flow field. Higher flow velocity helps to minimize these small effects, again at the cost of low efficiency. Showerhead reactors often have a lot of on-site flexibility like adjustable shower head height and hole configurations.

The planetary reactor has only one precursor inlet to the flow field in the center of the reactor. The injection nozzle is cooled to suppress gas-phase reactions, as is the ceiling and an entry section of the planetary wafer stage. The planetary reactor operates closer to the starved mode, with a high radial concentration gradient from inlet to outlet. The flow outlet in Figure 2.20 is a cylindrical collector ring with many holes to maintain a uniform flow across the planetary stage. The strategy to achieve uniformity is to rotate the wafers through the non-uniform flux field. This strategy requires 3D CFD numerical modeling as well as extensive experimental tuning.

2.4.3.2 Thermal Management

The reactor ceiling, precursor inlets and even walls may be actively cooled and the substrate is heated in the showerhead and planetary reactors. Convection is dominant in the heat transfer as it is in the mass transfer. The high gas velocities used to transport the precursor also produce significant cooling of the substrate below the susceptor or heater stage temperature. There will be significant thermal gradients in this reactor configuration, and a good deal of experimentation and engineering may be required to achieve uniform wafer temperature, particularly in the radial direction. Natural convection is suppressed by the high forced flow velocity. The substrate heater must provide a high heat flux because of the high convection rate to the gas flow and high radiation rate to the ceiling. However, because we do not have to heat an entire furnace to steady state, which can take a substantial time depending on the refractory and insulation level, the reactor can be quickly cycled between runs.

The pattern of holes in the showerhead reactor will change the local heat convection, so thermal effects may contribute to uniformity issues. The nozzle injector in the planetary reactor also has a big influence on the flow pattern. As a general statement, the uniformity is highly dependent on the details of the flow dynamics, which are, in turn, highly dependent on minute details of reactor geometry. Commercial CVD equipment manufacturers must have research engineering teams working to manage uniformity for each new production tool.

2.4.3.3 Quality

High substrate temperatures provide higher growth rate, but also can reduce film quality. The high temperature gradients in the reactor can cause problems with uniformly heating the substrate, and it may even warp. Once a substrate warps slightly then it will continue to warp further as the temperature gradients increase where it is no longer in contact with the susceptor and where it is closer to the cold ceiling. The cold walls and ceilings suppress parasitic deposition, but gas reactions can still be a problem, particularly at higher pressure. Plasma assisted deposition has a higher risk of sputtering molecules from the reactor walls onto the substrate.

2.4.3.4 Throughput

Clearly, the showerhead and planetary reactors hold fewer wafers than the tube reactors, so other strategies are used to provide reasonable throughput. The growth rate is maximized by maximizing substrate temperature and precursor flux. Cleaning and loading/unloading time are minimized through design. Most of the commercial showerhead reactors have robotic wafer loaders and a system platform with two, four or six reaction chambers on the same supply, vacuum and control system. This helps reduce foot print and increase throughput. With the robot loader, high heating and cooling rates, and integrated reactive gas cleaning the process cycle can be greatly shortened.

2.5 Liquid Precursor Delivery and Process Control

We have discussed the strategies used to achieve uniformity, quality and throughput for conventional CVD systems, and we have seen how control of the precursor flow and pressure is vital to controlling the deposition process. The traditional CVD system delivers reactant gasses or precursor vapors in a carrier gas from a liquid bubbler or solid evaporator. The number of new precursor chemicals, particularly for MOCVD, that are available as liquids has led to a new approach in process control through direct liquid injection to a vaporizer, aerosol generator or spray nozzle. This section describes the components, system design and control strategies for these liquid source reactors. In general, once a vapor is generated it can be supplied to any of the range of CVD reactor types, including hot and cold-wall, low or atmospheric pressure, tube or showerhead reactors. However, the aerosol and spray delivery systems must be integrated with the reactor design.

There are several issues with the conventional bubblers and solid evaporators. The operating principle of the bubbler is that the carrier gas will become saturated with the precursor at the pressure and temperature in the bubbler. In practice, even very high contact bubbling produces less than 100% saturation, and the vapor can very easily condense on the tube walls and fittings on the way to the reactor inlet. Another problem is that the liquid is often a solution with different vapor pressure for precursor solvent. The solvent will evaporate faster, meaning that the solution concentration will change throughout the deposition process. A pressure or flow controller cannot be placed down-stream of the vaporizer, so as the solution mixture changes the precursor flow rate will change without measurement or control. The final issue with the good old bubbler is that the flow and bubbler liquid temperature can take some time to stabilize during start-up, resulting in chemical wastage.

2.5.1 Liquid Injection Delivery Systems

This section describes the liquid injection approach and reviews the key components and design considerations for liquid delivery systems. A range of liquid delivery and vaporization systems are commercially available. The French company Jipelec offers the JetPulse system using very fast on/off valves that feed the liquid precursor to an atomizer and heated vaporization chamber.[15] Advanced Technology Materials, Inc (ATMI) is an American company that has developed a range of liquid and solid precursor delivery systems for CVD tools, including large, automated systems featuring programmable logic controllers (PLC) that can integrate with different tools in the fab lab.[16]

Some companies have developed specialist vapor supply equipment, *e.g.*, the Japanese company LINTEC Co. Inc., which offers a flow control and vaporization system for ferroelectric materials. Bronkhorst, in the Netherlands, supplies liquid, gas, pressure and continuous vaporization control components that researchers can use to build home-made systems.

The main idea of the liquid delivery system is to control the flow of precursor and carrier gas separately, up-stream of the vaporizer, and thus to precisely control the total precursor flow stream. The direct liquid injection approach is additionally aimed at minimal start-up settling time and maintenance purging time with the associated costs of chemical and throughput. Figure 2.21 shows a schematic diagram of a range of liquid delivery systems with insets of the MFC, liquid pump and

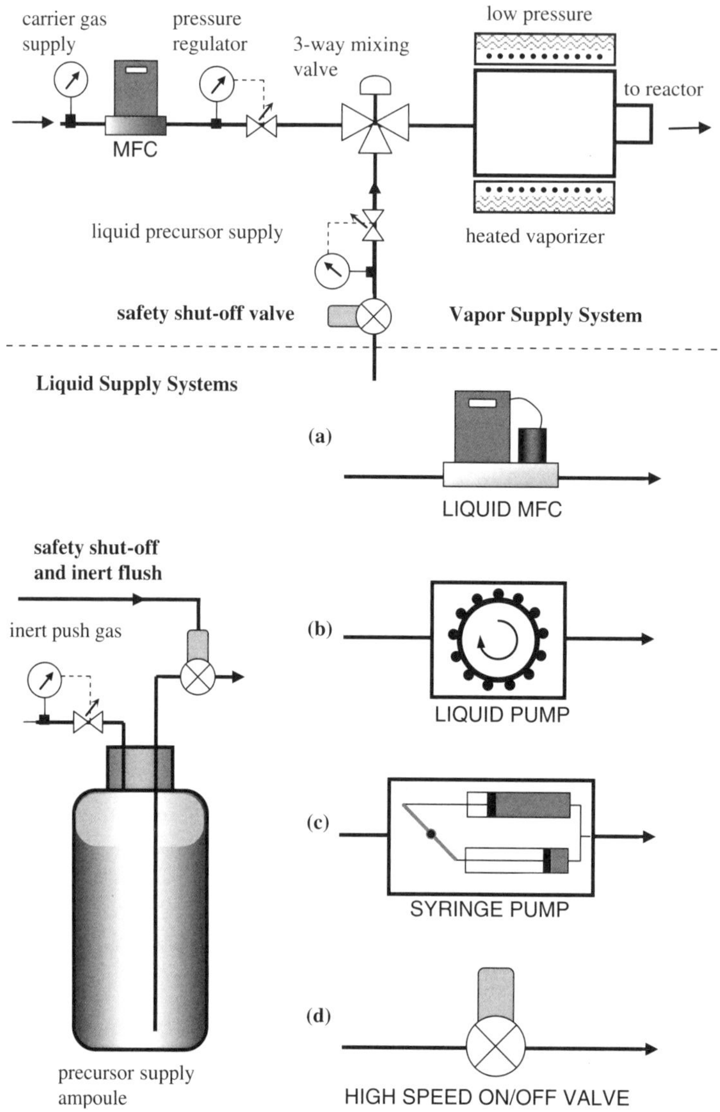

Figure 2.21 Liquid precursor delivery system components with three strategies for liquid delivery control: (*a*) direct liquid mass flow controller, (*b*) liquid gear pump or peristaltic pump, (*c*) mechanical syringe pump, (*d*) pulsed metering with high speed on/off valve.

high speed injector valve feed systems. The liquid precursor is mixed with the carrier gas in a mixing valve and flows into the heated vaporization chamber by means of the pressure difference between the delivery system and the low pressure of the up-stream reactor. The vaporization chamber may be called a *flash evaporator*. The vaporizer pressure and temperature are sufficient to quickly evaporate the precursor, but not high enough to risk decomposition reactions. The vaporizer can be as simple as a section of quartz tube wrapped with heat tape in a research experimental set-up. The flow dynamics in commercial vaporizers have been optimized to entrain, heat and shroud the liquid droplets so that they do not impinge on the heated walls or pass through any zones hot enough to cause homogeneous decomposition. Commercial vaporizers are typically specifically designed for certain precursors and supply rates. The promising "CEM" component by Bronkhorst combines the flow control, liquid-gas mixing valve and evaporation into a single in-line device to deliver vapor to the reactor.

In the simplest liquid supply system configuration, the precursor supply ampoule is pressurized with an inert gas which produces the liquid supply pressure. A pressure limiting valve must be installed on the gas supply to prevent overpressure of the ampoule. If liquid supply is connected directly to the low-pressure vaporizer, the low line pressure would cause in-line vaporization of the precursor. A needle valve can be used to try to control the liquid flow to the vaporizer to maintain the desired reactor pressure, but this rudimentary system would be prone to variation in the flow and would lack a measurement of the precursor flow rate. A load-cell or scale on the precursor supply ampoule can be used to estimate flow rate, but the gas and liquid connections can cause errors in the weight measurement. High-end chemical supply ampoules may be available with liquid level sensor, but the purpose is more for signaling supply replacement than process control.

A good degree of control can be achieved by using one of the flow control devices shown in Figure 2.21. The liquid mass flow controller (LMFC) works on the same temperature differential principle as the gas MFC. However, the flow tube is used as the measurement section and is usually heated directly with a temperature increase of less than 5 °C. The specific heat of the liquid precursor must be known, but calibration is relatively straightforward. LMFCs are available for aggressive chemicals.[17] Liquid gear pumps and peristaltic pumps operate on the positive volume displacement of liquid, and thus can be used to supply flow pressure as well as precise measurement and control of the flow rate. The range of flow rate depends on the displacement volume size and the pump shaft speed, which can usually be controlled with a PLC. Density of the precursor solution and concentration in the solvent are the only physical properties necessary for calibration of the pump. The syringe pump is also a positive displacement pump that can deliver a very smooth flow and pressure. The configuration shown in Figure 2.21(c) is a dual syringe pump that fills one syringe from the source while dispensing the other. Chemically resistant precision-controlled liquid pumps are expensive, and the complex internal shapes make cleaning and flushing a problem.

A high-speed on/off valve can be used to meter and control the liquid flow from a pressurized source. The valves are normally closed, and the desired flow is established by the length of time the valve is opened and the cycle frequency. The flow rate through the valve while open will be a function of the liquid pressure, properties and valve orifice size, but this can be calibrated relatively easily. The commercially available JetPulse® system from Jipelec, a division of qualiflow, has been engineered to smooth out the pressure pulses in the resulting vapor and carrier gas stream. This liquid delivery system is patent-protected, but the concept of the on/off valve may be one of the lowest-cost and most effective means of flow control and measurement for research scale systems. The Jipelec on/off valve system is also called "pulsed-MOCVD" in the literature.[18]

The safety, footprint, maintenance and control issues for liquid delivery systems are much the same as for gas supply systems discussed previously. The liquid supply system is installed in a vented cabinet with a monitor and control screen visible from the outside so precursor supply and operation can be observed without opening the cabinet. Line isolation and flushing must be integral functions of the liquid delivery system for safe maintenance. Flushed lines can be drained to the tool by-pass, or to an internal drain container in the liquid delivery cabinet. A good example of a liquid delivery system is the Unichem™ by ATMI, which has a very small fab lab footprint with a vertical cabinet configuration. A large chemical supply canister made of stainless steel or PTFE is loaded into the base of the cabinet and supplied with inert push gas. The bulk supply canister supplies smaller canisters and ampoules for different CVD tools in the fab lab and the dispensing, shut-off and flush functions can be controlled by the PLC and integrated with the various deposition tools. One bulk canister supplying several systems reduces footprint and maintenance down-time for re-loading ampoules. Another safety feature is unique quick connect (pressure sealing) connections for each type of chemical, with different size or shape fittings to prevent accidental installation of the wrong precursor. Notably, there are standards for semiconductor industry equipment that may serve as design and safety guides for research facilities building their own equipment.

2.5.2 Aerosol and Spray Reactors

One of the first improvements on the traditional bubbler was "Aerosol Assisted CVD."[19] Rather than passing the carrier gas through the liquid and relying on evaporation, the precursor is placed in the well of a nebulizer with an un-heated ultrasonic actuator in the base. Much like the familiar humidifier used in homes, the actuator throws a steady stream of micron and sub-micron sized droplets off of the liquid surface. The nebulizer can produce droplets of high vapor pressure liquids that would be very difficult to evaporate without risking thermal decomposition. These aerosol particles are then entrained in the carrier gas flow and carried to the heated vaporization chamber as in Figure 2.21 or into the AACVD reactor as in Figure 2.22(a). The aerosol generator reduces the problem of differential evaporation and settles to a steady operation in a fairly short time. The liquid supply rate can be measured indirectly by monitoring liquid consumption rate over a span of time, but it cannot be controlled in the nebulizer except indirectly by changing the ultrasonic power

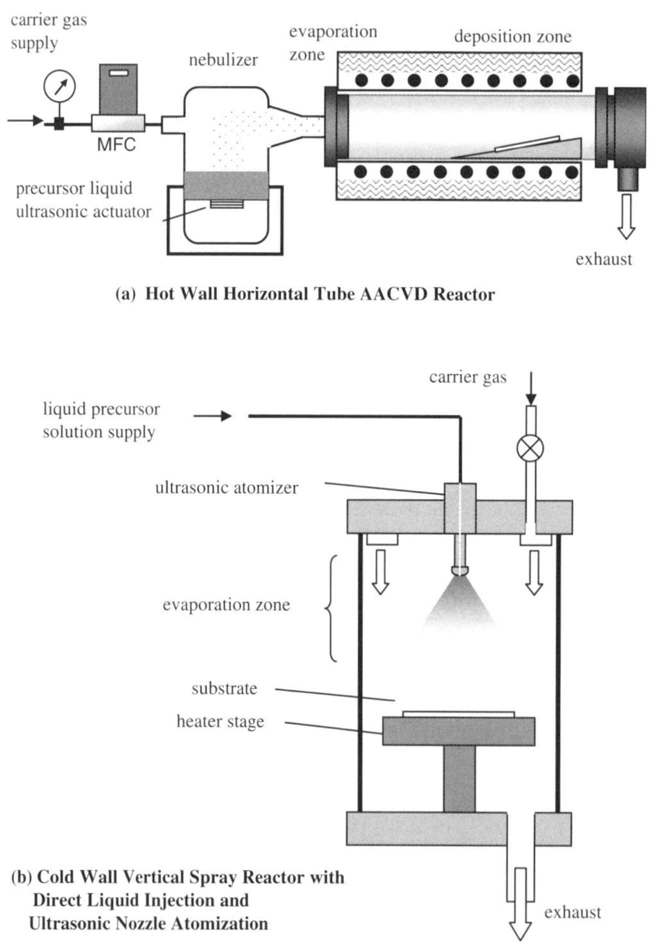

Figure 2.22 Ultrasonic atomization of liquid precursor using (*a*) a nebulizer and (*b*) an ultrasonic nozzle.

or frequency. The aerosol droplet size is also affected by changes in the nebulizer power and frequency. Aggressive chemicals may not be compatible with the nebulizer transducer materials.

The carrier gas needs to be at sufficient pressure and density to entrain the aerosol particles and transport them to the reactor. The aerosol transport from the nebulizer to the reactor needs to be a smooth, laminar flow path to prevent momentum impact on delivery tube walls. A great deal of engineering fluid mechanics has been done to manage aerosol transport in a range of other processing applications, and the same principles apply in AACVD. Aerosols are transported in a gas by viscous drag forces and follow streamlines when the aerosol momentum is in balance with the viscous forces. If the cross section of the flow duct decreases smoothly, as in a converging nozzle, then the flow accelerates, the forces on the aerosols increases and the aerosols are accelerated with the flow. If the flow decelerates, as in a diverging nozzle, then the drag force on the aerosols will reduce and they will continue along their previous trajectory path at the previous velocity. If the flow makes a bend, the particles can be thrown out of the flow as in a cyclone particle separator. If the reactor were at vacuum, the flow could be greatly accelerated in the supply duct, then the rapid gas expansion and pressure drop as it enters the reactor would essentially generate a high speed aerosol particle beam in the reactor.

The AACVD strategy is normally to organize the aerosol flow path so that the droplets have sufficient time to evaporate prior to reaching the substrate. Variations include a separate vaporization chamber at the entrance to the reactor, heated lines leading to the deposition zone in the reactor, or a long heated entrance length in a hot wall tube reactor (Figure 2.22). This is a challenging engineering approach in terms of uniformity, quality and process controls. At the time of writing, this author is not aware of any commercial AACVD reactor systems.

Direct liquid injection into the reactor using an ultrasonic nozzle is called "spray pyrolysis" when droplets are directed onto the substrate, and "liquid injection CVD with ultrasonic atomization" when the droplets evaporate before reaching the deposition zone. The reactor pressure in spray pyrolysis is usually high, and cold-wall reactors are used to prevent the droplets from evaporating. Obviously, the spray needs to be directed at the substrate, and may be directed or shaped by the nozzle configuration and shroud gas. Liquid injection CVD can supply either hot or cold-wall reactors, and the geometry can be horizontal or vertical. Pulsed liquid injection CVD with the spray into a vaporizer as supplied by Jipelec is currently being used in some new Aixtron tools. Very high vapor pressure liquids can be atomized directly into the reaction chamber, *e.g.*, in the reactor shown in Figure 2.22(b). Clearly, there is a risk that rogue large droplets can reach the substrate, particularly if the substrate temperature is low or the reactor pressure is higher than the liquid vapor pressure.

Ultrasonic atomizers, such as supplied by Sono-Tek, Inc., must be used to generate droplets into a vacuum chamber. Normal atomization relies on a large pressure difference across the atomization orifice, but the liquid droplets are actually produced through viscous interaction between the resulting thin liquid sheet and the gas into which the liquid is sprayed. The low density gas in CVD is not capable of producing the surface shear forces needed to overcome the surface tension in the liquid jet or sheet and break it up into droplets. The ultrasonic nozzle operates by setting up standing waves on a thin film of fluid that whets the atomizer tip. The ultrasonic waves throw off droplets with size on the order of the ultrasonic wavelength. Another important feature of the ultrasonic atomizer is that the droplets have very low velocity as they leave the atomizing surface. The ultrasonic atomizer nozzle produces a gentle mist as opposed to the strong spray associated with normal pressure atomization. This is also an important consideration in positioning the substrate for spray pyrolysis.

Conventional direct liquid injection CVD systems have a carrier gas and operate with a continuous flow of precursor. However, there is one process, pulsed-pressure CVD, which does not use a carrier gas, but uses discrete, timed injections of a set volume of precursor into a continuously evacuated reactor.[20] The precursor flash evaporates in the low pressure reactor, causing a sharp spike in reactor pressure. The evaporated vapor expands to fill the reactor very quickly, then the reactor is pumped back down to low pressure to start the cycle over. The consumption time is much faster than the residence time, so the system can have high efficiency. Research-scale pulsed-pressure MOCVD systems are available commercially from Sono-Tek, Inc.

2.6 Deposition Control by Surface Processes

The previous discussion focused on achieving the performance requirements of quality, uniformity and throughput through control of the exhaust system, precursor supply system, reactor flow dynamics and reactor thermal field. Low pressure, low temperature and flow field engineering were discussed as strategies for producing uniform thin films. There are two other requirements for modern electronic devices, conformality and epitaxy. Conformality, or step coverage, implies uniformity of thin films on 3D features and holes. Atomic layer deposition (ALD) produces conformal films by an alternating absorption–reaction process that involves formation of a monolayer of precursor on the substrate surface followed by reaction with a gas. Epitaxy means that the thin film has no grain boundaries or defects, and that the crystallographic structure is oriented to the substrate. Chemical beam epitaxy (CBE) is essentially an adaptation of the UHV MBE process to use gas and metalorganic precursor sources. CBE is a surface reaction limited process that requires a high degree of control of substrate temperature to meet quality and uniformity requirements. Details of these two processes are described elsewhere in this book. In this chapter we discuss the unique aspects of reactor design and process control needed to generate the conditions for the surface self-limiting process control in ALD and CBE.

2.6.1 ALD Reactors

All of the components, measurements, system engineering and process control topics covered in terms of CVD are applicable to ALD tools. ALD requires at least two precursors and an inert purge gas. The ALD process of surface monolayer formation is accomplished by supplying sufficient partial pressure of the precursor. The reactor chamber is flooded with the first half-reaction precursor vapor by simple on/off valve on a gas or liquid delivery system with carrier gas. The first precursor is then turned off and flushed from the reactor. The same process is repeated for the second half-reaction precursor followed by a purge gas flow to carry away the reaction products. Multi-layer, sealed and graded film structures are produced in the same deposition process by simply supplying different precursors in different sequences. This on/off precursor sequence is the reason ALD is called a "digital" process.[21–23]

ALD can be carried out at any pressure from atmospheric to molecular flow range, depending on the precursor physical properties, but normally systems are built to UHV standards to ensure cleanliness. A sufficiently high vapor partial pressure is required above the substrate to form a monolayer of reactant on the surface and into holes. This means that the reactor cannot be starved during the process, and ALD processing requires some degree of precursor wastage, which is a problem for cost of ownership. The primary strategy to increase conversion efficiency is to minimize the reactor volume. The showerhead reactor is the configuration of choice as the reactor volume can be minimized and the system is amenable to plasma and UV enhancement, and *in situ* surface metrology.

If the precursor vapor and reactant gas are allowed sufficient time of exposure to chemical rich vapor, then uniformity is assured. The ALD process does not need substrate manipulation to ensure uniformity. The relatively slow surface adsorption and saturation process, the purging steps and the fact that each cycle represents less than one monolayer of material mean that the engineering challenge in ALD systems is to increase the efficiency, throughput and availability while preserving quality. One of the key challenges in ALD reactors is the fluid dynamics of supply and

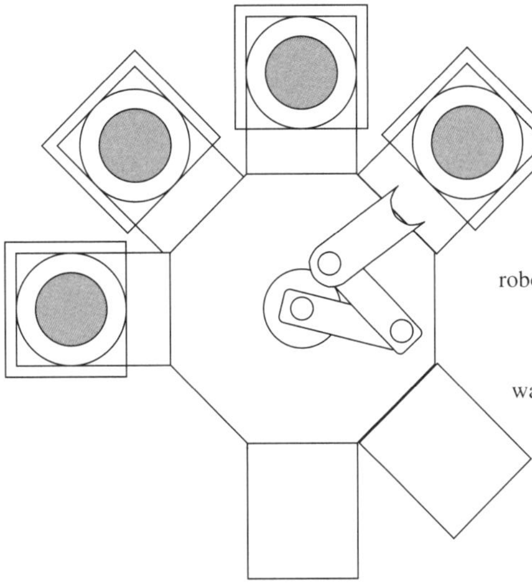

Figure 2.23 ALD reactors are clustered to increase wph (wafers per hour) while economizing on equipment and fab lab footprint.

purge to ensure that there are no "dead spots" in the flow, and that the fill and purge times are optimized. Reactive species also need to be isolated until mixing in the space above the substrate. This is a challenge because of the small ceiling height and large surface area. In fact, many of the ALD tool manufacturers have patented gas delivery systems that have been developed through extensive modeling and experimentation. Vapor delivery strategies are quite varied, but each can be developed into a reliable system. One approach is to use a square reactor volume and supply the gas through a slot along one side and exhaust from the opposite side. Other flow configurations are variations on the center fill with center exhaust below the wafer, and the showerhead with ring exhaust.

Single-wafer cluster platforms are the standard for the semiconductor industry. Figure 2.23 gives a representative sketch of the cluster reactor configuration with a robotic wafer handling system and wafer vacuum lock-loading stations that handle an entire production run and pre-heat the wafers. The cluster platform can economize on equipment and fab lab space by having common precursor supply, controls and exhaust for as many as eight single-wafer reactors. For example, the SMI Spin CVD®, the Aviza Celsior™, the Vesta Technology IRIS™ALD and the Genus LYNX2® are all flexible cluster platforms that have been engineered for minimum footprint, maximum reliability, uniformity and throughput.[24] The platform can be expanded by adding more single-wafer reactor modules. Flexible process control allows for in-house process development to take advantage of the continually emerging range of new ALD chemicals. Modules are usually designed to accept different size wafers, and to allow variable temperature range operation. By contrast, the Planar-ALD system for manufacturing flat panel displays is a large batch platform, and Planar work with customers to develop new materials and products using their industrial scale machines. Oxford Instruments, Picosun, and Kurt J. Lesker manufacture single wafer research-scale ALD reactors.

2.6.2 CBE Reactors

CBE makes use of the same UHV reactor technology as MBE and is also called MOMBE, gas-source MBE, and UHV-MOVPE. CBE takes advantage of high vapor pressure chemical sources that thermally decompose to supply the desired element for incorporation into the film. Hybrid CBE-MBE systems have effusion cells supplying molecules from pure solid sources as well as chemical gas sources. The UHV processing has the advantages of superior uniformity, no reactor viscous flow patterns and elimination of gas-phase reactions. A large number of individual sources can be used with source rates controlled to produce complex doped materials. However, the local growth rates, dopant incorporation, phase and composition depend on precursor flux rates in a complex way that is difficult to understand and directly control. CBE deposition is kinetically controlled through surface chemical reactions between source gases and the substrate. As a result, growth is dependent on the substrate temperature profile and geometric orientation relative to the source. CBE is a subject of interest and research, including kinetic surface reaction modeling, Monte Carlo modeling of the injector velocities and trajectories and experimental material science with *in situ* analysis using LEED or RHEED.[25–27]

Run-to-run uniformity in UHV processing is dependent on the reproducibility of the source fluxes. In CBE the source control includes the mass flow rate, thermal cracking and beam focus. The key technology for CBE is the gas supply to a UHV reactor and the injector, including heating and beam control.[28,29] The industry leaders in MBE products and research systems provide vapor and gas injection components. For example, RIBER and Varian produce CBE gas injectors, gas distribution systems and UHV reactor platforms with capability for MBE, CBE and combinations of sources.

The schematic diagram in Figure 2.24 shows a simplified UHV deposition system with a set of solid, liquid and gas sources. The reactor operates in the molecular flow regime, usually with the mean free path longer than the source–substrate distance. Clearly, very high pumping rates and low base pressure are required of the exhaust system. The reactor mass transport is modeled as ballistic trajectory of source molecules from the injector "nozzle." The injector is not a nozzle in the same sense as a pressure nozzle, but serves the same purpose of focusing the gas molecules into a desired pattern through collision with the directional plates. The injector may have heated molybdenum plates or baffles to crack thermally stable precursor gasses as they enter the deposition environment.

The gas injector is the key to process control in CBE.[30] The gas injector must be bakeable for the UHV de-gassing pre-deposition process. Commercially available gas injectors, for example from RIBER, have automatically controlled valves to allow selective injection of several different sources, either singly or mixed in the nozzle. Figure 2.24 shows the operating principle of the gas injector. For reactive gas sources, a MFC could be used on the supply line up-stream of a leak valve, but the extremely low pressures are outside of the operating range of MFCs. Liquid precursor sources for CBE are usually just heated slightly with the ampoule in a water bath, and the resulting vapor is transported at the supply pressure, P_s, in heated lines to prevent condensation. Again, MFCs are not appropriate for these vapors. Gas sources that are of the same family can be premixed by using MFCs at the supply pressure, then fed to the injector, but it is better to meter each gas individually, then mix them in the injector nozzle.

CBE gas supply rather relies on pressure control. The supply pressure is stepped down to the metering pressure, P_1, by a pressure controller. A fixed leak orifice with very low conductance then delivers the gas to the nozzle at the reactor vacuum, P_R. Equation (2.4) can be used to measure the gas flow rate, Q, for a known conductance, C_L, and a measured pressure difference. Because the reactor pressure is so much smaller than the delivery pressure, the simple calibration equation can be programmed into the injector PLC and flow measured and controlled by controlling the pressure. The injector exhaust is connected to the vacuum system and quickly evacuates the lines for post processing flushing, de-gassing and safety shut-off. The injector vent is also a by-pass connection to the vacuum system and is used during start-up to stabilize the gas flow through the leak orifice prior to deposition. The control valve can stop and start the gas flow with very fast response time. This gas supply feature provides the capability for doping and for depositing distinct multi-layers.

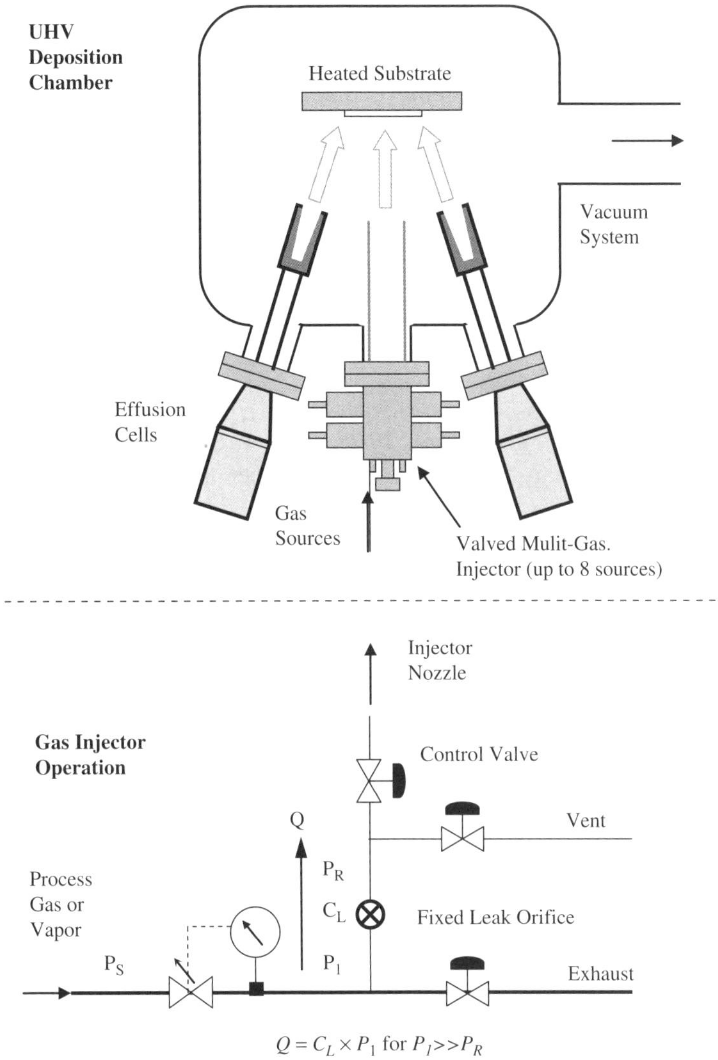

Figure 2.24 CBE apparatus to deliver a controlled flux of an ever growing number of precursors to a UHV deposition chamber.

In contrast to the up-stream control of gas sources, the solid sources are controlled by shutters and crucible temperature. The effusion cell flux can not be measured during the deposition. Rather, a retractable flux gage is used to measure the beam equivalent pressure directly and related back to the effusion cell temperature.

References

1. R. Beccard, *Compound Semiconductor* 2005, available at www.aixtron.com. (April 2007).
2. K. K. Schuegraf ed., *Handbook of Thin-Film Deposition Processes and Techniques*, Noyes Publications, New Jersey, 1988.
3. M. L. Hitchman and K. F. Jensen eds., *Chemical Vapor Deposition, Principles and Applications*, Academic Press, London, 1993.

4. M. Ohring, Chapter 2, Vacuum Science and Technology in *Materials Science of Thin Films Deposition and Structure*, Academic Press, San Diego, 2002.
5. L. I. Maissel and R. Glang, eds., Chapter 1 Vacuum Evaporation in *Handbook of Thin Film Technology*, McGraw-Hill, New York, 1970.
6. A. Roth, *Vacuum Technology*, North-Holland, Amsterdam, 1990.
7. D. G. Coronell and K. F. Jensen, *J. Electrochem. Soc.*, 1992, **139**(No. 8), 2264.
8. C. R. Kleijn, R. Dorsman, K. J. Kuijlaars, M. Okkerse, H. van Santen, *Journal of Crystal Growth*, 2007, **303**, 362.
9. R. L. Mahajan, *Adv. in Heat Transfer*, 1996, **28**, 339.
10. H. Schlicting, *Boundary Layer Theory*, McGraw-Hill, New York, 1979.
11. F. P. Incropera and D. P. DeWitt, *Fundamentals of Heat and Mass Transfer*, John Wiley & Sons, Inc. New York, 1996.
12. D. G. Coronell and K. F. Jensen, *J. Electrochem. Soc.*, 1992, **139**(No. 8), 2264.
13. R. de Blank, F. Huussen, E. Oosterlaken, P. Zagwijn, T. Claasen-Vujcic, A. Hasper, "Vertical Batch Furnaces: Hotter than ever," ASM white paper available at www.asm.com. (April, 2007).
14. M. Dauelsberg, C. Martin, H. Protzmann, A. R. Boyd, E. J Thrush, J. Käppeler, M. Heuken, R. A. Talalaev, E. V. Yakovlev, A. V. Kondratyev, "Modelling and process design of III-Nitride MOVPE at near-atmosphere pressure in Close Coupled Showerhead and Planetary Reactors," available at www.aixtron.com. (April, 2007).
15. J. P. Sénateur, F. Weiss, O. Thomas, R. Madar, A. Abrutis, *US Patent 9 308 38, 1999; EU Patent 9 400 858*, 1999.
16. J. F. Roeder, T. H. Baum, S. M. Bilodeau, G. T. Stauf, C. Ragaglia, M. W. Russell and P. C. Van Burskirk, *Adv. Mater. Opt. Electron.*, 2000, **10**, 145.
17. Y. Senzaki, A. K. Hochberg and J. A. T. Norman, *Adv. Mater. Opt. Electron.*, 2000, **10**, 93.
18. G. Garcia, J. Caro, J. Santiso, J. A. Pardo, A. Figueras and A. Abrutis, *Chem. Vap. Deposition*, 2003, **9**(No. 5), 279.
19. C. Y. Xu, M. J. Hampdensmith and I. T. Kodas, *Chemistry of Materials*, 1995, **7**(8), 1539.
20. S. Krumdieck and R. Raj, *Chem. Vap. Deposition*, 2001, **7**(No. 2), 85.
21. R. G. Gordon, D. Hausmann, E. Kim and J. Shepard, *Chem. Vap. Deposition*, 2003, **9**(No. 2), 73.
22. M. Leskela and M. Ritala, *Thin Solid Films*, 2002, **409**(1), 138.
23. H. Kim, *J Vac. Sci. Technol. B*, 2003, **21**(6), 2231.
24. O. Sneh, R. B. Clark Phelps, A. R. Londergan, J. Winkler and T. E. Seidel, *Thin Solid Films*, 2002, **402**(1–2), 248.
25. V. M. Donnelly and A. Robertson Jr., *Surf. Sci.*, 1993, **293**, 93.
26. R. Bellman and R. Raj, *Vacuum*, 1997, **48**(No. 2), 165.
27. M. R. Leys, *J. of Cryst. Growth*, 2000, **209**, 225.
28. J. L. Benchimol, F. Alexandre, B. Lamare and P. Legay, *Prog. Crystal Growth and Charact.*, 1996, **33**, 473.
29. H.-S. Lee, K. Nishimura, Y. Yagi, M. Tachibana, N. J. Ekins-Daukes, Y. Ohshita, N. Kojima and M. Yamaguchi, *J. of Cryst. Growth*, 2005, **275**, e1127.
30. C. R. Abernathy, *Mat. Sci. Engr.*, 1995, **R14**, 203.

CHAPTER 3

Modeling CVD Processes

MARK D. ALLENDORF,[a] THEODORE. M. BESMANN,[b] ROBERT J. KEE[c] AND MARK T. SWIHART[d]

[a] Sandia National Laboratories, PO Box 969 MS 9291, Livermore, CA 94551-0969, USA; [b] Oak Ridge National Laboratory, PO Box 2008 MS 6063, Oak Ridge, TN 37831-6063, USA; [c] Division of Engineering, Colorado School of Mines, Golden, CO 80401, USA; [d] Department of Chemical and Biological Engineering, State University of New York, Buffalo, NY 14260-4200, USA

3.1 Introduction

The modeling of CVD systems is in some ways a mature field, resting on scientific foundations in the fields of fluid dynamics, thermodynamics, gas-phase kinetics and surface science. Much of the theory and methods used to model the chemically reacting flows occurring in CVD systems are an outgrowth of decades-long efforts to understand combustion processes. Although combustion typically lacks the element of surface chemistry, the complex flows and interactions with chemical reactions at elevated temperatures bear many similarities to processes that also occur during CVD. As a result, it is possible to utilize computational tools and theoretical approaches originally developed to understand combustion of hydrocarbon fuels. In some ways, CVD processes are simpler than combustion. In particular, CVD reactors most often operate in the laminar (*i.e.*, low Reynolds number) regime, in which viscous flow dominates and turbulent mass transport does not occur. This means that many commercial software packages can be used, and since such flows can be simulated with precision and relatively minimal computational resources (in contrast with turbulent flows), the mass transport and fluid dynamics are essentially a solved problem. In addition, many CVD systems operate at sufficiently low temperatures that gas-phase chemistry does not occur, which greatly simplifies the modeling process. Many CVD systems that do operate at temperatures high enough to cause gas-phase precursor decomposition often have less complex gas-phase chemistry than combustion processes, due to the absence of oxygen and consequent lack of radical-chain mechanisms that lead to ignition–extinction phenomena and chemical instabilities.

That is not to say that CVD processes are simple. Unlike many combustion processes, CVD reactors often have quite complex geometries, necessitating two- and even three-dimensional computational fluid dynamics (CFD) modeling. An even more serious problem is that the

Chemical Vapour Deposition: Precursors, Processes and Applications
Edited by Anthony C. Jones and Michael L. Hitchman

Published by the Royal Society of Chemistry, www.rsc.org

thermodynamics, kinetics and transport properties of the species involved are far less well understood than hydrocarbon systems. As a result, major assumptions are often made to make modeling a given precursor system possible. Such assumptions most often concern the chemical reactions at the surface leading to deposit formation. Global reaction chemistries involving one or perhaps a few chemical reactions are often used, even in situations in which the gas phase species interacting with the surface are known. The most sophisticated treatments of CVD surface chemistry are found for relatively simple systems (involving deposition of only a single element, such as silicon) or for materials such as diamond for which experiments, theory and similarities to gas-phase systems produced a high degree of understanding. Unfortunately, these represent a very small percentage of the CVD chemistries in use today. In fact, the situation is becoming progressively worse, since new precursors systems are constantly under development to lower deposition temperatures and improve the quality of deposits.

Most CVD modeling to date has focused on predicting growth rates. This is because control of layer thickness and uniformity is critical in many applications, particularly in the electronics industry, but also for optical materials and coatings on glass. However, both the composition and microstructure of deposits can be critical to the intended application. For example, amorphous deposits are often desirable for many electronics applications, since grain boundaries represent sources of defects. Alternatively, for thermal barrier coatings, columnar growth is desirable to produce weakness in the direction parallel to the substrate so that stresses due to thermal expansion mismatch with the substrate are relaxed. Generally, one wants equiaxed grains in a ceramic or metal coating, or if they are columnar they should have random orientation. Minimization of impurities such as carbon, which is a component of many precursors, is essential to the performance of not only electronic devices but also of ceramics and MEMS (micro-electro-mechanical systems). Prediction of composition is elusive in many cases for two reasons. First, most CVD processes operate at temperatures too low for thermodynamic equilibrium to be achieved and, second, the complexity of the surface processes involved makes it very difficult to identify rate-controlling steps. Predicting the phase and microstructure of deposits extends modeling from the purely molecular to much larger length scales in the meso- and even macro-scale. Such calculations are computationally intensive, particularly if multiple models at differing length scales are required (see ref. 1 for a review of multi-length-scale CVD models).

Despite the wealth of scientific understanding underlying many aspects of CVD, modeling any specific CVD chemistry can be a major challenge. Two particular hurdles are faced in most efforts to develop practical, robust process models. First, data of a fundamental nature are often lacking: thermodynamic and transport properties of gas-phase species, mechanisms and rate constants for gas-phase processes and, most difficult of all, rate constants for surface processes. Second, data useful for testing and validating models are frequently either unavailable or were obtained from reactors of such complexity as to be virtually useless for developing kinetic models. It is not uncommon to find reports in the literature lacking critical information, such as flow rates or temperature profiles, which are necessary for comparing model predictions with measured quantities. Serious efforts to develop models useful beyond a very specific reactor often, therefore, require an extensive data gathering effort, requiring both experimental and computational resources. This is not to say that less detailed models, incorporating only mass transport or empirically obtained global chemistry, cannot be useful. However, the problem with such models is that their generality can be very limited. Although they may predict growth rates accurately in reactors of one design and/or scale, they may be completely inaccurate in other cases. Consequently, considerable effort continues to be devoted by researchers in the CVD community to expand databases and provide growth-rate data using experimental facilities that are readily susceptible to computational modeling.

Obviously, to do justice to this large and diverse subject would require an entire book, not just a single chapter. Therefore, the objective here is to introduce the reader to critical issues in CVD modeling and to the techniques used to address them. The naturally cursory treatment is buttressed by references to much more detailed descriptions provided elsewhere. Fortunately, in most cases, textbooks and review articles exist that cover many of the important topics in much greater detail. The principal topics covered here are: (1) equilibrium thermodynamic modeling; (2) reacting-flow modeling; (3) theoretical approaches to predicting gas-phase thermochemistry and kinetics; (4) surface chemistry; and (5) particle formation and growth. The latter is an important subtopic within CVD, since homogeneous nucleation often occurs in CVD reactors and must be controlled to avoid defects in films. Additionally, CVD-like methods are in use on an industrial scale to manufacture powders of various types. This chapter considers only thermally driven CVD processes; reviews of plasma CVD process modeling are available elsewhere.[2]

3.2 Thermodynamic Modeling of CVD

3.2.1 Application of Thermochemical Modeling to Chemical Vapor Deposition

Thermochemical modeling of a CVD process is relatively easy as compared to developing a full computational fluid dynamics (CFD) description coupled with reaction kinetics for a geometrically complex system. As such, a computational thermochemical study should be performed before embarking on the development of any new CVD process or material. The results of this kind of analysis can provide important information about whether the phases of interest are thermochemically allowed to form from a proposed precursor system. It can also indicate whether secondary phases can form and give some idea as to the maximum theoretical efficiency of the process. All of this information is predicated on reaching chemical equilibrium in a system, which is the fundamental assumption of thermochemical analysis. Although the presumption of chemical equilibrium is not realistic, given the relatively short residence time of precursors in CVD reactors, reactions will proceed toward equilibrium to a sufficient extent that thermodynamic modeling is still very useful for gaining process insights. In addition, it is possible to constrain equilibrium calculations to provide a more realistic result, for example, by eliminating a phase from consideration when it is known that kinetic or steric conditions will prevent its formation even when it is thermochemically permitted.

3.2.2 Thermochemistry of CVD

The thermodynamic modeling of chemical vapor deposition processes has been performed at least since the early 1970s, and a search of relevant papers between 1972 and 2006 yielded 335 citations. Some of the earliest work, like that of Wong and Robinson,[3] Ban,[4a] Besmann and Spear,[4b] and Madar *et al.*,[5] used the first computer-based free-energy minimization programs such as SOLGASMIX.[6] This now common, but very useful tool is generally applied to CVD processes under development as exemplified recently by Varanasi, *et al.* for the CVD of yttria-stabilized zirconia (YSZ),[7] Perez, *et al.*[8] for preparing iron aluminide coatings on steels, and Chaussende *et al.* for growing SiC single-crystal materials.[9] Chemical kinetic and mass-transport phenomena that could effect phase formation are not considered in strictly thermochemical calculations, and thus they may not always accurately predict the phases that actually form. Yet, without a phase being thermochemically allowed to form it would be difficult to obtain the material, which would be metastable if deposited.

The level of sophistication in utilizing thermochemical analysis varies widely. Approaches range from simple calculations to determine if changes in heats of reaction (ΔH_{rxn}) are positive (no reaction) or negative (deposition is possible) for the most relevant chemical reactions (*e.g.*, see ref. 10) to global Gibbs free-energy minimization that considers all possible gaseous species and condensed phases, as well as potential complex solid solution/defect structures in the deposited phases (*e.g.*, see ref. 7). The thermochemical concept is based on whether governing reactions are thermochemically favored. For example, the simple model CVD reaction:

$$AB_3(g) + 1.5C_2(g) = A(s) + 3BC(g) \tag{3.1}$$

will proceed to the right and deposit the desired "A" phase if the change in ΔH_{rxn} is negative. Solely knowing the ΔH_{rxn} of the single reaction, however, can often be inadequate as it gives no indication of whether competing reactions would yield more negative ΔH_{rxn} values and thus be more favorable. In addition, as three of the species are gaseous, their vapor pressures, and therefore their activities, also govern the thermochemistry of the reaction. These thermochemical concepts are explained in more detail in several excellent texts[11–14] and will be considered briefly here.

The most comprehensive thermochemical approach for assessing a CVD system is to determine the Gibbs free-energy change in a deposition reaction (ΔG°_{rxn}) for the system as the precursors are computationally allowed to react and come to equilibrium. To determine ΔG°_{rxn} requires a summation of the Gibbs free energies of formation (ΔG°_{f}) for constituents at the temperature of interest, defined as:

$$\Delta G^{\circ}_{f} = \Delta H^{\circ}_{f}(298\,\text{K}) + \int_{298\text{K}}^{T} \Delta C_p \text{d}T - T\Delta S^{\circ}(298\text{K}) - \int_{298\text{K}}^{T} (C_p/T)\text{d}T \tag{3.2}$$

where $\Delta H^{\circ}_{f}(298\,\text{K})$ is the standard heat of formation at 298 K, C_p is the heat capacity, T is absolute temperature and $S^{\circ}(298\,\text{K})$ is the standard entropy at 298 K. Thus ΔG°_{rxn} can be written using the law of mass action as:

$$\Delta G^{\circ}_{rxn} = \sum \Delta G^{\circ}_{f,\text{products}} - \sum \Delta G^{\circ}_{f,\text{reactants}} = -RT\ln(\Pi a_{\text{products}}/\Pi a_{\text{reactants}}) \tag{3.3}$$

where a is the activity of the phases and species, and for gaseous species (assuming the gas is ideal) the activity is defined as the partial pressure, p, in bar. Thus for the reaction of Equation (3.1) we can write ΔG°_{rxn} as:

$$\Delta G^{\circ}_{A} + 3\Delta G^{\circ}_{BC} - \Delta G^{\circ}_{AB3} - 1.5\Delta G^{\circ}_{C2} = -RT\ln(a_A p^3_{BC}/p_{AB}p_C{}^{1.5}) \tag{3.4}$$

where the product "A" is a pure material whose activity is by definition unity.

A relatively simple example of computing the conditions for deposition of a single phase is the CVD of SiC from $SiCl_4$ and CH_4. The overall reaction is:

$$SiCl_4(g) + CH_4(g) = SiC(s) + 4HCl(g) \tag{3.5}$$

Determining the ΔH°_{rxn} for the reaction requires having standard heat of formation, ΔH°_{f}, for each of the constituents. Using the FactSage[15] computational package and associated database, and assuming all components are in their standard state (unit activity, 1 bar pressure) and a constant temperature of 1200 °C, one can calculate the value of ΔH°_{rxn}, which is 296.7 kJ mol^{-1}. Viewing the system simplistically this positive value for ΔH°_{rxn} indicates the reaction shown in Equation (3.5) will *not* proceed to the right and form SiC.

The determination that ΔH°_{rxn} is positive, however, does not necessarily mean that SiC cannot be deposited. The most accurate approach to determining whether desired phases will form requires computing the minimum total Gibbs free energy (G) for the system and thus the resultant activities of all possible species, expressed as:

$$G = \sum_j \left(\sum_i n_i^j \right) G^j \tag{3.6}$$

where n is the number of moles of species i in phase j.

Table 3.1 shows an example of the results of a Gibbs free energy minimization calculation, again for the deposition of SiC from the tetrachloride and methane. To include consideration of all possible gaseous species and condensed phases requires use of nonlinear mathematical routines that can find the minimum system free energy, and thus all activities, which for ideal gases are their partial pressures. It was assumed that the temperature was 1200 °C, the total pressure was 1 bar (CVD is an open system and as such pressure can be kept constant), and an initial mole of each of the reactants were used.

Several things are quickly apparent that would not have been evident from a simple determination of whether a single reaction forming SiC from the reactants had a negative value of ΔH°_{rxn}. First, although ΔH°_{rxn} is positive as noted above, the overall Gibbs free energy under equilibrium conditions is negative, in this case –1796 kJ mol^{-1} (a value provided elsewhere in the calculational output), so that some SiC is expected to form. Second, single-phase SiC is not formed, but rather carbon (as graphite) is predicted to co-deposit with SiC, and in even a greater quantity. Third, the deposition process is relatively inefficient, with approximately one-third of the $SiCl_4$ precursor remaining unreacted.

In practice, the SiC deposition system described above usually includes significant amounts of hydrogen added to suppress carbon formation. Repeating the calculation with a hydrogen : silicon atomic ratio of 20 : 1 results in almost a two-thirds reduction in the amount of carbon predicted to form. Under actual experimental conditions carbon is not detectable in the coatings at all, as shown by the work of Fischman and Petuskey[16] and others. Thus, thermochemical calculations can be misleading. Experience indicates that carbon formation from methane is kinetically hindered in this case and that high hydrogen concentrations help improve efficiency.

3.2.3 Consideration of Non-stoichiometric/Solution Phases

The example of the deposition of SiC is relatively simple as the condensed phases are stoichiometric, no significant solid solutions exist and at the temperature of interest there are no liquids or liquid solutions. As technological systems grow in complexity there is a greater need to deposit multi-component coatings and films that have significant homogeneity ranges (non-fixed stoichiometry) and solid solutions. Some important examples are the ceramic high-temperature superconductors such as $YBa_2Cu_3O_{7-x}$,[17] $Al_{1-x}In_xN$ and $Ga_{1-x}In_xSb$ semiconductor layers for optoelectronic devices,[18,19] and yttria-stabilized zirconia (YSZ) for thermal barrier and fuel cell applications.[7]

The thermochemical solution concept is well established, with simple to complex solution models described in basic thermochemical texts.[11,12,14] The simplest model is an ideal solution where the components are treated as mixing randomly with no interactions (no bonding energetics or short-range order). The Gibbs free energy for an ideal solution is expressed as:

$$\begin{aligned} G^\circ &= \sum n_i G_i^{\circ j} \\ G^{id} &= RT \sum n_i \ln(n_i) \end{aligned} \tag{3.7}$$

where the first value is the sum of the Gibbs standard free energy for the constituent species in the solution and the second equation is the ideal mixing contribution, with the sum of the two

Table 3.1 Edited FactSage calculational output for the equilibrium state from input of 1 mol each of $SiCl_4$ and CH_4 at 1 bar and 1200 °C. Indicated are the initial conditions, the composition of the gas phase at equilibrium, and the equilibrium condensed phases with amounts of SiC and carbon (graphite) which are stable with other phases not stable.

T = 1200.00 C
P = 1.00000E + 00 bar
V = 4.15020E + 02 dm3

STREAM CONSTITUENTS	AMOUNT/mol
$SiCl_4$	1.0000E + 00
CH_4	1.0000E + 00

	EQUIL AMOUNT	MOLE FRACTION	FUGACITY
PHASE: gas_ideal	mol		bar
HCl_FACT53	1.4585E + 00	4.3044E-01	4.3044E-01
H2_ELEM	1.2462E + 00	3.6780E-01	3.6780E-01
SiCl4_FACT53	5.3028E-01	1.5650E-01	1.5650E-01
SiCl3_FACT53	7.3674E-02	2.1744E-02	2.1744E-02
SiHCl3_FACT53	4.2739E-02	1.2614E-02	1.2614E-02
SiCl2_FACT53	3.5019E-02	1.0335E-02	1.0335E-02
CH4_FACT53	1.2923E-03	3.8140E-04	3.8140E-04
SiH2Cl2_FACT53	5.4030E-04	1.5946E-04	1.5946E-04
H_FACT53	2.6068E-05	7.6935E-06	7.6935E-06
Cl_FACT53	2.4688E-05	7.2862E-06	7.2862E-06
C2H2_FACT53	9.4177E-06	2.7795E-06	2.7795E-06
SiH3Cl_FACT53	1.9940E-06	5.8848E-07	5.8848E-07
CH3Cl_FACT53	1.2623E-06	3.7253E-07	3.7253E-07
C2H4_FACT53	1.1361E-06	3.3530E-07	3.3530E-07
SiCH3Cl3_FACT53	7.3965E-07	2.1829E-07	2.1829E-07
CH3_FACT53	6.5419E-07	1.9307E-07	1.9307E-07
SiCl_FACT53	7.9638E-08	2.3504E-08	2.3504E-08
Cl2_ELEM	7.6048E-08	2.2444E-08	2.2444E-08
C2HCl_FACT53	5.9758E-09	1.7636E-09	1.7636E-09
C2H6_FACT53	T 4.6976E-09	1.3864E-09	1.3864E-09
SiH4_FACT53	2.9984E-09	8.8493E-10	8.8493E-10
CH2Cl2_FACT53	4.7342E-10	1.3972E-10	1.3972E-10
TOTAL:	3.3883E + 00	1.0000E + 00	1.0000E + 00
	mol		ACTIVITY
C_graphite(s)_ELEM	6.8094E-01		1.0000E + 00
SiC(s2)_FACT53	3.1774E-01		1.0000E + 00
SiC(s)_FACT53	0.0000E + 00		8.5474E-01
C_diamond(s2)_ELEM	0.0000E + 00		5.1994E-01
Si(s)_ELEM	0.0000E + 00		6.4433E-03

Cp_EQUIL J.K–1	H_EQUIL J	S_EQUIL J.K–1	G_EQUIL J	V_EQUIL dm3
4.01597E + 02	-3.56863E + 05	9.77000E + 02	-1.79613E + 06	4.15020E + 02

Mole fraction of system components:

	gas_ideal
C	4.6064E-01
Si	7.8569E-02
C	1.5156E-04
H	4.6064E-01

The cutoff limit for phase or gas constituent activities is 1.00E-10
Data on 1 constituent marked with 'T' are extrapolated outside their valid temperature range

providing the system Gibbs free energy. The ideal mixing term is the excess entropy resulting from randomly mixing the solution constituents. Where there are significant interactions between species and therefore an energetic contribution to the Gibbs free energy an excess energy term needs to be included. A common formalism for these excess terms is termed the Redlich-Kister formulation, where for a binary solution system A-B:

$$G_{AB}^{ex} = n_A n_B \sum_k L_{k,AB}(n_A - n_B)^k \tag{3.8}$$

in which L is an expansion coefficient in k that can also be temperature dependent. In a "regular" solution k equals zero, giving a single energetic value or interaction energy. At one time it was believed that all metal alloy solutions were regular solutions.[11] Now, better fits to metal alloy thermochemical behavior take into account specific energetics and are represented by expansions in multiple compositional and temperature dependent terms. In general, the total Gibbs free energy for a non-ideal solution is described as:

$$G = G^\circ + G^{id} + G^{ex} \tag{3.9}$$

A fundamental problem with the regular solution representation is that where there are significant interaction energies between species they will cause some short-range order, and therefore the assumption that the species randomly mix is not correct. The model, however, works well when the interaction energies are not large, and thus descriptions of metallic solutions have been particularly successful. This issue is more important for salts and chalcogenides, where the interaction energies are more significant. The problem has been addressed in several ways, including approaches such as the quasichemical,[20] compound energy formalism[21] and associate species models.[22]

A set of more complex calculations including solid solutions is demonstrated for the deposition of YSZ from metal-organic precursors carried in a solvent, in an example performed using the Thermo-Calc software.[23] The object of this investigation was to determine optimum conditions for depositing 8% yttria-stabilized zirconia, although the entire compositional region was explored. In this case the database available with Thermo-Calc did not include a representation of ZrO_2–$YO_{1.5}$ solid solutions, so that solution information (type of solution model and interaction parameters) had to be included manually. The representation and thermochemical values for the system constituents were adopted from Du *et al.*[24]

The overall reaction for the CVD process to prepare YSZ is:

$$\begin{aligned} &n_Y Y(C_{11}H_{19}O_2)_3 + n_{Zr}(C_{11}H_{19}O_2)_4 + 250(n_Y + n_{Zr})C_4H_4O \\ &\quad + 0.5 n_O O_2 \leftrightarrow (ZrO_2 : YO_{1.5}) + \text{byproducts} \end{aligned} \tag{3.10}$$

in which the metal-organic precursors are carried in the tetrahydrofuran (C_4H_4O) solvent. With the composition information and the thermochemical values for the various species, phases and solid solutions it is possible to explore CVD conditions to identify likely successful parameters for deposition of single-phase YSZ of the desired composition. Figure 3.1 is an example of a CVD diagram of the deposition temperature *versus* input oxygen that indicates the conditions under which specific phases can form. A result of the use of organic species is the potential for carbon co-deposition with the YSZ phase; the calculated boundary indicating where carbon is and is not predicted to form is shown in Figure 3.1. Experimental efforts successfully used the computed diagram to determine conditions for deposition of single-phase material.[23] Thus, this example demonstrates how diagrams derived from basic thermochemical information can direct conditions for efficient deposition of desired phases.

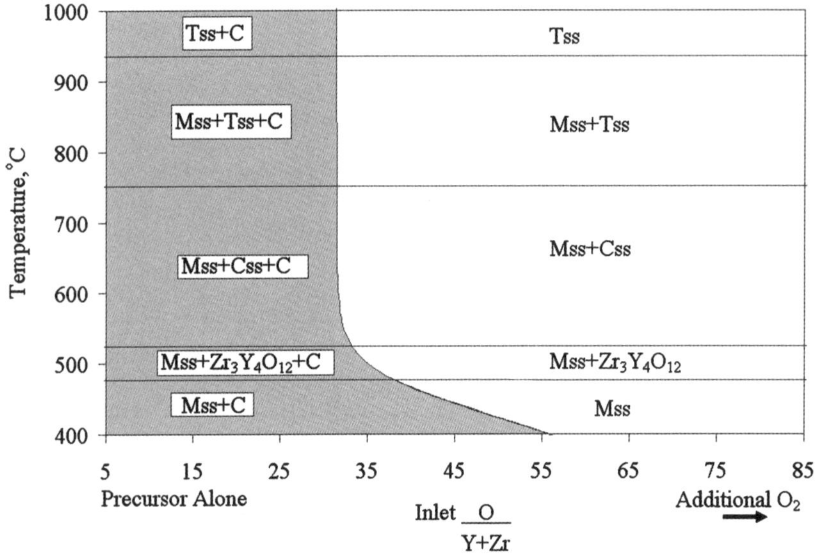

Figure 3.1 Computed CVD phase diagram for ZrO_2–$YO_{1.5}$. Note that the oxygen inherent in the precursor and solvent fix the minimum oxygen introduced in the system. (T_{ss} is tetragonal solid solution; M_{ss} is monoclinic solid solution; C_{ss} is cubic solid solution; C is carbon.).

Thermochemical calculations can also be useful for understanding deposition mechanisms and establishing maximum yields. Calculations performed with the constraint that no condensed phases can form potentially provide information about the gas phase above a substrate before the deposit forms. This has been explored, for example, for boron deposition,[25] SiC coatings[16,26] and aluminide coatings.[8] The investigation of SiC deposition from SiH_4 and C_2H_2 in a hydrogen environment illustrates the use of equilibrium calculations to identify potentially important gas-phase species. In addition to the expected stable species, the calculations included thermochemical values for 37 organosilicon species computed by first-principles quantum-chemistry methods. Figure 3.2 is a plot of species mole fraction with all condensed phases eliminated from the calculations. The results indicate that, under low pressure and relatively low temperatures, the formation of organosilicon radicals is favored, while radicals containing only silicon and hydrogen are not. The propensity for forming these radical species (Figure 3.2) leads to relatively low-temperature deposition and potential homogeneous nucleation, both of which are noted in experimental observations.

The work of Goujard *et al.* is a good illustration of how thermochemical equilibrium calculations can be used to determine coating composition and yield.[27] In this work the Si-B-C system was investigated for applications related to oxidation protection of carbon/carbon and carbon/silicon carbide composites. Because of uncertainties in key thermochemical values, it was necessary to perform a critical assessment of the thermochemical data for some species and phases to determine the most appropriate values. Also included was a solution model of the wide homogeneity of boron carbide (extending from $B_{10}C$ to B_4C). The precursor system was methyltrichlorosilane (MTS, CH_3SiCl_3) and BCl_3 in hydrogen. Figure 3.3 is an example of the predicted equilibrium yield, defined as the mole fraction of material formed at equilibrium divided by input boron, silicon or carbon plotted as a function of the MTS/BCl_3 fraction. From the results it is apparent that for this system SiC forms in relatively high concentrations even at low MTS/BCl_3 fraction, while the boron carbide phase is a minor constituent except at values of MTS/BCl_3 fraction less than 0.5.

Equilibrium thermochemical modeling is much less successful when applied to low-temperature processes. At high temperatures chemical kinetics are generally rapid due to the exponential dependence of reaction rates on temperature. High reaction rates decrease or eliminate the effect of

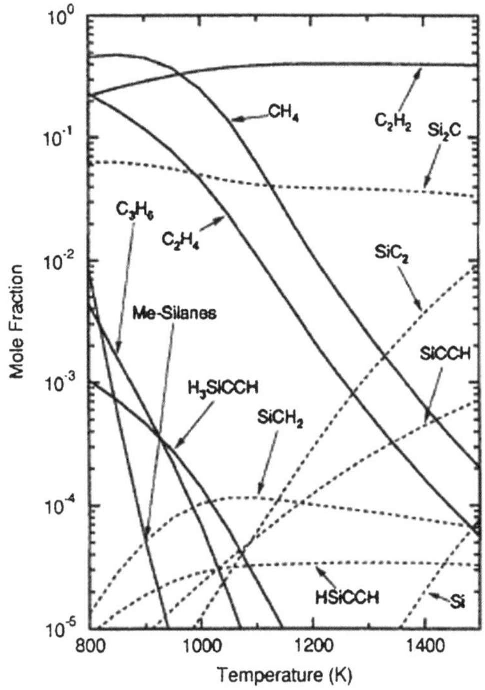

Figure 3.2 Computed equilibrium mole fractions of gaseous species in the SiH_4–C_2H_2 system. Initial conditions: pressure = 0.01 bar; number of moles: $Si_2H_6 = 1.0$, $C_2H_4 = 11.0$. The line labeled "Me-silanes" is the sum of the mole fractions for the $SiH_{4-n}(CH_3)_n$, $n = 1$–;4 species. Solid lines are stable species and dashed lines are radicals. (Reprinted with permission from ref. 26.)

individual reaction rates on the approach to equilibrium. However, at the low deposition temperatures used to deposit materials for microelectronics, for example, chemical-kinetics dominate and it is possible to deposit phases far from equilibrium. This is apparent in the often amorphous morphology of oxides deposited when the temperature is too low to ensure adequate species mobilities to form structures with long-range order. For example, SiO_2 and Ta_2O_5 layers deposited at low temperature form amorphous films.[28,29] Unfortunately, there are no firm guidelines with regard to temperatures or other conditions that govern whether deposited systems are near or far from equilibrium. A rough rule of thumb is to consider temperatures approaching 1000 °C as likely to form crystalline deposits and be governed by equilibrium thermochemistry, whereas deposition of films, particularly oxides, in the range of 500 °C or lower will likely be amorphous and potentially far from equilibrium.

3.2.4 Thermochemical Equilibrium Software Packages

The calculations just described were performed with the FactSage[15] or Thermo-Calc[30] software packages using their supplied databases. There are several other high-quality, very versatile software systems available for performing sophisticated thermochemical calculations, including generating plots of various output values such as partial pressures, activities, compositions, species quantities, as well as other types of information including phase diagrams and predominance diagrams. Other available packages include Thermosuite,[31] MTDATA,[32] PANDAT,[33] HSC,[34] and MALT.[35] The advent of relatively fast personal computers allows almost all of this type of nonlinear solver software to run on relatively standard machines, typically with a Windows interface. The selection of which package is most appropriate for an application or organization will likely be

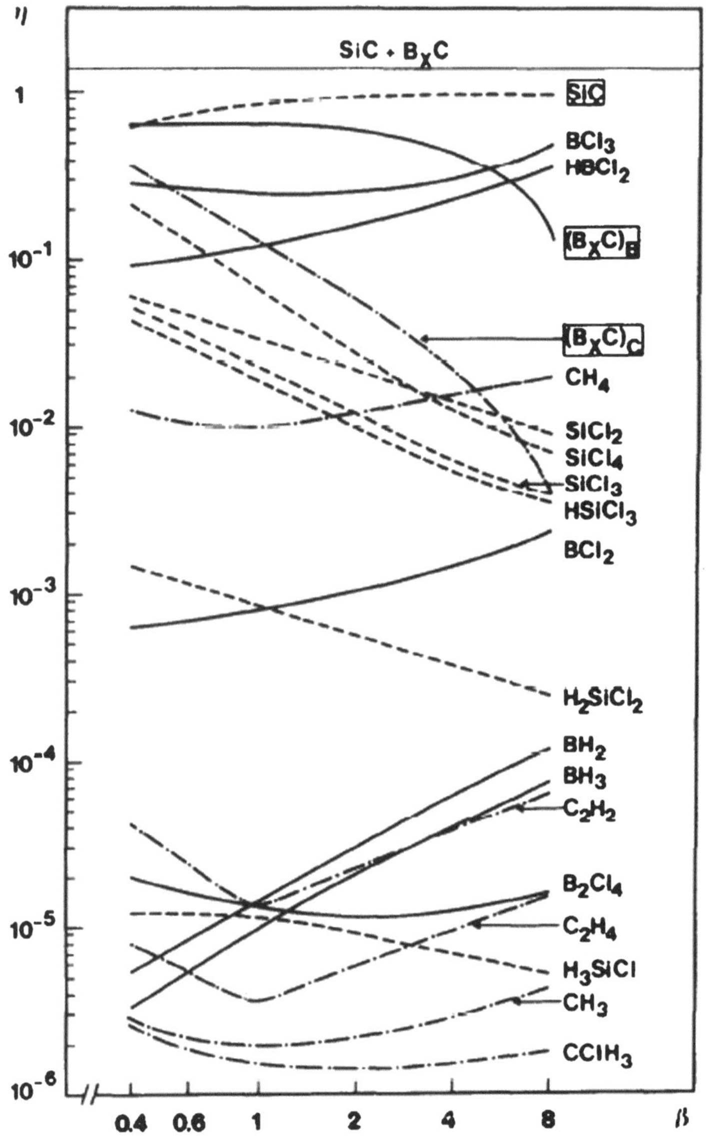

Figure 3.3 Equilibrium yields for phases in the boron carbide system. Yields are defined as the fraction of species/phase formed compared to the base element input to the system (η) of the different gaseous and solid species at $T = 1127\,^{\circ}C$, total pressure = 0.395 bar, H_2/MTS = 20 *versus* the MTS/BCl_3 (β) variable. The species phases are defined as —(B),– – – (Si),–·– (C) (containing species); η for $B_xC(s)$ is presented related to both input boron and carbon. (Reprinted with permission from ref. 27.)

determined by cost, the applicability of the available databases to the problem of interest, and personal preference with regard to the interface.

3.2.5 Thermochemical Data and Databases

Commercial equilibrium software packages are generally accompanied by thermochemical databases for a wide variety of chemical systems. The computational engines in the software automatically obtain from the databases the values needed to perform the calculations. This presents the user of the software with two critical issues. First, does the database provided contain values for all the species and phases of interest? Not only do all possible stoichiometric phases for a chemical system need to be included, but also any solid and possibly liquid solution phases likely to be important. Solid solutions need to be represented by specific solution/defect models, and these can be relatively complex. Thus the user must assure that these phases are available in the databases the software accesses and are properly considered. For the ZrO_2–Y_2O_3 example discussed above, data obtained from sources other than the Thermo-Calc supplied databases were necessary to properly consider the solution phases.

A second issue concerning thermochemical databases is their accuracy and reliability. Most commercial databases have been assessed, which means the data included in the database have been critically evaluated with regard to the source methodology (experimental or computational) used to obtain the data and accuracy. In addition, the data for a species or phase must be consistent with information for related species and phases that reside in the database. That is, calculations performed with the data and that for other species or phases must result in the appropriate relationships between the phases and species (*e.g.*, phase equilibria, activities and partial pressures). Users of commercial databases need to ensure that the data they are using have been assessed. In addition, the use of data from more than one source can be problematic in that the values may be consistent within the database, but not consistent between databases. Checking a set of data used in calculations against known behavior, such as by reproducing experimental phase equilibria, will help ensure that the information is consistent and will give accurate results.

Thermochemical data have been compiled for several decades, and among the best known compendia are the *NIST-JANAF Thermochemical Tables*[36] and *Thermochemical Data of Pure Substances*.[37] While most common substances are included in these compilations, it is not unusual for critical phases or species needed for thermochemical calculations to be absent from these tabulations. It would therefore be necessary to perform a literature search to locate measured and published values. Currently, with advances in first-principles modeling, data for some systems have been determined computationally, although this is much more likely for gaseous species than for condensed phases.

Databases are also available from commercial sources: the Scientific Group Thermodata Europe (SGTE)[32] is a well-established source of thermodynamic data and has a continuing program to assess systems to improve values and incorporate new species and phases. The Japanese database in MALT[35] is more limited than SGTE, with a focus on providing values for practical problems in industry. Many of the databases available with the commercial software packages are often entirely from outside sources such as SGTE, but may contain additional values from the supplier's work. This is particularly true for FactSage[15] and Thermo-Calc.[30]

A solution to the problem of missing thermochemical values is to resort to relatively simple estimation techniques, which in many cases can give sufficiently accurate values. Kubaschewski *et al.*[12] have presented an extensive discussion of estimation techniques that are extremely useful. For example, heat capacities of constituent oxides in complex oxide systems can be linearly summed to give very good representations of the heat capacity relationship. Enthalpies of formation in similar systems often exhibit linear relationships with atomic number.

3.3 Reactor Modeling

3.3.1 Chemically Reacting Fluid Flow

Broadly speaking, CVD is a process in which gas-phase precursors react to form a solid film at a surface. Usually a high-value thin film is the desired result. The primary objective of this section is to discuss fluid-mechanical and molecular-transport aspects of CVD, and their relationships to reaction chemistry.

CVD processes typically seek to grow a film or coating that is spatially uniform. In some cases, such as a semiconductor wafer, the deposition surface is flat at the macroscopic length scale of the wafer (*i.e.*, the wafer diameter of around 300 mm). However, at the micro-scale (*i.e.*, length scales of a micron and smaller) uniformity may be required in depositing films within trenches or vias. In other cases, the process must deliver a uniform film on a relatively large but complex-shaped part such as a turbine blade.

There is no single design rule for developing a CVD process and the reactor to implement it. Designing a CVD process depends on several important considerations. The standard state of precursor chemicals may be gaseous, liquid or solid. The process may be batch or continuous. Growing thin films for semiconductor devices, for example, is usually a batch process, operating on one, or more than one, wafer at a time. However, some applications, such as applying anti-reflective coatings to large glass sheets, are usually a continuous process in which the glass moves through the CVD reactor. Process pressure is another important consideration, ranging from vacuum conditions to atmospheric pressure or greater. As with most chemical processes, CVD is greatly influenced by temperature, both in the gas phase and at the deposition surface.

3.3.2 Rate Controlling Processes

Chemically reacting fluid flow is a balance between convective transport, diffusive transport and chemical reaction. Optimal process and reactor design usually depends on identifying and accommodating rate-limiting processes. Most CVD processes operate at atmospheric pressure or below. At higher pressures convective transport tends to be dominant. As pressure decreases toward vacuum conditions, diffusive processes become dominant because diffusion coefficients are generally proportional to the inverse of pressure. Reduced pressure usually leads to more uniform films on complex shapes, including microscopic features. However, because of reduced gas-phase collision frequency at the deposition surface, deposition rates are also reduced. In contrast, deposition rates can usually be increased by increasing the pressure, but convective fluid transport becomes increasingly important relative to diffusive transport. In this case, controlling the boundary-layer behavior at the deposition surface is important to achieving uniform deposition.

Temperature, especially at the deposition surface, is perhaps the most important consideration in CVD processes. Increasing temperature generally increases chemical reaction rates. All other factors being equal, increased reaction rates lead to higher deposition rates, which can be desirable. However, all other factors are not equal. The film's chemical composition may depend greatly on temperature. Furthermore, a wide range of film microstructures and morphologies can result that depend on growth conditions. As temperature increases, the deposited material may vary from being amorphous, to polycrystalline, to a single-crystal epitaxial film. Further, temperature can have a strong influence on the grain size of polycrystalline films. Owing to convective and diffusive transport, the substrate temperature affects the temperature of the gas-phase boundary layer adjacent to the deposition surface. The gas-phase temperature, in turn, affects gas-phase reaction rates. Some, but not all, CVD processes depend on gas-phase reaction prior to the surface reactions that ultimately deposit the desired film. For example, the parent precursors that initially enter the reactor may need to react in the gas phase to produce surface-active reaction products. As a consequence of all these considerations, there are many constraints on process temperature that control the required properties of the resulting product.

3.3.3 General Conservation Equations

Gas flow within CVD reactors is nearly always laminar. A combination of relatively low velocities and often reduced pressure lead to low Reynolds numbers. Thus, in the design and analysis of CVD

processes, it is unnecessary to consider turbulence. The reacting flow within a CVD reactor is described by the Navier–Stokes equations (conservation of mass and momentum), together with conservation equations for species and thermal energy. For a general and detailed derivation, one may refer to Kee *et al.*[38] These equations are stated in general vector form as:
Mass continuity:

$$\frac{\partial p}{\partial t} + \nabla \cdot (p\mathbf{V}) = 0 \tag{3.11}$$

Momentum:

$$\rho \frac{D\mathbf{V}}{Dt} = \mathbf{f} - \nabla p + \nabla \cdot \mathsf{T}' \tag{3.12}$$

Species continuity:

$$\rho \frac{DY_k}{Dt} = -\boldsymbol{\nabla} \cdot \mathbf{j}_k + \dot{\omega}_k W_k \tag{3.13}$$

Thermal energy:

$$\rho c_p \frac{DT}{Dt} = \frac{Dp}{Dt} + \boldsymbol{\nabla} \cdot (\lambda \boldsymbol{\nabla} T) - \sum_{k=1}^{K} c_{pk} \mathbf{j}_k \cdot \boldsymbol{\nabla} T - \sum_{k=1}^{K} h_k \dot{\omega}_k W_k \tag{3.14}$$

Equation of state:

$$\rho = \frac{p}{RT} \frac{1}{\sum Y_k / W_k} \tag{3.15}$$

Generally speaking, these equations represent balances between convective transport (left-hand sides) and diffusive transport and volumetric sources (right-hand sides). As written here, the left-hand sides of the transport equations are written in compact form using the substantial-derivative operator, which incorporates convective transport. The operator includes explicit temporal variations $\partial/\partial t$ as well as convective transport *via* the velocity field. The substantial derivative operator for a scalar variable (*e.g.*, temperature T) is written as:

$$\frac{DT}{Dt} \equiv \frac{\partial T}{\partial t} + \mathbf{V} \cdot (\boldsymbol{\nabla} T) = \frac{\partial T}{\partial t} + (\mathbf{V} \cdot \boldsymbol{\nabla}) T \tag{3.16}$$

The substantial derivative of a vector (*e.g.*, velocity $\mathbf{V}$) is written as:

$$\frac{D\mathbf{V}}{Dt} \equiv \frac{\partial \mathbf{V}}{\partial t} + (\mathbf{V} \cdot \boldsymbol{\nabla}) \mathbf{V} \tag{3.17}$$

In non-cartesian coordinates, care must be taken to expand the second term as:

$$(\mathbf{V} \cdot \boldsymbol{\nabla}) \mathbf{V} \equiv \frac{1}{2} \boldsymbol{\nabla} (\mathbf{V} \cdot \mathbf{V}) - [\mathbf{V} \times (\boldsymbol{\nabla} \times \mathbf{V})] \tag{3.18}$$

The independent variables are time t and the spatial coordinates. Dependent variables include the mass density p, velocity vector ($\mathbf{V}$), pressure (p), temperature (T), and the species mass fractions (Y_k). The momentum equation includes body forces $\mathbf{f} = \rho \mathbf{g}$, which in CVD reactors are the result of buoyancy caused by density variations associated with temperature and composition variations. In addition to the forces associated with the pressure gradient, the momentum equations also involve the divergence of the deviatoric stress tensor T'. The deviatoric stress tensor relates the fluid strain rates to the viscous stresses *via* the velocity field. Written out in cylindrical coordinates, this tensor is:

$$\mathsf{T}' = \begin{pmatrix} 2\mu\frac{\partial u}{\partial z} + \kappa\nabla\cdot\mathbf{V} & \mu\left(\frac{du}{dr} + \frac{dv}{dz}\right) & \mu\left(\frac{1}{r}\frac{\partial u}{\partial\theta} + \frac{\partial w}{\partial z}\right) \\ \mu\left(\frac{du}{dr} + \frac{dv}{dz}\right) & 2\mu\frac{\partial v}{\partial r} + \kappa\nabla\cdot\mathbf{V} & \mu\left(\frac{dw}{dr} - \frac{w}{r} + \frac{1}{r}\frac{dv}{d\theta}\right) \\ \mu\left(\frac{1}{r}\frac{\partial u}{\partial\theta} + \frac{\partial w}{\partial z}\right) & \mu\left(\frac{dw}{dr} - \frac{w}{r} + \frac{1}{r}\frac{dv}{d\theta}\right) & 2\mu\left(\frac{1}{r}\frac{\partial w}{\partial\theta} + \frac{v}{r}\right) + \kappa\nabla\cdot\mathbf{V} \end{pmatrix} \tag{3.19}$$

where u, v, and w are the axial, radial and circumferential components, respectively, of the velocity vector and μ and κ are the fluid's dynamic and bulk viscosities. According to Stokes' hypothesis, the bulk viscosity is usually taken as $\kappa = 2\mu/3$.

The species conservation equations balance convective transport, diffusive transport and the production (or consumption) of species *via* gas-phase chemical reactions. The variable W_k represents the molar production rate of species k by chemical reaction. CVD processes can often involve many elementary reactions, with rates depending on temperature, pressure and composition. The species diffusive mass flux vector is stated as:

$$\mathbf{j}_k = \rho Y_k \mathbf{V}_k \tag{3.20}$$

where $\mathbf{V}_k$ is the diffusion-velocity vector for the k-th species. The diffusion velocity may be written as:

$$\mathbf{V}_k = \frac{1}{X_k \overline{W}} \sum_{j \neq k}^{K} W_j D_{kj} \nabla X_k - \frac{D_k^T}{\rho Y_k} \frac{1}{T} \nabla T \tag{3.21}$$

The ordinary multicomponent diffusion coefficient matrix D_{kj} and the thermal diffusion coefficients D_k^T are determined from the binary diffusion coefficients using kinetic theory. The mole fractions are represented as X_k, the molecular weights are W_k, and the mean molecular weight is $\overline{W}$.

Transport properties (viscosity, thermal conductivity and diffusion coefficients) are determined from kinetic theory and the underpinning theory and methodology is well understood.[39–41] However, species-specific parameters are needed before individual species properties can be evaluated. The parameters include the potential-well depth and collision diameter, as well as dipole moment and polarizability. CVD processes often use chemical species for which the needed parameters are not known or catalogued. Thus, without specific experiments to measure properties, the analyst must often rely on estimation techniques.[40,42]

As written in Equation (3.14), the thermal-energy equation is restricted to ideal-gas mixtures. The specific heat capacity is represented as c_p. The first term on the right-hand side of Equation (3.14), which is often negligible, represents the contribution to thermal energy of pressure–velocity interactions. The second term, which represents the conduction of heat through the gas, involves the mixture thermal conductivity λ_k. The third term represents the transport of thermal energy *via* diffusive mass fluxes in a varying temperature field. The last term represents the contribution to thermal energy by chemical reactions. The species enthalpies are written as h_k.

3.3.4 Boundary and Initial Conditions

For any given reactor, the reactor geometry must be specified. Solving the system of partial differential equations requires appropriate boundary and initial conditions. For transient problems, the field of all dependent variables must be specified at some initial time. For steady-state problems, initial conditions are not needed, but the boundary conditions can be complex.

Generally speaking, inlet and outflow conditions must be specified. Temperature (or some other thermal condition such as a specified heat flux) must be specified at the reactor walls. For CVD reactors, special care is needed at the deposition surfaces. The species mass balance at these surfaces can be written as:

$$\mathbf{n} \cdot [\rho Y_k(\mathbf{V}_k + \mathbf{u})] = \dot{s}_k W_k, \quad (k = 1, \ldots, K_\mathrm{g}) \tag{3.22}$$

where $\mathbf{n}$ is the unit outward-pointing normal vector that defines the spatial orientation of the surface. This equation states that the convective and diffusive species fluxes of the K_g gas-phase species are balanced by the reaction of these species *via* heterogeneous chemistry at the deposition surface. When net mass is exchanged between the gas phase and the deposition surface there is a non-zero fluid velocity normal to the deposition surface. This reaction-induced Stefan velocity $\mathbf{u}$ is evaluated as:

$$\mathbf{n} \cdot \mathbf{u} = \frac{1}{\rho} \sum_{k=1}^{K_\mathrm{g}} \dot{s}_k W_k \tag{3.23}$$

The expression for the Stefan velocity is easily obtained from the interfacial mass balance, Equation (3.22), by summing over all K_g species, noting that the mass fractions must sum to unity and that mass conservation requires that the sum of the diffusive fluxes must vanish:

$$\sum_{k=1}^{K_\mathrm{g}} \rho Y_k \mathbf{V}_\mathrm{k} = 0 \tag{3.24}$$

For chemically inert portions of the reactor walls, Equation (3.22) still applies. However, the reaction rate $\dot{s}_k$ and the Stefan velocity both vanish. The surface reaction rates $\dot{s}_k$ are usually the result of several elementary heterogeneous reactions that involve both gas-phase and surface-adsorbed species. Note that the mass balance at the surface [*i.e.*, Equation (3.22)] directly includes only the gas-phase species. In general, however, the heterogeneous reaction mechanism involves gas-phase, surface and bulk species. For a steady-state process, the surface state must be stationary. That is, the net production rates of surface-adsorbed species must vanish:

$$\dot{s}_k = 0, \quad (k = 1, \ldots, K_\mathrm{s}) \tag{3.25}$$

where K_s is the number of surface species. The net production rate of bulk species (*i.e.*, species underneath the deposition surface) represents the deposition rate. That is, the growth rate G (measured in thickness per unit time) can be represented as:

$$G = \sum_{k=1}^{K_\mathrm{b}} \frac{\dot{s}_k W_k}{\rho_\mathrm{b}} \tag{3.26}$$

where K_b is the number of bulk species and ρ_b is the mass density of the deposited film. A much fuller discussion of gas-phase, surface and bulk species, together with heterogeneous reaction chemistry, has been given by Kee *et al.*[38]

3.3.5 Computational Solution

Although the complete system of partial differential equations is highly nonlinear, stiff and generally complex, it is solvable computationally. In fact, high-quality commercial software for solving such chemically reacting flow problems is readily available (*e.g.*, FLUENT, www.ansys.com). These software packages handle complex three-dimensional reactor geometries, as well as elementary or global reaction chemistry.

Evidently, from the full system of conservation equations, one must handle multicomponent thermodynamic properties, transport properties and reaction chemistry. As the chemical processes increase in complexity, so too do the requirements for handling relatively large systems of chemical species and reaction mechanisms. Software packages such as CHEMKIN and CANTERA are designed specifically for this purpose. CHEMKIN is FORTRAN-based software that was developed at Sandia National Laboratories to provide general capabilities to represent multicomponent thermodynamics, transport and reaction chemistry in chemically reacting flow simulations. The underlying theory has been documented by Kee *et al.*[38] Commercially supported implementations of CHEMKIN are now available (www.reactiondesign.com). CANTERA is object-oriented software written in C++. The software was developed by David Goodwin at Caltech and is freely available as shareware.[43]

Most computational fluid dynamics (CFD) software packages that are designed to solve chemically reacting flow problems have user interfaces that enable the incorporation of complex reaction chemistry, both in the gas phase and at surfaces. Several commercial offerings include interfaces to CHEMKIN, and some are also incorporating CANTERA interfaces.

3.3.6 Uniform Deposits in Complex Reactors

CVD processes are implemented in reactors that may be geometrically complex, including provisions for introducing gaseous chemical precursors and removing exhaust gases. Thus, thermal and chemical conditions can vary at different positions of the reactor walls. For example, some portions of the walls may be insulated while others are controlled to achieve a desired temperature. Deposition may occur on some surfaces, while other portions are chemically inert to inhibit deposition or other heterogeneous chemistry. The fluid flow is generally three-dimensional. However, because spatially uniform deposits are usually desired, the reactor design and operating conditions are developed to deliver a lower-dimensional result. Consider, for example, deposition on a flat semiconductor wafer. The deposit is "one-dimensional" in the sense that the deposited film thickness is the same everywhere on the wafer surface. Thus, the designer is challenged to develop a three-dimensional reactor that delivers a one-dimensional result.

3.3.7 Reactor Design

3.3.7.1 Historical Perspective

Figure 3.4 illustrates a highly simplified account of CVD reactor development for depositing films on semiconductor wafers. As illustrated in Figure 3.4(a), early CVD reactors were often implemented in a flow channel with a heated wafer on the channel floor. A boundary-layer model of such reactors was developed by Coltrin *et al.*[44,45] This model was the first to incorporate elementary reactions into a CVD mechanism. Because of the boundary-layer development, deposition thickness varied from the leading edge to the trailing edge of the wafer. Assuming transport-limited growth, the deposition rate would be higher at the leading edge, where the boundary-layer thickness is smaller. However, it is not necessarily the case that deposition rate is highest on the upstream portions of the wafer. For example, when homogeneous reactions of the precursors are needed to produce surface-active species, deposition rates could be higher on downstream sections. This is

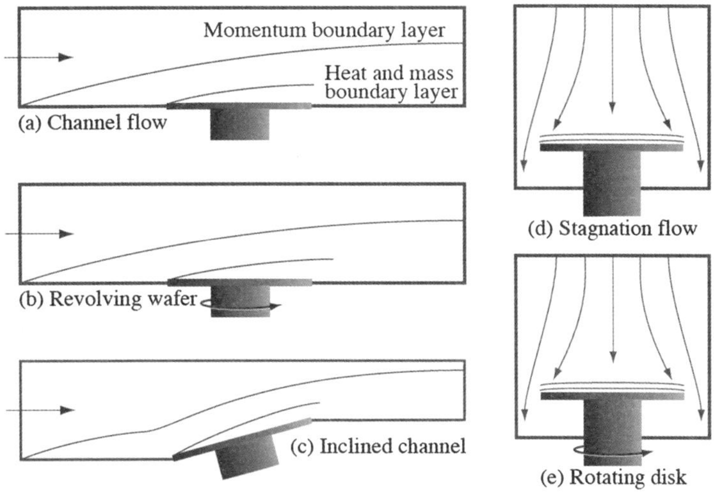

Figure 3.4 Simplified description of the evolution of channel-based and stagnation-based CVD reactors.

because the gas-phase reaction kinetics may require a certain residence time at elevated temperature to deliver appropriate levels of the surface-active species. In other situations, the deposition may be rate-limited by surface chemistry. In this case, the wafer temperature alone is the most important factor affecting growth rate. Under these circumstances, the fluid flow has a relatively small effect on the deposit uniformity, which is governed primarily by maintaining uniform wafer temperature.

Assuming the growth rate is limited by fluid-mechanical transport or gas-phase reaction, there can be benefits to slowly revolving the wafer on the channel floor (Figure 3.4b). The wafer revolution serves to continuously exchange the upstream and downstream portions of the wafer. If the deposition rate varies nearly linearly along the channel length, revolving the wafer results in a nearly uniform deposit thickness. If the rotation rate is relatively small, then the channel flow can be reasonably represented as a two-dimensional boundary-layer flow [Equations (3.18) and (3.19)]. However, if the rotation rate becomes too large, a complex three-dimensional flow develops.

Figure 3.4c illustrates another approach that seeks to limit thickness variations in the deposited film. Again assuming transport-limited growth, controlling boundary-layer thickness serves to control growth rate. By inclining the channel floor (or alternatively inclining the upper channel wall), the flow over the wafer must accelerate. The result is that the boundary-layer growth is suppressed. Consequently, the deposition thickness is more uniform than it would be without the restriction in the channel width. Combinations of channel geometry and wafer rotation could also be implemented.

In some sense the stagnation flow illustrated in Figure 3.4(d) represents a limiting case of the inclined channel. Here, the deposition surface is oriented perpendicular to the primary flow direction. This turns out to be an especially advantageous situation. In 1911, K. Heimenz showed that the stagnation flow situation could be formulated and solved as a one-dimensional ordinary-differential-equation boundary-value problem. A very important outcome of his analysis is that the boundary-layer thickness is uniform, independent of position on the stagnation surface. When the deposition rate is transport limited, this is an extremely desirable property for a CVD reactor. All modern semiconductor fabrication facilities employ many stagnation-flow reactors. The immense impact that this mathematical result of 1911 has had on the modern semiconductor-processing industry is remarkable. Of course, at the time, Heimenz could have not even begun to contemplate the implications of his work for future technological development and manufacturing.

In 1921, T. von Kármán developed a one-dimensional analysis for the rotating-disk problem as illustrated in Figure 3.4(e). Like Heimenz, von Kármán's primary motivation was to find practical solutions to complex fluid mechanics problems for certain limiting circumstances. As with the stagnation-flow problem, the similarity solution reveals that the boundary-layer thickness is uniform everywhere on the rotating disk.[46] Rotating disk reactors are also widely used in commercial CVD for semiconductor processes, usually for opto-electronic applications.

Kee *et al.*[38] provide a detailed derivation and discussion of the stagnation-flow and rotating-disk problems. Although perhaps not recognized at the time of the original derivations, both problems are described by the very same system of equations. These equations, written for the axisymmetric situation, can be summarized as the following system of ordinary differential equations:

Mass continuity:

$$\frac{d(\rho u)}{dz} + 2\rho V = 0 \tag{3.27}$$

Radial momentum:

$$\rho u \frac{dV}{dz} + \rho(V^2 - W^2) = -\Lambda_r + \frac{d}{dz}\left(\mu \frac{dV}{dz}\right) \tag{3.28}$$

Circumferential momentum:

$$\rho u \frac{dW}{dz} + 2\rho V W = \frac{d}{dz}\left(\mu \frac{dW}{dz}\right) \tag{3.29}$$

Thermal energy:

$$\rho u c_p \frac{dT}{dz} = \frac{d}{dz}\left(\lambda \frac{dT}{dz}\right) - \sum_{k=1}^{K} \rho Y_k V_k c_{pk} \frac{dT}{dz} - \sum_{k=1}^{K} h_k W_k \dot{\omega}_k \tag{3.30}$$

Species continuity:

$$\rho u \frac{dY_k}{dz} = -\frac{d}{dz}(\rho Y_k V_k) + W_k \dot{\omega}_k \quad (k = 1,\ K) \tag{3.31}$$

These steady-state equations have a single independent variable, the distance from the deposition surface z. The axial velocity is represented as u (which is independent of radius r) and the scaled radial velocity is written as $V = v/r$, where v is the actual radial velocity. The scaled circumferential velocity is written as $W = w/r$, where w is the actual circumferential velocity. The variable $\Lambda_r = (1/r)$ $(\mathrm{d}p/\mathrm{d}r)$ in Equation (3.28) is an eigenvalue that represents the radial pressure gradient. All other variables have the same meanings as in the full system of conservations equations.

The stagnation-flow and rotating-disk problems were derived originally assuming a semi-infinite domain above the surface. However, in a practical CVD reactor, precursor flow is usually introduced through a manifold that is parallel to the deposition surface. Such a reactor is illustrated in Figure 3.5. Maintaining similarity requires that the manifold introduces flow at uniform velocity, temperature and composition. To accomplish this, manifolds are typically implemented as a porous frit or a showerhead fabricated with an array of small holes.

Solving the system of equations requires boundary conditions at the inlet manifold and the deposition surface. At the inlet manifold, the axial velocity is specified and the radial velocity vanishes owing to a no-slip condition at the manifold surface. Further, the inlet temperature and composition must be specified. At the deposition surface, the radial velocity vanishes and the

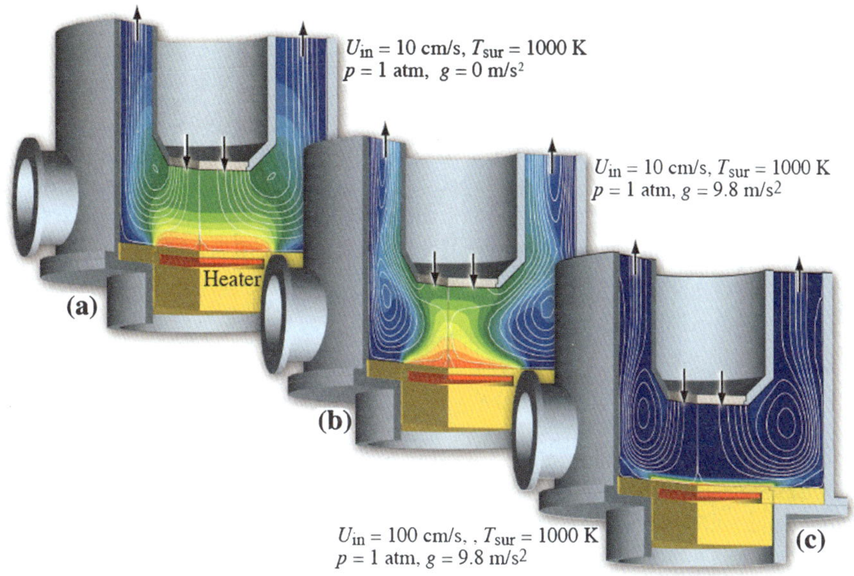

Figure 3.5 Computational solutions for stagnation-flow CVD reactors under different operating conditions.

temperature is specified. The boundary conditions for axial velocity and composition are the result of surface chemistry [*i.e.*, as stated in Equation (3.22) and following equations].

Considering the order of the system of equations, a keen observer will worry that there seem to be too many boundary conditions. The continuity equation is first-order in the axial velocity, while the other conservation equations are second order. The fact that two boundary conditions are specified for the axial velocity may appear to over-specify the problem. However, the value of Λ_r must be determined as an eigenvalue, which adds the extra degree of freedom needed to accommodate the specification of axial velocity at both boundaries. The system of equations is readily solved computationally. Solution algorithms are discussed in elsewhere.[38] Complex gas-phase and surface chemistry are easily incorporated, usually through software packages such as CHEMKIN or CANTERA.

Evidently, Equations (3.27)–(3.31) represent a boundary-value problem that is independent of radius r (except through scaled variables). This implies that the solutions are independent of radius, and are thus applicable for surfaces of indefinite radial extent. Of course, any actual reactor has a finite-radius deposition surface and is confined by reactor walls. Fortunately, it is both possible and practical to design a reactor that realizes the ideal stagnation-flow over most of the deposition surface.[47–50]

3.3.7.2 Practical Stagnation-flow Reactors

Figure 3.5 illustrates a possible reactor geometry, with downward inlet flow through a porous manifold and the deposition surface resting on a heater assembly. The exhaust flow exits upward through an annular region formed by the inlet assembly and the outer reactor walls. The colors in Figure 3.5 represent temperature contours and flow streamlines are shown as white lines. The two-dimensional solutions, which accommodate the actual reactor geometry, are computed using axisymmetric CFD software.

Figure 3.5(a) shows a solution for a low inlet velocity, but with gravity neglected. The low inlet velocity results in a relatively thick boundary layer. Importantly, it is seen that the boundary-layer thickness (as represented by the temperature contours) is nearly uniform over most of the heated deposition surface. Despite the fact that the flow is clearly two-dimensional and plainly does not satisfy the conditions for ideal similarity as it turns upward toward the exhaust annulus, even transport-limited deposition would be highly uniform.

Figure 3.5(b) uses the same boundary conditions as in Figure 3.5(a), but now with buoyant effects considered. With the heated deposition surface at the bottom, upward buoyant forces oppose the momentum of the downward flow directed towards the deposition surface. Under these flow conditions, buoyancy is important and causes a thermal plume to rise from the heated surface. Such natural-convective flow significantly alters the flow field, destroying the desired stagnation-flow similarity. The relative strength of buoyant convection can often be estimated in terms of a Reynolds number and a Grashof number. These dimensionless groups are defined as:

$$\mathrm{Re} = \frac{UL}{\nu}, \quad \mathrm{Gr} = \frac{g\beta\Delta T L^3}{\nu^2} \tag{3.32}$$

In these definitions U and L are characteristic velocity and length scales and ν is the fluid kinematic viscosity. The acceleration of gravity is g, the thermal expansion coefficient is β, and ΔT is a characteristic temperature difference. In a stagnation reactor, the characteristic velocity U may be the velocity through the inlet manifold, the characteristic length scale L may be the separation distance between the manifold and the deposition surface, and the characteristic temperature difference ΔT may be the difference between the inlet flow and the deposition surface. The relative importance of buoyancy is usually measured as the ratio $\mathrm{Gr}/\mathrm{Re}^2$. As this ratio increases, the likelihood of strong buoyant-driven flow increases. The exact value of the ratio depends on details of the reactor geometry.

It is usually possible to offset the potentially deleterious effects of buoyancy. Increasing inlet velocity, which increases Reynolds number, reduces the relative importance of buoyancy. Similarly, reducing the manifold-to-wafer distance L tends to suppress buoyancy. In some systems, it may be possible to orient the heated surface to face downward. In this case, the upward buoyant forces tend to stabilize the flow against the stagnation surface. The solution shown in Figure 3.5(c) uses an inlet velocity that is increased from 10 to 100 cm s^{-1}. The increased momentum of the inlet flow is sufficient to overcome the buoyant forces, leading to a stable stagnation flow. The boundary layer is also much thinner than the low-flow situation.

Beyond a relatively simple steady buoyant plume as shown in Figure 3.5(b), there can also be significantly more complex flow disruptions.[47,51–53] Fortunately, it is possible to predict such complex, often transient or even chaotic, flows with computational fluid dynamics models. Thus, there is a sound basis for the model-based design of CVD reactors.

Flow stability and deposition uniformity are often primary design considerations. However, there are also other important factors to be considered. One involves precursor flow that does not directly interact with the deposition surface. As can be seen in Figure 3.5, many of the streamlines that emanate from the inlet manifold turn toward the exhaust before entering the boundary-layer above the deposition surface. Thus, some of the precursor species that enter the reactor may leave without causing any deposition. This is especially the case for the relatively high flow rate and thin boundary layer represented in Figure 3.5(c). If the precursor chemicals are expensive, such flow bypass can increase process cost. However, in some sense, the unreacted flow is not entirely "wasted." The unreacted flow is indeed necessary to preserve the desired flat boundary layer above the deposition surface. Among other alternatives, the amount of unreacted flow can be reduced by reducing the separation distance between manifold and deposition surface.

3.4 Gas-phase Thermochemistry and Kinetics

The need for accurate gas-phase thermodynamic and kinetic data for the species involved in a CVD deposition mechanism cannot be overstated. Although the drive to lower deposition

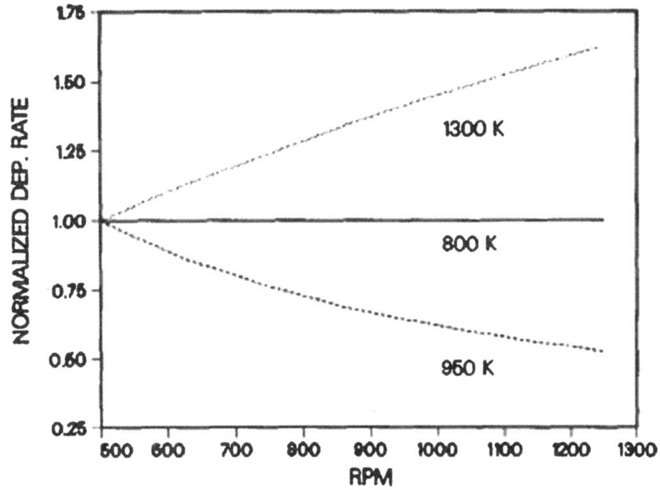

Figure 3.6 Relative silicon deposition rates in helium carrier gas in a rotating disk reactor a function of spin rate and temperature. Rates are normalized by the growth rate at 500 rpm. (Reprinted with permission from ref. 46).

temperatures minimizes or eliminates gas-phase chemistry in some cases [particularly in MOCVD and atomic layer deposition (ALD)[54]], many CVD processes employ temperatures that are more than sufficient to decompose the precursors and initiate complex subsequent reactions. High-temperature thermal methods for depositing many refractory materials used as wear-resistant coatings, structural ceramics, or thermal barrier layers,[55] single-crystal silicon carbide for electronic applications,[56] diamond deposition under some conditions,[57–61] and silicon deposition from silane,[62] fall into this category. Thus, the initial precursor may not be the actual growth species, and markedly different growth behaviors can occur depending on residence time, temperature and pressure. Deposition of silicon from silane provides an illustrative example.[46] In this case [(Figure 3.6); see additional discussion in Section 3.5.1], film growth rates in a rotating disk reactor can increase, decrease or remain constant with disk rotation rate, depending on the substrate temperature, which determines both the rate of surface reactions and the extent of gas-phase SiH_4 decomposition.

Heats of formation, enthalpies, and entropies as a function of temperature are the first requirement for modeling these complex chemistries. These data enable computation of: (1) chemical equilibria to predict stable species; (2) rate constants of unimolecular reactions in the high-pressure limit; and (3) reverse reaction rates through the equilibrium constant. Experimental efforts have not kept pace with the need for data relevant to new CVD chemistries. Fortunately, quantum-chemistry (QC) methods have reached the level of sophistication necessary to predict thermodynamic data for main-group compounds with accuracy comparable to or better than the available experimental values. This section provides a summary of methods capable of providing useful thermodynamic data and their limitations, as well as a sampling of the data now available in the literature.

3.4.1 *Ab Initio* Methods for Predicting Gas-phase Thermochemistry

This section will acquaint the reader with the most commonly used QC methods for predicting thermochemical properties of gas-phase molecules relevant to CVD modeling. The goal is to provide sufficient background information for the reader to judge the accuracy of predicted thermodynamic values. For those interested in the practical details of using QC codes, several good textbooks are available.[63,64] Reviews of QC methods as applied to the calculation of molecular thermochemistry have also been published.[65–67]

In contrast with the situation at the time of an earlier review,[68] there is now a great deal of thermodynamic data for CVD-relevant main-group compounds, including precursors and their decomposition products. Increased computer power and improvements in QC models enabled their use across a wide spectrum of molecules of varying size, chemical composition and electronic structure. In many cases, computed thermodynamic data are known to be more accurate than the best experimental data. Exceptions exist, of course, but at least for the precursors themselves and for closed-shell molecules (*i.e.*, those with no unpaired electrons) data obtained from the best methods (discussed below) can be considered quite reliable. Average deviations from experiment can be as little as $\pm 1.5\,\mathrm{kcal\,mol^{-1}}$. Thus, in the absence of experimental data, results obtained from quantum mechanics can be used to model many CVD processes with acceptable accuracy.

To understand the differences in these methods, a brief introduction to computational quantum chemistry is useful. In general, an individual QC computation consists of two components: first, a set of basis functions that comprise the electronic wave function and, second, the theoretical model used. Together these are sometimes referred to as the "model chemistry."

In principle, an infinite number of basis functions are required to completely describe the electronic structure of a molecule. Since this is obviously impractical, a decision must be made concerning the size of the basis set used in a calculation. Basis sets can take many different forms, but one of the most commonly used today are "Gaussian" basis sets, in which atomic orbitals are typically linear combinations of individual Gaussian functions termed "primitives." Gaussian functions are chosen because they can be efficiently integrated, resulting in shorter computation times. Gaussian basis sets are available from Internet sources[69,70] and are discussed in detail in the book by Hehre *et al.*[64] Basis-set size can be classified by the ζ number, which refers to the number of basis functions per atomic orbital. Thus, a "double-ζ" Gaussian basis set uses two Gaussian functions (sometimes referred to as "primitive functions") for each atomic orbital (*i.e.*, two for each s, p, d, *etc.*). In "split-valence" or "valence-multiple-ζ" basis sets, the core and valence-shell orbitals are treated separately. For split-valence Gaussian basis sets, the notation is L-$M_1M_2M_3$G, where L is the number of primitives composing each core orbital, the number of Ms gives the number of basis functions describing each valence-shell orbital, and the value of each M is the number of primitives composing a particular valence basis function. For example, the 6-31G basis set for carbon is composed of six primitives for the core 1s basis function and two basis functions to describe each of the 2s and 2p orbitals, for a total of nine basis functions. Of the two valence-shell basis functions, one is composed of three primitives while the other has only one. For calculations aimed at predicting molecular thermochemistry, it is advisable to choose a basis set of at least double-zeta or valence-double-ζ size. Composite methods (see below) such as G2 employ valence-triple-ζ basis functions to achieve their high accuracy. Although large basis sets of triple-ζ quality can yield highly accurate electronic energies, calculations employing basis sets of this size can be prohibitively "expensive" (*i.e.*, time consuming) because calculation times scale as N^M, where N is the number of basis functions and M is at least 4.[63]

The choice of computational method depends on the objective of the calculation. To determine molecular thermochemistry, a sequence of calculations is typically done in which the molecular geometry is first determined (a geometry optimization calculation). Vibrational frequencies are then calculated, since these are required input to statistical mechanical formulae used to obtain the heat capacity, entropy and enthalpy as a function of temperature. Finally, one or more "single-point calculations" are performed to determine the electronic energy at the optimized (and fixed) geometry.

3.4.1.1 Geometry Optimization and Frequencies

Typically, one of three methods is used to determine the molecular geometry and vibrational frequencies: Hartree–Fock/Self-consistent field theory (HF), second-order Møller–Plesset perturbation theory (MP2), or density functional theory (DFT). HF provides adequate accuracy in most cases, even though it does not include electron correlation. MP2 is used by some high-level composite methods (*e.g.* G2, see below), but is computationally more expensive. DFT using the B3LYP functional is probably the most widely used method today, since it provides geometries of accuracy equal to or better than HF and the most accurate frequencies. It is also computationally efficient and can thus be used to model large molecules such as organometallic precursors. A wide range of QC methods for predicting vibrational frequencies have been evaluated by Scott and Radom,[71] who provide scaling factors developed for low-frequency vibrations to correct systematic errors in both fundamental frequencies and low-frequency vibrations (which are common in CVD precursors because of the heavy atoms often present).[71]

3.4.1.2 Electronic Energies and the Calculation of Heats of Formation

Once the structure and vibrational frequencies of a molecule are known it is necessary to calculate the total electronic energy for the molecule at the optimized geometry. This value is used to calculate the heat of formation (ΔH_f°) and as a result should be as accurate as possible within the constraints of computational power and time. The raw electronic energy obtained from such calculations corresponds to the energy required to bring the electrons of the molecule from a distance of infinity to the atomic orbitals of the nucleus. It is usually reported in units of hartrees (1 hartree = 627.51 kcal mol^{-1}). This energy is converted into a heat formation by combining the electronic energy and the zero-point energy (obtained from the frequency calculation) with calculated electronic energies for the constituent atoms, from which one obtains the molecular heat of atomization at 0 K, ΣD_0:

$$\sum D_0 = \sum_i^n E_i(\text{atoms}) - [E_{\text{ab initio}}(\text{molecule}) + E_{\text{ZPE}}] \tag{3.33}$$

Referencing this energy against the experimental ΔH_f° (0 K) of the atoms in the gas phase yields the molecular ΔH_f° (0 K):

$$\Delta H_{\text{f,0 K}}^\circ = \sum_{\text{atoms}} \Delta H_{\text{f0, atoms}}^\circ - E_{\text{atomization}} \tag{3.34}$$

The methods considered most accurate and also most widely used for predicting electronic energies fall into three categories: (1) empirically corrected methods; (2) composite methods; and (3) density functional theory. Each of these is described below. The overarching concept is that in anything but a one-electron system the motion of an electron is affected by those of every other electron in the system. The energy associated with this is known as the electron correlation energy, and obtaining accurate values of the correlation energy has driven the development of QC methods.

Importantly, to calculate transition-state energetics, bond energies, or to obtain accurate heats of formation, generally the highest level of theory that is practical is desirable. Of course, the size of the molecule of interest may well limit this. A rough guide is that calculations employing DFT methods such as B3LYP can handle molecules containing up to ~50 non-hydrogen atoms, since efficient parallel implementations of these codes are now widely available. MP2 calculations are feasible for up to about 20 non-hydrogen atoms (again with parallel computing). Fourth-order MP perturbation theory and coupled cluster theory both scale as N^7, where N is the number of basis functions. These methods are therefore limited to rather small systems (≤ 10 non-hydrogen atoms). Thus, the choice of model chemistry must balance desired accuracy against the available computing power and time.

3.4.1.3 BAC Methods

Empirically corrected methods were developed in the 1980s to address the systematic errors resulting from finite basis sets and limitations of theory. At that time computing power severely limited the size of molecules that could be addressed by quantum mechanical methods. Bond additivity correction (BAC) calculations are a class of empirically corrected methods developed by C.F. Melius and co-workers that have been used extensively to predict thermochemistry for CVD systems.[72–75] The methods are based on the assumption that errors in electronic energies obtained from *ab initio* calculations are due to the finite size of the basis sets used and the application of limited electron correlation in the calculations. These errors are therefore systematic and can be corrected to achieve much more accurate heats of formation by applying various empirical corrections related to the elements and bonds in the molecule.

The BAC suite of methods consists of several levels of theory. The one most extensively applied is the BAC-MP4 method, which was the first to be developed. In this method, the molecular electronic energy is obtained from an *ab initio* electronic-structure calculation at the MP4(SDTQ) level of theory. Methods using MP2, (BAC-MP2), G2 theory (BAC-G2), and a hybrid method involving both density functional theory and MP2 have also been developed. These use a different approach for determining the empirical corrections to the *ab initio* electronic energy than the original BAC-MP4 method.[75]

The BAC-MP4 method has been used extensively to predict thermochemistry for main-group compounds, including compounds of boron,[76] silicon,[72,77–83] phosphorous,[84] indium,[85] tin[86] and antimony,[87] as well as halogenated hydrocarbons[88–90] and hydrocarbon intermediates.[74] Thermochemical data for group-III compounds derived from BAC-G2 method have also been published.[91,92]

3.4.1.4 Composite Methods

Composite methods simulate the effects of using large basis sets and high-order configuration interaction (CI) by using smaller basis sets and lower levels of theory coupled with empirical corrections, resulting in model chemistries that are more computationally efficient and accurate. Among the most successful and widely applied are the Gaussian-*n* methods. The objective of the original G1 method, which is rarely used today, is to achieve an estimate of the QCISD(T) energy (quadratic CI with single, double and triple excitations) using the computationally prohibitive 6-311+G(2df,p) basis set with diffuse-sp and 2df basis-set extensions, which is determined through an extensive series of electron-correlated calculations.[93] G2 theory raises the approximated level of theory to QCISD(T)/6-311+G(3df,2p)//MP2/6-31G(d) (for a review see ref. 94). The additional corrections included in G2 improve the predicted energies for ions, triplet-state molecules and hypervalent species (such as SO_2 and ClO_2). The average absolute deviation[i] in the 148 heats of formation in the G2/97 test set, a broad range of experimentally established heats of formation for compounds containing only the elements H–Cl, is 1.58 kcal mol^{-1}.[95]

More recently, the G3[66,96,97] and G4[98] methods were developed to address the deficiencies of G2, as well as provide a computationally more efficient method. G3 employs a different sequence of single-point energy calculations. In addition, smaller basis sets are used for the computationally intensive MP4 and QCISD(T) calculations in G3, and in G4 QCISD(T) is replaced by CCSD(T) to

[i] Note discussion below in Section 3.4.1.6, however, concerning the significance of this value.

Table 3.2 Comparison of average errors for various quantum-chemistry methods (kcal mol^{-1}).

Method	*Avg. error*	*Test set*	*Ref.*
G2	1.56	148 enthalpies (G2/97)	96
G3	0.94	148 enthalpies; G2/97	96
	1.19	270 enthalpies; G3/05	98
G4	0.80	270 enthalpies; G3/05	98
CBS-4	2.7	148 enthalpies (G2/97)	101
CBS-Q	1.2	148 enthalpies (G2/97)	101
BAC-MP4	1.25	93 enthalpies	75
BAC-G2	0.69	143 enthalpies	75
DFT/B3LYP	3.11	148 enthalpies (G2/97)	95

obtain the highest treatment of electron correlation. The (empirical) higher-level correction is also modified and corrections for atomic spin–orbit effects and core correlation are added. This yields a reduction of 0.62 kcal mol^{-1} in the average deviation (to 0.94 kcal mol^{-1}) for G3 relative to G2 using the G3/05 test set (see Table 3.2), including a decrease in the number of molecules with deviations greater than 2.0 kcal mol^{-1}, from 41 in G2 to only 9 in G3. In addition, computation times are shortened considerably. For example, the required CPU time for benzene is reduced by a factor of 1.9 and for $SiCl_4$ by a factor of 2.4. G4 improves significantly upon G3, reducing the average absolute deviation for the 454 molecules in the G3/05 test set from 1.13 to 0.83 kcal mol^{-1}. This method is also reported to reduce errors associated with non-hydrogen systems, which could be important for application to CVD systems. The test set includes molecules with as many as 12 heavy atoms (*e.g.*, C_6F_6) and thus its use should be feasible for CVD precursors of at least this size, although the use of CCSD(T) in the electronic energy calculation likely means that this is close to the upper size limit.

Another composite method useful for predicting molecular thermochemistry is the complete basis set (CBS) method. CBS takes a different approach from either BAC or G-n. Instead of relying on empirical corrections, the convergence trend in the electron correlation energy is extrapolated from the MP2 level to obtain an estimate of the energy that would be obtained in the complete basis set limit.[ii][99,100] The general intention is to provide accuracy comparable to G2 calculations, but at a lower cost, enabling application to larger molecules[99] by avoiding large basis-set calculations at the MP4 level. Three forms of CBS are available:[100] CBS-4M, which is the fastest of the three and is practical for molecules with up to 12 heavy atoms, an intermediate model CBS-QB3 and CBS-QCI/APNO, which is only practical for molecules with no more than six heavy atoms. Application of CBS-Q and CBS-4 to the G2 molecule test set yields energies whose mean deviation from accepted values is 1.2 and 2.7 kcal mol^{-1}, respectively.[101]

3.4.1.5 Density Functional Theory

DFT is now a ubiquitous tool for modeling molecular thermochemistry and kinetics because of its computational economy and relatively high accuracy. As such, it is frequently used to assist in the development of CVD models and is a good alternative to the first-principles all-electron methods discussed above. It is particularly advantageous for larger CVD precursors because DFT

[ii] The complete basis set limit is the result that would be obtained from a basis set that is complete in the sense that it provides full flexibility for describing the property in question and nothing will change upon addition of more basis functions of any kind. In contrast, the infinite basis set limit is the value of a property obtained using an infinitely large basis set. This is usually the same as the complete basis set limit, but it is possible to construct basis sets in such a way that even with infinitely many basis functions, the complete basis set limit won't be reached. Thus, the terms are not strictly speaking interchangeable.

computational time scales as N^3, where N is the number of atomic orbital basis functions used. In contrast, high-level methods such as MP4(SDTQ) scale as N^7 (for a discussion of MO scalings see ref. 63) Obviously, this is a severe penalty and, as remarked above, limits the use of the most accurate methods, such as G2 or G3, to relatively small molecules. There is a tradeoff with respect to accuracy when using DFT, however, and this has received much discussion in the literature.[63,71,102,103] A further disadvantage of DFT for calculating molecular thermochemistry is that, unlike the *ab initio* methods discussed above, there is no systematic path to improved accuracy *via* higher levels of theory, making it difficult to establish a convergence criterion. Therefore, although DFT predictions can be surprisingly accurate, they must be treated with some caution when applied to poorly characterized classes of molecules.

DFT is not a molecular orbital method, although the steps involved in setting up and running a calculation are very similar to those used for Hartree–Fock/self-consistent field theory. Thus, two basic decisions must be made. First, a basis set is used that functions mathematically in a very similar manner to the basis set used in a HF calculation. In fact, the same basis sets used for calculations using molecular-orbital theory can be used for DFT calculations. However, DFT does not attempt to optimize a molecular wave function by solving the Schrödinger equation. Instead, it approximates the solution by using an empirical "functional" (which converts a function into a number, in contrast to functions, which convert one number into another) of the electron density and its higher moments to determine the energy of the system. A key component of the functional is the "exchange-correlation energy," a sum of correction terms accounting for the fact that the electrons interact.[63,104] Thus, the second decision in performing a DFT calculation is to select a functional, which is analogous to a choosing the level of theory in a MO calculation.

Originally, DFT functionals depended only on the value of the electron density at a particular location, leading to the so-called local density approximation, but because of the high spatial non-uniformity of molecular systems these are not typically used (although they find extensive application to solid-state systems). Instead, functionals that also depend on the gradient of the electron density – known as "non-local" or "generalized gradient approximation" (GGA) functionals – are used, yielding accuracies that rival those of the more computationally intensive MO methods.

Today, the B3LYP functional is probably the most widely used to calculate molecular thermochemistry. Comparison of various GGA functionals shows that B3LYP can be very accurate for predicting thermochemistry for both organic and inorganic species.[95,102] B3LYP and other functionals have been used to predict thermochemistry for CVD precursors, including Si-H compounds,[105] Si/C/O/H species,[106] germanes,[107] metal carbonyls,[108] Ti/O/Cl species,[109] β-diketones,[110] indium compounds[111] and C/H/N compounds.[112]

DFT, and the B3LYP functional in particular, are also being successfully applied to the important topic of gas-phase kinetic mechanisms. Examples include GaN,[113,114] GaAs[115] and chalcogenides.[116,117] Of particular interest in this regard is recent work by Becke and co-workers in which they developed a new approach to account for electron exchange known as the "real-space correlation."[118] RSC yields atomization energies for the 222-molecule G3 test set with an average per-bond error of only 0.5 kcal mol^{-1}. This is comparable to G3, but far less expensive computationally. These authors also calculated barrier heights for a set of 70 reactions of various types and find that the real-space approach yields a mean absolute error of 1.4 kcal mol^{-1} without any reparameterization of the model. This result is highly encouraging and suggests the use of this method for CVD-related problems involving main-group elements.

3.4.1.6 *Uncertainties in Calculated Thermochemistry*

Uncertainties in thermochemistry derived from *ab initio* methods can be difficult to determine. Average deviations from experiment for various test sets (groups of molecules with established thermodynamic properties) are reported for G2, G3, and various CBS methods, as well as various DFT functionals (Table 3.2). However, these errors are not random, but are a function of the specific molecule and the degree to which the employed model chemistry is applicable. For example, the average deviation in the heat of formation predicted by G2 for hydrocarbons in the G2/97 test set is only 1.29 kcal mol^{-1}. In contrast, the heat of formation for CF_4 is too negative by 5.5 kcal mol^{-1}, while that of SiF_4 is too positive by 7.1 kcal mol^{-1}. Consequently, errors must be evaluated on a case-by-case basis. In general, one can say that the most widely used approaches, including perturbation theory, DFT and composite methods such as G2, are well suited to closed-shell systems with no low-lying electronic excited states, and to radicals with doublet ground states (*i.e.*, a single unpaired electron) such as CH_3. In these cases it is quite possible to obtain deviations from experiment of less than 2 kcal mol^{-1}. However, molecules with high-spin ground states (*e.g.* triplets), low-lying electronic states and transition metals will require higher levels of theory to achieve accurate results. An extensive comparison of mean and absolute errors for various MO- and DFT-based methods is given in ref. 63.

3.4.2 Sources of Gas-phase Thermodynamic Data

The following sources offer compilations of gas-phase thermochemical data obtained from experiments and/or QC modeling. Users should be cautious, as data do not always correspond to accepted values. This is particularly true of the *JANAF Thermochemical Tables*[36] (found online in the NIST Webbook), which have not been completely updated. Additional sources can be found in Chapter 13 of ref. 119.

- NIST Webbook: http://webbook.nist.gov/
- SGTE database: http://thermodata.free.fr/ Free database of condensed-phase inorganic binaries.
- SGTE web site: http://www.sgte.org/ Primary SGTE site with various databases available for a fee.
- Materials Processing Database: http://www.ipt.arc.nasa.gov/databasemenu.html. Database published by personnel at NASA/Ames Research Center, Mountain View, CA.
- Ivanthermo: http://www.ihed.ras.ru/thermo/ Database compiled by investigators at the Russian Academy of Sciences with links to a Windows version of the database and thermodynamic equilibrium software; see also ref. 120.
- Alexander Burcat database for combustion: http://garfield.chem.elte.hu/Burcat/burcat.html Database of combustion-related species compiled by A. Burcat (Israel Institute of Technology). Contains data for many species relevant to CVD.
- Thermodynamics Research Laboratory, University of Illinois: http://tigger.uic.edu/~mansoori/Thermodynamic.Data.and.Property_html A listing of many useful databases.
- Thermodynamics Resource: http://www.ca.sandia.gov/HiTempThermo/ Compiled by M. D. Allendorf and C. F. Melius (primarily main-group compounds with a few transition-metal species).
- Thermochemical and Chemical Kinetic Data for Fluorinated Hydrocarbons: http://www.cstl.nist.gov/div836/836.03/papers/NistTNThermo.html NIST database for combustion of fluorinated hydrocarbons. Includes both thermodynamic data and a kinetic mechanism.

3.4.3 Modeling Precursor Pyrolysis

When gas-phase chemistry is important in CVD processing it is often the breaking of bonds within the precursor that initiates the reaction. This process, called pyrolysis, is typically driven by so-called unimolecular reactions, in which chemical bonds in the precursor are thermally activated by collisions with bath-gas molecules, causing them to break. Subsequent reactions between the reaction fragments and precursor may accelerate the process through radical-driven chain reactions. However, in many CVD chemistries, chain-branching reactions,[121] which drive the ignition and combustion of hydrocarbons, are absent unless oxygen is a reactant and there is hydrogen in the precursor system. Typically, the overall rate constant for gas-phase chemistry in CVD processes is closely linked to the initiating reaction. In that case, one can estimate the extent of precursor decomposition based on residence time in the reactor and the reaction pre-exponential factor.[122]

There are numerous examples of CVD processes in which gas-phase chemistry plays a critical role. A very important one is silicon growth from silane.[46,123,124] Another example is diamond growth, in which not only the identity of the precursor but also the flux of radicals to the surface determine whether diamond or graphite is formed.[125,126] High growth temperatures also lead to extensive gas-phase chemistry in the epitaxial growth of silicon carbide,[127–129] gallium nitride[114,130] and tin oxide.[131] In addition to creating new species to interact with the growth surface, gas-phase reactions can also lead to homogeneous nucleation of clusters and particles (Section 3.6).

3.4.3.1 Transition State Theory of Unimolecular Reactions

Since the activation energies of pyrolysis reactions are typically much higher than any of the subsequent reactions of their decomposition products and can thus be rate-limiting, it is essential to have accurate rate constants for these reactions. For many CVD precursors these rates have not been measured, requiring a modeling approach to estimate the rate constant. Reactions involving the products can often be estimated by comparison with analogous chemistry (often involving hydrocarbons, for which there are copious data), so the key task of the CVD modeler in this regard is the prediction of the initial pyrolysis step. Fortunately, unimolecular reactions have been extensively studied and theoretical approaches for predicting their rates are well developed. Several useful textbooks are available, including those by Holbrook *et al.*,[132] Gilbert and Smith,[133] and Steinfeld *et al.*[134] This section provides a brief introduction to these concepts and the reader is referred to the more detailed treatments that are available.

Transition state theory (TST), initially formulated by Eyring, Evans and Polyani, and Wigner in the 1930s to predict the rates of chemical reactions[134] postulates that there is a unique location along a reaction coordinate at which a "transition state" (TS) exists. Once a molecule absorbs enough energy to attain the transition state, it proceeds irreversibly to products. From the point of view of the CVD modeler, the objective is to determine the geometry and thermodynamics of the transition state so that the results of TST can be used to predict the reaction rate. A good review of computational methods for modeling potential energy surfaces of chemical reactions[135] is available.

The most straightforward use of TST is to calculate the so-called "high-pressure limit" for a unimolecular reaction. In this case, the population of energy levels in the molecule is assumed to be at equilibrium and the rate is independent of pressure. It can thus be calculated without knowledge of the rates of collisional energy transfer, which are required to predict the low-pressure limit and intermediate "falloff" regimes (discussed below). The high-pressure limit is given by:

$$k_{\text{uni}}^{\infty} = \frac{k_{\text{B}}T}{h}\frac{Q^{\dagger}}{Q}\exp(-E_0/k_{\text{B}}T) \tag{3.35}$$

Here it is assumed that collisional energy transfer (by which bonds in the molecule accumulate sufficient energy to cross the activation barrier) is so high that a Boltzmann distribution of energies is produced, so that the reaction is limited by the rate at which molecules cross the activation barrier. The high-pressure limit has several useful features for modeling CVD chemistry. First, it represents the upper limit of the unimolecular reaction rate. If decomposition is limited by this rate (which can be the case when chain-branching reactions are absent) this enables an estimate of the extent of gas-phase reaction.[122] Second, since no knowledge of the energy transfer process is required, only the structure of the reactant and transition state (specifically, their moments of inertia) with their vibrational frequencies are required to calculate the rate. Finally, increasing molecular size and, hence, the number of vibrational frequencies, as well as decreasing their frequencies, shifts the unimolecular rate closer to the high-pressure limit. Since CVD precursors (especially organometallic compounds) are often relatively large molecules with heavy atoms and low vibrational frequencies, it is often not a bad approximation to assume that precursor pyrolysis is in the high-pressure limit.

The following brief derivation illustrates the connection between thermodynamic quantities of the transition state and partition functions. The former can be estimated from knowledge of bond strengths, especially in the case of simple bond-cleavage reactions. Partition functions can be accurately calculated if the structures of the reactant and transition state are known, with their moments of inertia and vibrational frequencies, all of which can be obtained from QC calculations.

Equation 3.35 can be rewritten in thermodynamic terms *via* the relationship between the partition functions Q and the equilibrium constant (in concentration units; (see ref. 134 for the derivation):

$$k = \frac{k_{\mathrm{B}}T}{h}\exp\left(\Delta S_0^{\dagger}/R\right)\exp\left(-\Delta H_0^{\dagger}/RT\right) \tag{3.36}$$

in which the correspondence with the familiar Arrhenius equation is evident:

$$k = A\mathrm{e}^{-E_{\mathrm{a}}/RT} \tag{3.37}$$

The activation energy can be related to the equilibrium constant through the Gibbs–Helmholtz equation and the ideal gas law, yielding:

$$E_{\mathrm{a}} = \Delta H_0^{\dagger} + RT - P\left(\Delta V_0^{\dagger}\right) = \Delta H_0^{\dagger} + RT \tag{3.38}$$

(for a unimolecular reaction, there is no change in the number of moles and thus $\Delta V_0^{\dagger}$ is zero).

The Arrhenius prefactor is:

$$A = \mathrm{e}\frac{k_{\mathrm{B}}T}{h}\exp\left(\Delta S_0^{\dagger}/R\right) \tag{3.39}$$

An excellent discussion of methods of estimating high-pressure rate constants from thermodynamic data and other empirical information, without the use of QC methods, is found in Benson's classic text.[136]

3.4.3.2 Types of Transition States

Two basic categories of unimolecular reactions exist: those with so-called "tight" transition states (Figure 3.7) and those with "loose" transition states (Figure 3.8). Tight transition states are found in reactions that require some structural rearrangement of the molecule, such as an isomerization or elimination reaction. In contrast, loose transition states correspond to the breaking of a single bond within the molecule resulting in the loss of an atom or molecular fragment. Examples of each for some reactions relevant to CVD are given below.

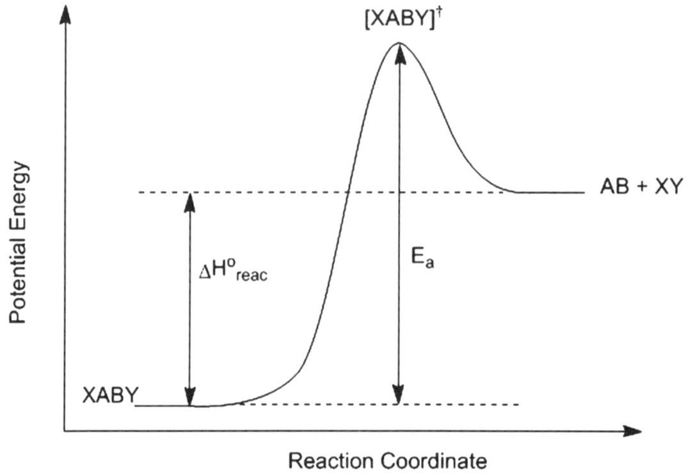

Figure 3.7 Potential energy diagram illustrating a tight transition state for a unimolecular elimination reaction, such as $SiH_4 \rightarrow SiH_2 + H_2$.

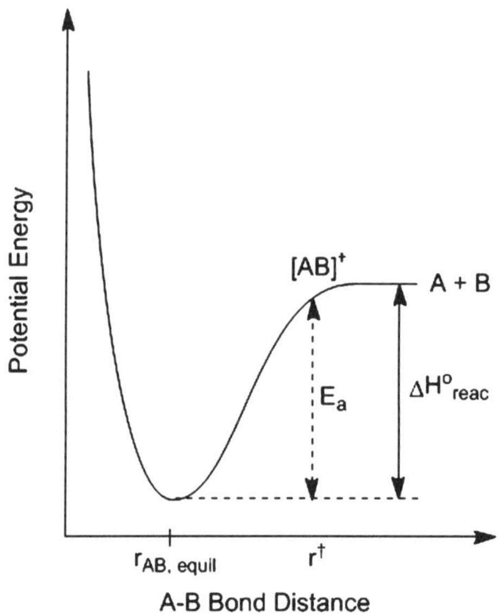

Figure 3.8 Potential energy diagram illustrating a loose transition state for a reaction such as $CH_3SiCl_3 \rightarrow SiCl_3 + CH_3$.

Tight transition states:

$$SiH_4 \leftrightarrow SiH_2 + H_2$$

$$CH_3SiCl_3 \leftrightarrow CH_2SiCl_2 + HCl$$

$$Si(OC_2H_5)_4 \leftrightarrow Si(OH)(OC_2H_5)_3 + C_2H_4$$

Loose transition states:

$$CH_3SiCl_3 \leftrightarrow CH_3 + SiCl_3$$

$$TiCl_4 \leftrightarrow TiCl_3 + Cl$$

$$C_4H_9SnCl_3 \leftrightarrow C_4H_9 + SnCl_3$$

The potential energy coordinate for a reaction having a tight TS is shown schematically in Figure 3.7. The reaction is characterized by a well-defined activation barrier E_a separating the reacting molecule from the products, with the TS (designated by †) located at the top of the barrier. This barrier is in addition to any thermodynamic barrier that might exist. The TS is located at a saddle point in the potential energy surface. QC methods can locate the geometry of the transition state by searching for a stationary point on the potential energy surface that has one imaginary vibrational frequency (these result from negative force constants). Although QC packages now include routines that search for such modes, the identification of transition states remains an art, in part because its geometry is often far from obvious and relatively flat potential energy surfaces in some cases make convergence difficult. An additional problem is that vibrational frequencies decrease substantially as bonds lengthen. CVD precursors involving atoms below the first row of the periodic table commonly have low-frequency vibrations ($\leq 600\,cm^{-11}$) and as these decrease approaching the TS the harmonic oscillator approximation becomes increasingly inaccurate. Both factors contribute to inaccuracies in computed temperature-dependent rate constants, as well as convergence difficulties. As a result, the uncertainty in computed activation energies can be a factor of two larger than for ground-state heats of formation.

In contrast, reactions with loose transition states have no activation barrier other than the heat of reaction ΔH_{reac} (Figure 3.8). In this case, the reaction coordinate corresponds to the simple stretching of a chemical bond until the fragments are sufficiently far apart that they separate, typically 2.5–3× the bond length. In this case, the absence of a well-defined TS geometry means that QC methods cannot be used to locate the TS and compute the required inputs to $Q^{\dagger}$. Instead, variational transition state theory (VTST) must be used to calculate the rate.[133] In this method, the rate is calculated at several locations along the reaction coordinate until the minimum is found. VTST defines this minimum as an upper limit to the true rate. Inputs to the calculation include the frequencies of the reactant and product fragments and the bond energy, both of which can be determined by QC methods. The moments of inertia along the reaction coordinate must also be known, but these can be calculated as a function of bond length using available codes (*e.g.*, the ChemRate code developed by Tsang[137] or the CHIMERA code developed by Korkin *et al.*;[138] both have routines that do this).

Notably, if one or both of the fragments formed in the reaction is a rotatable group rather than an atom (*e.g.*, dissociation of CH_3 from CH_3SiCl_3), the rocking vibrations corresponding to these groups in the intact molecule undergo a transition to a hindered rotor (sometimes described as a torsional vibration) and eventually to free rotors at sufficiently large separation distances. These must be treated separately from vibrations. Hindered rotors are characterized by a rotation barrier,[136] which in the ground-state molecule is on the order of $2–15\,kcal\,mol^{-1}$ and can be calculated as a function of bond length using QC methods. While tedious, this can be done with reasonable accuracy. The problem for modeling reactions involving rotors is that the statistical mechanics treatment is both complex and relatively imprecise. The method of Pitzer and Gwinn is often used,[139] but more elaborate approaches can be used *in situ*ations where highly accurate rate constants are desired.[140]

3.4.3.3 Collisional Energy Transfer, the Low-pressure Limit and Pressure fall-off Curves

Although high-pressure rate constants are relatively straightforward to predict, assuming one knows the energy and geometry of the transition state, it is often the case that at pressures and temperatures typical of CVD processes, the unimolecular rate constant is not k_{∞} but some lower

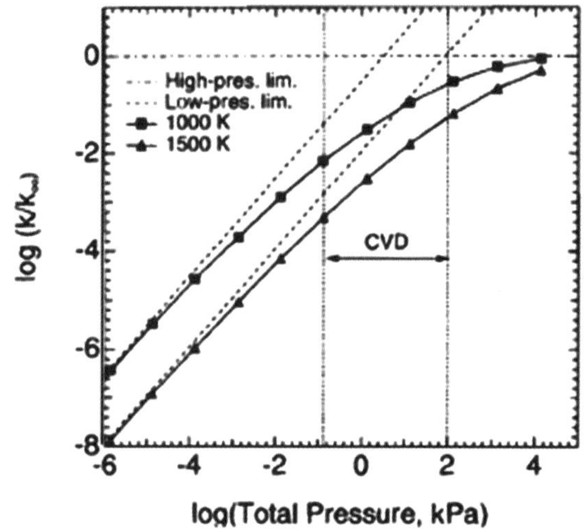

Figure 3.9 Fall-off curves for the pressure-dependent gas-phase decomposition of $TiCl_4$. (Reprinted with permission from ref. 145.)

value determined by both the pressure and the temperature. As shown in Figure 3.9 for the example of $TiCl_4$ decomposition, a simple single-channel unimolecular reaction, there is a transition between the high-pressure limit (independent of pressure) and the low-pressure bimolecular limit (linearly dependent on pressure). The intermediate region connecting the two is referred to as the "fall-off" regime and is determined by the rate of collisional energy transfer. Both experiments and modeling show that the rate constant for several important CVD precursors is in the fall-off regime. The extent of this effect can be quite substantial, as is evident in Figure 3.9, and neglect of it can result in large errors in the predicted precursor decomposition rate.

Pressure-dependent rate constants can be predicted at various levels of sophistication, ranging from QRRK theory to RRKM theory. When multiple product channels are accessible, RRKM coupled with a full master-equation treatment must be implemented to account for collisional energy transfer.[132,133,141,142] This approach has been used to model the pressure dependence of CVD precursor decomposition in a few cases, including SiH_4,[123,124] Si_2H_6,[143] CH_3SiCl_3[144] and $TiCl_4$,[145] as well as most hydrocarbons of interest in CVD.[iii] Its use is necessary to model complex situations such as Si_2H_6 decomposition, in which multiple product channels are possible. Computer codes are available to model these cases.[146]

The rate of energy transfer between bath-gas molecules and the precursor is required to predict the pressure dependence. Under the most favorable circumstances, this can be determined by fitting the results of, for example, RRKM calculations to measured rate constants. The treatment of SiH_4 by Moffat *et al.* is a good example of this approach.[124] Here, experimental data are available from several sources, allowing SiH_4 decomposition rate constant across a wide range of pressures and temperatures to be predicted. For most compounds of interest to CVD, however, no experimental rate data are available. Thus, some estimate of the energy transfer rate must be made. Prior to the development of RRKM theory, a strong collision model was assumed, in which collisions transfer large amounts of energy.[133] This approach is not accurate for light bath gases and leads to rates of energy transfer that are much higher than observed (by more than an order of magnitude). The

[iii] See the NIST kinetic database at http://kinetics.nist.gov/kinetics/index.jsp for a compilation of rate constants for hydrocarbons.

strong collision model can be improved by assuming a "collision efficiency," in which only a fraction of the collisions results in energy transfer, known as the "weak collision" model. Such approaches were initially implemented for mathematical convenience and not fidelity to physical reality.

Fortunately, several models have been developed to quantify rates of energy transfer. These models do not require detailed knowledge of the collision itself, only the mean energy transferred, a quantity that can be obtained either from experiment or with reasonable accuracy by first-principles calculations. Examples include the exponential-down (the most widely used model), biased random walk and ergodic models.[132,133] Compilations of experimentally measured energy transfer rates are also available, including a review of data for large polyatomic molecules by Oref and Tardy[147] (see also refs. 132,133 and references therein for additional reviews).

Although the RRKM/master equation approach can provide very accurate results, it is tedious to implement, requiring calculations at every temperature and pressure of interest. Since it is often desirable to model a CVD process across a broad range conditions, this is inconvenient at best. Furthermore, CVD reactors are typically not uniform in temperature, requiring expressions describing the temperature dependence at a given pressure to be incorporated into the model. Since it is computationally impractical to perform an RRKM calculation at every location within a reactor simulation, various practical approaches to modeling fall-off behavior have been developed – a problem addressed first by Lindemann and Hinshelwood. Today, the methods of Troe and co-workers are widely used and are accurate for thermally activated reactions proceeding through a single potential well and typical of many CVD precursor decompositions.[148] More recently, alternative approaches for multi-channel reactions with multiple potential wells have been developed, such as the damped pseudopotential approach of Venkatesh.[149]

3.5 Mechanism Development

The level of detail required for an effective CVD model depends on two factors: (1) the kinetic regime in which the process operates and (2) the purpose(s) of the model. The former determines whether chemical kinetics must be included (as opposed to mass and heat transport only). The latter determines the extent to which detailed knowledge of individual chemical reaction rates is required. Factors to be considered include the breadth of operating conditions over which the model must function (*e.g.*, a large temperature range, a single pressure *vs.* a wide pressure range, *etc.*), the goal of the growth process (*e.g.*, slow growth for highly ordered or epitaxial films, *vs.* fast growth rates for thick coatings or on-line glass coating), deposit composition and morphology. It is important to have a clear understanding of both factors so that unnecessary work is avoided. For example, there would be no point in developing a detailed gas-phase model if such chemistry does not influence the growth process. Furthermore, the modeling effort should result in a predictive capability with sufficient robustness for the purposes of the user is obtained. The remainder of this section illustrates these points using examples of actual CVD processes, in particular on the growth of silicon from silane. A review of the overall process of developing an experimentally validated model of CVD growth has been published by Lengyel and Jensen.[150]

3.5.1 Kinetic Regimes

In general, a given CVD process can operate in as many as three kinetic regimes, depending on the process conditions: surface rate-limited (Regime I), gas-phase rate limited (Regime II) and transport limited (Regime III). Which regime is operative depends on a competition between the rates of chemical reactions and transport. This is illustrated in Figure 3.6 for the case of silicon deposition from silane in a rotating disk reactor (discussed above). In this example, Regime I, the temperature (800 K) too low for gas-phase SiH_4 decomposition to occur, so growth is limited by the reaction of

SiH_4 with the surface. As seen in the figure, increasing the mass transport rate to the surface, by increasing the disk rotation rate has no effect. In Regime II (950 K), the gas phase is sufficiently hot that there is extensive SiH_4 decomposition, forming radical intermediates such as SiH_2 that are much more reactive with the surface than SiH_4. Increasing the mass transport rate in this regime actually decreases the growth rate by reducing the gas-phase residence time and thus the extent of gas-phase SiH_4 decomposition. Growth is therefore limited by the rate of gas-phase precursor pyrolysis. In Regime III (1300 K), the rates of both gas-phase and surface reactions are very fast and no SiH_4 reaches the surface. The growth rate is limited by the transport rate of reactive intermediates to the surface, as determined by the disk rotation rate. Although the temperatures at which these three regimes are operative vary greatly from one precursor to another, the behavior is characteristic of CVD processes in general, which exhibit growth controlled by gas-phase or surface kinetics at low temperatures ($E_a > \sim 2\,\text{kcal mol}^{-1}$) and mass-transport-limited growth at high temperatures ($E_a = 0$–$2\,\text{kcal mol}^{-1}$). A useful and more general way to describe these regimes is the Damkoller number (Da), a non-dimensional quantity that compares the characteristic times of transport and chemical reaction.[119] Large Da correspond to transport-dominated growth, while $\text{Da} < 1$ indicates chemical-kinetic control. Given a general expression for the growth rate and an understanding of the dominant transport mechanism (convection or diffusion), one can map process parameters onto the rate-controlling growth processes and thereby determine how to adjust growth rates to optimal values.

3.5.2 Global *versus* Elementary Mechanisms

A mechanism describing a CVD process can be as simple as a single reaction that converts the precursor into a solid material, or it can have hundreds of reactions describing the elementary chemistry occurring in both the gas phase and on the surface. It is worth drawing a distinction between these. For example, a "global" mechanism for silicon deposition could be $SiH_4 \rightarrow$ Si(solid) + $2H_2$. This mechanism contains no information concerning possible gas-phase chemistry or the details of the heterogeneous chemistry that converts silane into elemental silicon. Most likely it never occurs as written, but this approach has been applied to many CVD systems. Alternatively, a complete elementary reaction mechanism for this process might include these reactions among others:

Gas phase:

$$SiH_4(+M) \leftrightarrow SiH_2 + H_2(+M) \tag{3.40}$$

$$SiH_4 + SiH_2(+M) \leftrightarrow Si_2H_6(+M) \tag{3.41}$$

$$SiH_4 + H \leftrightarrow SiH_3 + H_2 \tag{3.42}$$

Surface:

$$SiH_4 + Si(s) \leftrightarrow Si(b) + SiH_2(s) + H_2 \tag{3.43}$$

$$SiH_2(s) \leftrightarrow Si(s) + H_2 \tag{3.44}$$

$$H(s) + H(s) \leftrightarrow H_2 + 2Si(s) \tag{3.45}$$

In this case, reactions shown by Equations (3.40)–(3.42) are elementary reactions, while the surface reactions given in Equations (3.43)–(3.45) may or may not be. Obviously, this type of mechanism has the potential to convey much more information about the CVD process. The problem comes in determining the rate constants for these reactions, which in many cases are unknown, especially for the surface reactions. This presents a dilemma for model development. A global mechanism, properly chosen to reproduce observed reaction orders, may be attractive because of its simplicity and the corresponding rate constant is relatively straightforward to determine. However, the resulting kinetic expression may not be useful beyond a limited range of deposition conditions or sufficiently robust to extend to reactor geometries other than the one for which it was determined. Alternatively, a detailed mechanism may contain so many unknowns that a large number of approximations must be made, limiting its accuracy. The choice of which approach to use must be made based on the extent of kinetic data available and the level of chemical detail necessary meet the purposes of the model. Before incorporating additional mechanistic detail into a CVD model, the question should always be asked: does this detail provide additional understanding or a more broadly applicable model, or does it simply expand the number of unknown parameters that must be fit to a limited body of experimental data?

3.5.3 Gas-phase Chemistry

Until the 1980s, it was not widely accepted that gas-phase chemical reactions often occur during CVD. Many investigations since that time demonstrated that this chemistry is quite prevalent and is responsible for not only growth-rate behaviors such as that illustrated in Figure 3.6, but also deposit composition, impurity formation and homogeneous particle nucleation. In general, much more is now known at the elementary chemistry level regarding gas-phase processes than about the corresponding heterogeneous reactions. This is because it is much easier to probe gas-phase reactions experimentally and because sophisticated theoretical approaches are available (Section 3.4). As a result, the large body of the thermodynamic and kinetic data available makes it relatively easy to develop detailed gas-phase models. Some simple principles can be applied in constructing these models for thermally driven CVD. Decomposition of methyltrichlorosilane (MTS), a widely used silicon carbide precursor, provides a suitable example to illustrate these points and summarize key steps in the process of developing a gas-phase mechanism. Both the thermochemistry[77,78,151] and the kinetics[129,144,152] of this system have been investigated in detail.

As a first step in developing a gas-phase mechanism it is necessary to determine the rate of precursor unimolecular decomposition. As remarked upon earlier, gas-phase CVD chemistry is usually initiated by pyrolysis of the precursor, even when oxygen is used as a reactant, because the precursor bonds are generally weaker than those of small-molecule reactants or carrier gases that might be used, such as H_2, N_2, NH_3, O_2 or HCl. Thus, knowledge of all bond energies within the precursor system is needed to determine which species are likely to decompose and thus avoid time-consuming TST or RRKM calculations for stable reactants. Table 3.3 gives typical bond energies, illustrating that M–H and M–C bonds are often the weakest, while M–halide or M–O bonds are significantly stronger. To illustrate for MTS:

$$CH_3SiCl_3 \leftrightarrow CH_3 + SiCl_3 \ (\Delta H^{\circ}_{298K} = 96.7\,\text{kcal mol}^{-1}) \tag{3.46}$$

$$CH_3SiCl_3 \leftrightarrow CH_3SiCl_2 + Cl \ (\Delta H^{\circ}_{298K} = 114.0\,\text{kcal mol}^{-1}) \tag{3.47}$$

$$CH_3SiCl_3 \leftrightarrow CH_2SiCl_2 + HCl \ (\Delta H^{\circ}_{298K} = 81.5\,\text{kcal mol}^{-1}) \tag{3.48}$$

$$CH_3SiCl_3 \leftrightarrow CH_2SiCl_3 + H \ (\Delta H^{\circ}_{298K} = 102.2\,\text{kcal mol}^{-1}) \tag{3.49}$$

Although the reaction shown in Equation (3.48) has the lowest reaction enthalpy, one can guess that its tight transition state will lead to a significant activation energy and possibly cause the reaction shown in Equation (3.46) to be the dominant channel; this is borne out by RRKM calculations.[144] Hydrogen, the typical carrier gas, and HCl, a major reaction product, have strong bonds ($>100\,\text{kcal mol}^{-1}$) so their decomposition is not a factor. These reactions are also in their

Table 3.3 Bond dissociation enthalpies for representative main-group compounds and precursors used in CVD.

Bond	*Dissociation enthalpy (kcal mol^{-1})*	*Bond*	*Dissociation enthalpy (kcal mol^{-1})*
H–H	104	HO–H	119
H_2B–H	105	O–O	119
Cl_2B–Cl	118	H–F	136
H_2Al–H	86	F–F	38
Cl_2Al–Cl	119	H–Cl	103
H_2Ga–H	82	Cl–Cl	58
Cl_2Ga–Cl	100	$(CH_3)_2B$–CH_3	105
H_2In–H	71	$(CH_3)_2Al$–CH_3	84
Cl_2In–Cl	88	Cl_2Al–CH_3	85
H_3C–H	105	$(CH_3)_2Ga$–CH_3	77
Cl_3C–Cl	70	$(CH_3)_2In$–CH_3	65
H_3Si–H	93	$(C_2H_5)_2In$–C_2H_5	58
Cl_3Si–Cl	111	CH_3–CH_3	90
H_3Ge–H	85	$(CH_3)_3Si$–CH_3	94
Cl_3Ge–Cl	93	Cl_3Si–CH_3	97
H_3Sn–H	75	$(CH_3)_3Ge$–CH_3	80
Cl_3Sn–Cl	84	$(CH_3)_3Sn$–CH_3	71
H_2N–H	109	$(CH_3)Cl_2Sn$–CH_3	70
N–N	226	$(CH_3)_2Sb$–CH_3	60
H_2P–H	82	$(C_2H_5)HSb$–C_2H_5	57
H_2As–H	76	Cl_2Sb–CH_3	60
H_2Sb–H	67		
Cl_2Sb–Cl	80		

Compiled from literature sources (298 K; kcal mol^{-1})

bimolecular limit, so their decomposition rates are very slow relative to larger molecular species whose reaction rates are closer to k^{∞}.

Once the most likely precursor(s) to decompose is identified, the second step is to determine corresponding products so that the necessary reaction chemistry can be included. Equilibrium calculations can be very helpful in this regard,[151,153] but chemical intuition must also guide mechanism development as well. In the case of MTS decomposition, as well as other main-group organometallic precursors, it is important to note that bonds within radical fragments formed by breaking one of the precursor bonds are often much weaker than those in the original molecule, leading to fast decomposition and formation of more stable intermediates. Main-group compounds typically follow a "high-low-high" trend in their bond energies.[iv] Therefore, if sufficient thermal energy is available to break the first bond, the second will follow immediately, as shown by Equation (3.50):

$$SiCl_3 \leftrightarrow SiCl_2 + Cl \ (\Delta H^{\circ}_{298K} = 68.8\,\text{kcal mol}^{-1}) \tag{3.50}$$

$$CH_3 + H_2 \leftrightarrow CH_4 + H \ (\Delta H^{\circ}_{298K} = -0.6\,\text{kcal mol}^{-1}) \tag{3.51}$$

$$SiCl_3 + HCl \leftrightarrow SiCl_4 + H \ (\Delta H^{\circ}_{298K} = -8.2\,\text{kcal mol}^{-1}) \tag{3.52}$$

These reactions, (3.50)–(3.52), are the source of radicals (reactive molecules) that can accelerate MTS decomposition, but because there are no significant chain-branching reactions in this system,

[iv] See, for example, ref. 77 for a discussion of this point.

their concentration remains tied to the MTS pyrolysis rate. Note that methyl radicals are an exception to the rule just discussed and typically do not decompose further. Instead, they undergo abstraction to create CH_4, which along with $SiCl_2$, $SiCl_4$ and HCl constitute the stable (at deposition temperatures) primary products of MTS decomposition.

If hydrocarbon species are formed, the third step is to incorporate well-developed chemical mechanisms for hydrocarbon pyrolysis and/or oxidation into the model. In the case of MTS decomposition, which typically occurs in the presence of hydrogen carrier gas, CH_3 radicals are quickly converted into CH_4 [reaction in Equation (3.51)]. Sources of data for this chemistry are listed below. These mechanisms can be quite large (hundreds of reactions), so it may be desirable to reduce their size to increase computational speed. Sensitivity analysis, which identifies reactions whose rate constants have the greatest impact on model predictions, and reaction-path analysis, which can be used to determine the primary pathways for production and consumption of individual species, are effective tools in this regard,[121] and programs such as Chemkin and Cantera include software for computing sensitivity coefficients. An example of the application of these tools to titanium carbide CVD can be found in ref. 154.

3.5.4 Sources of Gas-phase Kinetics Information

Kinetic data for gas-phase hydrocarbon oxidation and pyrolysis are available from various sources, primarily reviews published in *J. Phys. Chem. Ref. Data*. However, the best location to find a thorough compilation of such data is the NIST kinetics database web site (listed below). Unfortunately, this site does not have extensive holdings for reactions of compounds containing other main-group elements or organometallic compounds. Individual literature sources must be consulted for these:

- http://kinetics.nist.gov/kinetics/index.jsp NIST kinetics database.
- http://www.cstl.nist.gov/div836/ckmech/ Halocarbon mechanisms (including HFC), silane oxidation and a hydrocarbon combustion mechanism.
- http://www.me.berkeley.edu/gri_mech/ GRI-Mech hydrocarbon oxidation mechanism, widely used for modeling natural gas (CH_4 and C_2H_6) oxidation.
- http://www-cmls.llnl.gov/?url=science_and_technology-chemistry-combustion. Mechanisms for hydrocarbon oxidation.
- Materials Processing Database: http://www.ipt.arc.nasa.gov/databasemenu.html.Database published by personnel at NASA/Ames Research Center, Mountain View, CA. Contains kinetic mechanisms for dichlorosilane, trichlorosilane, and dimethylaluminium hydride.

3.5.5 Surface Chemistry

Although it is certainly logical to include surface chemistry in a CVD model, it is often much harder to write chemically reasonable reactions involving surfaces than for the corresponding gas-phase portion of the deposition mechanism. This is because the physical structure of the surface is often unknown; epitaxial growth processes on oriented single-crystal surfaces are typically used only in microelectronics fabrication. Potential reaction pathways are, therefore, numerous and rate constants are unknown and very difficult to determine, either experimentally or theoretically. In general, two approaches to address this problem are feasible. In the first, a schematic mechanism is constructed in which the individual reactions are in some sense chemically reasonable, but do not necessarily correspond to actual elementary processes. In this case, the details of surface structure are often ignored (*e.g.*, the actual surface plane is usually not specified) and rate constants are obtained by fitting experimental data, often with assumptions concerning the kinetic limits within which the process operates. Alternatively, mechanisms attempting to capture the elementary chemical steps have been developed, using rate constants obtained from either experiment or derived from computational modeling using first-principles methods. This approach became feasible with the advent of accurate QC codes and fast computers capable of handling systems with relatively large numbers of atoms. We illustrate both approaches here. An example of the first, applied to TiN growth, is discussed in detail below and recent examples from the literature describing the second approach are reviewed. Note that empirical approaches to estimating rate constants for elementary surface reaction have been developed.[155] The reader is also referred to several good reviews of CVD surface chemistry that have been published.[57,156–158]

Some general principles must be observed in the development of surface mechanisms. First, the mechanism must reproduce the deposit stoichiometry. This is straightforward in the case of solids composed of a single element, such as silicon or a metal, but is not always obvious in the case of non-molecular solids such as Si_3N_4. Second, most mechanisms are intended to reproduce the steady-state growth rate, rather than time-dependent phenomena such as the initiation or nucleation processes. Thus, a steady-state analysis of the rate equations should yield a system of rate equations that is neither over- nor under-determined. While this may seem obvious, a series of apparently "elementary" surface reactions that appear reasonable from a chemical point of view may not have this property. Third, the resulting steady-state rate equation should reproduce known reactant dependencies. All three of these principles are illustrated in the first example below: TiN CVD from $TiCl_4$ and NH_3 mixtures.

3.5.5.1 *Mechanism Construction: Surface Site Formalism*

The development of surface-kinetic CVD mechanisms frequently involves the identification of surface "sites" upon which species adsorb and react. Although it is not always clear whether this approach is based on physical reality, it has proven effective for modeling several CVD processes, including Si, diamond, Si_3N_4, TiN, SnO_2 and GaAs. This concept is used by the Surface Chemkin suite of routines,[119,159] for example.

In constructing a mechanism with this software, the user may include any of the following components:

- phases
- surface sites
- surface species
- bulk species

Phases correspond to actual materials being deposited and may include some or all of the other three components. Sites can be identified (named) such that they correspond to what is envisioned physically to exist on the surface. For example, mechanisms of diamond growth can include sp^2 and sp^3 carbon leading to graphite or diamond.[160] Alternatively, sites can be simply named "open" and correspond to an available location at which adsorbed species may reside. Surface species typically correspond to atoms or groups of atoms residing in the top-most layer of the deposit and can be created by reactions including adsorption, desorption and surface diffusion. Finally, bulk species can be created when a surface site is "covered" by adsorption of a gas-phase molecule:

$$SiH_4 + Si(s) \rightarrow SiH_4(s) + Si(b) \tag{3.53}$$

Note that surface sites are conserved, a general feature of this type of surface mechanism. The invocation of surface species and sites is not a requirement for a CVD mechanism. As discussed in

Section 3.5.2, a purely global process could be written that does not explicitly involve surface chemistry at all:

$$SiH_4 \rightarrow Si(b) + 2H_2 \tag{3.54}$$

In the absence of surface rate constants, the inclusion of detailed surface reactions may, in fact, only result in additional unknown parameters and thus a fitting exercise.

The reaction shown in Equation (3.53) as written is irreversible. However, the reverse rate constant could be calculated *via* the equilibrium constant (microscopic reversibility assumption) if the reaction thermochemistry is known. In the case of a simple adsorption reaction such as this (whose rate is often described by a sticking coefficient) the adsorption energy can be determined from an experiment such as temperature-programmed desorption.

In general, however, individual surface reactions are very difficult to observe experimentally, particularly at the temperatures typical of CVD. In addition, theoretical approaches for predicting these rates are a relatively recent innovation and are also computationally intensive, often requiring an expert user with access to parallel computing resources. Thus, the number of CVD mechanisms for which surface rate constants are known from experiment or even by theoretical prediction is quite limited. Typically, rate constants for the interaction of gas-phase molecules with the surface are cast in the form of sticking coefficients, for which measurements are available in some cases (*e.g.*, SiH_4 and SiH_2 sticking on Si are well characterized). Sticking coefficients for radicals are often assumed to be 1.0 (reasonable in many cases). Other reactions involving only surface and/or bulk species are either assumed to be fast or are obtained by fitting experimental data.

3.5.5.2 Titanium Nitride: an Example of a Schematic Model

The schematic modeling approach is particularly appealing in cases where a body of experimental data exists from which rate constants can be derived. Although the individual reactions do not necessarily represent the actual chemistry occurring on the surface, they do represent what *must* be happening in an overall sense. Another attractive feature of this approach is that specialized computational tools and expertise are not necessary, unlike cluster or periodic boundary-layer DFT approaches. Essentially, the user requires only the tools of standard chemical engineering reaction analysis.

To illustrate this approach we describe a model for titanium nitride (TiN) deposition from $TiCl_4$/NH_3 mixtures developed by R. S. Larson.[161] There are essentially five steps to the process: (1) define representative surface species; (2) write reactions using these species that reproduce the overall deposition stoichiometry; (3) assume the steady state condition for surface species and solve for the overall deposition rate; (4) obtain values for the constituent rate constants by fitting the rate expression to available experimental data; and (5) validate the model by comparing with data sets not used in the fitting exercise.

For TiN, the overall stoichiometry is written as:

$$6TiCl_4(g) + 8NH_3(g) \rightarrow 6TiN(b) + 24HCl(g) + N_2(g) \tag{3.55}$$

in which the notation (b) indicates a bulk (solid) species and (g) a gas-phase species. Experimentally, it is observed that the deposition rate is either zero-order or slightly negative order in $TiCl_4$ and second-order in NH_3. Gas-phase complex formation is possible, but temperatures are too low for unimolecular decomposition to occur. The titanium atoms undergo a change in oxidation state from +4 to +3, a complicating factor that necessitates the formation of N_2 gas in the mechanism. Clearly, a series of reactions must occur in which Ti–Cl bonds are successively broken and replaced by Ti–N bonds, with a similar set of reactions occurring to replace the N–H bonds. Thermodynamically, it is logical that this should occur *via* the formation of gas-phase HCl.

Undoubtedly, there are multiple ways to describe this process. In Larson's mechanism, the surface site formalism is adopted in which the following species are defined:

Gas: $TiCl_4(g)$ $NH_3(g)$ $HCl(g)$ $N_2(g)$
Surface: $TiCl_3(s)$ $TiCl_2(s)$ $TiCl(s)$ $Ti(s)$ $Ti^*(s)$ $NH_2(s)$ $NH(s)$ $N(s)$ $N^*(s)$ $N^{**}(s)$
Bulk: $Ti(b)$ $N(b)$

In this notation, Ti–N bonds are not specifically indicated. Thus, $TiCl_3(s)$ is a surface species in which one of the original Ti–Cl bonds is replaced by a Ti–N bond, while in Ti(s) all four of the Ti–Cl bonds have been replaced. In Ti*(s), however, one of the new Ti–N bonds has been severed. The nitrogen-containing species are analogous, so that in N**(s) two of the N–Ti bonds have been broken. Ti*(s) is the immediate precursor to a bulk titanium, Ti(b), while N(s) is the species that becomes N(b). An N_2 molecule is formed by the reaction of two N**(s) species, breaking the last N–Ti bond to each.

A set of 14 reactions is then defined that fit naturally into five groups:

Titanium deposition:

1. $TiCl_4(g) + NH_2(s) + Ti^*(s) \rightarrow TiCl_3(s) + NH(s) + HCl(g) + Ti(b)$
2. $TiCl_4(g) + NH(s) + Ti^*(s) \rightarrow TiCl_3(s) + N(s) + HCl(g) + Ti(b)$

Nitrogen deposition:

3. $TiCl_3(s) + NH_3(g) + N(s) \rightarrow TiCl_2(s) + NH_2(s) + HCl(g) + N(b)$
4. $TiCl_2(s) + NH_3(g) + N(s) \rightarrow TiCl(s) + NH_2(s) + HCl(g) + N(b)$
5. $TiCl(s) + NH_3(g) + N(s) \rightarrow Ti(s) + NH_2(s) + HCl(g) + N(b)$

Surface condensation:

6. $TiCl_3(s) + NH_2(s) \rightarrow TiCl_2(s) + NH(s) + HCl(g)$
7. $TiCl_3(s) + NH(s) \rightarrow TiCl_2(s) + N(s) + HCl(g)$
8. $TiCl_2(s) + NH_2(s) \rightarrow TiCl(s) + NH(s) + HCl(g)$
9. $TiCl_3(s) + NH(s) \rightarrow TiCl(s) + N(s) + HCl(g)$
10. $TiCl(s) + NH_2(s) \rightarrow Ti(s) + NH(s) + HCl(g)$
11. $TiCl(s) + NH(s) \rightarrow Ti(s) + N(s) + HCl(g)$

Bond breaking:

12. $Ti(s) + N(s) \rightarrow Ti^*(s) + N^*(s)$
13. $Ti(s) + N^*(s) \rightarrow Ti^*(s) + N^{**}(s)$

N_2 liberation:

14. $2Ti(s) + 2N^{**}(s) + 2N(b) \rightarrow 2Ti^*(s) + N_2(g) + 2N(s)$

Note that each reaction conserves surface sites and for each arriving gas-phase species a bulk species is created by "burying" a surface species. The opposite occurs in Reaction 14, in which the recombination of unsaturated surface nitrogen atoms leads to desorption of gas-phase N_2 and the exposure of previously buried nitrogen atoms.

The next step is to assume that the surface species are reaction intermediates whose concentrations are at steady state. This allows a conservation equation to be written for each. For example, denoting the rate of reaction i by R_i, the steady-state equation for the species Ti*(s) is:

$$-R_1 - R_2 + R_{12} + R_{13} + 2R_{14} = 0 \tag{3.56}$$

Since the total concentrations of Ti and N species are both conserved, only eight of the ten such equations are independent. By rearranging one can write the eight equations in the following compact form:

$$\begin{aligned}R_1 + R_2 =& R_3 + R_6 + R_7 = R_4 + R_8 + R_9 \\ =& R_5 + R_{10} + R_{11} = 3R_{12} = 3R_{13} = 6R_{14} \equiv r\end{aligned} \tag{3.57}$$

$$\begin{aligned}R_3 + R_4 + R_5 =& R_1 + R_6 + R_8 + R_{10} \\ =& R_2 + R_7 + R_9 + R_{11} \equiv s\end{aligned} \tag{3.58}$$

The net production rates of the gaseous and bulk species can then be cast in terms of the shorthand quantities r and s as follows:

$$\mathrm{HCl(g)}: R_1 + R_2 + R_3 + \cdots + R_{11} = 3s = 4r$$

$$\mathrm{TiCl_4(g)}: -R_1 - R_2 = -r$$

$$\mathrm{NH_3(g)}: R_{14} = \frac{1}{6}r$$

$$\mathrm{N_2(g)}: R_{14} = \frac{1}{6}r$$

$$\mathrm{Ti(b)}: R_1 + R_2 = r$$

$$\mathrm{N(b)}: R_3 + R_4 + R_5 - 2R_{14} = s - \frac{1}{3}r = r$$

It can be seen by comparing with the reaction shown in Equation (3.55) that the 14 surface reactions combine to reproduce exactly the assumed stoichiometry of the overall reaction, provided that the surface species are at steady state.

The kinetic law for the process can now be derived by inserting expressions for the rates of the individual steps into Equations (3.57) and (3.58), assuming mass-action kinetics. The goal is to write the overall deposition rate r (or s, which is proportional to it) in terms of the *rate constants of the individual steps* and the concentrations $C_{\mathrm{TiCl_4}}$ and $C_{\mathrm{NH_3}}$ of the gas-phase reactants. To avoid introducing unnecessary complexity in the absence of kinetic data for these reactions, the rate constants within each of the five surface reaction groups are taken to be equal and are denoted by α, β, γ, λ and δ, respectively. Since the concentration of N(b) is a constant, it can be incorporated into the rate constant δ. The concentrations of the ten surface species can be eliminated using the eight conservation equations together with the condition that the total concentrations of Ti- and N-containing surface species are each equal to $\frac{1}{2}$ of the overall surface site density (ρ). After some algebraic manipulation, one obtains:

$$r = \frac{3\lambda}{4(\phi_n + 2)^2}\left\{\left[\frac{\lambda}{2\delta} + \rho^2\frac{(\phi_n + 2)\theta\phi_t\phi_n}{1 + \phi_t + \theta\phi_t\phi_n}\right]^{1/2} - \left(\frac{\lambda}{2\delta}\right)^{1/2}\right\}^2 \tag{3.59}$$

where $\phi_t = (5\alpha/3\gamma)C_{\mathrm{TiCl_4}}$, $\phi_n = (5\beta/4\gamma)C_{\mathrm{NH_3}}$, and $\theta = \gamma/5\lambda$. This is a complex expression, but it can be simplified in certain limiting cases that are also physically realistic. For example, if ϕ_t and ϕ_n are both arbitrarily small (as for fast surface condensation, with deposition rate-limited by precursor adsorption), then:

$$r \approx \frac{25}{384}\delta\left(\frac{\alpha\beta\rho^2}{\gamma\lambda}\cdot C_{TiCl_4}\cdot C_{NH_3}\right)^2 \quad (3.60)$$

and the overall reaction is second-order in each reactant. Alternatively, if ϕ_n is arbitrarily small but $\phi_t \gg 1$(*i.e.*, deposition rate-limited by nitrogen deposition), then:

$$r \approx \frac{3}{128}\delta\left(\frac{\beta\rho^2}{\lambda}\cdot C_{NH_3}\right)^2 \quad (3.61)$$

so that the reaction is still second-order in NH_3 but now zeroth-order in $TiCl_4$. This agrees reasonably well with the literature reports; a small negative order in $TiCl_4$ is actually observed, but this is attributed to competing gas-phase complex formation.[162,163]

One must now determine the values of the five rate constants to make the model capable of quantitatively predicting deposition rates. This must be done by fitting experimental data, since their schematic nature prohibits first-principles prediction or direct measurement. In this instance, only one set of data, obtained in a stagnation-flow reactor (SFR), was available for which the reaction conditions were fully specified. Two sets of published data lacked experimental details crucial to modeling (a not uncommon occurrence).[162,163] However, it proved feasible to extract rate constants using a set of data obtained in a rotating disk reactor.[164] However, since the experimental data sets exhibit a small negative reaction order with respect to $TiCl_4$, the following gas-phase reaction was added, having a third-order rate constant k and activation energy E, providing a path whereby gas-phase $TiCl_4$ is depleted and providing the possibility of a negative reaction order:

$$TiCl_4 + 2NH_3 \rightarrow TiCl_4\cdot 2NH_3 \quad (3.62)$$

The activation energy E was needed due to the existence of a substantial temperature difference between the deposition surface and the inlet showerhead. There were thus seven adjustable parameters to be determined by optimizing the fit to the experimental deposition rates over the entire range of $TiCl_4$ concentrations; Figure 3.10 shows the resulting fit.

Unsurprisingly, with such a large number of parameters, a good fit can be obtained. A more rigorous test is provided by using the resulting rate constants to predict the growth rates reported in the literature. This becomes possible if the assumption of differential reaction conditions (*i.e.*, zero reactant depletion) is made, which appears to be valid for at least one of the data sets. As seen in Figure 3.11, although the Larson model tends to underpredict the deposition rates reported by Srinivas *et al.*,[163] the discrepancy is at most a factor of 2.7. In contrast, the observed trends in the deposition rate with NH_3 concentration are predicted quite well (Figure 3.12). Even more revealing, however, is that the rate expressions reported in the literature are clearly inferior to Larson's model when used to predict the growth rates obtained from the SFR (Figure 3.10), departing from the measured deposition rates by more than a factor of ten in some cases. Thus, the schematic mechanism approach succeeds in producing a significantly more robust model. An additional useful result is that the fitted activation energy for the reaction show in Equation (3.61) is unrealistically large, which has the effect of confining the complexing reaction to the immediate vicinity of the substrate. This suggests that the negative reaction order in $TiCl_4$ may be due to a more complicated gas-phase process or even a heterogeneous reaction, thus providing an unexpected chemical insight.

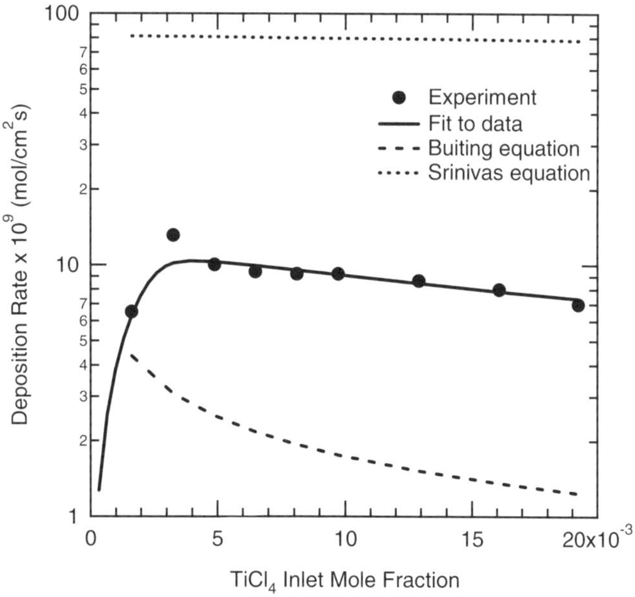

Figure 3.10 Optimized fit of TiN deposition data from a Sandia rotating disk reactor and comparison with predicted deposition TIN rates obtained for $TiCl_4/NH_3$ using published rate expressions. (Adapted from ref. 161.)

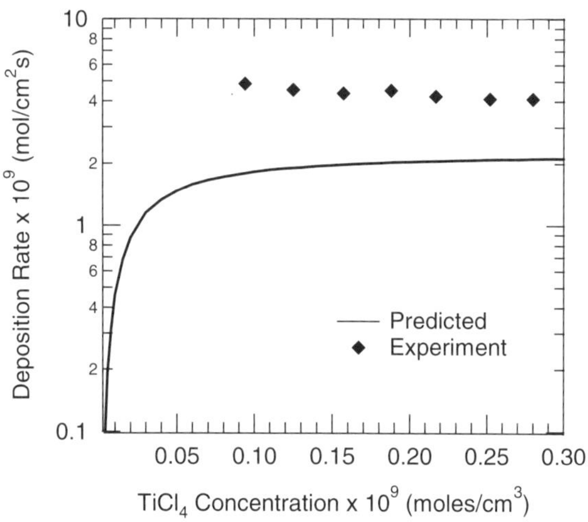

Figure 3.11 Predicted $TiCl_4$ dependence of the TiN deposition rate *vs.* observed deposition rates. Data are those of Srinivas *et al.*[163] (Adapted from ref 161.)

3.5.5.3 Mechanism Development Based on First-principles Modeling

Despite improvements in the sophistication of theoretical methods of modeling surface reactions and vast increases computational speed over the past decade, it is still quite difficult to predict rates for surface reactions from first principles. As a result, these methods are most often used to identify thermodynamically favorable pathways or to examine individual reactions that have not been experimentally characterized but which are thought to be rate-determining. Since the feasibility of

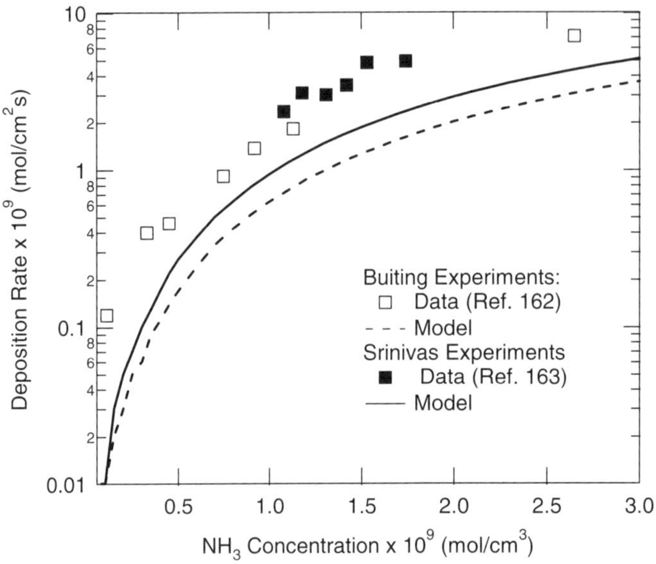

Figure 3.12 Predicted NH_3 dependence of the TiN deposition rate *vs.* deposition rates reported by Buiting *et al.*[162] and Srinivas *et al.*[163] (Adapted from ref. 161.)

this approach will no doubt increase as theoretical methods for locating reaction pathways improve, we provide an example from the literature in which a mechanism for CVD growth of aluminium is developed. An introduction to the application of the cluster approach for modeling semiconductor surfaces is found in ref. 165.

The cluster approach is the most commonly used technique, since it allows standard QC methods to be employed as they are for gas-phase molecules. It has been used to model reactions in several CVD systems, including Si_3N_4,[166] ZrO_2,[167] aluminium,[168] TiO_2[169] and GaAs.[170] Predictions of diamond growth based on cluster models have been particularly effective, largely due to the high level of understanding of carbon-based systems, as well as to the unique properties of this material.[171] There are, however, only a few examples of non-carbon-based CVD systems where extensive, if not complete, use of computational methods enable an accurate growth model to be developed.

Alternatively, the surface can be treated as a slab and modeled with plane-wave DFT methods employing periodic boundary-layer conditions (sometimes referred to as the supercell approach[172]). These methods, such as VASP,[173,174] ABINIT[175] and CASTEP,[176] are designed to treat surfaces and bulk materials. They are capable of predicting properties such as orientation-specific growth, defects, terraces and multiple phases that can be important in CVD, as well as molecular desorption energies and other reaction thermodynamics. Plane-wave DFT has seen minimal use in CVD modeling – in part because of lack of familiarity with solid-state physics theory on the part of potential users. In addition, however, such calculations are computationally expensive and often require parallel processing capabilities to complete in manageable time frames.

The primary concern in using a cluster instead of a slab to model surface is that non-local effects may not be captured. Unfortunately, the extent of non-local effects, which can be addressed by periodic DFT, has not been addressed fully for most materials of interest in CVD. Musgrave *et al.* have compared the predictions of a cluster model *vs.* periodic slab calculations, however, for the adsorption of NH_3 on Si(100)-(2×1) and found that non-local effects along the dimer row are significant, but across the trenches they are much less important.[177] They conclude that a three-trimer cluster is necessary to accurately model this surface ($Si_{21}H_{20}$) and that cluster calculations using DFT/B3LYP will reproduce experimental reaction and activation energies as well as strain effects. Interestingly, they also find a DFT/B3LYP calculation using the large 6-311++G(2d, p) basis set produces a more accurate activation energy for NH_3 dissociation than periodic-slab DFT calculations using the generalized gradient approximation.

The papers of Shimogaki, *et al.* and Yamashita *et al.* provide an illustration of how a cluster can be used to model deposition of aluminium films from dimethylalane (DMA, $[AlH(CH_3)_2]$.[178,179] This work is an unusually thorough treatment, including gas-phase, surface energies and vibrations, transition-state calculations, finite-difference modeling (CRESLAF code[180]), and comparison with experimental results. This example, rather than growth of diamond from hydrocarbons, was selected because it demonstrates what can be accomplished by combining QC predictions and CFD modeling with experimental measurements of growth rates and reaction products. In addition, the state of knowledge concerning the chemistry of aluminium MOCVD is much more representative of CVD systems in general than is diamond growth. These authors determined through experiments that the rate-limiting step is a surface reaction,[179,181] modeled the energetics using *ab initio* QC methods,[168,182] then validated the model by comparison with growth-rate measurements.[179] Realistically, this is probably the best route to a robust growth-rate model for CVD systems for which surface rate constants are unknown.

The basic elements of the QC-based cluster approach are as follows. First, the crystal face of interest must be selected. In the case of Al CVD,[168,182] the Al(111) surface was chosen. A cluster simulating this surface is then constructed, using the minimum number of atoms required to reproduce known material properties, in particular the local bonding arrangement at the surface. It is generally advisable to maintain the highest level of symmetry possible to minimize computational times. Figure 3.13 shows possible aluminium cluster sizes and configurations. Here, two-layer structures containing no more than ten atoms are sufficient to reproduce the experimentally measured average bond strength in bulk aluminium. However, a three-layer structure is necessary to have both three-fold surface sites available, so the 6-3-1 structure shown in Figure 3.14 was used to develop the surface reaction mechanism.

A level of theory sufficient to achieve the desired level of fidelity with available experimental data is then selected. Using this, a geometry optimization calculation is performed to relax the cluster geometry and minimize its energy. The resulting geometry should be in agreement with the known crystal structure; if not, a higher level of theory and/or larger basis set may be necessary. The DFT B3P86/LANL2DZP method was used in the investigation of Al CVD. DFT is often the method of choice for cluster calculations because it is computationally inexpensive and can handle systems containing large numbers of heavy atoms better than fully electron-correlated methods such as MP2. The energy obtained for the optimized structure is then used to calculate reaction energies for adsorption and other processes occurring on the surface.

Once the geometry of the bare cluster and its energy are established, possible adsorption sites can be identified. In the case of Al(111), there are four possible sites: top, bridging, and two different three-fold sites (Figure 3.14). Adsorption energies are then calculated for all species of interest, including the precursors and any decomposition products resulting from gas-phase reactions that are thought to play are role in the growth process. Atomic or molecular fragments that may form as result of surface reactions must also be considered. Yamashita *et al.* considered the following in their DFT calculations:[168,182] (1) dissociative precursor adsorption (dimethylaluminium hydride; DMH); (2) surface diffusion of adsorbates; (3) desorption of reaction products; and (4) dissociation of adsorbed methylaluminium to deposit aluminium, including the effect of steps on the activation energy. The resulting reaction mechanism (Table 3.4) includes both the enthalpies of individual adsorbates and the activation energies of reaction transition states.

In some cases, an initially intact molecule on the surface may be predicted to dissociate. This is the case for DMA (Reaction 1 in Table 3.4), which is predicted to decompose on the surface to form an adsorbed H atom and an $Al(CH_3)_2$ group. If so, this indicates that there is no energy barrier to

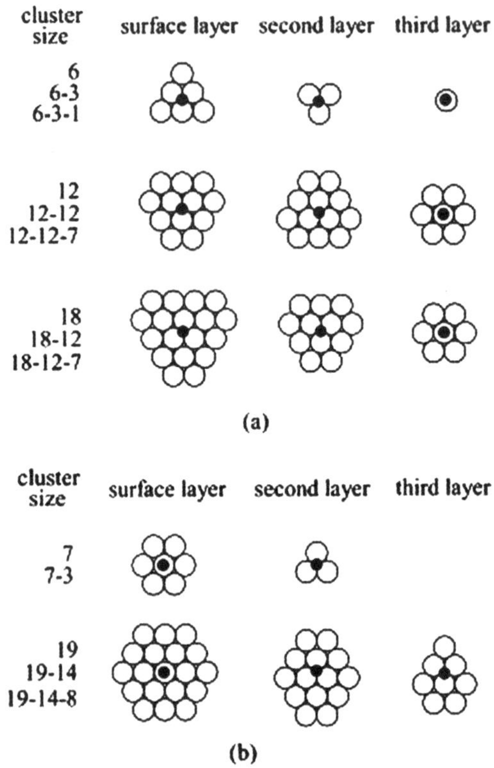

Figure 3.13 Cluster models for the Al(111) surface considered by Nakajima *et al.* to model CVD of aluminium. In (*a*) the centers of the three layers coincide with the three-fold surface sites. In (*b*) the centers coincide with terminal surface sites. Black dots indicate the center of the surface layer in each cluster. Numbers on the left-hand side correspond to the number of atoms in each layer. (Reprinted from ref. 168 with permission.)

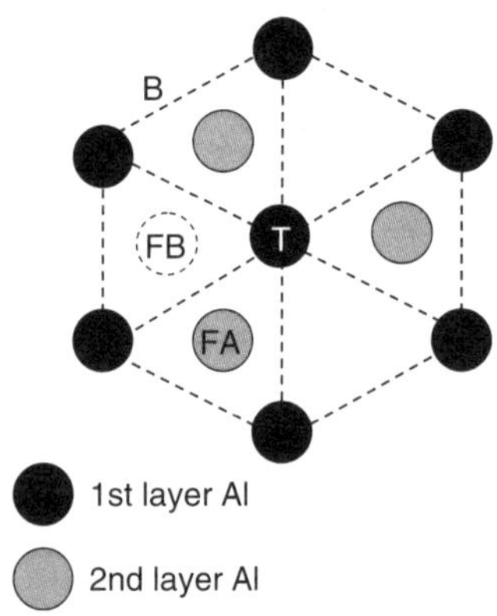

Figure 3.14 Adsorption sites on the 7–3 cluster model of the Al(111) surface. Adsorption T, B, F and FF represent the terminal, bridge, threefold A and threefold B sites, respectively. (Adapted from ref. 178.)

Table 3.4 Elementary reaction mechanism for aluminium CVD from dimethylaluminium hydride (DMAH)[178,179] with their estimated reaction rate constants. DMAH_m indicates DMAH monomer. AlD indicates the aluminium deposited on the surface. T, B, F and FF indicate the adsorption sites defined in Figure 3.14. O indicates an open site.

	Reactions
1.	DMAH_m + O(T) + O(F)=CH_3(T) + AlH(CH_3)(F)
2.	AlH(CH_3)(F) + O(FF)=H(FF) + AlH(CH_3)(F)
3.	TMA + O(B)=CH_3_AlH(CH_3)$_2$(B)
4.	CH_3_Al(CH_3)$_2$(B) + O(T)=AlH(CH_3)$_2$(B) + CH_3(T)
5.	AlH(CH_3)$_2$(B)=CH_3_Al(CH_3)(B)
6.	CH_3_Al(CH_3)(B) + O(T) + O(F)=Al(CH_3)$_2$(F) + CH_3(T) + O(B)
7.	CH_4 + O(FF)=H_ CH_3(FF)
8.	H_CH_3(FF) + O(T)=H(FF) + CH_3(T)
9.	H_2 + O(T)=H_H(T)
10.	H_H(T) + 2O(FF)=H_H(FF) + O(T)
11.	H_H(FF)=2H(FF)
12.	AlH(CH_3)(F) + O(T)=CH_3(T) + O(F) + AlD

the reaction. An additional factor to be considered is that surface coverage may affect adsorbate stability. For example, adsorbed species in adjacent sites on the Al(111) surface are destabilized relative to isolated adsorbates.

Once energies are available for surface species and transition states (at a minimum, as heats of formation at 0 K) it becomes feasible to compute reaction energies and identify which reaction pathways are most likely to occur. Numerous reaction pathways can potentially exist in any CVD mechanism, which complicates mechanism development. For this reason, it is always helpful to have experimental data that provide clues to which reactions are active. In the case of Al CVD, measurements of reactor outlet gas composition failed to detect CH_4, a logical product of the decomposition of DMA. This allowed the elimination of a substantial number of reaction pathways that might otherwise have been considered.

To compute rate constants at realistic temperatures, vibrational frequencies are needed for the cluster and adsorbed species. Computing frequencies for small molecular species is relatively straightforward, but for clusters containing a large number of atoms the calculation may be computationally so expensive that it becomes impractical. Accurate frequencies can be obtained using DFT/B3LYP, however, but a correction for systematic errors (which can be as large as 100 cm^{-1})[165] is necessary. Alternatively, new embedded-cluster methods,[165] such as the ONIOM method,[183,184] can reduce the computational cost by treating the surface as a two-region domain, in which the active site for chemical reaction is surrounded by a larger zone whose properties are treated at a lower level of theory. It is also possible to estimate vibrational frequencies using empirical approaches.[155,179] Regardless of the level of theory used, the calculated frequencies for a cluster model are only an approximation of the actual phonon spectrum of the solid, which ultimately limits their accuracy. Low-frequency vibrations characteristic of surface–adsorbate stretching and bending modes are particularly problematic (frequencies $< 200\,cm^{-1}$), since they make the greatest contribution to the adsorbate entropy and are also the most likely to be in error. For example, a 50 cm^{-1} error in vibrations with frequencies of order 100 cm^{-1} will cause a factor of 2 error *per vibration* in a rate constant at 500 K.[179] Thus, it is advisable to compare predicted frequencies with experimentally measured values (obtained from electron energy loss measurements, for example) to assess the magnitude of potential errors.

Once the reaction energetics and adsorbate frequencies are known, it then remains to apply TST, with an equation analogous to Equation (3.35) (but with partition functions corresponding to each reactant) to determine rate constants for these reactions, using the calculated enthalpy and entropy changes, as well as the predicted frequencies. Returning to the example of aluminium CVD, the results of the QC calculations enabled Shimogaki *et al.* to eliminate some pathways, reducing the number of required reactions to 12 and yielding a mechanism that predicts both deposition and the formation of the observed gas-phase products. The number of calculated rate constants was further reduced by assuming sticking coefficients for gas-phase molecules. Typically, adsorption processes are exothermic and proceed without an activation barrier, eliminating the need to identify a transition state. Reverse rate constants can be calculated through the equilibrium constant, assuming the adsorbate bond energy is known or predicted from theory. Note that QC calculations cannot be used to determine a sticking coefficient. Instead, Monte Carlo calculations must be performed to obtain rate constants for these reactions if experimental data are unavailable.[185] Alternatively, values can be estimated from analogous reactions, usually assuming values between 0.1 and 1.0 for sticking of radicals and lower values for intact precursors. Though relatively crude, this can yield a model in qualitative agreement with observations and is often a good starting point.

In our example, as in many other models of CVD chemistry, it is assumed that there is no surface coverage dependence and that Langmuir–Hinshelwood (LH) kinetics are operative, *i.e.*, that simple mass-action kinetics are in effect in which the rate of reactant diffusion is much faster than that of reaction. However, it has been shown that deviations from LH kinetics can occur if the rate of surface diffusion becomes comparable to or slower than surface reaction. The importance of this effect can be estimated using the following expression:[179]

$$k = \frac{1}{\frac{1}{k'} + \frac{1}{2\pi N_A D}} \tag{3.63}$$

in which k' is the surface reaction rate, D is the surface diffusion constant and N_A is Avogadro's number. If k' and $2\pi N_A D$ are of similar magnitude, diffusion cannot be neglected and mass-action kinetics can no longer be used. Instead, the rate of surface diffusion must be explicitly accounted for in the mechanism, requiring an activation energy and frequency for the diffusional hopping process to be determined. Atomistic methods, such as kinetic Monte Carlo or molecular dynamics, can then be used to calculate this rate.[186]

Despite the careful treatment described above, errors in calculated rate constants cannot be avoided, due to limitations in the accuracy of calculated energetics, vibrational frequencies and the cluster approximation itself. As a result of these errors, Shimogaki *et al.* found it necessary to adjust the activation energy of one surface reaction, as well as the sticking coefficient of DMA, by fitting to experimental data to achieve the desired level of accuracy. Fitting was accomplished by modeling deposition rates measured in a tubular CVD reactor using the boundary-layer code CRESLAF from the Chemkin software package.[159] The good agreement with the measured deposition rate profile, which included temperature effects due to the deliberate imposition of a temperature gradient in the reactor, suggests that the cluster approach successfully identified the key reactions, including the rate-limiting step. An important observation is that the reaction pathways examined were either barrierless, resulting in fast kinetics, or had large activation barriers, making them unimportant. Sensitivity analysis showed that DMA adsorption and TMA desorption are the key steps. Thus, the relatively accurate thermodynamics and activation barriers resulting from the cluster-QC approach narrowed considerably the number of reactions that need to be considered. This result motivated more detailed investigations to better understand these reactions so that model accuracy and robustness can be improved.

3.6 Particle Formation and Growth

3.6.1 Introduction

Formation of particles in CVD processes is often undesirable. Generally, the desired product of a CVD process is a thin film, deposited on a substrate. Any particles that are produced serve to consume precursors (reducing their utilization for film deposition), are potential sources of defects in the film (if they deposit on the surface), and can increase contamination and required maintenance of the CVD reactor and downstream equipment such as scrubbers and vacuum pumps. Particulate contamination is a leading cause of yield loss in semiconductor processing. As feature sizes in integrated circuits decrease, the critical defect size that leads to device failure decreases proportionally. The International Technology Roadmap for Semiconductors (ITRS)[187] predicts that from 2007 to 2018 the critical particle size (minimum size expected to cause a device failure) will decrease from 33 to 9 nm. As the critical particle size becomes smaller, and as improved cleanroom and gas purification technologies eliminate external sources of particles, nucleation of particles within the processing environment is becoming the most important source of particulate contamination. To maximize reactor throughput, one would often like to carry out a CVD process at the maximum deposition rate that does not lead to particle formation. Models can be useful in predicting how the onset of particle formation will vary with process conditions.

On the other hand, processes very similar to CVD are sometimes used to intentionally produce particles. This is often called Chemical Vapor Synthesis (CVS). A large-scale commercial example is the production of nickel powders *via* decomposition of nickel carbonyl.[188] Typical examples of this process are presented in the work of Winterer, Hahn and co-workers.[189–193] Variations on this process include methods in which a laser is used to heat the gas and initiate decomposition in a cold-wall reactor,[194–197] or methods where the precursors are delivered as liquid droplets that evaporate within the reaction chamber.[198] Energy to initiate the reaction can also be supplied by a flame, or a thermal or non-thermal plasma. Such processes are somewhat more complex than thermally driven processes. They are analogous to combustion CVD and plasma CVD processes that are likewise generally more complex than thermal CVD processes.

3.6.2 Modeling Approaches

In modeling particle formation, the transport phenomena, gas phase chemistry and surface chemistry involved are essentially the same as those involved in any CVD process. However, additional complications arise from (1) the need to describe *nucleation*, the process by which the smallest entities that will be treated as particles come to exist; (2) the need to describe a continuous size distribution of particles along with discrete chemical species; (3) the need to describe particle–particle interactions such as aggregation and sintering of particles; and (4) the need to incorporate additional driving forces for motion of particles, particularly thermophoresis. The goal of a particle formation model is to predict the particle concentration and size distribution as a function of time, position and/or process parameters. This particle formation model must be coupled to models of the fluid flow, heat transfer and chemical reactions that have already been described in this chapter. When little particle formation occurs, as in many cases where particle formation is undesirable, this coupling can be only in one direction. That is, one solves for the flow, temperature and concentrations fields in the absence of particle formation, and then subsequently models the particle formation treating the flow, temperature and concentration fields as known quantities. In other cases, where precursor consumption by particle formation is significant, the particle formation must be fully coupled to the rest of the problem, and the effect of particle formation on flow, temperature and concentration fields cannot be neglected.

In general, one can model flows containing particles from either an Eulerian perspective (a fixed viewpoint, with particles moving relative to the observer) or a Lagrangian perspective, in which one tracks particles through the flow. These perspectives are analogous to watching from the river bank as a boat goes past (Eulerian perspective) *vs.* riding in the boat (Lagrangian perspective). The Eulerian approach is more appropriate for cases in which particle generation takes place, because of difficulties in tracking particles that do not yet exist at the reactor inlet, and the importance of Brownian motion (diffusion) for very small particles, which is not easily accommodated in a Lagrangian framework. In addition, the fluid flow, heat transfer and species transport equations are almost always treated from an Eulerian perspective, and therefore this approach provides equations for the evolution of the particle size distribution that are similar in form to the equations describing the evolution of the temperature, velocity and concentration fields. From this point of view, the general starting point for describing the evolution of the particle size distribution is a particle population balance that is usually called the aerosol general dynamic equation:

$$\begin{aligned} &\frac{\partial n(v)}{\partial t} + \nabla \cdot \mathbf{V} n(v) - \nabla \cdot D(v) \nabla n(v) + \nabla \cdot \mathbf{V}_{\mathrm{th}} n(v) \\ &= \frac{1}{2} \int_0^v \beta(\bar{v}, v-\bar{v}) n(\bar{v}) n(v-\bar{v}) d\bar{v} - n(v) \int_0^\infty \beta(\bar{v}, v) n(\bar{v}) d\bar{v} \\ &- \frac{\partial}{dv}\left(n(v)\frac{dv}{dt}\right) + I(v^*)\delta(v-v^*) \end{aligned} \tag{3.64}$$

In Equation (3.64), $n(v)$ is the particle size distribution function, defined such that $n(v)\,\mathrm{d}v$ is the number of particles (per unit volume or per unit mass of aerosol) with particle volumes in the range from v to $v+\mathrm{d}v$. Our goal in a particle formation model is to predict $n(v)$ as a function of position, time, *etc.* The first term in Equation (3.64) represents changes in $n(v)$ with time, and is absent in steady state models. The other three terms on the left-hand-side represent particle transport by convection, diffusion and thermophoresis, respectively. $\mathbf{V}$ is the vector velocity of the gas in which the particles are suspended, $D(v)$ is the particle diffusion coefficient, which is a function not only of v, but also of temperature and gas composition, and $\mathbf{V}_{\mathrm{th}}$ is the particle velocity, relative to the gas, due to thermophoresis, which depends primarily on the local temperature gradient. The first two terms on the right-hand-side represent particle–particle coagulation; β is a coagulation coefficient, which is analogous to a second-order reaction rate constant for particle–particle collisions. It is a function not only of the sizes of the two colliding particles, but also of the gas composition, temperature, and pressure. The first of these two terms represents formation of particles of volume v through collision of all possible combinations of particles of volumes $\bar{v}$ and $v-\bar{v}$. The second coagulation term represents loss of particles of volume v by collision with particles of all sizes. The third term on the right-hand-side describes particle growth by physical or chemical deposition of gas-phase species. In this term, $\mathrm{d}v/\mathrm{d}t$ is the particle growth rate (volume per time), and all of the gas-surface chemistry occurring on the particle surface is incorporated into this term. The final term represents particle nucleation, where $I(v^*)$ is the rate of formation (particles per time per aerosol volume or mass) of incipient particles of volume v^*. This implicitly assumes a single minimum size that defines a particle. More complex models of the particle formation and evolution process can take into account additional variables such as particle surface area and composition. Addition of such variables makes the size distribution multidimensional (a distribution in both particle volume and surface area, for example). However, such treatments are beyond the scope of the present discussion.

The most common shape for an aerosol size distribution is approximately a log–normal function. A log–normal distribution can be written as:[199,200]

$$n(v) = \frac{N}{\sqrt{2\pi} v \ln \sigma_{\mathrm{g}}} \exp\left[-\frac{1}{2}\left(\frac{\ln v - \ln \bar{v}_{\mathrm{g}}}{\ln \sigma_{\mathrm{g}}}\right)^2\right] \tag{3.65}$$

Or, as a function of $\ln(v)$:

$$n(\ln v) = \frac{N}{\sqrt{2\pi}\ln\sigma_g}\exp\left[-\frac{1}{2}\left(\frac{\ln v - \ln\bar{v}_g}{\ln\sigma_g}\right)^2\right] \tag{3.66}$$

The integral of the expression in Equation (3.65) over all v gives the total particle concentration, N, as does the integral of the expression in Equation (3.66) with respect to $\ln v$. Equation (3.66) shows that $n(\ln v)$ is a normal (Gaussian) distribution in $\ln v$, with mean $\ln \bar{v}_g$ and standard deviation $\ln \delta_g$. The geometric mean volume, $\bar{v}_g$, is also the median volume (half of the particles are larger, half are smaller). The geometric standard deviation, δ_g, represents the width of the size distribution. It has a minimum value of 1 for a perfectly monodisperse size distribution (all particles have $v = \bar{v}_g$). The geometric mean and geometric standard deviation for any distribution, log–normal or not, are defined by:

$$\ln\bar{v}_g = \frac{1}{N}\int_0^\infty (\ln v)n(v)\mathrm{d}v$$

and:

$$\ln\sigma_g = \sqrt{\frac{1}{N}\int_0^\infty (\ln v - \ln\bar{v}_g)^2 n(v)\mathrm{d}v} \tag{3.67}$$

These are the statistics most commonly used to describe particle size distribution functions in aerosols. Figure 3.15 shows an example of a log–normal distribution, with $N = 10^8$ particles cm^{-3}, $\bar{v}_g = 10\,\text{nm}^3$, and $\sigma_g = 1.4$.

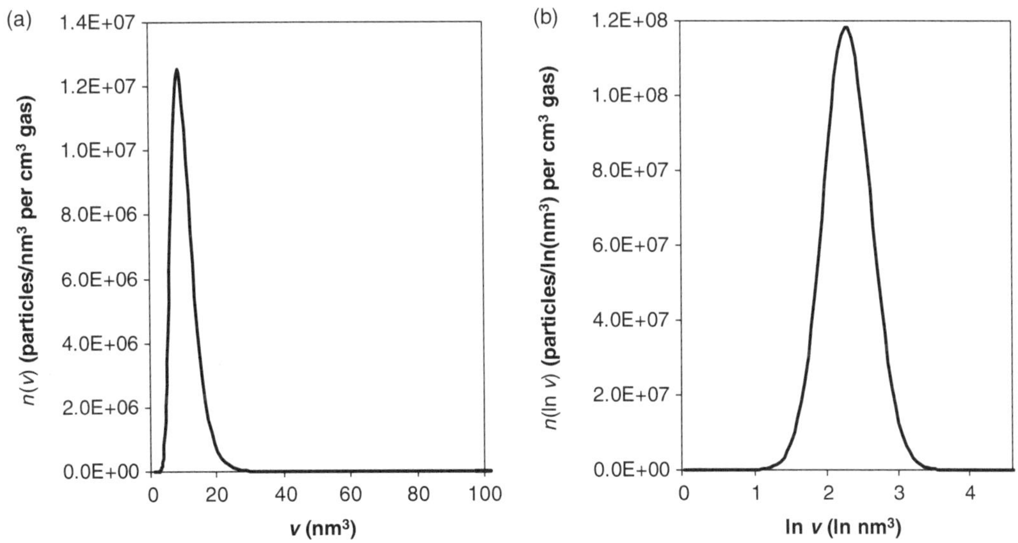

Figure 3.15 Example of a log–normal size distribution. In this example, $N = 10^8$ particle cm^{-3}, $\bar{v}_g = 10\,\text{nm}^3$, and $\sigma_g = 1.4$. Plotted in (*a*) as $n(v)$ *vs.* v, and in (*b*) as $n(\ln v)$ *vs.* $\ln v$.

As written above, the aerosol general dynamic equation is a nonlinear, partial integro-differential equation. It is almost never practical to apply this equation in the full form presented above, where the size distribution is a function of time and three spatial dimensions as well as of particle volume. In practice, the continuous size distribution is replaced by either a discrete distribution or an assumed functional form that is defined by a finite number of values at each point in space and time. Several strategies for doing this are briefly outlined below. There are inevitably trade-offs between the level of detail with which the particle size distribution can be described and the level of detail with which other aspects of the process, such as the reactor geometry or chemical kinetics, can be described while keeping the problem computationally tractable. Thus, here we briefly describe four approaches that provide increasing levels of flexibility and detail with regard to the particle size distribution, at correspondingly increasing computational cost.

3.6.2.1 Monodisperse Model

The simplest approximation to the particle size distribution is obtained by assuming that all particles are the same size. The two values that describe the size distribution are then the total particle concentration, N (per unit volume or mass of aerosol), and the particle size (diameter or volume). An equation for N can then be written as:

$$\frac{\partial N}{\partial t} + \nabla \cdot \mathrm{V}N - \nabla \cdot D\nabla N + \nabla \cdot \mathrm{V}_{\mathrm{th}}N = -\frac{1}{2}\beta N^2 + I \tag{3.68}$$

in which the diffusion coefficient D and coagulation coefficient β depend on the particle volume. I is the nucleation rate, which will generally be taken as zero throughout most of the computational domain. The representative particle volume, v_{r}, is governed by a similar equation:

$$\frac{\partial v_r}{\partial t} + \nabla \cdot \mathrm{V}v_r - \nabla \cdot D\nabla v_r + \nabla \cdot \mathrm{V}_{\mathrm{th}}v_r = \frac{\partial v}{\partial t} + \frac{1}{2}\beta v N \tag{3.69}$$

where the diffusion coefficient and coagulation coefficient are the same as those in the equation for N. As in the original form of the aerosol general dynamic equation, $\partial v/\partial t$ is the particle growth rate by chemical reaction or physical condensation of vapor-phase species. Thus, the monodisperse model adds only two additional equations to the set of equations describing the reactor. It is relatively easy to incorporate such a model into a detailed computational fluid dynamics (CFD) simulation of a reactor without adding unreasonable computational costs. This approach is also easily extended to include a simple description of the evolution of non-spherical particles by introducing a third equation for the particle surface area, which depends on the rate of coalescence of particles as well as the rates of coagulation and growth by deposition of gas-phase species. A good example of this latter approach is given by Kruis *et al.*[201]

3.6.2.2 Method of Moments (MOM)

An efficient and reasonably accurate means of approximately solving the aerosol general dynamic equation is the method of moments (MOM), which has been extensively used due to its relative ease of implementation and low computational cost.[202–204] The k-th moment of the particle size distribution function $n(v)$ is given by:

$$M_k = \int_0^\infty v^k n(v)\mathrm{d}v \tag{3.70}$$

The zeroth moment (M_0) represents the total particle concentration, the first moment (M_1) gives the total particle volume (volume of particles per unit volume or mass of aerosol) and the second moment (M_2) is related to the light-scattering intensity from particles. Using MOM, the problem can be reduced to a small set of moment equations, most commonly using just these first three moments. These equations are obtained by multiplying the aerosol general dynamic equation by v^k then integrating it over all particle sizes. A disadvantage of this method is that it requires that all terms in the moment equations be expressed as functions of the moments themselves. This closure of the moment equations is usually achieved by assuming the shape of the size distribution to be log–normal. Many experimental results suggest that aerosol size distributions are generally log–normal and the log–normal distribution has three parameters that are readily related to the first three moments of the size distribution.[205] Thus, in its most common implementation, this approach adds just three equations to the set of equations describing the process of interest. Solving these provides information not only on the particle concentration and average size, but also the width of the size distribution. Like the monodisperse model, this method can be incorporated into CFD simulation without increasing the computational cost dramatically. In fact, the monodisperse model can be considered a moment model in which only the first two moments are used. In this context, it can easily be modified to assume a size distribution of fixed width (geometric standard deviation) rather than assuming that all particles are the same size.

3.6.2.3 Quadrature Method of Moments (QMOM)

The quadrature method of moments (QMOM), first used in the field of aerosol dynamics by McGraw,[206] avoids using an assumed shape for the size distribution by approximating the moments of the size distribution by an n-point Gaussian quadrature. This solves the problem of obtaining closure of the moment equations, so the coagulation, growth, diffusion and thermophoretic terms can be expressed in their original forms. The moments of the size distribution are expressed as functions of abscissas and weights of the Gaussian quadrature:

$$M_k = \int_0^\infty r^k n(r)\mathrm{d}r \approx \sum_{i=1}^{Nq} f(r_i)\,\mathrm{Weights}_i \tag{3.71}$$

For a three-point quadrature approximation, six radial moments (M_0 to M_5) are required. Note that in QMOM, the moments are most often defined using the particle radius (r) rather than the volume. QMOM does not define or produce an explicit size distribution, but the six moments could be used with an assumed functional form for a size distribution with six degrees of freedom to produce one *a posteriori*.[207,208]

3.6.2.4 The Sectional Method (SM)

The sectional method approximates the continuous size distribution by a finite number of sections or bins within which one numerically conserved aerosol property is held constant.[209–212] This is computationally more expensive than the methods described above, but it is able to capture the details of early stages of particle formation where the assumption of a log–normal size distribution (for MOM) often breaks down. Using a finite element style formulation to discretize the size distribution in particle volume, the particle size distribution can be approximated as:

$$n(v) \approx \sum_{i=1}^{i_{\max}} n_i \phi_i \tag{3.72}$$

where the ϕ_i are basis functions that are equal to one within bin i (for $v_{i-1}<v<v_i$) and zero elsewhere; the n_i are the coefficients of the discretized particle size distribution, which are constant within each bin; and $i_{\max}$ is the number of bins used. Multiplying the GDE by ϕ_i and integrating over all v from 0 to ∞ gives an equation for the number of particles in the size range $v_{i-1} < v < v$, which converts the aerosol general dynamic equation into a set of $i_{\max}$ equations for the number of particles in each size range. The particle number concentration in each bin changes due to coagulation, nucleation and growth, as well as through transport processes (convection, diffusion and thermophoresis). Most often, the bins are evenly spaced on a logarithmic scale in volume, so that the width of a bin $(v_i - v_{i-1})$ is proportional to v_i. The number of additional equations added to the process model is equal to the number of bins used, which could range from ten to more than one hundred. Several variations on this approach have been employed, including moving sectional models, where the volume bins vary with time and/or position,[213–215] and discrete-sectional models in which very small particles are treated as discrete clusters of an integer number of atoms or monomers, while larger particles are treated with a sectional approach.[216–219]

3.6.3 Mechanisms of Particle Formation, Growth and Transport

In the previous section we described approaches to approximately solving the aerosol general dynamic equation without giving much thought to the physical origin of the terms in the equation. Here, we briefly consider the physicochemical origins of the various terms in the equations, and introduce means of estimating numerical values for parameters in the equations.

3.6.3.1 Convection and Diffusion

The transport of aerosol particles by convection and diffusion is essentially the same as the transport of molecules by these same mechanisms. Diffusion is generally only important for very small particles. Even a 10 nm particle has a diffusion coefficient that is more than 100× smaller than a typical gas molecule. The diffusion coefficient for a spherical solid particle in a gas can be computed from:[199]

$$D = \frac{k_B T}{3\pi\eta d} C_c(d) = \frac{k_B T}{3\pi\eta d}\left(1 + \frac{\lambda}{d}\left[2.34 + 1.05\exp\left(-0.39\frac{d}{\lambda}\right)\right]\right) \tag{3.73}$$

in which d is the particle diameter, k_B is Boltzmann's constant, η is the gas viscosity, and λ is the mean free path of gas molecules. The factor $C_c(d)$ is called the "slip correction factor." In the *continuum regime* of aerosol dynamics, d is significantly greater than λ, $C_c(d)$ is approximately 1, and the diffusion coefficient is inversely proportional to d. In the *free molecular regime*, d is significantly smaller than λ, $C_c(d)$ is inversely proportional to d, and the diffusion coefficient is inversely proportional to d^2. In either case, D is easily computed from the particle size and the known properties of the gas. For non-spherical particles, more complex variations of this expression are available. Often, this equation is applied for non-spherical particles using an effective diameter that does not correspond to any particular physical diameter.

3.6.3.2 Thermophoresis

In a temperature gradient, an aerosol particle is transported in the direction of decreasing temperature. For particles much smaller than the mean free path of the gas molecules, this simply results from the fact that gas molecules impinging on the particle from the region of higher temperature have higher average velocity than those impinging on the particle from regions of lower temperature. This results in greater momentum transfer from gas molecules to the particle on the higher temperature side of the particle than on the lower temperature side. Equating this thermophoretic force to the drag force on the particle yields a steady-state thermophoretic velocity [$\mathbf{V}_{th}$ in Equation (3.74)]. For particles smaller than the mean free path of the gas molecules, this velocity is independent of particle size, and is approximately given by:[199]

$$\mathbf{V}_{th} = \frac{-0.55\eta\nabla T}{\rho_g T} \tag{3.74}$$

For particles larger than the mean free path of the gas molecules, a temperature gradient within the particle can develop. In this case, more complex expressions for the thermophoretic velocity, which depend on particle size and the thermal conductivities of both the particle and the gas, are required.[199] Thermophoretic velocities can be substantial. It has been said that much of the success of thermal CVD processes in microelectronics can be attributed to thermophoresis, because in cold-wall CVD reactors the steep temperature gradients near the wafer surface create relatively high thermophoretic velocities that move particles away from the wafer and prevent deposition.

3.6.3.3 Coagulation

In the gas phase, whenever particles collide they stick together. If the temperature is sufficiently high, they will coalesce into a spherical particle, whereas at temperatures far below the melting point of the particles they will form loose agglomerates with a fractal structure. For small particles, the rate of coagulation is simply the rate at which the particles happen to collide as they move by Brownian diffusion. For particles larger than the mean free path of the gas molecules, the collision coefficient for particles of diameters d_1 and d_2 with corresponding diffusion coefficients D_1 and D_2 is:[200]

$$\begin{aligned}\beta(d_1, d_2) &= 2\pi(d_1 + d_2)(D_1 + D_2)\\ &= \frac{2k_B T}{3\eta}(d_1 + d_2)\left(\frac{C_c(d_1)}{d_1} + \frac{C_c(d_2)}{d_2}\right)\end{aligned} \tag{3.75}$$

For particles much smaller than the mean free path of the gas molecules, the particles behave like large gas molecules, and their collision rate is the product of the cross-sectional area for collision with their average relative thermal velocity, calculated from the kinetic theory of gases:[200]

$$\begin{aligned}\beta(d_1, d_2) &= \frac{\pi}{4}(d_1 + d_2)^2(\bar{c}_1^2 + \bar{c}_2^2)^{1/2}\\ &= \left(\frac{3k_B T}{\rho_p}\right)^{1/2}(d_1 + d_2)^2\left(\frac{1}{d_1^3} + \frac{1}{d_2^3}\right)^{1/2}\end{aligned} \tag{3.76}$$

In these equations, $\bar{c}$ is the mean thermal velocity of a particle, and ρ_p is the density of the particle, which is assumed to be independent of particle size. For the transition regime, where the particle diameter is comparable to the mean free path of the gas molecules, interpolating expressions developed by Fuchs and by Dahneke can be used,[200] or one can simply take the harmonic (geometric) mean of the values given by Equations (3.75) and (3.76). The most rapid coagulation occurs between particles of different sizes. In both Equations (3.75) and (3.76), there is a term involving $d_1 + d_2$ that is dominated by the larger particle, while the last term in each expression is dominated by the smaller particle. The product of these terms is largest for particles of different sizes. Physically, the larger particle provides a large cross-sectional area for collision, while the small particle has a high average thermal or diffusive velocity, increasing its probability of colliding with a large particle. For an aerosol whose size distribution is evolving only due to coagulation, this variation in coagulation coefficient with relative particle size causes the system to achieve a self-preserving size distribution, in which the geometric mean diameter increases and the total concentration decreases with further coagulation, but the shape of the distribution and the geometric standard deviation remain constant. The shape of this distribution is approximately log–normal with a geometric standard deviation near 1.45.[220]

3.6.3.4 Particle Growth

In CVD and CVS processes, aerosol particles grow by the same mechanisms as in conventional CVD film growth. Thus, one can apply the same surface chemistry mechanisms and models for particle growth that are applied for film deposition. Very small particles may be more reactive than a growing film, due to high curvature, larger numbers of edge and vertex atoms, *etc*. However, it is very rare to have any information on reaction kinetics as a function of particle size, so these effects are not usually considered. A key difference between film growth and particle growth is that the surface area of the aerosol is a function of time and position. If one has an aerosol of volume distribution $n(v)$, then the total surface area (per unit volume or mass of gas) is simply obtained by integrating over this distribution:

$$A = \int_0^\infty A(v)n(v)\mathrm{d}v = \int_0^\infty \left(36\pi v^2\right)^{1/3} n(v)\mathrm{d}v \tag{3.77}$$

If one is using a sectional model for the aerosol dynamics, then this integral becomes a sum over size bins. For a moment model, it can usually be evaluated analytically from the moments. In much of the aerosol literature, particle growth is treated as condensation from supersaturated vapor, both because this is the simplest mechanism of particle growth and because it is of great practical importance in atmospheric aerosols.[199,200] However, this is usually not a very realistic treatment for particle growth in CVD and CVS processes where growth is likely to occur from molecular precursors. Another simplified approach is to neglect growth by chemical reaction or condensation altogether, and attribute all particle growth to coagulation. This approach can be effective if the vapor phase precursor is highly reactive, so that the precursor molecules can be treated as particles that always react upon colliding with each other or with actual particles. In such highly reactive systems, the precursor molecules are rapidly consumed and the evolution of the particle size distribution is mostly governed by coagulation. Thus, getting the details of the gas to particle conversion process correct is not essential.

3.6.3.5 Particle Nucleation

Computing the particle nucleation rate, the final term in Equation (3.64), is often the most challenging aspect of modeling particle formation, particularly in systems where particle formation is to be avoided. For intentional particle synthesis processes, where most of the precursor molecules are converted into particles, and where particle concentrations are high, the final size distribution is often dominated by coagulation, and thus a crude model of nucleation may be adequate. However, in systems where a small fraction of the precursors is converted into particles and particle concentrations are too low for coagulation to dominate on the time scales of interest, nucleation can be the most important component in determining the particle concentration and size distribution. Classical theories of homogeneous nucleation of aerosol particles are based on condensation of a supersaturated vapor. In these theories, there is a critical cluster size that is the least stable (has the highest Gibbs energy). Smaller particles (clusters) tend to evaporate, while larger clusters tend to grow. The size of this critical cluster depends on the supersaturation (ratio of the partial pressure of the condensing species to its vapor pressure) and the surface tension of the aerosol droplet (usually assumed to be the same as the bulk surface tension).[200] As was the case for particle growth, this is not usually an appropriate description for CVD and CVS processes, but it may produce acceptable results if the final size distribution is mostly determined by coagulation.

An alternative approach to homogeneous nucleation theory is to write a detailed mechanism of gas-phase chemical reactions that leads to nucleation. Such a mechanism includes sequences of reactions that lead to formation of larger and larger molecules. A critical size is defined, and any molecules that reach the critical size are considered particles, rather than molecules. The nucleation rate is simply the rate of formation of molecules of the critical size, computed using the reaction mechanism. For example, one can imagine an oversimplified version of this for silicon particle nucleation from silane:

$$SiH_4 \leftrightarrow SiH_2 + H_2$$
$$SiH_4 + SiH_2 \leftrightarrow Si_2H_6$$
$$Si_2H_6 + SiH_2 \leftrightarrow Si_3H_8$$
$$Si_3H_8 + SiH_2 \leftrightarrow Si_4H_{10}$$
$$Si_4H_{10} + SiH_2 \leftrightarrow \text{particle}$$

In this simplified example, the nucleation rate in the aerosol general dynamic equation would be taken as the rate of the final reaction. Of course, this simplified example produces linear silane polymers, not silicon particles, and a more realistic model would have to include elimination of hydrogen, formation of rings and polycyclic clusters, *etc.* Such mechanisms can quickly become unmanageably large. Estimation methods are needed to provide approximate rate parameters for the reactions in such schemes, because the number of reactions can quickly become too large for one to carry out detailed calculations or experiments for all of the reaction rate parameters. In analogy with classical nucleation theory, one expects that there will be a critical size, below which it is thermodynamically favorable for clusters to decompose, and above which it is favorable for them to grow. The critical size used in a kinetic model like this should be at (or above) that critical size, which will depend on reaction conditions as well as the properties of the clusters. This example of silicon particle nucleation is one for which some of the most detailed kinetic models have been constructed, ranging from tens to thousands of reactions.[221–226] Development of more general approaches to describing such chemical nucleation processes remains an open and active area of research for which no single approach is likely to succeed.

3.6.4 Particle Formation: Modeling Examples

Before concluding our discussion of particle formation and growth, we mention a few examples from the literature. Nijhawan and co-workers[227] presented a model of particle transport in a low-pressure parallel-plate-CVD reactor during thermal CVD of polysilicon from silane. In this geometry, the gas flow near the center of the wafer is nearly a stagnation-point flow. They coupled a model of particle transport to a two-dimensional axisymmetric computational fluid dynamics model of the flow and temperature fields. By doing so, they were able to compute efficiencies of particle transport to the wafer and deposition patterns of particles on the reactor walls. This allowed them to develop reactor design criteria for directing particles to the exhaust, preventing deposition on the wafer and reducing contamination of the reactor walls. In this system, most particles are present in a thin sheath above the heated wafer surface where the thermophoretic velocity, due to the temperature gradient between the inlet and wafer, balances the convective velocity of the gas toward the wafer.

Talukdar and Swihart[228] presented a model of silicon nanoparticle formation under isothermal, plug flow conditions, in which they coupled a detailed chemical kinetic model for nucleation (about 1100 gas phase and 100 surface reactions) with three different approaches to the aerosol dynamics: the method of moments, a quadrature method of moments and a sectional method. The three methods produced nearly identical results for the particle number concentration and average particle size as a function of residence time. However, at short residence times, the sectional approach predicted a bimodal particle size distribution, with some very small particles being generated by ongoing nucleation, and a larger particle mode produced by surface growth and agglomeration. They also compared the computational cost of the different approaches. Even the moment model, which added only three equations to a model with more than 130 equations for species concentrations, more than doubled the solution time. The sectional method, with 100 particle size bins, increased the solution time by about an order of magnitude relative to the method of moments. As shown in this study, the aerosol dynamics portion of a coupled chemically reacting flow and aerosol dynamics simulation may dominate the computational cost, even when the chemically reacting flow portion is quite complex.

Girshick *et al.*[229] and Nijhawan *et al.*[204] coupled a detailed chemical kinetic model of silicon nanoparticle nucleation to a one-dimensional stagnation-point flow model of a parallel plate-CVD reactor, using the method of moments to describe the aerosol dynamics. Comparisons of these models with experimental measurements of particle formation during polysilicon CVD from silane suggested that the nucleation model overpredicted the nucleation rate, but the overall model gave results in good qualitative agreement with experiment.

Kommu *et al.*[216,217] coupled a discrete-sectional treatment of the aerosol dynamics to two- and three-dimensional computational fluid dynamics simulations of particle formation and transport in a commercial reactor for silicon epitaxy from trichlorosilane. This is an example of a model with a great deal of geometric detail, and a detailed discretization of the particle size distribution, coupled with a very simple description of the chemical reaction kinetics, consisting of a single gas-phase precursor decomposition reaction and a single deposition reaction. Likewise, a single, first-order reaction was used to model particle nucleation. This is an appropriate approach when particle transport, as opposed to generation, is of greatest interest, and when the deposition itself is mass-transfer limited.

The above examples have focused mostly on situations where particle formation is to be avoided. However, there is an even larger literature on modeling of processes where particle formation is the goal. One example is the work of Tsantilis *et al.*, who modeled flame synthesis of titania nanoparticles from titanium tetraisopropoxide in a premixed methane–oxygen flame.[230] They used a moving sectional aerosol dynamics model to investigate the impact of three different mechanisms of particle formation and growth on the final particle size distribution. This was coupled to an experimentally measured flame temperature profile. They found that a model that included surface growth (reaction of gas-phase precursors with aerosol particles) gave the best agreement with experiment at short reaction times. At longer times, the predicted aerosol size distribution was governed by coagulation and was nearly independent of the initial particle formation mechanism.

3.6.5 Summary

Both unwanted and intentional particle formation in CVD processes and by CVD-related methods are of significant technological interest. There are well-developed approaches to modeling of particle formation, but these are less mature than many other aspects of CVD reactor modeling. Although particle transport and coagulation are well understood, prediction of particle nucleation and growth by surface reaction remain major challenges for most material systems. Although not discussed in detail here, the prediction of particle sintering is also a significant challenge that is especially important for intentional nanoparticle synthesis, where fully-sintered, spherical particles are often desirable. Good introductions to aerosol dynamics are available in several textbooks,[199–200,220] and a few modeling examples have been briefly described. These examples, and references therein, should provide a convenient introduction to the literature of this field.

References

1. C. R. Kleijn, R. Dorsman, K. J. Kuijlaars, M. Okkerse and H. Van Santen, *J. Cryst. Growth*, 2007, **303**, 362.
2. C. Cavallotti, M. di Stanislao and S. Carra, *Prog. Cryst. Growth. Char. Mat.*, 2004, **48–9**, 123.
3. P. Wong and M. Robinson, *J. Amer. Ceram. Soc.*, 1970, **53**, 617.
4. (a) V. S. Ban, *J. Cryst. Growth*, 1972, **17**, 19; (b) T. M. Besmann and K. E. Spear, *J. Electrochem. Soc.*, 1977, **124**, 786.
5. R. Madar, F. Weiss and R. Fruchart, *J. Cryst. Growth*, 1978, **45**, 37.
6. G. Erikkson, *Chem. Scr.*, 1975, **8**, 100.
7. V. G. Varanasi, T. M. Besmann and T. J. Anderson, *J. Electrochem. Soc.*, 2005, **152**, C7.
8. F. J. Perez, M. P. Hierro, J. A. Trilleros, M. C. Carpintero, L. Sznchez, J. M. Brossard and F. J. Bolivar, *Intermetallics*, 2006, **14**, 811.
9. D. Chaussende, E. Blanquet, F. Baillet, U. Magali and G. Chichignoud, *Chem. Vap. Dep.*, 2006, **12**, 541.
10. S. Liu and J. Q. Mang, *Surf. Rev. Lett.*, 2005, **12**, 727.
11. K. Hack, *The SGTE Casebook Thermodynamics at Work*, The Institute of Materials, Herzogenrath, 1996.
12. O. Kubaschewski, C. B. Alcock and P. J. Spencer, *Materials Thermo-Chemistry*, Pergamon, 1993.
13. E. B. Smith, *Basic Chemical Thermodynamics*, Clarendon Press, Oxford, 1990.
14. J. R. Warn and A. P. Peters, *Concise Chemical Thermodynamics*, Imperial College Press, London, 1996.
15. C. Bale, P. Chartran, S. A. Degterov, G. Eriksson, K. Hack, R. Ben Mahfoud, J. Melancon, A. D. Pelton and S. Petersen, *Calphad*, 2002, **26**, 189.
16. G. S. Fischman and W. T. Petuskey, *J. Am. Ceram. Soc.*, 1985, **68**, 185.
17. C. Vahlas and T. M. Besmann, *J. Am. Ceram. Soc.*, 1992, **75**, 2679.
18. J. Kosikova, J. Leitner, P. J. K. Melichar, K. Jurek, I. Drbohlav and J. Stejskal, *Semicond. Sci. Technol.*, 2001, **16**, 759.
19. A. Koukitu, Y. Kumagai and H. Seki, *Phys. Stat. Sol.*, 2000, **180**, 115.
20. A. D. Pelton, S. A. Degterov, G. Eriksson, C. Robelin and Y. Dessureault, *Metal. Mater. Trans. B*, 2000, **31B**, 651.
21. M. Hillert, *J. Alloys Comp.*, 2001, **320**, 161.
22. T. M. Besmann and K. E. Spear, *J. Am. Ceram. Soc.*, 2002, **85**, 2887.
23. V. G. Varanasi, T.M. Besmann, R.L. Hyde, A. Payzant and T.J. Anderson, *J. Alloys Compd.*, in press. DOI: 10.1016/j.allcom.2008.02.103.
24. Y. Du, Z. P. Jin and H. PY, *J. Amer. Ceram. Soc.*, 1992, **75**, 3040.
25. H. Rebenne and R. Pollard, *J. Electrochem. Soc.*, 1985, **132**, 1932.
26. M. D. Allendorf, *J. Electrochem. Soc.*, 1993, **140**, 747.
27. S. Goujard, L. Vandenbulcke and C. Bernard, *Calphad*, 1994, **18**, 385.
28. J. Locquet, C. Marchiori, M. Sousa and J. Fompeyrine, *J. Appl. Phys.*, 2006, **100**, 14.
29. M. Schumacher, P. K. Baumann and T. Seidel, *Chem. Vap. Dep.*, 2006, **12**, 99.
30. B. Sundman, B. Jansson and J. Anderson, *Calphad*, 1985, **9**, 153.
31. B. Cheynet, P. Y. Chevalier and P. Fischer, *Calphad*, 2002, **26**, 167.

32. A. T. Dinsdale, *Calphad*, 1991, **15**, 317.
33. S. L. Chen, F. Zhang, S. Daniel, F. Y. Xie, X. Y. Yan, Y. A. Chang, R. Schmid-Fetzer and W. A. Oates, *JOM*, 2003, **55**, 48.
34. L. Wu and N. J. Themelis, *CIM Bull.*, 1988, **81**, 97.
35. H. Yokokawa, S. Yamauchi and R. Matsumoto, *Calphad*, 2002, **26**, 155.
36. M. W. Chase, C. A. Davies, J. R. Downey, D. J. Frurip, R. A. McDonald and A. N. Szverud, *J. Phys. Chem. Ref. Data*, 1998, **27**, 1.
37. I. Barin and F. Sauert, *Thermochemical Data of Pure Substances*, VCH, Weinheim, 1989.
38. R. J. Kee, M. E. Coltrin and P. Glarborg, *Chemically Reacting Flow: Theory and Practice*, John Wiley & Sons, Hoboken, 2003.
39. J. O. Hirschfelder, C. F. Curtiss and R. B. Bird, *MolecularTheory of Gases and Liquids*, John Wiley & Sons, New York, 1954.
40. R. B. Bird, W. E. Stewart and E. N. Lightfoot, *Transport Phenomena*, John Wiley & Sons, 2002.
41. G. Dixon-Lewis, *Proc. Royal Soc*, 1968, **A304**, 111–135.
42. R. A. Svehla, NASA Report R-132, 1962.
43. D. G. Goodwin, www.cantera.org, (7, 2008).
44. M. E. Coltrin, R. J. Kee and J. A. Miller, *J. Electrochem. Soc.*, 1984, **131**, 425.
45. M. E. Coltrin, R. J. Kee and J. A. Miller, *J. Electrochem. Soc.*, 1986, **133**, 1206–1213.
46. M. E. Coltrin, R. J. Kee and G. H. Evans, *J. Electrochem. Soc.*, 1989, **136**, 819.
47. C. Houtman, D. B. Graves and K. F. Jensen, *J. Electrochem. Soc.*, 1986, **133**, 961–970.
48. C. R. Kleijn, *Thin Sol. Films*, 1991, **206**, 47.
49. L. Raja, R. Kee, R. Serban and L. Petzold, *J. Electrochem. Soc.*, 2000, **147**, 2718.
50. H. Van Santen, C. R. Kleijn and H. E. A. Van den Akker, *J. Cryst. Growth*, 2000, **212**, 311.
51. W. G. Breiland and G. Evans, *J. Electrochem. Soc*, 1991, **138**, 1807–1816.
52. S. Joh and G. H. Evans, *Numer. Heat Transfer*, 1997, **31**, 867–879.
53. H. V. Santen, C. R. Kleijn and H. E. A. V. D. Akker, *J. Cryst. Growth*, 2000, **212**, 311–323.
54. S. D. Elliott, *Top. Appl. Phys.*, 2007, **106**, 73.
55. W. S. Rees Jr., ed., *CVD of Nonmetals*, VCH, Weinheim, 1996.
56. D. J. Larkin, *MRS Bull.*, 1997, **22**, 36.
57. J. Wei and J. T. Yates, *Crit. Rev. Surf. Chem.*, 1995, **5**, 1.
58. H. M. Liu and D. S. Dandy, *Diam. Relat. Mat.*, 1995, **4**, 1173.
59. S. Lee, Z. Lin and X. Jiang, *Mat. Sci. Eng. R*, 1999, **25**, 123.
60. M. Asmann, J. Heberlein and E. Pfender, *Diam. Relat. Mat.*, 1999, **8**, 1.
61. S. Ferro, *J. Mat. Chem.*, 2002, **12**, 2843.
62. M. L. Hitchman and K. F. Jensen (eds.), *Chem Vap Dep Academic*, London, 1993.
63. C. J. Cramer, *Essentials of Computational Chemistry*, Wiley, Chichester, 2003.
64. W. J. Hehre, L. Radom, P. v. R. Schleyer and J. A. Pople, *Ab initio Molecular Orbital Theory*, Wiley, New York, 1986.
65. J. Cioslowski (ed.), *Quantum-Mechanical Prediction of Thermochemical Data*, Kluwer, Dordrecht, 2001.
66. L. A. Curtiss and K. Raghavachari, *Theor. Chem. Acc.*, 2002, **108**, 61.
67. L. A. Curtiss, P. C. Redfern and D. J. Frurip, in *Rev. Comp. Chem.*, K. B. Lipkowitz and D. B. Boyd eds., Wiley, New York, 2000, p. 147.
68. C. F. Melius, M. D. Allendorf and M. E. Colvin, in *Fourteenth Int. Conf. Chem. Vap. Dep.*, M. D. Allendorf and C. Bernard eds., The Electrochemical Society Proceedings Series, Paris, 1997, p. 1.
69. Environmental Molecular Sciences Laboratory Basis Set Exchange, https://bse.pnl.gov/bse/porta/ (7, 2008).

70. K. L. Schuchardt, B. T. Didier, T. Elsethagen, L. Sun, V. Gurumoorthi, J. Chase, J. Li and T. L. Windus, *J. Chem. Inf. Model.*, 2007, **47**, 1045.
71. A. P. Scott and L. Radom, *J. Phys. Chem.*, 1996, **100**, 16502.
72. P. Ho, M. E. Coltrin, J. S. Binkley and C. F. Melius, *J. Phys. Chem.*, 1985, **89**, 4647.
73. P. Ho and C. F. Melius, *J. Phys. Chem.*, 1990, **94**, 5120–5127.
74. C. F. Melius, in *Chemistry and Physics of Energetic Materials*, S. N. Bulusu ed., Kluwer Academic Publishers, Dorderecht, 1990, p. 21.
75. C. F. Melius and M. D. Allendorf, *J. Phys. Chem.*, 2000, **104**, 2168.
76. M. D. Allendorf and C. F. Melius, *J. Phys. Chem. A*, 1997, **101**, 2670.
77. M. D. Allendorf and C. F. Melius, *J. Phys. Chem.*, 1992, **96**, 428.
78. M. D. Allendorf and C. F. Melius, *J. Phys. Chem.*, 1993, **97**, 720.
79. M. D. Allendorf and C. F. Melius, *J. Phys. Chem. A*, 2002, **106**, 6370.
80. M. D. Allendorf, C. F. Melius, P. Ho and M. R. Zachariah, *J. Phys. Chem.*, 1995, **99**, 15285.
81. P. Ho, M. E. Coltrin, J. S. Binkley and C. F. Melius, *J. Phys. Chem.*, 1986, **90**, 3399.
82. P. Ho and C. F. Melius, *J. Phys. Chem.*, 1995, **99**, 2166.
83. C. F. Melius and P. Ho, *J. Phys. Chem.*, 1991, **95**, 1410–1419.
84. P. C. Glaude, H. J. Pitz, W. J. Westbrook and C. K. Gann, *R. G. Proc. Comb. Inst.*, 2000, **28**, 1749.
85. A. J. Skulan, I. M. B. Nielsen, C. F. Melius and M. D. Allendorf, *J. Phys. Chem. A*, 2006, **110**, 281.
86. M. D. Allendorf and C. F. Melius, *J. Phys. Chem. A*, 2005, **109**, 4939.
87. A. J. Skulan, I. M. B. Nielsen, C. F. Melius and M. D. Allendorf, *J. Phys. Chem. A*, 2006, **110**, 5919.
88. R. Berry, D. Burgess, M. Nyden, M. Zachariah, C. Melius and M. Schwartz, *J. Phys. Chem.*, 1996, **100**, 7405.
89. R. Berry, C. Ehlers, D. Burgess, M. Zachariah, M. Nyden and M. Schwartz, *Theochem J. Mol. Str.*, 1998, **422**, 89.
90. M. Zachariah, P. Westmoreland, D. Burgess and W. Tsang, *J. Phys. Chem.*, 1996, **100**, 8737.
91. M. D. Allendorf, C. F. Melius and C. Bauschlicher Jr., *J. Phys. Fr. IV*, 1999, **9**, 23.
92. M. D. Allendorf, C. F. Melius, B. Cosic and A. Fontijn, *J. Phys. Chem. A*, 2002, **106**, 2629.
93. J. A. Pople, M. Head-Gordon, D. J. Fox, K. Raghavachari and L. A. Curtiss, *J. Chem. Phys.*, 1989, **90**, 5622.
94. L. A. Curtiss and K. Raghavachari, in *Quantum Mechanical Electronic Structure Calculations with Chemical Accuracy*, S. R. Langhoff (ed.), Kluwer Academic, Netherlands, 1995, p. 139.
95. L. A. Curtiss, P. C. Redfern, K. Raghavachari and J. A. Pople, *J. Chem. Phys.*, 1997, **106**, 1063.
96. L. A. Curtiss, K. Raghavachari, P. C. Redfern, V. Rassolov and J. A. Pople, *J. Chem. Phys.*, 1998, **109**, 7764.
97. L. A. Curtiss, P. C. Redfern and K. Raghavachari, *J. Chem. Phys.*, 2005, **123**, 124107.
98. L. A. Curtiss, P. C. Redfern and K. Raghavachari, *J. Chem. Phys.*, 2007, **126**, 084108.
99. J. W. Ochterski, G. A. Petersson and J. A. Montgomery Jr., *J. Chem. Phys.*, 1995, **104**, 2598.
100. G. A. Petersson, in *Quantum-Mechanical Prediction of Thermochemical Data*, J. Cioslowski ed., Kluwer Academic, Dordrecht, 2001, p. 99.
101. G. A. Petersson, D. K. Malick, W. G. Wilson, J. W. Ochterski, J. A. Montgomery and M. J. Frisch, *J. Chem. Phys.*, 1998, **109**, 10570.
102. K. Raghavachari, B. B. Stefanov and L. A. Curtiss, *Mol. Phys.*, 1997, **91**, 555.
103. G. Zhang and C. B. Musgrave, *J. Phys. Chem. A*, 2007, **111**, 1554.
104. W. Koch and M. C. Holthausen, *A Chemist's Guide to Density Functional Theory*, Wiley-VCH, Weinheim, 2001.

105. H. W. Wong, J. C. A. Nieto, M. T. Swihart and L. J. Broadbelt, *J. Phys. Chem. A*, 2004, **108**, 874.
106. T. B. Casserly and K. K. Gleason, *Plasma Processes Polym.*, 2005, **2**, 669.
107. L. M. Wang and J. S. Zhang, *J. Phys Chem A*, 2004, **108**, 10346–10353.
108. I. Hyla-Kryspin and S. Grimme, *Organometal.*, 2004, **23**, 5581.
109. R. H. West, G. J. O. Beran, W. H. Green and M. Kraft, *J. Phys. Chem. A*, 2007, **111**, 3560.
110. J. R. B. Gomes and M. A. V. Ribeiro da Silva, *J. Phys. Chem. A*, 2006, **110**, 13948.
111. B. H. Cardelino, C. E. Moore, C. A. Cardelino, D. O. Frazier and K. J. Bachmann, *J Phys. Chem. A*, 2001, **105**, 849–868.
112. L. Catoire and M. T. Swihart, *J. Prop. Power*, 2002, **18**, 1242.
113. D. Moscatelli, P. Caccioppoli and C. Cavallotti, *Appl. Phys. Let.*, 2005, **86**, 091106.
114. D. Moscatelli and C. Cavallotti, *J. Phys. Chem. A*, 2007, **111**, 4620.
115. C. Cavallotti, I. Lengyel, M. Nemirovskaya and K. F. Jensen, *J. Cryst. Growth*, 2004, **268**, 76–95.
116. C. Cavallotti, D. Moscatelli and S. Carra, *J. Phys. Chem. A*, 2004, **108**, 1214–1223.
117. C. Cavallotti, D. Moscatelli, M. Masi and S. Carra, *J. Cryst. Growth*, 2004, **266**, 363–370.
118. R. M. Dickson and A. D. Becke, *J. Chem. Phys.*, 2005, **123**, 111101.
119. R. J. Kee, M. E. Coltrin and P. Glarborg, *Chemically Reacting Flow: Theory and Practice*, John Wiley & Sons, Hoboken, 2003.
120. L. V. Gurvich, I. V. Veyts and C. B. Alcock, *Thermodynamic Properties of Individual Substances*, CRC Press, Boca Raton, 1994.
121. J. Warnatz, U. Maas and R. W. Dibble, *Combustion Physical and Chemical Fundamentals, Modeling and Simulation, Experiments, Pollutant Formation*, Springer, Heidelberg, 2006.
122. M. D. Allendorf and C. F. Melius, *Surf. Coat. Tech.*, 1998, **108–109**, 191.
123. A. Dollet, S. de Persis and F. Teyssandier, *Phys. Chem. Chem. Phys.*, 2004, **6**, 1203.
124. H. K. Moffat, K. F. Jensen and R. W. Carr, *J. Phys. Chem.*, 1991, **95**, 145.
125. F. G. Celii and J. E. Butler, *An. Rev. Phys. Chem.*, 1991, **42**, 643.
126. S. J. Harris, A. M. Weiner and T. A. Perry, *J. Appl. Phys.*, 1991, **70**, 1385.
127. M. D. Allendorf and R. J. Kee, *J. Electrochem. Soc.*, 1991, **138**, 841.
128. J. Meziere, M. Ucar, E. Blanquet, M. Pons, P. Ferret and L. Di Cioccio, *J. Cryst. Growth*, 2004, **267**, 436.
129. G. D. Papasouliotis and S. V. Sotirchos, *J. Electrochem. Soc.*, 1994, **141**, 1599.
130. J. R. Creighton and G. T. Wang, *J. Phys. Chem. A*, 2005, **109**, 10554.
131. A. G. Zawadzki, C. J. Giuna and R. G. Gordon, *J. Phys. Chem.*, 1992, **96**, 5364.
132. K. A. Holbrook, M. J. Pilling and S. H. Robertson, *Unimolecular Reactions*, Wiley, Chichester, 1996.
133. R. G. Gilbert and S. C. Smith, *Theory of Unimolecular and Recombination Reactions*, Blackwell, Oxford, 1990.
134. J. I. Steinfeld, J. S. Francisco and W. L. Hase, *Chemical Kinetics and Dynamics*, Prentice Hall, Englewood Cliffs, 1989.
135. H. B. Schlegel, *J. Comp. Chem.*, 2003, **24**, 1514.
136. S. W. Benson, *Thermochemical Kinetics*, John Wiley & Sons, New York, 1976.
137. V. Mokrushin and W. Tsang, ChemRate, http://www.nist.gov/kinetics/chemrate/chemrate.html (7, 2008).
138. K. P. Novoselov, D. B. Shirabaikin, S. Y. Umanskii, A. S. Vladimirov, A. K. Minushev and A. A. Korkin, *J. Comp. Chem.*, 2002, **23**, 1375.
139. K. S. Pitzer and W. D. Gwinn, *J. Chem. Phys.*, 1942, **10**, 428.
140. R. Schmid and D. Basting, *J. Phys. Chem. A*, 2005, **109**, 2623.
141. J. A. Miller and S. J. Klippenstein, *J. Phys. Chem. A*, 2006, **110**, 10528.
142. M. J. Pilling and S. H. Robertson, *Annu. Rev. Phys. Chem.*, 2003, **54**, 245.

143. H. K. Moffat, K. F. Jensen and R. W. Carr, *J. Phys. Chem.*, 1992, **96**, 7683.
144. T. H. Osterheld, M. D. Allendorf and C. F. Melius, *J. Phys. Chem.*, 1994, **98**, 6995.
145. F. Teyssandier and M. D. Allendorf, *J. Electrochem. Soc.*, 1998, **145**, 2167.
146. J. R. Barker, *Int. J. Chem. Kinet.*, 2001, **33**, 232.
147. I. Oref and D. C. Tardy, *Chem. Rev.*, 1990, **90**, 1407.
148. R. G. Gilbert, K. Luther and J. Troe, *Ber. Bunsenges. Phys. Chem.*, 1983, **87**, 169.
149. P. K. Venkatesh, *J. Phys. Chem. A*, 2000, **104**, 280.
150. I. Lengyel and K. F. Jensen, *Thin Sol. Films*, 2000, **365**, 231.
151. M. D. Allendorf, *J. Electrochem. Soc.*, 1993, **140**, 747.
152. C. Raffy, E. Blanquet, M. Pons, C. Bernard, C. F. Melius and M. D. Allendorf, *J. Phys. IV*, 1999, **9**, 205.
153. M. D. Allendorf and A. M. B. van Mol, in *Precursor Chemistry of Advanced Materials*, R. A. Fischer ed., Springer, Berlin, 2005, p. 1.
154. S. de Persis, F. Teyssandier, A. H. McDaniel and M. D. Allendorf, *Chem Vap Dep*, 2002, **8**, 63.
155. Y. F. Wang and R. Pollard, *J. Electrochem. Soc.*, 1995, **142**, 1712.
156. J. G. Ekerdt, Y. M. Sun, A. Szabo, G. J. Szulczewski and J. M. White, *Chem. Rev.*, 1996, **96**, 1499.
157. S. M. Gates, *Chem. Rev.*, 1996, **96**, 1519.
158. H. N. Waltenburg and J. T. Yates, *Chem. Rev.*, 1995, **95**, 1589.
159. M. E. Coltrin, R. J. Kee and F. M. Rupley, *Int. J. Chem. Kinet.*, 1991, **23**, 1111.
160. M. Frenklach and H. Wang, *Phys. Rev. B*, 1991, **43**, 1520.
161. R. S. Larson and M. D. Allendorf, in *Thirteenth Int. Conf. Chem. Vapor Dep.*, T. M. Besmann, M. D. Allendorf, M. Robinson and R. K. Ulrich eds., The Electrochemical Society Proceedings Series: Pennington, Los Angeles, 1996, Vol. **96-5**, p. 41.
162. M. J. Buiting, A. F. Otterloo and A. H. Montree, *J. Electrochem. Soc.*, 1991, **138**, 500.
163. D. Srinivas, J. T. Hillman, W. M. Triggs and E. C. Eichman, in *Advanced Metallization for ULSI Applications*, V. V. S. Rana, R. V. Joshi and I. Ohdomari eds., Materials Research Society, 1992, Vol. **V-07**, p. 319.
164. J. S. Custer and P. M. Smith, personal communication, 1995.
165. K. Raghavachari and M. D. Halls, *Mol. Phys.*, 2004, **102**, 381.
166. A. A. Bagatur'yants, K. P. Novoselov, A. A. Safonov, J. V. Cole, M. Stoker and A. A. Korkin, *Surf. Sci.*, 2001, **486**, 213.
167. W. Chen, D. W. Zhang, J. Ren, H.-L. Lu, J.-Y. Zhang, M. Xu, J.-T. Wang and L.-K. Wang, *Thin Sol. Films*, 2005, **479**, 73.
168. T. Nakajima, T. Tanaka and K. Yamashita, *Surf. Sci.*, 2000, **444**, 99.
169. Z. Hu and C. H. Turner, *J. Am. Chem. Soc.*, 2007, **129**, 3863.
170. C. Cavallotti, I. Lengyel, M. Nemirovskaya and K. F. Jensen, *J. Cryst. Growth*, 2004, **268**, 76.
171. C. B. Musgrave, S. J. Harris and W. A. Goddard, *Chem. Phys. Lett.*, 1995, **247**, 359.
172. H. Simka, B. G. Willis, I. Lengyel and K. F. Jensen, *Prog. Cryst. Growth. Char. Mat.*, 1997, **35**, 117.
173. G. Kresse and J. Furthmuller, *Comput. Mat. Sci.*, 1996, **6**, 15.
174. G. Kresse and J. Furthmuller, *Phys. Rev. B*, 1996, **54**, 11169.
175. X. Gonze, J. M. Beuken, R. Caracas, F. Detraux, M. Fuchs, G. M. Rignanese, L. Sindic, M. Verstraete, G. Zerah, F. Jollet, M. Torrent, A. Roy, M. Mikami, P. Ghosez, J. Y. Raty and D. C. Allan, *Comput. Mat. Sci.*, 2002, **25**, 478.
176. V. Milman, B. Winkler, J. A. White, C. J. Pickard and M. C. Payne, *Int. J. Quant. Chem.*, 2000, **77**, 895.
177. Y. Widjaja and C. B. Musgrave, *Surface Science*, 2000, **469**, 9–20.
178. M. Sugiyama, H. Itoh, J. Aoyama, H. Komiyama and Y. Shimogaki, *Jap. J. Appl. Phys. 1*, 2000, **39**, 1074.

179. M. Sugiyama, T. Nakajima, T. Tanaka, H. Itoh, J. Aoyama, Y. Egashira, K. Yamashita, H. Komiyama and Y. Shimogaki, *Jap. J. Appl. Phys. 1*, 2000, **39**, 6501.
180. M. E. Coltrin, H. K. Moffat, R. J. Kee and F. M. Rupley, Sandia National Laboratories Report, SAND93-0478 1993.
181. H. Komiyama, Y. Shimogaki and Y. Egashira, *Chem. Eng. Sci.*, 1999, **54**, 1941.
182. T. Nakajima and K. Yamashita, *Bull. Chem. Soc. Japan*, 2001, **74**, 2279–2283.
183. K. Morokuma and T. Kerdcharoen, *Chem. Phys. Lett.*, 2002, **355**, 257.
184. M. Svensson, S. Humbel, R. D. J. Froese, T. Matsubara, S. Sieber and K. Morokuma, *J. Phys. Chem.*, 1996, **100**, 19357.
185. C. C. Battaile and D. J. Srolovitz, *An. Rev. Mat. Res.*, 2002, **32**, 297.
186. A. F. Voter, F. Montalenti and T. C. Germann, *An. Rev. Mat. Res.*, 2002, **32**, 321.
187. The International Technology Roadmap for Semiconductors, www.itrs.net, International Sematech, 2005.
188. V. Paserin, S. Baska, A. Zaitsev, J. Shu, F. Shojai and W. Nowosiadly, *J. Nanosci. Nanotech.*, 2008, **8**, 4049.
189. J. U. Brehm, M. Winterer and H. Hahn, *J. App. Phys.*, 2006, **100**, 064311.
190. H. Sieger, M. Winterer, H. Muehlenweg, G. Michael and H. Hahn, *Chem. Vap. Dep.*, 2004, **10**, 71.
191. S. Seifried, M. Winterer and H. Hahn, *Chem. Vap. Dep.*, 2000, **6**, 239.
192. S. Klein, M. Winterer and H. Hahn, *Chem. Vap. Dep.*, 1998, **4**, 143.
193. H. Hahn, *Nanostruct. Mat.*, 1997, **9**, 3.
194. W. R. Cannon, S. C. Danforth, J. H. Flint, J. S. Haggerty and R. A. Marra, *J. Am. Ceram. Soc.*, 1982, **65**, 324.
195. W. R. Cannon, S. C. Danforth, J. S. Haggerty and R. A. Marra, *J. Am. Ceram. Technol.*, 1982, **65**, 330.
196. Y. He, X. Li and M. T. Swihart, *Chem. Mater.*, 2005, **17**, 1017.
197. X. Li, Y. He, S. S. Talukdar and M. T. Swihart, *Langmuir*, 2003, **19**, 8490.
198. H. Zhang and M. T. Swihart, *Chem. Mater.*, 2007, **19**, 1290.
199. W. C. Hinds, *Aerosol Technology: Properties, Behavior, and Measurement of Airborne Particles*, John Wiley & Sons, New York, 1999.
200. J. H. Seinfeld and S. N. Pandis, *Atmospheric Chemistry and Physics: From Air Pollution to Climate Change*, Wiley, New York, 1998.
201. F. E. Kruis, K. A. Kusters, S. E. Pratsinis and B. Scarlett, *Aerosol Sci. Technol.*, 1993, **19**, 514.
202. K. W. Lee and H. Chen, *Aerosol. Sci. Soc*, 1984, **3**, 327.
203. S. E. Pratsinis and K. S. Kim, *J. Aerosol. Sci.*, 1989, **20**, 101.
204. S. Nijhawan, P. H. McMurry, M. T. Swihart, S.-M. Suh, S. L. Girshick, S. A. Campbell and J. E. Brockmann, *J. Aerosol Sci.*, 2003, **34**, 691.
205. S. E. Pratsinis, *J. Colloid Interface Sci.*, 1988, **124**, 416.
206. R. McGraw, *Aerosol Sci. Technol.*, 1997, **27**, 255.
207. R. McGraw, S. Nemesure and S. E. Schwartz, *J. Aerosol Sci.*, 1998, **29**, 761.
208. R. B. Diemer and J. H. Olson, *Chem. Eng. Sci.*, 2002, **57**, 2211.
209. F. Gelbard and J. H. Seinfeld, *J. Colloid Interface Sci.*, 1980, **76**, 541.
210. M. J. Hounslow, R. L. Ryall and V. R. Marshall, *AIChE J.*, 1988, **34**, 1821.
211. M. Kostoglou and A. J. Karabelas, *J. Colloid Interface Sci.*, 1994, **163**, 420.
212. S. Kumar and D. Ramkrishna, *Chem. Eng. Sci.*, 1996, **51**, 1311.
213. J. D. Litster, D. J. Smit and M. J. Hounslow, *AIChE J.*, 1995, **41**, 591.
214. S. Kumar and D. Ramkrishna, *Chem. Eng. Sci.*, 1996, **51**, 1333.
215. Y. P. Kim and J. H. Seinfeld, *J. Colloid Interface Sci.*, 1990, **135**, 185.
216. S. Kommu, B. Khomami and P. Biswas, *Chem. Eng. Sci.*, 2004, **59**, 345.

217. S. Kommu, B. Khomami and P. Biswas, *Chem. Eng. Sci.*, 2004, **59**, 359.
218. J. D. Landgrebe and S. E. Pratsinis, *J. Colloid Interface Sci.*, 1990, **139**, 63.
219. J. J. Wu and R. C. Flagan, *J. Colloid Interface Sci.*, 1988, **123**, 339.
220. S. K. Friedlander, *Smoke, Dust and Haze: Fundamentals of Aerosol Dynamics (2nd Ed)*, Oxford University Press, 2000.
221. A. Yuuki, Y. Matsui and K. Tachibana, *Jap. J. Appl. Phys.*, 1987, **26**, 747.
222. U. V. Bhandarkar, M. T. Swihart, S. L. Girshick and U. R. Kortshagen, *J. Phys. D: Appl. Phys.*, 2000, **33**, 2731.
223. M. T. Swihart and S. L. Girshick, *J. Phys. Chem. B.*, 1999, **103**, 64.
224. M. Frenklach, L. Ting, H. Wang and M. J. Rabinowitz, *Israel J. Chem.*, 1996, **36**, 293.
225. H.-W. Wong, X. Li, M. T. Swihart and L. J. Broadbelt, *J. Phys. Chem. A.*, 2004, **108**, 10122.
226. C. J. Guinta, R. J. McCurdy, J. D. Chapple-Sokol and R. G. Gordon, *J. Appl. Phys.*, 1990, **67**, 1062.
227. S. Nijhawan, P. H. McMurry and S. A. Campbell, *J. Vac. Sci. Technol. A.*, 2000, **18**, 2198–2206.
228. S. S. Talukdar and M. T. Swihart, *J. Aerosol Sci.*, 2004, **35**, 889–908.
229. S. L. Girshick, M. T. Swihart, S.-M. Suh, M. R. Mahajan and S. Nijhawan, *J. Electrochem. Soc.*, 2000, **147**, 2303–2311.
230. S. Tsantilis, H. K. Kammler and S. E. Pratsinis, *Chem. Eng. Sci.*, 2002, **57**, 2139–2156.

CHAPTER 4

Atomic Layer Deposition

MIKKO RITALA AND JAAKKO NIINISTÖ

Laboratory of Inorganic Chemistry, University of Helsinki, P.O. Box 55, Helsinki, FI-00014 Finland

4.1 Introduction

Atomic layer deposition (ALD) is a special variant of chemical vapour deposition (CVD) technology used to grow high-quality thin films through alternate self-limiting surface reactions (Figure 4.1). ALD has gained considerable interest in recent years as a thin film deposition method to overcome many technological problems facing, especially, the semiconductor industry both now and in the near future. In integrated circuit manufacturing the simultaneously decreasing feature sizes, increasing aspect ratios, complicated designs and adaptation of new materials call for new methods that enable highly controlled deposition of very thin and conformal films, requirements perfectly matched with ALD.

ALD technology was invented and developed to production use for flat panel displays as early as in the 1970s,[1] but for wider application it has suffered from a relatively low throughput. This limitation is no longer a severe one for many applications, as the required film thickness is often below 10 nm. Indeed, Intel announced in 2007 that ALD is involved in "the biggest change to computer chips in 40 years," *i.e.* replacement of SiO_2 based gate dielectrics by HfO_2 based high-permittivity (high-k) dielectrics and polysilicon gates with metal gates in complementary metal-oxide silicon devices.[2] Besides applications in microelectronics, ALD offers a wide range of possibilities in many other areas, such as optics and optoelectronics, micro-electromechanical systems, catalysis, magnetic recording head technology, and protective and antireflective coatings as well as in the field of nanotechnology.

The present chapter describes the basic principles, including the benefits and limitations of ALD. Successful application of ALD relies on two factors: precursor chemistry and reactor design, and these topics will form the core of this chapter. Finally, current and possible application areas of ALD are briefly reviewed.

4.2 Basic Features of ALD

ALD technology was developed and patented some 30 years ago by Suntola and co-workers in Finland.[3] The purpose was to develop thin film electroluminescent (TFEL) displays where ALD,

Chemical Vapour Deposition: Precursors, Processes and Applications
Edited by Anthony C. Jones and Michael L. Hitchman

Published by the Royal Society of Chemistry, www.rsc.org

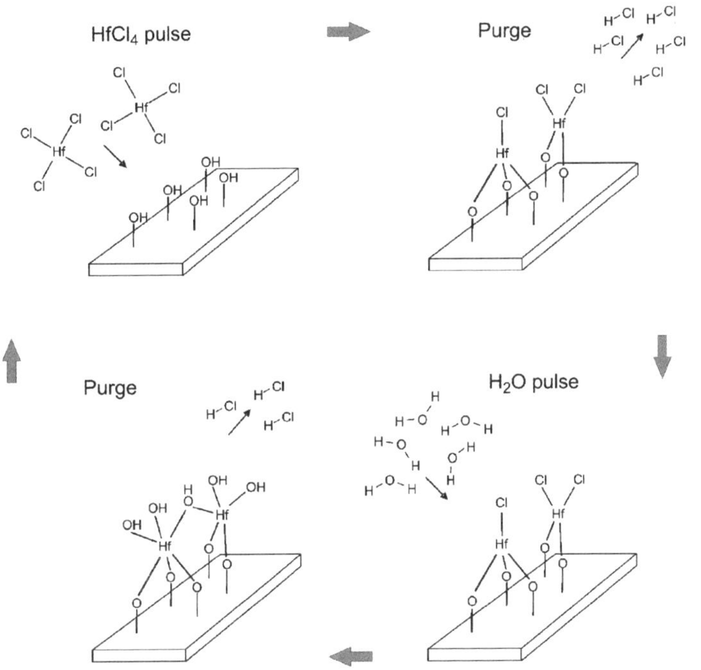

Figure 4.1 Schematic illustration of an ALD cycle of HfO_2 process where precursors, $HfCl_4$ and H_2O, are alternately pulsed and separated by inert gas purging.

then known as atomic layer epitaxy (ALE) was needed for depositing the demanding electroluminescent and dielectric layers that are subjected to high electric fields (Section 4.5).[1] TFEL display production was the first industrial application of ALD and, indeed, successful production still continues.[4] As mentioned, current interest in ALD stems mainly from the semiconductor industry, where the continuous shrinking of device size and the need to use complex geometries has demanded alternative deposition methods to grow, for example, high permittivity gate and capacitor dielectrics. The strength of the ALD technology lies in its capability to produce high-quality, dense and pinhole-free films on large surface areas with excellent uniformity and conformality as well as with thickness and composition control at an atomic level.

4.2.1 Principle and Characteristic Features of ALD

ALD processes and their applications have been frequently reviewed,[5–9] most extensively and recently by Puurunen[10] as well as by Ritala and Leskelä.[11] ALD is a variant of the CVD method but, unlike CVD, ALD relies on sequential and saturating surface reactions of the alternately applied precursors. The precursor pulses are separated by inert gas purging or evacuation of the reaction chamber to avoid gas-phase reactions between the precursors. Growth proceeds in a cyclic manner, enabling easy thickness control. The basic principle of ALD is shown in a simplified manner in Figure 4.1, where one ALD cycle of HfO_2 deposition process is presented. At first the exposure of the substrate surface to the gaseous $HfCl_4$ and its chemisorption on the available surface sites (–OH groups) leaves the surface saturated with the metal precursor. In the example shown the chemisorption actually involves an exchange reaction with the surface groups, releasing HCl. After inert gas purging of the excess precursor and HCl by-product, the surface is exposed to

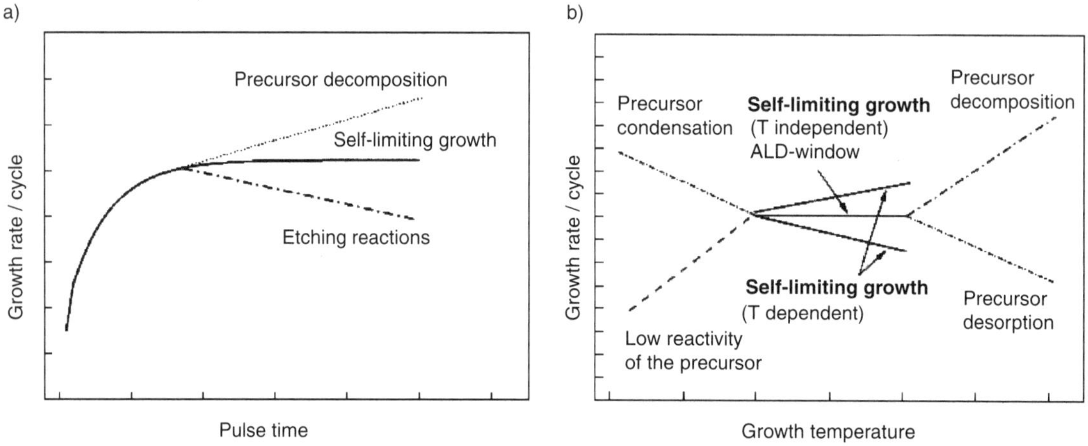

Figure 4.2 ALD and non-ALD type of behaviour of growth rate *vs.* (*a*) precursor pulse time and (*b*) temperature curves. Possible factors limiting the self-limiting growth are indicated in both parts.

the non-metal precursor, H_2O. The surface reaction produces the desired HfO_2, releases the rest of the Cl ligands and recreates the –OH groups. After inert gas purging the surface is ready for the next ALD cycle.

The alternate exposure of the precursors alone does not qualify the process as ALD. As important is that all the reactions are saturative because only such saturation ensures that the *film growth is self-limiting*. Under self-limiting conditions the same amount of film is deposited on all the surfaces irrespective of the dose received, provided the dose is high enough to maintain saturation.

A self-limiting ALD-type process is verified by observing the saturation of the growth rate with pulse time (Figure 4.2a, solid line). In other words, the pulse time must be long enough to reach the saturation, and precursor decomposition must be avoided as this would lead to a CVD type growth. Ideally, ALD growth proceeds by one atomic layer per cycle, but in practise, due to steric hindrances and sometimes possibly also due to a limited number of reactive surface sites, the growth rate per cycle usually is only a small fraction of a monolayer (ML) thickness, typically less than 0.5 ML. As the growth proceeds in a cyclic manner, and the purging periods take some time, the ALD technique is rather slow, but for applications such as high-k dielectric deposition where very thin films are needed this is not a critical issue. The limitations of ALD are discussed in more detail in Section 4.2.2.

Often, but not always, a temperature region with a constant deposition rate, also known as an ALD window, is observed (Figure 4.2b).[9,11,12] The ALD window is not a requirement for an ALD-type growth mode, but it is a desirable feature that improves the reproducibility of the film growth. Especially if a ternary material or nanolaminate structure is to be deposited, overlapping ALD windows of the constituent binary processes offer a good starting point. The commonly observed growth rate *vs.* temperature curves in ALD processes are shown and explained in Figure 4.2(b).

Owing to the self-limiting growth mode, ALD provides unique film deposition capability. The key for a successful ALD process relies on the matching precursor chemistry: fast and complete surface reactions give advantages no other deposition technique can fulfil. Table 4.1 summarizes the beneficial features ALD offers. As mentioned, the film thickness can be controlled precisely simply by changing the number of deposition cycles without controlling the dose of the precursor. As long as the dose is large enough to saturate the surface, perfectly uniform and conformal films can be deposited onto large area substrates and complex structures with high aspect ratios (Figure 4.3). The composition can be controlled at an atomic level, and, in many cases, pure films with low impurity

Table 4.1 Characteristic features of ALD and the resulting benefits.

Characteristic feature of an ALD process	*Practical advantage*
Self-limiting growth process	Precise film thickness control by the number of deposition cycles
	No need to control reactant flux homogeneity
	Excellent uniformity and conformality
	Large-area and batch capability
	Dense, uniform, homogeneous and pinhole-free films
	Atomic level composition control
	Good reproducibility and straightforward scale-up
Surface exchange reactions by separate dosing of reactants	Low thermal budget for high-quality film material
	No gas-phase reactions, allowing the use of highly reactive precursors
Often wide temperature windows	Multilayer structures can be deposited, doping straightforward

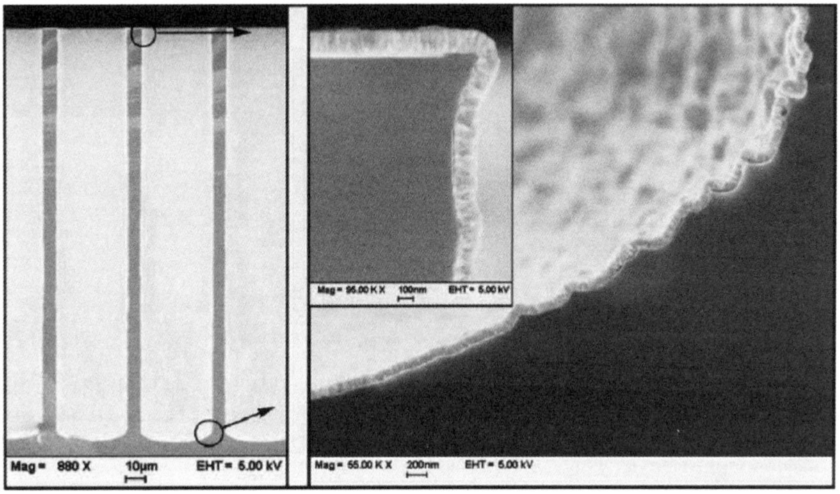

Figure 4.3 A Ru film deposited by ALD from $RuCp_2$ and O_2, demonstrating excellent conformality in terms of both trench structure and microroughness. (Copyright T. Aaltonen, University of Helsinki.)

levels can be obtained over a wide deposition temperature range. The possibility of tailoring the composition offers straightforward doping and deposition of structures with several layers (Figure 4.4). Reproducibility and straightforward scale-up are industrially important consequences of the forgiving nature of the self-limiting growth mechanism.

Table 4.2 lists the materials that have reportedly been deposited by ALD. For references, see more detailed reviews.[10,11] One can conclude that the material selection is broad – yet many important materials are still to be added.

4.2.2 Limitations of ALD

After learning all the benefits of ALD, the question arises as to why the use of ALD is still so limited. This section discusses the limitations of ALD, starting from the most important, its slowness. A critical look is also made at the numerous ways how real ALD processes often deviate from the ideal one, thereby causing process and material specific limitations.

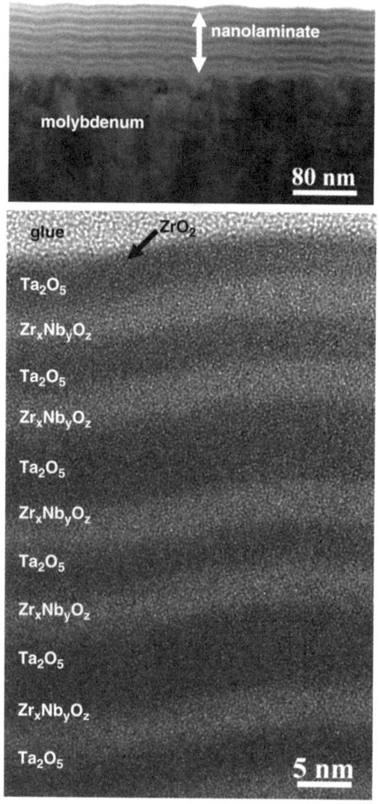

Figure 4.4 Representative transmission electron microscope images of Ta_2O_5-$Nb_xZr_yO_z$ nanolaminates where the Ta_2O_5 layer thickness has been varied on purpose. The roughness arises from the conformal growth on the rough molybdenum underlayer. (Reprinted from ref. 191 with permission from Elsevier.)

4.2.2.1 *Low Effective Deposition Rate*

The low effective deposition rate (thickness-increment per time-unit) is the major limitation of ALD that has retarded its wider use over the years. Deposition rates are process and reactor dependent but 100 nm h^{-1} is quite a common value for good ALD processes. A record rate of 2 μm h^{-1} has been achieved in GaAs ALE, where each cycle quite exceptionally deposits a full monolayer and an uncommon rotating substrate holder reactor (Section 4.4) was used.[13] Though modern devices require increasingly thinner films, low throughput is still widely considered as the main problem of ALD. Large batch processing can be used to improve the throughput and is used in TFEL display production, for example.[11] ALD tools for semiconductor applications are mainly single wafer tools, however, though a batch reactor for up to 150 wafers is available too.[14,15]

The effective deposition rate is directly related to the growth rate per cycle and inversely related to the cycle time. The growth rate per cycle may be maximised by optimizing precursor combinations and deposition temperatures without sacrificing the film quality requirements. The cycle time, in turn, depends largely on the reactor design (Section 4.4). However, if the reactions do not proceed rapidly, the chemistry may become the limiting factor for the cycle time.

The effective deposition rate has been increased by sacrificing to some extent the true ALD conditions, either by using such short cycle times that the reactions are not fully saturating or by accepting some precursor decomposition, leading to higher growth rate per cycle (see below).

Table 4.2 List of materials deposited by ALD. The table attempts to cover all materials whose ALD has been reported. Notably, the quality and applicability of the processes may vary widely. Of the films with three or more elements, mostly those are included that exhibit a crystalline phase with all these elements either in an as-deposited state or after post-deposition annealing. This leaves out several mixtures and nanolaminates studied. Doped transparent conductors and phosphors are included, however.

Materials	
Oxides	
Dielectric	Al_2O_3, TiO_2, ZrO_2, HfO_2, Ta_2O_5, Nb_2O_5, Sc_2O_3, Y_2O_3, MgO, B_2O_3, SiO_2, GeO_2, La_2O_3, CeO_2, PrO_x, Nd_2O_3, Sm_2O_3, EuO_x, Gd_2O_3, Dy_2O_3, Ho_2O_3, Er_2O_3, Tm_2O_3, Yb_2O_3, Lu_2O_3, $SrTiO_3$, $BaTiO_3$, $PbTiO_3$, $PbZrO_3$, Bi_xTi_yO, Bi_xSi_yO, $SrTa_2O_6$, $SrBi_2Ta_2O_9$, $YScO_3$, $LaAlO_3$, $NdAlO_3$, $GdScO_3$, $LaScO_3$, $LaLuO_3$, $Er_3Ga_5O_{13}$
Conductors/semiconductors	In_2O_3, In_2O_3:Sn, In_2O_3:F, In_2O_3:Zr, SnO_2, SnO_2:Sb, ZnO, ZnO:Al, ZnO:B, ZnO:Ga, RuO_2, RhO_2, IrO_2, Ga_2O_3, V_2O_5, WO_3, W_2O_3, NiO, FeO_x, CrO_x, CoO_x, MnO_x
Other ternaries	$LaCoO_3$, $LaNiO_3$, $LaMnO_3$, $La_{1-x}Ca_xMnO_3$
Nitrides	
Semiconductors/dielectric	BN, AlN, GaN, InN, SiN_x, Ta_3N_5, Cu_3N, Zr_3N_4, Hf_3N_4
Metallic	TiN, Ti-Si-N, Ti-Al-N, TaN, NbN, MoN, WN_x, WN_xC_y
II-VI compounds	ZnS, ZnSe, ZnTe; CaS, SrS, BaS; CdS, CdTe, MnTe, HgTe
II-VI based TFEL phosphors	ZnS:M (M = Mn, Tb, Tm); CaS:M (M = Eu, Ce, Tb, Pb); SrS:M (M = Ce, Tb, Pb)
III-V compounds	GaAs, AlAs, AlP, InP, GaP, InAs
Fluorides	CaF_2, SrF_2, MgF_2, LaF_3, ZnF_2
Elements	Ru, Pt, Ir, Pd, Rh, Ag, W, Cu, Co, Fe, Ni, Mo, Ta, Ti, Al, Si, Ge
Others	La_2S_3, PbS, In_2S_3, Cu_xS, $CuGaS_2$, Y_2O_2S, WS_2, TiS_2, SiC, TiC_x, TaC_x, WC_x, $Ca_x(PO_4)_y$, $CaCO_3$, $Ge_2Sb_2Te_5$

However, all approaches leading to deviations from self-limiting growth conditions involve a risk of reduced process control and uniformity.

4.2.2.2 Impurities

Impurities from precursor molecules are always incorporated to some extent in ALD films, which is characteristic of chemical techniques – in contrast to PVD techniques. Another source of impurities is the carrier gas. The actual residual concentrations are highly process dependent and their significance application dependent; passive materials such as protective layers are apparently more tolerant to impurities than active layers. Good ALD processes result in impurity contents below the detection limits of most surface analytical tools, *i.e.* in the range of 0.1 at.%, while some other processes may leave a few at.% of impurities, yet the films show reasonably good materials properties. Hydrogen is a common residue but because it can be analysed by only a few special techniques (elastic recoil detection analysis, nuclear reaction analysis) it too often remains unreported.

4.2.2.3 Limited Materials Selection

This has been one factor limiting the broader use of ALD. Several important materials, like Si, Ge, SiO_2, many metals, metal silicides, multicomponent oxide superconductors, ferroelectrics and chalcogenides, and so on, lack efficient processes. Some of these materials are listed in Table 4.2 among reported ALD materials but the processes studied so far do not seem to be production worthy. There is a clear need for precursor chemistry innovation in finding proper precursor combinations. However, some materials, like electropositive metals, may remain beyond the capabilities of thermal ALD. With the aid of plasmas the material selection has been substantially expanded, including also the electropositive metals Ti and Ta, but with concerns specific to plasma and radicals, such as surface damage creation and limited penetration into the highest aspect ratio structures.

4.2.2.4 Deviations from "Ideal ALD Growth"

Several unwanted characteristics are more or less frequently observed in ALD but as they are not common to all ALD processes, they should not be considered as limitations of the method itself but as process-specific deviations from an "ideal ALD growth". One may depict the ideal ALD process as completely saturating and self-limiting, depositing an exact monolayer in each cycle over a reasonably wide temperature range, fully preserving the substrate surface smoothness, and releasing by-products that are readily removed from the reactor.

As already noted, the growth rate is usually not an exact monolayer even if the process otherwise fully follows the ALD principle. The reason for the less than 1 ML $cycle^{-1}$ growth rate is a too low density of precursor molecules in the chemisorption layer (Figure 4.1) because of either steric hindrance between the precursors or a low number of reactive adsorption sites, such as hydroxyl groups. Still, film build-up can take place layer-by-layer, it just takes more than one cycle to complete one monolayer. Therefore, the less than 1 ML $cycle^{-1}$ growth rate alone is rarely a problem, an exception perhaps being epitaxial superlattices where monolayer-by-monolayer growth would allow the most accurate thickness control.

During the very first cycles, ALD film is deposited on the starting surface, which is either a substrate or an underlying film, while during the rest of the process the growth occurs on the film material itself. If the starting surface differs chemically from the film surface, the growth rate is not necessarily constant but may change quite substantially as the surface becomes covered by the film. Usually the growth rate is slower at the beginning, during the nucleation period, and it takes a certain number of cycles before a linear correlation between film thickness and number of ALD cycles becomes established. This kind of retarded nucleation has been observed in the technologically important application of ALD of high-*k* oxides on hydrogen-terminated silicon surfaces (Figure 4.5).[16] This surface contains only few, if any, hydroxyl groups that are the favourable reaction sites in ALD of oxides. Indeed, on silicon oxide surfaces the growth is much more linear from the very beginning (Figure 4.5). Besides complicating thickness control, the retarded nucleation may also lead to poor morphology, as discussed below.

The deposition rate may also be temperature dependent (Figure 4.2b). This is due to the temperature dependent density of precursor molecules in the saturated chemisorption layer, which, in turn, is affected by either the temperature dependent density of reactive sites or the temperature dependent structure of chemisorbed precursor molecules. Nevertheless, usually, the temperature dependence of the growth rate is so weak that requirements for temperature control and uniformity of the reaction chamber are not as strict as in CVD.

The growth is often not truly self-limiting because the precursors also thermally self-decompose. The higher the temperature the higher the contribution arising from self-decomposition (Section 4.3.3.1). In fortunate cases the decomposition product is the same as the one to be deposited by ALD, *e.g.* metal alkoxides are often used to deposit corresponding oxides by exchange reactions with water, but they also self-decompose to the same oxide if the temperature is too high. As far as the self-decomposition proceeds with only a modest rate it is likely to be a surface reaction limited rather than diffusion limited process, thus maintaining reasonably well good uniformity and conformality. When these two conditions (decomposition results in the desired deposit material and is surface reaction limited) are valid it is highly application dependent as to what extent

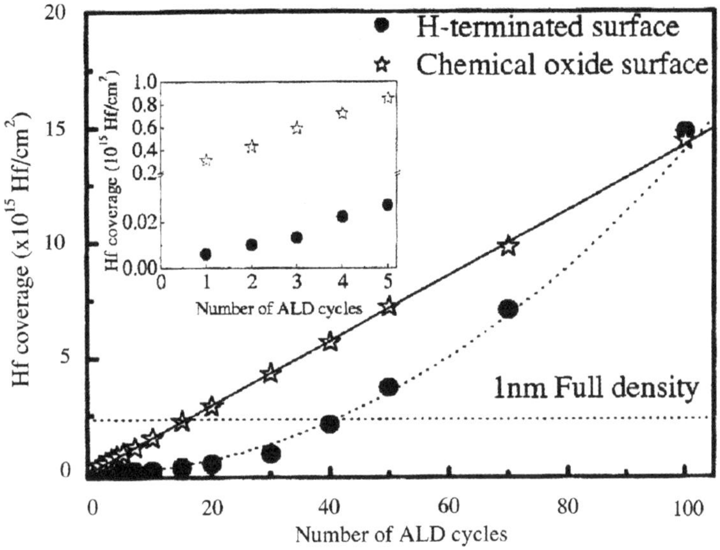

Figure 4.5 Hafnium coverage, measured by *in situ* MEIS (medium energy ion scattering), as a function of the number of ALD cycles in the $HfCl_4$-H_2O process on hydrogen terminated silicon and on silicon with a chemical oxide layer. The dotted line indicates the Hf-coverage corresponding to a uniform 1 nm thick HfO_2 layer with a bulk density. Inset: very first cycles in detail. (Reprinted with permission from ref. 16. Copyright 2007, American Institute of Physics.)

self-decomposition may be acceptable – for protective coatings it is of less concern than for gate oxides or optical films, for example.

Surface reactions may also remain somewhat short of saturation because of slow kinetics. Increasing exposure time could drive the reactions to saturation but with impractically long process times. Another option is, of course, to accelerate the reactions by increasing the deposition temperature, but with an increasing risk of precursor decomposition. Properties of the resulting films in comparison with the requirements in a given application again determine whether the process will be useful despite the non-saturated surface reactions.

Instead of exactly copying the substrate surface morphology ALD growth has often been observed to cause roughening of the surface at the nanometre level, though still showing excellent conformality on any larger scale. Such a roughening seems to be characteristic to polycrystalline films;[16–18] amorphous as well as epitaxial films usually retain the smoothness of the substrate surface. The dominant factor with respect to the surface morphology of amorphous and epitaxial films is the minimization of the surface area, which favours atomically smooth films because either the surface energy is isotropic (amorphous films) or the crystal orientation is dictated by the substrate (epitaxial films). With polycrystalline films the crystal growth of the originally more or less randomly oriented nuclei adds an important contribution. Before growing into contact with each other, crystals originating from the nuclei grow independently. The newly added atoms are able to diffuse over some distance to find energetically favoured sites. Each crystallite attempts to reach its equilibrium shape determined by the relative surface energies of the differently oriented surfaces. Though the equilibrium is usually probably not reached the attempt is already enough to tilt the crystal surfaces with respect to the substrate plane, in other words cause roughening of the film surface. The effect is further emphasized by the retarded nucleation discussed above: the lower the nucleation density the rougher the film becomes because the crystallites can grow independently for a longer time (Figure 4.6).[17,18]

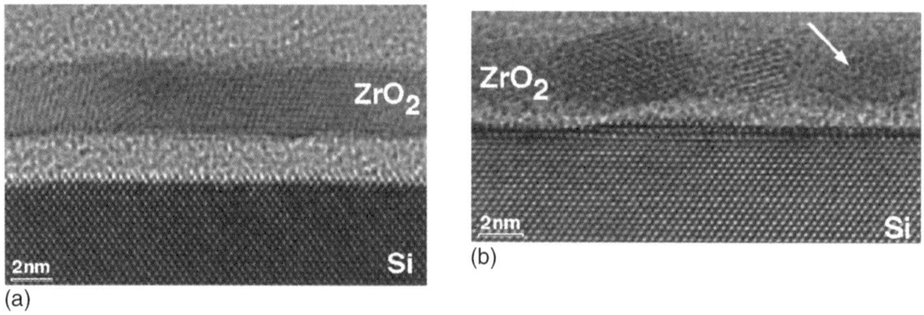

Figure 4.6 TEM images of ZrO_2 films grown on (*a*) oxidized and (*b*) hydrogen-terminated silicon from $ZrCl_4$ and H_2O at 300 °C. (Reprinted with permission from ref. 18. Copyright 2006, American Institute of Physics.)

Many reaction by-products are polar molecules and as such show reactivity toward the surface. If the by-products readsorb on the surface they can block adsorption sites from the precursor molecules and thereby decrease the growth rate. Furthermore, if the by-product readsorption does not occur uniformly across the substrate it can lead to thickness non-uniformity. Therefore, it is likely to have more severe effect in cross-flow reactors than in perpendicular flow reactors (Section 4.4) because in the former different substrate areas can experience substantially different exposures to by-products. If film growth begins only near the leading edge of the substrate, this part is exposed to only a small amount of by-product molecules while the trailing edge will receive a much higher amount. The effect is further emphasized by the fact that in cross-flow reactors the by-products travel in front of the precursor pulse that is producing them.[11] In contrast, in properly designed perpendicular flow reactors the film growth begins everywhere at about the same time, and hence the effect of by-product readsorption should be more uniform too.

In the most severe cases reactive by-products can etch the film material or the underlying material. Pitting of copper films has, for example, been observed when $TiCl_4$- and WF_6-based nitride processes have been applied on copper.[19] Exposure to only the metal halides did not cause the pitting, so it could be concluded that hydrogen halides released in reactions with ammonia were responsible for the pitting. Another possible reason for film non-uniformity is slow etching of the film material by the precursor molecules themselves (Section 4.3.1). In the worst case the etching may fully prevent film growth. It is also possible that the precursor molecule reacts directly with the underlying material (Section 4.3.1). For a thorough discussion on factors affecting film uniformity see ref. 20.

Deviations from ideal ALD are occasionally created on purpose to increase the throughput of the process. As noted above, decomposition of the precursor adds a CVD component to the process, and if the decomposition occurs in surface reaction rather than diffusion limited mode it does not destroy the film conformality but increases the growth rate per cycle and thereby throughput. Yet, the alternate supply with the second precursor may have beneficial effects on film purity and morphology, for example, thus favouring the use of this kind of ALD-like, pulsed-CVD approach instead of pure CVD. A good example is deposition of tungsten films by an alternate supply of WF_6 and SiH_4 (or Si_2H_6 or B_2H_6): several monolayers, up to 10 Å, are deposited in one cycle due to limited decomposition of the reducing agent.[21,22]

Another way of increasing throughput is to run the ALD process under non-saturative conditions. Total saturation can often be reached only by giving enough time for the surface reactions, not by increasing the partial pressure of the precursor. Furthermore, increasing the saturation from say 90% to 100% may take more time than reaching the first 90%. Thus it may be beneficial to accept the saturation as being only 90% because then the exposure time can be substantially shortened. In addition, because there are fewer precursors to be removed, purging can be completed

Table 4.3 Basic differences between ALD and CVD methods.

ALD	*CVD*
Precursors react separately on the substrate	Precursors react at the same time on the substrate, possibly also in the gas phase
Precursors highly reactive	Precursors less reactive
No precursor decomposition	Precursors can decompose
Exact control of the precursor dose is not necessary	Precursor dose must be carefully controlled
Deposition rates low (up to few $nm\,min^{-1}$)	Deposition rates high (up to $\mu m\,min^{-1}$)
Surface controlled	Process parameter controlled

faster. Such an approach is called limited optimized reactions ALD (LORA).[23] Up to ten-fold improvements in effective growth rate have been reported. The applicability of LORA is apparently highly dependent on the reactor design, however. While good uniformity may be reached with perpendicular-flow reactors, cross-flow reactors (Section 4.4) may in the worst case lead to division of the substrate to an upstream growth-area and a downstream non-growth-area.

4.2.3 Comparison of ALD and CVD

As stated before, ALD is a variant of the well-established CVD technique. After describing the principle as well as benefits and limitations of ALD it is useful to summarize the differences between these methods. Table 4.3 outlines the main differences.

4.3 Precursor Chemistry

The success of ALD is built on chemistry. The unique characteristics of ALD can be achieved and benefited from only with precursors that provide the self-limiting film growth through saturative surface reactions. Speed and cost-effectiveness of the process, in turn, require that the surface reactions proceed rapidly to saturation.

The requirements for ALD precursors are largely similar to those for CVD precursors. However, some important differences arise from the unique features of the ALD process. The self-limiting growth requires that the precursors do not thermally decompose on their own but only react with each other on the surface. The alternate supply of the precursors in turn ensures that there is no risk of gas-phase reactions, and hence more reactive combinations can be chosen than in CVD. In fact, high reactivity leading to rapid saturation of the alternate surface reactions is a key for optimizing the productivity of ALD.

Section 4.3.1 discusses the ALD precursor requirements. Section 4.3.2 gives some successful examples of various ALD precursor types. A survey of ALD process characterization methods (Section 4.3.3) completes the discussion on ALD chemistry.

4.3.1 ALD Precursor Requirements

Table 4.4 summarises the requirements for ALD precursors. They can be divided into two groups, mandatory and "nice-to-have". As will be seen, often not all of the requirements can be fulfilled at the same time, and then a compromise must be looked for by sacrificing those in the latter group.

Volatility is an obvious demand as ALD is a vapour phase method. A vapour pressure of 0.1 Torr in the source conditions, which often involve elevated temperatures, can be considered

Table 4.4 Summary of ALD precursor requirements.

Essential properties	*Desirable but not necessary*
Volatility	Unreactive volatile by-products
Fast and complete reactions	Inexpensive
No self-decomposition	Easy to synthesize and handle
No etching of the film or substrate material	Nontoxic and environmentally friendly
No dissolution into the film or substrate	
Sufficient purity	

as a minimum requirement. In practice, gases and high vapour pressure liquids are preferred because they are easy to handle and can be delivered with high fluxes. High fluxes are important because with fast surface reactions the precursor transportation becomes the limiting factor for the pulse time needed to saturate the surfaces. Obviously, the importance of high fluxes, and thereby high vapour pressure sources, increases with increasing surface area to be coated. In any event, with proper source designs (Section 4.4) solids can be used quite readily too, a good example being the use of low vapour pressure solid $ZnCl_2$ in the deposition of ZnS by batch processes in manufacturing TFEL displays. An additional risk of solid sources, especially with very fine particle size solids, *e.g.* $HfCl_4$, is particle transportation from the source to the substrates, potentially leading to particulate defects in the films. With proper source design the problem seems to have been solved, however, and the $HfCl_4$–H_2O process for HfO_2 gate dielectric has been proved ready for production.[24,25]

The key feature of ALD, the self-limiting film growth, relies on saturative surface reactions. All reactions that could lead to deviation from the saturation must be avoided. Precursor molecules must be chosen so that both they themselves and the surface species they form upon adsorption and reaction as intermediates are stable against self-decomposition at the chosen growth temperature. Thermally stable metal halides are in this respect the best choices while metal compounds with organic ligands are often, though not always, limited to temperatures below 300 °C. However, as noted in Section 4.2.2, in some applications with less strict demands on uniformity and thickness control, a minor contribution from decomposition may be accepted as a non-ideality provided that the decomposition product is pure and the same as obtained from the exchange reactions. This is often the case with metal alkoxides; both reaction with water and decomposition result in the same oxide. High productivity requires that the reactions are saturated rapidly in each step. The reactions should also be complete to produce pure films. Therefore, aggressively reacting precursors must be looked for. Owing to the separate dosing of the precursors, they do not meet in the gas phase, thereby automatically eliminating all detrimental gas-phase reactions. For the same reason, it is high reactivity towards the surface species formed by the other precursor that is crucial, rather than a high reactivity towards the other precursor molecule itself. In most cases these two reactivities are closely correlated but sometimes there may be substantial differences. In any case, the desire for highly reactive precursor combinations distinguishes ALD from CVD where usually only moderate reactivity can be tolerated because of the risk of gas-phase reactions. Thermodynamically, ALD net reactions should have as negative a Gibbs free energy change as possible. An illustrative example is the $ZrCl_4$–H_2O process with a net reaction of:

$$ZrCl_4(g) + 2H_2O(g) \rightarrow ZrO_2(s) + 4HCl(g) \quad (\Delta G = -148 \text{ kJ mol}^{-1} \text{ at } 300\,^{\circ}\text{C}) \tag{4.1}$$

This has been found to be a good ALD process, but in CVD water is usually formed *in situ* from CO_2 and H_2 to avoid the gas-phase reactions.[26]

Precursor molecules should not etch the film material as this is likely to have a negative effect on film uniformity, and in the worst case destroys the whole process. This was observed in attempts to deposit Nb_2O_5 from $NbCl_5$,[27] Mo from $MoCl_5$[28] and WO_3 from WF_6,[29] for example:

$$3NbCl_5(g) + Nb_2O_5(s) \rightarrow 5NbOCl_3(g) \quad (4.2)$$

$$MoCl_5\,(g) + Mo(s) \rightarrow MoCl_{5-x}(g) \quad (4.3)$$

$$2WF_6(g) + WO_3(s) \rightarrow 3WOF_4(g) \quad (4.4)$$

In the case of tungsten oxide the problem was circumvented by replacing WF_6 as a precursor with WOF_4, which does not have the etching capability, and similar approaches should apply to other etching cases as well. Etching to a lesser extent is possible too, so that film growth still proceeds but not in an ideal manner. For example, with Ta_2O_5 a reaction analogous to that shown in Reaction (4.2) becomes significant only above about 300 °C.[30,31] Of course the precursors should not etch the substrate or any underlying film material either. Notably, the etching problem is highly emphasized in ALD compared to CVD because in CVD there is always the competing pathway for film deposition through reactions with the other precursor. In ALD the other precursor is not simultaneously present and hence the etching can proceed without competition.

Yet another unwanted, but usually less detrimental, reaction is an exchange reaction, *e.g.* an exchange of cations of the metal precursor and the underlying oxide material:[32,33]

$$3TiO_2(s) + 4AlCl_3(g) \rightarrow 2Al_2O_3(s) + 3TiCl_4(g) \quad (4.5)$$

$$3ZnO(s) + 2Al(CH_3)_3(g) \rightarrow Al_2O_3(s) + 3Zn(CH_3)_2(g) \quad (4.6)$$

These reactions usually self-terminate after a few cycles when a continuous product layer has formed, and are therefore usually less harmful, but also more difficult to observe, than the etching reactions. The main complications caused by exchange reactions lie in the composition control of multicomponent films and thickness control of multilayer structures.

By-products released from the precursors in the ALD process should be unreactive so that they can be readily removed from the reactor. Reactive by-products, like hydrogen halides formed from the metal halides, may in the worst case etch the film or the underlying materials, as already noted in Section 4.2.2 for the growth of TiN and WN on copper where the hydrogen halides caused pitting of the copper film.[19] Other possible harmful reactions of by-products are reactive site blocking by readsorption (Section 4.2.2) and corrosion of the reactor and pump materials.

In general, precursors should not dissolve into the film or the underlying material. If the dissolution is irreversible it leads to film contamination, and if it is reversible an uncontrolled process results. The latter was observed in attempts to deposit copper using elemental zinc as a reducing agent for CuCl.[34] Once copper nuclei had formed (Reaction 4.7), zinc started to dissolve into these nuclei, forming brass (Reaction 4.8). After ending the zinc supply the reverse process occurred, *i.e.* zinc outdiffused from the brass. As this occurred also during the CuCl pulse, the two precursors were available in the reactor at the same time, leading to non-uniform and poorly controlled film growth (Reaction 4.9):

$$CuCl(ads) + 0.5Zn(g) \rightarrow Cu(s) + 0.5ZnCl_2(g) \quad (4.7)$$

$$Cu(s) + Zn(g) \leftrightarrow Cu(Zn)(s) \quad (4.8)$$

$$CuCl(g) + 0.5Zn(g) \rightarrow Cu(s) + 0.5ZnCl_2(g) \quad (4.9)$$

In contrast, in some cases precursor dissolution may be crucial to initiating film growth. Many polymer substrates lack functional surface groups that are considered necessary for the nucleation of the films, yet Al_2O_3 films have been deposited on all tested polymers using $Al(CH_3)_3$ and water as precursors.[35] Studies with a quartz crystal microbalance (QCM) concluded that $Al(CH_3)_3$ dissolves into the near-surface region of polymers and thereby becomes available even if no chemisorption or surface reaction had occurred. Growth of various noble metals with molecular oxygen as another reactant is an example where one reactant – atomic oxygen formed by dissociating O_2 – dissolves into the subsurface layers of the film and then reacts with the subsequently dosed metal precursor (Section 4.3.3.2).[36]

The "desirable but not necessary" requirements given in Table 4.4 are common to CVD and obvious in nature. When it comes to expense, its inverse correlation with the production scale must be underlined. No precursor candidate should be neglected at a research stage only because of its current price as this may drop quite dramatically if the compound becomes adopted into wider user.

As already noted, the nice-to-have requirements often must be sacrificed. For example, $Al(CH_3)_3$ is a pyrophoric compound and thus needs special care in handling, but at the same time this high reactivity is one of the key factors for the $Al(CH_3)_3$–H_2O process being probably the best ALD process developed to date. Other beneficial properties of $Al(CH_3)_3$ are its liquid state, convenient vapour pressure, high thermal stability, benign methane released as a by-product in reactions with water, and low price, which is largely due to its use in large quantities in polymerization catalysts.

4.3.2 Precursor Types

This section gives an overview of precursors used and studied in ALD. Examples are given of successful representatives of each type of precursors. For a more thorough survey of ALD processes see, for example, references 10 and 11.

4.3.2.1 Non-metal Precursors

Hydrides are by far the most widely used non-metal precursors. These include, for example, H_2O, H_2O_2, H_2S, H_2Se, H_2Te, NH_3, PH_3, AsH_3 and HF. Apart from most of them being quite toxic, the hydrides meet well the ALD precursor requirements. Even if some of them may have limited stability at elevated temperatures, this is not of a concern as the decomposition products are volatile too. Alkylated derivatives of the hydrides have also been used, *e.g.* $RAsH_2$, as the alkylation decreases the toxicity. *tert*-butylamine ($^tBuNH_2$) has been studied because it has somewhat more reducing power than ammonia. Unfortunately, the alkylation of hydrides seems to lead also to an increased risk of carbon contamination. Nevertheless, fully alkylated compounds like Et_2Te and alkylamides $(Me_2N)_3P$ and $(Me_2N)_3As$ have also been explored.

Chalcogenide elements (S, Se, Te) have been used too in reactions with elemental zinc and cadmium. By contrast diatomic gaseous elements such as O_2, H_2 and N_2 are usually too inert for use in ALD; nitrogen is in fact often used as an inert carrier gas. In some cases O_2 and H_2 have shown sufficient reactivity, however. Molecular oxygen becomes reactive in noble metal processes where the metal catalyses its dissociation to atomic oxygen. Also, some metal iodides react with O_2, but probably only after dissociation of one of the iodide ligands. H_2 has been explored quite widely as a reducing agent for metal deposition but only in a few cases has this been successful.

While O_2, H_2 and N_2 have usually been found inert in thermally activated ALD, they have been the most common sources for radicals in radical and plasma enhanced ALD processes. The unreactivity of these molecules without plasma activation means that with pulsed plasmas there is no need for purging after the plasma is turned off. In fact, the radical source gas flow can be constantly on, also during the metal precursor pulse. Avoidance of one purge period makes the ALD cycle shorter and thereby increases productivity. Ammonia has been used as a radical precursor, too.

Ozone (O_3) is a quite unique precursor compound unmatched with other elements, and thereby gives additional freedom in oxide deposition. Ozone shows high reactivity towards most metal precursors. Originally, ozone was used mainly with metal precursors that do not react efficiently with water, in particular with β-diketonates. More recently it has also been used with compounds that do react with water – even the successful $Al(CH_3)_3$–H_2O process is now being challenged with $Al(CH_3)_3$–O_3 and $Al(CH_3)_3$–O processes.[37] The reason for replacing water with ozone and oxygen radicals is the slow purging of water because of the strong interaction of the polar water molecule with oxide surfaces. This has become a concern especially in low temperature deposition and with high aspect ratio structures. With oxygen radicals Al_2O_3 can be deposited even at room temperature from $Al(CH_3)_3$.[38] However, ozone and oxygen radicals share the problem of sometimes being too oxidizing for the underlying material. This limits their use in gate oxide deposition on silicon where the formation of interfacial SiO_2 should be minimized.

Notably, the alternative non-metal precursors lead to quite different chemistries. This becomes most obvious by considering the fate of the metal precursor ligands. With hydrides, the ligands are simply protonated without any fragmentation, net reactions being, for example:[39,40]

$$2Al(CH_3)_3(g) + 3H_2O(g) \rightarrow 2Al_2O_3(s) + 6CH_4(g) \tag{4.10}$$

$$Ti(OCH(CH_3)_3)_4(g) + 2H_2O(g) \rightarrow TiO_2(s) + 4(CH_3)_3CHOH(g) \tag{4.11}$$

$$Sr(thd)_2(g) + H_2S(g) \rightarrow SrS(s) + 2Hthd(g) \tag{4.12}$$

By contrast, with ozone and oxygen radicals, the reaction is more complicated, combustion like, and involves more ligand fragmentation,[41] though details of these reactions remain still largely unexplored:

$$\begin{aligned}2Al(CH_3)_3(g) + nO_3(g) \rightarrow\ & 2Al_2O_3(s) + CO_x(g) \\ & + H_2O(g) + \text{possible other by-products}\end{aligned} \tag{4.13}$$

$$2Al(CH_3)_3(g) + nO(g) \rightarrow 2Al_2O_3(s) + CO_x(g) + H_2O(g) + \text{possible other by-products} \tag{4.14}$$

$$\begin{aligned}Ce(thd)_4(g) + nO_3(g) \rightarrow\ & CeO_2(s) + CO_x(g) \\ & + H_2O(g) + \text{possible other by-products}\end{aligned} \tag{4.15}$$

Because of the less clean removal of the ligands, there is an increased risk of contamination, especially if the film material is reactive toward the by-products. For example, many oxides have been found to contain carbonate when deposited with ozone.[42] In favourable cases, however, the films obtained with ozone have been purer than those obtained with water.[43,44] Few details are known about reactions employing radicals other than oxygen.

In some cases metal compounds have also been used as precursors for non-metals. One example is the use of metal alkoxides as both oxygen and metal sources in reactions with metal chlorides:[45,46]

$$ZrCl_4(g) + Ti(O^iPr)_4(g) \rightarrow ZrTiO_4(s) + 4^iPrCl(g) \tag{4.16}$$

Reaction (4.16) is an idealized reaction; *in situ* reaction mechanism studies have revealed that propene and HCl are also formed as by-products.[47] The chemistry becomes even more complicated when tris(*tert*-butoxysilanol) is reacted with, for example, $Al(CH_3)_3$; to explain the self-limiting growth of nearly stoichiometric SiO_2 with a rate as high as 12 nm cycle^{-1} a unique mechanism involving growth and cross-linking of siloxane polymer chains was proposed.[48]

Another example of metal compounds as non-metal precursors is the use of TiF_4 and TaF_5 as fluorine sources for CaF_2, SrF_2, MgF_2 and LaF_3.[49,50] In these processes metal β-diketonates were used as metal precursors and the net reaction, presumably, was:

$$TiF_4(g) + 2Ca(thd)_2(g) \rightarrow 2CaF_2(s) + Ti(thd)_4(g) \tag{4.17}$$

4.3.2.2 Metal Precursors

Compared with non-metals, the options for metal precursors are much more numerous. Figure 4.7 gives examples of the most common types of metal precursors.

Elements are ideal precursors as they do not introduce any impurities into the films. Unfortunately, only zinc, cadmium and mercury are volatile at reasonably low temperatures. Zinc and cadmium have been used in ALD of chalcogenides but no reports on mercury exist.

Metal halides meet well the requirement of thermal stability and usually also show high reactivity with non-metal hydrides, water in particular, although this does not apply to every case. Only a few ($TiCl_4$, $SnCl_4$, $SiCl_4$) metal halides are liquid, however, and not all metals have volatile enough halides. Potential disadvantages of metal halides include formation of corrosive hydrogen halides as by-products, halide residues in the films and in some cases etching reactions may also occur.

Chlorides are the most often used metal halide ALD precursors. Good processes exist for oxides with, for example, $AlCl_3$, $TiCl_4$, $ZrCl_4$, $HfCl_4$, $TaCl_5$, $SnCl_4$ and $InCl_3$ as precursors, although with the last two the reactions are not as fast with the others. Nitrides have been deposited from $AlCl_3$, $TiCl_4$, $TaCl_5$, $NbCl_5$, $MoCl_5$, GaCl and $GaCl_3$. Ammonia has been used as the nitrogen precursor but, compared to water, higher deposition temperatures have usually been needed to achieve reasonably low chlorine residue levels. In addition, with transition metal nitrides the growth rates have been quite low and there have been incompatibility issues while depositing on copper films (see above). For sulfide deposition, $ZnCl_2$ was exploited in one of the first ALD processes, and later also $CdCl_2$, $InCl_3$. $GaCl_3$, CuCl, $MnCl_2$ (evaporated around 500 °C) and $TiCl_4$ have been used. One of the big disappointments has been the poor reactivity of $SiCl_4$: in the growth of SiO_2 very long exposure times have been needed.[51] The reaction can be catalysed by amines, however, but this works only at low temperatures close to room temperature.[51,52] However, $SiCl_4$ has been found to

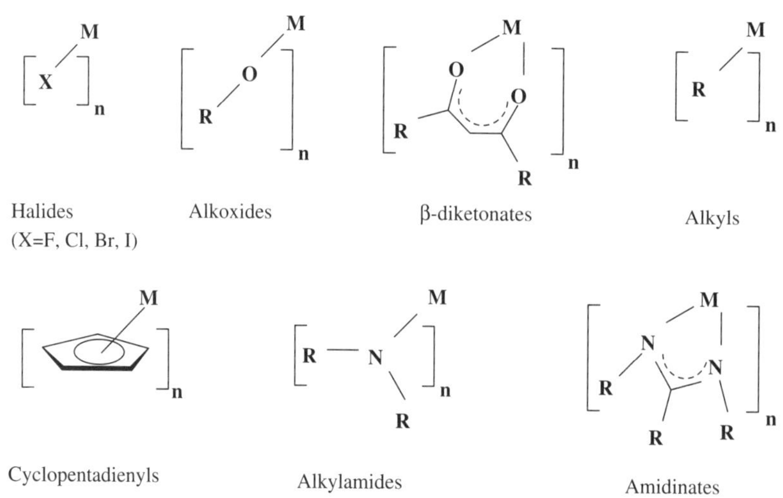

Figure 4.7 Examples of precursor types applied in ALD. R represents alkyl groups such as methyl or ethyl.

serve as a much more efficient precursor in the growth of mixed oxides. In fact, this behaviour seems common to all silicon precursors: while the growth of binary SiO_2 is usually very sluggish, mixed oxides up to quite high silicon contents can be deposited.

Metal iodides used in ALD include TiI_4 for both oxide and nitride, and ZrI_4, HfI_4, SnI_4, TaI_5, NbI_5 and CoI_2 for oxides. Both water and molecular oxygen have been used as oxygen precursors. With O_2 the growth starts usually only at higher temperatures, and therefore it seems that the metal iodide must first partially decompose to become reactive enough toward O_2. In any event, once the reactions start they lead to very pure films. Notably, in the metal iodide–oxygen processes neither precursor contains hydrogen, thereby offering a hydrogen-free alternative to other ALD oxide processes that almost all use hydrogen-containing precursors, *i.e.* water and/or metal precursors with organic ligands.

Of the metal fluorides WF_6 is by far the most studied. It has been used for depositing metallic tungsten with either silanes or boranes as reducing agents,[53] and for tungsten nitride with ammonia alone or with triethylboron as a coreactant.[19,54] The latter approach eliminates the incompatibility problem with copper. As noted above, WF_6 can not be used for WO_3 deposition because of the etching reaction but with the oxyfluoride WOF_4 this is possible.[29] TiF_4 and TaF_5 have been used as metal precursors for oxides and, as noted above, as fluoride precursors for metal fluorides.

Metal alkoxides have served as good precursors for several oxides. Well behaving ALD binary oxide processes have been based on $Ti(OMe)_4$, $Ti(OEt)_4$, $Ti(O^iPr)_4$, $Al(OEt)_3$, $Al(OPr)_3$, $Ta(OEt)_5$ and $Nb(OEt)_5$; and silicon alkoxides have been used for mixed oxides. Many of these alkoxides are liquids and thermally stable up to around 250 °C, at least. However, several other alkoxides do not show sufficient stability for ALD. Among these are zirconium and hafnium alkoxides that have been explored quite widely because of the wide interest in ZrO_2 and HfO_2. Besides the simple alkoxides, various donor functionalized alkoxides[55] have also been tested for zirconium and hafnium but the results have not been entirely promising; the compounds either decompose readily or if they are stable their reactivity is quite modest. Nevertheless, encouraging results like stability up to 350 °C and a high growth rate of 0.9 Å $cycle^{-1}$ have been reported for $Hf(mp)_4$ (mp = 3-methyl-3-pentoxide = OCMeEt).[56] In addition with, for example, the otherwise problematic lead, promising results have been achieved using an amine donor functionalized alkoxide.[57] $SrTa_2(OEt)_{10}(dmae)_2$ (dmae = dimethylaminoethoxide) is a successful example of a double metal alkoxide that carries the two metals in a fixed ratio into the oxide film.[58]

Metal β-diketonates form a very versatile group of precursors as a volatile compound can be found for nearly every metal.[59] Of the various ligands 2,2,6,6-tetramethyl-3,5-heptanedione (thd) is the most common, but the simpler and less expensive acetyl acetone (acac) is used whenever possible, *i.e.* when the resulting compound is volatile and stable enough. Metal β-diketonates have been used mainly for the growth of oxides,[60,61] but also for sulfides, in particular alkaline earth metal sulfides, fluorides, copper and noble metals. In addition, manganese and lanthanide β-diketonates have been used for doping of ZnS, CaS and SrS films for TFEL displays.[59] Interestingly, though metal β-diketonates react readily with H_2S, they do not usually react efficiently with water and thus require ozone for oxide deposition.

Metal alkyls of Group 12–14 elements have been widely studied in ALD. These compounds usually exhibit reasonably high volatility and high reactivity, and they are also often liquids.[62] The reactivity is often so high that the metal alkyls are pyrophoric and must be carefully protected from air and moisture. Some metal alkyl precursors have turned out to be very successful in oxide and sulfide ALD, the most well-known example being trimethylaluminium [$Al(CH_3)_3$];[63] with water, ozone and oxygen radicals Al_2O_3 can be grown in a highly controlled manner over a wide temperature range. Replacement of one methyl ligand with an alkoxide ligand converts the pyrophoric $Al(CH_3)_3$ into non-pyrophoric $Al(CH_3)_2(OR)$, though with a substantially lower thermal stability.[64] Dimethyl- and diethyl-zinc compounds also react readily with both water and hydrogen sulfide. In any event, metal alkyls have been explored most comprehensively for the epitaxy (ALE) of III-V compound semiconductors, in particular GaAs.[65] Methyl and ethyl compounds have again been the most common ones. The growth has not necessarily been purely self-limiting, however, but involved partial pyrolysis of the metal alkyls, thereby complicating the growth control and causing carbon contamination, which leads to relatively high p-type carrier concentrations.

Metal cyclopentadienyls broaden the selection of organometallic precursors from the quite limited number of metals exhibiting sufficiently stable alkyl compounds. In these compounds at least one of the ligands is a five-carbon cyclopentadienyl ring ($C_5H_5^-$; Cp) or, as it is often the case, its substituted derivative $C_5R_xH_{5-x}^-$. Cyclopentadienyls of hafnium and zirconium (*e.g.* $ZrCp_2Cl_2$, $ZrCp_2Me_2$) have proved potential precursors for the corresponding high-*k* oxides in combination with either water or ozone,[66–68] and they have been modified further by adding alkoxides ligands, *e.g.* $(MeCp)_2HfMe(OMe)$.[69,70] Strontium and barium cyclopentadienyls, $Sr(C_5{}^iPr_3H_2)_2$, $Sr(C_5Me_5)_2$ and $Ba(C_5{}^tBu_3H_2)_2$, in turn opened the possibility of depositing these metal oxides and, more importantly, multicomponent oxides like $SrTiO_3$ and $BaTiO_3$ with water based chemistry.[71,72] This is important because of the high tendency of these elements to form carbonates when ozone is used as an oxygen source with the metal β-diketonates. The same strontium and barium compounds have also been used for sulfide deposition.[73] With rare earth oxides the cyclopentadienyls have also been important in opening routes for water based chemistry as compared with the older metal β-diketonate–ozone processes.[74] Other cyclopentadienyls used for oxides include $MgCp_2$ and $Ni(CpMe)_2$. For noble metals $RuCp_2$, $Ru(CpEt)_2$, Ir(CpEt)(CHD) (CHD = cyclohexadiene) and $(CpMe)PtMe_3$ have been employed with either molecular oxygen or ammonia or with hydrogen plasma. Manganese cyclopentadienyls have in turn been used for depositing ZnS:Mn TFEL display phosphors.

Metal alkylamides offer a wide range of volatile metal compounds, yet only a few have been found useful in ALD. This is mostly because of their limited thermal stability. Zirconium and hafnium alkylamides, $M(NMe_2)_4$, $M(NMeEt)_4$ and $M(NEt_2)_4$, are stable up to about 250–300 °C in the case of Zr and somewhat higher in the case of Hf,[75,76] and have been quite widely studied as precursors for the corresponding oxides with water, ozone and oxygen radicals as the other reactant. Tantalum alkylamides show stability comparable to zirconium and hafnium but titanium alkylamides seem to be too unstable for use in ALD.[77,78] Other alkylamides studied in ALD include dimethylamides of aluminium, gallium, germanium, antimony and tungsten, and alkylimido-alkylamides of tungsten and molybdenum, for example.

Metal silylamides complement the metal selection of alkylamides but they also suffer from limited thermal stability. In addition, decomposition leads to silicon incorporation into the films, as observed with lanthanum and praseodymium oxides, for example. However, quite promising results have been achieved with $Bi[N(SiMe_3)_2]_3$ in growth of multicomponent oxides Bi-Ta-O and Sr-Bi-Ta-O, though only at temperatures below 200 °C.[79] This is important considering the problems in finding bismuth precursors for ALD.

Metal acetamidinates are the newest group of ALD precursors and cover a broad range of elements.[80] Promising results have been reported for metals (Cu, Fe, Co, Ni) even with molecular hydrogen as a reducing agent. In addition, several oxides have been deposited. Full validation of these compounds is still waiting as, for example, scale-up results have not been reported yet.

4.3.2.3 Reducing Agents

Reduction of a metal is needed in ALD growth of metal and transition metal nitride films but it has been surprisingly difficult to find effective reducing agents for thermal ALD. Silanes and boranes have been used in the growth of tungsten[53,81,82] but otherwise their use has not been reported, perhaps because of their hazardous nature. Substituted silanes and boranes, however, have turned out to be quite inefficient. For copper deposition many organic compounds, like alcohols, aldehydes and carboxylic acids, have been tested but seemingly with only limited success.

Several ALD noble metal processes have been developed using O_2 as the other precursor.[36,83,84] Although in these processes the noble metals are reduced, O_2 can hardly be considered as a reducing agent; the reaction mechanism is more properly described as rather complicated oxidative decomposition of the organometallic precursor.

In transition metal nitride processes the common nitrogen precursor ammonia often has enough power to reduce the metal from its higher oxidation state in the metal precursor to the formal oxidation state +III in the desired nitride phase MN, *e.g.* $Ti^{IV}Cl_4 \rightarrow Ti^{III}N$. However, if the metal can also form a dielectric nitride phase with a higher oxidation state, the reducing power of ammonia may be insufficient. For example, the $TaCl_5 + NH_3$ process results in dielectric Ta_3N_5.[85] Elemental zinc, in contrast, can reduce tantalum to TaN.[85] Zinc also has beneficial effects in other transition metal nitride processes, though not as clear as with TaN. Unfortunately, concerns of zinc contamination prohibit the use of zinc in microelectronics. *tert*-Butylamine, allylamine[86] and trimethylaluminium[87] can also reduce Ta(V), but they leave carbon and aluminium residues into the films. Triethylboron has been found to have a beneficial effect in the growth of WN_xC_y from WF_6 and NH_3, and though the main benefit probably arises from an elimination of HF release, triethylboron may also have a reducing effect in the process.[54] In the growth of TaC and WC, diethylsilane and trimethyl- or triethyl-aluminium have been used as combined carbon sources and reducing agents.[25]

Hydrogen radicals produced by plasma discharges from H_2 are by their nature highly reducing, and thereby largely overcome the problems met in conventional thermal ALD.[8] Even highly electropositive metals such as titanium, tantalum[88] and aluminium[89] have been deposited with hydrogen radicals but the processes are apparently not trivial and have probably not been adapted to a wider use. H_2-N_2 and NH_3 plasmas, in turn, have been widely used in ALD of transition metal nitrides.[8]

4.3.3 Characterization of ALD Chemistry

Many approaches exist for characterizing ALD chemistry (Table 4.5). However, ALD shares the common dilemma of surface chemistry: the more details one wants to obtain, the further one must go from the real process conditions, with an increasing concern about how representative the results are. Indeed, most ALD chemistry characterization has been done under typical ALD reactor conditions (Section 4.4), *i.e.* around a pressure of 1 Torr and that is where also this section is limited to. Besides experiments specifically probing the chemistry, information obtained indirectly from film growth experiments will also be discussed. A broader review covering also the other approaches mentioned in Table 4.5 can be found in ref. 11.

4.3.3.1 Film Growth Experiments

The ultimate goal of ALD film growth experiments is to identify the optimal conditions for a reproducible growth of high quality films. This is done by varying the experimental parameters and observing their effect on growth rate and film properties. In ALD the growth rate is given in units of Å $cycle^{-1}$ or nm $cycle^{-1}$ and is usually obtained by dividing the resulting film thickness by the number of deposition cycles applied.

Growth rate vs. *precursor dose*. Usually, the first task in ALD film growth experiments is to examine whether the film growth proceeds in the self-limiting manner as characteristic of ALD. This is done by fixing all other parameters (growth temperature, dose of the other precursor, purge times and number of ALD cycles) and varying a dose of one precursor. Most commonly, this is carried out by varying the pulse (*i.e.* exposure) time, but the partial pressure of the precursor may be varied too. If the growth is self-limiting, the growth rate saturates to a constant level beyond a certain dose or pulse time (Figure 4.2a). When the growth reactions are favourable the saturation is

Table 4.5 Summary of methods used to characterize ALD processes.

Method	*Advantages*	*Limitations*
Film growth experiments	Focus on the overall goal; optimization of the film growth process	Limited chemical information; interpretation requires additional data
In situ measurements under real growth conditions:	Relevancy to the actual film growth processes	Difficulties in sampling
• microgravimetry	Direct measurement of (relative) surface mass changes	Interpretation requires certain assumptions of the surface species
• mass spectrometry	Identification of volatile by-products at various stages of an ALD cycle	Cracking and reactions during ionization complicate interpretation
• optical methods	Sensitive observation of changes in surface termination	Chemical interpretation is difficult
Measurements after an inert transfer from the growth conditions to the analysis chamber	Availability of numerous applicable surface analytical techniques ensures thorough characterization	Possible changes during the sample transfer
Reactions under high vacuum conditions	Availability of numerous applicable surface analytical techniques ensures thorough characterization	Representative nature is questionable because of differences in pressures (pressure gap) and reaction times
Reactions on high surface area substrates	The large amount of products makes it possible to use routine chemical techniques, like IR, NMR and elemental analysis	Representative nature is questionable because of very long reaction times
		Possible changes during the sample transfer

achieved rapidly, but with less aggressive chemistry longer exposure times may be needed. Naturally, rapid saturation is beneficial for throughput.

If the growth rate does not saturate but continues to increase with a steady slope with increasing metal precursor pulse time, precursor decomposition is likely taking place (Figure 4.2a). A slope that increases with temperature gives further evidence for decomposition. Under these conditions the growth is not true ALD but involves a CVD component from the decomposition reaction. Two regions are then distinguished from the growth rate *vs.* pulse time curves: with small pulse times the desired exchange reactions dominate while with longer pulses the precursor decomposition dominates. The lack of saturation does not necessary mean that the process is useless, however, but special care is needed to deal with the decomposition contribution. In essence the process is a mixed ALD-CVD, where the ALD part is usually maximised by using pulse times shorter than that corresponding to the inflection point between the two regions.

Sometimes the growth rate may show a maximum at certain exposure time and then start to decrease. This is an indication of a weak tendency of the precursor to etch the film. If the etching reaction is more favourable, the film growth may be fully prevented.

A similar experiment is repeated with the other precursor. Most non-metal precursors do not cause a steady increase of growth rate even if they were decomposing because their decomposition products are usually volatile. Thus, for non-metal precursors the outcome of this experiment is usually a nicely saturating growth rate *vs.* pulse time curve.

Purge times. In true ALD processes the purge periods must be long enough to completely separate the precursors in the gas phase. Proper purge times are found by varying them and finding the shortest value that still results in a constant growth rate and uniform film; with purge times that are too short the pulses overlap, resulting in increased growth rate and less uniform films.

For subsequent work the shortest possible value is usually not chosen, though, but some safety margin is left.

Growth rate vs. *temperature*. In principle the above experiments should be repeated at every growth temperature of an interest. This is highly time consuming, however, and most often the temperature effect is probed by using a set of pulse and purge times found to result in self-limiting growth at one temperature. The resulting growth rate *vs.* temperature curve may adopt different shapes (Figure 4.2b) but usually three regions are distinguished. The self-limiting ALD region is most often found at intermediate temperatures. In this region the growth rate may be temperature independent but, unlike as often thought, this is not a necessary requirement for the growth to be self-limiting ALD. The growth may decrease or increase with temperature, but then the self-limiting temperature range may be hard to distinguish without the above described growth rate *vs.* dose experiments. The decrease of growth rate occurs if the density of reactive surface sites (intermediates) is the limiting factor for the growth rate and decreases with increasing temperature. This has quite often been observed in ALD of oxides where surface hydroxyl groups are often the preferred reaction sites for the adsorbing metal precursors. The concentration of the hydroxyl groups is limited by a dehydroxylation process:

$$2\text{–OH(s)} \rightarrow \text{–O – (s)} + H_2O(g) \tag{4.18}$$

Because of the heterogeneity of the surface, dehydroxylation does not occur sharply at any certain temperature but increases quite smoothly with temperature. An increase in growth rate in the temperature region corresponding to the self-limiting growth is less common but has been observed with some metals.[80,83,84] With noble metal processes using O_2 as the other precursor the increase is likely due to increased formation of subsurface oxygen (Section 4.3.3.2).

At the lower temperature side of the self-limiting range the growth rate usually decreases rapidly while going towards lower temperatures (Figure 4.2b). This is commonly taken as an indication of kinetic limitation, *i.e.* the given pulse times are not long enough for the reactions to proceed to completion. In principle, longer exposure times could lead to completion of the reactions, thereby returning the process to the self-limiting conditions. The required times may be impractically long, however. Sometimes the growth rate may also increase towards lower temperatures because of condensation of low vapour pressure precursors to adsorption layers thicker than a monolayer.

The upper limit for the temperature range of self-limiting growth is usually set by the metal precursor decomposition. If decomposition occurs on the substrate it leads to an increase in growth rate (Figure 4.2b), but if it occurs already before the substrate a decrease of the growth rate may occur because of a reduced precursor flux arriving on the surface. Another possible reason for a decrease of the growth rate with increasing temperature is desorption of the adsorbed precursors, but, apart from elemental precursors, little evidence exists for this mechanism.

Nucleation. A plot of film thickness *vs.* the number of ALD cycles reveals details of the nucleation behaviour. If the plot is linear and goes through the origin, nucleation is likely unproblematic and occurs without delay. However, if the curve is nonlinear at low cycle numbers, or if the lowest cycle numbers have not been studied and the linear extrapolation of the measurement points does not go through the origin, the nucleation appears to be retarded for some reason (Figure 4.5). To reveal the fine details, one should not actually make the plot with film thickness but with the amount of atoms deposited (atoms cm^{-2}), as measured with, for example, MEIS or Rutherford backscattering spectroscopy. This is because at the nucleation stage the film is often non-continuous. Another approach for following the nucleation is to measure the surface coverage of substrate and film atoms *vs.* the number of ALD cycles with low energy ion scattering. The longer it takes to completely cover the substrate, *i.e.* to reach full coverage of the film, the more retarded and/or island-like the nucleation is.

4.3.3.2 In Situ Characterization

As will be described in Section 4.4, most thermal ALD processes are run at pressures around 1 Torr, and plasma enhanced processes around 10 mTorr. These pressures rule out many common surface analytical techniques from *in situ* characterization of the ALD chemistry. Therefore, microgravimetry, mass spectroscopy and optical techniques have been most often used.

A quartz crystal microbalance (QCM) is a widely used microgravimetry tool for ALD process characterization.[30,31,39] QCM can be used at least up to 450 °C provided that temperature variations are carefully eliminated or compensated for.[90] Therefore, direct exposure of QCM to cool gases just fed into the reactor must be avoided. This can be done by placing the QCM after the substrates so that the gases are warmed up while flowing across the substrates before they reach the QCM (Figure 4.8).

When carefully calibrated, QCM provides absolute mass changes occurring during each step in an ALD process. However, relative mass changes are already enough for studying the chemical

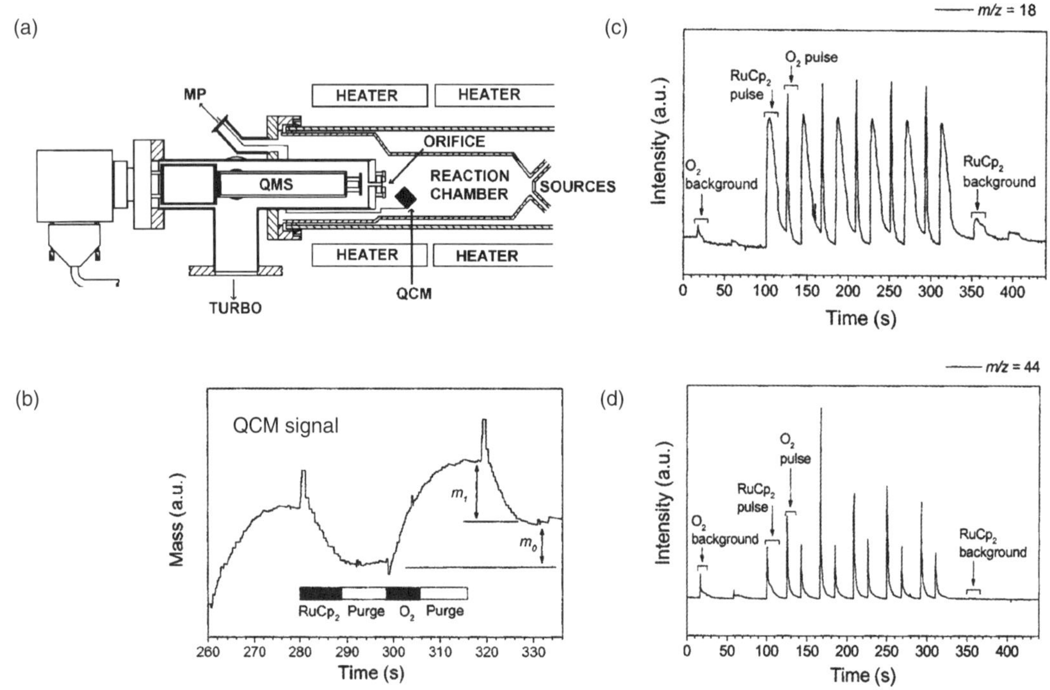

Figure 4.8 (*a*) A flow type ALD reactor with a quartz crystal microbalance (QCM) and a differentially pumped quadrupole mass spectrometer (QMS) for *in situ* reaction mechanistic studies. Precursors are transported with the carrier gas to the reaction chamber from sources (not shown) on the right-hand side and pumped out by the mechanical pump (MP). A small fraction of the flow goes through the orifice and the QMS chamber to the turbomolecular pump. (Reprinted with permission from ref. 91. Copyright 2001, American Chemical Society.) As an example, QCM signal (*b*) and two masses [*m*/*z* 18, H_2O (*c*); *m*/*z* 44, CO_2 (*d*)] in the Ru ALD process with $RuCp_2$ and O_2 as precursors are shown. In the QCM data m_0 is the mass change during one complete ALD cycle and m_1 the mass change during the $RuCp_2$ pulse. The background signals in the QMS data arise when several pulses of one precursor are given without the other precursor in between, hence representing the contribution arising directly from the precursor rather than from the surface reactions with the other precursor. (Reproduced by permission of The Electrochemical Society from ref. 36.)

reactions. The mass change during a full ALD cycle (m_0) is proportional to the molar mass of the film material (M_0) while the mass change during a given precursor pulse (m_1) is proportional to the molar mass of species added or removed (M_1), hence $m_0/m_1 = M_0/M_1$. The species involved in the reaction during the precursor pulse can be studied by comparing the measured m_0/m_1 ratio to M_0/M_1 ratios calculated for various suggested reactions. For example, in the $TiCl_4$–D_2O process for TiO_2, the adsorbing $TiCl_4$ may, in principle, react with 0–4 hydroxyl groups, losing the corresponding number of its chlorine ligands. Each proposed mechanism leads to a unique M_0/M_1 ratio (Table 4.6), thus enabling one to distinguish the dominant route. In practice, the reaction mechanism changes quite smoothly with temperature: at low temperatures two ligands are lost upon adsorption while at higher temperatures molecular adsorption seems to occur (Figure 4.9).[91] Naturally, the mechanism deduced is an average of all the elementary reactions occurring.

Figure 4.8 gives an example of Ru process with $RuCp_2$ and O_2 as precursors.[83] At the beginning of the $RuCp_2$ pulse, the mass first increases but soon starts to decrease and the overall mass change is negative, indicating that more material is removed than added. During the oxygen pulse the mass

Table 4.6 Possible reactions in the TiO_2 ALD process with $TiCl_4$ and D_2O as precursors; n is the number of chlorine ligands released during the $TiCl_4$ pulse. The m_0/m_1 ratios have been calculated as $m_0/m_1 = M(TiO_2)/[M(TiCl_4) - nM(DCl)]$ and are directly comparable with the m_0/m_1 ratios determined with QCM, with m_0 being the mass change during one complete ALD cycle and m_1 the mass change during the $TiCl_4$ pulse.

N	*Reactions*	m_0/m_1
0	$TiCl_4$ (g) → $TiCl_4$ (s) $TiCl_4$ (s) + $2D_2O$ (g) → -TiO_2 (s) + 4DCl (g)	0.42
1	-OD (s) + $TiCl_4$ (g) → -O-$TiCl_3$ (s) + DCl (g) -O-$TiCl_3$ (s) + $2D_2O$ (g) → $(\text{-O-})_2$Ti(OD) (s) + 3DCl (g)	0.53
2	2(-OD) (s) + $TiCl_4$ (g) → $(\text{-O-})_2TiCl_2$ (s) + 2DCl (g) $(\text{-O-})_2TiCl_2$ (s) + 2 D_2O (g) → $(\text{-O-})_2Ti(OD)_2$ (s) + 2DCl (g)	0.70

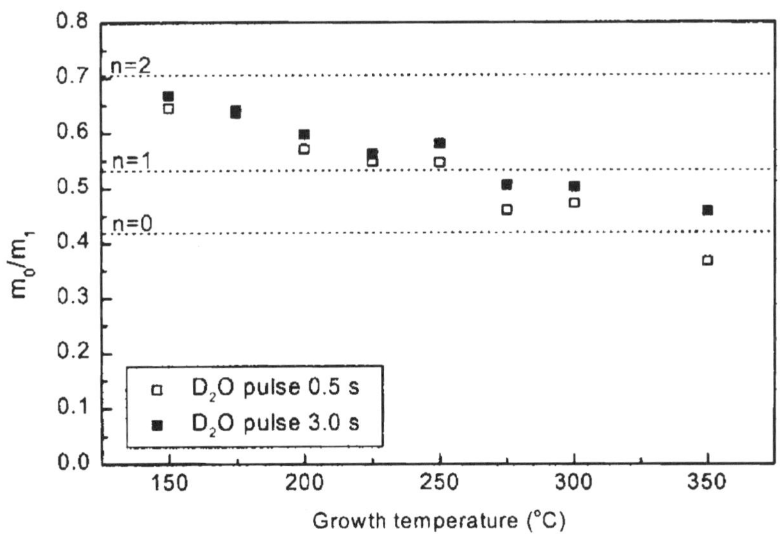

Figure 4.9 Temperature dependence of m_0/m_1 ratios determined with QCM for the $TiCl_4 + D_2O \rightarrow TiO_2$ ALD process with 0.5 and 3.0 s D_2O pulses (*cf.* Table 4.6). (Reprinted with permission from ref. 91. Copyright 2001, American Chemical Society.)

increases so much that the total mass change during a complete ALD cycle is positive, as it must be for a deposition process. This behaviour is quite unique to noble metal ALD processes; in other ALD processes a mass increase is seen during the metal precursor pulse and a mass decrease has been more common than a mass increase during the non-metal precursor pulse. In understanding the different behaviour of noble metal ALD processes mass spectrometry has proven to be an extremely valuable supplement to QCM, as will be explained next.

Mass spectrometry provides information about gas phase species – precursors and by-products – in the reactor during the various stages of an ALD cycle. Because pressures in typical ALD reactors are too high for the commonly used quadrupole mass spectrometers (QMS), differential pumping is needed (Figure 4.8). Rapid and representative sampling is ensured by placing the QMS in an immediate vicinity of the reaction chamber and by using orifices instead of capillaries for the pressure reduction from the reaction chamber to the QMS chamber. Careful heating of the sampling port and ion source is essential to avoid condensation of low vapour pressure compounds.

Figure 4.8(c) and (d) shows traces of the main by-products H_2O (*m*/*z* 18) and CO_2 (*m*/*z* 44), respectively, formed in the reactions between $RuCp_2$ and O_2 in the same Ru ALD process as analysed with QCM.[83] Formation of CO_2 and H_2O indicates that the cyclopentadienyl ligands are oxidatively decomposed. As these by-products are formed also during the $RuCp_2$ pulse, oxygen must remain on the surface after the O_2 pulse. From the growth rate and the relative ratios of CO_2 and H_2O liberated during the $RuCp_2$ and O_2 pulses one could estimate that the amount of oxygen left on the surface after the O_2 pulse and reacting with the incoming $RuCp_2$ is equal to 3 ML. Indeed, from surface science studies it is known that a ruthenium surface dissociates O_2 and at elevated temperatures the resulting oxygen atoms can penetrate into the subsurface region.

Figure 4.10 shows the proposed reaction mechanism for the $RuCp_2$–O_2 process. The surface left after the previous O_2 pulse contains atomic oxygen both at the surface and in the subsurface region.

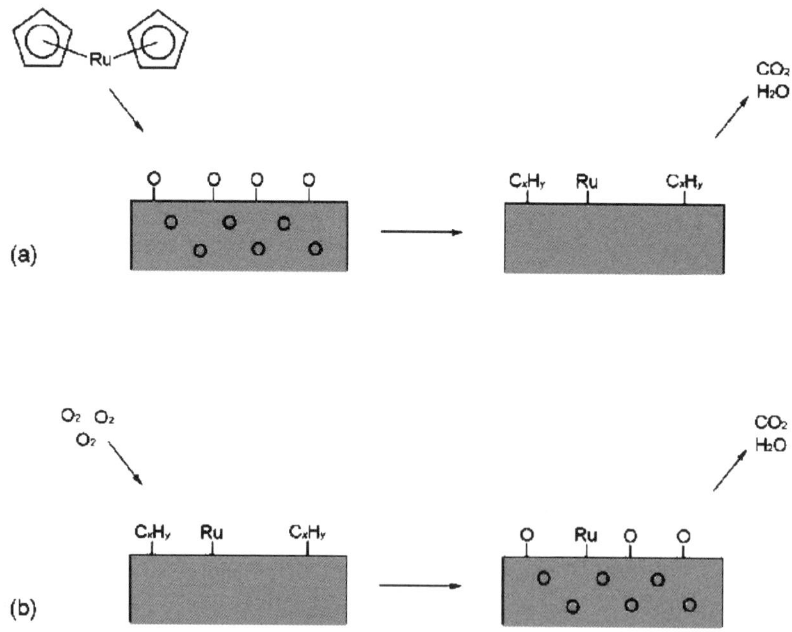

Figure 4.10 Proposed reactions taking place on the ruthenium surface (*a*) during the $RuCp_2$ pulse and (*b*) during the O_2 pulse. (Copyright T. Aaltonen 2005, University of Helsinki.)

Once $RuCp_2$ adsorbs on the surface, the oxygen atoms react with the ligands, forming CO_2 and H_2O. There is not enough oxygen to oxidize the ligands completely, and consequently some hydrocarbon residues remain. These become oxidized during the following O_2 pulse, which also reforms the oxygen subsurface region. The mass of oxygen consumed from the surface during the $RuCp_2$ pulse is higher than the mass of Ru atoms and ligand residues added, hence explaining the observed mass decrease that is uncommon to ALD processes other than noble metals.

Optical techniques do not have pressure limitations but provide only limited chemical information. Vibration spectroscopies (IR, Raman) are in this respect an important exception but their sensitivity is usually not high enough to detect monolayers on planar surfaces. However, under carefully optimized conditions, FTIR has been successfully employed in characterizing ALD growth of various oxides.[92,93] In addition, FTIR has been used quite extensively in analyzing ALD surface reactions on high surface area substrates (Section 4.3.3.3). Another useful method is optical emission spectroscopy that has been used in characterizing plasma species in plasma enhanced ALD processes.[94]

4.3.3.3 High Surface Area Substrates

High surface area substrates have been used in ALD both because of the interest in coating them for various applications, like catalysts, and to enable studies on ALD chemistry with techniques that are not sensitive enough on planar surfaces. Basically, two kinds of porous substrates have been used: relatively thin porous layers and membranes, and thicker beds of powders.

The thin porous layers already provide a high enough surface area, and thereby sufficient quantities of reaction intermediates so that FTIR can be used much more readily than with planar substrates.[53] The measurements can be performed either *in situ* inside the reactor or *ex situ* after inert transfer. Both transmission and reflection mode can be used. From the appearance and disappearance of the characteristic peaks of surface intermediates one can follow the surface reactions.

Porous powders like silica and alumina may have specific surface areas as high as hundreds of $m^2 g^{-1}$. Therefore, the amounts of surface intermediates formed in the ALD reactions increase to such high levels that techniques that are not commonly considered as surface analytical methods can be used.[95,96] NMR and elemental analysis have been the most common of these. Elemental analysis gives, for example, the ligand-to-metal ratio in the adsorbate, and thereby reveals how many ligands a precursor molecule has lost while adsorbing on the surface. NMR, on the other hand, distinguishes and quantitatively measures differently bound hydroxyl groups before and after the reaction, for instance.

A limitation of the porous and high surface area substrates is that longer exposure times are inevitably needed to saturate the surface reactions. With thin porous layers and membranes the exposure times have been minutes and with powders an hour has typically been needed compared with second or even subsecond scales on planar surfaces. In an ideal ALD process the difference in the exposure times should not matter. However, as is often the case, the chemistry is not fully ideal but involves some undesired side reactions, most often precursor self-decomposition. Such non-saturative reactions naturally become more pronounced the longer the exposure times. Therefore, the decomposition onset temperatures obtained from experiments with planar and porous substrates may be quite different, the latter giving a lower onset temperature.

4.4 ALD Reactors

Similar to ALD chemistry, ALD reactors also share many similarities with their CVD counterparts, yet they also have some important differences that reflect the unique features of ALD. One is the need to supply the precursors alternately hundreds or even thousands of times in each deposition process. Reasonable throughput requires the cycling to be carried out as rapidly as possible, thereby putting strict demands on both the source and reaction chamber design. In contrast, thanks to the self-limiting growth mechanism, less attention needs to be devoted to uniform distribution of the precursors across the substrates. It must be emphasized that this relaxation is strictly valid only when the process is truly self-limiting, in other words for ideal ALD. If there is any CVD component involved, the reactor design must take this into account, thus approaching CVD reactors. Finally, because growth rates in ALD are typically only weakly temperature dependent, temperature uniformity is usually not as critical in CVD.

This chapter focuses mostly on the unique features of ALD reactors. Chapter 2 of this book covers the components that are in common with CVD reactors. For detailed descriptions of ALD reactors see, for example, refs. 11,15 and 97–101.

4.4.1 Operation Pressure

ALD processes have been run over a wide pressure range from UHV to atmospheric, and even supercritical fluid has been tested, though with limited success so far.[102] In practice, however, most reactors seem to be of the flow-type where a continuous flow of inert gas is used for precursor transportation and purging, and the pressure is commonly in the range 1–10 Torr. Purging with inert gas is used because that is faster than evacuating the reactor between each precursor pulse. Most often nitrogen or argon is used as the inert gas, with flow rates ranging from 0.5 slm in small research reactors to some tens of slm in the largest production reactors. With such high flow rates the inert gas serves potentially as the largest source of impurities in the process. Reactive residual contents must be at least on a ppm level, and even much lower in the most sensitive processes.

Plasma and radical assisted ALD processes are often though not always run in higher vacuum (*e.g.* 1–100 mTorr range) than thermal ALD because of more favourable plasma generation and longer lifetime of radicals under the reduced pressure. To maintain the low pressure, usually there is no constant flow of inert gas for precursor transportation. Also, the purging with inert gas may be replaced by simple evacuation, though purge pulses are still quite common.

Truly high vacuum conditions have been quite rarely used in ALD. Sometimes molecular beam epitaxy reactors have been operated in the ALD mode. Another reason for using high vacuum conditions is to study the ALD surface chemistry with various surface characterization techniques that are not applicable in the pressure range of common ALD type reactors (Section 4.3.3).

4.4.2 Precursor Sources with Valving System

The basic requirement of ALD, the alternate supply of the precursors, requires a separation of the precursor source and the reaction chamber by a mechanical valve or some other rapid valving system. For fast cycling, all cold spots, dead spaces and turbulences in the source lines and the reaction chamber must be carefully eliminated.

ALD sources have conventionally been classified as external or internal, depending on whether they are placed in the same vacuum chamber as the substrates. They have also been called gas/liquid and solid sources, respectively. This division is not clear, however, as solids are nowadays increasingly supplied from external sources.[103,104] Therefore, the valving system and the vapour pressure of the compound being supplied perhaps best describe the differences between the two main ALD source types: mechanically valved high vapour pressure sources and inert gas valved (see below) low vapour pressure sources. Though the choice of source type follows the vapour pressure of the precursor, this distinction is not always clear either, and for many compounds both kinds of source types have been used in different reactors.

High vapour pressure sources are nearly always external, outside the reaction chamber, and a mechanical pulsing valve separates the source from the reaction chamber. If the precursor needs to be heated to reach the desired vapour pressure, source lines and valves must also be heated to avoid condensation. The whole source–valve assembly may be enclosed into a separate oven.

High vapour pressure sources can further be divided into vacuum draw type and conventional bubblers. In the vacuum draw the pressure of the precursor in the source is higher than the reactor pressure on the other side of the pulsing valve, and thus the vapour is simply drawn from the source into the reactor once the valve is opened. Vacuum draw is therefore applicable for all gases and for those liquids that have vapour pressure higher than the reactor pressure. No carrier gas is used to drive the precursor into the reactor, but further transportation inside flow-type reactors usually employs carrier gas. In bubblers, an inert carrier gas flows through or over the source compound. Bubblers can therefore also be used with precursors whose vapour pressure is not high enough for vacuum draw but still high enough so that no extensive heating of the source is needed. Though somewhat more complicated, bubblers are often used also with liquids whose vapour pressure would allow their delivery by the simpler vacuum draw as well. Bubblers become the choice over the vacuum draw especially when there is a need to increase the delivery rate for large surface area applications.

Pulsing of the high vapour pressure sources is relatively simple with fast solenoid or pneumatic valves. Short response times below 20 ms, reliable operation for millions to tens of millions of cycles, no particle or other kind of contamination and consistent flow performance are key requirements for ALD valves.[105] Pneumatically driven ultrahigh-purity diaphragm valves seem to be the best at meeting semiconductor application purity requirements. High operation temperature is of course also very important with less volatile precursors. Special pneumatic valves can be operated up to 200–250 °C, which can be considered as the present upper limit for the mechanically valved sources.

Mechanical valves also enable simple control of the amount of precursor being delivered in each pulse. This is done by placing two valves in series and leaving a known volume between them (Figure 4.11).[106] First the valve next to the source is opened to fill the volume with the precursor vapour with a pressure controlled by the source temperature. Then the first valve is closed and the second one opened to empty the volume into the reaction chamber.

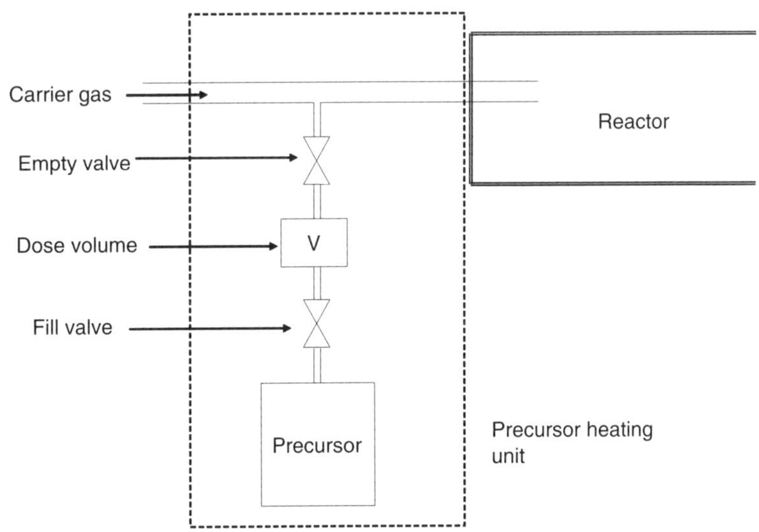

Figure 4.11 Schematic of a source setup with a precursor vessel, two on-off valves and a precursor dose controlling volume enclosed in a heating system, *e.g.* oven, maintaining the components at a common temperature. During operation the dose volume is first filled by opening only the fill valve, and then emptied by opening only the empty valve. (According to ref. 76.).

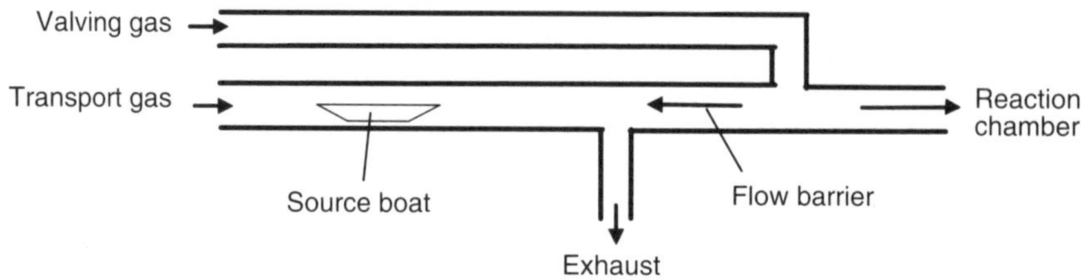

Figure 4.12 Schematic of an inert gas valving system.

Special low vapour pressure sources are unavoidably needed with low volatility precursors that do not reach a high enough vapour pressure at temperatures achievable with the mechanically valved sources. For example, $ZnCl_2$, a precursor used in the first commercial application of ALD, *i.e.* thin film electroluminescent displays, needs to be heated around 350 °C to reach a sufficient vapour pressure. Even the newest high temperature valves can not be used at this high temperature. The low vapour pressure precursors are placed inside the reactor in a region that is heated to a temperature giving the desired vapour pressure, typically 0.1–1 Torr. Alternatively, the heated source may also be placed outside the reactor and connected to the chamber with heated lines and an appropriate hot valving system.

Pulsing of low vapour pressure sources is not trivial because of the high temperatures involved. A major breakthrough in ALD technology was the development of an inert gas valving system (Figure 4.12).[11,97–99] The valving action is realized by directing two inert gas flows into each source tube, for example by employing coaxial tubes. One is a transport gas flowing over the source and the other one a valving gas. The valving gas is fed into the source line at a point between the source and the reaction chamber. When the source is in its off-state, the transport gas flow is off and the valving gas on. At the point of its entry into the source line the valving gas divides into two parts: one purges the reaction chamber and the other flows toward the source setting a diffusion barrier that prevents precursor molecules entering the reaction chamber. The source is turned on simply by switching the valving gas flow off and the transport gas on. This breaks the diffusion barrier and lets a pulse of precursor molecules to enter the reaction chamber. While the valves employed for switching the gas flows remain at room temperature, the valving action itself can take place at any temperature, often it is the reaction chamber temperature. Careful control of conductances around the valving point is obviously critical for the inert gas valving system.

While inert gas valved sources were first developed for truly low vapour pressure precursors, they are quite extensively used also with precursors that from vapour pressure point of view could be supplied from mechanically valved sources too. This choice is often made with research reactors that are already equipped with the inert gas valving system; once available, these sources are very convenient to operate, in research at least. Inert gas valving may also be combined with mechanical valves to back-up them in case of minor leakage.[104]

Liquid injection has been increasingly studied as an alternative way for ALD precursor delivery. Particularly attractive liquid injection delivery systems are those that inherently operate in the pulsed, drop-on-demand mode. One example is the TriJet device that is based on similar technology as used in automobile fuel injectors and has been widely explored especially in the related atomic vapour deposition method.[23] Liquid injection delivery can be used with both liquid precursor compounds and solutions containing the precursor dissolved into an appropriate, preferably inert solvent. The liquid is injected to an evaporation zone where it is completely vaporized and transported with an inert gas to the reaction chamber. In most systems evaporation and further transportation occur in an immediate sequence and probably to some extent also simultaneously. In an alternative device[107] the evaporation and transportation are separated time wise: first the liquid is evaporated into a storage volume that is then rapidly emptied to deliver the precursor vapour upon demand. Monitoring the pressure inside the storage volume enables one to accurately calculate and control the precursor dose.

The benefit of liquid injection delivery is that the precursor experiences elevated temperature only for a very short time, a few seconds at maximum. This is critical if the precursor does not fully meet the requirement of being thermally stable against self-decomposition and may degrade if heated for prolonged times in the source. It is also possible to control accurately the volume of liquid injected in each pulse. Furthermore, especially from a production point of view it is important that the liquid injection sources are less labour intensive than internal sources. Potential disadvantages include reliability issues, like clogging of the injector, residual particle formation and, in the case of solutions, participation of the solvent in growth reactions.

Avoidance of particles on the films is critical for high process yields in semiconductor applications, for example. Some solid precursors consist of very fine particles and are therefore highly potential particle emitters, though particles may form from liquid residues too. Effective particle filtering is therefore necessary. Mechanical, electrostatic and inertial trap filters have been suggested.[103,104] Mechanical filters, such as sinters and membranes, may suffer from low flow conductances, however, which increase the precursor pulse rise and decay times and thereby lengthen the ALD cycle time. These problems may be avoided by doing the filtering before the valving point.

Commonly, precursors are transported to the reaction chamber along separate lines, one for each precursor. This avoids deposition on the source line walls and the consequent cleaning procedures. The closer to the substrate the lines merge, the less precursor becomes wasted. However, the region next to the merging point of the source lines, *i.e.* the point where the film growth starts, is often disturbed by various kinds of inlet effects. Hence there should be some dummy area between the merging point and the substrate. The inlet effects are often clear in the cross-flow reactors but may be hidden in the perpendicular-flow reactors (see below).

4.4.3 Reaction Chamber

ALD reactors are usually of either the cross-flow (also called as a flow-channel or travelling-wave reactor) or perpendicular-flow type (Figure 4.13). Perpendicular-flow reactors may employ a showerhead or a simpler gas dispersing device to ensure uniform gas distribution. Showerheads may either have common channels for the two precursors so that the ALD growth occurs already inside the showerhead or there may be separate channels for the precursors so that the growth occurs only outside the showerhead. In the simplest case the perpendicular-flow reactor may just have precursor inlets on top of the chamber and the uniformity control is left to the self-limiting ALD growth mechanism. However, if the top wall with the inlet port is brought very close to the substrate and susceptor (Figure 4.13), the reactor is better described as a radial-flow channel reactor, rather than perpendicular-flow reactor, and the benefits and concerns related to the cross-flow reactor apply, though in a radial symmetry.

The benefit of the cross-flow reactor over the perpendicular-flow reactor is the speed with which the cycling can be performed, and hence better throughput. The speed benefit arises from the geometry of the flow channel between the substrate and the top wall, or another substrate as is often the case with two or multiple substrates. The height of the channel (distance between the opposite surfaces) can be made very small, down to a millimetre range. Because of the close proximity of the substrate and the adjacent wall, precursor molecules make multiple hits with the surfaces while being transported over the substrate along the flow channel. The multiple-hit conditions increase the probability of a molecule finding an open adsorption site as well as the rate with which the substrate surface becomes saturated with the precursor. Therefore, (1) precursor utilization efficiency is increased and (2) exposure time shortened compared to more open reaction

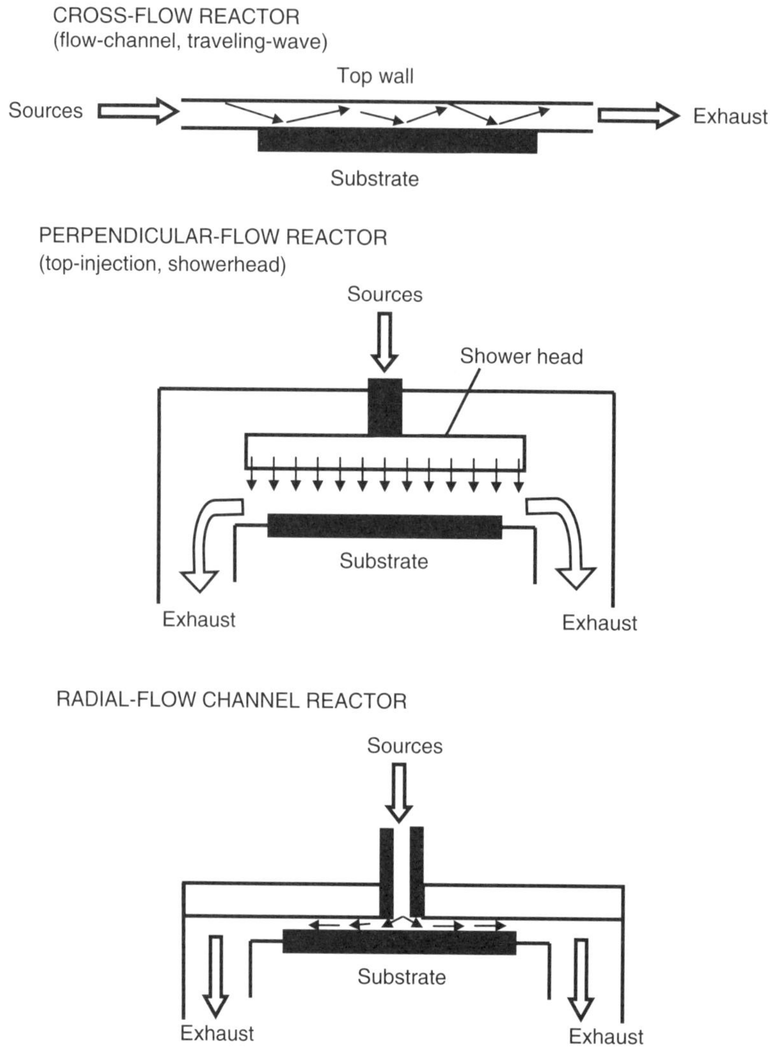

Figure 4.13 Common reactor chamber designs in flow-type ALD reactors.

chamber configurations; with good chemistry and a high enough precursor partial pressure 0.1 s can be long enough to saturate the surface.

The small volume of the flow channel, together with the plug-like flow conditions in the most commonly used pressure range, also ensures very rapid purging of the reactor, under favourable conditions 0.1 s suffices for completing the purging. So, with rapidly proceeding chemistry one may complete an ALD cycle in less than 0.5 s. With large substrates larger amounts of precursor need to be transported, but cycle times below 1 s have been reported, *e.g.* 475 ms for Al_2O_3 deposition from $Al(CH_3)_3$ and O_3 onto 300 mm wafers.[25]

The disadvantage of the cross-flow reactor is its sensitivity to various kinds of non-idealities like precursor self-decomposition and by-product readsorption (section 4.2.3), both leading to thickness non-uniformity. By-product readsorption can be highly pronounced in cross-flow reactors because the by-products travel in front of the precursor pulse that is producing them; at the front of the pulse precursor molecules find open adsorption/reaction sites on the surface, react with them and release by-product molecules that thereby replace the precursor molecules in the flow.[11] The effect becomes further emphasized if the growth starts immediately before the substrate and hence different parts of the substrate are subjected to largely different doses of by-product.

Recognizing the widespread use of showerhead reactors in CVD, it is natural that the perpendicular-flow reactor is more tolerant to the non-idealities that add a CVD kind of contribution to the ALD process. An optimum result may be expected when the reactor is carefully designed so that each site on the substrate surface receives the precursor pulse front at the same time without being first exposed to the by-products. The injector piece unavoidably increases the reaction chamber volume to be purged, however. To minimize the volume the injector piece can be brought closer to the substrate than commonly used in CVD.

4.4.4 Batch Reactors

Single wafer reactors are preferred in semiconductor industry but due to the low effective deposition rates of ALD these have obvious limitations in throughput. Therefore, semi-batch reactors (*e.g.* four wafers) and multiple parallel process modules in a single process tool have been suggested. On the other hand, large-batch tools with up to 150 wafers with 200 or 300 mm diameter[14,15,25] or 50 wafers with 300 mm diameter[108] are also commercially available, though with significantly longer cycle times. In non-semiconductor applications large-batch reactors are common, *e.g.* 44 substrates measuring 40×50 cm or 82 substrates of 16×27 cm size.[11,109] In these reactors, substrates are packed back-to-back so that the growth surfaces from the adjacent substrate pairs form a flow channel similar to the single wafer cross-flow reactors (Figure 4.14). Thus, there are multiple parallel flow channels, and the cycling can be carried out nearly as fast as in single wafer reactors; cycle times have been as short as 2–4 s in the 44 substrate reactor.[25] Obviously, the perpendicular-flow design can not be employed in large batch processing. In protective coating applications, like those applied on jewellery (Section 4.5), quite massive batch processing is necessary to make the process cost-effective (Figure 4.15).

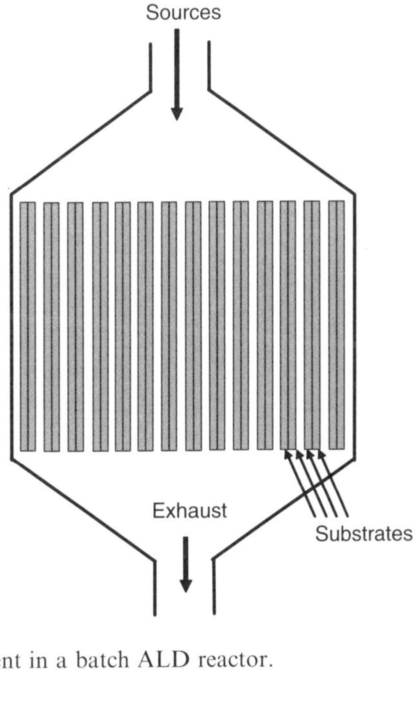

Figure 4.14 Substrate placement in a batch ALD reactor.

Figure 4.15 Example of a batch ALD reactor with 2000 silver jewellery pieces being coated simultaneously. (Copyright Beneq Oy, Finland.)

4.4.5 Wall Temperature

As the capability to rapidly change the gas composition in the reaction chamber is central to an efficient ALD process, any slow adsorption–desorption processes on the walls of the reaction chamber and the source lines must be excluded. This is done by employing hot wall reactors. Often the temperature is the same or nearly the same as that of the substrate, and film grows also on the walls. Frequent cleaning of the walls is therefore needed. A warm wall configuration has been suggested as an alternative, in which the walls would be at an intermediate temperature that is below the onset temperature of the film growth but high enough to ensure rapid desorption processes. Such a reactor could avoid the need for frequent cleaning of the wall deposits but appropriate temperatures can hardly be found for all processes.

Many ALD reactors are based on a design where the reaction chamber and its heating elements are enclosed in a larger vacuum chamber with cold walls.[101] This is a practical solution, especially with large substrates and batches. In addition, it gives freedom to the reaction chamber designer because the reaction chamber does not need to withstand atmospheric pressure.

4.4.6 Plasma Sources

Plasma sources may be located either remotely, upstream of the substrates,[110] or so that the substrates become immersed in the plasma (Figure 4.16). Accordingly, different processes may be distinguished and named,[111] the main difference being how intense an ion bombardment the substrate surface experiences. Immersion of the substrate into the plasma is called direct plasma ALD, and is usually realized by placing the substrate on an electrode involved in plasma generation. Remote plasma configurations may further be divided to two groups depending on which species reach the substrate. If only radicals but not ions nor electrons reach the substrate, the process is called radical enhanced ALD (REALD). The intermediate case where the plasma is remote but still a small flux of charged species reaches the substrate is called remote plasma ALD. For example, at a pressure of about 10 mTorr, an ion flux of about $2 \times 10^{14}\,cm^{-2}\,s^{-1}$ has been

a)

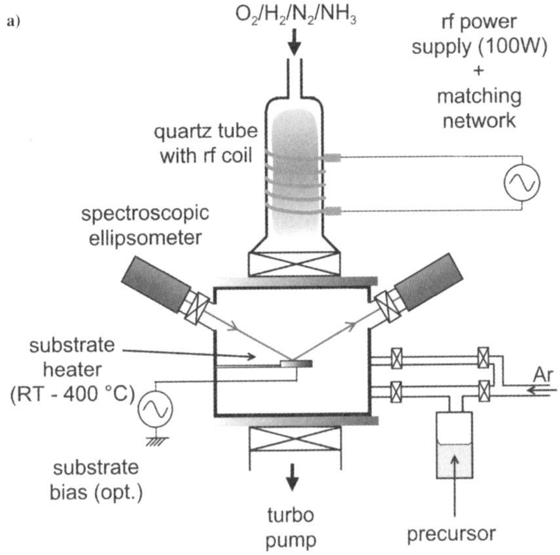

b)

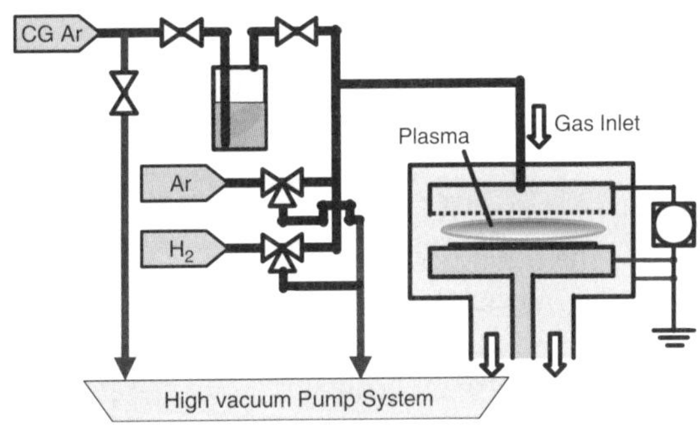

Figure 4.16 (*a*) Schematics of a remote plasma ALD reactor with an inductively coupled RF plasma source. Only the non-metal precursor gases are fed through the discharge region while metal precursors are introduced directly into the growth chamber. One high vapour pressure source is shown. The ellipsometer allows real-time measurement of thickness and optical properties of the growing film. (Copyright S. B. S. Heil and W. M. M. Kessels, Eindhoven University of Technology.) (*b*) Schematics of a direct plasma ALD reactor with a capacitively coupled RF-discharge. The plasma is pulsed so that it is on only during the non-metal precursor pulse. One bubbler source for metal precursor is shown. (Copyright C. Hoßbach, Dresden University of Technology.)

estimated to arrive on a substrate placed 30 cm from an inductively coupled H_2–N_2 plasma source.[112] Under these conditions the electron temperature close to the substrate is around 3 eV, which means that the plasma in the downstream region can still be considered as an active plasma.

No matter what kind of a plasma source is used, only non-metal precursors may be subjected to the plasma discharge because metal precursors could be fragmented to non-volatile products, thereby destroying the self-limiting growth mechanism. Applying plasmas to batches is obviously very difficult.

The most commonly used remote plasma source in ALD is the inductively coupled RF (radio-frequency) plasma (Figure 4.16a). To supply radicals in pulses as needed in an ALD process, the plasma discharge is either pulsed or kept constantly on but isolated from the reaction chamber by a high conductance valve. To avoid rapid recombination of plasma species outside the discharge these reactors are usually operated in sub-Torr pressure range. As already noted, a combination of pulsed plasmas and inert radical source gases (H_2, N_2, O_2) allows one to skip the purge step after the plasma pulse. The shortened cycle time makes the process faster than thermal ALD.

For operating the remote plasma configuration in the pressure range of a few Torr that is commonly needed for the inert gas valving, a surfatron microwave source has been integrated into a research scale reactor.[113] In this source the plasma discharge is ignited and maintained by a travelling surface wave on the inner wall of a quartz tube. The wave is launched by the surfatron placed outside the ALD reactor and runs into the reactor along the tube. The length of the discharge column can be adjusted by the power and gas flow rates so that the discharge extends close to the substrates, yet leaving them in the downstream position. In this way the distance the radicals need to travel can be made short enough so that REALD can be accomplished also in the inert gas valved flow-type reactors despite their relatively high pressure. However, scale-up to large substrates may be problematic.

Capacitively coupled RF plasma is the dominant direct plasma in ALD (Figure 4.16b). The substrate is placed on one electrode, usually the ground electrode, and the opposing surface is the powered electrode. The powered electrode may also be a showerhead.

4.4.7 Other Reactor Configurations

While the cross-flow and perpendicular-flow reaction chambers dominate, several alternative configurations have been reported too, though mostly for special purposes. The problem of coating of high aspect ratio structures was addressed by a flow-through reactor configuration (Figure 4.17).[114] As the name implies, the reactor may be used only with such substrates that have holes running through them from one face to the other. In the flow-through reactor all the precursor molecules are forced to flow through the substrate while in conventional reactors transportation into the pores occurs only by diffusion of some molecules from the main stream flowing along the substrate surface. Therefore, the flow-through approach both speeds up the process and increases substantially the precursor utilization efficiency.

Another reactor type possibly useful with high aspect ratio structures, in this case also close-ended, is that employing a fill-in–hold–pump-down approach. Rather than having a continuous flow of precursors through the reaction chamber, the chamber is isolated for a certain time between feeding in the precursor and pumping it out. The static exposure gives enough time for saturation to occur also at the bottom of the deep trenches without wasting precursor while waiting for the saturation to be completed. Because of the long exposure times, batch processing is preferable for

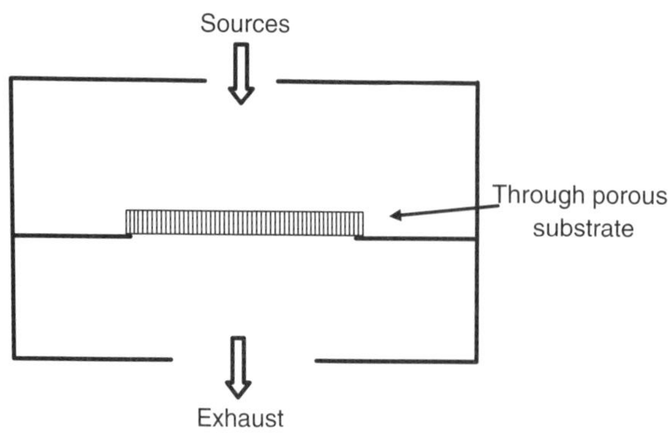

Figure 4.17 Flow through reactor design for coating samples with pores extending through the substrate.

improving productivity. Though this reactor type was first suggested in 1970s[3] it still has not entered wider use.

Besides pulsing the precursors onto a stationary substrate, an ALD process can also be realized by moving the substrate alternately to two constant precursor fluxes, with exposure to vacuum in between to remove excess precursor and by-products. This is performed most simply by rotating the substrate holder,[1,3,97,98,115,116] though other kinds of movements have been used, too.[117,118] In fact, a rotation in a high vacuum chamber with evaporation sources for elemental zinc and sulfur was used in the early ALD experiments.[1,3,97,98] While this approach avoids issues related to valving the precursors, it has other concerns, in particular mixing and reaction of the two precursors in the chamber. In the flow-type embodiment a buffer gas flow is used to avoid the mixing.[116] With this kind of a reactor the ALD cycle has been completed in 0.5 s. Despite the good research results in ALE of III-V compounds, no commercial reactors employing this concept have appeared yet.

Powders, when processed in any larger amounts like 1 cm^3, represent an extreme case both in terms of surface area to be coated and conformality. Early applications of ALD on powders dealt with silica and alumina catalyst supports that consist of porous powders with specific surface areas of several hundreds of $m^2 g^{-1}$.[95,96] However, as the focus was on modifying the surface composition by applying only a few ALD cycles, the reactor requirements were not as severe as in complete coating of powders. The powder was placed into a vessel with a porous bottom plate and the gases were flowed through the powder bed without any stirring. To saturate all the surfaces long exposure times of up to hours were necessary even with small amounts of powder. However, coating of non-porous powders with a continuous film is, reactor wise, more challenging, even if the surface area may be substantially less than with the porous powders. This is because of a risk of fusing the particles together with the coating if the process is performed without constant movement of the powder bed. Various reactors have been developed to overcome this problem. In one approach, the reactor vessel is constantly shaken by a mechanical vibration,[119] while another reactor involves a rotary porous tube as a holder for the particles and the balancing of gravity and centrifugal forces (<1 G) agitates the particles.[120] Yet another approach is to use a fluidized bed, but this does not allow one to use static exposures, which might be necessary to increase the utilization of less reactive precursors. The high surface area of powders leads to high precursor consumption and thereby emphasizes the importance of the precursor utilization efficiency.

4.4.8 Process Control Devices

One of the attractive features of the ALD method is that the self-limiting growth mechanism inherently provides good process control, including film uniformity, thickness control and repeatability. Therefore, it has not been that common to add separate process control devices. As seen in section 4.3.3, quadrupole mass spectrometry, quartz crystal microbalance and optical methods have been used in studying the ALD chemistry. These techniques can be used also for process control, as demonstrated with mass spectrometry, though with quite long cycle times (>60 s).[121] Effluent stream monitoring by optical emission spectroscopy was also shown to be sensitive for monitoring the process tool condition even if it could not distinguish separate pulses.[122] Pressure measurements in various points of the reactor, in precursor delivery systems in particular, are a simple way of monitoring the reactor condition, like integrity of seals in pulsing valves.[104]

4.5 Applications of ALD

The increase in application areas where ALD is considered and studied has matched the rapid growth of ALD research in general. Yet commercial applications of ALD are still quite limited.

This is largely because ALD is a relatively slow method and as such is often more expensive than the alternatives. Consequently, ALD appears to become adopted into commercial use only when the other methods fail. This is well exemplified by the existing commercial applications discussed below. Notably, however, it is quite hard to assess how broadly ALD is truly in production use already today: sometimes press-releases may be premature, sometimes the commercial use of ALD may be kept as a trade secret.

This section begins with a brief survey of applications where ALD is already in production, proceeding in more or less chronological order. This is followed by an overview of areas where ALD is under active research, with a possibility of production use in the near future.

4.5.1 Thin Film Electroluminescent Displays (TFELs)

TFELs (Figure 4.18) served as the original motivation for developing ALD technology, including both chemistry and reactors.[1,123–127] In the mid-1970s the concept of the TFEL display was known but the existing thin film deposition methods did not allow their production in a reliable and cost-effective manner. This was largely due to the high electric field (1–2 $MV\,cm^{-1}$) to which the insulator-luminescent-insulator three-layer structure is exposed in operation of the display. The high electric field makes the large area device sensitive to breakdown-causing defects, like pinholes. ALD turned out to enable a high yield in TFEL display manufacturing and has been in commercial use since the early 1980s. The thin film materials produced by ALD for the TFEL display are ZnS:Mn as the luminescent layer, Al_2O_3 or Al_xTi_yO as the insulators and Al_2O_3 as the passivation and protective layers. It must be emphasised that, unlike sometimes thought, the films in this first application of ALD are not atomically thin but at least 200 nm each, with the luminescent layer even exceeding one micrometer. Various other ALD materials and processes were also developed and explored over the years in TFEL display research.[11,127] Nevertheless, a major problem is still a lack of efficient blue emitting luminescent material. Commercial TFEL displays are therefore either monochrome or multicolour but not full-colour. However, the ALD research on TFEL display materials, particularly on insulators,[127] has served as a valuable basis for bringing ALD into new application areas.

4.5.2 Magnetic Heads

The second commercial application of ALD appeared around the turn of the millennium in thin film magnetic heads that are used to read and write data into hard disks. Because of the aggressive downscaling of the magnetic heads, the previously used sputtering method was rapidly approaching its capability limit in depositing thin insulating gap layers on non-planar surfaces. This created an urgent need for a low temperature deposition process of conformal insulator with good dielectric strength. ALD technology could, fortunately, reply to this request with the already existing nearly ideal $Al(CH_3)_3$–H_2O process for Al_2O_3[128] and rapidly became adopted into production.[129,130] It could even be stated that ALD saved the magnetic head industry.[130] Besides the insulating gap layers, ALD could obviously also find other use in magnetic heads but the required materials are such metals that have been lacking good ALD processes.

4.5.3 Microelectronics

At the end of the 1990s, intensive research also began into applying ALD in microelectronics, an area that bears the largest potential for high volume commercialization and thereby rapidly became dominant in the field of ALD. In fact, worldwide interest in ALD as a deposition method in the past decade has largely been driven by the need for highly conformal, ultrathin films for

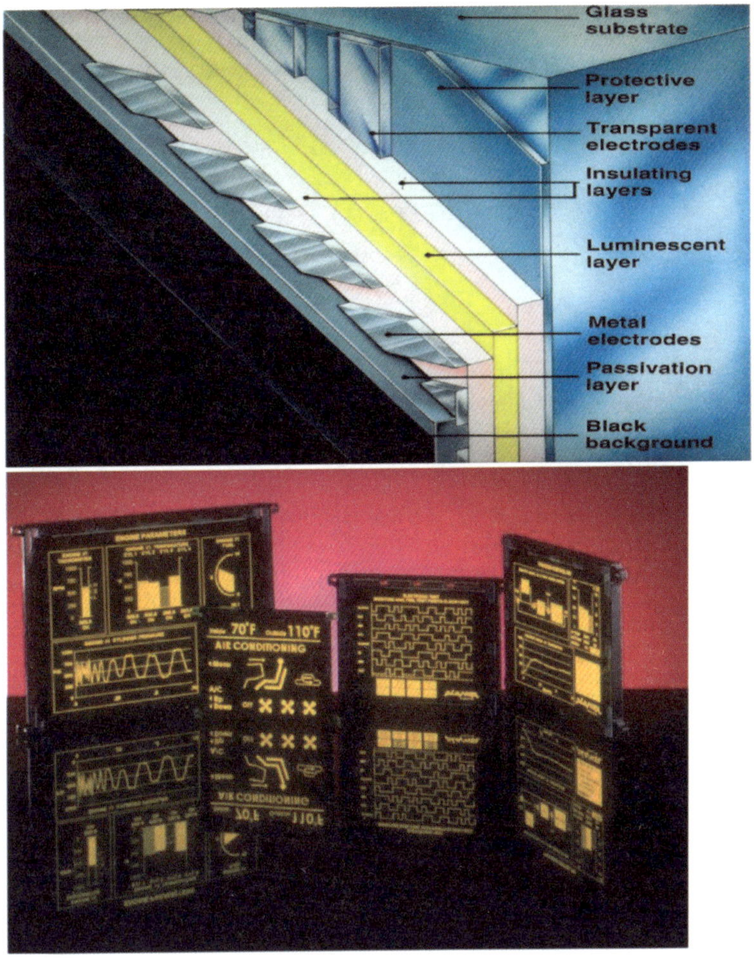

Figure 4.18 Schematics and photograph of TFEL displays. In the production of these displays ALD is used to deposit the luminescent (ZnS:Mn, about 1 μm thick), insulating (Al_2O_3 or Al_xTi_yO, about 200 nm each) as well as protective (Al_2O_3) and passivating (Al_2O_3) layers. In operation of the display a pixel is turned on by applying about 200 V ac voltage to the electrodes crossing at the pixel. The high ac field across the insulating-luminescent-insulating layer stack forces electrons to move back-and-forth across the ZnS:Mn luminescent layer. During this travel the electrons gain high energy from the electric field and impact excite the manganese ions. De-excitation results in light emission. The protective layer prevents sodium outdiffusion from soda lime glass while the passivating layer is against ambient. (Copyright Planar Systems Inc.)

dimensionally down-scaled semiconductor devices. Especially, high-*k* materials as alternative gate oxides in metal-oxide-semiconductor field effect transistors (MOSFET) as well as capacitor dielectrics in dynamic random access memories (DRAM) are the main application areas for ALD, as recognized by the semiconductor industry.[131] ALD is currently the leading candidate as the method to produce the films for these applications. ALD has been employed in DRAM production since early 2000, and has very recently been adopted, after a rather long period of intensive research, for the production of microprocessors.[2] In addition, the ALD of metals and nitrides for electrodes and interconnects has been studied widely. These three mainstream semiconductor applications, capacitor, gate-stack and interconnects (Figures 4.19–21 below) have expanded

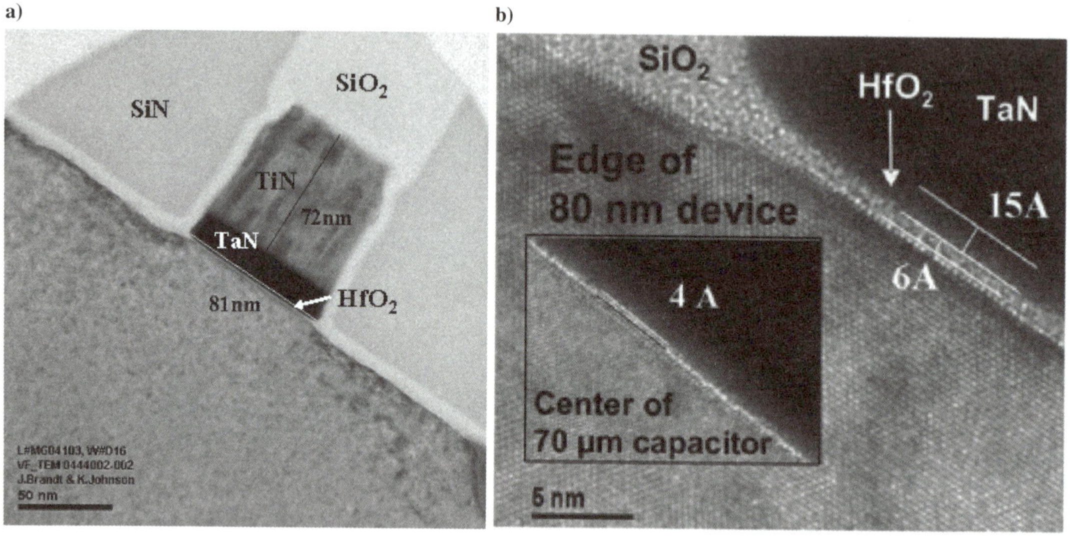

Figure 4.19 (*a*) Cross sectional TEM-image of an 80 nm transistor with ALD-HfO_2 gate dielectric and TaN/TiN gate electrode made by PVD. (Courtesy of Lars-Åke Ragnarsson at IMEC and Kevin D. Johnson at Intel.) (*b*) High-resolution TEM image of the edge of the transistor. Inset: the centre of the large area capacitor. (Reprinted with permission from L.-Å. Ragnarsson *et al.* *IEEE Trans. Electron Device*, 2006, **53**, 1657. Copyright 2006, IEEE.)

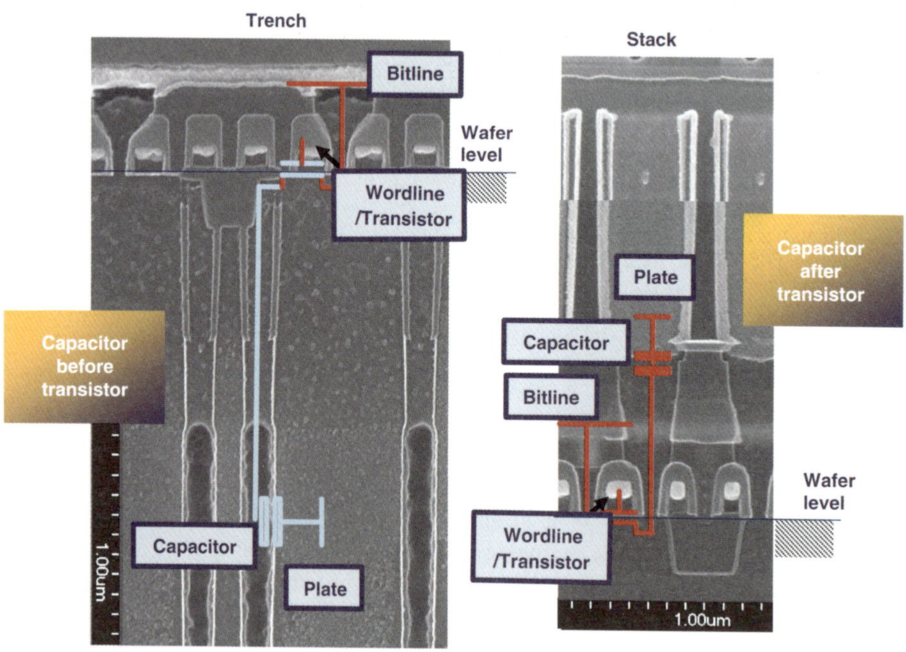

Figure 4.20 Cross-sectional images of trench and stack DRAM capacitors. In the trench structure the capacitor is buried into the wafer and is made before the transistor. In the stack structure the capacitor is prepared above and after the transistors.

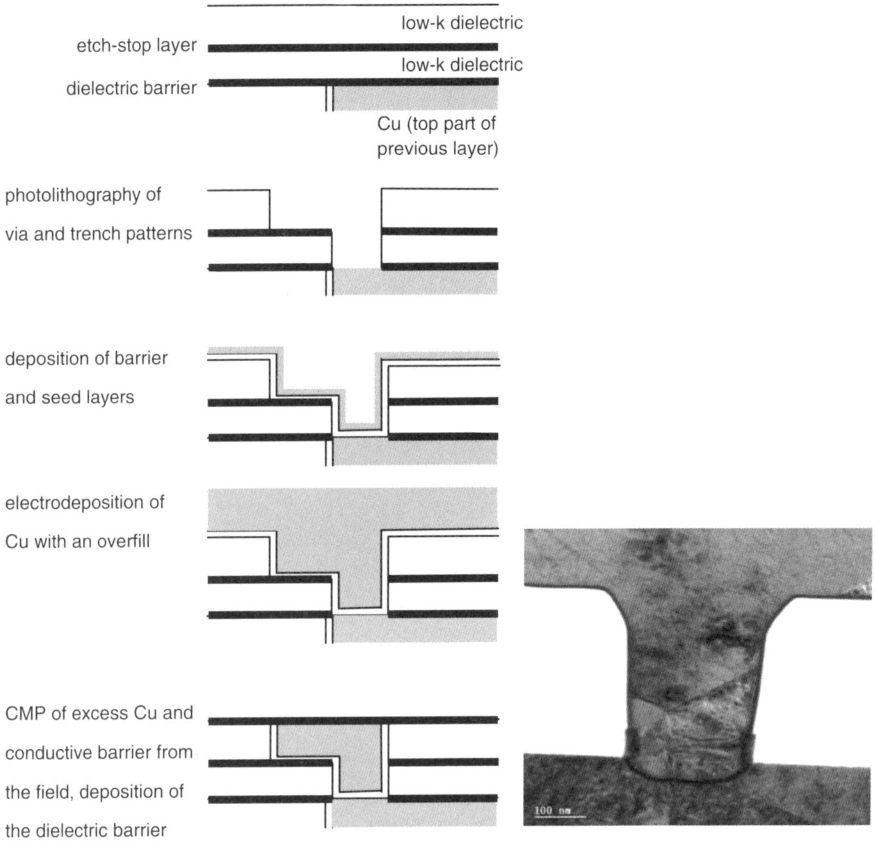

Figure 4.21 Dual-damascene process for the fabrication of copper interconnects. Dielectric films are deposited first and then patterned to open trenches and vias for copper wires. After barrier and seed layer deposition, copper is electrochemically overplated into the trenches and vias, and finally the excess is removed by chemical mechanical polishing (CMP). The TEM image shows an ALD WC_xN_y barrier in a via. (The TEM image is reprinted from ref. 140. Copyright 2002, with permission from Elsevier.)

ALD research activities from only a couple of dozens of research groups in the mid-1990s[132] to the current extensive worldwide research efforts in industry and academia.

The semiconductor industry has been following for several decades the so-called Moore's law, *i.e.* a steady exponential increase in the number of transistors on a silicon chip. To follow this scaling, the gate oxide layer in MOSFET, first SiO_2 and then nitrided SiO_2 has been downscaled to a thickness of only a few monolayers. However, this thin layer can no longer effectively prevent tunnelling resulting in high leakages. Therefore, the silicon oxide based gate insulator material must be replaced by a higher permittivity material. The search for such an alternative high-*k* material has been extensive and most ALD publications in the past ten years have focused on different aspects of depositing various high-*k* gate dielectrics.[133] To date, it seems that hafnium oxide based materials, containing silicon and nitrogen (HfSiON), have been selected as a short-term solution. However, the research effort continues to seek even better alternatives. In addition, three-dimensional transistor structures and replacement of silicon channel with germanium or compound semiconductors, maybe even carbon nanotubes and graphene, lead to integration issues of their own kind.

Owing to the incompatibility of high-*k* oxides with polysilicon gate, metal gates have been widely studied for MOSFET applications (Figure 4.19). A dual metal gate approach will most likely be used, one with a work function matched for pMOS and another one for nMOS.[134] This will lead to quite complicated process flow, especially if a so-called gate last or gate replacement approach needs to be chosen instead of the current gate first approach.[8] In the gate last approach the gate metal needs to be deposited into a trench structure, thus calling for good conformality of ALD. Accurate thickness control of the metal is also important in the preparation of the gate stacks. Many alternatives for the metal gates have been explored but working solutions have not been revealed. Because of the problems finding appropriate work functions, due especially to Fermi level pinning, alternative approaches to tuning the work function have also been sought. One way to tune the work function is to employ very thin dielectric capping layers between the high-k oxide and metal gate.[135]

High-k materials are needed for memory applications as well. The capacitance of a DRAM capacitor can be increased by decreasing the SiO_2 dielectric thickness, increasing the effective surface area and/or introducing a high-k dielectric. Storage capacitors can be divided in two types, stacked capacitors and trench capacitors (Figure 4.20), of which the latter offers the highest density, but the manufacturing process complexity may become an obstacle. Trench capacitors are constructed into high aspect ratio trenches to increase the surface area and thus the effective capacitance density. The area can further be increased by widening the trench profile (bottle-shape trenches) and by roughening the sidewalls of the trenches. ALD-Al_2O_3 has been evaluated as a replacement for silicon oxide and nitride-based insulators but higher permittivity material, such as oxides of zirconium and hafnium, are needed in the near future. As the trench aspect ratio is expected to increase up to $\sim$95 : 1 by the end of the decade, ALD is probably the only viable technique for such depositions.[131] Stacked capacitors have now been introduced into the sub-100 nm technology CMOS nodes and metal–insulator—metal capacitors with high-k materials, such as Ta_2O_5, are applied. Potential long-term solutions are based on ultrahigh-k dielectrics, such as perovskites. However, although the conformality requirements favour ALD as the deposition method, complex chemistry to produce a dielectric layer with a permittivity value of over 100, *e.g.* (Ba,Sr)TiO_3, sets challenges.[136] The high aspect ratios of the DRAM capacitor structures (trench capacitors in particular) make it necessary to also explore ALD for the capacitor electrodes. In trench capacitors TiN is currently used due to its high thermal stability and relatively low resistivity. In stack DRAM structures, noble metals are employed, such as Ru and Ir.

The main driving force for the adoption of ALD for interconnects is the shrinking geometries, which result in increasing aspect ratios. Modern copper based interconnects are made by the dual-damascene process (Figure 4.21). ALD has in the first place been explored for depositing the barrier layers that need to be made as thin as possible, a few nanometres only, to maximize the space available for copper. While being ultrathin, the barrier should prevent interdiffusion of copper and surrounding insulators and also serve as an adhesion promoter for Cu interconnect. In many cases, obtaining this thin continuous film is challenging. The rapid closure of growing nuclei is needed despite the sensitive surface chemistry in the initial stages of the growth. For a diffusion barrier, ALD TaN_x[137,138] and WC_xN_y[54,139,140] seem to be the most promising solutions. On top of the diffusion barrier, ideally significantly below 10 nm thick, a seed layer for Cu electroplating is required. A viable ALD process for Cu film growth has not been found, however, due to a lack of appropriate chemistry. The requirements for copper seed include also high purity, good adhesive properties towards the barrier and naturally high conformality and uniformity. The alternative candidates for copper seed are mostly noble metals, among which Ru has gained considerable interest recently. Ruthenium can also be employed as a combined barrier and seed.[141] Tungsten, on the other hand, has been grown by ALD as a seed for tungsten plug fill by CVD.[81,82]

Besides these three mainstream applications, emerging applications for ALD include RF and decoupling capacitors as well as non-volatile memories such as FeRAM, Flash and phase change memories.

4.5.4 Protective Coatings

The latest reported commercial application of ALD came from a quite surprising field, jewellery.[142] ALD Al_2O_3 proved effective in preventing silver tarnishing at thicknesses as low as about 10 nm. This low thickness is important because it leaves the visual appearance of the silver objects essentially unchanged. A robust process, inexpensive precursors and batch processing with 2000 valuable pieces (Figure 4.15) ensure the cost effectiveness of the process. Finnish companies Kalevala Koru and Lapponia Jewelry have proven the process in their production since 2006. Obviously, the effective sealing by conformal ALD coatings could also be used in providing chemical protection for many other objects. Promising results have been obtained in the preventing corrosion of stainless steel, for example.[143] There is a concern, however, about the erosive load that is nearly always present to some extent too. Because it is not economic to make the ALD films very thick their lifetime is limited under erosion. The most apparent solution would be to combine an ALD film with a thicker coating that would protect against the erosion while the ALD film below would seal the substrate against chemical attack.

4.5.5 Solar Cells

Besides reported commercial applications, ALD has been studied in several other areas. One of the first was solar cells. CdTe and CdS films were successfully deposited for CdTe based solar cells.[144,145] Promising results were also achieved in replacing CdS buffer layer in $Cu(In,Ga)Se_2$ solar cells with ALD made ZnO, ZnS and In_2S_3.[146–149] In addition, high quality transparent conducting oxides were developed.[11,150–153] However, it appears that in these conventional thin film solar cells ALD could not bring any clear benefit compared to other thin film deposition methods and thus could not compensate for its slowness and price. By contrast, in nanostructured solar cells, like the semiconductor nanocomposite made by infiltrating $CuInS_2$ into the pores of nanostructured TiO_2[154] and nanowire based dye-sensitized solar cells,[155,156] ALD obviously better distinguishes itself in providing a conformal coating of the highly demanding three-dimensional structures.

4.5.6 Optical Applications

In optics ALD seems to follow a similar development path as in solar cells: though good results have been obtained with conventional thin film devices ALD has not been able to replace the currently used faster methods, but in three-dimensional components more opportunities exist. Film thickness control requirements in certain areas of optics are stricter than anywhere else, thus in principle favouring, but also challenging, ALD. Multilayer interference coatings have been made with ALD for wavelength ranges from IR to visible and soft and hard X-rays,[157–160] but attempts to commercialize ALD for multilayer optics have so far failed as far as is known. In three-dimensionally structured optical components development is still going on and ALD has been explored in the preparation of novel optical devices from, for example, nanogratings[161] and porous silicon.[162] Inverse opal[163–166] (Figure 4.22) and woodpile[167] photonic crystals, inductive grid filters,[168] erbium-doped waveguides[169] and ultrahigh-resolution Fresnel zone plates for X-ray microscopy[170] are further examples of the exploitation of the conformality of ALD in three-dimensionally structured optical components.

4.5.7 Heterogeneous Catalysts

Heterogeneous catalysts were one of the first application areas ALD was targeted to.[95,96] The catalysts consist of high surface area porous powder supports, typically silica and alumina, whose

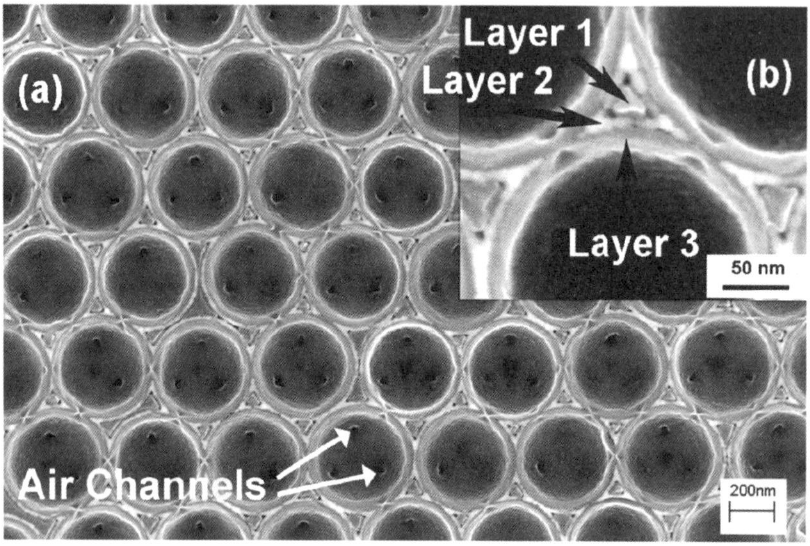

Figure 4.22 Ion milled cross section of the inverse opal photonic crystal (PC) with three-layer structure. The PC was fabricated by infiltration of opal template consisting of 466 nm silica spheres with 10 nm ZnS:Mn (layer 2) and 26 nm TiO_2 (layer 1), removing the silica template with HF, and backfilling with 10 nm TiO_2 (layer 3). (Reprinted with permission from ref. 166. Copyright 2006, American Institute of Physics.)

surfaces are modified with additives to achieve the desired catalytic performance. Usually, there is no need to deposit continuous films; instead, simply to tailor the surface chemistry with one or a few ALD cycles only. The high specific to surface area, typically in the range of hundreds of $m^2 g^{-1}$, requires very long exposure times to transport enough precursors for saturation of the whole surface. The results achieved with ALD made catalysts have been comparable and sometimes even somewhat better than those achieved by conventional methods. However, no such clear benefit has been found that would have justified the higher cost of ALD compared to impregnation, for example. Therefore, in heterogeneous catalysts ALD seems to remain in the role of a valuable research tool, allowing controlled preparation of catalysts with well-defined composition.

4.5.8 Coatings on Powders

Besides the surface modification of porous powders, ALD has also been used in encapsulating compact powders with continuous films.[119,120,171] If the ongoing work on upscaling the coating process to large batches turns out to be successful and economically feasible, potential application areas are numerous.

4.5.9 Photocatalysts

ALD has been shown to produce photocatalytically active TiO_2 films.[172] Titanium dioxide is the most widely used photocatalyst in air and water purification as well as in self-cleaning and sterilizing coatings. Many methods exist for the preparation of TiO_2 photocatalysts, making it hard for ALD to compete. Nevertheless, based on its excellent conformality and possibility to tailor the composition at atomic layer level ALD may find use in preparation of tailored three-dimensional photocatalyst structures.[173]

4.5.10 Coatings on Polymers

ALD films, Al_2O_3 in particular, have also been deposited on polymers at temperatures below 100 °C.[174] With radical enhanced ALD, Al_2O_3 has even been deposited at room temperature.[175] One of the main motivations for studies on the low temperature deposition on polymers has been for barrier coatings against water and oxygen permeation. A low water vapour transmission rate of 1.7×10^{-5} g m^{-2} per day at 38 °C was achieved with 25 nm Al_2O_3 deposited at 120 °C from $Al(CH_3)_3$ and water on polyethylene naphthalate.[176] This value approaches the challenging requirement of $< 10^{-5}$ g m^{-2} per day in organic light emitting diodes (OLEDs). Al_2O_3 films seem to degrade over time with exposure to water, however.[177] Combination of Al_2O_3 with SiO_2 into multilayer structures improves the stability and results in a lower water vapour transmission rate than either of the multilayer constituents alone.[177]

4.5.11 Micro-electro-mechanical Systems (MEMS)

MEMS nearly always contain three-dimensional structures whose conformal coating is an obvious application area for ALD. Yet so far there have been limited publications on ALD in MEMS applications.[178] Nevertheless, for example, wear resistant,[179] lubricating,[180] anti-stiction[181] and charge dissipating[182] coatings have been deposited by ALD on MEMS components.

4.5.12 Nanotechnology

Last but definitely not least, nanotechnology is an application area where ALD research is currently increasing the most rapidly. As is common to nanotechnology in general, commercial applications seem still quite distant, though rapid entries would not be that surprising either. As in many other areas, ALD benefits greatly from its perfect conformality and accurate thickness control. Various nanofibres and nanotubes have been coated by ALD, like carbon nanotubes,[183,184] Ge/Si[185] and ZnO nanowires,[155,186,187] self-assembled polymer fibres,[188] cellulose fibres,[189] metal wires made by templating with porous membranes[173] and electrospun ceramic and polymer nanowires.[190,191] In addition, many nanoporous membranes have been coated by ALD, like anodic alumina,[173,192] γ-Al_2O_3[193] and track-etched polycarbonate.[194] Other nanostructured objects coated with ALD include porous silicon,[195,196] porous anodic alumina films,[197,198] aerogels,[199] synthetic opal,[163–166]

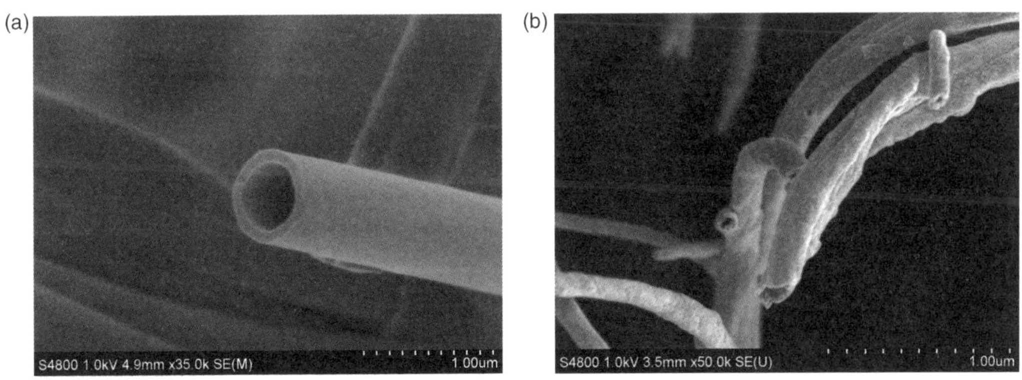

Figure 4.23 Hollow nanotubes made by coating electrospun polymer nanofibres (*a*) and cellulose nanofibres in paper (*b*) by ALD of Al_2O_3 and TiO_2, respectively, and removing the template fibres by calcination in air. [(*a*) Reprinted from ref. 191. Copyright 2007, with permission from Elsevier. (*b*) Copyright M. Kemell 2005, University of Helsinki.]

spheres[188] and objects originating from nature, *e.g.* diatoms,[200] viruses[201] and nanostructured butterfly wings.[202] If the substrate is removed after the coating, a hollow replica is formed from the ALD deposited film. Such a templating process has been used to prepare nanotubes with electrospun nanofibres (Figure 4.23a),[190] anodic alumina[197,198] and cellulose fibres[189] (Figure 4.23b) as templates, hollow nanospheres from polymer nanospheres[188] and inverse opals from synthetic opals.[163–166] Another way to form tubes from wire templates is to rely on the Kirkendall effect by reacting the wire with the coating; a hollow core is formed when the outdiffusion of the core material is faster than the indiffusion of the coating material.[203] For a more comprehensive overview of using ALD in nanotechnology, see the review by Knez *et al.*[204]

4.6 Conclusions

Its unique self-limiting growth mechanism gives ALD features unmatched by any other method: perfect conformality and uniformity over large areas, including large batch processing, and easy and accurate thickness control down to an atomic layer level. However, ALD is a slow method and even if this can be compensated for by large batch processing the speed has served as the main limitation for its wider use. In addition, relatively few processes fulfil all the requirements of ALD chemistry and thereby fully reach the ideal ALD characteristics. Indeed, the success of ALD is built very much on chemistry, not forgetting reactors that need to correct non-idealities in the chemistry. Another highly important requirement of reactor design is the need to perform an ALD cycle in as short a time as possible. While designing both ALD chemistry and reactors, the special features of ALD need to be thoroughly perceived.

ALD has been used commercially since early 1980s in electroluminescent display manufacturing. Currently, the semiconductor industry has become an area with the largest application potential. Other areas where ALD has been reported to be in production are magnetic heads and jewellery. In addition, there are many emerging application areas with great potential for commercialization, like MEMS devices, three-dimensionally structured solar cells and optics components, protective coatings, coated powders and various branches of nanotechnology.

After 30 years, from its development to a production proven method, ALD is now experiencing a rapid increase of interest. This is likely to lead to new processes and applications at an increasing rate.

References

1. T. Suntola and J. Hyvärinen, *Annu. Rev. Mater. Sci.*, 1985, **15**, 177.
2. M. T. Bohr, R. S. Chan, T. Ghani and K. Mistry, *IEEE Spectrum*, 2007, **44**(10), 29.
3. T. Suntola and J. Antson, *U.S. Patent 4058430,* 1977.
4. A. Pakkala, *AVS 4th International Conference on Atomic Layer Deposition*, Helsinki, Finland, 2004, presentation available on CD-ROM.
5. L. Niinistö, M. Ritala and M. Leskelä, *Mater. Sci. Eng. B*, 1996, **41**, 23.
6. M. Leskelä and M. Ritala, *Thin Solid Films*, 2002, **409**, 138.
7. M. Leskelä and M. Ritala, *Angew. Chem. Int. Ed.*, 2003, **42**, 5548.
8. H. Kim, *J. Vac. Sci. Technol. B*, 2003, **21**, 2231.
9. L. Niinistö, J. Päiväsaari, J. Niinistö, M. Putkonen and M. Nieminen, *Phys. Status Solid. A*, 2004, **201**, 1443.
10. R. L. Puurunen, *J. Appl. Phys.*, 2005, **97**, 121301.
11. M. Ritala and M. Leskelä, in *Handbook of Thin Film Materials*, ed. H. S. Nalwa, Academic Press, New York, 2002, **vol. 1**, pp. 103–159.
12. T. Suntola, *Appl. Surf. Sci.*, 1996, **100/101**, 391.

13. A. Dip, G. M. Eldallal, P. C. Colter, N. Hayafuji and S. M. Bedair, *Appl. Phys. Lett.*, 1993, **62**, 2378.
14. R. de Blank, G. J. Snijders, S. Beulens, L. Vandezande, R. Wilhelm and A. Hasper, *Electrochem. Soc. Proc.*, 2003, **14**, 225.
15. E. Granneman, P. Fischer, D. Pierreux, H. Terhorst and P. Zagwijn, *Surf. Coat. Technol.*, 2007, **201**, 8899.
16. K. B. Chung, C. N. Whang, H. S. Chang, D. W. Moon and M.-H. Cho, *J. Vac. Sci. Technol. A*, 2007, **25**, 141.
17. J. Ihanus, M. Ritala, M. Leskelä, T. Prohaska, R. Resch, G. Friedbacher and M. Grasserbauer, *Appl. Surf. Sci.*, 1997, **120**, 43.
18. M. Copel, M. Gibelyuk and E. Gusev, *Appl. Phys. Lett.*, 2000, **76**, 436.
19. K.-E. Elers, V. Saanila, P. J. Soininen, W.-M. Li, J. T. Kostamo, S. Haukka, J. Juhanoja and W. F. A. Besling, *Chem. Vap. Deposition*, 2002, **8**, 148.
20. K.-E. Elers, T. Blomberg, M. Peussa, B. Aitchison, S. Haukka and S. Marcus, *Chem. Vap. Deposition*, 2006, **12**, 13.
21. T. Ostead, G. C. D'Couto, S.-H. Lee, P. Wongsenakaum, J. Collins and K. Levy, *Solid State Technol.*, 2002, **45**(9), 51.
22. S.-H. Kim, E.-S. Hwang, B.-M. Kim, J.-W. Lee, H.-J. Sun, T. E. Hong, J.-K. Kim, H. Sohn, J. Kim and T.-S. Yoon, *Electrochem. Solid-State Lett.*, 2005, **8**, C155.
23. M. Schumacher, P. K. Baumann and T. Seidel, *Chem. Vap. Deposition*, 2006, **12**, 99.
24. E. J. Shero, C. G. Wang, M. Verghese, J. Reed, E. Liu and G. Wilk, *AVS 7th International Conference on Atomic Layer Deposition*, San Diego, United States, 2007, presentation available on CD-ROM.
25. S. Haukka, *ECS Transactions*, 2007, **3**, 15.
26. M. Ritala and M. Leskelä, *Appl. Surf. Sci.*, 1994, **75**, 333.
27. K.-E. Elers, M. Ritala, M. Leskelä and E. Rauhala, *Appl. Surf. Sci.*, 1994, **82/83**, 468.
28. M. Juppo, M. Vehkamäki, M. Ritala and M. Leskelä, *J. Vac. Sci. Technol. A*, 1998, **16**, 2845.
29. P. Tägtsröm, P. Mårtensson, U. Jansson and J.-O. Carlsson, *J. Electrochem. Soc.*, 1999, **146**, 3139.
30. J. Aarik, A. Aidla, K. Kukli and T. Uustare, *J. Cryst. Growth*, 1994, **144**, 116.
31. J. Aarik, K. Kukli, A. Aidla and L. Pung, *Appl. Surf. Sci.*, 1996, **103**, 331.
32. J. Skarp, *U.S. Patent 4486487*, 1984.
33. J. W. Elam and S. M. George, *Chem. Mater.*, 2003, **15**, 1020.
34. M. Juppo, M. Ritala and M. Leskelä, *J. Vac. Sci. Technol. A*, 1997, **15**, 2330.
35. C. A. Wilson, R. K. Grubbs and S. M. George, *Chem. Mater.*, 2005, **17**, 5625.
36. T. Aaltonen, A. Rahtu, M. Ritala and M. Leskelä, *Electrochem. Solid-State Lett.*, 2003, **6**, C130.
37. S.-C. Ha, E. Choi, S.-H. Kim and J. S. Roh, *Thin Solid Films*, 2005, **476**, 252.
38. A. Niskanen, K. Arstila, M. Ritala and M. Leskelä, *J. Electrochem. Soc.*, 2005, **152**, F90.
39. A. Rahtu, T. Alaranta and M. Ritala, *Langmuir*, 2001, **17**, 6506.
40. A. Rahtu and M. Ritala, *Chem. Vap. Deposition*, 2002, **8**, 21.
41. S. B. S. Heil, P. Kudlacek, E. Langereis, R. Engeln, M. C. M. van de Sanden and W. M. M. Kessels, *Appl. Phys. Lett.*, 2006, **89**, 131505.
42. A. Kosola, M. Putkonen, L.-S. Johansson and L. Niinistö, *Appl. Surf. Sci.*, 2003, **211**, 102.
43. S.-C. Ha, E. Choi, S.-H. Kim and J. S. Roh, *Thin Solid Films*, 2005, **476**, 252.
44. H. B. Park, M. Cho, J. Park, S. W. Lee, C. S. Hwang, J.-P. Kim, J.-H. Lee, N.-I. Lee, H.-K. Lee and S.-J. Oh, *J. Appl. Phys.*, 2003, **94**, 3641.
45. M. Ritala, K. Kukli, A. Rahtu, P. I. Räisänen, M. Leskelä, T. Sajavaara and J. Keinonen, *Science*, 2000, **288**, 319.
46. A. Rahtu, M. Ritala and M. Leskelä, *Chem. Mater.*, 2001, **13**, 1528.

47. A. Rahtu and M. Ritala, *Langmuir*, 2002, **18**, 10046.
48. D. Hausmann, J. Becker, S. Wang and R. G. Gordon, *Science*, 2002, **298**, 402.
49. T. Pilvi, M. Ritala, M. Leskelä, M. Bischoff, U. Kaiser and N. Kaiser, *Appl. Opt.* 2008, **47**, C271.
50. T. Pilvi, K. Arstila, M. Ritala and M. Leskelä, *Chem. Mater.*, 2007, **19**, 3387.
51. J. W. Klaus, O. Sneh, A. W. Ott and S. M. George, *Surf. Rev. Lett.*, 1999, **6**, 435.
52. J. W. Klaus, O. Sneh and S. M. George, *Science*, 1997, **278**, 1934.
53. J. W. Klaus, S. J. Ferro and S. M. George, *Thin Solid Films*, 2000, **360**, 145.
54. W. Koh, D. Kumar, W.-M. Li, H. Sprey and I. J. Raaijmakers, *Solid State Technol.*, 2005, **48**(6), 54.
55. A. C. Jones, H. C. Aspinall, P. R. Chalker, R. J. Potter, T. D. Manning, Y. F. Loo, R. O'Kane, J. M. Gaskell and L. M. Smith, *Chem. Vap. Deposition*, 2006, **12**, 83.
56. W. Cho, K.-S. An, T.-M. Chung, C. G. Kim, B.-S. So, Y.-H. You, J.-H. Hwang, D. Jung and Y. Kim, *Chem. Vap. Deposition*, 2006, **12**, 665.
57. G. W. Hwang, H. J. Lee, K. Lee and C. S. Hwang, *J. Electrochem. Soc.*, 2007, **154**, G69.
58. M. Vehkamäki, M. Ritala, M. Leskelä, A. C. Jones, H. O. Davies, T. Sajavaara and E. Rauhala, *J. Electrochem. Soc.*, 2004, **151**, F69.
59. M. Tiitta and L. Niinistö, *Chem. Vap. Deposition*, 1997, **3**, 167.
60. J. Päiväsaari, M. Putkonen and L. Niinistö, *Thin Solid Films*, 2005, **472**, 275.
61. M. Leskelä, K. Kukli and M. Ritala, *J. Alloys Compd.*, 2006, **418**, 27.
62. G. B. Stringfellow, *Organometallic Vapor-Phase Epitaxy: Theory and Practice*, Academic Press, 1999, pp. 572.
63. G. S. Higashi and C. G. Flemming, *Appl. Phys. Lett.*, 1989, **55**, 1963.
64. W. Cho, K. Sung, K.-S. An, S. S. Lee, T.-M. Chung and Y. Kim, *J. Vac. Sci. Technol. A*, 2003, **21**, 1366.
65. M. Ozeki, *Mater. Sci. Rep.*, 1992, **8**, 97.
66. J. Niinistö, M. Putkonen, L. Niinistö, K. Kukli, M. Ritala and M. Leskelä, *J. Appl. Phys.*, 2004, **95**, 84.
67. J. Niinistö, M. Putkonen, L. Niinistö, S. L. Stoll, K. Kukli, T. Sajavaara, M. Ritala and M. Leskelä, *J. Mater. Chem.*, 2005, **15**, 2271.
68. J. Niinistö, M. Putkonen, L. Niinistö, K. Arstila, T. Sajavaara, K. Kukli, M. Ritala and M. Leskelä, *J. Electrochem. Soc.*, 2006, **153**, F39.
69. J. Niinistö, M. Putkonen, L. Niinistö, F. Song, P. Williams, P. N. Heys and R. Odedra, *Chem. Mater.*, 2007, **19**, 3319.
70. K. Kukli, J. Niinistö, A. Tamm, J. Lu, M. Ritala, M. Leskelä, M. Putkonen, L. Niinistö, F. Song, P. Williams and P. N. Heys, *Microelectr. Eng.*, 2007, **84**, 2010.
71. M. Vehkamäki, T. Hatanpää, T. Hänninen, M. Ritala and M. Leskelä, *Electrochem. Solid-State Lett.*, 1999, **2**, 504.
72. M. Vehkamäki, T. Hänninen, M. Ritala, M. Leskelä, T. Sajavaara, E. Rauhala and J. Keinonen, *Chem. Vap. Deposition*, 2001, **7**, 75.
73. J. Ihanus, T. Hänninen, T. Hatanpää, T. Aaltonen, I. Mutikainen, T. Sajavaara, J. Keinonen, M. Ritala and M. Leskelä, *Chem. Mater.*, 2002, **14**, 1937.
74. J. Päiväsaari, J. Niinistö, P. Myllymäki, C. Dezelah, M. Putkonen, M. Nieminen, L. Niinistö and C. H. Winter, *Top. Appl. Phys.*, 2007, **15**.
75. K. Kukli, M. Ritala, T. Sajavaara, J. Keinonen and M. Leskelä, *Chem. Vap. Deposition*, 2002, **8**, 199.
76. D. M. Hausmann, E. Kim, J. Becker and R. G. Gordon, *Chem. Mater.*, 2002, **14**, 4350.
77. W. J. Maeng and H. Kim, *Electrochem. Solid-State Lett.*, 2006, **9**, G191.
78. J. W. Elam, M. Schuisky, J. D. Ferguson and S. M. George, *Thin Solid Films*, 2003, **436**, 145.
79. M. Vehkamäki, T. Hatanpää, M. Ritala and M. Leskelä, *J. Mater. Chem.*, 2004, **14**, 3191.

80. B. S. Lim, A. Rahtu and R. G. Gordon, *Nature Mater.*, 2003, **2**, 749.
81. M. Yang, H. Chung, A. Yoon, H. Fang, A. Zhang, C. Kenpfler, R. Jackson, J. S. Byun, A. Mak, M. Eizenberg, M. Xi, M. Kori and A. K. Sinha, *Proc. of ULSI XVII*, 2002, 655.
82. S.-H. Kim, N. Kwak, J. Kim and H. Sohn, *J. Electrochem. Soc.*, 2006, **153**, G887.
83. T. Aaltonen, P. Alen, M. Ritala and M. Leskelä, *Chem. Vap. Deposition*, 2003, **9**, 45.
84. T. Aaltonen, M. Ritala, V. Sammelselg and M. Leskelä, *J. Electrochem. Soc.*, 2004, **151**, G489.
85. M. Ritala, P. Kalsi, D. Riihelä, K. Kukli, M. Leskelä and J. Jokinen, *Chem. Mater.*, 1999, **11**, 1712.
86. P. Alén, M. Juppo, M. Ritala, M. Leskelä, T. Sajavaara and J. Keinonen, *J. Mater. Res.*, 2002, **17**, 107.
87. P. Alén, M. Juppo, M. Ritala, T. Sajavaara, J. Keinonen and M. Leskelä, *J. Electrochem. Soc.*, 2001, **148**, G566.
88. S. M. Rossnagel, A. Sherman and F. Turner, *J. Vac. Sci. Technol. B*, 2000, **18**, 2016.
89. Y. J. Lee and S.-W. Kang, *J. Vac. Sci. Technol. A*, 2002, **20**, 1983.
90. A. Rahtu and M. Ritala, *Appl. Phys. Lett.*, 2002, **80**, 521.
91. R. Matero, A. Rahtu and M. Ritala, *Chem. Mater.*, 2001, **13**, 4506.
92. M. M. Frank, Y. J. Chabal and G. Wilk, *Mat. Res. Soc. Symp. Proc.*, 2003, **745**, N.2.41.
93. Y. Wang, M.-T. Ho, L. V. Goncharova, L. S. Wielunski, S. Rivillon-Amy, Y. J. Chabal, T. Gustafsson, N. Moumen and M. Boleslawski, *Chem. Mater.*, 2007, **19**, 3127.
94. S. B. S. Heil, P. Kudlacek, E. Langereis, R. Engeln, M. C. M. van de Sanden and W. M. M. Kessels, *Appl. Phys. Lett.*, 2006, **89**, 131505.
95. M. Lindblad, S. Haukka, A. Kytökivi, E.-L. Lakomaa, A. Rautiainen and T. Suntola, *Appl. Surf. Sci.*, 1997, **121/122**, 286.
96. S. Haukka, E.-L. Lakomaa and T. Suntola, *Stud. Surf. Sci. Catal.*, 1998, **120**, 715.
97. T. S. Suntola, A. J. Pakkala and S. G. Lindfors, *U. S. Patent. 4389973*, 1983.
98. T. S. Suntola, A. J. Pakkala and S. G. Lindfors, *U. S. Patent. 4413022*, 1983.
99. T. Suntola, in *Handbook of Crystal Growth*, ed. D. T. J. Hurle, Elsevier, Amsterdam, 1994, p. 601.
100. M. Ylilammi, *J. Electrochem. Soc.*, 1995, **142**, 2474.
101. J. I. Skarp, P. J. Soininen and P. T. Soininen, *Appl. Surf. Sci.*, 1997, **112**, 251.
102. D. Barua, T. Gougousi, E. D. Young and G. N. Parsons, *Appl. Phys. Lett.*, 2006, **88**, 92904.
103. J. Kesälä, *U.S. Patent 6699524*, 2001.
104. B. J. Aitchison, J. Maula, T. Lang, P. Kuosmanen, K. Härkönen, M. Sonninen, *U.S. Patent Appl., 2004/0124131.*
105. W. Glime and T. Seidel, *Semicond. Int.*, 2005, **28**(9), 76.
106. D. M. Hausmann, E. Kim, J. Becker and R. G. Gordon, *Chem. Mater.*, 2002, **14**, 4350.
107. P. Meneghini, A. Shajii, W.R. Entley, D. Smith, S.P. Nagarkatti, B. McKinney, *AVS 6th International Conference on Atomic Layer Deposition*, Seoul, Korea, 2006, presentations available on CD-ROM.
108. Y. Okuyama, C. Barelli, C. Tousseau, S. Park and Y. Senzaki, *J. Vac. Sci. Technol. A*, 2005, **23**, L1.
109. *European Semiconductor*, April 1999, p. 69.
110. S. B. S. Heil, J. L. van Hemmen, C. J. Hodson, N. Singh, J. H. Klootwijk, F. Roozeboom, M. C. M. van de Sanden and W. M. M. Kessels, *J. Vac. Sci. Technol. A*, 2007, **25**, 1357.
111. J. L. van Hemmen, S. B. S. Heil, J. H. Klootwijk, F. Roozeboom, C. J. Hodson and M. C. M. van de Sanden, *J. Electrochem. Soc.*, 2007, **154**, G165.
112. S. B. S. Heil, E. Langereis, F. Roozeboom, M. C. M. van de Sanden and W. M. M. Kessels, *J. Electrochem. Soc.*, 2006, **153**, G956.

113. A. Niskanen, A. Rahtu, T. Sajavaara, K. Arstila, M. Ritala and M. Leskelä, *J. Electrochem. Soc.*, 2005, **152**, G25.
114. M. Ritala, M. Kemell, M. Lautala, A. Niskanen, M. Leskelä and S. Lindfors, *Chem. Vap. Deposition*, 2006, **12**, 655.
115. S. M. Bedair and N. A. El-Masry, *Appl. Surf. Sci.*, 1994, **82/83**, 7.
116. M. A. Tischler and S. M. Bedair, *Appl. Phys. Lett.*, 1986, **48**, 1681.
117. A. Usui and H. Sunakawa, *Jpn. J. Appl. Phys.*, 1986, **25**, L212.
118. A. Koukitu, H. Nakai, A. Saegusa, T. Suzuki, O. Nomura and H. Seki, *Jpn. J. Appl. Phys.*, 1988, **27**, L744.
119. L. F. Hakim, S. M. George and A. W. Weimer, *Nanotechnology*, 2005, **16**, S375.
120. J. A. McCormick, B. L. Clouter, A. W. Weimer and S. M. George, *J. Vac. Sci. Technol. A*, 2007, **25**, 67.
121. L. Henn-Lecordier, W. Lei, M. Anderle and G. W. Rubloff, *J. Vac. Sci. Technol. B*, 2007, **25**, 130.
122. J. P. Loo, *AIP Conf. Proc.*, 2005, **788**, 187.
123. T. Suntola, J. Antson, A. Pakkala and S. Lindfors, *SID 80 Digest*, 1980, **11**, 108.
124. T. Suntola, *Mater. Sci. Rep.*, 1989, **4**, 261.
125. M. Leskelä and L. Niinistö, in *Atomic Layer Epitaxy*, ed. T. Suntola, M. Simpson, Blackie, Glasgow, 1990, p. 1.
126. R. Törnqvist, *Displays*, 1992, **13**, 81.
127. M. Leskelä, W.-M. Li and M. Ritala, *in Electroluminescence, Semiconductors, Semimetals*, 1999, **64**, 413.
128. A. Paranjpe, S. Gopinath, T. Omstead and R. Bubber, *J. Electrochem. Soc.*, 2001, **148**, G465.
129. A. Devasahayam, *Solid State Technol.*, 2005, **48**(9), Suppl. S16.
130. M. Kautzky, R. Lamberton, S. Chakravarty, L. Stearns, A. Kumar, J. Dolejsi, S. McKinlay, L. Colaianni and C. Perrey, *AVS 3th International Conference on Atomic Layer Deposition*, San Jose, United States, 2003, presentations available on CD-ROM.
131. *International Technology Roadmap for Semiconductors, Front end processes*, 2005, http://public.itrs.net. accessed August 2007.
132. L. Niinistö, *Curr. Opin. Solid State Mat. Sci.*, 1998, **3**, 147.
133. M. Ritala, in *High-k Gate Dielectrics*, ed. M. Houssa, Institute of Physics Publishing, Bristol, United Kingdom, 2004, pp. 17–64.
134. G. D. Wilk, R. M. Wallace and J. M. Anthony, *J. Appl. Phys.*, 2001, **89**, 5243.
135. L. Pantisano, T. Schram, B. O'Sullivan, T. Conard, S. De Gendt, G. Groeseneken, P. Zimmerman, A. Akheyar, M. M. Heyns, S. Shamuilla, V. V. Afanas'ev and A. Stesmans, *Appl. Phys. Lett.*, 2006, **89**, 113505.
136. M. Vehkamäki, T. Hatanpää, M. Ritala, M. Leskelä, S. Väyrynen and E. Rauhala, *Chem. Vap. Deposition*, 2007, **13**, 329.
137. P. Ho, R. Rajagopalan, H. Chong, H. Chung and J. Yu, *Semicond. Int.*, 2004, **27**(6), 61.
138. J. Gelatos, L. Chen, H. Chung, R. Thakur and A. Sinha, *Solid State Technol.*, 2003, **46**(2), 44.
139. W.-M. Li, M. Tuominen, S. Haukka, H. Sprey and I. J. Raaijmakers, *Solid State Technol.*, 2003, **46**(7), 103.
140. S. Smith, W.-M. Li, K.-E. Elers and K. Pfeifer, *Microel. Eng.*, 2002, **64**, 247.
141. T. P. Moffat, M. Walker, P. J. Chen, J. E. Bonevich, W. F. Egelhoff, L. Richter, C. Witt, T. Aaltonen, M. Ritala, M. Leskelä and D. Josell, *J. Electrochem. Soc.*, 2006, **153**, C37.
142. S. Sneck, *in NanoTechnology in Northern Europe conference*, Helsinki, 2007. www.beneq.com/download.php/ accessed August 2007.
143. R. Matero, M. Ritala, M. Leskelä, T. Salo, J. Aromaa and O. Forsen, *J. Phys. IV*, 1999, **9**, Pr8–493.

144. J. Skarp, E. Anttila, A. Rautiainen and T. Suntola, *Int. J. Sol. Energy*, 1992, **12**, 137.
145. T. Suntola, *MRS Bull.*, 1993, **45**(10).
146. S. Chaisitsak, T. Sugiyama, A. Yamada and M. Konagai, *Jpn. J. Appl. Phys.*, 1999, **38**, 4989.
147. A. Shimizu, S. Chaisitsak, T. Sugiyama, A. Yamada and M. Konagai, *Thin Solid Films*, 2000, **361–362**, 193.
148. E. B. Yousfi, T. Asikainen, V. Pietu, P. Cowache, M. Powalla and D. Lincot, *Thin Solid Films*, 2000, **361–362**, 183.
149. D. Hariskos, S. Sprieng and M. Powalla, *Thin Solid Films*, 2005, **480–481**, 99.
150. T. Asikainen, M. Ritala and M. Leskelä, *J. Electrochem. Soc.*, 1995, **142**, 3538.
151. H. Viirola and L. Niinistö, *Thin Solid Films*, 1994, **251**, 127.
152. V. Lujala, J. Skarp, M. Tammenmaa and T. Suntola, *Appl. Surf. Sci.*, 1994, **82/83**, 34.
153. A. Yamada, B. Sang and M. Konagai, *Appl. Surf. Sci.*, 1997, **112**, 216.
154. M. Nanu, J. Schoonman and A. Goossens, *Adv. Funct. Mater.*, 2005, **15**, 95.
155. M. Law, L. E. Greene, A. Radenovic, T. Kuykendall, J. Liphardt and P. Yang, *J. Phys. Chem. B*, 2006, **110**, 22652.
156. A. B. F. Martinson, J. W. Elam, J. T. Hupp and M. J. Pellin, *Nano Lett.*, 2007, **7**, 2183.
157. H. Kumagai, K. Toyoda, K. Kobayashi, M. Obara and Y. Iimura, *Appl. Phys. Lett.*, 1997, **70**, 2338.
158. M. Ishii, S. Iwai, H. Kawata, T. Ueki and Y. Aoyagi, *J. Cryst. Growth*, 1997, **180**, 15.
159. D. Riihelä, M. Ritala, R. Matero and M. Leskelä, *Thin Solid Films*, 1996, **289**, 250.
160. F. H. Fabreguette, R. A. Wind and S. M. George, *Appl. Phys. Lett.*, 2006, **88**, 13116.
161. J. J. Wang, X. Deng, R. Varghese, A. Nikolov, P. Sciortino, F. Liu, L. Chen and X. Liu, *J. Vac. Sci. Technol. B*, 2005, **23**, 3209.
162. V. Kochergein and H. Foell, *Mater. Sci. Eng. R*, 2006, **52**, 93.
163. J. S. King, D. P. Gaillot, E. Graugnard and C. J. Summers, *Adv. Mater.*, 2006, **18**, 1063.
164. A. Rugge, J. S. Becker, R. G. Gordon and S. H. Tolbert, *Nano Lett.*, 2003, **3**, 1293.
165. J. S. King, C. W. Neff, C. J. Summers, W. Park, S. Blomquist, E. Forsythe and D. Morton, *Appl. Phys. Lett.*, 2003, **83**, 2566.
166. J. S. King, E. Graugnard and C. J. Summers, *Appl. Phys. Lett.*, 2006, **88**, 081109.
167. J. H. Lee, W. Leung, J. Ahn, T. Lee, I.-S. Park, K. Constant and K.-M. Ho, *Appl. Phys. Lett.*, 2007, **90**, 151101.
168. K. Jefimovs, J. Laukkanen, T. Vallius, T. Pilvi, M. Ritala, T. Meilahti, M. Kaipiainen, M. Bavdaz, M. Leskelä and J. Turunen, *Microel. Eng.*, 2006, **83**, 1339.
169. K. Solehmainen, M. Kapulainen, P. Heimala and K. Polamo, *IEEE Photonics Technol. Lett.*, 2004, **16**, 194.
170. K. Jefimovs, J. Vila-Comamala , T. Pilvi, J. Raabe, M. Ritala, C. David, *Phys. Rev. Lett.*, 2007, **99**, 264801.
171. J. D. Ferguson, A. W. Weimer and S. M. George, *Appl. Surf. Sci.*, 2000, **162–163**, 280.
172. V. Pore, A. Rahtu, M. Leskelä, M. Ritala, T. Sajavaara and J. Keinonen, *Chem. Vap. Deposition*, 2004, **10**, 143.
173. M. Kemell, V. Pore, J. Tupala, M. Ritala and M. Leskelä, *Chem. Mater.*, 2007, **19**, 1816.
174. C. A. Wilson, R. K. Grubbs and S. M. George, *Chem. Mater.*, 2005, **17**, 5625.
175. A. Niskanen, K. Arstila, M. Ritala and M. Leskelä, *J. Electrochem. Soc.*, 2005, **152**, F90.
176. P. F. Garcia, R. S. McLean, M. H. Reilly, M. D. Groner and S. M. George, *Appl. Phys. Lett.*, 2006, **89**, 31915.
177. A. A. Dameron, S. M. George, P. F. Garcia, R. S. McLean, *AVS 7th International Conference on Atomic Layer Deposition*, San Diego, United States, 2007, presentations available on CD-ROM.
178. C. R. Stoldt and V. M. Bright, *J. Phys. D: Appl. Phys.*, 2006, **39**, R163.

179. T. M. Mayer, J. W. Elam, S. M. George, P. G. Kotula and R. S. Goeke, *Appl. Phys. Lett.*, 2003, **82**, 2883.
180. T. W. Scharf, S. V. Prasad, M. T. Dugger, P. G. Kotula, R. S. Goeke and R. K. Grubbs, *Acta Mater.*, 2006, **54**, 4731.
181. N. D. Hoivik, J. W. Elam, R. J. Linderman, V. M. Bright, S. M. George and Y. C. Lee, *Sens. Actuators A*, 2003, **103**, 100.
182. C. F. Herrmann, F. W. DelRio, D. C. Miller, S. M. George, V. M. Bright, J. L. Ebel, R. E. Strawser, R. Cortez and K. D. Leedy, *Sens. Actuators A*, 2007, **135**, 262.
183. A. Javey, H. Kim, M. Brink, Q. Wang, A. Ural, J. Guo, P. McIntyre, P. McEuen, M. Lundstrom and H. Dai, *Nature Mater.*, 2002, **1**, 241.
184. D. B. Farmer and R. G. Gordon, *Nano Lett.*, 2006, **6**, 699.
185. J. Xiang, W. Lu, Y. Hu, H. Yan and C. M. Lieber, *Nature*, 2006, **441**, 489.
186. J. Hwang, B. Min, J. S. Lee, K. Keem, K. Cho, M.-Y. Sung, M.-S. Lee and S. Kim, *Adv. Mater.*, 2004, **16**, 422.
187. K. Keem, D.-Y. Jeong, S. Kim, M.-S. Lee, I.-S. Yeo, U.-I. Chung and J.-T. Moon, *Nano Lett.*, 2006, **6**, 1454.
188. R. H. A. Ras, M. Kemell, J. de Wit, M. Ritala, G. ten Brinke, M. Leskelä and O. Ikkala, *Adv. Mater.*, 2007, **19**, 102.
189. M. Kemell, V. Pore, M. Ritala, M. Leskelä and M. Linden, *J. Am. Chem. Soc.*, 2005, **127**, 14178.
190. E. Santala, M. Kemell, T. Pilvi, M. Ritala and M. Leskelä, *Nanotechnology in Northern Europe Conference*, Helsinki 2006, Abstracts, p. 104.
191. M. Leskelä, M. Kemell, K. Kukli, V. Pore, E. Santala, M. Ritala and J. Lu, *Mater. Sci. Eng. C*, 2007, **27**, 1504.
192. J. W. Elam, D. Routkevitch, P. P. Mardilovich and S. M. George, *Chem. Mater.*, 2003, **15**, 3507.
193. M. Pan, C. Cooper, Y. S. Lin and G. Y. Meng, *J. Membr. Sci.*, 1999, **158**, 235.
194. G. Triani, P. J. Evans, D. J. Attard, K. E. Prince, J. Bartlett, S. Tan and R. P. Burford, *J. Mater. Chem.*, 2006, **16**, 1355.
195. C. Dücsö, N. Q. Khanh, Z. Horváth, I. Bársony, M. Utriainen, S. Lehto, M. Nieminen and L. Niinistö, *J. Electrochem. Soc.*, 1996, **143**, 683.
196. M. Kemell, M. Ritala, M. Leskelä, E. Ossei-Wusu, J. Carstensen and H. Föll, *Microel. Eng.*, 2007, **84**, 313.
197. M. S. Sander, M. J. Côte, W. Gu, B. M. Kile and C. P. Tripp, *Adv. Mater.*, 2004, **16**, 2052.
198. M. Daub, M. Knez, U. Goesele and K. Nielsch, *J. Appl. Phys.*, 2007, **101**, 09J111.
199. J. Biener, T. F. Baumann, Y. Wang, E. J. Nelson, S. O. Kucheyev, A. V. Hmza, M. Kemell, M. Ritala and M. Leskelä, *Nanotechnology*, 2007, **18**, 055303.
200. D. Losic, G. Triani, P. J. Evans, A. Atanacio, J. G. Mitchell and N. H. Volecker, *J. Mater. Chem.*, 2006, **16**, 4029.
201. M. Knez, A. Kadri, C. Wege, U. Gösele, H. Jeske and K. Nielsch, *Nano Lett.*, 2006, **6**, 1172.
202. J. Huang, X. Wang and Z. L. Wang, *Nano Lett.*, 2006, **6**, 2325.
203. H. J. Fan, M. Knez, R. Scholz, K. Nielsch, E. Pippel, D. Hesse, M. Zacharias and U. Gösele, *Nature Mater.*, 2006, **5**, 627.
204. M. Knez, K. Nielsch and L. Niinistö, *Adv. Mater.*, 2007, **19**, 3425.

CHAPTER 5

Basic Chemistry of CVD and ALD Precursors

MOHAMMAD AZAD MALIK AND PAUL O'BRIEN

School of Chemistry, The University of Manchester, Oxford Road, Manchester, M13 9PL, UK

5.1 Introduction

In chemical vapor deposition (CVD) processes thermal decomposition of a precursor is the key step that produces a thin-film deposit, and ideally the ligands associated with the precursor are cleanly lost into the gas phase. However, there are problems as the ligands may fragment and lead to impurities in the films. A rational development of precursors for specific applications is thus only possible with insight into the molecular decomposition pathways.

Studies on precursor decomposition mechanisms really need to address vapor phase reactions and subsequent surface reactions of the precursors or relevant fragments. Techniques used in such work include GC-MS and matrix-isolation FTIR spectroscopy (MI-IR).

Systematic studies comparing the performance of precursors under similar conditions remain scarce. This chapter highlights some relevant examples of relationships between precursors and materials and analyzes the contribution of synthetic chemistry to precursor design for CVD or ALD processes with respect to oxides, chalcogenides, pnictides, heterometallic and metallic compounds.

5.2 Precursor Requirements for CVD

5.2.1 Precursor Purity

It is desirable to have precursors with a high level of purity to prevent contamination of the thin film with undesirable side products. Purification techniques and the methods used during precursor preparation are therefore very important.

Chemical Vapour Deposition: Precursors, Processes and Applications
Edited by Anthony C. Jones and Michael L. Hitchman

Published by the Royal Society of Chemistry, www.rsc.org

5.2.2 Precursor Reactivity

The precursor needs to react at the substrate surface to give the desired thin film. Substrate surfaces are generally heated to a temperature of the order of several hundred °C. The precursor should be reactive at these temperatures. However, one potential problem is that the precursor can react in the vapor phase throughout the reactor, as opposed to reacting only on the substrate surface. In general such reactions should be minimized to stop the resulting decomposition of the precursor in the reaction chamber. Moreover, reactions in the gas phase potentially generate particles that can pollute the film. A precursor is needed that will react on the substrate surface, yet being not so reactive as to react in the vapor phase under the delivery conditions. A level of thermal stability is an important parameter as liquid injection systems, which are increasingly used for CVD, contain heated evaporators where the precursor may decompose, leading to blockage of the flow and contamination.

5.2.3 Volatility

Volatility is a prerequisite for conventional CVD. It is affected by intermolecular forces (van der Waals interactions, π-stacking or hydrogen bonds etc.) these are influenced by molecular weight, geometry and, for solids, lattice structure. Ideally, a precursor should be readily volatile at a temperature well below that of its decomposition.

It is possible to reduce the pressure inside a reactor to compensate for lower volatility in precursors; however, a reduction in growth rate on the substrate may occur. It is also possible to raise the evaporator temperature to compensate for lower volatility in a precursor but the precursor may then decompose prematurely.

5.2.4 Stability in Air

It is desirable for a potential precursor to be stable under normal storage conditions to avoid the need for specialist storage.

5.2.5 Toxicity

A low degree of toxicity for a precursor is also desirable, avoiding the need for specialist storage and special procedures during thin film growth.

5.2.6 Synthesis

To be valuable for the use in CVD thin layer growth, any precursor must be able to be synthesized in quantities of at least several grams, with as few synthetic steps as possible. The synthesis should be adaptable to larger scale production without major problems.

5.2.7 Environment and Cost

As environmental issues are becoming more important it is advisable that new precursors should be as non-toxic as possible and preferably recyclable. Because different deposition techniques are available, the price of the CVD precursor must be as low as possible.

5.3 Metal Chalcogenides

The use of chemical routes for the deposition of semiconducting materials is not new[1] but the growth of II-VI films only became practical with Manasevit's use of volatile metal alkyls such as dimethylcadmium (Me_2Cd) and dimethylzinc (Me_2Zn) in a two-component mixture with H_2S, H_2Se or Me_2Te using hydrogen as the carrier gas.[2] A series of films including ZnS, ZnSe, CdS, CdSe, and CdTe were grown using this method.

The deposition of metal chalcogenides from metal alkyl precursors proceeds as shown in Reactions (5.1) and (5.2):

$$M(CH_3)_2 + H_2S \rightarrow (CH_3)M{-}SH + CH_4 \tag{5.1}$$

$$(CH_3)M{-}SH \rightarrow MS + CH_4 \tag{5.2}$$

The above reaction might appear as a clean route to the chalcogenide but soon it was realized that a gas-phase reaction between the metal alkyl and the chalcogenide took place in the mixing chamber with intermediates "$CH_3M(SH)$" to less volatile oligomers, which nucleate before reaching the substrate. This can result in the deposition of polycrystalline material of non-uniform morphology.

To overcome the limitations in this method the following strategies were employed:

1. Use of alternative chalcogenide precursors;
2. Metal alkyl adducts – to reduce volatility;
3. Single source precursors – to reduce pre-reaction.

5.3.1 Alternative Chalcogenide Precursors

There are problems with some compounds used in conventional MOCVD. Metal alkyls are pyrophoric, and can produce oxide dust of a hazardous particle size; most of the hydrides are permanent gases of considerable toxicity. Extensive efforts have been directed towards the development of alternative precursors for conventional MOCVD in attempts to control volatility, reduce toxic hazard, improve the quality of the grown layers and reduce the temperature required for growth.

Alkyl chalcogenides can be used as alternative chalcogen sources. The chemistry of alkyl chalcogenide precursors has been described in several reviews[3–7] and will be discussed only briefly. MeSeH,[8] Et_2Se,[9,10] ^tBuSeH[11]and $^tBu_2Se$[12] have been used but require higher deposition temperatures due to their thermal stability. Similarly, dialkyltellurides, alkyl(allyl)tellurides and diallyltellurides have been used[13] but they are not very stable on storage.[14] A detailed investigation of the decomposition mechanism of tellurium precursors has underlined the importance of homolytic Te–C bond scission and the involvement of surface-bound alkyl or ally1 radicals during the film growth process.[15] The formation of radicals in $R_2Te/CdMe_2$ mixtures has been shown to lead to alkyl group scrambling, *e.g.*, $^iPr_2Te$ can be partly converted into thermally more stable (and hence less suitable) derivatives such as Me_2Te_2 and $MeTe^iPr$.

5.3.1.1 Metal Alkyl Adducts as Precursors

Adducts of zinc and cadmium alkyls with dioxan,[16–18] thioxan,[18] triethylamine[19–24] or triazine[19] have been used as precursors in MOCVD and can successfully inhibit pre-reaction of the chalcogenide and metal alkyls.[25,26] Adducts as precursors have several advantages, including:

1. The vapor pressure of the metal alkyl is reduced in the adduct, *e.g.*, eliminating the necessity to cool bubblers containing dimethylzinc.
2. The chances of a homogeneous prereaction are considerably reduced.
3. An improved quality of films due to the purification of the alkyl moiety during the preparation of the adduct.

Numerous adducts have been studied as potential precursors. However, by far the most useful seems to be the triethylamine adduct of dimethylzinc, as first reported by Thiele in the 1960s. Several groups have now shown that this compound is useful both in the deposition of chalcogenides[19–24] and in the p-type doping of III-V materials with zinc.[27,28] A useful feature of this compound is that the 1 : 1 species is a eutectic mixture and the stoichiometry of the precursor is consequently self-limiting.

Crystal growth results indicate that there is a marked difference in the extent to which each adduct inhibits homogeneous prereaction with the Group 16 hydride during growth, a surprising observation in view of the similar, and largely dissociated, nature of the adducts in the vapor phase.

Despite the increasing use of such compounds in MOCVD, little is known about the nature of the reactive chemical species present in the gas phase in the MOCVD reactor. IR studies indicate that many of the adducts of dimethylcadmium and dimethylzinc are close to fully dissociated in the vapor phase.[29] Totally dissociative vaporization is common for organometallic species; one of the simplest examples is trimethylindium, which is a tetramer in the solid state[30] and a monomer in the vapor phase.[31] The adduct of dimethylcadmium with the chelating ligand tetramethylethylenediamine shows some evidence of association at room temperature, but under MOCVD conditions any association is likely to be extremely limited.[32]

The effectiveness of the adducts in inhibiting homogeneous reaction in MOCVD reactors had been interpreted[19] in terms of the blocking the homogeneous reaction *via* an initial step involving the formation of an addition compound (adduct) between the Group 12 source and the metal alkyl which is inhibited by excess of a "stronger" Lewis base. The reactions suggested are shown in Reactions (5.3–5.5):

$$ZnMe_2 + H_2S \rightarrow ZnMe.EH \tag{5.3}$$

$$ZnMe_2.EH_2 \rightarrow ZnE + \text{volatile products } (CH_4 \text{ etc.}) \tag{5.4}$$

$$ZnMe_2 + L \rightarrow ZnMe_2.L \tag{5.5}$$

where E = chalcogen and L = Lewis base. Suppression of the homogenous reaction in the system Me_2Zn/H_2Se/pyridine at pyridine to Zn ratios as low as 0.05 : 1[33] suggests that the Lewis bases react not so much with the metal dialkyls but with reaction intermediates such as $[MeM(SH)]_n$. Dialkyl metals of Group 12 are weak Lewis acids with Lewis acidity decreasing in the sequence M = Zn > Cd > Hg. Replacement of alkyl groups with sulfur, which is more electronegative element, will increase the Lewis acidity of the metal and hence its tendency to bind additional ligands or form aggregates of higher nuclearity.

The oligomeric structures of alkylzinc thiolates [RZn(SR′)] such as $[MeZn(S^tBu)]_5$, a pentamer (Figure 5.1a),[34,35] $[MeZn(S^iPr)]_8$[36] an octamer and $[EtZn(SEt)]_{10}$ a decamer[37] provide good indicators of what other species formed in the gas-phase nucleation process might look like.[38] Pyridine reacts with the pentameric $[MeZn(S^tBu)]_5$[35] to give dimeric adducts $[MeZn(S^tBu)(Py)]_2$ (Figure 5.1b), showing how these ligands can block the polymerization of thiolato intermediates. All of the dimers have similar structures, zinc is four-coordinate, the methyl and amine groups are monodentate and

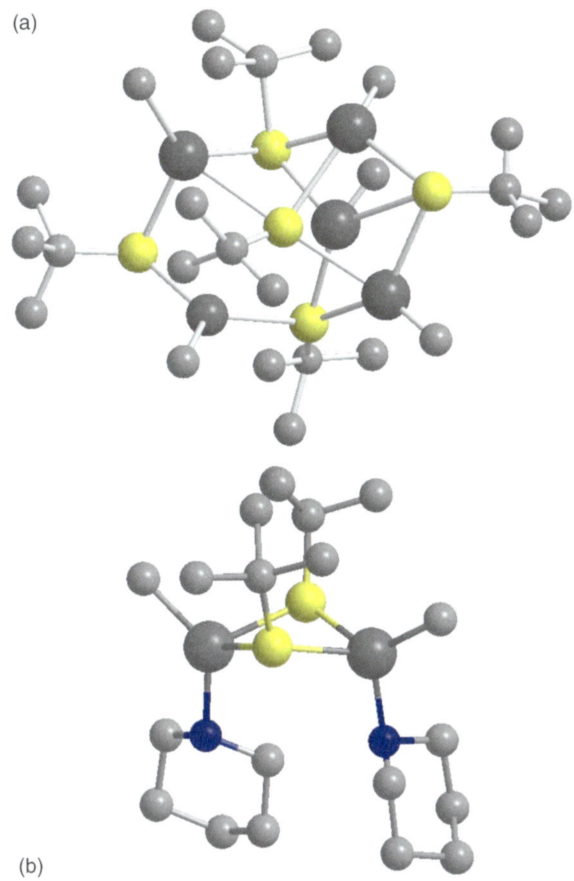

Figure 5.1 Structure of (*a*) $[MeZn(S^tBu)]_5$; (*b*) $[MeZn(S^tBu)(Py)]_2$.

the thiolates form bridges. The crystallographic structures of several adducts were investigated and the four-coordinate structure of the dimethylzinc adduct with hexahydrotrimethyltriazine[39] is shown in Figure 5.2.

5.3.2 Single-molecule Precursors

5.3.2.1 II-VI Materials

Chalcogenide Complexes. Zinc or cadmium chalcogenide complexes are the simplest candidates as single-molecule precursors for zinc or cadmium chalcogenides. The chemistry of divalent Zn and Cd with a chalcogen-containing ligand such as thiolates is generally typified by the formation of polymeric structures with a tetrahedral metal center and such compounds are practically involatile.[40] Some structural modifications such as cadmium complexes with pyridinethione and 2,3-sulfanylbenzothiazole resulted in polymeric chains.[41,42] However, many such chalcogen-containing complexes decompose to the corresponding II-VI material. The lack of volatility means that they are not generally useful as precursors for LP-MOCVD. One approach to the preparation of precursors is to modify the properties of the thiolate by forming adducts.

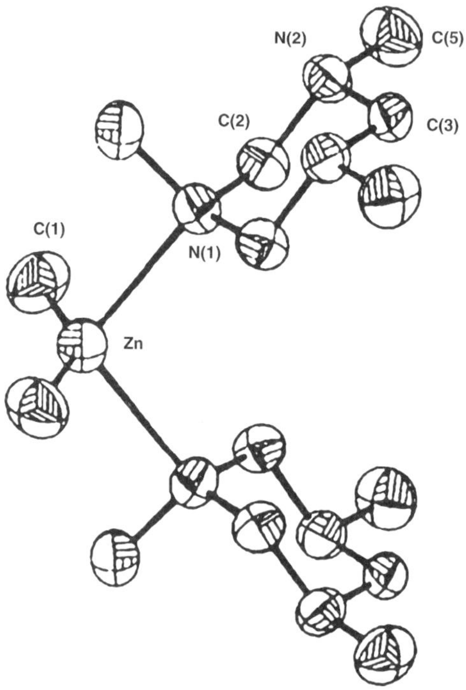

Figure 5.2 Structure of dimethylzinc.hexahydrotrimethyltriazine adduct.

A novel series of precursors prepared with 1,2-bis(diethylphosphino)ethane (depe), $M(ER)_2$ (M = Zn, Cd, Hg: E = S, Se, Te), has been reported by Steigerwald and co-workers for the deposition of II-VI materials.[43] Complexes containing one or two mole equivalents of the phosphine have been isolated. The 1 : 2 species are polymeric and the 1 : 1 complexes dimers; crystal structures have been reported for the complexes $[Cd_2(SeC_6H_5)_4(depe)]_n$ and $[Hg(SeC_6H_5)_2(depe)]_2$. Decomposition of such compounds in high-boiling-point solvents, such as 4-ethylpyridine, leads to the deposition of sub-nanometer clusters of CdSe from $[Cd_2(SeC_6H_5)_4(depe)]_n$. Cadmium selenide was also obtained as the result of solid-state pyrolysis of the adduct in a vacuum sealed tube. Similarly, decomposition of bischalcogenato complexes [44,45] gives rise to metal chalcogenides.

Hampden-Smith and co-workers have prepared a compound of stoichiometry $[Zn(SEt)Et]_{10}$ by the insertion of sulfur into the Zn–C bond of diethylzinc.[46] Although this decameric thiolate possesses a similar arrangement of zinc and sulfur atoms to that found in wurtzite, pyrolysis of the material at 250 °C leads to predominantly cubic ZnS; aerosol assisted CVD does, however, lead to the hexagonal form.[46]

Adduct formation does not provide a solution to the involatility of chalcogenide complexes. The volatility issue was then addressed by using bulky chalcogenide ligands to reduce the molecularity of the resulting complexes. Mercury proves, as is often the case, to be an exception, and Bradley and Kunchur[47] reported that the mercury complex of *tert*-butyl thiol has only weak intermolecular interactions between its molecular $Hg(SR)_2$ units, even in the solid state. A series of low-coordination metal complexes have been prepared with 2,4,6-tri-iso-propylbenzenethiol (tipt), initially pioneered by Dilworth *et al.*[48] Bochmann *et al.*[49–52] have now further extended and developed such chemistry and produced a range of precursors for II-VI materials based on 2,4,6-tri-*tert*-butylphenylchalcogenolate. The general preparation method of one such complex is shown in

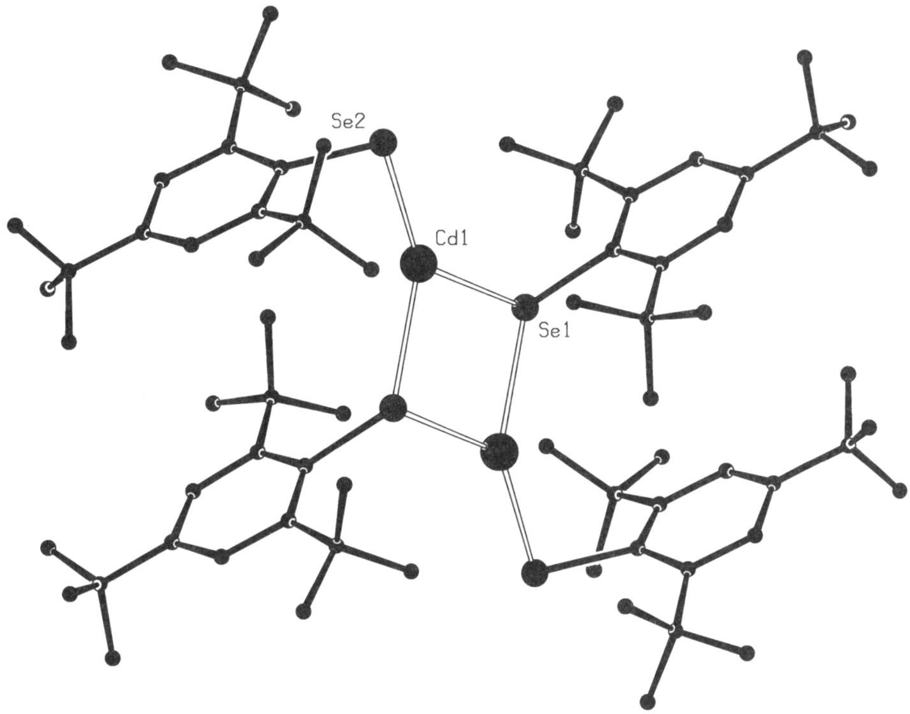

Figure 5.3 Dimeric structure of super mesityl complex $[Cd(SeC_6H_2{}^tBu_3\text{-}2,4,6)]$.

Reactions (5.6)–(5.8), and the structure of the product is given in Figure 5.3:

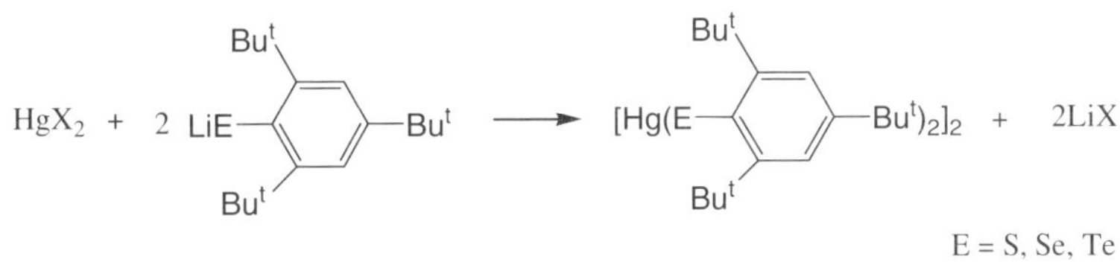

(5.6)

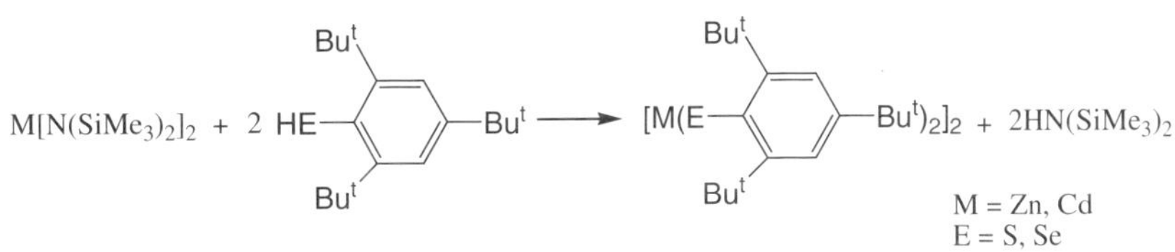

(5.7)

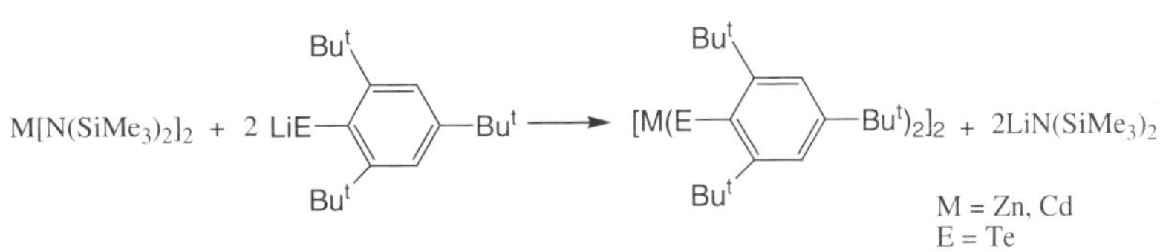

(5.8)

These compounds have been used to deposit thin films of the metal sulfides or selenides, in preliminary low-pressure growth experiments. One problem with such ligands is that steric bulk is achieved by the incorporation of large numbers of carbon atoms and carbon incorporation into thin films grown from such precursors remains a possibility. The compounds are essentially dimeric, even in the vapor phase, and can be used in low-pressure MOCVD for the deposition of II-VI materials.[53,54] For example, $[Cd(SC_6H_2Bu^t_3\text{-}2,4,6)_2]_2$ gave polycrystalline CdS thin films in low-pressure MOCVD at a substrate temperature of 450 °C (10^{-2} Torr). During the deposition diaryl sulfide was detected as a residual pyrolysate. The mercury analogues readily decompose *via* a reductive elimination path to form atomic Hg and diaryldichalco-genides, which may indicate that these compounds are more suitable for photo-assisted rather than thermal MOCVD.[51] Mixed alkyl complexes[55] with aryl chalcogenates RMSeR′ ($SeR' = SeC_6H_2Pr^i_3\text{-}2,4,6$; M = Zn, R = Me, Et, Pr, Pr^i; M = Cd, R = Me) have also been prepared but have not been used in deposition studies.

Another series of precursors involve bulky silicon-based systems of stoichiometry $M[ESi(SiMe)_3]_2$ (M = Zn, Cd, Hg; E = S, Se or Te) and these have been used by Arnold and co-workers to deposit a range of chalcogenides.[56,57] The most detailed work has been reported on the tellurides: thin films of the tellurides have been deposited by low-pressure MOCVD.[58] The tellurium-containing ligand, $HTeSi(SiMe_3)_3$, is termed HSitel and this reagent is potentially generally useful in the preparation of metal tellurolates.[59–61] Metal complexes of Sitel are generally prepared as illustrated in Reaction (5.9). Their decomposition in MOCVD proceeds *via* an elimination path [Reaction (5.10)]:

$$M[N(SiMe_3)_2]_2 + 2HTeSi(SiMe_3)_3 \rightarrow [M\{TeSi(SiMe_3)_3\}_2]_2(s) + 2HN(SiMe_3)_2 \quad (5.9)$$

$$[M\{TeSi(SiMe_3)_3\}_2]_2(g) \rightarrow (MTe)n + Te\{Si(SiMe_3)_3\}_2 \quad (5.10)$$

where M = Zn, Cd, Hg. Zinc telluride was deposited at temperatures between 250 and 350 °C onto quartz, silicon, InAs and GaSb substrates. The cadmium precursor (1.4 Torr, 290 °C) showed the deposition of hexagonal phase and indicated that the films may have been of better stoichiometry than those of ZnTe.

Dithio-/diselenophosphinato Complexes. Thiophosphinato complexes are another class of chalcogen-containing compounds that may be useful as precursors. Takahashi[62] deposited cadmium sulfide using dimethylthiophosphinates $M(S_2PMe_2)_2$ (M = Cd or Zn). Cadmium sulfide normally shows n-type conduction, due to non-stoichiometry, and doping with phosphorus would lead to highly compensated semi-insulating material. Evans and Williams reported[63] that highly orientated sulfide films could be grown using the dimethylthiophosphinates as precursors.

CdS and ZnS films were deposited from $[Cd(S_2P^iBu_2)_2]_2$ and $[Zn(S_2P^iBu_2)_2]_2$, respectively, using a home made cold-wall low-pressure reactor.[64] The cadmium complex has a classic C_i symmetric two-step ladder (or "chair") structure (Figure 5.4) of type 23 for the central eight-membered $Cd_2S_4P_2$ ring.[65] This geometry contrasts with the C_2 symmetric "boat" conformation observed for the closely related isopropoxide complex $Cd_2[(i\text{-}C_3H_7O)_2PS_2]_4$.[66] The geometry at cadmium is severely distorted tetrahedral, with angles ranging between 78.82(6) and 128.63(7)°, the acute angle being associated with the bite of the chelating ligand. X-Ray analysis of the zinc complex shows it to adopt a geometry for the $Zn_2S_8P_4$ core that is virtually identical with that seen in the cadmium and zinc complexes with *O,O*-diisopropylphosphordithioate.[66] It has previously been postulated that a reason for a change from a C_i symmetric chair to a C_2 symmetric boat conformation for the central eight-membered metallocyclic ring for phosphordithioates/dithiocarbamates is a consequence of the packing requirements of the differently sized R groups.

A series of metal complexes of bis(dialkylselenophosphinyl)selenide, $[(R_2PSe_2)Se]$, with the general formula $[M(R_2PSe_2)_n]$ (M = Zn^{II}, Cd^{II}, Pb^{II}, In^{III}, Ga^{III}, Cu^{I}, Bi^{III}, Ni^{II}; R = ^{i}Pr, Ph) and $Mo^V{}_2O_2Se_2(Se_2P^iPr_2)_2]$ have been synthesized and used for the deposition of metal selenide thin

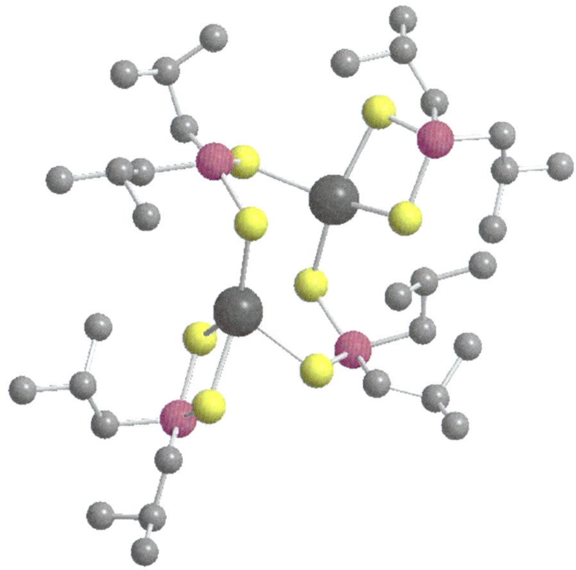

Figure 5.4 Molecular structure of $[Cd(S_2P^iBu_2)_2]_2$ showing its "chair" conformation.

films by CVD.[67,68] The compound $[R_2PSe_2]_2Se$ (R=^{i}Pr, Ph) was prepared by the reaction of NEt_3 with $^iPr_2PCl$ or $^iPr_2PCl$ and $HSiCl_3$ in cold toluene [Reactions (5.11) and (5.12)]:

$$R_2PCl + HsiCl_3 + NEt_3 \xrightarrow[6h,\ rt]{Toluene} R_2PSiCl_3 + (HNEt_3)Cl \quad (5.11)$$

$$R_2PSiCl_3 + 3Se \xrightarrow[20h,\ reflux]{Toluene} [R_2PSe_2]_2Se + Si_2Cl_6 \quad (5.12)$$

This compound was then reacted with different metal salts in methanolic solutions to give the metal complexes as precipitates that were subsequently recrystallized from dichloromethane or toluene before further analysis. The complexes are soluble in chloroform, toluene or dichloromethane. All are stable at room temperature under open atmosphere for months, making them potentially useful as single-source precursors for the deposition of metal selenide thin films. Solid state structures of several of these complexes were determined by single-crystal X-ray analysis.

The copper complex is based on the tetramer $[Cu_4(Se_2P^iPr_2)_4]$ (Figure 5.5). None of the ligands are chelating, all four ligands are bridging. One selenium is bonded to phosphorus and copper only whilst the other is bonded to two copper atoms and one phosphorus atom in each ligand. Each Cu is bonded to three Se atoms, giving overall a cubane structure. In an analogous reaction the molybdenum complex $[Mo_2O_2Se_2(Se_2P^iPr_2)_2]$ is formed. In a dimeric structure each ligand chelates to one Mo atom and the two Mo atoms are doubly bridged by two selenium atoms (Figure 5.6). The geometry around Mo is a distorted square pyramidal.

The indium complex shows the expected $[In(^iPr_2PSe_2)_3]$ molecule (Figure 5.7a). All three diselenophosphinate ligands are chelating to form three four-membered rings (Se-P-Se-In) with In–Se bond distances of 2.729(6)–2.768(6) Å. The geometry on indium is trigonally distorted octahedral due to the restricted bite angle of the chelating diselenophosphinate ligand.

The structure of $[Ga(^iPr_2PSe_2)_3]$ (Figure 5.7b) has a four-coordinate gallium center in contrast to the six-coordinate indium. Until relatively recently[69] all reports on structurally characterized pairs of analogues In/Ga tris(chelates) with (O,O′) or (S, S′) ligands have been shown to be

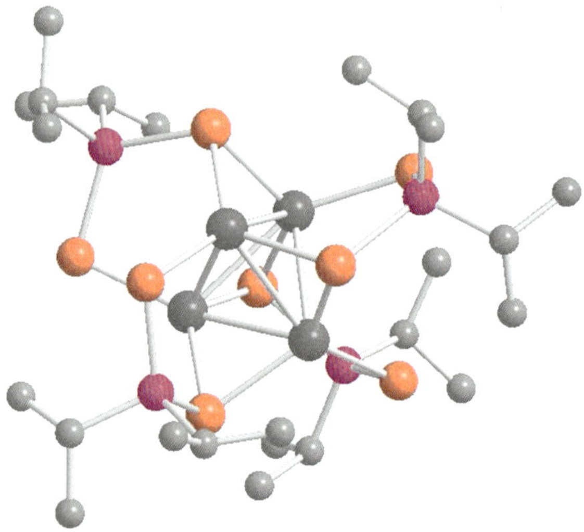

Figure 5.5 Tetrameric structure of $[Cu(Se_2P^iPr_2)]_4$.

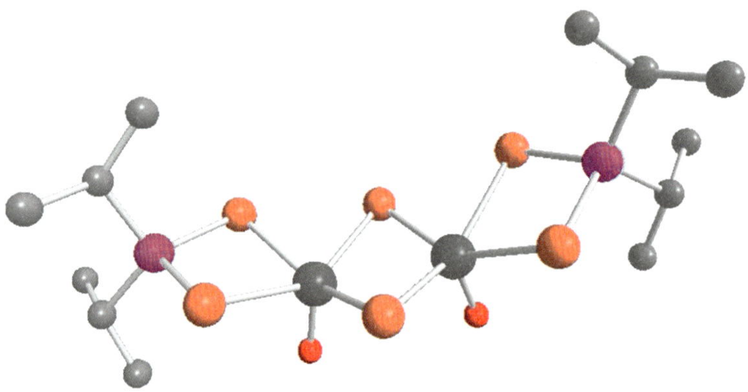

Figure 5.6 Dimeric structure of $[Mo_2O_2Se_2(Se_2P^iPr_2)_2]$.

isostructural.[70,71] The first such structure of a gallium complex was reported for tris(di-isobutyldithiophosphinato)gallium(III), $[Ga(^iBu_2PS_2)_3]$.[9] The structure consists of one chelating and two pendant di-isopropyldiselenophosphinate ligands in a distorted tetragonal geometry. The complex $[Ga(Ph_2PSe_2)_3]$ showed a similar coordination.

Dialkylthio-/diselenocarbamato. Dialkyldithio-/diselenocarbamato metal complexes with the general formula ($M(E_2CNR_2)_2$ (symmetrical) or $M(E_2CNR_1R_2)_2$ (unsymmetrical) R = alkyl, E = S, Se; M = Zn, Cd) are precursors used for the deposition of II/VI materials thin films. All of these compounds are stable crystalline solids with some volatility. Their solid state structures have been determined by X-ray crystallography. Most of them, *e.g.* $Zn(S_2CNR_2)_2$, with R = Me, Et, or ^iPr[72–74] and in $Cd(S_2CNR_2)_2$ (Figure 5.8), $Zn(Se_2CNEt_2)_2$, $Cd(Se_2CNEt_2)_2$, $Zn(S_2CNMeR)_2$, with R=Et, ^nPr, ^iPr, or ^nBu, have dimeric structures;[75–78] the metal atom is five-coordinate with a geometry between trigonal bipyramid and a tetragonal pyramid. Saunders *et al.*[79] were the first to use diethyldithiocarbamates $M[S_2CNEt_2]_2$ (M = Cd, Zn) for the deposition of ZnS, CdS and $Zn_xCd_{1-x}S$ thin films by LP-MOCVD. The same precursors were used to

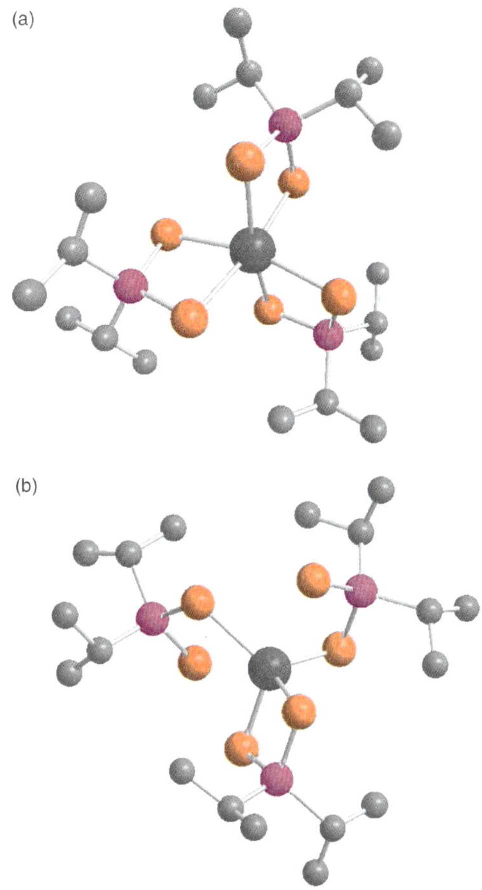

Figure 5.7 Structure of (*a*) $[In(^{i}Pr_2PSe_2)_3]$; (*b*) $[Ga(^{i}Pr_2PSe_2)_3]$.

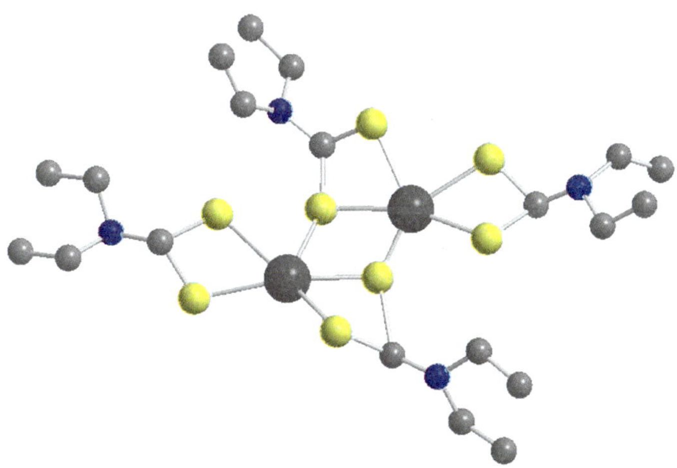

Figure 5.8 Dimeric structure of $Cd(S_2CNEt_2)_2$.

grow heterostructures by organometallic vapor phase epitaxy (OMVPE).[80,81] Frigo *et al.* deposited good quality CdS or $Zn_xCd_{1-x}S$ prepared from an equimolar mixture of the precursors $Zn[S_2CNEt_2]_2$ and $Cd[S_2CNEt_2]_2$ on (100)-oriented GaAs or InP substrates[82] by LP-MOCVD. The quality of ZnS films deposited from $Zn[S_2CNEt_2]_2$ was poor.

The deposition by MOCVD of ZnS usually gives the hexagonal α-ZnS wurtzitic phase but Nomura *et al.* deposited the cubic β-ZnS zinc blende phase epitaxially on (111)-Si by OMVPE, using $Zn[S_2CNEt_2]_2$ in a cold-wall horizontal reactor.[83] The depositions conducted under low pressure (10^{-2} Torr) produced hexagonal α-ZnS of poor crystallinity and morphology whereas depositions with the introduction of a carrier gas (N_2) gave good quality (111)-oriented β-ZnS. Fainer *et al.* used $Cd[S_2CNEt_2]_2$ with He or H_2 as a carrier gas in LP-MOCVD[84] and plasma-enhanced MOCVD (PE-MOCVD)[85] to grow CdS on SiO_2, Al_2O_3 (sapphire), (111)-Si and (111)-InP.[84]

Bis(dialkyldithio-/selenocarbamato)-cadmium/zinc compounds have the advantage of stability for years but the volatility of the precursors could be low and the quality of the films was often poor, and for the selenides Se contamination was a problem. However, O'Brien *et al.* have developed some novel air-stable unsymmetrical precursors based on bis(methyl-alkyl-di-seleno-thio-carbamato)zinc or -cadmium, which decompose cleanly in MOCVD to selenides or sulfides.[86–88] Volatilization experiments have shown that the unsymmetrically substituted $[Zn(S_2CN(Me)Et)]_2$, $[Zn(S_2CN(Me)\text{-}^nPr)]_2$ and $[Zn(S_2CN(Me)^nBu)]_2$ sublime well above their melting points, whereas $[Zn(S_2CNEt_2)_2]_2$ sublimes as a solid.[78] ZnS was grown on glass from $[Zn(S_2CN(Me)^nBu)]_2$ by LP-MOCVD at 450 °C in a cold-wall reactor.[78] Deposition studies on $Cd[S_2CNEt_2]_2$ and $Cd(SCNMe^nBu)_2$ showed that the methyl butyl derivative was more volatile[89] and the films grown from this precursor on InP at 500 °C gave polycrystalline CdS.

A series of other unsymmetrical dithio- and diselenocarbamates were also synthesized for use as single source precursors for the deposition of thin films. Bis(n-hexyl(methyl)dithio/selenocarbamato)cadmium/zinc (Figure 5.9) proved to be the best unsymmetrical derivatives for the growth of chalogenides.[90]

Since dithio- and diseleno-carbamates and their derivatives are used so widely in the deposition of thin films and nanoparticles a mechanistic study of their decomposition was carried out by O'Brien *et al.*[88] Wold *et al.*[91] studied the decomposition products of $Zn(S_2CNEt_2)_2$ using GC-MS, and their reported deposition path shows clean elimination of ZnS from the precursor [Reactions (5.13) and (5.14)]. However, the proposed decomposition route is somewhat different to the step-by-step fragmentation observed in the EI-MS of the compound [Reaction (5.15)]. This difference

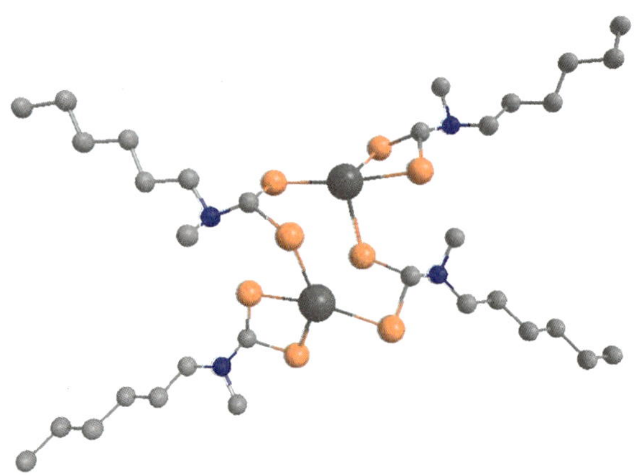

Figure 5.9 Structure of $[Cd(Se_2CNMe^nHex)_2]_2$.

can be attributed to inherent differences between the two techniques:

$$Zn(S_2CNEt_2)_2 \rightarrow ZnS + EtNCS + Et_2NCS_2Et \quad (5.13)$$

$$Et_2NCS_2Et \rightarrow Et_2NH + CS_2 + C_2H_4 \quad (5.14)$$

$$Zn(S_2CNEt_2)_2 \xrightarrow{-S_2CNEt_2} Zn(S_2CNEt_2) \xrightarrow{-SCNEt_2} ZnS \quad (5.15)$$

In contrast, the analogous diethyl-diselenocarbamates have been shown to be poor sources for the deposition of ZnSe or CdSe films. Under similar reaction conditions (10^{-3}–10^{-4} Torr, 370–420 °C) the diethyl-diselenocarbamate precursors give films of the metal selenide heavily contaminated with selenium.[77] However, mixed alkyl-diselenocarbamates complexes [Reaction (5.16)] have been used successfully to deposit thin films of CdSe or ZnSe:[92,93]

$$Me_2M + M(Se_2CNEt_2)_2 \rightarrow 2MeM(Se_2CNEt_2) \quad M = Cd \text{ or } Zn \quad (5.16)$$

ZnSe films were deposited from $Zn(Se_2CNMe^nHex)_2$, which has two different alkyl substituents on the nitrogen. This complex is monomeric in the solid phase[86] (Figure 5.10), in contrast to the analogous diethyl-diselenocarbamates and the mixed alkyl-diselenocarbamates complexes, which are both dimers. All of the dithiocarbamates prepared can be used to deposit CdS or ZnS by MOCVD but the diselenocarbamates show a different pattern of behavior; $M(Se_2CNEt_2)_2$ (M = Zn, Cd) deposit films heavily contaminated with selenium whereas $M(Se_2CNRR')_2$ (M = Zn, Cd) and $EtZnSe_2CNEt_2$ deposit the metal selenide. It was of interest to study the decomposition of the compounds by GCMS and EI-MS to determine why the latter are successful precursors; and also to determine plausible decomposition pathways for comparison with the dithiocarbamato complexes. Plausible schemes for the decomposition of the precursor have been proposed (Scheme 5.1).

Among several limiting factors in the deposition of chalcogenides by these compounds, the volatility of the complexes in particular influences the results of the deposition. Some attempts to obtain more volatile dithiocarbamate complexes have been made. Dithiocarbamate complexes derived from trimethylpropylenediamine were synthesized[94,95] by a conproportionation reaction. This complex showed a different structure from the dimers and is a weakly bonded polymer (Figure 5.11). The compound is apparently more volatile than the simpler dithiocarbamates and has successfully been used in deposition experiments to deposit CdS thin film on glass or GaAs substrates at 350 °C.[94]

The related class of *N*-alkyldithiocarbamato compounds $[M(S_2CNHR)_2]$ have potential as precursors. Very little was reported[96] about these complexes due to the supposition that they are not very stable and were also expected to be less volatile, and consequently less suitable as single-source CVD precursors. Recently, O'Brien *et al.*[97] have synthesized a series of complexes of cadmium or zinc with varying alkyl chain lengths, $[M(S_2CNHR)_2]$ (M = Cd(II) Zn(II); R = C_2H_5, C_4H_9, C_6H_{13}, $C_{12}H_{25}$).

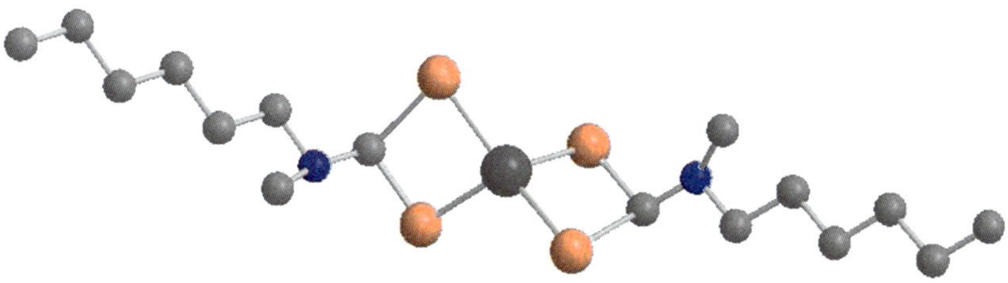

Figure 5.10 Structure of $[Zn(Se_2CNMe^nHex)_2]$.

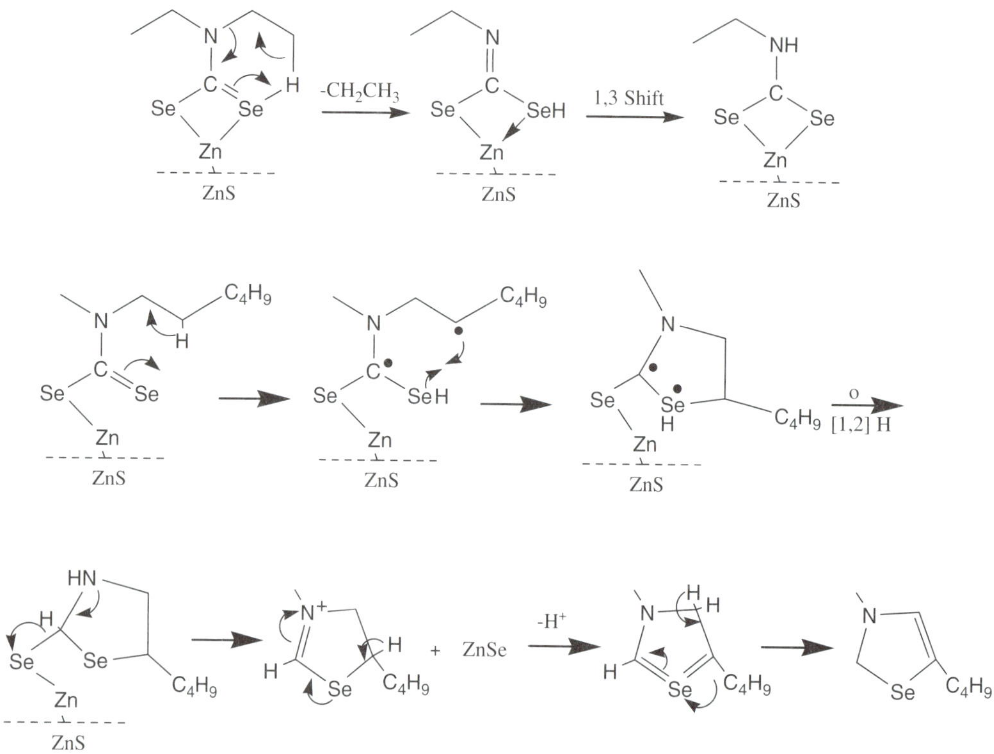

Scheme 5.1 Decomposition mechanism of bis(diethyldiselenocarbamato)zinc(II) and of bis[hexyl(methyl)diselenocarbamato]zinc(II).

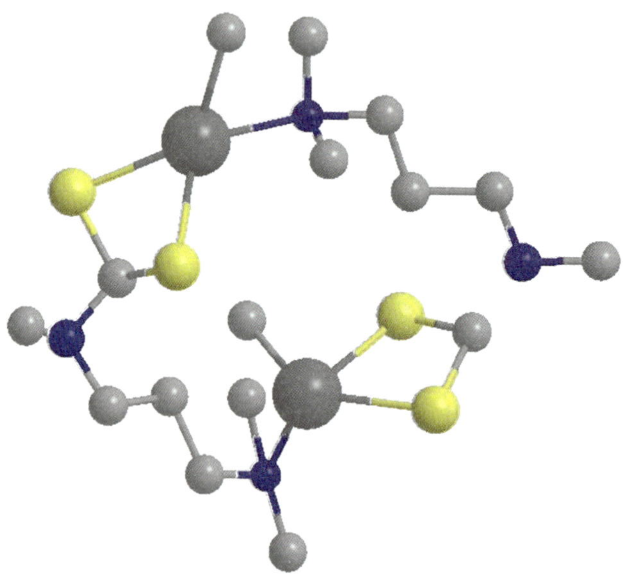

Figure 5.11 Structure of $[MeCd(MeN(CH_2)_3NMe_2)]_n$.

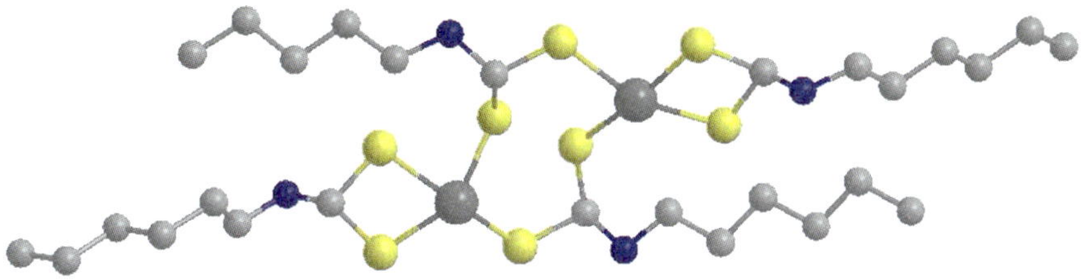

Figure 5.12 Structure of $[Zn(S_2CNHHex)_2]_2$.

The condensation of the primary amines with carbon disulfide in the presence of metal salts produced $[M(S_2CNHR)_2]$ [M = Cd(II) or Zn(II)] in reasonable purity with good yield. Most of these complexes were only sparingly soluble in organic solvents. The zinc complexes were comparatively stable and more soluble. The compound $[Zn(S_2CNHHex)_2]$ was recrystallized from dichloromethane to give good quality crystals used for an X-ray study (Figure 5.12).[97]

The cadmium complexes decomposed to a cadmium sulfate complex with pyridine and water, which crystallized from pyridine. Stable bis(*N*-alkyldithiocarbamato)cadmium(II) complexes have been prepared by metathesis between lithium *N*-alkyldithiocarbamato salts and cadmium chloride in neutral aqueous solution.[98] The *N*-alkyldithiocarbamato complexes are less stable than their corresponding *N*,*N*-dialkyldthiocarbamato complexes due to the presence of the acidic hydrogen at the nitrogen.

Thermogravimetric analysis (TGA) was conducted to evaluate the physical behavior of *N*-alkyldithiocarbamato complexes of cadmium and zinc.[97] The cadmium compounds decompose in an unresolved two-step mechanism. In contrast, the zinc compounds decomposed in a single step. Surprisingly, the decomposition behavior of the compounds improved as the chain length was increased, as indicated by the final residue of compounds. The final remaining residue from $[M(S_2CNHR)_2]$ (M = Zn(II), Cd(II); R = Hex, Dodecyl) are very close to the values calculated for the mass percentage of bulk metal sulfide. In contrast, short substituted alkyl chain compounds such as Et or butyl afford a significant large remaining residue, indicating incomplete decomposition or impurities present after the decomposition of precursor.

Mixed Alkyl/Dithio- or Diseleno-carbamates. These compounds were first prepared by Noltes[99] using an insertion reaction[100] [Reaction (5.17)]:

$$2MeZn(NEt_2) + 2CS_2 \rightarrow [MeZn(S_2CNEt_2)_2] \quad (5.17)$$

Conproportionation [Reaction (5.18)] also provides a convenient synthesis for these compounds:[119–121]

$$Me_2Zn + Zn(S_2CNEt_2)_2 \rightarrow [MeZn(S_2CNEt_2)_2] \quad (5.18)$$

The compounds are dimers[100] in the solid state and the parent dimeric structure has been confirmed for a wide range of compounds where R = Me, Et, Bu^t or Me_3CCH_2, M = Zn or Cd, E = S or Se, and R′ = Me or Et.[101,102] X-ray single crystal structures of neopentylcadmium and diethyldiselenocarbamate complexes are shown in Figure 5.13(a) and (b).[103]

An interesting application of the conproportionation reaction is the preparation of a mixed species such as methylcadmium/methylzinc diethyldiselenocarbamate, which is useful for the

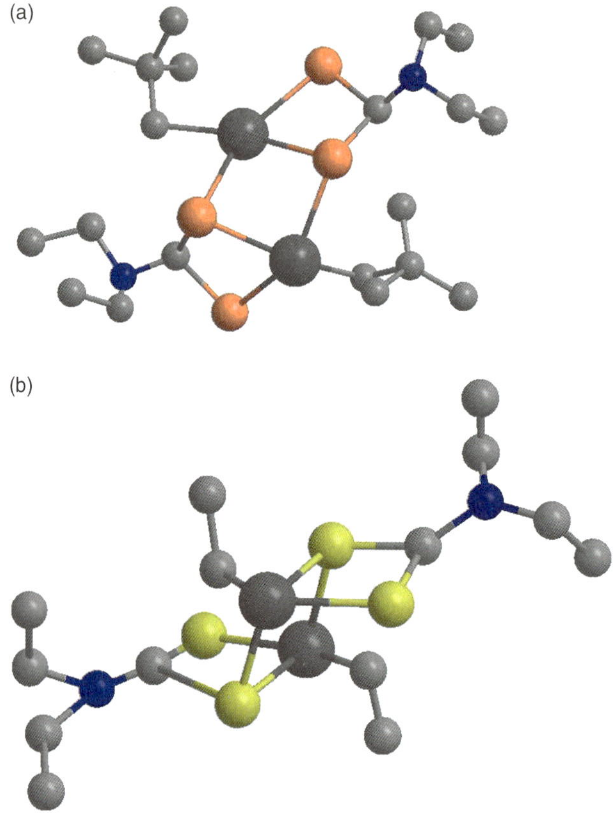

Figure 5.13 Structure of (*a*) $[NpCdSe_2CNEt_2]_2$; (*b*) $[EtZnSe_2CNEt_2]_2$.

deposition of thin films of ternary solid solutions of $Cd_{0.5}Zn_{0.5}Se$. Thus, the reaction of Me_2Zn with $Cd(Se_2CNEt_2)_2$ gave $Me_2CdZn(Se_2CNEt_2)_2$ as shown in Reaction (5.19):

$$(CH_3)_2Zn + Cd(Se_2CNEt_2)_2 \rightarrow [(CH_3)_2CdZn(Se_2CNEt_2)_2] \quad (5.19)$$

A polycrystalline $Cd_{0.5}Zn_{0.5}Se$ layer, of which the bandgap energy was estimated to be 2.1 eV, was deposited on a glass substrate by low-pressure MOCVD (450 °C, 0.2 Torr).

The mixed aggregate showed similar dimeric molecular units, $[RM(Se_2CNEt_2)]_2$, to other alkylmetal dithio- and diselenocarbamates. In the solid-state structure, the cadmium and zinc atoms were modeled as randomly occupying the metal sites.[93] Many of these mixed alkyl diseleno- and dithiocarbamate compounds have been used to deposit thin films of metal chalcogenides by low-pressure MOCVD (10^{-2}–10^{-3} Torr). Growth rates are generally of the order of 0.1 μm h^{-1}.[104] The dithiocarbamate complexes derived from trimethylpropylenediamine were also synthesized[94,95] by a conproportionation reaction. The structure of the polymer complex is shown in Figure 5.11. The compound is apparently more volatile than the simpler dithiocarbamates.

Xanthate Complexes. Cheon *et al.*[105] have reported the deposition of NiS by both a thermal and photochemical CVD using $[Ni(S_2COCHMe)_2]$. Recently, O'Brien *et al.*[106] deposited NiS thin films on glass by aerosol assisted chemical vapor deposition(AACVD) using single source precursors of the type $[Ni(S_2COR)_2]$ ($R = C_2H_5$ or C_3H_7). The metal alkyl xanthate compounds $M(S_2COCHR)_2$ ($M = Ni$ and $R =$ ethyl or isopropyl) were prepared by stoichiometric reaction of potassium alkyl xanthate and metal salts following the literature method,[107–109] which involved the reaction of an aqueous solution of the corresponding metal salt with the potassium xanthate aqueous solution. The metal xanthate forms rapidly and precipitates. TGA of $[Ni(C_2H_5OCS_2)_2]$ carried out between 25 and 500 °C at 10 °C min^{-1} under N_2. $[Ni(C_2H_5OCS_2)_2]$ showed a rapid mass loss of 67% at 155–200 °C. The $Ni(C_3H_7OCS_2)_2$ grown under similar conditions showed a 71% mass loss between *ca.* 167 and 202 °C. Single-step decomposition was observed for both compounds. Thin films of NiS were deposited from $[Ni(C_2H_5OCS_2)_2]$ at 200, 250 and 300 °C. SEM studies show that the morphology of the films changes with temperature. At 200 °C NiS films with rod-like morphologies are observed. These structures are approximately 500 nm long and 250 nm wide, with uniform coverage on the substrate. At 250 °C the films have a twinned cube shape with interpenetrating structures. This shape is very uniform across the substrate with particles sizes of 400–500 nm. There is a further change of morphology at 300 °C, where granular close-to spherical particles are observed. The coverage is non-uniform, with clustered growth predominant on the substrate.

Growth from $[Ni(C_3H_7OCS_2)_2]$ at 200 °C differs from that from ethyl xanthate at a similar temperature.[2] At 200 and 250 °C the films deposited could be described as granular, similar to those grown at 250 °C from $Ni(C_2H_5OCS_2)_2$. There is a distinct increase in grain size as the temperature of deposition is increased from 200 to 300 °C. However, films grown at 300 °C are similar to those grown from the ethyl derivative at this temperature. A clustered growth of spherical particles is predominant on the substrate.

Nanocrystalline ZnS and CdS thin films were deposited onto SiO_2 substrates by CVD at temperatures between 200 and 450 °C from $[M(C_3H_7OCS_2)_2]$ (M = Zn, Cd) as single-source precursors.[110] Thermal decomposition and fragmentation of $[M(C_3H_7OCS_2)_2]$ was also investigated by thermal analyses and mass spectrometry.

Barrecal *et al.* have investigated the properties of $M(S_2COCHR)_2$ (M = Cd and R = ethyl or isopropyl) as potential single-source molecular precursors for the CVD of CdS thin films.[111] The structure of bis(*O*-isopropyldithiocarbonato)cadmium(II) $[Cd(C_3H_7OS_2)_2]_n$ is shown in Figure 5.14.[112]

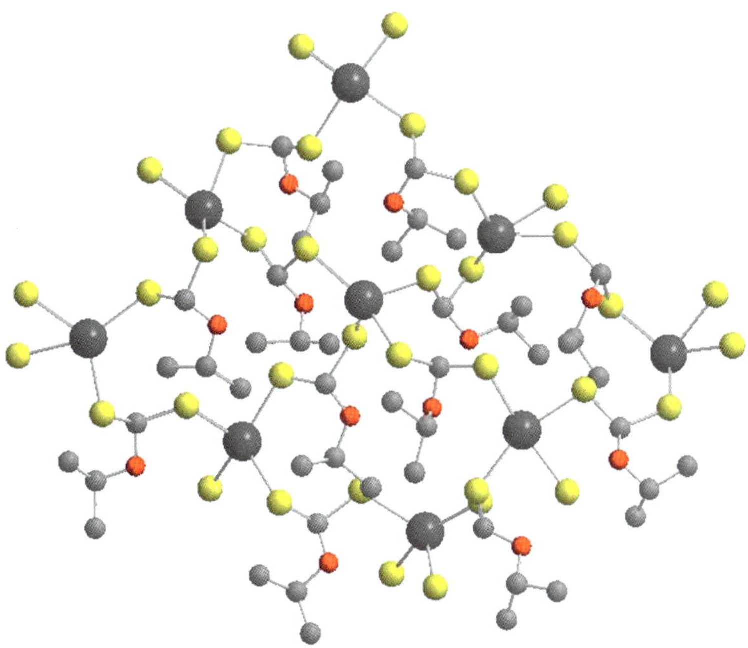

Figure 5.14 Structure of $[Cd(C_3H_7OS_2)_2]_n$.

Two distinct motifs are known for $[Cd(S_2COR)_2]$, a square-planar geometry with R = CH_2CH_2OMe, and weak Cd–S interactions above and below the square plane,[113] and tetrahedrally coordinated Cd, *i.e.*, when R = Et[114] and R = Bu.[115] The reported structure of $[Cd(S_2COC_3H_7)_2]$ also conforms to this motif.[116]

Monothiocarbamato Complexes. Bis(diethylmonothiocarbamato)cadmium(II) has shown a reasonable volatility despite its polymeric structure[117] and was used to deposit films of CdS by LP-MOCVD at 300–400 °C.[118] Compounds with the general formulae $M(SOCR)_2$(tmeda) (tmeda = *N,N,N,N*-tetramethylethylenediamine, R = methyl[119] or *tert*-butyl[120] were used to grow CdS, ZnS and $Cd_xZn_{1-x}S$ by AACVD. $Zn(SOCCH_3)_2$(tmeda) and $Cd(SOCCH_3)_2$(tmeda) are isostructural, monomeric complexes in the solid state, with the metal atom in a distorted tetrahedral coordination environment, consisting of two nitrogens and two sulfurs.[119]

Dichalcogenoimidodiphosphinato Complexes. Dichalcogenoimidodiphosphinate anions (Figure 5.15) were first synthesized by Schmidpeter *et al.* in the 1960s.[121]

In 1995, Woollins and co-workers initiated the chemistry of the selenium analogue.[122] Much of the early development of the coordination chemistry of these ligands with both main group[123] and transition metals[123,124] focused on the phenyl derivatives. In 2004 it was demonstrated that metal complexes incorporating the more volatile *iso*-propyl ligand are excellent precursors for the production of various binary metal selenides by CVD techniques.[125–127]

Imino-diisopropylphosphineselenides, are prepared by the oxidative insertion of elemental selenium.[128–130] The cadmium imino-bis(diisopropylphosphine selenide) compound $Cd[N(SeP^iPr_2)_2]_2$ (Figure 5.16) was first synthesized by Woollins *et al.*[131] from diisopropylchlorophosphine *via* a

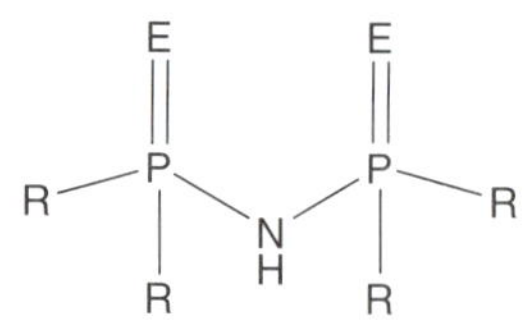

Figure 5.15 Dichalcogenoimidodiphosphinate anions (E = O, S, Se; R = methyl, isopropyl, butyl, phenyl).

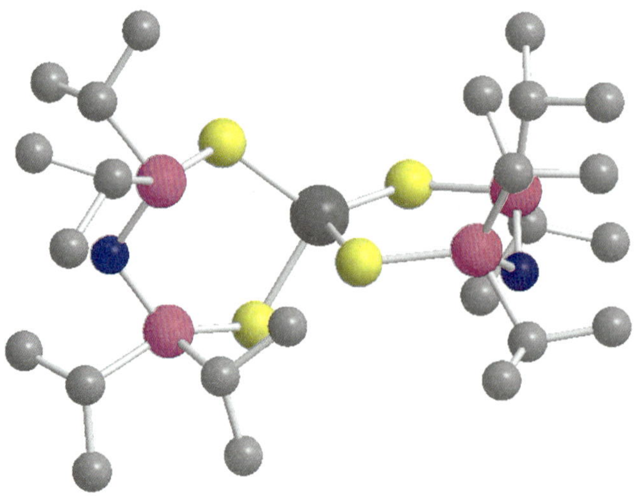

Figure 5.16 Structure of $Cd[N(SeP^iPr_2)_2]_2$.

two-step strategy. Improved yields for $Cd[N(SeP^{i}Pr_2)_2]_2$ above those reported[132,133] can be afforded by utilizing $CdCl_2$/NaOMe conditions rather than metal carbonates [Reaction (5.20–5.22)]:

$$NH(SiMe_3)_2 + {}^{i}Pr_2PCl \xrightarrow[\text{stirring(3 hours)}]{50-70^{\circ}C} NH(P^{i}Pr_2)_2 + 2ClSiMe_3 \tag{5.20}$$

$$NH(P^{i}Pr_2)_2 \xrightarrow[\text{reflux}]{E=S,\ Se} NH(EP^{i}Pr_2)_2 \tag{5.21}$$

$$2NH(EP^{i}Pr_2)_2 \xrightarrow[\text{MeOH}]{\text{NaOMe}} M[(EP^{i}Pr_2)_2N]_2 + 2NaCl \tag{5.22}$$

Imino-bis(dialkylphosphineselenide) complexes have been used as CVD precursors.[132,133] The $NH(SeP^{i}Pr_2)_2$ ligand is more thermally stable than bulky selenolate ligands, such as $[SeSi(SiMe_3)_3]^-$, and thermolysis of its complexes produces cleaner products with reduced contamination caused by undesired ligand degradation reactions.[134,135] $[M\{(EP^{i}Pr_2)_2N\}_2]$ (M = Cd, Zn and E = S, Se) and $[M\{(SePPh_2)_2N_2\}]$ (M = Cd, Zn) complexes have been used as precursors for zinc/cadmium selenide films by LP-MOCVD[132,133,136] at 400–500 °C. The suitability of the precursors were determined by TGA at atmospheric pressure; they all showed clean sublimation without any residues, as is desirable in precursors for MOCVD studies. All of these precursors have dimeric structures.[136] The complex $[MeCd\{(SeP^{i}Pr_2)_2N\}]_2$ was prepared by conproportionation of Me_2Cd and $Cd[(SeP^{i}Pr_2)_2N]_2$ in anhydrous toluene.[137] Its structure (Figure 5.17)[137] was determined by X-ray crystallography and consists of dimeric molecular units; each diselenoimidodiphosphinate chelates to one cadmium atom and bridges to the next. Each cadmium is four-coordinate and bound to three selenium atoms and one carbon. The compound is suitable for the deposition of CdSe films by low pressure CVD.

Aerosol-assisted chemical vapor deposition (AACVD) of CdTe has been carried out using $Cd[(TeP^{i}Pr_2)_2N]_2$ at substrate at 375–475 °C.[138] The Te analogue of $Cd[(SeP^{i}Pr_2)_2N]_2$ could not be synthesized by direct reaction of $NH(P^{i}Pr_2)_2$ with tellurium. An alternative approach involved metallation of $NH(PR)_2$ with NaH, prior to reaction with tellurium, which facilitates the preparation of $Na[N(TePR_2)_2]$ (R=Ph, ^{i}Pr).[139,140] This reagent was then used in metathetical reactions

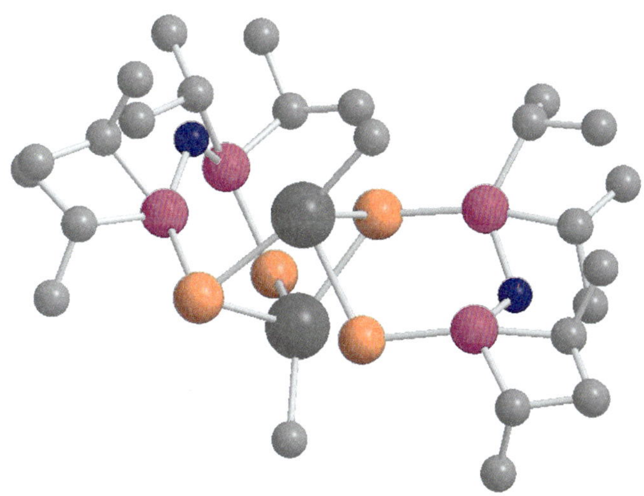

Figure 5.17 Structure of $[MeCd\{(SeP^{i}Pr_2)_2N\}]_2$.

$$\mathrm{Na[(P^{i}Pr_2)_2N]} \quad + \quad \mathrm{2Te} \xrightarrow[\text{TMEDA}]{\text{Hot Toluene}} \mathrm{(tmeda)Na[N(TeP^{i}Pr_2)_2]}$$

$$\mathrm{MX_2} \; + \; \mathrm{2\,(tmeda)Na[N(TeP^{i}Pr_2)_2]} \xrightarrow[\text{-2 NaX,-tmeda}]{\text{Hot Toluene}} \mathrm{M[(TeP^{i}Pr_2)_2N]_2}$$

M = Cd(1), X = I; M = Hg(2), X= Cl, tmeda = tetramethylethanediamine

Scheme 5.2 Synthetic scheme for $M[(TeP^iPr_2)_2N]_2$ (M = Cd, Hg).

with metal halides to generate homoleptic complexes of the type $M[(TeP^iPr_2)_2N]_2$ (M=Cd, Hg) (Scheme 5.2).[140]

Although crystalline samples of these metal complexes can be handled in air for short periods, extended exposure to moist air results in decomposition, especially for powdered samples. Consequently, these precursors have to be handled under the inert atmosphere of a glove box. The TGA of both complexes showed a single-step weight loss at 360 and 340 °C for $Cd[(TeP^iPr_2)_2N]_2$ and $Hg[(TeP^iPr_2)_2N]_2$, respectively. The residues obtained tentatively correspond to CdTe for the tellurium derivative (found: 22%, calcd. 21%) and HgTe for the mercury derivative (found: 22%, calcd. 27%).

XRD of the films grown shows the formation of cubic CdTe between 425 and 475 °C. At low deposition temperature (375 °C), a mixture of hexagonal tellurium and cubic cadmium telluride is observed. SEM images reveal that the growth temperatures do not have a profound effect on the morphologies of films. Surface analysis by XPS of films deposited at 475 °C showed the growth of Te-rich films.

AACVD of $Hg[(TeP^iPr_2)_2N]_2$ resulted in deposition of hexagonal tellurium,[138] which may be due to reductive elimination of mercury at higher temperatures.[141] Previously, mercury chalcogenide compounds have been shown to produce R_2E_2 (E = S, Se, Te) and Hg under CVD conditions rather than HgE.[142,143] It is known that the anionic ligand $[(TeP^iPr_2)_2N]_2$ is readily oxidized to the ditelluride $(TeP^iPr_2NiPr_2PTe_2)_2$ (which can be viewed as R_2E_2, where R = $TeP^iPr_2NiPr_2P$ and E = Te).[144] Thus, reductive elimination of mercury with concomitant formation of this ditelluride is a feasible pathway for the decomposition of $Hg[(TeP^iPr_2)_2N]_2$. Subsequent degradation of this ditelluride to give hexagonal Te films may account for the current observations. To confirm such a decomposition pathway, pyrolysis of the mercury precursor was carried out at 500 °C. The black powder obtained was investigated by XRD, which confirmed the presence of hexagonal Te along with cubic HgTe.

5.3.2.2 Group III-VI Materials

Group 13 and 16 elements form several types of compounds, M_2E_3 (M = Ga, In; E = S, Se, Te): a zinc-blende type (M = Ga) and a defect spinel (M = In) and also a solid ME (with the same layered structure as in M_2E_3).[145] The sesquichalcogenides are direct gap semiconductors with wide bandgaps like the II-VI compounds. Consequently, III-VI compounds are alternatives to II-VI materials as potential optoelectronic and photovoltaic materials. However, their polytypism and the variety of stoichiometries accessible are more severe problems than in II-VI materials, and so III-VI compounds have not been studied extensively.

Thiolates. The preparation of III-VI thin layers has essentially been initiated by the use of single-source systems.[146,147] Indium and gallium thiolato complexes are typically involatile polymeric aggregates similar to those formed by Group 12 thiolates.[148] Nomura *et al.* successfully prepared dimeric volatile alkylindium alkane thiolates by introducing steric bulk both at the alkyl group bound to the indium center and to the thiolate moieties.[149] These compounds are typically liquids at room temperature and can be distilled under reduced pressure ($>10^{-3}$ Torr). The deposition of the two types of indium sulfide was controllable by the number of thiolate ligands bound to the indium atom; dialkylindium monothiolates gave InS and monoalkylindium dithiolates afforded β-In_2S_3 at 300 °C under static pyrolysis conditions. Deposition of III-VI materials by MOCVD is considerably complicated by the fact that there are a plethora of known phases with varying compositions of M_nE_x. Nomura's group used $^nBuIn(S^iPr)_2$ as a source for indium sulfide deposition. Highly orientated films of β-In_2S_3 with a strongly preferred (103) growth direction were deposited on Si(111) and quartz substrates at 300–400 °C.[150] Polycrystalline sulfur-deficient films of In_6S_7 were obtained when the substrate temperature was raised to 450 °C; the films were both conductive and photoresponsive. However, orthorhombic InS layers could be deposited at lower temperatures using $Bu_2In(SPr)$ as a source, but the presence of trace amounts of oxygen in the carrier gas caused the partial oxidation of the InS layers thus deposited and the result was a highly conductive film of indium oxide doped with sulfur by MOCVD even at 275 °C.[151] This process leads to a procedure for the preparation of transparent and conducting indium oxide layers by MOCVD at the lowest reported growth temperature yet achieved.

Barron and co-workers have deposited a novel metastable cubic GaS phase from the $[^tBuGaS]_4$ cube, at temperatures in the region of 400 °C; in this case it is tempting to draw a correlation between the cubic phase observed and the cubane structure of the precursor as shown in Figure 5.18.[152] Another, perhaps simpler, interpretation of this observation is that the films deposited are extremely sulfur deficient, and the deposition also led to an amorphous gallium-containing material containing *ca.* 20% sulfur.

In related work the dimeric indium thiolate $[^tBu_2In(S^tBu)]_2$ was used to deposit a polycrystalline, tetragonal high-pressure form, β-In_2S_3.[152] The authors believe that cubic GaS layers are useful as passivation layers for GaAs wafers. The gallium sulfide cubane can be prepared by the reaction of Ga^tBu_3 with H_2S followed by static pyrolysis at 45 °C.[153] Workers at Kodak have reported one of the most thorough studies of the deposition of these materials. Gysling *et al.* deposited thin films of a novel cubic InSe (300–250 °C), and the known hexagonal In_2Se_3 (470–530 °C), on GaAs(100) by a modified MOCVD method using a sonicated spray evaporator using $Me_2In(SePh)$ and $In(SePh)_3$ as precursors. The structures of the precursors are unknown and the spray MOCVD process (as in ref. 111) was developed to deal with the relatively low volatility of the source material.[154]

Thiocarbamates. O'Brien and co-workers have grown thin films from alkyl metal dithiocarbamates[155] and dialkyldichalcogenocarbamates.[156,157] Films of indium sulfide have been grown from dimethyl-, diethyl- or dineopentyl-indium diethyldithiocarbamates ($R_2InS_2CNEt_2$ with R = Me, Et, or Np), which are air-sensitive compounds.[155] The films were grown on (100)-GaAs

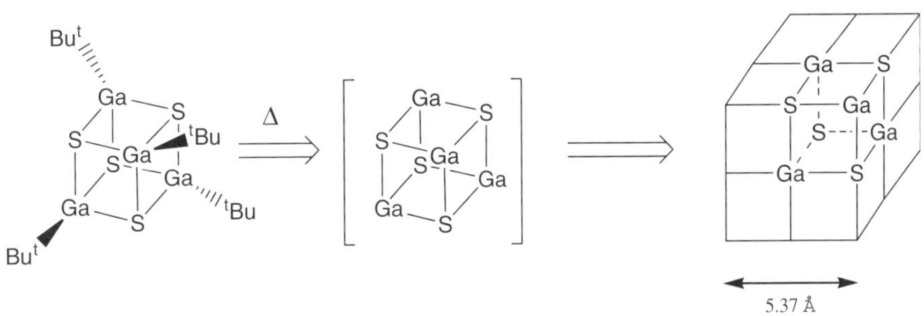

Figure 5.18 Cubic structure of $[^tBuGaS]_4$.

substrates, using a cold-wall, low-pressure reactor. The methyl complex deposits orthorhombic InS and monoclinic In_6S_7 phases between 425 and 400 °C, whereas growth at 325 °C results in cubic β-In_2S_3. In contrast, the ethyl compound deposits monophasic, crystalline β-In_2S_3 over the range 350–400 °C. The ethyl complex shows a different behavior than the other two for β-H elimination and the deposition of single phase cubic β-In_2S_3 over the entire temperature range. These results show the role played by the change of alkyl groups. They suggest that the amount of carbon incorporation, which relates to the β-H elimination process, might definitely influence the film's composition. Experiments with related gallium precursors were less conclusive. Indium tris-complexes of unsymmetrically substituted alkylmethyldichalcogeno-carbamates ($In[E_2CN(Me)R]_3$) have been used in experiments where E = S with R = n-butyl or n-hexyl,[157] and E = Se with R = n-hexyl.[156] As for the zinc and cadmium bis-complexes of the same ligands[66,78] these compounds are air-stable. In_2S_3 films were grown from $In[S_2CN(Me)^nBu]_3$ or $In[S_2CN(Me)^nHex]_3$, at 450–500 °C on glass, GaAs or InP by LP-MOCVD.[157] Cubic α-In_2Se_3 was the predominant phase, sometimes mixed with some hexagonal β-In_2Se_3 phase when the n-butyl derivative was used. The deposition of predominately α-In_2Se_3 regardless of growth conditions shows a clean gas phase decomposition process. These results are very similar to those of Bessergenev *et al.* with indium isopropyl xanthate ($In(S_2CO^iPr)_3$),[158] which has a similar InS_6 core. Similarly, thin films of cubic α-In_2Se_3 have been grown from $In[Se_2CN(Me)^nHex]_3$.[156] These results are interesting with regard to the deposition of $CuInS_2$ and $CuInSe_2$ using complexes of the type $M[E_2CN(Me)R]_n$, where R = ^{n}Bu and ^{n}Hex; M = Cu (n = 2) and M = In (n = 3).[159]

Xanthates and Monothiocarbamates. $In(S_2CO^iPr)_3$ has proved a volatile complex from which thin films of cubic α-In_2S_3 could be grown at temperatures as low as 210 °C. Films deposited on glass showed high (111)-orientation.[158] Thin films of In_2S_3 have been produced from $In(SOCNEt_2)_3$[160] and $In(SOCN^iPr_2)_3$[161] on borosilicate glass substrates, by LP-MOCVD. $In(SOCN^iPr_2)_3$ proved both more volatile and more efficient for film delivery than $In(SOCNEt_2)_3$. The former led to the tetragonal β-phase at temperatures as low as 350 °C, and the latter to the cubic β-phase at 400 °C and above. Hampden-Smith and co-workers have taken the facile elimination of thioacetic anhydride from polyether adducts of Group 2 metal thioacetates (as a pathway to deposit high-purity, stoichiometric metal sulfides[162] and applied it to adducts of gallium and indium thioacetates.[163,164] Two gallium complexes, $Ga(SCOMe)_2(CH_3)(dmp)$ and $Ga(SCOMe)_3(dmp)$ (dmp = 3,5-dimethylpyridine) and an indium complex, $[Hdmp]^+[In(SCOCH_3)_4]^-$,[164] have been used, by AACVD, to prepare thin films of the metal sesquisulfides. The respective proposed decomposition pathways for the gallium and indium complexes are shown in Reactions (5.23–5.25):

$$2Ga(SCOMe)_2(CH_3)(dmp) \rightarrow 2\ dmpy + [MeC(O)SC(O)Me] + 2(Me_2CO) + Ga_2S_3 \quad (5.23)$$

$$2Ga(SCOMe)_3(dmp) \rightarrow 2\ dmpy + 3[MeC(O)SC(O)Me] + Ga_2S_3 \quad (5.24)$$

$$\begin{aligned} 2[Hdmp]^+[In(SCOMe)_4] &\rightarrow 2HSOCMe + 2(dmp)\,In(SOMe)_3 \\ &\rightarrow 2(dmpy) + 3[MeC(O)SC(O)Me] + Ga_2S_3 \end{aligned} \quad (5.25)$$

Dichalcogenoimidodiphosphinates. $\{M(\mu\text{-}Te)[N(^iPr_2PTe)_2]\}_3$ (M = In, Ga) have been employed to deposit M_2E_3 films onto glass and Si(100) substrates at deposition temperatures of 325–475 °C.[165] The indium precursor gave cubic In_2Te_3 exclusively, whereas the gallium complex generated a mixture of cubic Ga_2Te_3, monoclinic GaTe and hexagonal Te. Mass spectrometric studies indicate that fragmentation of the indium precursor to give In_2Te_3 is accompanied by the formation of $In[N(^iPr_2PTe)_2]_2^+$ and $[N(^iPr_2PTe)_2]_2^-$.

These complexes are structurally different from the previously utilized homoleptic Cd and Sb complexes of the $[N(^iPr_2PTe)_2]_2$ ligand. Instead of the octahedral complex that one might expect, both complexes contain a central preformed M_3Te_3 (M = In, Ga) ring, with each tetrahedral metal center further coordinated by a $[N(^iPr_2PTe)_2]_2$ ligand (Figure 5.19).[166] Scheme 5.3 shows the general synthesis.

TGA of these complexes showed them to decompose in a single step between 300 and 375 °C for the indium complex and 280–360 °C for the gallium complex. Both precursors afford a significantly

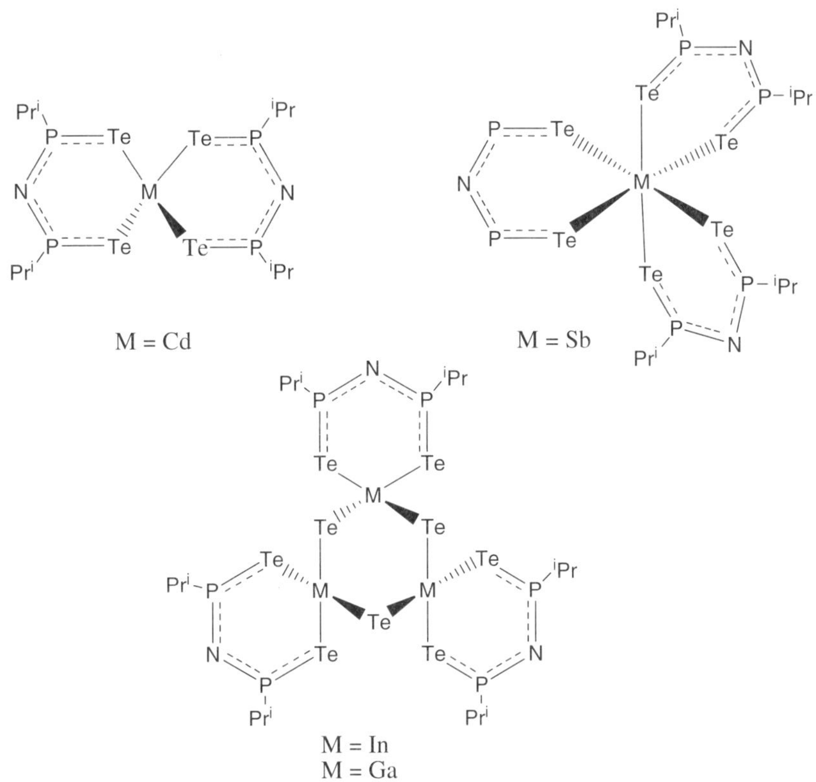

Figure 5.19 Structures of $Cd[N(^iPr_2PTe)_2]_2$, $Sb[N(^iPr_2PTe)_2]_3$ and $[M(\mu\text{-}Te)[N(^iPr_2PTe)_2]]_3$ (M = In, Ga).

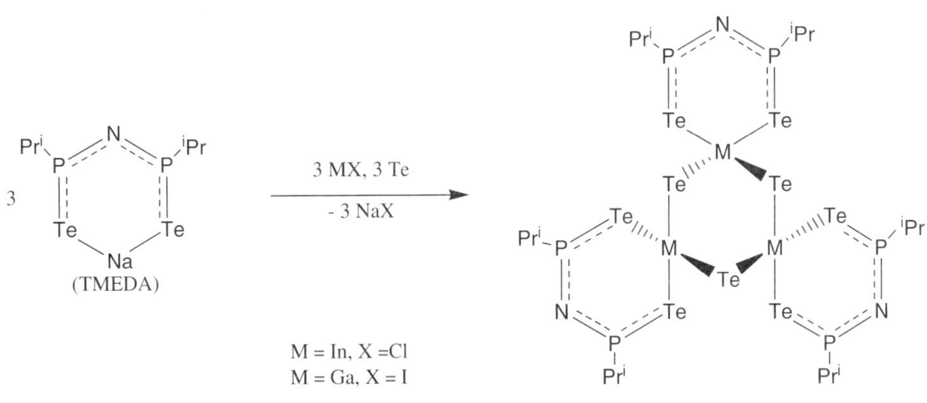

Scheme 5.3 General synthesis of tellurium derivatives.

large residue (59% and 58%, respectively), indicating incomplete decomposition or the presence of impurities after the decomposition of precursor.

5.4 Metal Pnictides

5.4.1 III-V Materials

The best known metal pnictides are Group 13-15 compounds such as GaAs, InP, $Al_xGa_{1-x}As$ etc.

There are two approaches to growing these materials; the conventional approach, in which separate Group III and V precursors are used, and the single-source precursors. Each approach has inherent advantages and disadvantages, but it is worth noting that many of the original precursors introduced by Manasevit are still favored today. The single-source precursor approach has yet to make a significant impact on the growth of III-V materials, especially where low temperature growth is important.

Metal-organic precursors traditionally employed have been those which are readily available commercially, and which have suitable vapor pressures. These include the volatile Group III trialkyls trimethylgallium (Me_3Ga), trimethylaluminium (Me_3Al) and trimethylindium (Me_3In), in combination with the Group V hydride gases arsine (AsH_3) and phosphine (PH_3). The films grown from these sources showed a very low carbon contamination. This observation is attributed to the large quantity of active atomic hydrogen released by the pyrolysis of AsH_3 or PH_3, which causes the clean removal of carbon-containing fragments from the growth surface.[167] A mechanism has been proposed[168] for the growth of GaAs from Me_3Ga and AsH_3 (Scheme 5.4).

Most of the methyl (CH_3) radicals react with AsH_x species to form a methane molecule that is easily removed from the surface as stable gas. However, a small proportion becomes more strongly adsorbed and subsequently decompose, leading to carbon incorporation in the films in an electrically active form (*i.e.* as a p-dopant) at an arsenic surface site. On the basis of infrared data under UHV conditions, it has been proposed[169] that carbon incorporation proceeds *via* the dehydrogenation of adsorbed methyl radicals to give strongly bound carbene-like species ($=CH_2$). Further dehydrogenation of these species leads to carbon incorporation in the GaAs layers.

The concentration of carbon incorporated in GaAs grown from Me_3Ga/AsH_3 is extremely low (*ca.* 0.001 atomic ppm) and the layers are high purity.[170] Similarly, high purity InP, essentially free from carbon impurities, can readily be grown from Me_3In and PH_3. There is little reason to look for alternative precursors as far as purity of the films is concerned. However, the toxicity of these gases has stimulated research into safer liquid replacements for Group V hydride gases, and similar reasons have prompted chemists to investigate single source precursors to III-V materials.

5.4.1.1 Alternative Group V Sources

AsH_3 and PH_3 are extremely toxic gases, stored in high pressure cylinders, and their use needs to be carefully managed. A body of work has appeared[171] aimed at developing safer liquid alternatives that disperse more slowly in the atmosphere in the case of accidental release. Trialkylarsine compounds, such as trimethylarsine (Me_3As) and triethylarsine (Et_3As), give heavily carbon contaminated GaAs layers due to the absence of active AsH_x species necessary for carbon removal. Therefore, arsenic precursors that contain one or more hydrogen atoms such as the ethylarsines (Et_2AsH, $EtAsH_2$) or *tert*-butylarsine ($^tBuAsH_2$) are better. The most successful liquid arsenic

$$Ga(CH_3)_x + AsH_x \longrightarrow Ga(CH_3)_{x-1} + AsH_{x-1} + CH_4\uparrow$$

Scheme 5.4 Proposed mechanism for the growth of GaAs from Me_3Ga and AsH_3.

source is $^tBuAsH_2$, with a convenient vapor pressure and pyrolysis at a lower temperature than AsH_3 (50% pyrolyzed at 425 °C compared with 575 °C for AsH_3). The more efficient pyrolysis allows the growth of GaAs at lower V/III ratios and the increased concentration of active AsH_x species on the substrate reduces carbon contamination.[167] The pyrolysis of $^tBuAsH_2$ may proceed by homolytic fission of the As–C bond to give [C_4H_9] and [AsH_2]. Subsequent radical disproportionation, recombination and exchange reactions then produce the fragments such as C_4H_8, C_4H_{10} and C_8H_{18}.[172] In an alternative mechanism,[173] two competing decomposition pathways were identified. The dominant route was proposed to be an intramolecular hydrogen transfer in $^tBuAsH_2$, leading to the elimination of C_4H_{10} and formation of [ASH] species. However, at substrate temperatures > 350 °C this is accompanied by a minor decomposition route involving the β-hydride elimination of C_4H_8 and the formation of AsH_3. In addition to safety considerations, there are good technological reasons for seeking a replacement for PH_3. The high thermal stability of PH_3 (only 50% decomposed at 700 °C) in the presence of less thermally stable AsH_3 leads to problems of composition control in the growth of quaternary alloys such as InGaAsP. The trialkylphosphines (*e.g.*, Me_3P, Et_3P) are not useful, as they are more thermally stable than PH_3 and would in any case lead to increased carbon contamination.

The most successful alternative phosphorus source is to date is *tert*-butylphosphine, ($^tBuPH_2$),[171] which is a liquid with a convenient vapor pressure (184.9 mbar at 10 °C) suitable for a wide range of MOVPE applications. Its intrinsic toxicity is lower than that of PH_3 and it pyrolyzes at a significantly lower temperature (50% pyrolyzed at 450 °C), probably by homolytic phosphorus–carbon bond fission. The reduced thermal stability of $^tBuPH_2$ relative to PH_3 allows the growth of InP at lower V/III ratios and leads to big improvements in the uniformity of InGaAsP. As well as these advantages, $^tBuPH_2$ has a more favorable gas phase chemistry than other RPH_2 precursors, which can prereact with Me_3In, even at room temperature to liberate methane and deposit a white solid [probably an $(MeInPR)_n$ polymer]. In contrast, $^tBuPH_2$ undergoes little or no homogenous reaction and this effect may be due to the large steric hindrance of the bulky *tert*-butyl group, inhibiting the formation of gas-phase intermediates, such as [$Me_3InPH_2{}^tBu$], that are likely precursors to the polymeric $(MeInPR)_n$ deposit.[167]

5.4.1.2 III-V Adducts

Several reviews have been published on adduct related complexes for III-V materials.[174–176] Benz *et al.* used III-V adducts as substitutes for highly reactive group III trialkyls[177] and Moss and Evans used adducts as precursors, but they synthesized them from the components directly in the reactor to avoid side reactions.[178] Constant and Maury proposed the use of Lewis acid-base adducts of the general formulae $R_2GaCl.ER'_3$ ($E = As^V$, P^V; R and R′=Me or Et) to overcome the problem of high reactivity and toxicity of conventional dual sources.[179] Films of GaP[180,181] and InP[181] have been grown from the corresponding $[Et_2M\text{-}P(Et)_2]_3$ complexes (M = Ga, In) – two cyclic trimeric molecules with covalent metal–phosphorus bonds.[182] Epitaxial growth of GaAs has been achieved using $R_2GaCl.AsEt_3$ (R = Me, Et).[183] On the same kinds of substrate, metal-organic vapor phase epitaxy (MOVPE) of GaAs from $(C_6F_5)Me_2Ga.AsEt_3$ and $[Et_2GaCl.AsEt_2]_2CH_2$ was attempted. The former gave epilayers on (111)-GaAs in the range 600–700 °C. Whilst with the latter epitaxial growth was observed on both types of substrate in the lower temperature range 500–625 °C.[184] These precursors belong to two series of compounds of general formulae $(C_6F_5)_{3-x}Me_xGa.AsEt_3$ ($x = 0$ or 2) and $[ClR_2Ga.AsEt_2]_2CH_2$ (R = Me, Et),[185] whose thermal decompositions have been systematically investigated by *in situ* mass spectrometric analysis of the vapor in a cold-wall CVD reactor, with helium and hydrogen as carrier gases.[186,187] In these studies, the role of the relative stabilities of the central M–E bond with respect to peripheral M–ligand and E–ligand bonds was investigated through two different approaches.[174,188] The first approach involves the use of rather fragile Lewis acid-base dative bonds, whose strength can be tuned by means of the more or less electron-donating nature of alkyl groups, and varied admixture of halogen (Cl) or pseudo-halogen groups (C_6F_5). There are several drawbacks to this approach, such as loss in volatility and surface mobility. The second approach uses complexes with covalent M–E bonds to give a single source precursors, as developed by Cowley and Jones.[189]

5.4.1.3 *Single Source*

The chemistry of complexes involving a direct bond between Group 13 and 15 elements was pioneered by Coates,[190,191] and extensively studied by Cowley and Jones,[192–194] Maury[195,196] and Wells.[197–199] These compounds were prepared, notably by the Cowley group, to investigate their potential as precursors for the CVD of III/V semiconducting materials. Many of the complexes prepared were used to grow thin films of materials such as GaAs and InP.[200] Figure 5.20 shows the structure of typical complexes.

Cowley and Jones have focused on the design and synthesis of organometallics that feature σ-bonding between Group III and V elements.[189] Their work on GaAs was reviewed in 1994 as an illustration of the concept of a single-source precursor.[175] They successively tested three tetranuclear precursors: $[Me_2Ga(\mu\text{-As}^tBu_2)]_2$,[189,192,201–203] $[Et_2Ga(\mu\text{-As}^tBu_2)]_2$[175] and $Ga(As^tBu_2)_3$.[203] The first two complexes both have a preformed (M_2E_2) ring in their solid-state structures.

Preparation of the precursor[194] involves the reaction of di–*tert*–butylarsine and trimethylgallium. This method involves the preparation of a Grignard reagent and its reaction with arsenic trichloride to obtain $^tBu_2AsCl$, which is then hydrogenated by $LiAlH_4$ to give di–*tert*-butylarsine.[204] A slightly modified method was reported by O'Brien *et al.*,[205,206] where instead of reacting trimethylgallium and di-*tert*-butylarsine at 57 °C for 4 days as reported by Cowley *et al.*[194] the reactants were dissolved in diethyl ether and refluxed at 40 °C for 3 days. Figure 5.21 shows the X-ray single-crystal structures of $[^tBu_2AsGaMe_2]_2$[205] and $[^tBu_2AsInEt_2]_2$.[206]

$[Me_2Ga(\mu\text{-As}^tBu_2)]_2$ was used to deposit GaAs thin films by LP-CVD at a precursor temperature of 145 °C and deposition temperature of 450–700 °C using H_2 or He as the carrier gas. The films deposited were arsenic-deficient.[202] Substitution of ethyl groups for methyl is known to lower carbon contamination. Epitaxial films were obtained from $[Et_2Ga(\mu\text{-As}^tBu_2)]_2$ over the temperature range 400–500 °C but showed low Hall mobilities due to the presence of Ga islands caused by

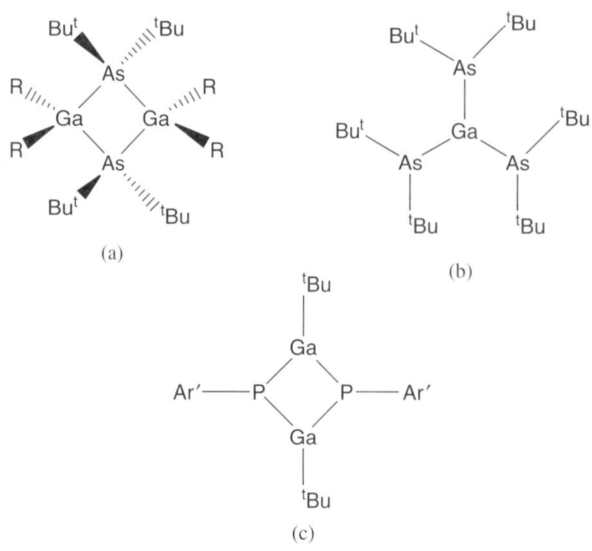

Figure 5.20 Structure of single source precursors for GaAs (*a*, *b*) and GaP (*c*).

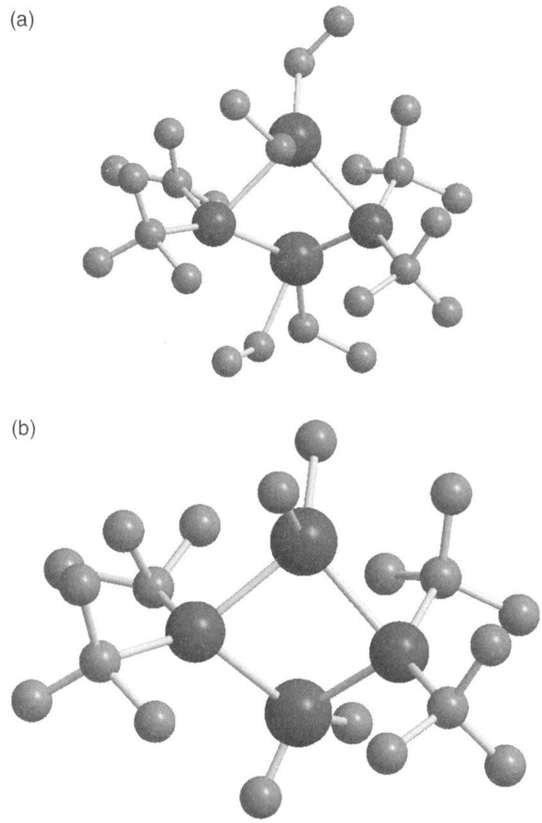

Figure 5.21 Structures of (*a*) [$^tBu_2AsInEt_2]_2$; (*b*) [$^tBu_2AsGaMe_2]_2$.

highly reactive ethylgallium species.[175] The use of $Ga(As^tBu_2)_3$, which contains an excess of arsenic and no gallium–carbon bonds, produced films with carbon levels about two-orders of magnitude lower than that of the methyl derivative. The excess of arsenic in the precursors gave rise to n-type films, while they were p-type with previous precursors.[203] Pyrolysis[202,207] and temperature-programmed desorption[203] studies suggest two modes of decomposition for these precursors: As–C bond hydrolysis to form ^{t}Bu radicals, and β-H elimination from ^{t}BuAs moieties followed by CH_3 . . . H coupling at Ga to form isobutene and methane.

Trimeric hexanuclear complexes such as $[Me_2Ga(\mu\text{-}AsMe_2)]_3$ and $[Me_2Ga(\mu\text{-}As^iPr_2)]_3$ failed to give GaAs films because of the facile loss of tetraalkyldiarsine, as shown by pyrolysis studies.[207,208] $[^nBu_2Ga(\mu\text{-}As^tBu_2)]_2$ has been used to grow epitaxial GaAs, either by ultrahigh vacuum metal-organic chemical vapor deposition (MOCVD)[209] or by spray pyrolysis.[210] Films of InP have been prepared by Cowley *et al.*[189,192] and Bradley and co-workers,[211] independently, from the same tetranuclear molecule, $[Me_2In\text{-}P^tBu_2]_2$. Cowley *et al.* carried out the deposition studies at between 450 and 700 °C in a cold-wall reactor with H_2 or He as the carrier gas. Bradley *et al.* used an MBE reactor and found that stoichiometric growth was only possible at 480 °C with a simultaneous flux of dissociated phosphine. Indium-rich films were obtained at lower temperature. $[Me_2In(\mu\text{-}Sb^tBu_2)]_3$ was used to grow films of InSb on (100)-Si wafers in a hot-wall reactor at 450 °C.[212] The deposition of AlSb films from $[Et_2AlSb(SiMe_3)_2]_2$ or $[^iBu_2AlSb(SiMe_3)_2]_2$ on (100)-Si or polycrystalline Al_2O_3 by LP-MOCVD has been investigated. Optimal deposition temperatures were found to be in the range 375–425 °C for $[Et_2AlSb(SiMe_3)_2]_2$, and 425–475 °C $[^iBu_2AlSb(SiMe_3)_2]_2$. Silicon contamination was observed at higher temperatures.[213]

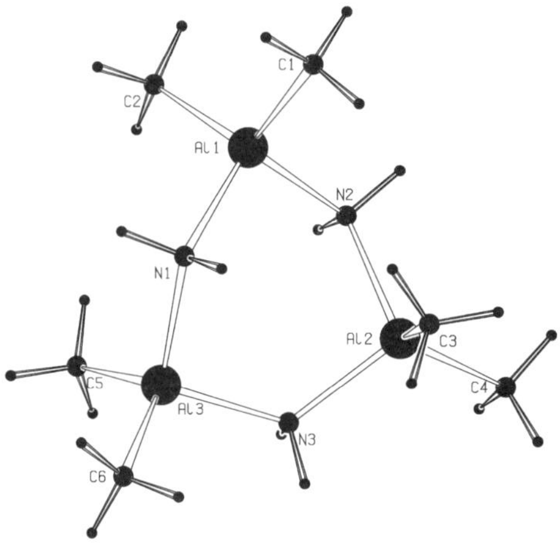

Figure 5.22 Structure of $[Me_2AlNH_2]_3$.

Growth temperatures for AlN and GaN were considerably lowered by using single source precursor.[214] AlN films, with no detectable carbon have been grown by low pressure MOCVD at 400–800 °C from the trimeric precursor $[Me_2AlNH_2]_3$ (Figure 5.22).

The relatively low levels of carbon in the AlN films can be attributed to the removal of $[CH_3^{\cdot}]$ radicals by the NH_2 groups bonded to the Al center In contrast, $[Me_3Ga.NH_3]$ and $[Me_2GaNH_2]_3$ pyrolyze with cleavage of the relatively weak gallium–nitrogen bond to deposit Ga metal rather than GaN.[215] A series of other single-source precursors have been used for AlN or GaN, which include $[Me_2AlNHR]_2$ (R = ^{i}Pr, ^{t}Bu), $[M(NR_2)_3]$ (M = Al, Ga; R = Me, Et), $[Et_2M(N_3)]$ (M = Al, Ga) and $[(Me_2N)_2Ga(N_3)]_2$.[216] However, although these precursors allow the growth of AlN or GaN at low/moderate temperatures (400–800 °C) they generally have only very low vapor pressures ($\ll$1 mbar) at room temperature, thereby requiring the heating of source and reactor lines and the use of high vacuum MOCVD equipment. Also, organometallic azides [*e.g.*, $Et_2M(N_3)$] are of unknown stability and are possibly hazardous. It has been proposed[217] that the amine-stabilized organometallic azides (Figure 5.23) are more stable than conventional azides. These compounds have been successfully used for the growth of AlN, GaN and InN at $<$600 °C.[217]

5.5 Metals

There has been a considerable interest in the CVD of metals and a good understanding of CVD processes has been developed for W, Al and Cu (see Ch. 7). Hampden-Smith and Kodas have discussed metal deposition by CVD processes in their book[218] and have also reviewed further developments.[219]

5.5.1 Precursor for Metals

Precursors for CVD of metals can be broadly classified into three types: inorganic-metal precursors, metal-organic precursors, which contain organic ligands but do not contain any carbon–metal bonds, and organo-metallic precursors, which contain organic ligands with carbon–metal bonds.

Traditional methods of precursor delivery to CVD reactors often rely on evaporation of the precursor, which is limited by the equilibrium vapor pressure. Substantial vapor pressures are required to allow convenient delivery of precursors into CVD reactors and to achieve high

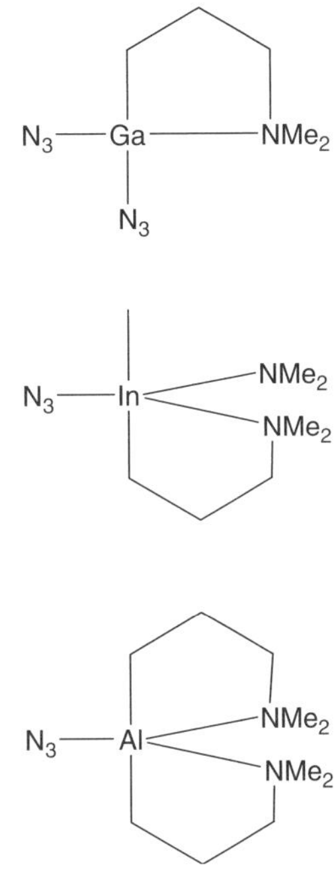

Figure 5.23 Monomeric amine stabilized Group 13 organometallic azides used for the low-temperature growth of GaN, InN and AlN.

deposition rates. The factors that effect the vapor pressure of CVD precursors, such as aggregation, are discussed under metal oxides (Section 5.6.1, Ch. 8 Section 8.2).

Another way to increase the vapor pressure is to increase the temperature of the precursor but the precursor often starts to decompose, resulting in irreproducible delivery rates and the transport of "unknown" species. This problem is found, for example, with Cu precursors, where vapor pressures are typically in the 10–100 mTorr range. However, this limitation can be avoided by using alternative delivery methods such as aerosol delivery or liquid delivery. These approaches rely on evaporation of the precursor from aerosol droplets to a carrier gas or from a solution sprayed onto a hot surface, therefore allowing the precursor to be kept at or even below room temperature until it is rapidly vaporized. While this approach has been used for various materials[220–222] but it has not been used widely for metals.[223–225]

The decomposition pathway for a precursor determines the purity of the deposited film and the deposition rate. In some cases, the reaction pathway can be modified by changing the nature of the ligands to give a pathway that deposits higher purity material at higher rates and, if desired, at lower temperatures, as discussed below. Importantly, rates are often limited by the precursor vapor pressure, which suggests that the synthesis of precursors with higher vapor pressures is an important goal. This is particularly crucial for classes of CVD precursors that already decompose at low temperatures but which have relatively low vapor pressures. This includes most of the precursors for Cu, Pt, Pd, Ru, Co, Au, Ag and many of other metals.

5.5.2 CVD of Metals

Inorganic precursors such as metal halides do not produce pure metal films below approximately 600 °C and generally require the presence of a reducing agent such as H_2 or SiH_4 to avoid halide contamination. Furthermore, they are often solids at the source temperature (there are several important exceptions: WF_6, IrF_6, $SnCl_4$ and $TiCl_4$), which often makes delivery into the reactor more difficult to control compared to liquid precursors. However, given these drawbacks, high-purity films can be deposited at high rates, and the precursors are generally inexpensive, easily purified and commercially available. Therefore, metal halides are often the optimum precursors within the constraints described above, as is currently the case for WF_6. In applications such as solar and wear resistant coatings, where low cost is often the highest priority and high deposition temperatures can be tolerated, such precursors are used.

In contrast, metal-organic and organometallic precursors have the advantages (over inorganic precursors) that they often thermally decompose at lower temperatures, usually have higher vapor pressures and are more often liquids, which can simplify precursor delivery. They have the disadvantage that impurity incorporation (C, O, F, P) is common, and to obtain pure films they often require a reducing agent such as H_2 or H_2O, or an oxidizing agent such as O_2. The problem of impurity incorporation is particularly difficult for reactive metals, such as Cr and Ti, for which no good organometallic or metal-organic precursors currently exist.[226,227] However, there are many examples where high purity metal films have been deposited from such precursors without any co-reactant (such as a reducing agent). This approach requires a decomposition mechanism where the organic ligands are removed from the reaction chamber intact or in which a reaction pathway exists to form volatile products that desorb easily from the surface. Cases where this occurs are the disproportionation of copper(I) compounds,[228–233] β-hydride elimination of $^iBu_3Al$, homolysis of Au–C bonds in $MeAuPMe_3$[234–236] and the use of precursors such as $AlH_3.NR_3$ which do not contain metal–carbon bonds.[237]

The well-known β-hydride elimination is an example of a mechanism that cleaves the metal–carbon bond. However, because other organic radicals can be liberated (*e.g.*, isobutyl radicals on an Al surface), alternative reactions can occur that lead to impurity incorporation (*e.g.*, β-methyl group elimination in the case of Al CVD from $^iBu_3Al$).[238] For this reason, alternative precursors for the CVD of Al have been sought, and the Lewis base adducts of AlH_3 provide a viable method to deposit high-purity aluminium films from precursors that are unlikely to produce carbon contaminated films because the precursor does not contain aluminium–carbon bonds. Rough guidelines can be derived for choosing the organic portion of a organometallic precursor. As the metal–carbon bond order increases in the order metal alkene $\approx$ metal alkyne $\leq$ metal alkyl $<$ metal(η^3-allyl) $<$ metal carbonyl $<$ metal cyclopentadienyl $<$ metal arene in organometallic compounds the chances for carbon incorporation are increased due to the increased strength of M–C bonding. For example, CVD of Fe using $Fe(CO)_5$ gives rise to relatively pure films as compared to $Fe(\eta^5\text{-}C_5H_5)_2$ (in the absence of H_2).[239]

In general, the more noble the metal and the lower it is in a given group in the periodic table the more easily it is reduced, the less stable the M–X bonds, and the lower likelihood of carbon incorporation. For example, $Ni(CO)_4$ gives rise to purer metal films than other transition metal carbonyl compounds earlier in the first row of the transition metal series in the periodic table. Furthermore, films derived from Pd, Pt, Cu and Ag cyclopentadienyl complexes have lower carbon incorporation than films derived from cyclopentadienyl complexes of metals earlier in the periodic table. However, these trends can be modified by the details of the reaction mechanism, which varies depending on the metal and ligands involved. In the absence of an appropriate reaction pathway, organometallic and metal-organic precursors often give rise to impurity incorporation, particularly carbon and oxygen, as is apparent from a review of the CVD of many transition metals. Impurity incorporation can be avoided in certain cases by the addition of a reducing agent (*e.g.*, H_2) or an oxidizing agent to remove the organic ligands. The reduction of metal carbonyl compounds by H_2 produces pure metal films due to Fischer–Tropsch type catalysis chemistry.[240] Metal β-diketones are popular choices as precursors mainly because the fluorinated derivatives exhibit high vapor pressures and are generally capable of high transport rates. However, the presence of a reducing agent such as H_2 is generally required to avoid extensive contamination in the film; for example, the CVD of high-purity copper films has been achieved by reduction of $Cu(hfac)_2$ with H_2. For reactive metals such as Ti, though, even the use of a hydrogen plasma does not provide sufficiently reducing conditions to form pure metal films from organometallic precursors such as Ti(η^5-C_5H_5)(η^7-C_7H_7).

An example of the use of an oxidizing agent such as oxygen is in the CVD of Ir from (η^5-C_5H_5)Ir(l,5-COD).[241] The formation of pure metal films in the presence of O_2 is only likely to occur where the metal forms an oxide that is thermodynamically unstable with respect to formation of the metal at the deposition temperature, such that only the carbon-containing ligands are oxidized to CO. Reactions catalyzed by the metal surface, such as Wacker-type processes, are likely under these conditions. [240]

5.6 Metal Oxides

The successful deposition of complex multicomponent oxides with uniform properties over a large area requires that the precursor chemistry is well understood with, as far as possible, precursors having matched thermal properties and reasonable vapor pressures. This contribution of properties allows vaporization without decomposition or condensation and clean decomposition on the substrates without incorporation of carbon from the ligands or solvent. Conventional CVD with the use of bubblers has proved difficult to develop for the deposition of oxides, primarily because the precursors have low vapor pressures, which reach only 1 Torr at temperatures well in excess of 100 °C. Prolonged heating of the bubbler to achieve sufficient vapor pressure often results in the slow decomposition of the precursor and a changing vapor pressure with time. This makes run to run reproducibility very difficult. However, several modified CVD technologies based on liquid injection, (Ch. 1, Section 1.5) or the use of aerosol formation have been developed. These require that the precursor can be dissolved in a suitable solvent to make a stable solution. Here the precursor only sees elevated temperatures for a short time as it passes from point of injection (or point of evaporator) to the deposition zone within the CVD kit.

The most common precursors used for the deposition of oxides thin films include metal alkoxides, metal carboxylates, β-diketonates and an oxidant. Owing to the low vapor pressure and low thermal stability of many oxide precursors, liquid injection MOCVD is now widely used for the deposition of metal oxide films. In this method, the precursor is dissolved in relatively inert solvent, usually an ether, *e.g.*, tetrahydrofuran, or a hydrocarbon, *e.g.*, toluene, heptane or nonane. The precursor solution is stored at room temperature until required. Liquid injection MOCVD has the requirement that precursors must be soluble and stable for long periods in the chosen solvent and must not react with other precursors in the same solution. In addition, the presence of a single heated evaporator at a fixed temperature makes it important that co-precursors evaporate at similar temperatures, otherwise it will be difficult to control the stoichiometries of complex oxides. To optimize the compositional uniformity it is also important that the various precursors deposit oxide in a similar temperature regime.

5.6.1 Precursors

The molecular structure of a metal complex has a crucial bearing on its physical properties, and on its effectiveness as an MOCVD or ALD precursor. Key structural factors in oxide precursor chemistry are discussed below, and there is further detailed discussion on oxide precursor chemistry in Chapter 8 (Section 8.2).

5.6.1.1 Metal Alkoxides

Generally, alkoxides have higher vapor pressures than the β-diketonates but are less stable, which can lead to lower deposition temperatures; however, they are more reactive to water and oxygen and hence can easily suffer degradation over time. To control the volatility and aggregation the following approaches have been developed.

Methods of Controlling the Volatility and Aggregation. *Charge and polarity:* Ionic molecules tend to be involatile because they must overcome strong intermolecular electrostatic forces to vaporize. The same principle applies to polar molecules, which tend to be less volatile due to the presence of electrostatic forces. Therefore, when considering CVD precursor, a molecule is sought that is covalently bonded, and is not polar.

Molecular size: In general, the larger an atom or molecule is the more polarizable and hence the less volatile it is. Logically, therefore, the smaller the ligands surrounding a metal center the greater that complex's volatility will be. It is somewhat surprising, therefore, that one common method for increasing a molecule's volatility is by the introduction of bulky groups. This can be explained by the fact that the larger groups minimize intermolecular interaction by steric hindrance. It is in this way that the incorporation of bulky groups into a molecule can increase its volatility. A compromise must be found between effectively isolating the metal center and avoiding ligands that are too large and, therefore, give rise to a more polarizable molecule.

CF_3 groups: The introduction of a CF_3 group into a given molecule can appreciably increase its volatility. In general, the presence of these groups in a precursor will also have little or no adverse effect on the CVD results, unless the substrate temperature is so high that the C–F bonds are broken, at which point fluorine atoms may become a major contaminant of the thin films. The effect of CF_3 groups on a molecule's volatility can be explained by the various properties of this chemical group. C–F bonds are very stable and extremely polar and the CF_3 group is not particularly bulky. This high bond polarity means that CF_3 groups often behave as self-repellent electron rich balls. It is important to differentiate between the positive effect of the CF_3 group's polarity and the undesirable effect of dipolarity within a molecule. The CF_3 group is an electronegative entity, "pulling" negative charge towards itself. The rest of the molecule is left with a slightly positive charge which is spread out among the entire molecule and is extremely small in relation to the concentrated negative charge on the CF_3 group.

Hence, the molecule contains a polar group, but no dipole and hence has an increased volatility.

Bulky ligands: The pioneering studies by Bradley and co-workers[242] describe the relationship between the molecular structure of metal alkoxides and their physical properties such as degree of association and volatility. It was realized that, to inhibit oligomerization in metal alkoxides containing large, highly positively-charged metal atoms, bulky sterically demanding ligands such as *tert*-butoxide must be employed. This facilitates the preparation of the volatile mononuclear alkoxides $[Zr(O^tBu)_4]$ and $[Hf(O^tBu)_4]$, which have been successfully used for the MOCVD of ZrO_2[243] and HfO_2.[244] However, these precursors contain unsaturated four-coordinate metal centers and the *tert*-butoxide ligand undergoes a catalytic hydrolytic decomposition reaction in the presence of trace water. These complexes are therefore extremely air- and moisture-sensitive, which limits their shelf-life and makes them difficult to handle and use in MOCVD, especially for solution based liquid injection delivery.

Oxoalkoxide: Oxoalkoxides are less susceptible to hydrolysis than alkoxides.[245] The low valence of metals such as Ba or lanthanides associated with large metal radii makes oxoalkoxides more stable than $[M(OR)_x]_n$ oligomers or their alcohol solvated analogs.[246] This has been well established for yttrium and lanthanide isopropoxides. The square pyramidal pentanuclear $M_5(\mu_5\text{-}O)(\mu_3\text{-}OR)_4(\mu\text{-}OR)_4(OR)_5$ (M = Y, Ln) complex with the electron-rich oxide ion encapsulated (Figure 5.24) appears as a basic building block for oxophilic trivalent (Ln, In, Al, Fe) and divalent (Sr, Ba) metals with alkoxides but also with other ligands.[246] In contrast to OR ligands, they can assemble up to six metals and ensure high coordination numbers despite the lack of an appropriate Lewis base.[245,247] Oxo ligands can modify physical properties and reactivity patterns. Despite a high nuclearity,

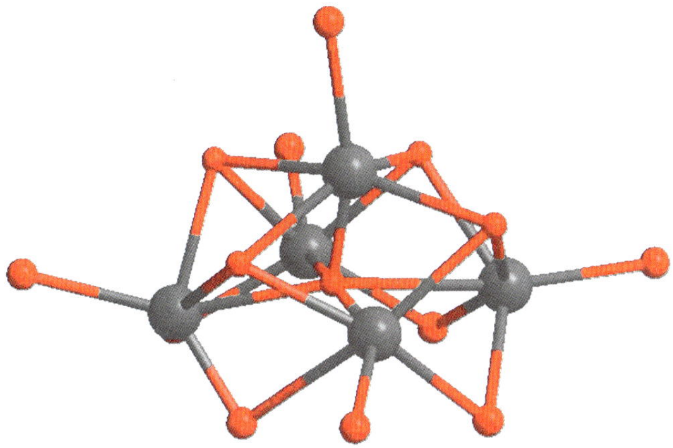

Figure 5.24 Structure of $M_5(\mu_5\text{-}O)(\mu_3\text{-}OR)_4(\mu\text{-}OR)_4(OR)_5$ (M = Y, Ln).

oxoalkoxides are quite soluble in hydrocarbons due to their peripheral OR ligands. Although hydrolytic susceptibility decreases with increasing the number of oxo ligands, derivatives such as $M_5(\mu_5\text{-}O)(\mu_3\text{-}OR)_4(\mu\text{-}OR)_4(OR)_5$ remain highly moisture sensitive since the lanthanides are only six-coordinate. Metal alkoxides or oxoalkoxides based on elements other than silicon are highly reactive with reagents having labile hydrogens.

Alcohols: Modification of metal alkoxides with diols or polyols can optimize rheology[248] or introduce functional sites (non-deprotonated OH) for controlling the stoichiometry of heterometallics.[245] Polyols provide cost-effective access to alkoxides starting from oxides such as Ti, Al, Si, Pb, Bi, *etc*. Solubility depends on the deprotonation of the polyol. Complete deprotonation affords soluble species, partial deprotonation leaves residual hydroxyls available for intra- or intermolecular hydrogen bonding. Comparison of structural features of functional alkoxides can provide guidelines for molecular design. For instance, pinacolate ligands ($OCMe_2CMe_2O$) are less bulky than two isopropoxides and have a better bridging ability than $OC_2H_4NMe_2$ but a better chelating ability than OC_2H_4OMe. Constrained polyols such as triethanolamine can also depolymerize insoluble, late transition metal alkoxides although its deprotonation by divalent metals is only partial.[244] Tetradentate triethanolaminates and/or networks of H-bonds are a hurdle for facile elimination and crystallization at low temperatures but they can act as templates for porosity of oxides up to high temperatures.[245] Other tripodal ligands such as tris(hydroxymethyl)ethane have been used for reducing the reactivity of Ti, Zr and Nb alkoxides and optimization of properties of ceramics such as PZT or PNM.[249]

Surface capping agents: Surface capping or surface controlling agents (SCA) act as "polymerization-lockers;" they decrease nuclearity and favor colloids by hydrolysis. β-Diketones have been the primary choice for such purposes and for the stabilization of oxide colloids. Acetylacetone, the most attractive in view of ceramic yield, was shown to be easily modified[250] or to degrade oxo aggregates in mild conditions, depending on the solvent.[251] Trioctylphosphine oxide (TOPO) or 2-pyrrolidone can be alternatives for stabilization of oxide colloids.[252]

Bidentate ligands: Another, strategy for inhibiting oligomerization in metal alkoxides, as well as increasing the coordination number of the highly positively-charged central metal atoms, is to incorporate bidentate donor-functionalized alkoxide ligands.[253] Alkoxide ligands such as dimethylaminoethoxide (dmae) [$OCH_2CH_2NMe_2$] or methoxy ethoxide [OCH_2CH_2OMe] contain an [NR_2] or [OMe] donor group that can increase the coordinative saturation of the complex and minimize intermolecular aggregation. For example, the reaction of dmae with $[Ta(OEt)_5]_2$ and $[Nb(OEt)_5]_2$ leads to the monomeric complexes [$Ta(OEt)_4$(dmae)] and [$Nb(OEt)_4$(dmae)],[254] which have significantly higher vapor pressures than the dinuclear parent alkoxides. The sterically hindered ligand 1-methoxy-2-methyl-2-propanolate (mmp; $OCMe_2CH_2OMe$) is particularly effective at shielding large, highly positively-charged metal centres.[253] This is because in addition to containing the [OMe] donor function it also contains two [Me] groups on the α-carbon closest to the metal center. This inhibits alkoxide polymerization and facilitates the formation of mononuclear metal alkoxide complexes, and the mmp ligand has been used to prepare a series of complexes, $[Zr(OBu^t)_2(mmp)_2]$, $[Hf(OBu^t)_2(mmp)_2]$, $[Zr(mmp)_4]$ and $[Hf(mmp)_4]$, which are six-coordinate octahedral monomers in the solid state[255] (Chapter 8, Figure 8.7).

5.6.1.2 *β-Diketonates*

β-Diketonate complexes of metals are known for their volatility and, therefore, have been used to separate metals by fractional sublimation and gas chromatography.[256–259] In the 1970s they were utilized for the first time in CVD work.[260,261] The thermal stability and volatility of the β-diketonate-type metal chelates make them suitable for metal-organic chemical vapor depositions (MOCVD).[262–265] The first generation of CVD oxide precursors was based on β-diketonates, generally 2,2,6,6-tetramethylheptane-3,5-dionates, $M(thd)_n$, which are solids or, for early transition metals, on $M(OR)_n$ or $M(OR)_{n-x}$ $(\beta\text{-dik})_x$ species. In view of the aggregation of 2,2,6,6-tetramethylheptane-3,5-dionates of Group 2 metals, $[M(thd)_2]_x$ (Sr, $x=3$; Ba, $x=4$), a major challenge has been to find a low melting point Ba source with sufficient vapor pressure and thermal-cycling stability. Oligomerization in the solid and in the vapor can be controlled by the steric bulk of ligands, manipulation of Lewis acid–base reactions by intramolecular donor sites (donor-functionalization approach) or by additional ligands.

5.6.1.3 *Alkylamides*

The fundamental chemistry of metal alkylamide complexes such as $[Ti(NMe_2)_4]$, $[Zr(NEt_2)_4]$ and $[Hf(NEt_2)_4]$ was established many years ago by Bradley and co-workers.[266,267] More recently it has been shown that these metal alkylamide complexes are good MOCVD precursors for the deposition of both metal nitride[268] and metal oxide thin films. It was shown in the mid-1990s that $[Zr(NEt_2)_4]$ can be used for the MOCVD of high quality ZrO_2 films in the temperature range 500–580 °C.[269] More recently, the liquid alkylamide complex $[Hf(NEt_2)_4]$ has been used to deposit HfO_2 by conventional (*i.e.* "bubbler-based') MOCVD.[270,271] The levels of residual carbon and nitrogen were found to decrease with increasing oxygen flow rates and increasing substrate temperatures, and, at 450 °C, stoichiometric polycrystalline HfO_2 containing few residual impurities was deposited.

5.6.2 CVD of Metal Oxides

5.6.2.1 *CVD of Dielectric and Ferroelectric Oxides*

Application of the basic chemical principles described above has allowed the selection and molecular design of a large range of volatile precursors for the MOCVD of metal oxide thin films. A great variety of oxide films have been deposited by MOCVD, using these precursors, including dielectric oxides such as Ta_2O_5, TiO_2, ZrO_2, HfO_2 and the lanthanide oxides and silicates, as well as ferroelectric oxides such as $SrTiO_3$, $Pb(Zr,Ti)O_3$ and $Pb(Sc,Ta)O_3$. The CVD of these oxides is described in detail in Chapter 8 (Sections 8.3 and 8.4), and so only a summary of some commonly used precursors is given (Table 5.1).

Table 5.1 Precursors used for the deposition of dielectric and ferroelectric metal oxide films by MOCVD.

Metal oxide	*Precursors*	*Reference*
Ta_2O_5	$[Ta(OEt)_5]$	272
	$[Ta(OEt)_4(thd)]$	273
	$[Ta(OEt)_5(dbm)]$	274
	$[Ta(OEt)_4(dmae)]$	275
TiO_2	$[TiCl_4]$	276,277
	$[Ti(O^iPr)_4]$	278,279
	$[Ti(NO_3)_4]$	280,281
	$[Ti(O^iPr)_2(tboac)_2]$	282
ZrO_2, HfO_2	$[M(acac)_4]$, $[M(thd)_4]$	283,284
	$[M(tfac)_4]$	285
	$[M(NEt_2)_4]$	269–271
	$[M(NO_3)_4]$	280–281
	$[M(O^tBu)_4]$	286,287
	$[M(O^tBu)_2(mmp)_2]$, $[M(mmp)_4]$	255,288–290
PrO_x	$[Pr(thd)_3]$	291
	$[Pr(hfac)_3(diglyme)]$	291
	$[Pr(mmp)_3]$/3 tetraglyme	292,293
Gd_2O_3	$[Gd(thd)_3(tetraglyme)]$	294
	$[Gd(mmp)_3]$/3 tetraglyme	292,295
La_2O_3	$[La(acac)_3]$	55
	$[La(thd)_3]$	296,297
	$[La(mmp)_3]$/3 tetraglyme	292,298
Nd_2O_3	$[Nd(F_3C_5H_4O_2)_3]$	299
	$[Nd(thd)_3]$	300
	$[Nd(mmp)_3]$/3 tetraglyme	292,301
$PrSi_xO_y$	$[Pr\{N(SiMe_3)_2\}_3]$	292,293
$LaSi_xO_y$	$[La\{N(SiMe_3)_2\}_3]$	292
$SrTiO_3$	$[Sr(thd)_3]]/[Ti(O^iPr)_2(thd)_2]$	302
	$[Sr_2Ti_2(^iPrO)_8(thd)_4]$	303
$Pb(Zr,Ti)O_3$	$[Pb(thd)_2]/[Zr(thd)_4]/[Ti(O^iPr)_2(thd)_2]$	304
	$[Pb(thd)_2]/[Zr(O^iPr)_3(thd)]/[Ti(O^iPr)_2(thd)_2]$	305
$Pb(Sc,Ta)O_3$	$[Pb(thd)_2]/[Sc(thd)_3]/[Ta(OMe)_4(thd)]$	306

Abbreviations: thd=2,2,6,6-tetramethylheptane-3,5-dionate, dbm=1,3-diphenylpropane-1,3-dionate, acac=pentane-2,4-dionate, tfac=1,1,1-trifluoropentane-2,4-dionate, tboac=*t*-butylacetoacetate, dmae=2-dimethylaminoethanolate, mmp=1-methoxy-2-methyl-2-propanolate. (See also Table 8.2, Chapter 8 for ligand nomenclature.)

5.6.2.2 *CVD of MgO*

Sung and co-workers[307] have synthesized MgO thin films from methyl magnesium *t*-butoxide[308] without using an added oxidizing agent. The films were of high purity but could only be deposited at the high temperature of 800 °C. The alkoxide is easy to synthesize but is moisture sensitive, and hence is difficult to work with. Matthews *et al.*[309] have reported the successful preparation of a homoleptic, monomeric magnesium β-ketoiminate complex and used it as Single Source Precursor (SSP) for the deposition of MgO thin films at 450 °C; however, the synthesis of this precursor is difficult, with low yields (58%), and the films are highly contaminated with carbon. Films grown from the mixed ligand alkoxide/β-diketonate complex $[Mg(thd)(OMe)(MeOH)]_4$ were also contaminated with carbon.[310]

Cluster complexes of higher nuclearities have shown improved volatility as CVD precursors. The hexameric diethylcarbamatomagnesium cluster (Figure 5.25) was used as SSP for MgO thin films.

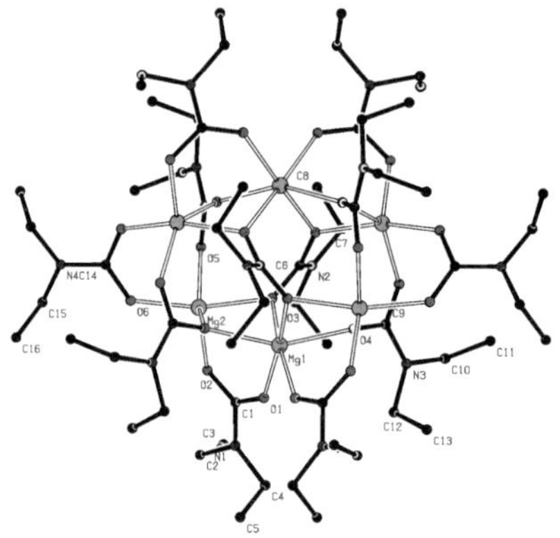

Figure 5.25 Structure of the cluster hexameric diethylcarbamatomagnesium.

The carbamato cluster is easily synthesized in one step from inexpensive starting materials [Reaction (5.26)]. Addition of CO_2 to a dry solution of $MgBu_2$ and Et_2NH yields the complex in 68% yield, by a CO_2 insertion reaction into the Mg–N bond of a diethylaminomagnesium intermediate:[110(b)]

$$6MgBu_2 + 12Et_2NH + 12CO_2(\text{Toluene}) \rightarrow Mg_6(O_2CNEt_2)_{12} + 12BuH \qquad (5.26)$$

Whilst dry reaction conditions are required to allow the formation of the carbamato cluster in preference to the potential polymeric hydroxo and carbonato hydrolysis by-products, the hexamer is quite stable to hydrolysis once it has been formed. The shelf-life of the solid precursor extends from several weeks in air to several months when stored under a dry, inert atmosphere. A TGA study showed the complex to have a reasonable volatility.

GC-MS studies of the hexamer show the formation of a trinuclear species[311,312] that decomposes to give smaller, final fragments that include CO_2 and C_2H_4N, and $Mg(OH)_2$, which is likely to have been formed by the hydration of hygroscopic MgO nanoparticles.[313] $Mg(OH)_2$ will readily dehydrate to MgO above 300 °C.[314] The presence of the highly volatile carbonaceous by-products CO_2 and C_2H_4N shows that the main mode of organic ligand fragmentation is to volatile molecules, suitable for the deposition of MgO.

5.6.2.3 *CVD of $MgAl_2O_4$*

$[MgAl_2(O^iPr)_8]$, $[MgAl_2(OBu^t)_8]$, and $[MgAl_2(O^tBu)_4H_4]$ can be used as single source precursors to deposit $MgAl_2O_4$ but at different temperatures. The hydride precursor, $[MgAl_2(O^tBu)_4H_4]$ produces crystalline $MgAl_2O_4$ at temperatures as low as 450 °C, whereas 600 and 550 °C are necessary to produce films from $[MgAl_2(O^iPr)_8]$ and $[MgAl_2(O^tBu)_8]$ respectively.[315]

It has been reported that vapor transport of $[MgAl_2(O^iPr)_8]$ poses problems, because it tends to oligomerize to other species, such as $[MgAl_2(O^iPr)_8]_2$ and $[Mg_2Al_3(O^iPr)_{13}]$, which are less volatile;[316] this is caused by ageing as a freshly prepared supply had no such problem. Nevertheless, the molecular framework based on bulkier *tert*-butoxide ligands in $[MgAl_2(O^tBu)_8]$ is thermally more robust and can be used for the deposition of stoichiometric spinel films; however, the increased molecular weight of the compound and bulkier alkoxy groups seem to hamper its volatility. Although alkoxy ligands with increased steric profile are expected to increase the volatility by reducing the nuclearity, this effect is not apparent in the case of $[MgAl_2(O^tBu)_8]$ because the overall framework in the three monomeric precursors is comparable. In addition, the increased thermal stability necessitates high decomposition temperatures. In the context of precursor properties, $[MgAl_2(O^iPr)_8]$ (liquid) is more suitable than $[MgAl_2(O^tBu)_8]$ (solid).

Kim *et al.* have synthesized the alkyl-modified alkoxides $[MgAl_2(O^iPr)_4Me_4]$ and $[MgAl_2(O^tBu)_4Me_4]$[316b] by replacing the terminal alkoxide ligands present on aluminium atoms with methyl groups. Both the precursors are much more volatile than $[MgAl_2(OR)_8]$ compounds; $Mg[(O^iPr)_2AlMe_2]_2$ can be vapor-transported at room temperature, whereas $Mg[(O^tBu)_2AlMe_2]_2$ needs to be heated to 60 °C. The isopropoxide precursor, $[MgAl_2(O^iPr)_4Me_4]$, showed better transport properties due to its lower molecular weight; however, the spinel films obtained were largely amorphous at 600 °C and needed annealing at 900 °C in air for crystallinity. The simple Mg-Al isopropoxide complex $[MgAl_2(O^iPr)_8]$ is sufficiently volatile and provides crystalline deposits of $MgAl_2O_4$ at 600 °C and is to be preferred.

Veith *et al.* have replaced terminal alkoxide ligands on Al centers by hydride ligands[317] to obtain the hydride modified MgAl *tert*-butoxide, $[MgAl_2(O^tBu)_4H_4]$. As a consequence of the drastic reduction in the molecular weight, an enhanced volatility is observed for the modified precursors ($[MgAl_2(O^tBu)_8]$, 100 °C; $[MgAl_2\text{-}Me_4(O^tBu)_4]$, 60 °C; $[MgAl_2H_4(O^tBu)_4]$, 45 °C). Both alkyl and hydride modifications are based on the abstraction of β-hydrogen (by an incipient carbanion) from the alkoxy group to eliminate methane ($[MgAl_2Me_4(O^tBu)_4]$) or dihydrogen $[MgAl_2H_4(O^tBu)_4]$) and iso-butene.

This principle in organometallic chemistry has already been exploited for the deposition of ZnO and MgO from methyl zinc alkoxide and methylmagnesium alkoxide, respectively.[318] Gas-phase pyrolysis of the precursors $[MgAl_2(O^iPr)_8]$, $[MgAl_2(O^tBu)_8]$ and $[MgAl_2(O^tBu)_4H_4]$ was performed. Structural characterization of the precursors confirms the presence of an appropriate Mg : Al ratio, in all three cases, for the formation of a stoichiometric spinel phase. However, they show different vapor pressure characteristics depending upon their molecular weight and nature of the ligands. Online mass spectrometric (MS) analysis of the residual gases gives clues for a possible decomposition profile of individual precursors. The major species detected in the CVD of $[MgAl_2(O^iPr)_8]$ are fragments of isopropyl alcohol and acetone [Reaction (5.27)]:

$$\begin{aligned} MgAl_2(O^iPr)_8 \rightarrow MgAl_2O_4 + {}^iPrOH + Me_2C = O \\ + (Me)HC = CH_2 + H_2 + H_2O \end{aligned} \tag{5.27}$$

This was confirmed by recording the mass spectral patterns of isopropyl alcohol and acetone under identical conditions followed by a differential peak analysis of the MS data recorded during the CVD of $[MgAl_2(O^iPr)_8]$. Further, the results observed correlate to the mass spectral data recorded during the CVD of other isopropoxide precursors.[319,320] The observation of other species (water, propene) in the mass spectra is because products eliminated during the first step of decomposition of the molecular precursor (primary reaction) undergo further fragmentation (secondary reactions) to produce different species whose composition and concentration depend on the process temperature. For example, isopropyl alcohol can decompose on a hot aluminium oxide surface, subject to the local temperature, through dehydration or dehydrogenation reactions, to produce propene or acetone, respectively.[321] These reactions are obviously accompanied by the formation of water, molecular hydrogen and isopropene. These species can either react with the precursor (*e.g.*, hydrolysis of the precursor by H_2O, formed in the dehydration of ^iPrOH) or incorporate in the film material (*e.g.*, elemental carbon formed by the reduction or pyrolysis of the organic compounds). Therefore, for an efficient decomposition process, a precisely defined temperature window should be coupled with an efficient removal of the gaseous by-products from the deposition zone.

In contrast to metal isopropoxides, the decomposition of metal *tert*-butoxides is relatively simple, with *tert*-butyl alcohol, isobutene and water as the major gas-phase products, as found in the case of $[MgAl_2(OBut)_8]$. The origin of water lies in the dehydration of ^tBuOH to produce isobutene and water. As a result, the overall amount of iso-butene observed in the decomposition of

$[MgAl_2(OPr^i)_8]$

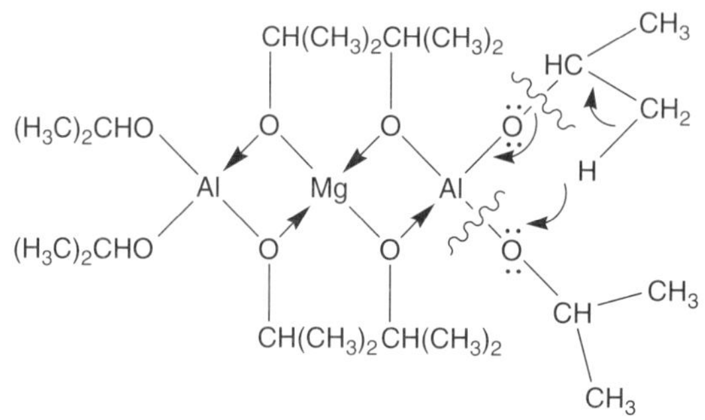

$[MgAl_2(OBu^t)_8]$

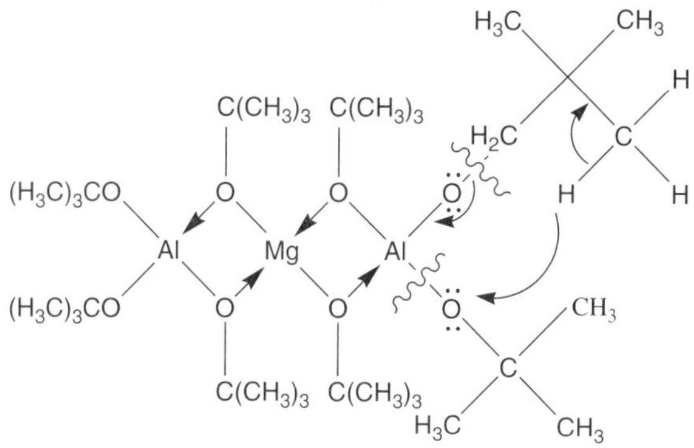

$[MgAl_2(OBu^t)_4H_4]$

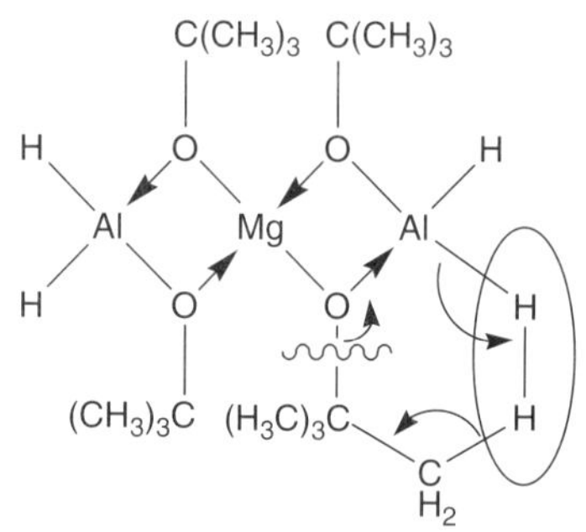

Scheme 5.5 Proposed ligand elimination mechanisms in the decomposition of $[MgAl_2(OPr^i)_8]$, $[MgAl_2(OBu^t)_8]$ and $[MgAl_2(OBu^t)_4H_4]$ by thermolysis.

$[MgAl_2(O^tBu)_8]$ is less than that found in $[MgAl_2(O^tBu)_4H_4]$ because it contains only four as against the eight -O^tBu groups in $[MgAl_2(O^tBu)_8]$ [Scheme 5.5; Reactions (5.28) and (5.29)]:

$$MgAl_2(O^tBu)_8 \rightarrow MgAl_2O_4 + {}^tBuOH + Me_2C{=}CH_2 + CH_4 + H_2O \tag{5.28}$$

$$MgAl_2(O^tBu)_4H_4 \rightarrow MgAl_2O_4 + Me_2C{=}CH_2 + H_2 \tag{5.29}$$

The decomposition mechanism of $[MgAl_2(O^tBu)_4H_4]$ is based on β-hydrogen elimination in which one of the protons of the *tert*-butyl moiety migrates to the hydride ligand present on the Al centers to facilitate the formation of molecular hydrogen, whilst the rest of the organic frame work is eliminated as isobutene molecules.[317] For [$MgAl_2(O^iPr)_8$] and [$MgAl_2(O^tBu)_8$], the protonation of one of the alkoxy oxygen atoms (alcohol formation) and the C–O bond fission (alkene elimination) are competitive reactions, whereas in [$MgAl_2(O^tBu)_4H_4$] the presence of terminal hydride ligands compels a β-hydride elimination. Formation of *t*-butyl alcohol is not observed, because this would require the abstraction of oxygen bonded to two electropositive metal centers and therefore is energetically unfavorable. Nevertheless, if the elimination of isobutene is the first step in the decomposition reaction, a metal-bound hydroxy group will be left in the precursor framework after the migration of a protic hydrogen to the alkoxide oxygen with a concomitant elimination of C_4H_8 molecule [Scheme 5.5; Reactions (5.27–5.29)]. Such an intermediate can react with other molecules to produce either water or *tert*-butyl alcohol by condensation reactions. Plausible ligand elimination mechanisms active in the decomposition of [$MgAl_2(O^iPr)_8$], [$MgAl_2(O^tBu)_8$], and [$MgAl_2(O^tBu)_4H_4$] are represented in Scheme 5.5. The molecular structures are drawn on the basis of single-crystal X-ray diffraction data reported for the precursor molecules.[317,322]

5.7 Chemistry of ALD Precursors

In ALD gaseous precursors are introduced sequentially into the reactor in alternate pulses, and chemisorb onto the substrate, and reacting there only. Inert gas is purged between each pulse to avoid gas-phase reactions. Under ideal conditions, the substrate–precursor reactions are self-limiting and the surface is saturated with precursor fragments at the end of each pulse. Growth is therefore slow, but with the advantage of atomic-level control of the film thickness. (Ch. 4)

There has been a rapid increase in interest in ALD due to ever decreasing device dimensions and increasing aspect ratios in integrated circuits (IC) for silicon based microelectronics. The attractions are the low processing temperatures and control of film thickness, in many cases of the order of a few nanometres; the main drawback is the low deposition rate.

ALD has been described in several reviews focusing on different areas: catalysts,[323] nanotechnology[324] and electronic and optoelectronic materials.[325,326] A review by Ritala *et al.* covers the principles of the ALD method and the surface chemistry involved.[327]

5.7.1 ALD Precursors

During the last two decades a series of new precursors have been studied for ALD process. These compounds have been metals, halides, nitrates or inorganic chelates and organometallic compounds. Metal alkyls and cyclopentadienyl-type compounds have been studied systematically.

5.7.1.1 General Requirements for ALD Precursors

ALD involves vapor delivery, solid and liquid precursors must be volatile at the operating temperature and pressure, and, if heating is required to obtain sufficient vapor pressure, thermal stability of the precursor over a prolonged time is necessary. Some other general requirements for ALD precursors include:

1. Sufficient volatility at the deposition temperature;
2. Precursors must adsorb or react with the surface sites;
3. Sufficient reactivity towards the other precursor, *e.g.*, H_2O;
4. No etching of the substrate or the growing film;
5. Availability at a reasonable price;
6. Non-toxicity and easy handling.

5.7.1.2 In Situ Measurements

In situ studies of the ALD process give useful information about the reaction mechanisms and the nature of precursors. In a typical ALD flow-type reactor the pressures are 1–10 mbar, which limits the selection of possible *in situ* characterization methods. The quartz crystal microbalance (QCM) has successfully been used to examine several ALD processes for oxides.[328,329] A quadrupole mass spectrometer (QMS)-ALD reactor[330,331] and a QMS-QCM-ALD combination have been used.[329] The QCM signal shows the relative weight of the adsorbed species, while the QMS gives information on the gaseous species released in the surface reaction.[332] *In situ* studies have verified the importance of surface groups in ALD growth. The measurements show, for example in the case of oxide films, how many ligands are released when the metal precursor reacts with the surface –OH groups, *i.e.*, the mechanism of the adsorption step.

Several oxide depositions involving water as an oxygen source have been studied using a QCM. The precursors studied include titanium chloride[333,334] titanium iodide,[335] titanium ethoxide,[336] titanium isopropoxide,[337] trimethylaluminium,[331,338] vanadyl isopropoxide,[339] tantalum iodide,[340] and diethylzinc.[341] QMS or QMS-QCM combinations have been used in studies of reactions of water with titanium chloride,[342] titanium isopropoxide,[343] and titanium, niobium and tantalum ethoxides.[344]

Optical techniques[345] used include surface photo-absorption (SPA) and reflectance difference spectroscopy. A variation from the SPA technique is termed the incremental dielectric reflection (IDR) method, which has been used in analysis of growth of TiO_2 by ALD.[346]

Besides examining the form in which the precursor is adsorbed on the surface, *in situ* studies can also give information on several other aspects, like reaction by-product re-adsorption and possible etching processes. The latter is important especially when growing niobium or tantalum oxide from pentachlorides.

5.7.1.3 Types of ALD Precursors

The volatile metal containing precursors that have been used in ALD are divided into four main categories: (1) halides, (2) alkoxides, (3) β-diketonate complexes and (4) organometallics (Figure 5.26). Other compounds such as amido complexes, metal nitrates, carboxylates and isocyanates have also been used.[347,348]

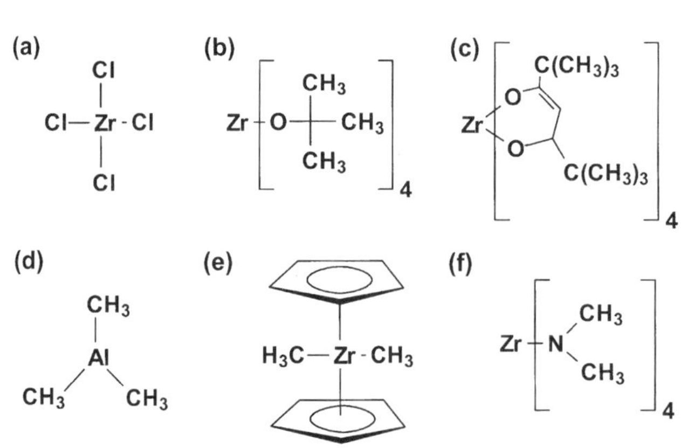

Figure 5.26 Examples of precursors for ALD depositions of oxide films. Volatile (*a*) halides, (*b*) alkoxides, (*c*) β-diketonates, (*d*) organometallics, (*e*) organometallic cyclopentadienyl-type compounds and (*f*) amido complexes have been exploited.

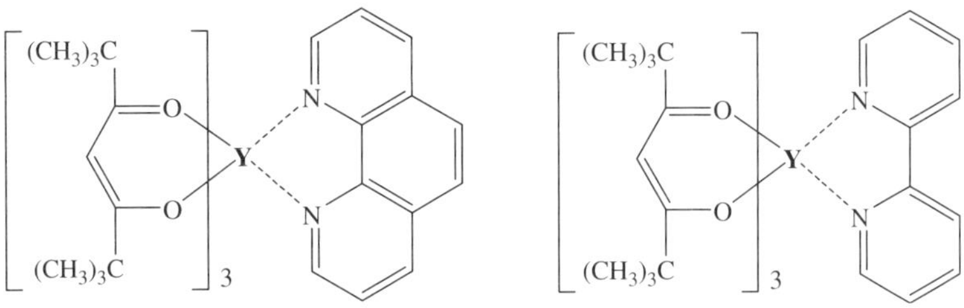

Figure 5.27 $Y(thd)_3$ adducts with (*a*) 1,10-phenanthroline and (*b*) 2,2′-bipyridyl.

5.7.1.4 β-Diketonates

In ALD, β-diketonates were first utilized in the 1980s as dopant precursors for ZnS[349]and other sulfide thin films.[350,351] Their use in metal oxide depositions[352,353] with water, hydrogen peroxide or oxygen requires much higher deposition temperatures. In addition, β-diketonates oligomerize due to being coordinatively unsaturated or they may react with moisture and become less volatile.[354–357] One way to protect them against oligomerization and room temperature reactions with moisture is to form adducts with neutral molecules (Figure 5.27).[358–361]

These adducts may dissociate during the heating process to release the neutral precursor molecule, which volatilizes separately instead of a single molecule, depending on the strengths of the bonds.[358,362] Adduct precursors dissociate at the deposition temperatures, producing original precursor.[363] Metal sulfide films have been deposited from [$Ca(thd)_2$(tetraen)] or [$Ce(thd)_3$(phen)] (thd = 2,2,6,6-tetramethyl-3,5-heptanedione), where the growth process appears to be similar to that of the unadducted precursor.[341,364] Similar results have also been obtained in the case of the deposition process for Y_2O_3 oxide films from [$Y(thd)_3$(phen)], [$Y(thd)_3$(bipy] (Figure 5.27) and O_3 as precursors.[365]

5.7.2 ALD Precursors for Oxides

β-Diketonates have been used with ozone to grow rare earth metal oxide films.[365,366] The resulting films are polycrystalline and contained only small amounts of carbon. [$Mg(thd)_2$] can be combined with hydrogen peroxide, and MgO film is formed at 325–425 °C at a growth rate of 0.10–0.14 Å cycle^{-1}.[367] In oxide deposition processes, water has most often been the oxide source, but ozone has been used when a more reactive source is needed.[365,366] Trimethylaluminium and water[368–371] and trimethylaluminium and ozone[372] have been used as precursors for the deposition of Al_2O_3.

Molecular oxygen has also been used as the oxygen precursor. TiO_2 has been grown from titanium iodide and oxygen at 200–350 °C,[373] and ZrO_2 from *tert*-butoxide and oxygen at 300–400 °C.[374] It seems that the metal precursor should be slightly thermally unstable so that oxygen is reactive enough to form the oxide at relatively low temperature.

Water or hydrogen peroxide may be too oxidative or reactive and, therefore, a water-free process based on alkoxides has been developed.[375] The metal alkoxide serves as both metal and oxygen source while another metal compound, typically a metal chloride, acts as the other metal source. The major benefit of not using separate oxygen precursors is that the substrate shows lower tendency to oxidation. This point can be especially important when thin oxide layers are to be deposited directly on silicon without creating an interfacial silicon oxide layer. Alkoxide precursors such as tetra-*tert*-butoxide,[376] tetrakis-dimethylaminoethoxide (DMAE), and mixed ligand complexes of DMAE with *iso*-propoxide and *tert*-butoxide have been studied for the deposition of zirconium oxide films. Vanadyl tri-isopropoxide [$VO(O^iPr)_3$] and water were used to deposit vanadium oxide, which has application for electrochemical batteries.[377]

$[Si(OEt)_4]$ or $[Si(OBu)_4]$ with metal halides have been used to deposit metal silicates, which are useful gate dielectric materials. Exchange reactions between metal halides and alkoxides have further been studied with tantalum oxide and titanium-zirconium oxide.[378,379] The growth proceeds without water, but water can be added as a supplementary oxygen source. The precursors $[Hf(NR_2)_4]$ and $[(^tBuO)_3\ SiOH]$[380,381] have been used for the deposition of high-*k* materials such as $Ba_{1-x}Sr_xTiO_3$ and ALD processes have been developed for barium and strontium titanates,[345,382] using Sr and Ba cyclopentadienyl compounds as precursors, together with titanium isopropoxide and water. Excellent conformality was achieved in the ALD STO films and the films were crystalline as-deposited at 325 °C.[382]

In situ studies on different titanium precursors in ALD of TiO_2 at 250 °C revealed an interesting and quite unexpected difference in reaction mechanisms; while $TiCl_4$ lost one and $Ti(OEt)_4$ two ligands in reactions with surface hydroxyl groups when arriving on the surface, less than 10% of the ligands in $Ti(OEt)_4$ were released at this stage. This observation suggests that $Ti(OEt)_4$ adsorbs molecularly rather than undergoing exchange reactions with surface hydroxyl groups. However, as the other two cases indicate, there clearly should be hydroxyl groups on the TiO_2 surface at this temperature for $Ti(OEt)_4$ to react with it. Therefore, another explanation for the observed difference is an adsorption of ethanol formed during the $Ti(OEt)_4$ pulse such that it becomes released from the surface only when replaced by water during the next pulse. Separate experiments verified the ethanol adsorption on TiO_2 under these conditions. Above 250 °C, the behavior of $TiCl_4$ resembled that of $Ti(OEt)_4$, whereas $Ti(O^iPr)_4$ started to decompose.[383] The decomposition of $Ti(OEt)_4$ commenced beyond 300 °C.

ZrO_2 and HfO_2 films have been grown by ALD using the halides $ZrCl_4$,[384] ZrI_4,[385] $HfCl_4$ and HfI_4.[386–389] Although good crystalline films of ZrO_2 and HfO_2 were obtained, chloride contamination and the risk of particle transport to the substrate are the problems with Zr and Hf halides.

The use of high purity metal alkoxides potentially offers significant process advantages over halide precursors, and $[Zr(O^tBu)_4]$,[390,391] $[Zr(OBut)_2(dmae)_2]$,[392,393] $[Zr(OPr^i)_2(dmae)_2]$[393] and $[Zr(dmae)_4]$ (dmae $= OCH_2CH_2NMe_2$)[393] have been used but none of the complexes gave self-limiting growth and the growth rate increased with the increase in pulse length. HfO_2 films have been deposited from $[Hf(mmp)_4]$ and H_2O in the temperature range 275–425 °C on borosilicate glass and Si(100) substrates,[394] although film growth was not entirely self-limiting, with the oxide growth rate increasing with increasing precursor pulse length. This is probably due to some thermal decomposition of the precursor. Similarly, the heteroleptic Hf precursor $[Hf(OBut)_2(mmp)_2]$ also failed to give fully self-limiting growth of HfO_2.[395]

In contrast to alkoxide precursors, Zr- and Hf-alkylamides, $[M(NR_2)_4]$ have recently been shown to give fully self-limiting film growth.[396,397] The conformality, or step coverage, of HfO_2 films deposited from $[Hf(NR_2)_4]$ is much higher than the conformality of films deposited from halide precursors, probably due to a higher surface mobility of $[Hf(NR_2)_x]$ species relative to $[HfCl_x]$ species. This property of $[Hf(NR_2)_4]$ precursors has been utilized in the deposition of highly conformal HfO_2/Al_2O_3 multilayers for DRAM capacitor structures.[398] In an effort to find a more thermally stable precursor for the ALD of HfO_2, $[Hf(ONEt_2)_4]$ has been investigated;[399] MOCVD studies had indicated that it may have a higher thermal stability than $[Hf(NMe_2)_4]$.[400] The HfO_2 films were deposited using alternate pulses of $[Hf(ONEt_2)_4]$ and H_2O at substrate temperatures between 250 and 350 °C on borosilicate glass and Si(100) substrates.

Relatively few precursors are available for the ALD of rare earth oxides. Metal β-diketonates have been investigated,[401] but ozone is required to obtain acceptable growth rates and no ALD growth window could be found. Recently, metal amidinates $[M(R\text{–}R'AMD)_n]$ (M = Ti, V, Mn, Fe, Co, Ni, Cu, Ag, La; R=Pr^i, Bu^t; R' = Me, Bu^t) have been proposed as a general class of precursors for the ALD of metal and metal oxide films,[402] and $LaAlO_3$ has been deposited by ALD using $[La(R\text{–}R'AMD)_3]$ with $[AlMe_3]$.[403] The lanthanide alkyl silylamides $Ln[N(SiMe_3)_2]_3$ are also volatile and relatively stable,[404] and have therefore been investigated as precursors in ALD.[292] PrO_x

films have recently been deposited over a range of substrate temperatures using alternate pulses of $Pr[N(SiMe_3)_2]_3$ and H_2O,[292] although the presence of Si in the PrO_x film indicates that the $[N(SiMe_3)_2]$ ligand is decomposing during the ALD process.

GdO_x films have also been grown by ALD using alternate pulses of $Gd[N(SiMe_3)_2]_3$ and H_2O. Growth was not self-limiting, even at low temperature.[292]

5.7.3 ALD of Metals

The first ALD thin films were deposited in the 1970s using elemental zinc and sulfur, reacting to form ZnS at 250–450 °C.[405] Although using metals as ALD precursors seem to be the most straightforward method of producing compound thin films, this type of process is limited by the generally low vapor pressure of elemental metals. Therefore only zinc and cadmium have been used as metal sources in ALD processes. However, some metalloids, *e.g.*, selenium and tellurium, can be used in this way.

5.7.3.1 Metal Amidinates

Recently, metal amidinates $[M(R\text{–}R'AMD)_n]$ (M = Ti, V, Mn, Fe, Co, Ni, Cu, Ag, La; R = Pr^i, Bu^t; R′ = Me, Bu^t) have been used as a general class of precursors for the ALD of metal and metal oxide films,[402] with molecular hydrogen or water as the second precursors. $LaAlO_3$ has been deposited by ALD using $[La(R\text{–}R'AMD)_3]$ with $[AlMe_3]$.[403] Metal amidinates are structurally analogous to the β-diketonates, except that the ligand is coordinating through nitrogen, making these compounds more reactive than the corresponding chelates coordinating through oxygen.

5.7.3.2 Halides

Metal halides, especially chlorides, have been widely applied in ALD for deposition of oxide, sulfide and nitride films.[345] Most of the halides are solids, but in ALD the solid precursors are not such a problem because it is only crucial that the precursor dose is large enough to saturate the surface, but the flux need not necessarily be constant or homogeneous.

Several metal halide precursors have been used in ALD processes, together with water as an oxygen source. However, the suitability of a particular metal halide for ALD depositions has been found to depend on the metal. Halide contamination and liberation of HX (X = F, Cl, Br, I) during the deposition process of the film can cause problems of corrosion and etching.[406–408] Oxyhalides have been used as more volatile inorganic precursors. Tungsten oxyfluoride (WO_xF_y) and H_2O have been used as precursors in the deposition of WO_3[409] while CrO_2Cl_2 and CH_3OH have been used in the deposition of chromium oxide.[410]

Although halides have long been known as good ALD precursors, not much attention was initially given to fluorides, bromides or iodides. More recently, fluorides and iodides have received more attention, but also new chloride precursors have been taken into use. BCl_3 is a new precursor, which has been introduced quite recently. BCl_3 reacts with ammonia and forms BN film in ALD mode.[411] Shimogaki *et al.*[412] reported on the first chloride adduct precursor in their flow modulation CVD process. They used a tantalum chloride thioether adduct and ammonia to deposit TaN films. The important benefit of this adduct is that, unlike $TaCl_5$ itself, it is a liquid. Tungsten hexafluoride has been examined as a precursor for tungsten metal and tungsten nitride films.

WF_6 can be transformed into W with disilane.[413] WF_6–Si_2H_6 process involves an exchange reaction during both reaction steps; the growth rate is high (2.5 Å $cycle^{-1}$) and the process window is wide (150–330 °C). The films were, however, amorphous and rather resistive. WF_6 is also a suitable precursor in combination with ammonia for nitride films.[414,415] The stoichiometry of the films may vary but the formation of polycrystalline W_2N has been reported. Volatile, *in situ* prepared tungsten oxofluorides are precursors for tungsten oxide, WO_3.[416] Polycrystalline films can be grown even at 200 °C. WF_6 could not be used as a precursor for oxide films because of etching effects.

The use of iodides as metal precursors has been studied in the deposition of titanium nitride,[417] and oxides of titanium,[335,418] zirconium[419,420] and tantalum.[340] Oxide formation reaction is enhanced if H_2O_2 is used instead of water. The reaction is self-limiting at lower temperatures, but at higher temperatures metal iodides slightly decompose. Reaction mechanism studies show that in the case of TiO_2 the situation is complicated because of the structural change (anatase–rutile) of the oxide, which changes the growth mechanism.[418] The TiI_4–H_2O_2 process has even facilitated epitaxial growth at low temperature (−40 °C) on single crystal sapphire and MgO substrates.[421] Oxide films grown from the iodides are iodine-free and the properties equal to films deposited with chloride–water ALD processes. In the TaI_5–H_2O_2 process, etching takes place above 350 °C, which is 50 °C higher than in the $TaCl_5$–H_2O process.[340]

5.7.4 Organometallic Precursors

Amongst organometallic compounds, metal alkyls and cyclopentadienyl derivatives have been the most widely investigated in ALD.

5.7.4.1 *Metal Alkyls*

Metal alkyls are an important group of compounds that have been exploited for ALD. Volatile metal alkyls were first used to deposit III–V semiconductors containing Ga, Al, In, As and P. In many cases (Al, Zn), alkyls behave almost ideally in making oxide and sulfide films, but in some cases (Ga, In) the results have not been very promising due to complex and unfavorable surface chemistry. GaAs has been deposited using Me_3Ga[422,423] or Et_3Ga[424] with AsH_3. The films obtained were contaminated by carbon due to the decomposition of trialkylgallium. Because of the low reactivity of AsH_3 higher deposition temperatures are required, which cause the decomposition of trialkylgallium. Similarly, indium alkyls, mainly Me_3In, have been used to deposit InAs,[425,426] InP[427,428] and In_2O_3[429] thin films.

Me_4Sn or Et_4Sn in combination with N_2O_4 were utilized to grow transparent SnO_2 thin films.[430] The conductivity of the films increased by almost two orders of magnitude by doping the films using BF_3 pulses. Me_2Zn and Et_2Zn, have been used as precursors for ZnS,[431–433] ZnSe[432,434,435] and ZnTe[436] thin films with H_2S, H_2Se and H_2Te as chalcogen sources, respectively. Zinc alkyls have also been used for the deposition of ZnO thin films, by using H_2O as the second reactant, and doped with boron[437,438] aluminium[439–441] or gallium[442] with B_2H_6, Me_3Al or Me_3Ga, respectively, to increase the conductivity of the films. Similarly, Et_4Pb with H_2S has been used for PbS thin films.[439]

Several III–V materials such as AlAs,[443–445] AlP[446,447] and AlN[448–450] have been deposited from aluminium metal alkyls, including Me_3Al, Me_2AlH and Et_3Al, with AsH_3, PH_3 or NH_3 as the pnictide source. Al_2O_3 thin films were grown with Me_3Al, Me_2AlCl,[451] $Me_2Al\ H$[452] or Et_3Al as aluminium source. Trimethylaluminium has been used with H_2O,[453–457] H_2O_2,[458–460] NO_2,[461] N_2O,[462] O_2 plasma[463,464] or O_3[465] as oxygen source. Deposition from Me_3Al with H_2O has a wide temperature window (100–500 °C) with a reasonably high deposition rate of up to 1.2 Å $cycle^{-1}$. At very low deposition temperatures hydroxyl impurities remain in the film. Different oxygen sources, such as H_2O_2, O_2 or O_3, have been investigated to reduce the impurities in the films.

Bismuth oxide cannot be grown from triphenylbismuth and H_2O_2, but a ternary Bi–Ti–O material could be deposited from this compound.[466] In the as-deposited films, bismuth may be partially metallic, and the ternary oxide phase is formed only upon annealing in air at 500 °C. Further annealing at 700 °C resulted in the crystalline dielectric $Bi_2Ti_2O_7$ phase, but it was not possible to obtain bismuth titanates with higher bismuth content.

5.7.4.2 Cyclopentadienyls

Cyclopentadienyl-based metal complexes (Figure 5.28) were first synthesized in the 1950s[467,468] and their reactions have been exploited with a series of elements.[469] Metallocenes are often thermally stable and volatile compounds and hence useful as precursors[470–473] for thin film growth in a CVD process. Highly reactive cyclopentadienyl compounds can be more suitable precursors for ALD process[474] in which high reactivity can be controlled by sequential pulsing of the precursors.

More recently, cyclopentadienyl compounds have been used in deposition of both alkaline earth titanates and sulfide films.[475] However, for the heavier alkaline earth metals, simple Cp_2Sr and Cp_2Ba compounds do not posses sufficient thermal stability for a controlled ALD process. Bulkier ligands such as in $[(C_5{}^iPr_3H_2)_2Sr]$ $[(C_5Me_5)_2Ba]$ and $[(C_5{}^tBu_3H_2)_2Ba]$ have been used with $[Ti(OCH(CH_3)_2)_4]$ and H_2O to deposit $SrTiO_3$[476,477] and $BaTiO_3$[477] thin films. SrS and BaS thin films have been deposited from $[(C_5{}^iPr_3H_2)_2Sr]$, $[Cp_2Sr]$ and $[Cp_3Ba]$ using H_2S as the sulfur source.[478]

MgO thin films have been deposited from Cp_2Mg and H_2O[479–481] with a film deposition rate almost ten-times higher than the traditional β-diketonate/ozone process.[482] Cp_2Mg/H_2O has also been used with Me_3Al/H_2O for the deposition of $MgAl_2O_4$ thin films.

Cp_3Sc and Cp_3Y can be used in ALD process but the thermal stability of the corresponding lanthanum and cerium compounds is not high enough for controlled film growth. The Cp_3Sc/H_2O process gives a high Sc_2O_3 deposition rate of 0.75 Å $cycle^{-1}$ at 250–350 °C[483] and the deposited films contain only a very small amount of carbon and hydrogen impurities. Cp_3Y/H_2O and $(C_5MeH_4)_3Y/H_2O$ give uniform and stoichiometric Y_2O_3 films below 400 °C with a deposition rate of 1.2–1.6 Å $cycle^{-1}$, which is 5–7 times higher than for the corresponding β-diketonate/O_3 process.[365] Cp_3Ce and $(C_5MeH_4)_3Ce$ have been investigated for the deposition of CeO_2 films but they seem to be too unstable for a controlled ALD deposition process; however, the bulkier $(C_5Me_4H)_3Ce$ has been successfully used to dope SrS thin films.[484,485] Cp_3Pr/H_2O gave only non-uniform films of praseodymium oxide due to its thermal instability. $\{[C_5H_4(SiMe_3)]_2LuCl\}_2$ deposited Lu_2O_3 thin films[483] with some decomposition.

Although there are several volatile cyclopentadienyl-type compounds of zirconium and hafnium only a few compounds have been used as ALD precursors. Cp_2ZrCl_2 was the first cyclopentadienyl compound used as a zirconium source[487] in the growth of ZrO_2 thin films. The same precursor has been explored to coat porous silica powders in the preparation of high-surface-area catalysts by the

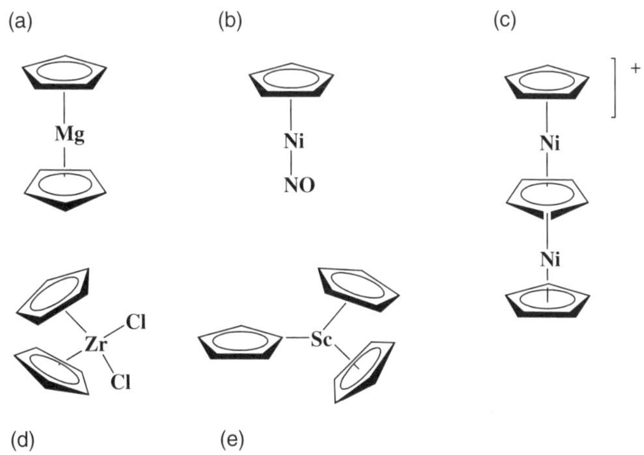

Figure 5.28 Examples of simple metallocene structures: (*a*) parallel sandwich, (*b*) half-sandwich, (*c*) multi-decker sandwich, (*d*) bent/tilted sandwich and (*e*) compounds with differently bonded cyclopentadienyl ligands.

ALD method.[488] Owing to the low reactivity of the precursor towards mild oxygen sources, such as H_2O or O_2, O_3 had to be used as an oxygen source.[487] Cp_2ZrCl_2 is reasonably thermally stable and uniform films were obtained below 400 °C.[487] It seems that ALD processes based on zircocene dichloride suffer similar contamination problems as other halide precursors, as 0.1–0.3 wt% chlorine contamination was observed below 300 °C in binary ZrO_2.[487] Chloride contamination in the films increased even more when yttria-stabilized zirconia (YSZ) films were deposited by adding $Y(thd)_3/O_3$ pulses in this process.[489]

ZrO_2 and YSZ film depositions have also been investigated from Cp_2ZrMe_2,[487,490] which is more reactive than Cp_2ZrCl_2 and has been successfully used as an ALD precursor with H_2O for the deposition of ZrO_2. Currently, there is also increasing interest in processing of HfO_2 thin films by ALD. Because of the similar ionic radius and analogous chemistry between zirconium and hafnium, organometallic hafnium compounds behave almost identically to the corresponding zirconium precursors.[491] In addition to the conventional halide,[492–494] nitrate[495] and amide[496,497] precursors, organometallic cyclopentadiene compounds have also been used to produce insulating HfO_2 thin films.[491] ALD growth of HfO_2 has been achieved by using either Cp_2HfCl_2 or Cp_2HfMe_2 as the hafnium source.

Cp_2Mn and $(C_5MeH_4)Mn(CO)_3$ have been used as manganese sources for doping ZnS thin films to produce yellow-emitting thin-film electroluminescent devices.[498]

Metallic Ru films have been deposited by using Cp_2Ru,[499,500] $(C_5EtH_4)_3Ru$[501] or $(C_8H_8)_3Ru$ as metal sources with O_2 or air, which decomposes the organic ligand but does not oxidize metallic ruthenium. $(C_5MeH_4)PtMe_3$ and O_2 were used as precursors[500,502] for the deposition of platinum films. Metallic nickel films were deposited in two steps.[503] In the first step NiO thin films were deposited using one ALD cycle of Cp_2Ni and H_2O and in the second step the NiO film was then reduced to metallic nickel by pulsing hydrogen radicals into the reactor.

Cp_2Bi[504] and Cp_2Pb[486] have been used for the preparation of bismuth- and lead-oxide thin films. Owing to the low reactivity of these complexes, ozone is required to obtain completely oxidized films.[513] Similarly, H_2O_2 with Ph_3Bi gave metallic bismuth films.[504]

No effective ALD processes have been found so far for silicon dioxide. In all reports, long pulse times have been required to saturate the reactions. An interesting report[383] discussed the deposition of SiO_2 from *N*,*N*-di-*t*-butyl-1,3-diaza-2-silacyclopent-4-en-2-ylidene and oxygen at 200 °C. The authors flash evaporated the solution and a deposition rate of 0.05 nm cycle^{-1} was obtained throughout the length (30 cm) of the deposition zone.

5.7.5 Non-metal Precursors

Hydrides (H_2O, H_2S, NH_3 and AsH_3) are the non-metal precursors usually used without any problems in terms of volatility and thermal stability. H_2O and H_2S have reasonable reactivity but NH_3 and AsH_3 have more limited reactivity.

Metal and nonmetal precursors are separated due to the sequential nature of ALD process, which makes it possible to control the reactivity and reactions of the metal precursor by selecting an oxidizing or reducing precursor. Metal alkoxides are a special case where no separate oxygen source is needed.[505–507] In the non-oxidizing regime, a second precursor (reductant) is also needed to deposit elemental metal films. Hydrogen is the most widely used reducing agent[508] but metallic zinc vapour[509,510] various silanes[511–513] and B_2H_6[514] have also been used. High deposition temperatures are required for molecular hydrogen but the reactivity can be increased by thermal or plasma cracking to atomic hydrogen.

Ammonia has been used for the deposition of nitride films such as TiN, Ta_3N_5, W_2N, NbN and WCN, where it serves both as the nitrogen source and the reducing agent. Other nitrogen containing compounds, such as $(CH_3)NNH_2$, $^tBuNH_2$ and CH_2CHCH_2NH, have also been studied.

H_2O, H_2O_2, N_2O_4,[430] N_2O,[461] O_2 or O_3 are used as oxidizing agents depending on the metal precursor. Water is the most commonly used oxygen source as it readily reacts with many metal halides, alkyls or alkoxides. H_2O_2 has been used when water is not successfully reacted.[515–518]

Alcohols have also been used as oxygen source, *e.g.*, Al_2O_3 has been deposited with a $AlCl_3$ and ROH precursor combination, where R is H, CH_3, CH_2OHCH_2OH, *t*-C_4H_9OH or *n*-C_4H_9OH.[519] The use of H_2O, H_2O_2, N_2O, CH_3COOH and O_2 as oxygen sources with β-diketonate-type compounds[520–523] gave highly carbon contaminated films due to their low reactivity.

5.7.6 Metal Nitrides

Metal alkylamides have been studied extensively as CVD precursors for nitride thin films. Recently, they have also been used in ALD process.[383] Tetrakis(dimethylamido)titanium [or tetrakis (ethylmethylamido)titanium]–ammonia processes have been shown to be self-limiting at 170–210 °C with high ammonia doses and long exposure times. Ti–Si–N films were grown on addition of silane in the process, and the Si content in the films can be varied up to 23 at.% by controlling the number of silane pulses.[382] An increased quantity of silane decreases the growth rate by blocking the growth. Ti–Si–N films showed excellent conformality and barrier properties against Cu diffusion, but are apparently resistive.[383]

t-Butylimidotris(diethylamidotantalum) with hydrogen radicals was used to grow tantalum nitride films.[524] The films were heavily contaminated with carbon (15 at.%) and were Ta rich. The aluminium precursor dimethyl(aluminium hydride)ethylenepiperidine has been employed in ALD of AlN films.[525]

ALD transition metal nitrides have been based on metal chlorides and ammonia. The best results have been obtained with metallic zinc as an additional reducing agent.[526,527] However, zinc contamination is a problem. Titanium iodide leads to a conducting nitride film that can be deposited without zinc.[415] Other reducing agents such as trimethylaluminium (TMA)[528] and 1,1-dimethylhydrazine as a nitrogen precursor have been used.[529] Low TMA doses clearly assist the formation of nitride films. Dimethylhydrazine reduces Ti, Nb and Mo chlorides and the corresponding low resistivity nitride films are formed at 400 °C and, in the case of TiN, even at 250–400 °C.[529] However, $TaCl_5$ cannot be reduced by dimethylhydrazine. The growth rate in the chloride–ammonia processes has been low, typically 0.2 Å $cycle^{-1}$. It is unknown whether this is due to a small number of reactive NH_x groups left on the surface after the NH_3 pulse, or due to adsorption site blocking by re-adsorbed reaction by-product HCl. Elers *et al.*[530] were, however, able to almost double the growth rate by *in situ* reducing titanium chloride with metallic titanium. The $TiCl_3$ species formed seemed to be able to utilize the active surface site more effectively than $TiCl_4$. The WF_6–NH_3 process mentioned above has resulted in a remarkably higher growth rate of 2.55 Å $cycle^{-1}$.[421] TiN films made by chloride–ammonia processes have shown good conformance and barrier properties up to 650 °C.

Transition metal nitride films can be deposited from alkyl amides and ammonia with high growth rates at low temperatures[382,383] but the alkyl amides are also quite prone to decomposition. TiN films deposited by these reactions have been amorphous and contained some carbon (4 at.%) and hydrogen (6 at.%) impurities.[531] The films have shown excellent conformality and negligible interdiffusion of Cu and/or Si at temperatures lower than 600 °C.[532]

5.7.7 Metal Films and Plasma-ALD

Copper films have been deposited by ALD using hydrogen as a reducing agent for CuCl and $Cu(thd)_2$.[533] Both processes seem to be substrate material-dependent, indicating that the substrate (Pt and Pd) actively participates or catalyzes the growth reaction. CuCl and also $MoCl_5$ can also be reduced to metal by zinc vapor.[534,535] The processes suffer, however, from a reversible dissolution–outdiffusion of zinc into the Cu and Mo films. The well-known solution (electroless deposition) chemistry for copper deposition where the reduction is carried out by organic compounds (formaldehyde, alcohols) has successfully been employed in ALD.[536] The Cu precursor used was 1,1,1,5,5,5-hexafluoroacetylacetonate hydrate, and low-resistivity (1.78–4.25 mV) films were deposited at 300 °C. Other thermal ALD metal processes studied to date include tungsten and nickel. In deposition of tungsten, the WF_6–Si_2H_6–process was employed.[413] Nickel deposition involves a two-step process where first NiO is grown from $Ni(acac)_2$ and ozone and then the oxide is converted into metal upon annealing in a hydrogen atmosphere at 260 °C.[537] In the deposition of nitride and metal films thermal energy is not necessarily enough to activate the reactions. Therefore, plasma enhanced growth has also been used in ALD. Thin films of inert refractory materials are used in semiconductor interconnect applications as diffusion barrier, seed and adhesion layers. Rossnagel *et al.*[538] deposited 100% conformal Ta and Ti films at 25–400 °C using a plasma enhanced ALD process, where $TaCl_5$ and $TiCl_4$ were reduced by atomic hydrogen generated upstream with an inductively coupled RF plasma discharge. Chlorine residues were below 3 at.% and the films deposited below 250 °C were amorphous. Further studies of Ti films with QCM system and several different substrate materials have revealed that the growth is saturated with increasing hydrogen plasma power and atomic hydrogen exposure time.[539]

Atomic hydrogen has been used in the epitaxial growth of silicon at 550–610 °C from dichlorosilane both on silicon (100) and (111) surfaces.[540,541] Plasma assistance has also been used in ALD growth of nitride and oxide films.

Tetrabutylimidotris(diethylamido)tantalum has been reduced to TaN by hydrogen radicals produced by a capacitively coupled RF plasma source.[529] The use of nitrogen radicals is also possible.[542] High-quality aluminium oxide films have been deposited by plasma enhanced ALD. After the aluminium precursor [(dimethylethylamine)alane] pulse, the reactor was pumped down and H_2 plasma was introduced. Then the aluminium layer was oxidized with O_2 plasma.[543] Al_2O_3 films 15-nm-thick were grown by repeating the process. The films showed very good electrical properties and the properties could be further improved by post-deposition oxygen-plasma annealing at 200 °C. The advantages of O_2 plasma have also been shown in the deposition of tantalum oxide films from ethoxide and aminoethoxide $[Ta(OEt)_4(OC_2H_4N(CH_3)_2)]$ zirconium oxide film from zirconium *t*-butoxide and aluminium oxide from TMA.[544,545] For example, the growth rate could be increased by 50–100% by using an O_2 plasma.

References

1. P. John, in *The Chemistry of the Semiconductor Industry*, S. J. Moss and A. Ledwith eds., Blackie, Glasgow, 1987, p. 98.
2. H. M. Manasevit, *J. Cryst. Growth*, 1981, **55**, 1.
3. G. B. Stringfellow, *Organometallic Vapour Phase Epitaxy: Theory and Practice*, Academic Press, New York, 1989.
4. D. J. Cole-Hamilton, *Chem. Br.*, 1990, 852.
5. J. B. Mullin and S. J. C. Irvine, *Prog. Cryst. Growth Charact.*, 1994, **29**, 217.
6. (a) L. M. Smith and J. Thompson, *Chemtronics*, 1989, **4**, 60; (b) J. B. Mullin, D. J. Cole-Hamilton, S. J. C. Irvine, J. E. Hails, J. Giess and J. S. Gough, *J. Cryst. Growth*, 1990, **101**, 1.
7. (a) P. O'Brien in: *Inorganic Materials*, D. W. Bruce, D. O'Hare eds., Wiley, Chichester, UK, 1992, 491; (b) P. O'Brien, *Chemtronics*, 1991, **5**, 61; (c) A. C. Jones, P. J. Wright and B. Cockayne, *Chemtronics*, 1988, **3**, 35.
8. S. Y. Hirata, S. Fujita and M. Isenuera, *J. Cryst. Growth*, 1990, **104**, 521.

9. (a) K. P. Giapis and K. F. Jensen, *J. Cryst. Growth*, 1990, **101**, 111; (b) K. P. Giapis, D. C. Lu, D. I Fotiadis and K. F. Jensen, *J. Cryst. Growth*, 1990, **104**, 629.
10. D. F. Foster, I. J. Patterson, L. D. James, D. J. Cole-Hamilton, D. N. Annitage, H. M. Yates, A. C. Wright and J. O. Williams, *Adv. Mater. Opt. Electron.*, 1994, **3**, 163.
11. K. Nishimura, Y. Nagao and K. Sakai, *J. Cryst. Growth*, 1993, **134**, 293.
12. (a) W. Kuhn, A. Naumov, H. Stanzl, S. Bauer, K. Wolf, H. P. Wagner, W. Gebhardt, U. W. Pohl, A. Krost, W. Richter, U. Dumichen and K. H. Thiele, *J. Cryst. Growth*, 1992, **123**, 605; (b) G. Sato, T. Numai, M. Hoshiyama, I. Suernune, H. Machida and N. Shimoyama, *J. Cryst. Growth*, 1995, **150**, 734.
13. J. E. Hails, S. J. C. Irvine, J. B. Mullin, D. V. Shenai-Khatkhate and D. J. Cole-Hamilton, *Mater. Res. Soc. Symp. Proc.*, 1989, **131**, 75.
14. J. E. Hails, S. J. C. Irvine and J. B. Mullin, *Mater. Res. Soc. Symp. Proc.*, 1990, **162**, 343.
15. (a) W. Bell, J. Stevenson, D. J. Cole-Hamilton and J. E. Hails, *Polyhedron*, 1994, **13**, 1253; (b) J. Stevenson, W. Bell, J. Ferry, D. J. Cole-Hamilton and I. E. Hails, *J. Organomet. Chem.*, 1993, **449**, 141; (c) J. E. Hails, D. J. Cole-Hamilton and W. Bell, *J. Cryst. Growth*, 1994, **145**, 596.
16. P. J. Wright, B. Cockayne, A. J. Williams, A. C. Jones and E. D. Orrell, *J. Cryst. Growth*, 1987, **84**, 552.
17. P. J. Wright, B. Cockayne and A. C. Jones, *Chemtronics*, 1988, **3**, 35.
18. B. Cockayne, P. J. Wright, A. J. Armstrong, A. C. Jones and E. D. Orrell, *J. Cryst. Growth*, 1988, **91**, 57.
19. P. J. Wright, P. J. Parbrook, B. Cockayne, A. C. Jones, E. D. Orrell, K. P. O'Donnell and B. Henderson, *J. Cryst. Growth*, 1989, **94**, 441.
20. P. J. Wright, B. Cockayne, P. J. Parbrook, A. C. Jones, P. O'Brien and J. R. Walsh, *J. Cryst. Growth*, 1990, **104**, 601.
21. H. M. Yates and J. O. Williams, *J. Cryst. Growth*, 1991, **107**, 387.
22. K. F. Jensen, A. Annapragada, K. L. Ho, J. S. Hiuh, S. Patnaik and S. Salim, *J. Physique, IZ*, 1991, **C2**, 243.
23. D. F. Foster, I. L. J. Patterson, L. D. James, D. J. Cole-Hamilton, D. N. Armitage, H. M. Yates, A. C. Wright and J. O. Williams, *Adv. Mater. Opt. Electron.*, 1994, **3**, 163.
24. J.-S. Huh, S. Patniak and K. F. Jensen, *J. Electron. Mater.*, 1993, **22**, 509.
25. P. J. Wright, B. Cockayne, A. C. Jones, E. D. Orrell, P. O'Brien and O. F. Z. Khan, *J. Cryst. Growth*, 1989, **94**, 97.
26. A. C. Jones, S. A. Rushworth, P. J. Wright, B. Cockayne, P. O'Brien and J. R. Walsh, *J. Cryst. Growth*, 1989, **97**, 537.
27. P. J. Wright, B. Cockayne, A. C. Jones and E. D. Orrell, *J. Cryst. Growth*, 1988, **91**, 63.
28. A. C. Jones, S. A. Rushworth, P. O'Brien, J. R. Walsh and C. Meaton, *J. Cryst. Growth*, 1993, **130**, 295.
29. O. F. Z. Khan, P. O'Brien, P. A. Hamilton, J. R. Walsh and A. C. Jones, *Chemtronics*, 1989, **4**, 2412.
30. E. I. Amma and R. E. Rundle, *J. Am. Chem. Soc.*, 1958, **80**, 4141.
31. G. E. Coates and R. A. Whitcombe, *J. Chem. Soc.*, 1956, **335**, 1204.
32. M. J. Almond, M. P. Beer, K. Hagen, D. A. Rice and P. J. Wright, *J. Mater. Chem.*, 1991, **1**, 1065.
33. P. J. Wright, B. Cockayne, P. J. Parbrook, E. P. Oliver and A. C. Jones, *J. Cryst. Growth*, 1991, **108**, 525.
34. (a) G. E. Coates and D. Ridley, *J. Chem. Soc.*, 1965, 1870; (b) G. W. Adamson, *Acta Crystallogr., B*, 1982, **38**, 462.
35. M. A. Malik, M. Motevaili, J. R. Walsh, P. O'Brien and A. C. Jones, *J. Mater. Chem.*, 1995, **5**, 731.

36. G. W. Adamson and H. M. M. Shearer, *J. Chem. Soc. Chem. Commun.*, 1969, 897.
37. D. H. Zeng, M. J. Hampden-Smith and E. N. Duesler, *Inorg. Chem.*, 1994, **33**, 5376.
38. M. Bochmann, *Chem. Vap. Depos.*, 1996, **2**, 85.
39. M. B. Hursthouse, M. Motevalli, P. O'Brien, J. R. Walsh and A. C. Jones, *Organometallics*, 1991, **10**, 3196.
40. I. G. Dance, *Polyhedron*, 1986, **5**, 1037.
41. M. B. Hursthouse, O. F. Z. Khan, M. Mazid, M. Motevalli and P. O'Brien, *Polyhedron*, 1990, **9**, 541.
42. O. F. Z. Khan and P. O'Brien, *Polyhedron*, 1991, **10**, 325.
43. (a) J. G. Brennan, T. Segrist, P. J. Carroll, S. M. Stuczynski, P. Reynders, L. E. Brus and M. L. Steigerwald, *J. Am. Chem. Soc.*, 1989, **111**, 4141; (b) J. G. Brennan, T. Siegrist, P. J. Carroll, S. M. Stuczynski, P. Reynders, L. E. Brus and M. L. Steigerwald, *Chem. Mater.*, 1990, **2**, 403.
44. K. Osakada and T. Yamamoto, *J. Chem. Soc., Chem. Commun.*, 1987, **1**, 117.
45. M. L. Steigerwald and C. R. Sprinkle, *J. Am. Chem. Soc.*, 1987, **109**, 7200.
46. (a) M. J. Hampden-Smith, personal communication; (b) D. Zeng, M. J. Hampden-Smith and E. N. Densler, *lnorg. Chem.*, 1994, **33**, 5376.
47. D. C. Bradley and D. N. Kunchur, *J. Chem. Phys.*, 1964, **8**, 2258.
48. P. J. Blower, J. R. Dilworth, J. P. Hutchinson and J. A. Zubieta, *J. Chem. Soc. Dalton Trans.*, 1985, 1533.
49. M. Bochmann, K. Webb, M. Harman and M. B. Hursthouse, *Angew. Chem., Int. Ed. Engl.*, 1990, **29**, 638.
50. M. Bochmann, K. J. Webb, M. B. Hursthouse and M. Mazid, *J. Chem. Soc., Dalton Trans.*, 1991, 2317.
51. M. Bochmann and K. J. Webb, *J. Chem. Soc., Dalton Trans.*, 1991, 2325.
52. M. Bochmann, G. C. Bwembya, R. Grinter, A. K. Powell, K. J. Webb, M. B. Hursthouse, K. M. Abdel Malik and M. A. Mazid, *Inorg. Chem.*, 1994, **33**, 2290.
53. M. Bochmann and K. J. Webb, *Mater. Res. Soc. Symp. Proc.*, 1991, **204**,149; *Chem. Abstr.*, 1992, **116**,3 3194h.
54. M. Bochmann, K. J. Webb, J. E. Hails and D. Wolverson, *Eur. J. Solid State lnorg. Chem.*, 1992, **29**, 155.
55. M. Bochmann, A. P. Coleman and A. K. Powell, *Polyhedron*, 1992, **11**, 507.
56. B. O. Dabbousi, P. J. Bonasia and J. Arnold, *J. Am. Chem. Soc.*, 1991, **113**, 3186.
57. P. J. Bonasi and J. Arnold, *Anorg. Chem.*, 1992, **31**, 2508.
58. J. Arnold, J. M. Walker, K. M. Yu, P. J. Bonasia, A. L. Seligson and E. D. Bourret, *J. Cryst. Growth*, 1992, **124**, 647.
59. P. J. Bonasia, D. E. Gindelberger, B. O. Dabbousi and J. Arnold, *J. Am. Chem. Soc.*, 1992, **114**, 5209.
60. P. J. Bonasia, G. P. Mitchell, F. J. Hollander and J. Arnold, *Inorg. Chem.*, 1994, **33**, 1797.
61. J. Arnold and P. J. Bonasia, *US Pat. 5157136, 20 Oct 1992*; *Chem. Abstr.*, 993, **118**, 213295s.
62. Y. Takahashi, R. Yuki, M. Sugiura, S. Motojima and K. Sugiyama, *J. Cryst. Growth*, 1980, **50**, 491.
63. M. A. H. Evans and J. O. Williams, *Thin Solid Films*, 1982, **87**, L1.
64. C. Byrom, M. A. Malik, P. O'Brien and A. White, *Polyhedron*, 2000, **19**, 211.
65. Y. Takahashi, R. Yuki, M. Sugiura, S. Motojima and K. Sugiyama, *J. Cryst. Growth*, 1980, **50**, 491.
66. S. L. Lawton and G. T. Kokotailo, *Inorg. Chem.*, 1969, **8**, 2410.
67. A. Adeogun, C. Q. Nguyen, M. Afzaal, M. A. Malik and P. O'Brien, *Chem. Commun.*, 2006, 2179.

68. C. Q. Nguyen, A. Adeogun, M. Afzaal, M. A. Malik and P. O'Brien, *Chem. Commun.*, 2006, 2182.
69. Jin-Ho Park, P. O'Brien, A. J. P. White and D. J. Williams, *Inorg. Chem.*, 2001, **40**, 3629.
70. (a) V. Garcia-Montalvo, R. Cea-Olivares, D. J. Williams and G. Espinosa- perez, *Inorg. Chem.*, 1996, **35**, 3948; (b) E. G. Zaitseva, I. A. Baidina, P. A. Stabnikov, S. V. Borisov and I. K. Igumenov, *Zh. Strukt. Khim.*, 1990, **31**, 184.
71. (a) K. Dymock, G. J. Palenok, J. Slezak, C. L. Raston and A. H. White, *J. Chem. Soc. Dalton. Trans.*, 1976, 28; (b) S. Bhattacharya, N. Seth, D. K. Srivastava, V. D. Gupta, H. Noth and M. Thomann-Albach, *J. Chem. Soc. Dalton. Trans.*, 1996, 2815.
72. H. P. Klug, *Acta Cryst.*, 1966, **21**, 536.
73. M. Bonamico, G. Mazzone, A. Vaciago and L. Zambonelli, *Acta Cryst.*, 1965, **19**, 898.
74. H. Miyame, M. Ito and H. Iwasaki, *Acta Cryst.*, 1979, **B35**, 1480.
75. A. Domenicano, L. Torelli, A. Vaciago and L. Zambonelli, *J. Chem. Soc. A*, 1968, **1**, 1351.
76. M. Bonamico and G. Dessy, *J. Chem. Soc. A*, 1971, 264.
77. M. B. Hursthouse, M. A. Malik, M. Motevalli and P. O'Brien, *Polyhedron*, 1992, **11**, 45.
78. M. Motevalli, P. O'Brien, J. R. Walsh and I. M. Watson, *Polyhedron*, 1996, **15**, 2801.
79. A. Saunders, A. Vecht and G. Tyrell, *Chem. Abstr.*, 1988, **108**, 66226h.
80. B. L. Druz, Y. N. Evtukhov and M. Y. Rakhlin, *Metallorg. Khim.*, 1988, **1**, 645.
81. B. L. Druz, A. I. Dyadenko, Y. N. Evtukhov, M. Y. Rakhlin and V. E. Rodionov, *Izv. Akad. Nauk. SSSR Neorg. Mater.*, 1990, **26**, 34.
82. D. M. Frigo, O. F. Z. Khan and P. O'Brien, *J. Cryst. Growth*, 1989, **96**, 989.
83. R. Nomura, T. Murai, T. Toyosaki and H. Matsuda, *Thin Solid Films*, 1995, **271**, 4.
84. N. I. Fainer, Y. M. Rumyantsev, M. L. Kosinova and F. A. Kuznetsov, in *Proc. XIV Int. CVD Conf. and EUROCVD11*, M. D. Allendorf, C. Bernard eds., Electrochemical Society, Pennington, NJ, 1997, **Vol. 97-25**, p. 1437.
85. N. I. Fainer, M. L. Kosinova, Y. M. Rumyantsev, E. G. Salman and F. A. Kuznetsov, *Thin Solid Films*, 1996, **280**, 16.
86. M. Chunggaze, J. McAleese, P. O'Brien and D. J. Otway, *J. Chem. Soc. Chem. Commun.*, 1998, 833.
87. P. O'Brien, D. J. Otway and J. R. Walsh, *Adv. Mater. CVD*, 1997, **3**, 227.
88. M. Chunggaze, M. A. Malik and P. O'Brien, *J. Mater. Chem.*, 1999, **9**, 2433.
89. P. O'Brien, J. R. Walsh, I. M. Watson, L. Hart and S. R. P. Silva, *J. Cryst. Growth*, 1996, **167**, 133.
90. B. Ludolph, M. A. Malik, P. O'Brien and N. Revaprasadu, *Chem. Commun.*, 1998, 1849.
91. R. D. Pike, H. Cui, R. Kershaw, K. Dwight, A. Wold, T. N. Blanton, A. A. Wernberg and H. J. Gysling, *Thin solid Films*, 1993, **224**, 221.
92. M. A. Malik and P. O'Brien, *Adv. Mater. Opt. Elec.*, 1994, **3**, 171.
93. M. B Hursthouse, M. A. Malik, M. Motevalli and P. O'Brien, *J. Mater. Chem.*, 1992, **9**, 949.
94. M. A. Malik, T. Saeed and P. O'Brien, *Polyhedron*, 1993, **12**, 1533.
95. M. A. Malik, M. Motevalli, T. Saeed and P. O'Brien, *Adv. Mater.*, 1993, **5**, 653.
96. V. M Agre, E. A Shugam and E. G. Rukhadze, *Tr. IREA*, 1967, **30**, 369.
97. A. A. M. Memon, M. Afzaal, M. A. Malik, C. Nguyen, P. O'Brien and J. Raftery, *Dalton Trans.*, 2006, 4499.
98. L. H. Van Poppel, T. H. Groy and M. T. Caudle, *Inorg. Chem.*, 2004, **43**, 3180.
99. J. G. Noltes, *Recl. Tran. Chim. Pays-Bas*, 1965, **84**, 126.
100. M. B. Hursthouse, M. A. Malik, M. Motevalli and P. O'Brien, *Organometallics*, 1991, **10**, 730.
101. M. A. Malik and P. O'Brien, *Chem. Mater.*, 1991, **3**, 999.
102. M. A. Malik, M. Motevalli, P. O'Brien and J. R. Walsh, *Organometallics*, 1992, **11**, 3436.
103. (a) I. Abrahams, M. A. Malik, M. Motevalli and P. O'Brien, *J. Organomet. Chem.*, 1994, **465**, 73; (b) M. A. Malik, M. Motevalli and P. O'Brien, *Acta Cryst*, 1996, **C52**, 1931.

104. M. A. Malik and P. O'Brien, *Adv. Mater. Opt. Electron.*, 1994, **3**, 171.
105. J. Cheon, D. S. Talaga and J. I. Zink, *Chem. Mater.*, 1997, **9**, 1208.
106. P. L. Musetha, N. Revaprasadu, M. A. Malik, P. O'Brien, *MRS Proc.*, 2005.
107. Z. Trávnícek, R. Pastorek, Z. Šindelár and R. Klicka, *Polyhedron*, 1995, **4**, 3627.
108. S. R. Rao, *Xanthate and Related Compounds*, Dekker, New York, 1971.
109. C. G. Sceney, J. O. Hill and R. J. Magee, *Thermo. Chim, Acta.*, 1973, **6**, 111.
110. (a) D. Barreca, A. Gasparotto, C. Maragno and E. Tondello, *J. Electrochem. Soc.*, 2004, **151**(6), G428–G435; (b) K.-C. Yang, C.-C. Chang, C.-S. Yeh, G.-H. Lee and S.-M. Peng, *Organometallics*, 2001, **20**, 126.
111. D. Barrecal, A. Gasparotto, C. Maragno, R. Seraglial, E. Tondello, A. Venzol, V. Krishnan and H. Bertagnolli, *Appl. Organometal. Chem.*, 2005, **19**, 59.
112. E. R. T. Tiekink, *Acta Cryst.*, 2000, **C56**, 1176.
113. (a) B. F. Abrahams, B. F. Hoskins, E. R. T. Tiekink and G. Winter, *Aust. J. Chem.*, 1988, **41**, 1117; (b) M. J Cox and E. R. T. Tiekink, *Rev. Inorg. Chem.*, 1997, **17**, 1.
114. (a) Y. Iimura, T. Ito and H. Hagihara, *Acta Cryst.*, 1972, **B28**, 2271; (b) C. K. Johnson, ORTEPII. Report ORNL-5138. 1976, Oak Ridge National Laboratory, Tennessee, USA.
115. H. M. Rietveld and E. N. Maslen, *Acta Cryst.*, 1965, **18**, 429.
116. D. W. Tomlin, T. M. Cooper, D. E. Zelmon, Z. Gebeyehu and J. M. Hughes, *Acta Cryst.*, 1999, **C55**, 717.
117. M. Chunggaze, M. A. Malik, P. O'Brien, A. J. P. White and D. J. Williams, *J. Chem. Soc., Dalton Trans.*, 1998, 3839.
118. M. Chunggaze, M. A. Malik and P. O'Brien, *Adv. Mater. Opt. Electron.*, 1998, **7**, 311.
119. M. Nyman, M. J. Hampden-Smith and E. Duesler, *Chem. Vap. Depos.*, 1996, **2**, 171.
120. M. Nyman, K. Jenkins, M. J. Hampden-Smith, T. T. Kodas, E. N. Duesler, A. L. Rheingold and M. L. Liable-Sands, *Chem. Mater.*, 1998, **10**, 914.
121. (a) A. Schmidpeter, R. Bohm and H. Groeger, *Angew. Chem., Int. Ed. Engl.*, 1964, **3**, 704; (b) A. Schmidpeter and K. Stoll, *Angew. Chem., Int. Ed. Engl.*, 1967, **6**, 252; (c) A. Schmidpeter and K. Stoll, *Angew. Chem., Int. Ed. Engl.*, 1968, **7**, 549.
122. P. Bhattacharyya, A. M. Z. Slawin, D. J. Williams and J. D. Woollins, *J. Chem. Soc., Dalton Trans.*, 1995, **1**, 2489.
123. For a review, see: C. Silvestru and J. E.Drake, *Coord. Chem. Rev.*, 2001, **223**, 117.
124. For a review, see: T. Q. Ly and J. D. Woollins, *Polyhedron*, 1998, **176**, 451.
125. M. Afzaal, D. Crouch, M. A. Malik, M. Motevalli, P. O'Brien, J.-H. Park and J. D. Woollins, *Eur. J. Inorg. Chem.*, 2004, **1**, 171.
126. D. J. Crouch, P. O'Brien, M. A. Malik, P. J. Skabara and S. P. Wright, *Chem. Commun.*, 2003, 1454.
127. M. Afzaal, D. J. Crouch, P. O'Brien, J. Raftery, P. J. Skabara, A. J. P. White and D. J. Williams, *J. Mater. Chem.*, 2004, **14**, 233.
128. I. G Dance, A. Choy and M. L. Scudder, *J. Am. Chem. Soc.*, 1984, **106**, 6285.
129. G. S. H. Lee, K. J. Fisher, D. C. Craig, M. Scudder and I. G. Dance, *J. Am. Chem. Soc.*, 1990, **112**, 6435.
130. S. L. Cumberland, K. M. Hanif, A. Javier, G. A. Khitrov, G. F. Strouse, S. M. Woessner and C. S. Yun, *Chem. Mater.*, 2002, **14**, 1576.
131. D. Cupertino, D. J. Birdsall, A. M. Z. Slawin and J. D. Woollins, *Inorg. Chim. Acta*, 1999, **290**, 1.
132. M. Afzaal, S. M. Aucott, D. Crouch, P. O'Brien, J. D. Woollins and J. H. Park, *Chem. Vap. Depos.*, 2002, **8**(5), 187.
133. M. Afzaal, D. Crouch, P. O'Brien and J. H. Park, *Mater. Res. Soc. Symp. Proc.*, 2002, 692.
134. K. Osakada and T. Yamamoto, *J. Chem. Soc. Chem. Commun.*, 1987, 1117.

135. J. G. Brennan, T. Siegrist, P. J. Carroll, S. M. Stuczynski, P. Reynders, L. E. Brus and M. L. Steigerwald, *Chem. Mater.*, 1990, **2**, 403.
136. M. Afzaal, D. Crouch, M. A. Malik, M. Motevalli, P. O'Brien, Jin-Ho Park and J. D. Woollins, *Eur. J. Inorg. Chem.*, 2004, **171**.
137. M. Afzaal, D. Crouch, M. A. Malik, M. Motevalli, P. O'Brien and Jin-Ho Park, *J. Mater. Chem.*, 2003, **13**, 639.
138. S. S. Garje, J. S. Ritch, D. J. Eisler, M. Afzaal, P. O'Brien and T. Chivers, *J. Mater. Chem.*, 2006, **16**, 966.
139. G. G. Briand, T. Chivers and M. Parvez, *Angew. Chem., Int. Ed.*, 2002, **41**, 3468.
140. T. Chivers, D. J. Eisler and J. S. Ritch, *Dalton Trans.*, 2005, 2675.
141. M. Green, G. Wakefield and P. J. Dobson, *J. Mater. Chem.*, 2003, **13**, 1076.
142. M. Bochmann, *Chem. Vap. Depos.*, 1996, **2**, 85.
143. Y. Okamoto and T. Yano, *J. Organomet. Chem.*, 1971, **29**, 99.
144. T. Chivers, D. J. Eisler, J. S. Ritch and H. M. Tuononen, *Angew. Chem., Int. Ed.*, 2005, **44**, 4953.
145. K. Maschke and F. Levy, Physics of Non-tetrahedrally Bonded Binary Compounds, in Landolt-Bornstein New Series, ed. O. Madelung, W. von der Ostem and U. Rossler, Springer Verlag, Berlin, 1983, Vol. 17f, ch. 9.7; A. MacKinnon. Physics of Non-tetrahedrally Bonded Ternary Compounds, in Landolt-Bornstein New Series, ed. O. Madelung, W. von der Ostem and U. Rossler, Springer Verlag, Berlin, 1985, Vol. 17h, ch. 10.1, p. 12–23.
146. R. Nomura, S. J. Inazawa, K. Kanaya and H. Matsuda, *Polyhedron*, 1989, **8**, 763.
147. R. Nomura, S. Fujii, K. Kanaya and H. Matsuda, *Polyhedron*, 1990, **9**, 361.
148. D. G. Tuck, in *Comprehensive Organometallic Chemistry*, ed. G. Wilkinson, F. G. A. Stone and E. W. Abel, Pergamon, Oxford. 1995.
149. R. Nomura, S. J. Inazawa, K. Kanaya and H. Matsuda, *Appl. Organomet. Chem.*, 1989, **3**, 195.
150. R. Nomura, K. Konishi and H. Matsuda, *Thin Solid Films*, 1991, **198**, 339.
151. R. Nomura, K. Konishi and H. Matsuda, *J. Electrochem. Soc.*, 1991, **138**, 631.
152. A. N. MacInnes, M. B. Power and A. R. Barron, *Adu. Mater. Opt. Electron.*, 1992, **1**, 229.
153. M. B. Power and A. R. Barron, *J. Chem. Soc., Chem. Commun.*, 1991, 1315.
154. H. J. Gysling, A. A. Wernberg and T. N. Blanton, *Chem. Mater.*, 1992, **4**, 900.
155. S. W. Haggata, M. A. Malik, M. Motevalli, P. O'Brien and J. C. Knowles, *Chem. Mater.*, 1995, **7**, 716.
156. P. O'Brien, D. J. Otway and J. R. Walsh, *Chem. Vap. Depos.*, 1997, **3**, 227.
157. P. O'Brien, D. J. Otway and J. R. Walsh, *Thin Solid Films*, 1998, **315**, 57.
158. V. G. Bessergenev, E. N. Ivanova, Y. A. Kovalevskaya, S. A. Gromilov, V. N. Kirichenko and S. V. Larionov, *Inorg. Mater.*, 1996, **32**, 592.
159. J. McAleese, P. O'Brien and D. J. Otway, *Chem. Vap. Depos.*, 1998, **4**, 94.
160. G. A. Horley, M. Chunggaze, P. O'Brien, A. J. P. White and D. J. Williams, *J. Chem. Soc., Dalton Trans.*, 1998, 4205.
161. G. A. Horley, P. O'Brien, J.-H. Park, A. J. P. White and D. J. Williams, *J. Mater. Chem.*, 1999, **9**, 1289.
162. K. Kunze, L. Bihry, P. Atanasova, M. J. Hampden-Smith and E. N. Duesler, *Chem. Vap. Depos.*, 1996, **2**, 105.
163. G. Shang, M. J. Hampden-Smith and E. N. Duesler, *Chem. Commun.*, 1996, 1733.
164. (a) G. Shang, K. Kunze, M. J. Hampden-Smith and E. Duesler, *Chem. Vap. Depos.*, 1996, **2**, 242; (b) G. Shang, M. J. Hampden-Smith and E. N. Duesler, *Chem. Commun.*, 1996, 1733.
165. S. S. Garje, M. C. Copsey, M. Afzaal, P. O'Brien and T. Chiversb, *J. Mater. Chem.*, 2006, **16**, 4542.

166. M. C. Copsey and T. Chivers, *Chem. Commun.*, 2005, 4938.
167. A. C. Jones, *Chemical Society Reviews*, 1997, 101.
168. T. F. Kuech and E. Veuhoff, *J. Cryst. Growth*, 1984, **68**, 148.
169. K. F. Jensen, D. I. Fotiadis and T. J. Mountziaris, *J. Crystal Growth*, 1991, **107**, 1.
170. E. J. Thrush, C. G. Cureton, J. M. Trigg, J. P. Stagg and B. R. Butler, *Chemtronics*, 1987, **2**, 62.
171. G. B. Stringfellow, *J. Cryst. Growth*, 1990, **105**, 260.
172. P. W. Lee, T. R. Omstead, D. R. McKennd and K. F. Jensen, *J. Cryst. Growth*, 1988, **93**, 134.
173. C. A. Larsen, N. I. Buchan, S. H. Li and G. B. Stringfellow, *J. Cryst. Growth*, 1988, **93**, 15.
174. F. Maury, *Adv. Mater.*, 1991, **3**, 542.
175. A. H. Cowley and R. A. Jones, *Polyhedron*, 1994, **13**, 1149.
176. P. O'Brien and S. Haggata, *Adv. Mater. Opt. Electron.*, 1995, **5**, 117.
177. K. W. Benz, H. Renz, J. Weidlein and M. H. Pilkuhn, *J. Electron. Mater.*, 1981, **10**, 185.
178. R. H. Moss and J. S. Evans, *J. Cryst. Growth*, 1981, **55**, 129.
179. A. Zaouk, E. Salvetat, J. Sakaya, F. Maury and G. Constant, *J. Cryst. Growth*, 1981, **55**, 135.
180. F. Maury, M. Combes, G. Constant, R. Carles and J. B. Renucci, *J. Phys.*, 1982, **1**(10), 347.
181. F. Maury, M. Combes and G. Constant, in *Proc. EUROCVD4*, J. Bloem, G. Verspui, L. R. Wolff, eds., Philips Centre for Manufacturing Technology, Eindhoven, The Netherlands, 1983, p. 257.
182. F. Maury and G. Constant, *Polyhedron*, 1984, **3**, 581.
183. A. Zaouk and G. Constant, *J. Phys.*, 1982, **C5**, 43.
184. F. Maury, A. E. Hammadi and G. Constant, *J. Cryst. Growth*, 1984, **68**, 88.
185. A. E. Hammadi, F. Maury, G. Muller, J. Bensoam and G. Constant, *Acad. Sci. Paris, Ser. II*, 1984, **299**, 1255.
186. F. Maury and A. E. Hammadi, *J. Cryst. Growth*, 1988, **91**, 97.
187. F. Maury and A. E. Hammadi, *J. Cryst. Growth*, 1988, **91**, 105.
188. F. Maury, in *Transformation of Organometallics into Common and Exotic Materials: Design and Activation*, (Ed: R. M. Laine), NATO ASI Series E, Martinus Nijhoff, The Netherlands 1988, **141**, 195.
189. A. H. Cowley, B. L. Benac, J. G. Ekerdt, R. A. Jones, K. B. Kidd, J. Y. Lee and J. E. Miller, *J. Am. Chem. Soc.*, 1988, **110**, 6248.
190. O. T. Beachley and G. E. Coates, *J. Chem. Soc.*, 1965, 3241.
191. O. T. Beachley, J. P. Kopasz, H. Zhang, W. E. Hunter and J. L. Atwood, *J. Organomet. Chem.*, 1987, **325**, 69.
192. A. H. Cowley and R. A. Jones, *Angew. Chem. Int. Ed. Engl.*, 1989, **28**, 1208.
193. A. M. Arif, B. L. Benac, A. H. Cowley, R. Greets, R. A. Jones, K. B. Kidd, J. M. Power and S. T. Schwab, *J. Chem. Soc. Chem. Comm.*, 1986, 1543.
194. A. M. Arif, B. L. Benac, A. H. Cowley, R. A. Jones, K. B. Kidd and C. M. Nunn, *New J. Chem.*, 1988, **12**, 553.
195. F. Maury, M. Combes, G. Constant, R. Carles and J. B. Renucci, *J. Phys.*, 1982, **45**, C1–347.
196. F. Maury and G. Constant, *Polyhedron*, 1984, **3**, 581.
197. C. G. Pitt, K. T. Higa, A. T. McPhail and R. L. Wells, *Inorg. Chem*, 1986, **25**, 2483.
198. R. L. Wells, A. P. Purdy, K. T. Higa, A. T. McPhail and C. G. Pitt, *J. Organomet. Chem.*, 1987, **325**, C7.
199. A. P. Purdy, R. L. Wells, A. T. McPhail and C. G. Pitt, *Organometallics*, 1987, **6**, 2099.
200. M. A. Malik, P. O'Brien, S. Norigar and J. Smith, *J. Mater.Chem.*, 2003, **13**, 2591.
201. J. E. Miller, K. B. Kidd, A. H. Cowley, R. A. Jones, J. G. Ekerdt, H. J. Gysling, A. A. Wernberg and T. N. Blanton, *Chem. Mater.*, 1990, **2**, 589.
202. J. E. Miller and J. G. Ekerdt, *Chem. Mater.*, 1992, **4**, 7.

203. J. G. Ekerdt, Y. M. Sun, M. S. Jackson, V. Lakhotia, K. A. Pacheco, S. U. Koschmieder, A. H. Cowley and R. A. Jones, *J. Cryst. Growth*, 1992, **124**, 158.
204. D. O'Conner and D. Phillips, *Time-Correlated Single Photon Counting*, Academic, London, 1984.
205. M. A. Malik, M. Afzaal, P. O'Brien, U. Bangert and B. Hamilton, *Matererial Science and Technology*, 2004, **20**, 959.
206. M. A. Malik, P. O'Brien and M. Halliwell, *J. Mater. Chem.*, 2005, **15**, 2.
207. J. E. Miller, M. A. Mardones, J. W. Nail, A. H. Cowley, R. A. Jones and J. G. Ekerdt, *Chem. Mater.*, 1992, **4**, 447.
208. A. H. Cowley, R. A. Jones, M. A. Mardones and C. M. Nunn, *Organometallics*, 1991, **10**, 1635.
209. J.-P. Lu, R. Raj and A. Wernberg, *Thin Solid Films*, 1991, **205**, 236.
210. A. A. Wernberg, D. J. Lawrence, H. J. Gysling, A. J. Filo and T. N. Blanton, *J. Cryst. Growth*, 1993, **131**, 176.
211. D. A. Andrews, G. J. Davies, D. C. Bradley, M. M. Faktor, D. M. Frigo and E. A. D. White, *Semicond. Sci. Technol.*, 1988, **3**, 1053.
212. A. H. Cowley, R. A. Jones, C. M. Nunn and D. L. Westmorland, *Chem. Mater.*, 1990, **2**, 221.
213. H. S. Park, S. Schulz, H. Wessel and H. W. Roesky, *Chem. Vap. Depos.*, 1999, **5**, 179.
214. A. C. Jones, C. R. Whitehouse and J. S. Roberts, *Chem. Vap. Depos.*, 1995, **1**, 65.
215. S. A. Rushworth, J. R. Brown, D. J. Houlton, A. C. Jones, V. Roberts, J. S. Roberts and G. W. Critchlow, *Adv. Mater. Opt Elec*, 1996, **6**, 119.
216. D. Neumeyer and J. G. Ekerdt, *Chem. Mater.*, 1996, **8**, 9.
217. A. Miehr, R. A. Fircher, Paper PDSP 5 Abstr. Book of 8th Int. Conf. On MOVPE, 9-13 June, 1996, Cardiff, UK.
218. *The Chemistry of Metal CVD* (Eds: T. T. Kodas, M. J. Hampden-Smith), VCH, Weinheim 1994.
219. M. J. Hampden-Smith and T. T. Kodas, *Chem. Vap. Depos.*, 1995, **1**(1), 8.
220. R. D. Pike, H. Cni, R. Kershaw, K. Dwight, A. Wold, T. N. Blanton, A. A. Wernberg and H. J. Gysling, *Thin Solid Films*, 1993, **224**, 221.
221. A. A. Wernberg and H. J. Gysling, *Chem. Mater.*, 1993, **5**, 1056.
222. H. J. Gysling and A. A. Wernberg, *Chem. Mater.*, 1992, **4**, 900.
223. C. Y Xu, M. J. Hampden-Smith and T. T. Kodas, *Adv. Mater.*, 1994, **6**, 745.
224. C. Roger, T. S. Corbitt, M. J. Hampden-Smith and T. T. Kodas, *Appl. Phys. Lett.*, 1994, **65**, 1022.
225. G. Blandenet, M. Court and Y. Legard, *Thin Solid Films*, 1981, **77**, 81.
226. R. M. Charatan and M. E. Gross, *Mater. Res. Soc.*, Mtg., Boston, MA, 1993.
227. C. Wyetzner, S. Komarov. C. Freel, M. Jones, A. F. Hepp, M. A. Fury, A. E. Kaloyeros, Mater. *Res. Soc. Mtg*., Boston, MA, 1993.
228. A. Jain, K. M. Chi, M. J. Hampden-Smith, T. T. Kodas, M. E Paffett and J. D. Farr, *J. Mater. Res.*, 1992, **7**, 261.
229. A. Jain. K. M. Chi, M. J. Hampden-Smith, T. T. Kodas, M. F. Paffett and J. D. Farr, *J. Electrochem. Soc.*, 1993, **140**, 1434.
230. J. A. T. Norman, B. A. Muratore, P. N. Dyer, D. A. Roberts and A. K. Hochberg, *J. de Phys. IV*, 1992, **1**, C2–271.
231. A. Jain, K. M. Chi, M. J. Hampden-Smith, T. T. Kodas, M. E. Paffett and J. D. Farr, *Chem. Mater.*, 1991, **3**, 995.
232. T. H. Baum and C. E. Larson, *Chem. Mater.*, 1992, **4**, 365.
233. S. K. Reynolds, C. J. Smart, E. F. Baran, T. H. Baum, C. E. Larson and P. J. Brock, *Appl. Phys. Lett.*, 1991, **59**, 2332.

234. P. F. Seidier, S. P. Kowalczyk, M. M. Banaszak, J. J. Ynrkas, M. H. Norcott and F. R. McFeely, *Mater. Res. Soc. Symp. Proc.*, 1993, **282**, 359.
235. M. M. Banaszek Holl, P. F. Seidler, S. P. Kowalcyk and E. R. McFeely, *Inorg. Chem.*, 1994, **33**, 510.
236. M. M. Banaszak Holl, P. F. Seidler, S. P. Kowalczyk and F. R. McFreely, *Appl. Phys. Lett.*, 1993, **62**, 1475.
237. W. L. Gladfelter, *Chem. Mater.*, 1993, **5**, 1372.
238. B. E. Bent, R. G. Nuzzo and L. H. Dubois, *J. Am. Chem. Soc.*, 1989, **111**, 1634.
239. R. Kaplan and N. Bottka, *Appl. Phys. Lett.*, 1982, **41**, 972.
240. (a) J. P. Collman, L. S. Hegedus, J. R. Norton and R. G. Finke, *Principles and Applications of Organotrunsition Metal Chemistry*, University Science Books, Mill Valley, CA, 1987; (b) *Appl. Phys. Lett.*, 1993, **62**, 1475.
241. J. B. Hoke, E. W. Stern and H. H. Murray, *J. Mater. Chem.*, 1991, **I**, 551.
242. D. C. Bradley, R. C. Mehrotra and D. P. Gaur, *Metal Alkoxides*, Academic Press, New York, 1978, Vol. **42**, 149.
243. Y. Takahishi, T. Kawae and M. Nasu, *J. Cryst. Growth*, 1986, **74**, 409.
244. (a) S. Pakswer and P. Skoug in *Thin Dielectric Films Made by Oxygen-Assisted Pyrolysis of Alkoxides*, J. Blocher and J. C. Withers ed., The Electrochemical Society, Los Angles, CA, 1970, 619; (b) D. C. Bradley, *Chem. Rev.*, 1989, **89**, 1317.
245. (a) L. G. Hubert-Pfalzgraf, *Coord.Chem.Rev.*, 1998, 178–180; (b) L. G. Hubert-Pfalzgraf, *J. Mater. Chem.*, 2004, **14**, 3113–180.
246. L. G. Hubert-Pfalzgraf, *Inorg.Chem.Commun*, 2003, **6**,102 and refs. therein; L. G. Hubert-Pfalzgraf, *New J. Chem*, 1995, **19**, 727.
247. H. W. Roesky, I. Haiduc and N. S. Hosmane, *Chem. Rev.*, 2003, **103**, 2579.
248. M. L. Calzada, A. Gonzalez, J. Garcia-Lopez and R. Jimenez, *Chem.Mater.*, 2003, **15**, 4783.
249. T. J. Boyle, T. M. Alam, D. Dimos, G. J. Moore, C. D. Buchheit, H. N. Al-Shareef, E. R. Mechenbier, B. R. Bear and J. W. Ziller, *Chem.Mater.*, 1997, **9**, 3187.
250. T. Kemmitt and M. Daghish, *Inorg.Chem.*, 1998, **37**, 2063.
251. B. Moraru, G. Kickelbick, M. Battistella and U. Schubert, *J. Organomet. Chem*, 2001, **636**, 27.
252. Z. Li, H. Chen, H. Bao and M. Gao, *Chem. Mater.*, 2004, **16**, 1391.
253. W. A. Herrmann, N. W. Huber and O. Runte, *Angew. Chem., Int. Ed. Engl.*, 1995, **34**, 2187.
254. A. C. Jones, T. J. Leedham, P. J. Wright, M. J. Crosbie, D. J. Williams, P. A. Lane and P. O'Brien, *Mater. Res. Soc. Symp. Proc.*, 1998, **495**, 11.
255. P. A. Williams, J. L. Roberts, A. C. Jones, P. R. Chalker, N. L. Tobin, J. F. Bickley, H. O. Davies, L. M. Smith and T. J. Leedham, *Chem.Vap.Deposition*, 2002, **8**, 163.
256. K. J. Eisentraut and R. E. Sievers, *J. Am. Chem. Soc.*, 1965, **87**, 5254.
257. C. A. Burgett and J. S. Fritz, *Talanta*, 1973, **20**, 363.
258. K. Utsunomiya and T. Shigematsu, *Anal. Chim. Acta.*, 1972, **58**, 411.
259. J. E. Schwarberg, R. E. Sievers and R. W. Moshier, *Anal. Chem.*, 1970, **43**, 1828.
260. L. Ben-Dor, R. Druilhe and P. Gibart, *J. Cryst. Growth*, 1974, **24–25**, 172.
261. L. Ben-Dor, E. Fischbein, I. Felner and Z. Kalman, *J. Electrochem. Soc.*, 1977, **124**, 451.
262. B. D. Fahlman and A. R. Barron, *Adv. Mater. Opt. Electron.*, 2000, **10**, 223.
263. A. F. Bykov, P. P. Semyannikov and I. K Igumenov, *J. Therm. Anal.*, 1992, **38**, 1477.
264. M. L. Hitchman, S. H. Shamlian, G. G. Condorelli and F. Chabert-Rocabois, *J. Alloys. Compd.*, 1997, **251**, 297.
265. A. E. Turgambaeva, V. V. Krisyuk, A. F. Bykov and I. K. Igumenov, *J. Phys.*, 1999, **IV**(9), 8.
266. D. C. Bradley and I. M. Thomas, *J. Chem. Soc.*, 1960, 3857.
267. D. C. Bradley and M. H. Gitlitz, *J. Chem. Soc. A*, 1969, 980.
268. R. Fix, R. G. Gordon and D. M. Hoffman, *Chem. Mater.*, 1991, **3**, 1138.

269. A. Bastianini, G. A. Battiston, R. Gerbasi, M. Porchia and S. Daolio, *J. Phys.*, 1995, **5**, C5–525.
270. Y. Oshita, A. Ogura, A. Hoshino, S. Hiiro and H. Machida, *J. Cryst. Growth*, 2001, **233**, 292.
271. Y. Ohshita, A. Ogura, A. Hoshino, S. Hiiro, T. Suzuki and H. Machida, *Thin Solid Films*, 2002, **406**, 215.
272. W. R. Hitchens, W. C. Krusell and D. M. Dobkin, *J. Electrochem. Soc.*, 1993, **140**, 2615.
273. M. J. Crosbie, P. J. Wright, D. J. Williams, P. A. Lane, J. Jones, A. C. Jones, T. J. Leedham, P. O'Brien and H. O. Davies, *J. Phys. IV, France*, 1999, **9**, Pr8–935.
274. P. A. Williams, A. C. Jones, P. J. Wright, M. J. Crosbie, J. F. Bickley, A. Steiner, H. O. Davies and T. J. Leedham, *Chem. Vap. Deposition*, 2002, **8**, 110.
275. A. C. Jones, *Chem. Vap. Deposition*, 1998, **4**, 169.
276. R. N. Goshtagore, *J. Electrochem. Soc.*, 1970, **117**, 529.
277. S. Hayashi and T. Hirai, *J. Cryst. Growth*, 1976, **36**, 157.
278. E. T. Fitzgibbons, K. J. Sladek and W. H. Hartwig, *J. Electrochem. Soc.*, 1972, **119**, 735.
279. M. Yokozawa, H. Iwasa and I. Teramoto, *Jpn. J. Appl. Phys.*, 1968, **7**, 96.
280. C. J. Taylor, D. C. Gilmer, D. G. Colombo, G. D. Wilk, S. A. Campbell, J. Roberts and W. L. Gladfelter, *J. Am. Chem. Soc.*, 1999, **121**, 5220.
281. R. C. Smith, T. Ma, N. Hoilien, L. Y. Tsung, M. J. Bevan, L. Colombo, J. Roberts, S. A. Campbell and W. Gladfelter, *Adv. Mater. Opt. Electron.*, 2000, **10**, 105.
282. R. Bhakta, F. Hipler, A. R. S. Devi, S. P. Ehrhart and R. Waser, *Chem.Vap. Depos.*, 2003, **9**, 295.
283. M. Pulver and G. Wahl, *Electrochem. Soc.*, 1977, **97–25**, 960.
284. A. C. Jones, T. J. Leedham, P. J. Wright, M. J. Crosbie, D. J. Williams, P. A. Lane and P. O'Brien, *Mater. Res. Soc. Symp. Proc.*, 1998, **495**, 11.
285. M. Balog and M. Schieber, *J. Crystal Growth*, 1972, **17**, 298.
286. B. J. Gould, I. M. Povey, M. E. Pemble and W. R. Flavell, *J. Mater. Chem.*, 1994, **4**, 1815.
287. S. Pakswer and P. Skoug, in *Thin Dielectric Oxide Films Made by Oxygen Assisted Pyrolysis of Alkxoxides*, eds. J. M. Blocher and J. C. Withers, The Electrochemical society, Los Angeles, 1970, p. 619.
288. A. C. Jones, *J. Mater. Chem.*, 2002, **12**, 2576.
289. S. Hori, K. Yamamoto, M. Asai, H. Miya and M. Niwa, *Jpn. J. Appl. Phys.*, 2003, **42**, 5176.
290. P. A. Williams, J. L. Roberts, A. C. Jones, P. R. Chalker, J. F. Bickley, A. Steiner, H. O. davies and T. J. Leedham, *J. Mater. Chem.*, 2002, **12**, 165.
291. R. L. Nigro, R. Toro, G. Malamdrino, V. Raineri and I. L. Fragalà, *Electrochem. Soc. Proc.*, 2003, **8**, 915.
292. A. C. Jones, H. C. Aspinall, P. R. Chalker, R. J. Potter, K. Kukli, A. Rahtu, M. Ritala and M. Leskelä, *J. Mater. Chem.*, 2004, **14**, 3101.
293. H. C. Aspinall, J. Gaskell, P. A. Williams, A. C. Jones, P. R. Chalker, P. A. Marshall, L. M. Smith and G. W. Critchlow, *Chem. Vap. Deposition*, 2004, **10**, 83.
294. J. McAleese, J. C. Plakatouras and B. C. H. Steele, *Thin Solid Films*, 1996, **286**, 64.
295. H. C. Aspinall, J. M. Gaskell, Y. F. Loo, A. C. Jones, P. R. Chalker, R. J. Potter, L. M. Smith and G. W. Critchlow, *Chem. Vap. Deposition*, 2004, **10**, 301.
296. Y. Shiokawa, R. Amino and A. Nomura M. Yagi, *J. Radioanal.Nucl.Chem.*, 1991, **152**, 373.
297. A. Weber and H. Suhr, *Mod. Phys. Lett.*, 1989, **B3**, 1001.
298. H. C. Aspinall, J. Gaskell, P. A. Williams, A. C. Jones, P. R. Chalker, R. J. Potter, L. M. Smith and G. W. Critchlow, *Chem. Vap. Deposition*, 2004, **10**, 13.
299. M. Langlet and R. D. Shannon, *Thin Solid Films*, 1990, **186**, 1.
300. S. Chavalier, G. Bonnet and J. P. Larpin, *Appl. Surf. Sci.*, 2000, **167**, 125.
301. Y. F. Loo, R. J. Potter, A. C. Jones, H. C. Aspinall, J. M. Gaskell, P. R. Chalker, L. M. Smith and G. W. Critchlow, *Chem. Vap. Deposition*, 2004, **10**, 306.

302. M. Kiyotoshi and K. Eguchi, *Appl. Phys. Lett.*, 1995, **67**, 2468.
303. C. Dubourdieu, H. Roussel, C. Jimenez, M. Audier, J. P. Senateur, S. Lhostis, L. Auvray, F. Ducroquet, B. J. O'Sullivan, P. K. Hurley, S. Rushworth and L. Hubert-Pfalzgraf, *Materials Science and Engineering B-Solid State Materials for Advanced Technology*, 2005, **118**, 105.
304. P. Schäfer and R. Waser, *Adv. Mater. Opt. Electron.*, 2000, **10**, 169.
305. A. C. Jones, T. J. Leedham, P. J. Wright, D. J. Williams, M. J. Crosbie, H. O. Davies, K. A. Fleeting and P. O'Brien, *J. Eur. Ceram. Soc.*, 1999, **19**, 1431.
306. A. C. Jones, H. O. Davies, T. J. Leedham, P. J. wright, P. A. Lane, M. J. Crosbie, D. J. Williams, J. C. Jones and C. L. Reeves, *Int. Ferroelectrics*, 2000, **30**, 19.
307. M. M. Sung, C. Kim, G. Kim and Y. Kim, *J. Cryst. Growth.*, 2000, **210**, 651.
308. S. Kim and S. Hishita, *Thin Solid Films*, 1996, **281–282**, 449.
309. J. S. Matthews, O. Just, B. Obi-Johnson and W. S. Rees, *Chem. Vap. Depos.*, 2000, **6**, 129.
310. H. O. Davies, A. C. Jones, T. J. Leedham, M. J. Crosbie, P. J. Wright, N. M. Boag and J. R. Thompson, *Chem. Vap. Depos.*, 2000, **6**, 71.
311. (a) M. R. Hill, A. W. Jones, J. J. Russell, N. K. Roberts and R. N. Lamb, *J. Mater. Chem.*, 2004, **14**, 3198; (b) K. C. Yang, C. C. Chang, C. S. Yeh, G.-H. Lee and S.-M. Peng, *Organometallics*, 2001, **20**, 126.
312. M. T. Caudle, R. A. Nieman and V. G. Young, *J. Inorg. Chem.*, 2001, **40**, 1571.
313. E. D. S. Lucas, A. Khaleel, A. Seitz, S. Fultz, A. Ponce, W. Li, C. Carnes and K. J. Klabunde, *J. Eur. Chem.*, 2001, **7**, 2505.
314. T. T. T. Yoshida, H. Yoshida, S. Takenaka, T. Funabiki, S. Yoshida and T. Murata, *J. Phys. Chem.*, 1995, **99**, 10890.
315. S. Mathur, M. Veith, T. Ruegamer, E. Hemmer and H. Shen, *Chem. Mater.*, 2004, **16**, 1304.
316. (a) J. A Meese-Marktscheffel, R. Fukuchi, M. Kido, G. Tachibana, C. M. Jensen and J. W. Gilje, *Chem. Mater.*, 1993, **5**, 755; (b) J. H. Boo, S. B. Lee, S. J. Ku, W. Koh, C. Kim, K. S. Yu and Y. Kim, *Appl. Surf. Sci.*, 2001, **169-170**, 581.
317. (a) S. Mathur, M. Veith, T. Ruegamer, E. Hemmer and H. Shen, *Chem. Mater.*, 2004, **16**, 1304; (b) M. Veith, A. Altherr and H. Wolfanger, *Chem. Vap. Dep.*, 1999, **5**, 87; (c) R. Winter, M. Quinten, A. Dierstein, R. Hempelmann, A. Altherr and M. Veith, *J. Appl. Crystallogr.*, 2000, **33**, 507.
318. J. Auld, D. J Houlton, A. C Jones, S. A. Rushworth, M. A Malik, P. O'Brien and G. W. Critchlow, *J. Mater. Chem.*, 1994, **4**, 1249.
319. M. Veith, S. Mathur, H. Shen, S. Hufner and M. Jilavi, *Chem. Mater.*, 2001, **11**, 4041.
320. M. Veith, S. Mathur, N. Lecerf, K. Bartz, M. Heintz and V. Huch, *Chem. Mater.*, 2000, **12**, 271.
321. C. C. Chang, W. H. Lee, T. Y. Her, G. H. Lee, S. M. Peng and Y. Wang, *J. Chem. Soc., Dalton Trans.*, 1994, 315.
322. M. Veith, S. Mathur and C. Mathur, *Polyhedron*, 1998, **17**, 1005.
323. S. Haukka, E. Lakomaa and T. Suntola, *Stud. Surf. Sci. Catal.*, 1999, **120**, 715.
324. M. Ritala and M. Leskelä, *Nanotechnology*, 1999, **10**, 19.
325. L. Niinistö, *Curr. Opin. Solid State Mater. Sci.*, 1998, **3**, 147.
326. L. Niinistö, *Proceedings of the CAS International Semiconductor Conference*, Sinaia, Romania, 2000, **1**, 33.
327. M. Leskelä and M. Ritala, *Thin Solid Films*, 2002, **409**, 138.
328. J. Aarik, A. Aidla, A. Jaek, A.-A. Kiisler and A.-A. Tammik, *Acta Polytechn. Scand., Chem. Technol.*, 1990, **195**, 201.
329. K. Kukli, J. Aarik, A. Aidla, H. Siimon, M. Ritala and M. Leskelä, *Appl. Surf. Sci.*, 1996, **112**, 236.

330. M. Ritala, M. Juppo, K. Kukli, A. Rahtu and M. Leskelä, *J. Phys. IV France*, 1999, **9**, Pr8–1021.
331. M. Juppo, A. Rahtu, M. Ritala and M. Leskelä, *Langmuir*, 2000, **16**, 4034.
332. A. Rahtu and M. Ritala, *Electrochem. Soc. Proc.*, 2000, **13**, 105.
333. J. Aarik, A. Aidla, H. Mandar and V. Sammelselg, *J. Cryst. Growth*, 2000, **220**, 531.
334. J. Aarik, A. Aidla, H. Mandar and T. Uustare, *Appl. Surf. Sci.*, 2001, **172**, 148.
335. K. Kukli, M. Ritala and M. Schuisky, *et al., Chem. Vap. Depos.*, 2000, **6**, 303.
336. J. Aarik, A. Aidla, V. Sammelselg, T. Uustare, M. Ritala and M. Leskelä, *Thin Solid Films*, 2000, **370**, 163.
337. J. Aarik, A. Aidla, T. Uustare, M. Ritala and M. Leskelä, *Appl. Surf. Sci.*, 2000, **161**, 385.
338. A. Rahtu, T. Alaranta and M. Ritala, *Langmuir*, 2001, **17**, 6506.
339. E. B. Yousfi, B. Weinberger, F. Donsanti, P. Cowache and D. Lincot, *Thin Solid Films*, 2001, **387**, 29.
340. K. Kukli, J. Aarik and A. Aidla, *et al., Chem. Mater.*, 2001, **13**, 122.
341. E. B. Yousfi, J. Fouache and D. Lincot, *Appl. Surf. Sci.*, 2000, **153**, 223.
342. R. Matero, A. Rahtu and M. Ritala, *Chem. Mater.*, 2001, **13**, 4506.
343. A. Rahtu and M. Ritala, *Chem. Vap. Depos.*, 2002, **8**, 21.
344. A. Rahtu, K. Kukli and M. Ritala, *Chem. Mater.*, 2001, **13**, 817.
345. M. Ritala and A. Leskelä, in *Handbook of Thin Film Materials*, H.S. Nalwa (ed.), Academic Press, San Diego, CA, 2001, **1**, 103.
346. A. Rosenthal, A. Tarre, P. Adamson, A. Gerst, A. Kasikov and A. Niilisk, *Appl. Surf. Sci.*, 1999, **142**, 204.
347. J. C. Badot, S. Ribes, E. B. Yousfi, V. Vivier, J. P. Pereira-Ramos, N. Baffier and D. Incot, *Electrochemical Solid-State Lett.*, 2000, **3**, 485.
348. K. Kobayashi and S. Okudaira, *Chem. Lett.*, 1997, 511.
349. M. Tammenmaa, M. Leskelä, T. Koskinen and L. Niinistö, *J. Less-Common Met.*, 1986, **126**, 209.
350. M. Leskelä, L. Niinistö, E. Nykanen, P. Soininen and M. Tiitta, *J. Less-Common Met.*, 1989, **153**, 219.
351. M. Tammenmaa, H. Antson, M. Asplund, L. Hiltunen, M. Leskelä and L. Niinistö, *J. Cryst Growth*, 1987, **84**, 151.
352. J. Aarik, A. Aidla, A. Jaek, M. Leskelä and L. Niinistö, *Appl. Surf. Sci.*, 1994, **75**, 33.
353. M. Tiitta and L. Niinistö, *Chem. Vap. Deposition*, 1997, **3**, 167.
354. P. Soininen, E. Nykanen, L. Niinistö and M. Leskelä, *Chem. Vap. Deposition*, 1994, **2**, 69.
355. V. Saanila, J. Ihanus, M. Ritala and M. Leskelä, *Chem. Vap. Deposition*, 1998, **4**, 227.
356. J. M. Buriak, L. K. Cheatham, R. G. Gordon, J. J. Graham and A. R. Barron, *Eur. J. Solid. State Inorg. Chem.*, 1992, **29**, 43.
357. M. L. Hitchman, S. H. Shamlian, D. D. Gilliland, D. J. Cole-Hamilton, S. C. Thompson, S. L. Cook and B. C. Richards, *Mater. Res. Soc. Symp. Proc.*, 1994, **33**, 249.
358. K. Timmer, S. L. Cook, C. Spee, *US Patent*, 1998, 5, 837, 321.
359. R. Gardiner, D. W. Brown, P. S. Kirlin and A. L. Rheingold, *Chem. Mater.*, 1991, **3**, 1053.
360. H. A. Meinema, K. Timmer, H. L. Linden and C. Spee, *Mater. Res. Soc. Symp. Proc.*, 1994, **335**, 193.
361. N. P. Kuzmina, L. I. Martynenko, Z. A. Tu, A. R. Kaul, G. V. Girichev, N. I. Giricheva, A. N. Rykov and Y. M. Korenev, *J. Phys. IV*, 1993, **C3**, 385.
362. R. G. Gordon, S. Barry, R. N. R. Broomhall-Dillard and D. J. Teff, *Adv. Mat. Opt. Electron.*, 2000, **10**, 201.
363. T. Hanninen, I. Mutikainen, V. Saanila, M. Ritala, M. Leskelä and J. C. Hanson, *Chem. Mater.*, 1997, **9**, 1234.

364. T. Leskelä, K. Vasama, G. Harkonen, P. Sarkio and M. Lounasmaa, *Adv. Mater. Opt. Electron*, 1996, **6**, 169.
365. M. Putkonen, T. Sajavaara, L. S. Johansson and L. Niinistö, *Chem. Vap. Deposition*, 2001, **7**, 44.
366. M. Nieminen, M. Putkonen and L. Niinistö, *Appl. Surf. Sci.*, 2001, **174**, 155.
367. T. Hatanpaa, J. Ihanus, J. Kansikas, I. Mutikainen, M. Ritala and M. Leskelä, *Chem. Mater.*, 1999, **7**, 1846.
368. D. E. Simon and C. G. James, *J. Mater. Chem.*, 2004, **14**, 3246.
369. L. G. Gosset, J. F. Damlencourt, O. Renault, D. Rouchon, P. H. Holliger, A. Ermolieff, I. Trimaille, J. J. Ganem, F. Martin and M. N. Semeria, *Journal of Non-Crystalline Solids*, 2002, **303**, 17–23.
370. A. C. Dillon, A. W. Ott, J. D. Way and S. M. George, *Surf. Sci.*, 1995, **322**, 230.
371. M. Ritala, M. Leskelä, J. P. Dekker, C. Mutsaers, P. J. Soininen and J. Skarp, *Chem. Vap. Depos.*, 1999, **5**, 7.
372. S. K. Kim, S. W. Lee, C. S. Hwang, Y. S. Min, J. Y. Won and J. Jeongd, *Journal of The Electrochemical Society*, 2006, **5**(153), F69–F76.
373. M. Schuisky, J. Aarik, K. Kukli, A. Aidla, J. Lu, A. Harsta, *Atomic Layer Deposition, Topical Conference, American Vacuum Society, Monterey*, 2001, May 14–15CA, USA.
374. J.P. Chang, Y.S. Lin, *Atomic Layer Deposition, Topical Conference, American Vacuum Society*, Monterey, May 14–15, 2001,CA, USA.
375. M. Ritala, K. Kukli and A. Rahtu, *et al., Science*, 2000, **288**, 319.
376. K. Kukli, M. Ritala and M. Leskelä, *Chem. Vap. Depos.*, 2000, **6**, 297.
377. J. C. Badot, S. Ribes and E. B. Yousfi, *et al., Electrochem. Solid State Lett.*, 2000, **3**, 485.
378. K. Kukli, M. Ritala and M. Leskelä, *Chem. Mater.*, 2000, **12**, 1914.
379. A. Rahtu, M. Ritala and M. Leskelä, *Chem. Mater.*, 2001, **13**, 1528.
380. R. G. Gordon, J. Becker, D. Hausman, *Atomic Layer Deposition, Topical Conference, American Vacuum Society*, Monterey, 2001, May 14–15, CA, USA.
381. J. S. Min, H. S. Park and S. W. Kang, *Appl. Phys. Lett.*, 1999, **75**, 1521.
382. M. Vehkamäki, T. Hanninen and M. Ritala, *Chem. Vap. Depos.*, 2001, **7**, 75.
383. J. S. Min, J. S. Park, H. S. Park and S. W. Kang, *J. Electrochem. Soc.*, 2000, **147**, 3868.
384. M. Ritala and M. Leskelä, *Appl. Surf. Sci.*, 1994, **75**, 333.
385. K. Kukli, K. Forsgren, J. Aarik, A. Aidla, T. Uutare, M. Ritala, A. Niskanan, M. Leskelä and A. Harsta, *J. Cryst. Growth*, 2001, **231**, 262.
386. M. Ritala, M. Leskelä, L. Niinistö, T. Prohaska, G. Friedbacher and M. Grasserbauer, *Thin Solid Films*, 1994, **250**, 72.
387. J. Aarik, A. Aidla, A.-A. Kiisler, T. Uustare and V. Sammelselg, *Thin Solid Films*, 1999, **340**, 110.
388. K. Forsgren, A. Harsta, J. Aarik, A. Aidla, J. Westlinder and J. Olsson, *J. Electrochemical Soc.*, 2002, **149**, F139.
389. M. Cho, J. Park, H. B. Park, C. S. Hwang, J. Jeong and K. S. Hyun, *Appl. Phys. Lett.*, 2002, **81**, 334.
390. D. C. Bradley, *Chem. Rev.*, 1989, **89**, 1317.
391. K. Kukli, M. Ritala and M. Leskelä, *Chem. Vap. Deposition*, 2000, **6**, 297.
392. R. Matero, M. Ritala, M. Leskelä, A. C. Jones, P. A. Williams, J. F. Bickley, A. Steiner, T. J. Leedham and H. O. Davies, *J. Non-Cryst. Solids*, 2002, **303**, 24.
393. A. C. Jones, P. A. Williams, J. L. Roberts, T. J. Leedham, H. O. Davies, R. Matero, M. Ritala and M. Leskelä, *Mater. Res. Soc. Symp. Proc.*, 2002, **716**, 145.
394. K. Kukli, M. Ritala, M. Leskelä, T. Sajavaara, J. Keinonen, A. C. Jones and J. L. Roberts, *Chem. Mater.*, 2003, **15**, 1722.

395. K. Kukli, M. Ritala, M. Leskelä, T. Sajavaara, J. Keinonen, A. C. Jones and J. L. Roberts, *Chem. Vap. Depos.*, 2003, **9**, 315.
396. K. Kukli, M. Ritala, T. Sajavaara, J. Keinonen and M. Leskelä, *Chem. Vap. Depos.*, 2002, **8**, 199.
397. D. M. Hausmann, E. Kim, J. Becker and R. G. Gordon, *Chem. Mater.*, 2002, **14**, 4350.
398. J. H. Lee, J. P. Kim, J. H. Lee, Y. S. Kim, H. S. Jung, N. I. Lee, H. K. Kang, K. P. Suh, M. M. Jeong, K. T. Hyun, H. S. Baik, Y. S. Chung, X. Liu, S. Ramanathan, T. Seidel, J. Winkler, A. Londergan, H. Y. Kim, J. M. Ha and N. K. Lee, *Proceedings of the Electron Devices Meeting*, held 9–11 Dec, 2002, San Francisco, IEDM, USA, 221–224.
399. K. Kukli, M. Ritala, M. Leskelä, T. Sajavaara, J. Keinonen, A. C. Jones and N. L. Tobin, *Chem. Vap. Depos.*, 2004, **10**, 91.
400. P. A. Williams, A. C. Jones, N. L. Tobin, P. R. Chalker, S. Taylor, P. A. Marshall, J. F. Bickley, L. M. Smith, H. O. Davies and G. W. Critchlow, *Chem. Vap. Deposition.*, 2003, **9**, 309.
401. M. Leskelä and M. Ritala, in *Handbook of Thin Film Materials*, H. S. Nalwa ed., Academic Press, New York, 2002, Vol. **1**, 103.
402. B. S. Lim, A. Rahtu, J.-S. Park and R. G. Gordon, *Inorg. Chem.*, 2003, **42**, 7951.
403. A. Rahtu, B. S. Lim and R. G. Gordon, *Abstract Book of ALD 2003 Conference*, 2003, San Jose, USA.
404. D. C. Bradley, J. S. Ghotra and F. A. Hart, *J. Chem. Soc., Dalton Trans.*, 1973, **1**, 1021.
405. T. Suntola, *Atomic Layer Epitaxy, Handbook of Crystal Growth*, Elsevier Science BV, Amsterdam, 1994.
406. J. Aarik, A. Aidla, T. Uustare and V. Sammelselg, *J Cryst Growth*, 1995, **148**, 268.
407. J. Aarik, A. Aidla, A. A. Kiisler, T. Uustare and V. Sammelselg, *Thin Solid Films*, 1999, **340**, 110.
408. J. Aarik, A. Aidla, H. Mändar and T. Uustare, *Appl. Surf. Sci.*, 2001, **172**, 148.
409. P. Tagtstrom, P. Martensson, U. Jansson and J. O. Carlsson, *J. Electrochem. Soc.*, 1999, **146**, 3139.
410. V. E. Drozd, A. A. Tulub, V. B. Aleskovski and D. V. Korol'kov, *Appl. Phys. Lett.*, 1994, **82/484**, 587.
411. J. D. Ferguson, S. M. George, *Atomic Layer Deposition, Topical Conference, American Vacuum Society*, Monterey, May 14–15, 2001, CA, USA.
412. Y. Shimogaki, H. Hamamura, R. Yamamoto, K. Y. Jun, I. Nishinaka, *Atomic Layer Deposition, Topical Conference, American Vacuum Society*, Monterey, May, 14–15, 2001. CA, USA.
413. J. W. Klaus, S. J. Ferro and S. M. George, *Thin Solid Films*, 2000, **360**, 145.
414. J. W. Klaus, S. J. Ferro and S. M. George, *J. Electrochem. Soc.*, 2000, **147**, 1175.
415. H. S. Sim, Y. T. Kim, H. Jeon, *Atomic Layer Deposition, Topical Conference, American Vacuum Society*, Monterey, May 14–15, 2001,CA, USA.
416. P. Tagstrom, P. Martensson, U. Jansson and J. O. Carlsson, *J. Electrochem. Soc.*, 1999, **146**, 3139.
417. M. Ritala, M. Leskelä, E. Rauhala and J. Jokinen, *J. Electrochem. Soc.*, 1998, **145**, 2916.
418. K. Kukli, A. Aidla and J. Aarik, *et al., Langmuir*, 2000, **16**, 8122.
419. K. Kukli, K. Forsgren, M. Ritala, M. Leskelä, J. Aarik and A. Harsta, *J. Electrochem. Soc.*, 2001, **148**, F227.
420. K. Kukli, K. Forsgren and J. Aarik, *et al., J. Cryst. Growth*, 2001, **231**, 262.
421. M. Schuisky, A. Harsta, A. Aidla, K. Kukli, A. A. Kiisler and J. Aarik, *J. Electrochem. Soc.*, 2000, **147**, 3319.
422. H. Yokoyama, M. Tanimoto, M. Shinohara and N. Inoue, *Appl. Surf. Sci.*, 1994, **82/83**, 158.
423. J. G. Reitl, H. M. Uridianyk and S. M. Bedair, *Appl. Phys. Lett.*, 1991, **59**, 2397.

424. J. P. Simko, T. Meguro, S. Iwai, K. Ozasa, Y. Aoyagi and T. Sugano, *Thin Solid Films*, 1993, **225**, 40.
425. W. G. Jeong, E. P. Menu and P. D. Dapkus, *Appl. Phys. Lett.*, 1989, **55**, 244.
426. B. Y. Maa and P. D. Dapkus, *Appl. Phys. Lett.*, 1991, **58**, 2261.
427. W. K. Chen, J. C. Chen, L. Anthony and P. L. Liu, *Appl. Phys. Lett.*, 1989, **55**, 987.
428. N. Pan, J. Carter, S. Hein, D. Howe, L. Goldman, L. Kupferberg, S. Brierley and K. C. Hsieh, *Thin Solid Films.*, 1993, **225**, 64.
429. A. W. Ott, J. M. Johnson, J. W. Klaus and S. M. George, *Appl. Surf. Sci.*, 1997, **112**, 205.
430. V. E. Drozd and V. B. Aleskovski, *Appl. Surf. Sci.*, 1994, **82/83**, 591.
431. A. Hunter and A. H. Kitai, *J. Cryst. Growth*, 1988, **91**, 111.
432. C. T. Hsu, *Thin Solid Films*, 1998, **335**, 284.
433. E. B. Yousfi, T. Asikainen, V. Pietu, P. Cowache, M. Powalla and D. Lincot, *Thin Solid Films*, 2000, **361–362**, 183.
434. M. Yokoyama, N. T. Chen and H. Y. Ueng, *J. Cryst. Growth*, 2000, **212**, 97.
435. H. Fujiwara, H. Kiryu and I. Shimizu, *J. Appl. Phys.*, 1995, **77**, 3927.
436. W. S. Wang, H. Ehsani and I. Bhat, *J. Electr. Mater.*, 1993, **22**, 873.
437. B. Sang, A. Yamada and M. Konagai, *Sol. Energy Mater. Sol. Cells*, 1997, **49**, 19.
438. Y. Yamamoto, K. Saito, K. Takahashi and M. Konagai, *Sol. Energy Mater. Sol. Cells*, 2001, **65**, 125.
439. A. W. Ott and R. P. H. Chang, *Mater. Chem. Phys.*, 1999, **58**, 132.
440. V. Lujala, J. Skarp, M. Tammenmaa and T. Suntola, *Appl. Surf. Sci.*, 1994, **82/83**, 34.
441. L. Stolt, J. Hedström and J. Skarp, *Conf Rec Photovoltaic Spec Conf.*, 1994, 250.
442. S. J. Yun, Y. S. Kim and S. H. K. Park, *Appl. Phys. Lett.*, 2001, **78**, 721.
443. S. M. Bedair, M. A. Tischler, T. Katsuyama and N. A. El-Masry, *Appl. Phys. Lett.*, 1985, **47**, 51.
444. S. Hirose, N. Kano, M. Deura, K. Hara, H. Munekata and H. Kukimoto, *Jpn. J. Appl. Phys.*, 1995, **34**, L1436.
445. M. Ozeki and N. Ohtsuka, *Appl. Surf. Sci.*, 1994, **82–83**, 233.
446. J. R Gong, S. Nakamura, M. Leonard, S. M. Bedair and N. A. El-Masry, *J. Electron. Mater.*, 1992, **21**, 965.
447. M. Ishii, S. Iwai, H. Kawata, T. Ueki and Y. Aoyagi, *J. Cryst. Growth*, 1997, **180**, 15.
448. T. M. Mayer, J. W. Rogers and T. A. Michalske, *Chem. Mater.*, 1993, **3**, 641.
449. D. Riihela, M. Ritala, R. Matero and M. Leskelä, *Thin Solid Films*, 1996, **289**, 1.
450. M. A. Khan, J. N. Kuznia, R. A. Skogman, D. T. Olson, M. Mac Millan and W. J. Choyke, *Appl. Phys. Lett.*, 1992, **61**, 2539.
451. K. Kukli, M. Ritala, M. Leskelä and J. Jokinen, *J. Vac. Sci. Technol. A.*, 1997, **15**, 2214.
452. R. Huang and A. H. Kitai, *J. Electron Mat.*, 1993, **22**, 215.
453. A.W. Ott, J. W. J. M Klaus Johnson, S. M George, *Thin Solid Films*, 1997, **292**,135.
454. K. Kukli, M. Ritala and M. Leskelä, *J. Electrochem. Soc.*, 1997, **144**, 300.
455. S. M. George, O. Sneh, A. C. Dillon, M. L. Wise, A. W. Ott, L. A. Okada and J. D. Day, *Appl. Surf. Sci.*, 1994, **82/84**, 460.
456. H. Kumagai, K. Toyoda, K. Kobayashi, M. Obara and Y. Iimura, *Appl. Phys. Lett.*, 1997, **70**, 2338.
457. A. Paranjpe, S. Gopinath, T. Omstead and R. Bubber, *J. Electrochem. Soc.*, 2001, **148**, G465.
458. J. F. Fan and K. Toyoda, *Appl. Surf. Sci.*, 1992, **60–61**, 765.
459. H. Kumagai and K. Toyoda, *Appl. Surf. Sci.*, 1994, **82–83**, 481.
460. J. F. Fan and K. Toyoda, *Mater. Res. Soc.Symp. Proc.*, 1993, **284**, 517.
461. V. E Drozd, A. P. Baraban and I. O. Nikiforova, *Appl. Surf. Sci.*, 1994, **82–83**, 583.
462. H. Kumagai, K. Toyoda, M. Matsumoto and M. Obara, *Jpn. J. Appl. Phys.*, 1993, **32**, 6137.

463. S. W. Choi, C. M. Jang, D. Y. Kim, J. S. Ha, H. S. Park, W. Koh and C. S. Lee, *J. Korean Phys. Soc.*, 2003, **2**, S975.
464. C. W. Jeong, J. S. Lee and S. K. Joo, *Jpn. J. Appl. Phys.*, 2001, **40**, 285.
465. J. Kim, K. Chakrabarti, J. Lee, K. Y. Oh and C. Lee, *Mat. Chem. Phys.*, 2003, **78**, 733.
466. M. Schuisky, K. Kukli, M. Ritala, A. Harsta and M. Leskelä, *Chem. Vap. Depos.*, 2000, **6**, 139.
467. T. J. Kealy and P. L. Pauson, *Nature*, 1951, **168**, 1039.
468. S. A. Miller, J. A. Tebboth and J. F. Tremaine, *J. Chem. Soc.*, 1952, 632.
469. N. J. Long, *Metallocenes: An Introduction to Sandwich Compounds*, Blackwell Science, Oxford, 1998.
470. S. Codato, G. Carta, G. Rossetto, G. A. Rizzi, P. Zanella, P. Scardi and M. Leoni, *Chem. Vap. Depos.*, 1999, **5**, 159.
471. A. Weber, H. Suhr, H. Schumann and R. D. Köhn, *Appl. Phys. A*, 1990, **51**, 520.
472. W. C. Yeh and M. Matsumura, *Jpn. J. Appl. Phys.*, 1997, **36**, 6884.
473. H. Suhr, *New J. Chem.*, 1990, **14**, 523.
474. D. K. Russell, *Chem. Vap. Depos.*, 1996, **6**, 223.
475. M. Vehkamäki, T. Hatanpaa, T. Hanninen, M. Ritala and M. Leskelä, *Electrochem. Solid State Lett.*, 1999, **2**, 504.
476. M. Vehkamäki, T. Hänninen, M. Ritala, M. Leskelä, T. Sajavaara, E. Rauhala and J. Keinonen, *Chem. Vap. Depo.*, 2000, **17**, 75.
477. M. Vehkamäki, T. Hatanpaa, T. Hanninen, M. Ritala and M. Leskelä, *Electrochem. Solid-State Lett.*, 1999, **2**, 504.
478. J. Ihanus, T. Hänninen, T. Hatanpaa, T. Aaltonen, I. Mutikainen, T. Sajavaara, J. Keinonen, M. Ritala and M. Leskelä, *Chem. Mater.*, 2002, **14**, 1937.
479. M. Putkonen, T. Sajavaara and L. Niinistö, *J. Mater Chem.*, 2000, **10**, 1857.
480. R. Huang and A. H. Kitai, *Appl. Phys. Lett.*, 1992, **61**, 1450.
481. R. Huang and A. H. Kitai, *J. Mater. Sci. Lett.*, 1993, **12**, 1444.
482. M. Putkonen, L. S. Johansson, E. Rauhala and L. Niinistö, *J. Mater. Chem.*, 1999, **9**, 2449.
483. M. Putkonen, M. Nieminen, J. Niinistö, T. Sajavaara and L. Niinistö, *Chem. Mater.*, 2001, **13**, 4701.
484. G. Harkonen, T. Kervinen, E. Soininen, R. Tornqvist, K. Vasama, M. Glanz, H. Schumann, US Patent, 6, 248, 605 (2001).
485. J. E. Lau, G. G. Peterson, D. Endisch, K. Barth, A. Topol, A. E. Kaloyeros, R. T. Tuenge and C. N. King, *J. Electrochem. Soc.*, 2001, **148**, C427.
486. G. Scarel, E. Bonera, C. Wiemer, G. Tallarida, S. Spiga, M. Fanciulli, I. L. Fedushkin, H. Schumann, Y. Lebedinskii and A. Zenkevich, *Appl. Phys. Lett.*, 2004, **85**, 630.
487. M. Putkonen and L. Niinistö, *J. Mater. Chem.*, 2001, **11**, 3141.
488. M. Kroger-Laukkanen, M. Peussa, M. Leskelä and L. Niinistö, *Appl. Surf. Sci.*, 2001, **183**, 290.
489. M. Putkonen, T. Sajavaara, J. Niinistö, L. S. Johansson and L. Niinistö, *J. Mater. Chem.*, 2002, **12**, 442.
490. M. Putkonen, J. Niinistö, K. Kukli, T. Sajavaara, M. Karppinen, H. Yamauchi and L. Niinistö, *Chem. Vap. Depos.*, 2003, **9**, 207.
491. J. Niinistö, M. Putkonen, L. Niinistö, *The American Vacuum Society Topical Conference on Atomic Layer Deposition*, August 4–6, 2003, San Jose CA USA.
492. M. Ritala, M. Leskelä, L. Niinistö, T. Prohaska, G. Friedbacher and M. Grasserbauer, *Thin Solid Films*, 1994, **250**, 72.
493. K. Forsgren, A. Harsta, J. Aarik, A. Aidla, J. Westlinder and J. Olsson, *J. Electrochem. Soc.*, 2002, **149**, F139.

494. H. B. Park, M. Cho, J. Park, S. W. Lee, C. S. Hwang, J. P. Kim, J. H. Lee, N. I. Lee, H. K. Kang, J. C. Lee and S. J. Oh, *J. Appl. Phys.*, 2003, **94**, 3641.
495. J. F. Conley Jr., Y. Ono, R. Solanki, G. Stecker and W. Zhuang, *Appl. Phys. Lett.*, 2003, **82**, 3508.
496. D. M. Hausmann, E. Kim, J. Becker and R. Gordon, *Chem. Mater.*, 2002, **14**, 4350.
497. K. Kukli, M. Ritala, T. Sajavaara, J. Keinonen and M. Leskelä, *Chem. Vapor. Depos.*, 2002, **8**, 199.
498. E. Soininen, G. Harkonen and K. Vasama, *3rd international Conference on the Science and Technology of Display Phosphors*, Huntington Beach, CA, USA, 1997.
499. T. Aaltonen, P. Alen, M. Ritala and M. Leskelä, *Chem. Vap. Depos.*, 2003, **9**, 45.
500. T. Aaltonen, A. Rahtu, M. Ritala and M. Leskelä, *Electrochem. Solid-State Lett.*, 2003, **6**, C130.
501. O. K. Kwon, S. H. Kwon, H. S. Park and S. W. Kang, *Electrochem. Solid-State Lett.*, 2004, **7**, C46.
502. T. Aaltonen, M. Ritala, T. Sajavaara, J. Keinonen and M. Leskelä, *Chem. Mater.*, 2003, **15**, 1924.
503. J. Chae, H. S. Park and S. Kang, *Electrochem. Solid State Lett.*, 2002, **5**, C64.
504. M. Schuisky, K. Kukli, M. Ritala, A. Harsta and M. Leskelä, *Chem. Vap. Depos.*, 2000, **6**, 139.
505. M. Ritala, K. Kukli, A. Rahtu, P. I. Raisanen, M. Leskelä and T. Sajavaara, *J. Keinonen Science*, 2000, **288**, 319.
506. K. Kukli, M. Ritala and M. Leskelä, *Chem. Mater.*, 2000, **12**, 1914.
507. A. Rahtu, M. Ritala and M. Leskelä, *Chem. Mater.*, 2001, **13**, 1528.
508. H. Kim, *J. Vac. Sci. Technol. A*, 2003, **21**, 2231.
509. M. Juppo, M. Vehkamaki, M. Ritala and M. Leskelä, *J. Vac. Sci. Tech. A*, 1998, **16**, 2845.
510. M. Juppo, M. Ritala and M. Leskelä, *J. Vac. Sci. Tech. A*, 1997, **15**, 2330.
511. J. W Klaus, S. J. Ferro and S. M. George, *Appl. Surf. Sci.*, 2000, **162–163**, 479.
512. J. W Elam, C. E. Nelson, R. K. Grubbs and S. M. George, *Surf. Sci.*, 2001, **479**, 121.
513. S. H. Lee, P. Wongsenakhum, L. Gonzale, J. Gao, L. Chan, J. Collins, K. Ashtiani, K. Levy, *Proceedings of American Vacuum Society Topical Conference on Atomic Layer Deposition*, 2002, Seoul, Korea.
514. M. Yang, H. Chung, A. Yoon, H. Fang, A. Zhang, C. Knepfler, R. Jackson, J. S. Byun, M. Eizenberg, M. Xi, M. Kori, A.K. Sinha, *Proceedings of the Advanced Metallization Conference*, 2001, 655.
515. K. Kukli, M. Ritala, M. Schuisky, M. Leskelä, T. Sajavaara, J. Keinonen, T. Uustare and A. Harsta, *Chem. Vap. Depos.*, 2000, **6**, 303.
516. M. Ritala, T. Asikainen and M. Leskelä, *Electrochem. Solid-State Lett.*, 1998, **1**, 156.
517. K. Kukli, K. Forsgren, J. Aarik, T. Uustare, A. Aidla, A. Niskanen, M. Ritala, M. Leskelä and A. Harsta, *J. Cryst. Growth*, 2001, **231**, 262.
518. K. Kukli, J. Aarik, A. Aidla, K. Forsgren, J. Sundqvist, A. Harsta, T. Uustare, H. Mandar and A. A. Kiisler, *Chem. Mater.*, 2001, **13**, 122.
519. L. Hiltunen, H. Kattelus, M. Leskelä, M. Mäkela, L. Niinistö, E. Nykänen, P. Soininen and M. Tiitta, *Mater. Chem. Phys.*, 1991, **28**, 379.
520. H. Molsa, L. Niinistö and M. Utriainen, *Adv. Mater. Opt. Electron.*, 1994, **4**, 389.
521. T. Hatanpa, J. Ihanus, J. Kansikas, I. Mutikaine, M. Ritala and M. Leskelä, *Chem. Mater.*, 1999, **11**, 1846.
522. M. Tiitta and L. Niinistö, *Chem. Vap. Depos.*, 1997, **3**, 167.
523. M. Ritala and M. Leskelä, *Atomic Layer Deposition (Handbook of Thin Film Materials)*, Academic, San Diego, 2002, pp. 103–159.

524. J. S. Park, M. J. Lee, C. S. Lee and S. W. Kang, *Electrochem. Solid State Let.*, 2001, **4**, C17.
525. J. Y. Kim, H. K. Kim, S. W. Seo, Y. Kim, Y. D. Kim, H. Jeon, *Atomic Layer Deposition, Topical Conference, American Vacuum Society*, Monterey, May 14–15, 2001, CA, USA.
526. M. Ritala, P. Kalsi, D. Riihela, K. Kukli, M. Leskelä and J. Jokinen, *Chem. Mater.*, 1999, **11**, 1712.
527. M. Ritala, M. Leskelä, E. Rauhala and P. Haussalo, *J. Electrochemical. Soc.*, 1995, **142**, 1670.
528. M. Juppo, P. Alen, M. Ritala and M. Leskelä, *Chem. Vap. Depos.*, 2001, **7**, 211.
529. M. Juppo, M. Ritala and M. Leskelä, *J. Electrochemical. Soc.*, 2000, **147**, 3377.
530. S. Haukka, K.-E. Elers and M. Tuominen, *Mat. Res. Soc. Symp. Proc.*, 2000, **612**, D6.4.1.
531. J. S. Min, H. S. Park, W. Koh and S. W. Kang, *Mater. Res. Soc. Symp. Proc.*, 1999, **564**, 207.
532. D. J. Kim, Y. B. Young, M. B. Lee, Y. H. Lee, J. H. Lee and J. H. Lee, *Thin Solid Films*, 2000, **372**, 276.
533. P. Martensson, Acta Univerisity Uppsala, *Comprehensive Summaries of Uppsala Dissertations from the Faculty of Science and Technology*, 1999, 421 and publications cited therein.
534. M. Juppo, M. Ritala and M. Leskelä, *J. Vac. Sci. Technol. A*, 1997, **15**, 2330.
535. M. Juppo, M. Vehkamäki, M. Ritala and M. Leskelä, *J. Vac. Sci. Technol., A*, 1998, **16**, 2845.
536. R. Solanki and B. Pathangey, *Electrochem. Solid State Lett.*, 2000, **3**, 479.
537. M. Utriainen, M. Kroger Laukkanen, L. S. Johansson and L. Niinistö, *Appl. Surf. Sci.*, 2000, **157**, 151.
538. S. M. Rossnagel, A. Sherman and F. Turner, *J. Vac. Sci. Technol. B*, 2000, **18**, 2016.
539. H. Kim, S. M. Rossnagel, *Atomic Layer Deposition, TopicalConference, American Vacuum Society*, Monterey, May 14–15, 2001, CA, USA.
540. Y. Satoh, K. Ikeda, S. Sugahara and M. Matsumura, *Jpn. J. Appl. Phys.*, 2000, **39**, 5732.
541. E. Hasunumu, S. Sugahara, S. Hoshina, S. Imai, K. Ikeda and M. Matsumura, *J. Vac. Sci. Technol., A*, 1998, 679.
542. S. W. Kang, *Atomic Layer Deposition, Topical Conference, American Vacuum Society*, Monterey, May 14–15, 2001, CA, USA.
543. C. W. Jeong, J. S. Lee and S. K. Joo, *Jpn. J. Appl. Phys.*, 2001, **285**, 285.
544. S. W. Choi, C. S. Lee, W. Koh, *et al.*, *Atomic Layer Deposition, Topical Conference, American Vacuum Society*, Monterey, May 14–15, 2001,CA, USA.
545. T. Doh, J. Koo, Y. Kim, J. Han, Y. Kim, H. Jeon, *Atomic Layer Deposition, Topical Conference, American Vacuum Society*, Monterey, May 14–15, 2001, CA, USA.

CHAPTER 6

CVD of III-V Compound Semiconductors

JAE-HYUN RYOU,[a] RAVI KANJOLIA[b] AND RUSSELL D. DUPUIS[c]

[a] Center for Compound Semiconductors and School of Electrical and Computer Engineering, Georgia Institute of Technology, 777 Atlantic Dr. NW, Atlanta, GA 30332-0250, USA; [b] SAFC Hitech, 1429 Hilldale Avenue, Haverhill, MA 01832, USA; [c] Center for Compound Semiconductors, School of Electrical and Computer Engineering, and School of Materials Science and Engineering, Georgia Institute of Technology, 777 Atlantic Dr. NW, Atlanta, GA 30332-0250, USA

6.1 Fundamentals of III-V Compound Semiconductors

Group III-V compound semiconductor material systems, consisting of Group III elements, such as aluminium (Al), gallium (Ga) and indium (In), and Group V elements, such as nitrogen (N), phosphorus (P), arsenic (As) and antimony (Sb), have semiconducting properties. As a result of the covalent bonding between the four nearest-neighbor adjacent atoms and the details of the energies of the allowed electronic states for the valence electrons, these materials exhibit an electronic band structure having a bandgap energy, which is a range of forbidden energies between the allowed energy states in the lower-energy valence band and in the upper-energy conduction band. Consequently, compounds consisting of alloys of Group III and Group V elements are classified as semiconductors, similar to the closely related Group IV materials such as silicon (Si) and germanium (Ge), and they are generally called III-V *compound semiconductors* as opposed to the single-element semiconductors, *e.g.*, Si and Ge, which are referred to as *elemental semiconductors*. Both column IV elemental and III-V compound semiconductors have covalent bonding as result of sharing electrons in the outer hybridized sp^3 orbitals of the tetrahedrally coordinated nearest-neighbor atoms; however, in the case of III-V semiconductors, a certain degree of charge transfer occurs, since the outer shells of the Group III and Group V atoms have a different number of valence electrons available, which creates an ionic character to the covalent bonding of III-V semiconductors. There are also other types of semiconductors, *e.g.*, Group II-VI compound semiconductors, Group IV alloy semiconductors, and Group IV compound semiconductors.

Chemical Vapour Deposition: Precursors, Processes and Applications
Edited by Anthony C. Jones and Michael L. Hitchman

Published by the Royal Society of Chemistry, www.rsc.org

Group II-VI compound semiconductors, such as HgCdTe (mercury cadmium telluride also known as MCT), ZnSe and ZnO, have a higher degree of ionic nature in the bonding. Typical Group IV alloy semiconductors are materials consisting of both Si and Ge (and sometimes including C, hence they are Si-Ge-C alloys). These alloy semiconductors are often called Group IV "compound" semiconductors but this is not an exactly correct expression, since the elements are alloyed randomly in their lattice sites, rather than each element having their own designated lattice site as is the case for the III-Vs. In contrast, SiC is a Group IV semiconductor having a specific stoichiometry and atomic lattice structure and it is a true "compound" semiconductor.

The history of III-V compound semiconductors (at least in the published literature) began with the identification in the early 1950s of the semiconducting properties of many III-V compounds and the study of their physical properties, *e.g.*, InSb, AlSb, GaSb, GaAs and InP were identified as semiconductors.[1] Since then, research and development efforts have been devoted to understand the fundamental properties and characteristics of III-V compound semiconductors and their structures and to develop useful devices with multiple applications and improved performance characteristics. As a result of these efforts over the past ~50 years, compound semiconductors have evolved into viable technological materials and play critically important roles in most of the photonic device applications and in some of electronic device applications today. Group III-V compound semiconductor technologies will not be able to replace the Si technology in many applications. Instead, they will be developed and positioned as technically important materials in many applications and will be employed in circuits in conjunction with Si-based integrated circuits without having to compete against and win over the technology. This section describes the characteristics and material properties of III-V compound semiconductors and compares them to the elemental semiconductor Si; this information will be important in the discussion of the applications of III-V compound semiconductors in Section 6.2.

6.1.1 Characteristics of III-V Compound Semiconductors

Group III-V compound semiconductors have various material, optical and electronic characteristics. Some of the characteristics are beneficial in device applications while others are not. The pros and cons of III-V compound semiconductors and their structures in comparison to Si are described below.

6.1.1.1 Electronic Band Structures

The most fundamental characteristics of III-V compound semiconductors that give them an advantage over Si in optoelectronic applications come from their electronic band structure and associated excess carrier transitions.[2] Many III-V semiconductors have a direct bandgap, while Si and Ge have indirect-bandgap energy-band structures. This characteristic makes many of the III-V materials efficient light emitters for photonic device applications. Figure 6.1 shows simplified energy (E) *versus* crystal momentum vector (k) diagrams of a direct-bandgap semiconductor and an indirect-bandgap semiconductor. In *direct-bandgap* semiconductors, the lowest minimum electron energy in the conduction band (CB) is located at the zone center in the first Brillouin zone ($k = 0$), and is aligned vertically in energy *vs.* k space at the maximum electron energy in the valence band (VB). In *indirect-bandgap* semiconductors, the lowest minimum energy electron state in the conduction band is *not* located at the zone center in the first Brillouin zone ($k \neq 0$), *i.e.*, it is *not* aligned near the maximum electron energy in the valence band. The transition of an electron between the CB and the VB requires energy conservation, that is, an upward transition (VB→CB) requires energy absorption, while a downward transition (CB→VB) accompanies energy emission, as well as conservation of crystal momentum vector, k. For direct-bandgap semiconductors, the transition

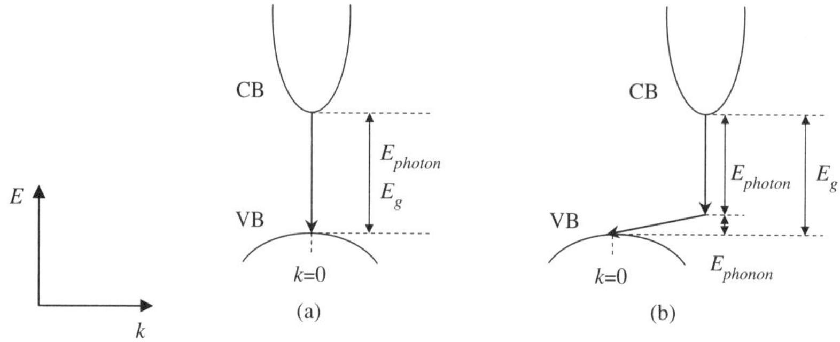

Figure 6.1 Energy (*E*) *versus* momentum vector (*k*) diagram of (*a*) direct bandgap and (*b*) indirect bandgap semiconductors (CB: conduction band; VB: valence band; E_g: bandgap; E_{photon}: photon energy; E_{phonon}: phonon energy).

energy is provided by the photon energy in the interaction between electrons and photons. The CB→VB and the VB→CB transitions in direct bandgap semiconductors do not require a change in momentum and hence can occur without a change in the *k*-value or the crystal momentum, a process that is commonly described as the involvement of a phonon – which is a lattice vibration with a relatively large crystal momentum, *k*, but little energy. In such transitions, the transition energy is conserved by either photon absorption or emission. The momentum change due to photon energy emission or absorption is negligible – strictly speaking the change in *k* is "nearly" zero, not completely zero – such a transition is represented by an "almost" vertical line and it is called *radiative recombination* in the case of the downward transition, when the electron in the CB recombines with the hole in the VB. In contrast, VB→CB and CB→VB transitions for electrons in an indirect-bandgap semiconductor require a significant change in both momentum and energy. The energy change is accommodated by both photons and phonons; however, the momentum change needs to be accommodated by one or more phonons in the crystal. Therefore, for the recombining carriers to satisfy both energy and momentum requirements for the transition, the carriers have to wait until the transition energy and momentum values are satisfied by the right *k*-value of one or more phonons. There are other transitions, which are non-radiative in nature, such as transitions *via* surface or bulk defects, and traps, as well as Auger recombination. Radiative recombination competes with such *non-radiative* recombination processes. The long waiting time for excess carriers to undergo radiative recombination in the indirect semiconductors, *i.e.*, the long radiative carrier lifetime, τ_r, makes the probability of radiative recombination of such carriers low – the transition of most carriers occurs *via* non-radiative recombination, meaning the non-radiative carrier lifetime, τ_{nr}, is smaller than τ_r. In contrast, τ_r is small in direct semiconductors, which makes the probability of the radiative recombination high. This direct-bandgap property enables most high-quality III-V compound semiconductors to function as "efficient" light emitters.

6.1.1.2 Crystal Structures

Elemental semiconductors, *e.g.*, Si, and the III-V semiconductors, have related but different crystal structures. A single *crystal* is formed when a *basis* arrangement of atoms is combined with a *lattice* that is a periodic array of points.[3] The crystal lattice can be thought of as an arrangement of identical geometrical set (or basis) of atoms with a given lattice parameter, or spacing, between each set. Elemental semiconductors have a diamond structure, which is a interpenetrating set of two FCC (face-centered cubic) lattices, each having a basis of two Si atoms that are near each other with the transitional relation $a = 1/4$, $b = 1/4$ and $c = 1/4$. In III-V compound semiconductors, two stable

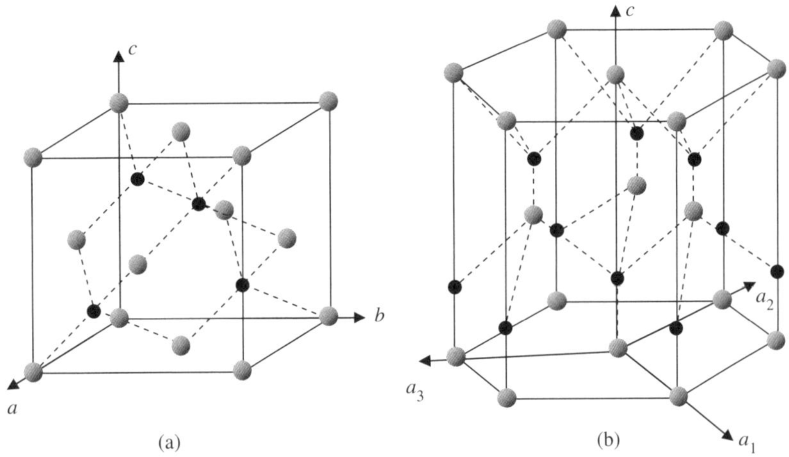

Figure 6.2 Crystal structures of (*a*) zinc blende and (*b*) wurtzite.

crystalline structures are often observed, depending on the specific materials (*e.g.*, ionicity, atom size) and thermodynamic parameters such as temperature and pressure: these two structures are called zinc-blende structure and wurtzite structure. Most III-As, III-P and III-Sb materials have a zinc-blende structure, which is similar to the diamond structure, that is an FCC lattice with a basis of two atoms; but, in this case, the basis consists of a pair composed of Group III and Group V atoms. III-N materials generally have a stable wurtzite structure. The wurtzite structures have an HCP (hexagonal close packed) lattice with a basis of a pair of Group III and Group V atoms that are separated by a $c = 3/16$ transitional relation. Since the zinc-blende structure and the wurtzite structure are based on the cubic and the hexagonal lattice, respectively, their indexing for crystallographic directions and planes are different. The cubic lattice follows *Miller indices* in the form of $(a\ b\ c)$ for planes and $[a\ b\ c]$ for directions and the hexagonal lattice follows *Miller–Bravais indices* in the form of $(a_1\ a_2\ a_3\ c)$ for planes and $[a_1\ a_2\ a_3\ c]$ for directions. In the Miller–Bravais indices of hexagonal lattice, only two basal plane indices are required for designation, since $a_1 + a_2 + a_3 = 0$. Therefore, $(a_1\ a_2\ a_3\ c)$ is often expressed as $(a_1\ a_2 \cdot c)$ (the dot between a_2 and c indicating that a_3 is omitted in the index). Also, using {} and $\langle\rangle$ instead of parenthesis ( ) and bracket [ ] for planes and directions indicates the "family" of equivalent planes and directions, respectively. For example, {111} includes the geometrically equivalent planes of (111), $(\bar{1}11)$, $(1\bar{1}1)$, $(11\bar{1})$, *etc.* Figure 6.2 shows the crystal structures of the unit cells of zinc-blende and wurtzite structures. They look very different in atomic arrangement but they are actually similar. Considering the close-packed planes, which are {111} and $\{00 \cdot 1\}$ for zinc blende and wurtzite, respectively, the atomic arrangements on the plane are the same; only the stacking sequence of the planes is different: stacking of the planes along the $\langle 111 \rangle$ direction in the zinc-blende structure follows the pattern ABCABC, while the stacking of $\{00 \cdot 1\}$ planes along the $\langle 00 \cdot 1 \rangle$ direction in the wurtzite structure follows the sequence ABAB . . .

6.1.1.3 High Degree of Freedom in Materials Selection

As the name indicates, III-V compound semiconductors consist of combination of Group III and Group V elements. Therefore, by choosing each element from Group III and Group V, twelve materials that have different properties can be formed – three elements in Group III and four elements in Group V [excluding boron (B), thallium (Tl), and bismuth (Bi) even though they are in Groups III or V in the periodic table], such as AlN, GaN, InN, AlP, GaP, InP, AlAs, GaAs, InAs,

AlSb, GaSb and InSb. They are the basic forms of the compound semiconductors and are called *binary* compound semiconductors. When the Group III and/or Group V are alloyed, *i.e.*, multiple elements from Group III and/or Group V elements are combined on the respective crystal sublattice, *ternary* and *quaternary alloy* compound semiconductors can be formed. For example, $Al_xGa_{1-x}As$, $In_xGa_{1-x}As$, $In_xGa_{1-x}P$ and $In_xGa_{1-x}N$ are technologically important ternary compound semiconductor material systems in device applications and $In_xGa_{1-x}As_yP_{1-y}$, $In_{0.53}(Al_xGa_{1-x})_{0.47}As$ and $In_{0.49}(Al_xGa_{1-x})_{0.51}P$ are important quaternary material systems. Therefore, in the selection of materials, III-V compound semiconductors provide various options. Different materials by nature have different materials properties in terms of their chemical, structural, mechanical, thermal, electrical and optical properties, such as the thermal conductivity, dielectric constant, refractive index, bandgap, lattice parameter, effective masses of carriers (electron and hole), *etc.* Such a large degree of freedom in material selection for compound semiconductors is critically beneficial in selecting the right properties of the materials that are required in various device applications. Among various properties of materials, the bandgap and lattice parameters are the properties of most interest, since they govern the operating wavelength of photonic devices and the growth feasibility of high-quality materials without the generation of defects. Figure 6.3 shows the bandgap energy *versus* lattice constant diagram of many III-V compound semiconductors. The darker black solid line indicates direct bandgap material, while the lighter grey line indicates an indirect bandgap. Binary compound semiconductors are represented as a point and ternary alloy semiconductors are represented as a line connecting two binary semiconductor end points. For example, $Al_xGa_{1-x}As$ is a line connecting the GaAs and AlAs binary materials and the bandgap energy and lattice parameter change as the mole fraction of Al in the alloy, x, (the Al composition) changes. Quaternary alloy semiconductors are represented by the area enclosed by binary compound elements and their associated ternary compounds. For example, $In_xGa_{1-x}As_yP_{1-y}$ is an area enclosed by GaAs, InAs, GaP, and InP binary materials and their related ternaries. The lattice parameter, a, changes linearly with mole fraction following Vegard's rule in the case of ternary $\mathrm{III}_{\mathrm{A}_x}\mathrm{III}_{\mathrm{B}_{1-x}}\mathrm{V}$ or $\mathrm{III}\mathrm{V}_{\mathrm{C}_x}\mathrm{V}_{\mathrm{D}_{1-x}}$:

$$a_{\mathrm{III}_{\mathrm{A}_x}\mathrm{III}_{\mathrm{B}_{1-x}}\mathrm{V}} = x a_{\mathrm{III}_\mathrm{A}\mathrm{V}} + (1-x) a_{\mathrm{III}_\mathrm{B}\mathrm{V}} \text{ or } a_{\mathrm{III}\mathrm{V}_{\mathrm{C}_x}\mathrm{V}_{\mathrm{D}_{1-x}}} = x a_{\mathrm{III}\mathrm{V}_\mathrm{C}} + (1-x) a_{\mathrm{III}\mathrm{V}_\mathrm{D}} \quad (6.1)$$

and quaternary alloy semiconductors $\mathrm{III}_{\mathrm{A}_x}\mathrm{III}_{\mathrm{B}_{1-x}}\mathrm{V}_{\mathrm{C}_y}\mathrm{V}_{\mathrm{D}_{1-y}}$:

$$\begin{aligned} a_{\mathrm{III}_{\mathrm{A}_x}\mathrm{III}_{\mathrm{B}_{1-x}}\mathrm{V}_{\mathrm{C}_y}\mathrm{V}_{\mathrm{D}_{1-y}}} =& xy a_{\mathrm{III}_\mathrm{A}\mathrm{V}_\mathrm{C}} + (1-x)y a_{\mathrm{III}_\mathrm{B}\mathrm{V}_\mathrm{C}} + x(1-y) a_{\mathrm{III}_\mathrm{A}\mathrm{V}_\mathrm{D}} \\ &+ (1-x)(1-y) a_{\mathrm{III}_\mathrm{B}\mathrm{V}_\mathrm{D}} \end{aligned} \quad (6.2)$$

The bandgap energy of ternary alloy compound semiconductors is not always linear with the change of mole fraction, x, as is shown in the parabolic curves for some of alloys in Figure 6.3. The bandgap energy, E_g, of the ternary alloy $\mathrm{III}_{\mathrm{A}_x}\mathrm{III}_{\mathrm{B}_{1-x}}\mathrm{V}$ as a function of mole fraction, x is:

$$E_g(x) = E_{g,\mathrm{III}_\mathrm{B}\mathrm{V}} + bx + cx^2 \quad (6.3)$$

If c is not zero, the bandgap change with mole fraction is not linear and it is called the *bowing parameter*. The bandgap energy (eV) is related to the wavelength [λ (μm)] by following expression:

$$E_g = 1.24/\lambda \quad (6.4)$$

Equation (6.4) is a basic consideration in the design of all the devices for certain photonic applications. For example, if a light emitting diode operating at ~870 nm is desired, the required semiconductor material has a 1.424 eV bandgap energy, and one possible choice is GaAs. In

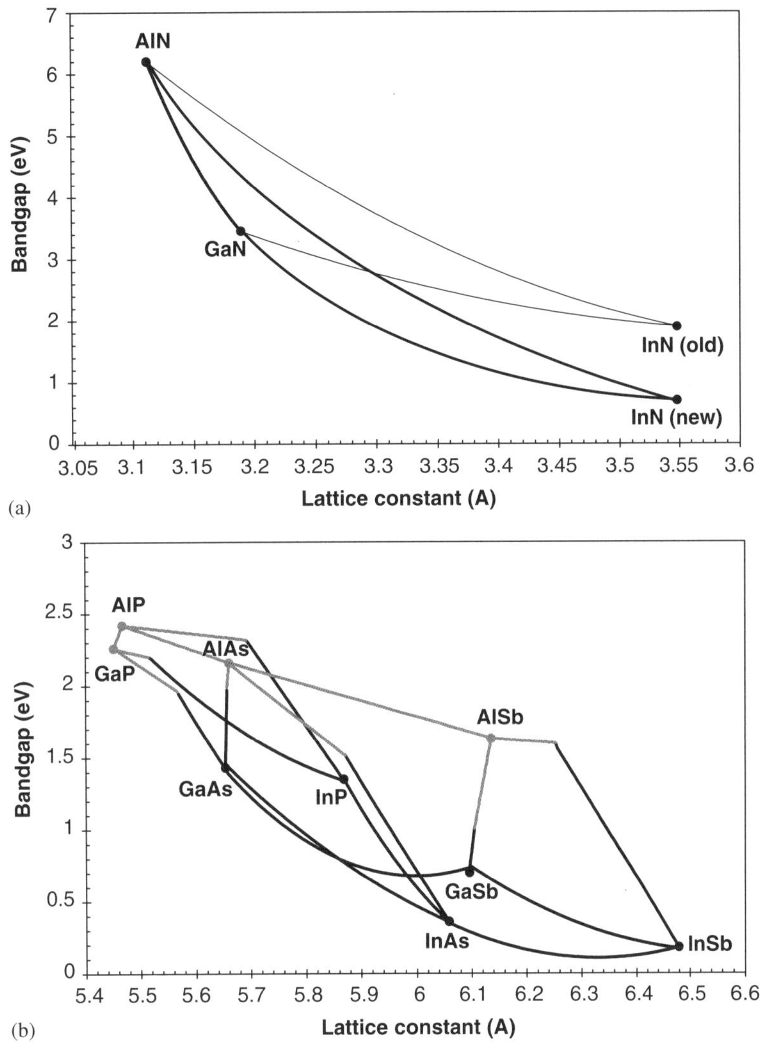

Figure 6.3 Energy *versus* lattice constant diagram of (*a*) wide bandgap materials and (*b*) narrow bandgap materials.

addition, if visible light emitters are needed, the optimal choice for the materials will be InAlGaP or InGaN alloys. In another example, if it is desired to fabricate a photodetector that operates in mid-IR (infrared) region with $\lambda > 1\,\mu m$, the absorption layer of the detector is required to have a bandgap energy less than 1.24 eV, and InGaAs is a viable choice. As shown in Figure 6.3, III-V compound semiconductors have a wide variation of bandgap energies, ranging from 0.1 eV ($\lambda = 12.4\,\mu m$) to 6.2 eV ($\lambda = 200$ nm), which covers from the mid-IR to the deep-UV (ultraviolet) spectral region. The bandgap energy of III-N materials is generally larger than that of III-As, III-P and III-Sb materials. This was particularly true until the bandgap of InN was more recently reported to be ~0.7 eV[4] (previously, the InN bandgap was believed to be ~1.9 eV). Also, the crystal structure of III-N semiconductors is different from other III-V materials. This large bandgap energy and different crystal structure are reasons why III-N materials are generally classified separately as "wide bandgap semiconductor materials."

For most semiconductor materials of interest for optoelectronic and electronic device applications, the material needs to be a single crystal with a low density of crystalline defects. To grow single-crystal layers for devices, a relatively thick single-crystal template or platform, called a *substrate*, is generally required. A deposition of a "single-crystal" film of a material upon a surface of a single-crystalline substrate is called *epitaxy* or *epitaxial growth*. Such a film is termed an *epitaxial layer*. In Greek, epi means "above" and taxis means "in ordered manner." If the film and substrate are composed of the same materials, the film is *homoepitaxial* and if the film and substrate are formed from different materials (with different lattice constants in many cases), the film is *heteroepitaxial*. For device applications, generally, the epitaxial layers that are grown are *lattice-matched* or *nearly-lattice-matched* to the substrate to avoid crystalline defect generation during the epitaxial growth. Therefore, the lattice constants of the epitaxial layers are generally chosen to be close to that of the substrate. The III-V single-crystal substrates that are readily available in large-area form are the binaries GaAs, InP, GaP and GaSb. For high-quality epitaxial layers to be grown on a GaAs substrate, the lattice constant of the materials in the structure need to be located close to the vertical line intersecting GaAs (having similar lattice constant to GaAs) in Figure 6.3. These materials include $Al_xGa_{1-x}As$, $In_{\sim0.49}Ga_{\sim0.51}P$, $In_{\sim0.47}Al_{\sim0.53}P$ and $In_{\sim0.47-0.49}(Al_xGa_{1-x})_{\sim0.53-0.51}P$, and are often called GaAs-based materials. By the same token, $In_{\sim0.53}Ga_{\sim0.47}As$, $In_{\sim0.52}Al_{\sim0.48}As$, $In_{\sim0.52-0.53}(Al_xGa_{1-x})_{\sim0.48-0.47}As$ and $In_xGa_{1-x}As_yP_{1-y}$ having lattice parameters close to that of InP, are called InP-based materials. $In_xAl_yGa_{1-x-y}N$ is a GaN-based material system. In this case, single-crystal sapphire (Al_2O_3) and/or SiC substrates are often used to first grow a thick GaN buffer layer, since crystal growth of bulk ingots of GaN for use as a source of thick substrates cannot be readily accomplished simply in the same way as GaAs or InP bulk crystals are grown. Relatively thick GaN substrates ($\sim$300–500 µm thick) have become available recently but they are not prepared by the slicing of wafers from a bulk crystal grown in the form of boule from a seed crystal as the case for "bulk" GaAs and InP substrates. Instead, a thick GaN layer (several hundreds of µm to a few mm thick) is grown on a (0001)-oriented sapphire wafer or other single-crystal substrate using a very high deposition rate by HVPE (hydride vapor phase epitaxy) and then the sapphire is removed. Individual substrate wafers are sliced from this thick heteroepitaxial structure. These "bulk" GaN substrates are often called "free-standing" GaN substrates. The growth of GaN on foreign substrates (*e.g.*, basal-plane sapphire) is a typical example of heteroepitaxy and various growth techniques have been developed (Section 6.6.4).

6.1.1.4 Heterostructures

We discussed various selections of materials for certain device applications in the previous section. For device applications, in addition to a high degree of freedom of materials selection, a structure consisting of multiple layers with different properties is critically beneficial. The structure consisting of multiple layers of III-V compound semiconductors is called an *epitaxial structure*. When the layers in the epitaxial structure are different materials, the structure is called a *heterostructure* as opposed to a *homostructure* that has layers of same materials (possibly with different electrical type and concentration). In addition, when the two layers that form the epitaxial structure are different materials *and* have different electrical conductivity types or doping concentrations to form a junction (typically, a *p-n* junction, but not limited to this case), the junction is called a *heterojunction*. By the same token, if the structure has the same basic semiconductor materials combined to form a junction, the junction is called a *homojunction*. The capability of implementing lattice-matched heterostructures in III-V compound semiconductors is a unique and advantageous characteristic of III-V compound semiconductors as compared to Si. Semiconductor heterostructures have had a tremendous impact on both science (*e.g.*, solid-state physics) and technology (*e.g.* electronics and optoelectronics) fields. The importance of the development of heterostructures was recognized by the Nobel Prize in Physics in 2000, which was awarded to Academician Zhores I. Alferov and Herbert Kroemer for their contribution of "developing semiconductor 'heterostructures' used in high-speed- and opto-electronics". Virtually all devices based on III-V compound semiconductor materials employ heterostructures[5] to improve the performance characteristics. Heterostructures utilize the different properties of different materials, including the bandgap, refractive index and electrical conductivity in an epitaxial structure grown on a substrate. In semiconductor laser structures, *e.g.*, for double heterostructures (DH) that have a laser active region sandwiched with wide-bandgap cladding layer materials, the performance characteristics such as the threshold current density, quantum efficiency and operating temperature limits are significantly improved. The typical DH lasers employ *N-i-P* heterostructures. Note that the capital letter *N* and *P* are used, as opposed to *n* and *p*, to indicate wide-bandgap *n*-type and *p*-type materials – such nomenclature using capitalized letters is often (but not always) used to indicate the wide-bandgap materials in the structure – and *i* is used to indicate the narrow-bandgap unintentionally doped (intrinsic) material. Wide-bandgap material creates the potential barrier and refractive index step at the interface of the wide-bandgap material and the narrow-bandgap material. Therefore, a heterostructure can facilitate carrier confinement in the active region and help confine the transverse optical mode in the active region. In bipolar junction transistor (BJT) structures, using an emitter material with a relatively wide bandgap improves the one-way carrier injection in an emitter-base junction carrier, which results in improvement of current gain. Such structures are classified as heterojunction bipolar transistor (HBT) structures to distinguish them from BJT with homostructures.

6.1.1.5 Oxide Formation

Silicon-based devices utilize the stable native oxide for MOS (metal-oxide-semiconductor) structures. In this case, SiO_2, the native oxide of silicon, is formed mainly by thermal oxidation of an exposed Si surface and can be used for the gate oxide for field control in field-effect transistors. For III-V compound semiconductors, the formation of a stable native oxide for a gate oxide is difficult and no reliable success has been achieved. The native oxides are not stable enough for many material systems and the interface between the oxide and the semiconductor has interface states.[6] This has hampered the application of III-V compound semiconductors to the MOS-based device structures. Epitaxially grown gate dielectrics on GaAs and GaN surfaces have been developed[7] but the actual applications are still limited. Stable native oxide layers can be formed for aluminium-containing alloy compound semiconductors such as AlGaAs and the native oxide of AlGaAs, produced by surface or lateral thermal oxidation in the presence of water vapor, is used for the current-guiding layer in both in-plane and surface-emitting laser structures.[8]

In summary, the major advantageous characteristics of III-V compound semiconductors are (1) most of them are efficient light emitters due to their direct bandgap nature; (2) various selections of materials can be made depending on the demand of the applications and the design requirements; and (3) implementing lattice-matched high-quality heterostructures is possible. In contrast, stable high-quality oxide formation for the use in MOS device structures is generally difficult in compound semiconductor systems.

6.1.2 Materials Properties of III-V Compound Semiconductors

This section summarizes the materials properties of important III-V compound semiconductors. For binary compounds, detailed parameters related to structural, electronic, optical and thermal properties are listed in Tables 6.1 and 6.2. More detailed properties are given in the references.[9]

6.2 Applications of III-V Compound Semiconductors

Direct bandgap nature of most III-V compound semiconductors enables us to use the compound semiconductors in efficient light emitting and detecting devices, *i.e.*, *photonic* or *optoelectronic*

Table 6.1 Basic structural parameters of binary III-V compound semiconductors.

Materials	*Crystal structure (stable phase)*	*Lattice constant (Å)*	10^{-22}*(atoms cm^{-3})*	*Density (g cm^{-3})*
AlN	wurtzite	$a=3.112$ $c=4.982$	9.98	3.255
GaN	wurtzite	$a=3.1892$ $c=5.185$	8.90	6.15
InN	wurtzite	$a=3.548$ $c=5.57$	6.81	6.4
AlP	zinc blende	$a=5.467$	4.9	2.40
GaP	zinc blende	$a=5.4505$	4.94	4.138
InP	zinc blende	$a=5.8687$	3.96	4.81
AlAs	zinc blende	$a=5.660$	4.41	3.70
GaAs	zinc blende	$a=5.65325$	4.42	5.3161
InAs	zinc blende	$a=6.0583$	3.59	5.667
AlSb	zinc blende	$a=6.136$	3.46	4.26
GaSb	zinc blende	$a=6.09593$	3.53	5.6137
InSb	zinc blende	$a=6.4794$	2.94	5.7747

Table 6.2 Electronic, optical and thermal parameters of binary III-V compound semiconductors.

Materials	*Bandgap, E_g (300 K) (eV)*	*Refractive index near E_g*	*Thermal expansion coefficient, α ($\times 10^{-6}$ K^{-1})*	*Thermal conductivity (W m^{-1} K^{-1})*
AlN	6.2 (direct)	2.1–2.2	$\alpha_a=4.15$, $\alpha_c=5.27$	2.85
GaN	3.39 (direct)	2.29	$\alpha_a=3.27$, $\alpha_c=5.59$	1.3
InN	0.6–0.7 (direct)	2.93	$\alpha_a=4.8$, $\alpha_c=2.9$	0.45
AlP	2.45 (indirect)	3.03	4.5	0.9
GaP	2.268 (indirect)	3.45	4.65	1.1
InP	1.34 (direct)	3.41	4.6	0.68
AlAs	2.14 (indirect)	3.18	4.9	0.8
GaAs	1.424 (direct)	3.65	5.73	0.55
InAs	0.356 (direct)	3.52	4.52	0.273
AlSb	1.63 (indirect)	3.4	4.0	0.57
GaSb	0.70 (direct)	3.82	7.75	0.32
InSb	0.18 (direct)	4.00	5.37	0.18

devices. Such devices include light-emitting diodes (LEDs) and laser diodes (LDs), which generate photons from electrical inputs, and photodetectors and solar cells, which generate electrical signals (detectors) or power (solar cells) from photon energy inputs. Optoelectronic devices are critical in optical data storage systems, display or illumination systems, and optical communication systems. For electronic devices, such as heterojunction bipolar transistors (HBTs) and heterostructure field-effect transistors (HFETs), III-V compound semiconductor materials are dominantly used for high-speed and high-power microwave and RF (radio-frequency) applications. Devices and systems based on compound semiconductors are important in current every-day life as well as in next-generation technologies. This section describes briefly the basic operating principles and important figures-of-merits of devices based on III-V compound semiconductors. The review papers or textbooks cited in each section offer further reading.

6.2.1 Photonic Device Applications

There are many currently used applications for III-V optoelectronic devices. For example, optical pickups in optical data storage systems[10] employ semiconductor lasers and detectors at 780 nm (AlGaAs materials) and 650 nm (InAlGaP materials) for CD (compact disc) and DVD (digital

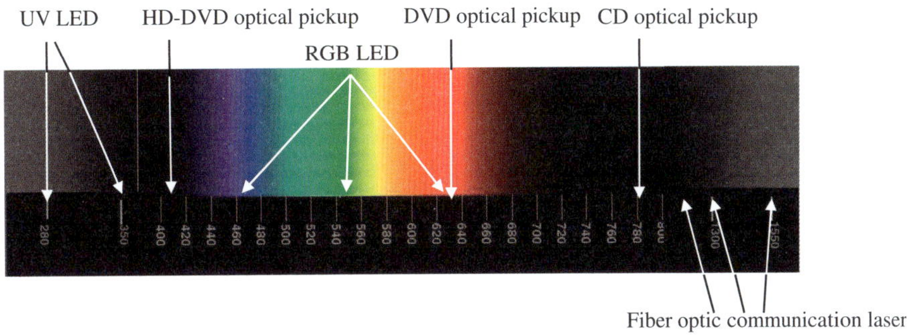

Figure 6.4 Examples of photonic and optoelectronic devices based on III-V compound semiconductors as a function of electromagnetic spectral wavelengths.

versatile disc), respectively. Next-generation optical data storage systems, the HD-DVD™ (high-density DVD) or Blu-ray Disc™ systems employ lasers emitting at 405 nm (InAlGaN materials) to increase the capacity of the disc to as high as 50 GB in the case of a dual-layer disc. Visible LEDs have been expanding their applications in color displays and illumination systems. High-brightness LEDs have been employed in traffic signal lights, automotive illuminating systems, various back-light applications including LCD and key pad back lights for mobile applications, *etc.* Silica-based optical fibers[11] have minimum attenuation and dispersion at 1550 and 1310 nm, respectively, and hence long-haul optical communication systems employ light emitters and detectors at those wavelengths based upon InGaAsP or InAlGaAs materials. Figure 6.4 shows several examples of photonic and optoelectronic devices based on III-V compound semiconductors and their operating wavelengths.

6.2.1.1 Light-emitting Diodes

Light-emitting diodes (LEDs)[12] utilize the processes of (1) carrier injection followed by (2) carrier radiative recombination for photon generation. The combined process is called *electroluminescence*. The generated photons need to be extracted from the devices for useful photon emission and this final process is (3) extraction process. Most epitaxial LEDs employ *P-i-N* DH structures. Wide-bandgap *p*-type layers (*P*) and *n*-type layers (*N*) supply the carriers (holes from *P* and electrons from *N*) under forward bias conditions into the unintentionally doped narrow-bandgap layer (*i*) where the radiative recombination occurs. The *i*, *P* and *N* layers are called the active layer, hole injection layer and electron injection layer, respectively. Wide-bandgap layers in the LED structures can act as carrier confinement layers for carriers to be confined in the active layer before radiative recombination and as window layers for generated photons not to be re-absorbed before the photon extraction from the device. The efficiency of an LED, η_{LED}, depends on the carrier injection efficiency (η_{in}), radiative recombination efficiency (η_{rad}) and extraction efficiency (η_{ex}) and is expressed by:

$$\eta_{LED} = \eta_{in}\eta_{rad}\eta_{ex} = \eta_{int}\eta_{ex} = \eta_{ext} \tag{6.5}$$

Here the radiative recombination efficiency is the ratio of radiative recombination rate (R_r) to the sum of radiative and non-radiative (R_{nr}) recombination rates:

$$\eta_{rad} = \frac{R_r}{R_r + R_{nr}} \tag{6.6}$$

The injection efficiency depends on the epitaxial structures and material qualities and the efficiency is generally high compared to other efficiencies. The extraction efficiency depends on epitaxial and device structures and packaging design. The product of injection efficiency and the radiative recombination efficiency is the LED internal quantum efficiency, η_{int}, and the products of all three efficiencies (or the product of the LED internal quantum efficiency and the extraction efficiency) is also called the LED external quantum efficiency, η_{ext}. The optical output power of an LED, P_{LED}, is expressed by:

$$P_{LED} = \eta_{in}\eta_{rad}\eta_{ex}\frac{h\nu}{q}I = \eta_{int}\eta_{ex}\frac{h\nu}{q}I = \eta_{ext}\frac{h\nu}{q}I \tag{6.7}$$

where I is a current (A), hv is the energy per photon and q is an elementary charge.

High-brightness visible LEDs consist of direct-bandgap semiconductor active layers. Visible LEDs based on AlGaAs and InAlGaP materials[13] cover the visible spectral region corresponding to red and yellow colors and the emission is limited to visible wavelengths longer than yellowish green due to the unavailability of wider direct-bandgap semiconductor active regions in these systems. High-brightness LEDs operating at shorter visible wavelengths, *i.e.*, in the violet, blue and green, were made possible by the emergence of the GaN-based InAlGaN materials having wide direct bandgaps.[14] The efficiencies of LEDs change with the operating wavelength and those of InAlGaP and InAlGaN based LEDs behave differently with the wavelength. The internal quantum efficiency decreases with decreasing wavelengths from red for InAlGaP and with increasing wavelength from violet for InAlGaN, both towards the green spectral region.

6.2.1.2 Laser Diodes

Semiconductor laser diodes (LDs),[15] similar to LEDs, rely on carrier injection followed by luminescence in the active region; however, the LD structures require one more component to create an *optical cavity* where photons have multiple reflections for optical gain. When the photons generated by carrier injection and radiative recombination travel along the cavity, photons can be multiplied by stimulated emission. When the increased optical gain overcomes the optical losses from the cavity itself and the mirrors, the lasing process begins and this condition is called laser *threshold* condition. When the current increases beyond the threshold condition, *i.e.*, beyond the *threshold current*, the light output begins to increase abruptly and lasing operation is achieved. The optical output power of LDs, P_{LD}, is expressed as follows:

$$P_{LD} = \eta_{d}\frac{h\nu}{q}(I - I_{th}); (I > I_{th}) \tag{6.8}$$

where I is a current (A), I_{th} is a threshold current (A), hv is the energy per photon, q is an elementary charge; η_d is a differential quantum efficiency and is related to internal quantum efficiency, optical confinement factor and cavity and mirror losses.

Depending on the direction of the optical cavity and the type of mirrors, LDs are classified as Fabry–Pérot (F-P) lasers, distributed feedback (DFB) lasers and vertical-cavity surface-emitting lasers (VCSELs). F-P lasers and DFB lasers are in-plane lasers that have the optical cavity formed along the epitaxial plane (normal to growth direction), while VCSELs[16] have the optical cavity parallel to growth direction (normal to the epitaxial plane). Figure 6.5 shows schematic structures of in-plane lasers and VCSELs and Table 6.3 gives a comparison of them. Fabry–Pérot lasers have mirrored surfaces typically formed by cleaving the crystal along a selected crystal plane. DFB lasers contain periodic optical gratings structures fabricated along the in-plane cavity. VCSELs have stacks of distributed Bragg reflector (DBR) layers designed to form a vertical optical cavity.

Most semiconductor LDs are used in optical storage and optical communication systems. For optical communication systems, 1310 and 1550 nm DFB lasers and 850 nm VCSELs are used for

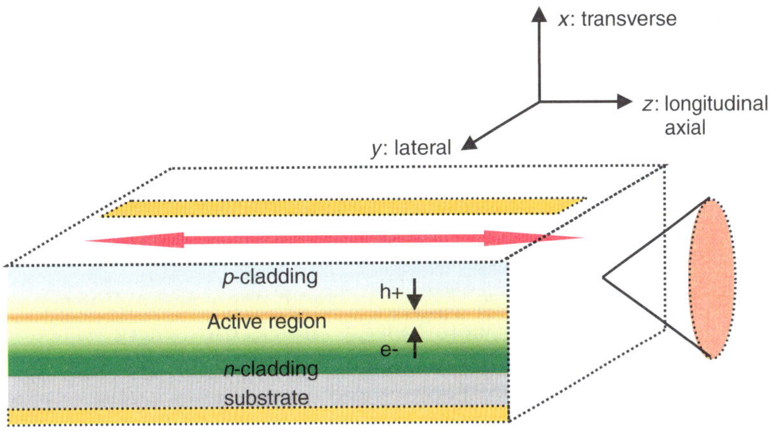

(a)

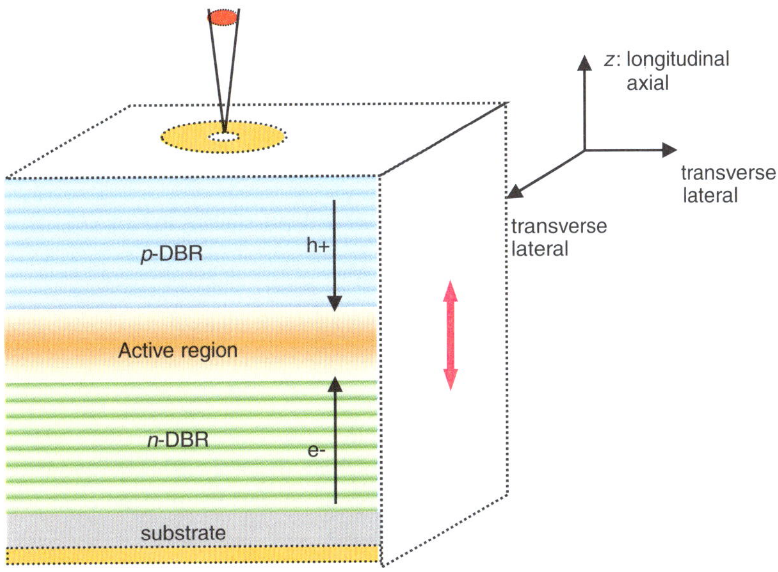

(b)

Figure 6.5 Schematic structures of (*a*) in-plane lasers (P-F lasers) and (*b*) vertical cavity surface emitting lasers (VCLELs).

Table 6.3 Comparison of in-plane lasers and VCSELs.

	VCSELs	*In-plane lasers*
Cavity (direction & length)	Along growth direction (vertical); short ~1 μm	Along the plane. Relatively long (hundreds of μm)
Mirror	DBR stack (reflectance > 99%)	Cleaved edge (reflectance ~30%) (not for DFB)
Carrier injection	Parallel to cavity Over DBR interface	Normal to cavity No interface barrier
Bean output	Top/bottom surface emitting Symmetrical circular	Edge emitting Elliptical

long-haul and short-haul systems, respectively. For optical storage, shorter-wavelength emitters are preferred to increase the storage density (the shorter the wavelength, the smaller the focused spot); hence, F-P lasers for optical storage have been developed toward shorter wavelength emitters, *e.g.*, the technology has moved from 780 nm (infrared) to 650 nm (red) and recently to 405 nm (ultraviolet).

6.2.1.3 Photodetectors

In photodetector devices,[17] energy conversion occurs in an opposite way to that exploited in light emitters, *i.e.*, energy is converted from photon energy into electrical energy. Therefore, the process of photodetection is essentially the reverse process of that used in light emitters. Photodetectors utilize the processes of (1) absorption of photons and generation of carriers, followed by (2) collection of the photogenerated carriers, resulting in the generation of a photocurrent. Major figures-of-merits for detectors are quantum efficiency and responsivity. The quantum efficiency of a detector, η_{det}, is defined by the number of carriers by photocurrent, I_{ph}, divided by the number of photons:

$$\eta_{det} = \frac{I_{ph}/q}{P_{in}/h\nu} \tag{6.9}$$

The responsivity, R (A W^{-1}), is expressed by the ratio of photocurrent, I_{ph}, to incident optical power, P_{in}:

$$R = \frac{I_{ph}}{P_{in}} = \frac{\eta q}{h\nu} \tag{6.10}$$

To efficiently collect the photocurrent, photodiodes are operated under reverse bias while light emitters are operated under forward bias to inject carriers. Figure 6.6 shows typical current–voltage (I-V) characteristics of diodes and the operating regime of various photonic devices, showing that light emitters and photodiodes are operated in the first and third quadrants, respectively.

There are several types of photodetectors and the primary ones are photoconductors, *p-i-n* (PIN, here capital letter designation is not related to wide bandgap materials) diodes, avalanche photodiodes (APDs) and metal-semiconductors-metal (MSM) photodiodes.

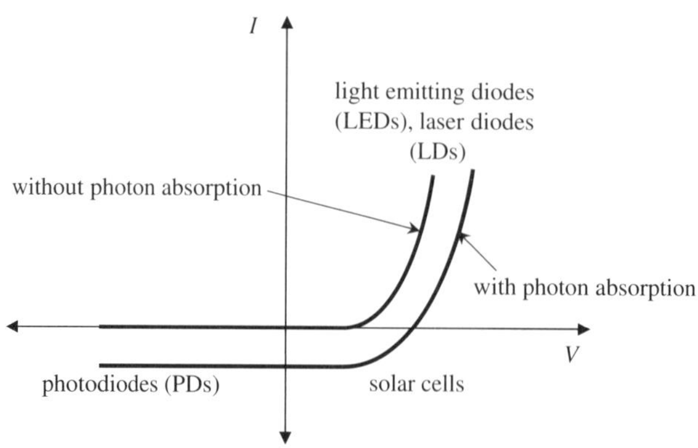

Figure 6.6 Current–voltage (*I-V*) characteristics of diodes and operating range of photonic devices.

6.2.1.4 Solar Cells

Solar cells[18] convert photon energy into electrical energy to generate DC power. The energy conversion is similar to that used in photodetectors, but the devices are not reverse-biased, unlike photodetectors. Instead, a load resistor is employed in a solar-cell circuit to deliver the power. Photons absorbed in the solar-cell structure create electron–hole pairs that diffuse toward the *n*-type and *p*-type contacts, which then flow in the external circuit, delivering power. Solar cells are operated in the fourth quadrant of Figure 6.6. Maximum power output, P_{m}, and power conversion efficiency, η_{power}, of a solar cell are given by:

$$P_{\mathrm{m}} = I_{\mathrm{m}} V_{\mathrm{m}} \tag{6.11}$$

$$\eta_{\mathrm{power}} = \frac{P_{\mathrm{m}}}{P_{\mathrm{in}}} \tag{6.12}$$

To improve the power output and power conversion efficiency, I_{SC} (short circuit current) and V_{OC} (open circuit voltage) must be increased and this can be achieved by decreasing the series resistance of a solar cell by using heavy doping on the *p*- and *n*-sides. In addition, the spectral response of the photon absorption coefficient of solar cell materials, minority carrier lifetime, surface recombination rate, absorption and reflection of photons on the surface, metal contact schemes are important considerations in solar cell design.

6.2.2 Electronic Device Applications

Unlike photonic devices, electronic devices based on III-V compound semiconductors[19] do not have to rely on the direct-bandgap nature of these materials. Instead, III-V electronic devices rely on the advantages derived from the availability of high-quality heterostructures for their operation in comparison to Si-based transistors. The use of heterostructures in electronic devices can make significant improvements in majority carrier transport and minority carrier injection efficiency. Devices based on III-V compound semiconductor heterostructures offer advantages in microwave, millimeter-wave and high-speed integrated circuit applications compared to similar Si-based devices.

6.2.2.1 Heterojunction Bipolar Transistors

Bipolar junction transistors (BJTs) that are mostly Si-based consist of homojunction, *npn* or *pnp* emitter-base-collector structures. Heterojunction bipolar transistors (HBTs)[20] employ a wide-bandgap emitter layer (both *Npn* or *Pnp* structures have been demonstrated) to increase emitter injection efficiency and hence to provide a higher current gain. In *Npn* HBT structures, the band offset at the emitter-base (E-B) interface can create a potential difference that facilitates the injection of electrons from the emitter to the base while retarding hole injection from the base to the emitter.[21] Also, the use of heterojunction E-B structures, which results in one-way preferential carrier injection from E→B, permits a lower emitter doping and higher base doping while not sacrificing the current gain. (In normal BJTs, the emitter doping should be higher than base doping to achieve a high current gain.) The heavy base doping results in a low base resistance and low junction capacitance (due to the light doping in the emitter), which are beneficial in achieving high power gain and high-frequency performance.

6.2.2.2 Heterostructure Field-effect Transistors

Silicon-based field-effect MOS transistors have a thin channel to confine the carriers that are controlled by the field, and, hence, the transport of free carriers in the channel is strongly affected by carrier scattering at the rough oxide–semiconductor interface. In contrast, in heterostructure field-effect transistors (HFETs),[22] free carriers (electrons) in the channel are confined by a smooth heterojunction interface and are also physically separated from the ionized donors (the channel is typically formed in an undoped layer), resulting in a significant increase of electron mobility and the saturation value of the electron drift velocity. Heterostructures consisting of *n*-type (modulated) doped wide-bandgap and undoped narrow-bandgap materials can also create the band bending and conduction-band offsets that result in formation of a two-dimensional electron gas (2-DEG),[23] *i.e.*, the channel is essentially a plane of electrons in narrow-bandgap undoped layer. Depending on the focus of the description of the device characteristics, physical mechanism and structural configuration of the HFETs, they are also called high-electron mobility transistors (HEMTs), modulation-doped FETs (MODFETs) and two-dimensional electron gas FETs (TEGFETs).

6.3 Fundamentals of CVD Processes for III-V Compound Semiconductors

Metal-organic chemical vapor deposition (MOCVD) technology has been developed to the point where it has become a major production and research tool in the field of compound semiconductor materials and devices through successful competition against other growth technologies. This section describes historical aspects of the development of MOCVD, which has become a major process technology for III-V epitaxial films, and the chemical reaction mechanisms and corresponding thermodynamics, hydrodynamics and kinetics of MOCVD growth to provide a fundamental understanding of the MOCVD process for III-V compound semiconductors.

6.3.1 Historical Overview of MOCVD Technology

Epitaxial growth by MOCVD was first reported in the scientific literature in 1968 by H. M. Manasevit, a chemist working at the North American Rockwell Corporation.[24,25] Similar processes and experimental results had been described prior to 1967 in the patent literature by other groups such as R. A. Ruehrwein with Monsanto, USA.[26] There were also several earlier reports on the preparation of compound semiconductors by metal-organic–hydride reaction process in a closed system, *i.e.*, mixing the metal-organic and hydride precursors at low temperature in a closed system and then applying heat to initiate the reaction.[27] Nevertheless, the significance of the early work of MOCVD growth in an "open-tube" reactor chamber by Manasevit *et al.*[28,29] cannot be underestimated and he is widely acknowledged as the inventor of the MOCVD technology that is being exploited today.[30] Interestingly, Manasevit's first paper on MOCVD, concerning the epitaxial growth of GaAs on insulating oxide crystalline substrates, *e.g.* single-crystal sapphire, did not describe the actual epitaxial process, which was simply referred to as "chemical vapor deposition."[24,31] In fact, in the late 1960s and early 1970s Manasevit and co-workers published papers describing the MOCVD growth of most of the III-V semiconductors – including GaN and AlN[32] – and the II-VI and IV-IV compounds, demonstrating the flexibility of this process.

From the early 1960s to late 1970s, various III-V materials technologies had been developed and had come to dominate the research and production of devices based on compound semiconductors, including (1) halide vapor-phase epitaxy (VPE) using Group V halides (*e.g.*, $AsCl_3$) and Group III metal elements (*e.g.*, Ga); (2) hydride VPE using Group V hydride (*e.g.*, AsH_3) and Group III trichlorides, *e.g.*, $GaCl_3$; (3) liquid-phase epitaxy (LPE) using Group III metal solutions (*e.g.* Ga melts with GaAs source material); and (4) molecular-beam epitaxy (MBE) using pure elemental sources (*e.g.*, Ga and As). In the development of the MOCVD epitaxial growth technology, Manasevit and co-workers explored the growth of various III-V, II-VI and IV-IV compounds over several years after 1968. However, much of Manasevit's work was "proof-of-concept" growth studies and he and his co-workers were not able to demonstrate "device-quality" materials that were comparable to those produced by other well-developed (at that time) III-V epitaxial growth technologies such as VPE and LPE. Also, MBE was being actively researched by a few groups as a next-generation epitaxial semiconductor growth technology, particularly at Bell Laboratories and IBM Research Laboratory. Consequently, there was not much interest in MOCVD – it was viewed as just "another" III-V materials technology – until R. D. Dupuis' work at Rockwell International in the late 1970s. In 1977, Dupuis *et al.* reported high-performance AlGaAs/GaAs solar cells and injection lasers grown by MOCVD,[33] demonstrating that MOCVD technology could perform at levels equal to the other III-V materials technologies. In 1978, Dupuis, collaborating with Holonyak *et al.*, at the University of Illinois, reported the first quantum-well semiconductor injection lasers operating continuous wave (CW) at room temperature,[34] clearly showing that the performance of MOCVD-grown materials and devices could, in fact, exceed that of alternative materials technologies. These results caused many groups to reconsider the exploration of MOCVD materials technology, beginning a rush to develop MOCVD as a major process technology for III-V epitaxial films.

6.3.2 Chemical Reactions in MOCVD Growth

MOCVD has also been called metal-alkyl vapor phase epitaxy (MAVPE), metal-organic VPE (MOVPE), organometallic CVD (OMCVD) and organometallic VPE (OMVPE). Generally, MO and OM, and CVD and VPE are mingled and interchanged in the terminology of this process. Manasevit first used the term "metal-organic"[35] (emphasizing the metal component) because that was the common term that chemists applied to the "metal alkyl" compounds at this time and "CVD" because he felt that the process could be broadly applied to "chemical vapor deposition" of many different materials, including polycrystalline and amorphous films on heterogeneous substrates. Therefore, the term "vapor-phase epitaxy" is a special case of the more general term "chemical vapor deposition." This section will use the more "generic" and historical name "metal-organic chemical vapor deposition" for this process. As its name implies, MOCVD is the method for the *deposition* of condensed phase (solid phase in most cases) from *vapor* phase *via* a *chemical* reaction employing *metal-organic* (MO) precursors. Therefore, the MOCVD process requires metal-organic *precursors* in chemical reaction for deposition. A precursor is a chemical compound that is used as an "input" to a chemical process to produce a desired product. Precursors are often referred to as "*sources*." Both Group III and Group V precursors can be metal-organic precursors; however, the growth of III-V compound semiconductors generally employs mixtures of Group III metal-organic and Group V hydride precursors in a carrier gas. Typical carrier gases are H_2, N_2, or mixture of H_2 and N_2, and the precursors are transported by one or more carrier gases and injected into an open-tube process chamber. The purity of precursors and the carrier gas(es) is critical in achieving high-quality materials in MOCVD growth.

The detailed chemical reactions occurring in MOCVD are very complicated; however, the process can be represented by a very simple formula for a reaction of III-V employing metal-organic precursors and hydride precursors, as in Equation (6.13):

$$R_3M(g) + EH_3(g) \rightarrow ME(s) + 3RH \uparrow (g) \tag{6.13}$$

where M is a Group III metal atom, *e.g.*, Ga, Al or In; R is an organic radical, typically CH_3 (methyl) or C_2H_5 (ethyl); E is a Group V atom *e.g.*, As, P or N; and H is atomic hydrogen. While this simplified net reaction ignores the intermediate reactions or additional by-product formation that may occur in the gas phase and on the surface, it provides a basic framework that can be used to describe the more complicated cases where more than one metal-organic and/or hydride precursors are involved, *e.g.*, for the growth of ternary and quaternary compound semiconductors. For example, the net reactions employed for the MOCVD growth of GaAs, $Al_xGa_{1-x}As$ and $In_xGa_{1-x}As_yP_{1-y}$ are given in Equations (6.14)–(6.16):

$$(CH_3)_3Ga(g) + AsH_3(g) \rightarrow GaAs(s) + 3CH_4 \uparrow (g) \tag{6.14}$$

$$\begin{aligned} x'(CH_3)_3Al(g) + (1 - x')(CH_3)_3Ga(g) + AsH_3(g) \\ \rightarrow Al_xGa_{1-x}As(s) + 3CH_4 \uparrow (g) \end{aligned} \tag{6.15}$$

$$\begin{aligned} x'(CH_3)_3In(g) + (1 - x')(CH_3)_3Ga(g) + y'AsH_3(g) + (1 - y')PH_3(g) \\ \rightarrow In_xGa_{1-x}As_yP_{1-y}(s) + 3CH_4 \uparrow (g) \end{aligned} \tag{6.16}$$

As indicated as x' and y', which differ from x and y, the ratio of precursors in the vapor phase does not have to be same as compositions in solids. The reaction for GaAs involves the input precursors, TMGa (trimethylgallium) and arsine (AsH_3), to form a solid-phase GaAs and the gas-phase by-product methane. This chemical reaction requires the decomposition of the input compounds into atomic elements and this decomposition occurs generally by thermal energy in the MOCVD. Decomposition of a compound using thermal energy is called "pyrolysis" or "pyrolytic decomposition." Pyrolytic decompositions and formation of by-products might involve many intermediate reactions. The chemical reactions for the MOCVD growth can be grouped as homogeneous gas-phase reactions and heterogeneous surface reactions. The homogeneous and heterogeneous reactions stem from homogeneous and heterogeneous nucleations of a phase. Nucleations for phase transitions or chemical reactions that occur between the same phases without the help of a heterogeneous interface are called "homogeneous." In chemical reactions for MOCVD growth, the pyrolysis of compounds in the vapor phase or the reaction between precursors (or decomposed products) in the vapor phase are considered homogeneous reactions. When the reactions limit the efficiency of the growth or form undesirable by-products that occur mainly in homogeneous reactions, the reactions are called "parasitic" reactions. In contrast, nucleations for phase transitions or chemical reactions that occur at the inter face between different phases are called "heterogeneous." The pyrolysis or the chemical reaction on solid semiconductor surface is considered a heterogeneous reaction. Besides the precursors used in the MOCVD growth, H_2 is also introduced as a carrier gas. More complete reaction models, including both precursors and carrier gas in the gas phase (homogeneous) and on the surface (heterogeneous), have been developed that include more than 40 individual intermediate reactions and by-products. Jensen *et al.* and Stringfellow *et al.* have reviewed the detailed reactions in the case of the growth of GaAs using TMGa and AsH_3 with H_2 carrier gas by MOCVD.[36] Such detailed dependence of reaction mechanisms on growth chemistry can help in the understanding of (1) the details of how the growth rate being affected by each step, such as decomposition, recombination, adsorption, surface reaction and desorption, and (2) how impurity incorporation being dependent on the growth parameters. Prior to describing the reaction chemistry, however, it needs to be noted that the actual chemical reactions and their kinetics are very complicated and that these reactions are not fully understood yet and can be significantly affected by many reactor and growth conditions, such as the growth reactor chamber geometry, temperature of susceptor (on which the substrates sit) and reactor chamber wall, pressure, residence time of precursors, *etc.*

The pyrolysis of TMGa can occur homogeneously (indicated as G for gas-phase reaction) to DMGa (dimethylgallium) and MMGa (monomethylgallium) as follows:

$$(G1)(CH_3)_3Ga \rightarrow (CH_3)_2Ga + CH_3 \tag{6.17}$$

$$(G2)(CH_3)_2Ga \rightarrow (CH_3)Ga + CH_3 \tag{6.18}$$

The pyrolysis of TMGa and DMGa can also occur heterogeneously (indicated as S for surface reaction) to MMGa (monomethylgallium) as follows:

$$(S1)(CH_3)_3Ga + S_G \rightarrow (CH_3)Ga^* + (CH_3)_2 \tag{6.19}$$

$$(S2)(CH_3)_2Ga + S_G \rightarrow (CH_3)Ga^* + CH_3 \tag{6.20}$$

S_G is the site for the adsorption of gallium species on the surface and * denotes the adsorbed species. The methyl radicals from pyrolysis can participate in subsequent reaction with arsine, hydrogen and other decomposed by-products in the gas phase to produce CH_4:

$$(G3)\, CH_3 + AsH_3 \rightarrow AsH_2 + CH_4 \tag{6.21}$$

$$(G4)\, CH_3 + H_2 \rightarrow H + CH_4 \tag{6.22}$$

$$(G5)\, CH_3 + (CH_3)Ga \rightarrow (CH_2)Ga + CH_4 \tag{6.23}$$

$$(G6)\, CH_3 + (CH_3)_3Ga \rightarrow (CH_2)(CH_3)_3Ga + CH_4 \tag{6.24}$$

MMGa is the most stable decomposition product of TMGa and the release of the remaining methyl radical is likely to occur on the surface *via* reaction with arsenic-containing species. Arsine is considered to be decomposed mostly heterogeneously on the surface *via* adsorption and subsequent release of H as follows:[37]

$$(S3)AsH_3 + S_A \rightarrow AsH_2^* + H_2 \tag{6.25}$$

$$(S4)AsH_2 + S_A \rightarrow AsH^* + H \tag{6.26}$$

$$(S5)AsH + S_A \rightarrow As^* + H \tag{6.27}$$

where S_A is the surface adsorption site for arsenic species. After the pyrolysis of precursors, GaAs growth can occur between gallium- and arsenic-containing species adsorbed on the surface as follows:

$$(S6)(CH_3)Ga^* + AsH^* \rightarrow GaAs + CH_4 + S_A + S_G \tag{6.28}$$

$$(S7)(CH_3)Ga^* + As^* \rightarrow GaAs + CH_3 + S_A + S_G \tag{6.29}$$

When hydrogen is used as a carrier gas, hydrogen participates in the chemical reaction, such as promoting the pyrolysis of TMGa by the attack of atomic H. Also, atomic hydrogen can adsorb on both types of surface (S_A and S_G) and desorb as H_2. Atomic hydrogen sometimes plays a critical role in the doping of III-V semiconductor materials, especially in *p*-type doping by magnesium (as discussed in Section 6.6.4). Vapor-phase adduct formation between TMGa (a Lewis acid) and AsH_3 (a Lewis base) can also occur:

$$(G7)(CH_3)_3Ga + AsH_3 \rightarrow (CH_3)_3Ga : AsH_3 \tag{6.30}$$

where ":" indicates the adduct bond. The stability of the adduct depends significantly on the temperature. With GaAs MOCVD growth, the adduct is unstable at typical growth temperatures; hence, it is not expected to play a role in the growth mechanism. However, for some MOCVD growth processes, such as the growth of AlN using TMAl and NH_3, the adduct is reported to be stable over a broad range of temperatures.

We have briefly reviewed the chemical reactions during the MOCVD growth of III-V semiconductors and considered the specific example of GaAs growth. In general, MOCVD growth of III-V semiconductor films can be complicated by homogeneous reactions in the vapor phase, precursor-dependent activation energies and pyrolysis efficiencies, and surface reactions that depend on substrate orientation.

6.3.3 Thermodynamics, Kinetics and Hydrodynamics of the MOCVD Process

Thermodynamic analysis can predict if the proposed chemical reaction will occur (energetically favorable) or not (energetically unfavorable). For chemical vapor deposition, thermodynamics can describe if the deposition of a solid will occur from the vapor phase under a given set of conditions in terms of the reaction "driving force." Consider the following simple chemical vapor deposition process (note that it does not necessarily reflect the actual chemical reaction in MOCVD, however):

$$\mathrm{Ga}(v) + \mathrm{As}(v) \leftrightarrows \mathrm{GaAs(s)} \tag{6.31}$$

The vapor/solid equilibrium is set up with the Gibbs free energy:[38]

$$G^{\mathrm{v}}_{\mathrm{Ga}} + G^{\mathrm{v}}_{\mathrm{As}} = G^{\mathrm{s}}_{\mathrm{GaAs}} \tag{6.32}$$

If a non-equilibrium situation is intentionally created for this system, for instance, $G^{\mathrm{v}}_{\mathrm{Ga}} + G^{\mathrm{v}}_{\mathrm{As}} > G^{\mathrm{s}}_{\mathrm{GaAs}}$, the system will try to restore the equilibrium condition. Intentional non-equilibrium is achieved by supplying reactants to the vapor phase to create a super-saturation condition and the deposition of a solid thin film will occur in the process of restoring equilibrium. Therefore, the difference of Gibbs free energy $\Delta G = G^{\mathrm{s}}_{\mathrm{GaAs}} - (G^{\mathrm{v}}_{\mathrm{Ga}} + G^{\mathrm{v}}_{\mathrm{As}})$ is the driving force for the deposition. The more negative ΔG is, the larger the driving force is. Thermodynamics governs the deposition reaction, more precisely it controls the direction of the reaction, and hence the "maximum" growth rate of the solid phase; however, it cannot tell the whole story of the deposition process. *Kinetics* also needs to be considered to understand the growth process and to describe the rate of reaction toward the equilibrium.[39] For Equation (6.32), the energy of $G^{\mathrm{v}}_{\mathrm{Ga}} + G^{\mathrm{v}}_{\mathrm{As}}$ is higher than $G^{\mathrm{s}}_{\mathrm{GaAs}}$, and a driving force for deposition exists; however, for the reaction to occur, the energy barrier of the overall reaction (which is related to intermediate state) needs to be overcome, which is called the activation free energy. The activation free energy is related to the rate constant, k:

$$k = \exp(-\Delta G^{*}/RT) \tag{6.33}$$

where ΔG^* is the activation free energy. Such thermodynamic and kinetic considerations can be applied to the detailed growth chemical reactions described in the previous section, including the pyrolytic decomposition reactions, to understand the growth process.[40] Since most MOCVD growth processes use a total growth chamber pressure of $P_{\mathrm{g}} > 10\,\mathrm{kPa}$, which is in a viscous-flow regime (as opposed to a molecular-flow regime, *e.g.*, $P_{\mathrm{g}} < 10\,\mathrm{Pa}$, where interactions between gases can be ignored), *hydrodynamics* is another important consideration in studying the MOCVD growth process. Hydrodynamics is extremely critical in designing the reactor chamber for an MOCVD epitaxial growth system.

With given conditions for the deposition, such as temperature, pressure, flow rates of precursors and carrier gases, reactor chamber geometry, *etc.*, the details of growth process are determined by thermodynamics, kinetics and hydrodynamics. In this section we discuss the effect of thermodynamics and kinetics first, and then consider the combined effects with hydrodynamics on the growth process and efficiency. For a typical MOCVD process in an open-flow system, the following steps occur:[41] (1) mass "transport" of input precursors by the flowing gas stream; (2) mass "transfer" (also called "diffusion" or "mass transport type II") of precursors and/or pyrolyzed precursors through the so-called "stagnant region" near the substrate surface (also called a boundary layer); and (3) surface processes including adsorption of precursors and/or pyrolyzed precursors, surface chemical reactions and desorption of by-products on the surface. The "boundary layer" is the region in the gas phase near the surface of the substrate where the gas velocity decreases from the more or less constant "bulk value" in the growth chamber to essentially zero at the substrate. Among various steps that occur in series for a certain overall process, the slowest reaction step (or intermediate step) determines or limits the kinetics of the whole process and this step is referred to as a rate-limiting or rate-determining step. In an open system as in a typical MOCVD growth chamber, mass transport is rarely a rate-limiting step, since it is generally fast enough and generally ignored in growth process analysis. Notably, a mass "transfer" step is often replaced with a mass "transport" step, and diffusion-limited and mass-transport-limited are used to refer to the mass transfer step in some literature; however, this does not refer to mass transport "by a flowing gas stream."

Kinetically, MOCVD processes can be modeled with rate-limiting steps and growth kinetics can be modeled as a function of temperature, flow dynamics, pressure, substrate orientation, *etc.* Figure 6.7 shows the rate-limiting step as a function of the substrate temperature during the growth. At low temperatures, the surface reaction is slow, since the rate of the surface reaction is sensitive to the temperature; therefore, it is surface-reaction-kinetically limited. In this case, the growth rate increases with increasing temperature and is affected by substrate orientation.[42] When the temperatures are high enough, the growth is limited by the mass transfer of precursors through a boundary layer. In these cases, the growth rate is relatively insensitive to the growth temperature and the substrate orientation. When the surface process rate is high, the vapor pressure on the surface becomes essentially the equilibrium value and the growth rate is then controlled by the partial pressure of precursors (generally of the Group III source) in the input gas stream near the boundary layer. When the temperature increases further, the growth rate decreases with increasing temperature. This can be due to the formation of "parasitic" homogeneous reaction(s),[43] enhanced deposition on the reactor growth chamber wall upstream of the substrate, and the desorption of

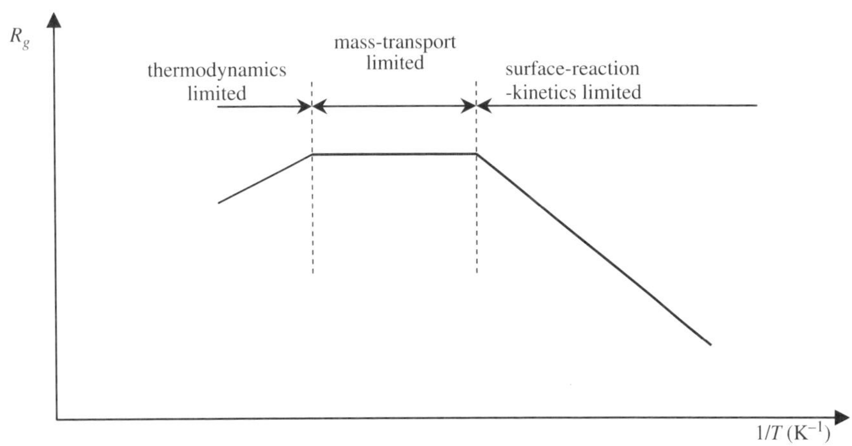

Figure 6.7 Rate-limiting step as a function of substrate temperature during the growth.

atoms from the solid surface. Those reactions become energetically favorable at high temperatures; for example, the driving force for Equation (6.32) changes in such a way that the reverse reaction becomes more significant. Therefore, in most cases, MOCVD growth at high temperatures is a thermodynamically limited process.

When the pressure of the growth changes, hydrodynamic effects need to be considered in the growth process. For very low total pressures ($P_g < 1$ kPa), the growth is entirely kinetically controlled, even at relatively high temperatures, resulting in a zero slope in the R_g *vs.* P curve. For pressures above 1 kPa, the growth rate is primarily controlled by mass-transfer, resulting in a $-\frac{1}{2}$ slope in a log R_g *vs.* log P_g curve. Growth in the pressure regime $P_{tot} < 1$ kPa is usually referred to as ultralow pressure MOCVD. At even lower pressures ($P_{tot} < 10$ Pa), the process is called ultrahigh vacuum (UHV) MOCVD and is sometimes referred to as "metal-organic molecular-beam epitaxy" (MOMBE) or "chemical-beam epitaxy" (CBE). Such low pressures are required so that molecules can traverse the space between the source "injector" or point of origin and the substrate surface without interacting with any other molecules.

6.3.4 Growth Mechanisms of the MOCVD Process in a Reactor Chamber

We have discussed the chemical reactions involved in the MOCVD growth, including intermediate reactions and the thermodynamic, kinetic and hydrodynamic aspects of the growth process. The reactions for the growth of epitaxial films typically occur in an open-tube system in a special vessel called a reactor growth chamber; the growth processes and mechanisms that occur in the chamber are discussed this section. The growth chamber can be divided into four "zones" to describe the growth processes: (1) precursor injection, (2) mixing, (3) boundary layer and (4) substrate surface. In the precursor injection zone, Group III and Group V precursors with the carrier gas are injected into the growth chamber. In this zone, the precursor gas stream experiences an abrupt change from the small cross-sectional area tubing in the gas manifold to the large cross-sectional area and volume growth chamber. Therefore, the uniform injection of precursors without creating turbulent flow is a critical goal in reactor chamber design. Generally, separately injected Group III and Group V precursors are mixed in the mixing zone where homogeneous gas-phase reactions, including the pyrolysis of precursors and possibly adduct formation, as well as the physical mixing of precursors can occur. When the precursor gas stream enters the boundary layer zone, precursors are transferred mainly *via* diffusion onto the surface. In this zone, homogeneous reactions, such as vapor-phase pyrolysis, can also occur, possibly with a higher efficiency than in the mixing zone due to the higher temperature generally present in this zone. Besides the diffusion of precursors onto the substrate surface, thermal convection can also occur, which can cause premature reactions and result in depleted precursor concentrations near the substrate surface. Such thermal convection-driven effects need to be minimized by careful reactor chamber design. For the three zones discussed so far, mass transport, which is related to hydrodynamics, and mass transfer *via* diffusion play an important role. On the substrate surface, where surface reaction kinetics are important, the adsorption of pyrolyzed precursors, surface diffusion of ad-atoms, chemical reactions and desorption of by-products occur to complete the crystal growth process.

The simplified understanding of the MOCVD growth process discussed earlier has been helpful in determining the choice of growth parameter and reactor designs in relation to the materials grown. Especially, with the advent of advanced computer modeling and the experimental verification of the general predictions of these models, it has recently become possible to use the results of computational fluid dynamics techniques to determine the most favorable operating regime for some reactor systems. However, for the study of specific materials and device parameters, the crystal grower has to explore the growth parameter space peculiar to the specific reactor and materials employed to determine the optimum conditions for the growth of epitaxial thin films.

6.4 MOCVD Reactor Systems

As described in the previous section, MOCVD growth involves very complex chemistry and the growth itself becomes even more complicated when it is influenced by reactor system conditions, especially the reactor chamber geometry. MOCVD reactor system design requires the integrated consideration of hydrodynamics, thermodynamics and kinetics. The importance of reactor design should not be underestimated, since it critically affects the quality of materials, including the (1) structural, electronic and optical qualities of material itself and (2) uniformity, controllability and run-to-run repeatability of the growth process. Therefore, several types of MOCVD reactor systems have been developed to improve not only the qualities of the materials grown by MOCVD but also growth efficiencies, repeatability, controllability and uniformity of the materials. In addition, MOCVD reactors are being developed for added safety features and high throughput capability, which are critical in production reactors.

The components of a general MOCVD reactor system can be grouped into several parts. In this section, we group a reactor system into three major components in terms of their functions and gas flow streams (from upstream to downstream): (1) the gas delivery system, (2) the reactor growth chamber and (3) exhaust system (Figure 6.8). Also, there is an additional critical component, even though it is not included in the gas flow stream, which is the reactor safety system. This section discusses the hardware issues of MOCVD reactor systems. Safety issues with MOCVD growth and apparatus for safety are discussed first, followed by the three major components of a typical

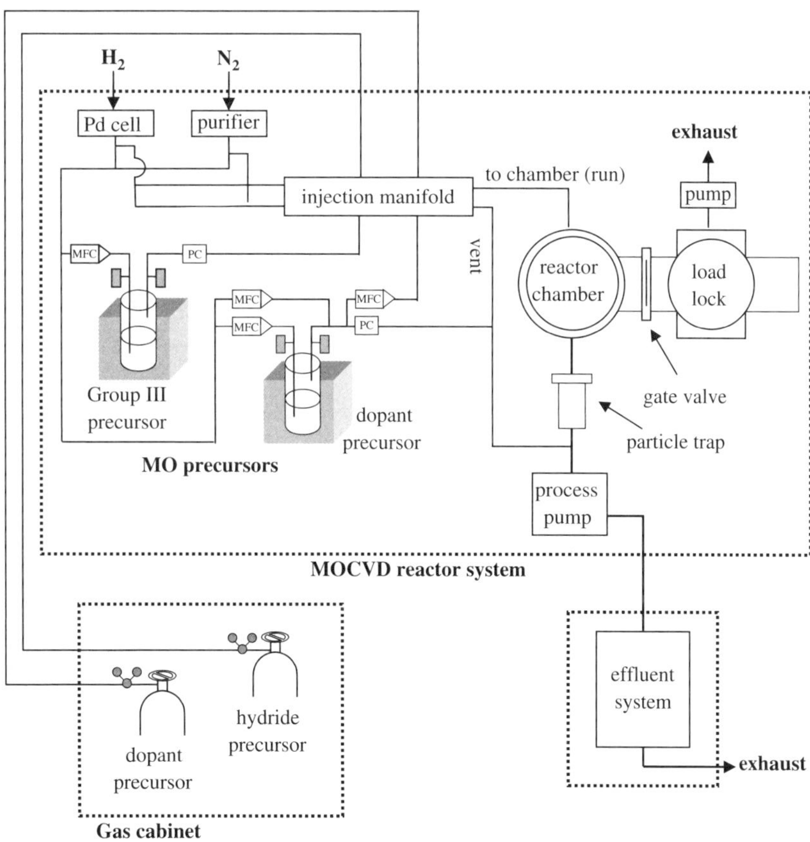

Figure 6.8 Schematic diagram of an MOCVD reactor system.

MOCVD system. Several types of growth chamber geometries are discussed in detail, since this design element is the heart of an MOCVD reactor system.

6.4.1 Safety Apparatus and System

MOCVD growth requires use of metal-organic and hydride precursors. Hydrides, such as arsine (AsH_3) and phosphine (PH_3), are particularly a concern due to their toxicity. Even ppm levels of exposure to these gases for a very short time can cause a lethal or life-threatening situation.[44] In addition to its toxicity, phosphine is pyrophoric. Furthermore, most metal-organic precursors are pyrophoric when exposed to air. Most MOCVD growth processes also employ a large amount of hydrogen as a carrier gas, which is flammable when the concentration is between 5% and 95%. Therefore, ensuring safety for the operation and maintenance of the MOCVD reactor system is the most critical element. Integrated safety apparatus and interlock systems in commercial MOCVD reactors in both laboratory and production environments have been described in the literature.[45]

6.4.1.1 Safety for Precursor and Effluent Transport

Potentially toxic, pyrophoric and/or flammable precursors and effluents in the MOCVD reactor systems in use today are transported in the form of vapor-phase streams using complex gas flow systems generally constructed of 316 L electropolished stainless steel[46] tubing, valves, flow meters, *etc.* Proper sealing (*i.e.*, achieving a high-level of leak-tightness) of the gas flow system is prerequisite to ensure the safe operation of these systems. In MOCVD systems, many joints between stainless-steel tubing sections and valves, *etc.* are required and welding, metal-face gasket sealing and O-ring sealing are used to join tubing and gas line components. Generally, welded joining is considered to be to be the safest (but also the least flexible) and any gas lines running outside the vented and interlocked reactor cabinets should be welded for safety. For added safety, double-walled welded stainless steel tubing (with monitoring of the space between inner wall and outer wall) is used especially used for extremely toxic high-pressure delivery tubing such as for arsine and phosphine gas sources. Metal-face gasket sealing is used for high-pressure tubing and components in the gas system located in the reactor cabinets. O-ring sealing is sometimes used for low-pressure tubing joints due to the ease of maintenance of these fittings. Reactor tube connections should be inspected by using a He-mass-spectrometer vacuum leak detector on a regularly scheduled basis as well as after installation and maintenance.

There are two general types of MOCVD precursors: condensed-phase precursors and compressed-gas precursors. Most metal-organic precursors are used in the condensed phase and these source cylinders or containers are located inside the MOCVD reactor cabinet as an integral part of gas handling system. Most hydride precursors are compressed gases and these are not generally stored in the MOCVD reactor enclosure. Instead, precursors in the form of compressed gases in high-pressure cylinders should be contained in a separate specially designed and interlocked gas cabinet. Both the MOCVD reactor system enclosure and the gas cabinets are individually vented to one or more exhaust fans to maintain a negative pressure inside the enclosures with respect to atmospheric pressure and these enclosures are also equipped with toxic gas sensors and air flow or pressure switches. The exhaust fan employed for venting the gas cabinets, reactor system enclosure and the exhaust of the reactor effluent scrubbing system should not to be confused with the exhaust line of the reactor chamber internal to the MOCVD system where the process effluent flows from the growth chamber to effluent scrubbing system. As described earlier, all critical gas tubing running between the gas cabinet and the MOCVD reactor system enclosure should be welded. Each gas cabinet is equipped with a gas control system consisting of a pressure regulator, an excess flow switch, several pneumatic isolation valves, and N_2 purge and vacuum/vent system. The pressure regulator is used to set the delivery pressure of a precursor and the N_2 purge and vacuum/vent system is used for "cycle" purge of the gas line tubing to remove residual precursors from the tubing for cylinder change or gas panel maintenance.

6.4.1.2 Effluent Treatment

The effluent of the MOCVD growth chamber that is carried in the exhaust line to the scrubber system is a mixture of particulates in a gaseous stream that consists of hydrogen and/or nitrogen, unreacted precursor vapors and volatile by-products. From a safety point of view, effluent materials are generally still toxic, pyrophoric and flammable and need to be treated to covert them into a non-toxic stable form of compounds before releasing them to the atmosphere and/or drain. As a part of the reactor effluent exhaust system, filters and scrubbers are incorporated for effluent treatment (Section 6.4.4).

6.4.1.3 Alarm and Interlock System

Several types of alarms and sensors are typically installed in the gas cabinets, the MOCVD reactor enclosure and the gas panel, and the MOCVD laboratory work area. Typical sensors are fire sensors (smoke or heat sensors), hydrogen detectors and toxic gas sensors for arsine, phosphine and ammonia. Toxic gas sensors generally have a sampling or sniffing port located in the exhaust duct from the MOCVD reactor system enclosure and the gas cabinets to the main exhaust fan. In addition, toxic gas sensors are used to monitor the laboratory "breathing" area outside of the gas cabinet and reactor enclosure to protect the operating personnel. Sensors are linked not only to an annunciator alarm panel to initiate both a visual and an audible alarm but also to a reactor interlock system to take automated safety measures according to the type of alarm that has occurred. For instance, in the MOCVD lab in the Center for Compound Semiconductors at Georgia Institute of Technology, arsine detection in one of the gas sensors located in the gas cabinet exhaust duct initiates the following measures: (1) annunciator visual (red flash) and audible (buzzer) alarm; (2) putting the MOCVD reactor that uses arsine in a safe condition; (3) shutting the arsine gas reactor input lines and the arsine cylinder valves off; and (4) automatically calling the Georgia Tech police department. Several levels of interlock can be designed, depending on the level of safety required, the specific reactor system and the laboratory configuration. In addition, most commercially manufactured MOCVD reactors have emergency shut off (EOS) buttons (or panic buttons) for manual emergency shut down when required. For ensured continuous operation of sensor monitoring systems and interlocks, all the sensors, alarms and interlocks need to have power sources that are backed up by a UPS (uninterrupted power supply).

6.4.1.4 Human Factors

In addition to the presence of a well-designed and complete safety hardware and software infrastructure for alarm and interlock operation, the most important factor in the safe operation of an MOCVD system is the human factor. All personnel involved in MOCVD growth should be aware of the hazardous and toxic nature of materials used in the process and the potential dangers if the system is not properly operated and maintained. MOCVD reactor operators should be trained to react to alarm or emergency situations according to the appropriate pre-designed plan. For routine reactor operation as well as for the maintenance of the reactor system, well-planned cautious approaches are absolutely required. Consequently, intensive training of the MOCVD growth system operators in terms of safety, material growth and reactor system maintenance is required. The use of procedures that the operator believes "should-be-OK" or the use of a "try-it-and-see" approach without considering the various unintended but possible outcomes of the trial are some of the most dangerous factors in MOCVD operations and maintenance. Such unpredictable and potentially unsafe practices are perhaps more dangerous than the hazards involved in the careful use of arsine as a precursor.

6.4.2 Gas Delivery System

The gas delivery system is designed for transporting and controlling the precursor mole fractions and carrier gases into the growth chamber and generally consists of stainless steel tubing, manifolds, purifiers, filters, automatic and manual valves, pressure controllers and mass flow controllers. The tubing and manifolds for gas delivery are typically made of Type 316L stainless steel that has electropolished internal surfaces to minimize internal wall out-gassing and are often heated by resistive heaters wrapped around the tubing or inserted into manifolds to prevent vapor-phase precursors condensing back into a liquid (or solid) phase. As described above, the gas delivery system needs to be leak-tight at levels detectable by a He mass spectrometer leak detector (*i.e.*, in the $\sim 10^{-10}$ atm-cc s^{-1} range). This is not only for safety but also to maintain the purity of the precursors, which is ultimately related to the quality of materials to be grown in the system. The design of the gas delivery system must also consider the time-dependent delivery of different precursors (for heterostructure growth) without experiencing flow or pressure transients from the required precursor switching or from residual gases caused by dead volumes in the gas piping system. Depending on the precursors and the carrier gas, the purification and control of transported gases can be different due to the different chemistry and phases of the precursors and other gases.

6.4.2.1 Carrier and Purge Gases

The carrier gas is the gas mixed with the precursor vapors for transporting the precursor into the growth chamber. The purge gas is used to purge some of gas piping or regions in the growth chamber, such as the optical port(s) used for *in situ* growth monitoring to prevent the part(s) from being exposed to precursor vapors and thus to minimize the deposition on or in the component. In most cases, hydrogen is used for carrier and purge gas while nitrogen or a mixture of hydrogen and nitrogen is used for the growth of some H-sensitive materials such as InGaN materials. The largest volumes of gas used in MOCVD system are the carrier and purge gases, which, for most laboratory- and production-scale MOCVD reactors, can range from the order of SLM (standard liter per min) to tens or even hundreds of SLM, depending on number of sources, growth chamber size and geometry. Therefore, maintaining of the purity of hydrogen and nitrogen as high as possible is critical. Hydrogen is commonly purified by passing atomic hydrogen through a specially designed palladium (Pd) alloy membrane cell heated at $\sim 400\,^{\circ}C$.[47] A heated Pd or Pd-Ag (silver) alloy surface can act as catalyst for the ionization and decomposition of hydrogen molecules (H_2). The Pd purifier element is designed so that that only monatomic hydrogen can diffuse through it, while other larger elements and molecules are blocked. After passing through the Pd element, the H atoms recombine into H_2. Ultrahigh purity (UHP) hydrogen is thus created at the "output side" of the Pd cell and delivered to the reactor system in this way. Sudden uncontrolled cooling of the Pd cell with hydrogen atoms trapped inside could cause cracks in the cell due to excessive stress. To avoid such a disastrous event, commercial Pd purifier systems are equipped with a nitrogen purge line to cause a slow hydrogen diffusion from the output side to the input side (which is then purged with UHP nitrogen) and, in the case of a power failure, the Pd cell heater is often backed up by a UPS. Carrier gas lines in the MOCVD reactor gas panel are generally divided into Group III lines for MO precursors and Group V lines for gas-phase precursors, with each line consisting of a "run" line that goes to the growth chamber and a "vent" line that goes to the reactor exhaust system (typically, directly to the low-pressure vacuum pump inlet). The carrier gas flow rate of the "run" line is actively controlled to define and stabilize the total flow rate of gas flowing in the growth chamber. The difference in pressures between the "run" and "vent" lines is maintained at zero using differential pressure controllers.

6.4.2.2 Metal-organic Precursors (Condensed Phase in Bubbler)

Most metal-organic precursors are in the liquid or solid phase near room temperature and atmospheric pressure. Because they generally have relatively high vapor pressures, the process of transporting the metal-organic precursors can employ the equilibrium vapor pressure of the condensed phase. When a gas (generally hydrogen or nitrogen) is introduced into the condensed-phase metal-organics, the gas will capture precursor vapors by "bubbling" through the liquid (or loosely packed solid) precursor so the cylinder containing a metal-organic precursor is commonly called a "bubbler." The gas bubbling though the precursor is sometimes called the carrier gas but in this section we use the term "bubbling gas" to avoid confusion with the main reactor carrier gas described above. Unless the gas flow rate is too high, the bubbling action sets up an equilibrium between the condensed phase and the vapor phase.[48] The equilibrium vapor pressure, p_{eq}, is normally given by:

$$\ln p_{eq,T} = -A/T + B(+C \ln T) \tag{6.34}$$

where T is the absolute temperature (K) and A, B and C are constants. The term in parenthesis is sometimes ignored. The properties of metal-organic precursors are discussed in detail in Section 6.5. Equation (6.34) shows that the equilibrium vapor pressure is strongly dependent on the temperature; therefore, source bubblers should be kept at a precisely controlled constant temperature. The molar flow rate of a metal-organic precursor, $Q_{metal-organic}$ (mol min^{-1}), is calculated from Equation (6.35):

$$Q_{metal-organic} = \frac{p_{(Eq,T)}}{P - p_{(Eq,T)}} \times \frac{F_{bubbling\_gas}}{C_{STP}} \tag{6.35}$$

where $F_{bubbling\_gas}$ is the flow rate of the gas flowing though the bubbler (in standard cubic centimeters per min, sccm), P is the bubbler pressure (Torr), $p_{Eq,T}$ is the vapor pressure of metal-organic precursors (Torr) and C_{STP} is a constant for the molar volume of an ideal gas at standard temperature (298.15 K) and pressure (760 Torr), *i.e.*, 22 406 cc mol^{-1}. Precise control of the molar flow rates of Group III precursors is crucial, since in most cases of MOCVD growth the Group III molar flow rates govern the growth rate and composition (solid mole fraction) of the epitaxial layers. Therefore, precise control of the gas flow rate, bubbler total pressure, and temperature of the precursor (and thus the bubbler itself) are important for the controllability and repeatability of the growth; the required precision control is carried out by the mass flow controllers, pressure controllers and isothermal baths, respectively. Figure 6.9(a) shows a schematic diagram of a bubbler and metal-organic precursor transport system.

6.4.2.3 Hydride Precursors (Liquefied Gas in Cylinder)

Most hydride precursor compounds are in a liquefied gas phase (at room temperature) contained in high-pressure cylinders and generally the gas is transported directly as a precursor. Group V hydride precursors employed in MOCVD are ultrahigh purity gases ($>$99.99995% min. for AsH_3 and NH_3 and $>$99.9999% min. for PH_3) contained in a specially inner-wall treated (for ultra-cleanliness) cylinders stored in a remotely located gas cabinet that is plumbed directly to reactors

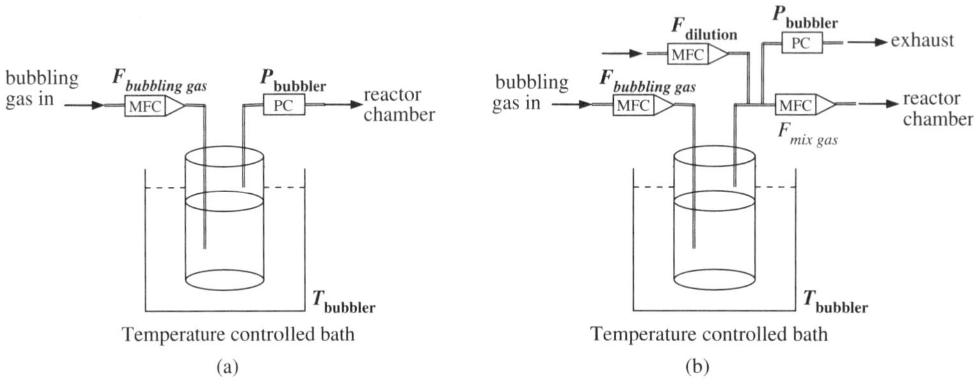

Figure 6.9 Schematic diagram of bubbler configuration of (*a*) a standard bubbler and (*b*) a double dilution bubbler.

through a pressure regulator that makes the pressure step-down from the cylinder pressure (typically, at pressures higher than 200 psig, depending on the gas) to the chosen delivery pressure (generally between 20 and 40 psig). The hydride delivery line is normally equipped with a specially constructed purifier system to remove any residual impurities from the gas such as oxygen and water vapor. The molar flow rate of a hydride precursor, Q_{hydride} (mol min^{-1}), is calculated from Equation (6.36):

$$Q_{\text{hydride}} = \frac{F_{\text{hydride}}}{C_{\text{STP}}} \times \frac{\text{Mixture\%}}{100} \tag{6.36}$$

where F_{hydride} is the flow rate of the gas from the cylinder (in sccm), which is regulated by electronic mass flow controllers.

6.4.2.4 *Dopant Precursors (Condensed Phase in Bubbler and Gas in Cylinder)*

Precursors for *n*- and *p*-type dopants can be in the condensed phase or the gas phase. For instance, Si_2H_6, SiH_4 and H_2S are in the gas phase (not necessarily liquefied at room temperature) and contained in high-pressure cylinders while Cp_2Mg, DEZn, TESn, CCl_4 and CBr_4 are condensed-phase sources typically contained in the bubblers. Generally, precursors need to be diluted to achieve the desired molar flow rate into the growth chamber. Such dilution can be carried out using additional carrier gas flows in separate "dilution lines" in addition to the "bubbler flow" for a precursor (Figure 6.9b). Typically, many applications for such dilution gas flow systems require such diluted mixtures having a precursor concentration generally in the part per million (ppm) range balanced with hydrogen. Similar concentrations can be achieved for gas-phase dopant precursors, *e.g.*, SiH_4 diluted with H_2 and stored in a cylinder. The molar flow rate of a dopant precursor, Q_{dopant} (mol min^{-1}), is calculated from one of following equations depending on the type of precursors:

$$Q_{\text{dopant,bubbler}} = \left(\frac{p_{(\text{Eq},T)}}{P - p_{(\text{Eq},T)}} \times \frac{F_{\text{bubbling\_gas}}}{C_{\text{STP}}}\right) \times \left(\frac{F_{\text{mix\_gas}}}{F_{\text{dilution}} + F_{\text{bubbling\_gas}}}\right) \tag{6.37}$$

$$Q_{\text{dopant,cylinder}} = \frac{F_{\text{dopant}}}{C_{\text{STP}}} \times \frac{\text{concentration in ppm}}{10^6} \tag{6.38}$$

where $F_{bubbling\_gas}$ is the flow rate of the gas flowing though the bubbler (sccm), $F_{dilution}$ is the flow rate of dilution gas after bubbling (sccm), F_{mix_gas} is the flow rate of diluted dopant precursor into growth chamber (sccm) and F_{dopant} is the flow rate of the diluted dopant gas from the cylinder (sccm).

6.4.2.5 Gas Transport/Switching Manifold

Every valve control line, mass-flow controller, pressure transducer and pressure controller input and output is routed to a computer-controlled analog/digital matrix, which allows complex switching instructions to be executed; most of the gas switching functions are performed by pneumatically actuated and/or solenoid valves. Precursor delivery is controlled by separate pressure-balanced switching manifolds, sometimes called injector blocks or injection manifolds. The switching manifold is a set of valves typically incorporated into a single machined stainless steel block, which can prevent transient flow and pressure conditions from affecting the growth in the growth chamber. In the switching manifold, the individual precursor gas flows are switched, combined and routed into either the growth chamber or the vent/exhaust line, while maintaining a differential pressure of nearly zero between the "run" and "vent" line so that the gas flow switching transient is minimized and does not strongly affect the growth process. Additional hydrogen or nitrogen carrier flows are added within each precursor transport line to the overall flows to aid with gas delivery through the remainder of the lines.

6.4.3 Growth Chamber

The introduction of precursors and carrier gas into the growth chamber of the reactor system and the actual growth processes that take place there include (1) flow injection (precursor mass transport), (2) mixing and diffusion (precursor mass transfer), (3) pyrolysis (homogeneous and heterogeneous reactions) and (4) adsorption of pyrolyzed precursors, surface diffusion of ad-atoms, chemical reaction and desorption of by-products (surface reactions). The growth chamber is a vessel made of quartz and/or stainless steel and includes gas flow injector, susceptor, heating element and chamber exhaust. The susceptor is located in the growth chamber where uniform and efficient growth occurs, considering the hydrodynamics of a given chamber geometry; the susceptor generally has machined surface depressions or "pockets" that can hold the semiconductor substrates during growth to prevent them moving around. The susceptor is generally made of graphite and can be SiC-coated. For improved uniformity of layers in terms of thickness, doping and alloy composition, the susceptor is usually rotated. Heating is used to apply the thermal energy to the susceptor and thus to the substrates. Resistive heating by graphite or molybdenum heating element(s), RF (radio-frequency) induction heating and IR (infrared) radiation heating are used, depending on the temperature range to be used and the reactor-chamber geometry. In many applications of production-type reactors, multiple identical wafers need to be grown in a single growth run to achieve a high-throughput, lower-cost epitaxial growth process. Large susceptors that can accommodate many wafers (susceptors with capacities of more than 40 two-inch dia. wafers are available as of the time of writing) are used, and independently controlled multiple-zone heaters are employed to ensure the lateral (or radial) temperature uniformity across the large susceptor.

There are basically two types of large-scale commercial growth chamber geometries for the MOCVD growth of III-V compound semiconductor materials – the description of these reactors depends on the primary flow direction of the input gases, which is determined by the design of the flow injector and the chamber exhaust. In this approach, the two types are horizontal and vertical reactors. In horizontal reactors, the gas phase precursors flow horizontally, in parallel to the susceptor and substrate, while in the vertical flow reactors the gas phase precursors flow vertically normal to the susceptor and substrate surface. Many commercial reactor chambers have more sophisticated features in the design of the gas flow path and, in many cases, they are designated by their own copyright-protected names.

All the commercial MOCVD reactor designs discussed below are supported by extensive numerical simulations of reactant chemistry and thermal profiles, allowing the calculation and optimization of many of the important performance parameters for advanced production reactors to be made without actually using the expensive tool to perform the experiments necessary to develop detailed predictions of the control of layer thickness, alloy composition, doping, *etc.*

6.4.3.1 Close-coupled Showerhead System

Modern MOCVD reactor systems offered by Thomas Swan Scientific Equipment Ltd.[49] are equipped with a close-coupled showerhead (CCS)™ reactor chamber system.[50] The precursors are dispersed uniformly across the entire area of the susceptor by a specially designed and constructed "showerhead," which is located close to the susceptor in a vertical-geometry reactor. The CCS is suited to the large-area deposition of uniform layers while allowing for easy scalability to larger chambers. In addition, the showerhead is constructed to enable precursors of Groups III and V to be separated right up to the point where they are uniformly injected into the chamber through a multiplicity of small tubes (over 4000 are used in a seven-inch dia. showerhead), which minimizes premature reactions between the precursors, leading to good chemical efficiency. The CCS design has been demonstrated to be a robust route to uniformity and scalability.

6.4.3.2 Horizontal Rotary Susceptor System

Horizontal rotary susceptor MOCVD systems[51] are offered by Taiyo Nippon Sanso.[52] In the horizontal rotary susceptor system, multiple substrates are placed on a radially symmetric susceptor that is slowly rotated [10–20 rpm (revolutions per minute]. The reactor system has a "triple-flow channel" gas injection system, which enables Group V (and carrier gas), Group III (and carrier gas) and purging gas flows to be injected separately. Generally, the Group V precursor channel is located close to the susceptor so that Group III precursor molecules diffuse though a Group V precursor flow layer while flowing down stream.[53] This flow configuration is claimed to be useful for improving the growth efficiency and the uniformity of the film. Also, while the outer reactor chamber is made of stainless steel, a gas flow restriction chamber liner (made of quartz) is carefully designed to achieve high-velocity gas flow with laminar flow conditions. For some reactor designs from Nippon Sanso, the substrates are placed in a "face-down" configuration using a specially designed inverted-geometry susceptor to prevent the precursor depletion effects induced by thermal convection and also to reduce the number of particles on the growing surface.

6.4.3.3 Planetary Reactor System

The Planetary Reactor®[54] was developed by Philips and is manufactured under exclusive license to Aixtron[55] to be used as its multiple-wafer MOCVD system technology platform. Planetary reactor systems are radial-flow horizontal reactor systems that can accommodate multiple wafers. The design features a central gas inlet with separate supplies of Group III and Group V precursor gases and substrates symmetrically located in an annular region around the inlet. In a Planetary reactor system, two rotational modes are employed to improve the layer uniformity: (1) the main susceptor is slowly rotated to average out any thermal imbalance induced by heating element and (2) the satellite susceptor discs that carry the substrates are individually slowly rotated in the opposite direction to ensure the uniformity of the layer properties by rotational averaging of the radial reactant depletion. Achieving laminar horizontal flow is critical in the precise control of the composition of ultrathin layers and heterojunctions and deposition rates at the monolayer level. The combination of this laminar-flow principle with the simultaneous multiple-substrate satellite disc rotation, known as gas foil rotation (GFR),[56] which utilizes a levitation of the satellite susceptor discs on a gas foil to provide momentum for rotation for all the individual satellite discs simultaneously, is claimed to ensure excellent deposition homogeneity regarding layer thickness, composition and doping over multiple wafers. In addition, the special reactor inlet valve, which allows the separation of certain gases, ensures a uniform outward radial flow and optimum distribution and efficient utilization of precursor gases.

6.4.3.4 Rotating Disk Reactor System

Rotating disk reactors (RDRs)[57] were originally commercially developed by Emcore Corporation and such reactors are now offered by Veeco Compound Semiconductor[58] and are marketed as TurboDisc® MOCVD systems. RDR reactor systems are vertical-geometry reactor systems equipped with a high speed rotating disk that operates at speeds of >1000 rpm. While other reactor designs employ rotating susceptors (mostly operating at low speeds <100 rpm), the salient feature of RDRs is the high-speed rotation of the susceptor. RDRs are designed to satisfy two fundamental requirements: (1) uniform precursor flow over the substrate and (2) a uniform substrate temperature. In a RDR, the input precursor concentrations are varied along the radius of the disk to take into account depletion effects of the precursor as it flows radially outward. RDRs are claimed to have inherent advantages in hydrodynamic symmetry and flow dynamics that enable growth to be laterally uniform, abruptly switchable and robust against variations in process parameters. In addition, the high rotating speed of susceptor provides a second mechanism affecting the gas flow since the rotating disk acts as a pump due to viscous drag at the gas/solid interface. In many cases, the reactors operate at a reduced pressure to maintain laminar flow and suppress buoyancy-driven convection.

6.4.4 Exhaust System

The exhaust system of a modern low-pressure MOCVD system serves two main functions: (1) removal of unreacted reactants and reaction by-products from the growth chamber and (2) maintaining a steady-state pressure condition in the chamber during growth. Typically, the primary elements in the exhaust system are one or more particle filters (to protect the vacuum pump from large particles), a pressure sensor connected to a computer-controlled variable-conductance vacuum valve (to control the reactor chamber total pressure) and a high-performance vacuum pump (typically an oil-free "dry" pump is used). As noted above, a process vacuum pump is required for modern MOCVD growth systems operating at sub-atmospheric pressures. Previously, oil-based process pumps with an oil back-streaming filter were dominantly used but they have been replaced with oil-free "dry" pumps that use the high-speed rotation of precision metal vanes having a small clearance between the vanes. Growth chamber pressure is regulated by a throttle valve system operating under feedback control using a pressure transducer attached to growth chamber. Gases exhausted from the growth chamber or vent line flow through the replaceable particle filter and the process pump into a scrubber system for effluent treatment.

There are several methods of MOCVD reactor effluent treatment: (1) utilizing the adsorption of reactants on the surface of an appropriate type of activated carbon; (2) using chemical reactions of appropriate solutions to form a stable oxides of reactants; and (3) using thermal energy (burning or heating) and chemical reactions to form a cracked or stable oxidized form of reactants.

6.5 Precursors – Synthesis, Purification, Analysis and Delivery

Precursors typically employed in epitaxial growth by MOCVD can be divided into the following sub-categories – Group III alkyls (or metal-organics), Group V hydrides and dopants. All these precursors share the common properties of reactivity, volatility, toxicity and/or pyrophoricity. These precursors must be of the highest purity to deposit materials with the desired optoelectronic performance. The selection of an optimal precursor for use in MOCVD technology is primarily based on the semiconductor alloy material being grown. For example, a precursor that decomposes at higher temperature can not be used for the alloy that itself melts below the decomposition temperature of the precursor or the resultant molecular species. Table 6.4 describes various precursors and their properties.

6.5.1 Synthesis and Purification

Most of alkyls of Groups II, III and V are synthesized by one of the following general chemical reaction schemes:[59]

$$MX_n + nR_3Al \rightarrow R_nM + nR_2AlX \tag{6.39}$$

$$MX_n + nRLi \rightarrow R_nM + nLiX \tag{6.40}$$

$$M_2Mg_5 + MeI \rightarrow 2Me_3M.ether + 2MeMgI + 3MgI_2(\text{Group III only}) \tag{6.41}$$

$$MX_n + nRMgX \rightarrow R_nM.ether + MgX_2 \tag{6.42}$$

Each method has its relative advantages and disadvantages. The reaction schemes shown above, with the exception of the trialkylaluminium (R_3Al) route, have the disadvantage of incorporating some concentration of unintended detrimental ether impurity in the product. The use of high boiling ethers, which form weak M–O bonds, can be used to mitigate this contamination. However, the highest purity precursors are achieved by employing proprietary adduct purification technology (Figure 6.10).[60]

In addition to the adduct purification technology, various distillation and sublimation techniques are employed to purify metal-organic precursors. Fractional columns for distillation are designed specifically for each product based on their physical properties and thermal stability. In some instances, chemical getters are also employed to selectively remove a detrimental impurity. While distillation technology works well for liquids, purification of solids requires sublimation, which is quite limited in effectiveness for the separation and removal of impurities. For example, to minimize oxygen impurities, care is taken not to select any raw materials and solvents that contain oxygen species.

The absolute volume of consumption of Group III-alkyls is ranked second to the volume of Group V sources consumed in MOCVD applications. Primary metal alkyls, like trimethylgallium (TMGa), trimethylindium (TMIn), trimethylaluminium (TMAl) and triethylgallium (TEGa), are produced annually in quantities of 100–5000 kg. Owing to the pyrophoric nature of these materials, coupled with the extreme sensitivities of the application toward impurities, specifically, the reaction toward oxygen and moisture, high integrity stainless steel plants are employed to manufacture these materials. Figure 6.11 shows such a modern TMGa production plant, which has a 10 metric ton annual capacity. This plant integrates all the steps of precursor production and packaging process, *i.e.*, from the formation of $GaCl_3$ to the production of TMGa, as well as TMGa purification and, finally, TMGa transfilling into smaller bubblers or into bulk containers. As epitaxial III-V materials have moved into main-stream production, the latest statistical process and quality control

Table 6.4 Properties of MOCVD precursors.

Precursor	*Acronym*	*Formula*	*Physical state*	*Main hazard*	*Vapor pressure data*	*Density* $(g\,cm^{-3})$
Group III						
Trimethylgallium	TMG	$(CH_3)_3Ga$	Liquid	Pyrophoric	$Log_{10}P(mmHg) = 8.07–1703/T(K)$	1.151
Triethylgallium	TEG	$(C_2H_5)_3Ga$	Liquid	Pyrophoric	$Log_{10}P(mmHg) = 8.08–2162/T(K)$	1.06
Trimethylaluminium	TMA	$(CH_3)_3Al$	Liquid	Pyrophoric	$Log_{10}P(mmHg) = 8.22–2134/T(K)$	0.752
Triethylaluminium	TEA	$(C_2H_5)_3Al$	Liquid	Pyrophoric	0.02 mmHg @20 °C; 0.14 mmHg@40 °C	0.835
Trimethylindium	TMI	$(CH_3)_3In$	Solid	Pyrophoric	$Log_{10}P(mmHg) = 10.52–3014/T(K)$	1.568
Group V						
Arsine		AsH_3	Gas	Toxic	$Log_{10}P(mmHg) = 6.569–762.4/T(K)$; bp = –62.5 °C	3.29 kg m^{-3} at 15 °C
Phosphine		PH_3	Gas	Toxic	$Log_{10}P(mmHg) = 6.984–754.1/T(K)$; bp –87.7 °C	1.45 kg m^{-3} at 15 °C
Ammonia		NH_3	Gas	Toxic	bp –33.35 °C, mp –77.7 °C	0.5967 g L^{-1}
tert-Butylarsine	TBA	$(t\text{-}C_4H_9)AsH_2$	Liquid	Toxic	$Log_{10}P(mmHg) = 7.5–1562.3/T(K)$	1.0
tert-butylphosphine	TBP	$(t\text{-}C_4H_9)PH_2$	Liquid	Toxic	$Log_{10}P(mmHg) = 7.586–1539/T(K)$	0.7
Unsym-dimethylhydrazine	DMHz	$(CH_3)_2NNH_2$	Liquid	Reactivity	$Log_{10}P(mmHg) = 8.19–1780/T(K)$	0.791
Trimethylantimony	TMSb	$(CH_3)_3Sb$	Liquid	Pyrophoric	$Log_{10}P(mmHg)=7.73–1709/T(K)$	1.528
Triethylantimony	TESb	$(C_2H_5)_3Sb$	Liquid	Pyrophoric	$Log_{10}P(mmHg)=7.90–2183/T(K)$	1.324
Tris(dimethylamino)antimony	TDMSb	$[(CH_3)_2N]_3Sb$	Liquid	Flammable	$Log_{10}P(mmHg) = 6.23–1734/T(K)$	1.3
Dopants						
Dimethylzinc	DMZ	$(CH_3)_2Zn$	Liquid	Pyrophoric	$Log_{10}P(mmHg) = 7.80–1560/T(K)$	1.386
Diethylzinc	DEZ	$(C_2H_5)_2Zn$	Liquid	Pyrophoric	$Log_{10}P(mmHg) = 8.28–2109/T(K)$	1.20
Bis(cyclopentadienyl) magnesium	BisMg	Cp_2Mg	Solid	Pyrophoric	$Log_{10}P(mmHg) = 10.0–3372/T(K)$	–
Carbon tetrachloride			Liquid		$Log_{10}P(mmHg) = 8.05–1807/T(K)$	1.59
Carbon tetrabromide		CBr_4	Solid		$Log_{10}P(mmHg) = 7.78–2346/T(K)$	3.42
Silane		SiH_4	Gas	Pyrophoric	$Log_{10}P(mmHg) = 7.06–680.4/T(K)$: bp –111 °C	1.35 kg m^{-3} at 15 °C
Diethyltelluride	DETe	$(C_2H_5)_2Te$	Liquid	Flammable	$Log_{10}P(mmHg) = 7.99–2093T(K$	1.599
Bis(cyclopentadienyl)iron		Cp_2Fe	Solid		$Log_{10}P(mmHg) = 10.27–3680/T(K)$	–

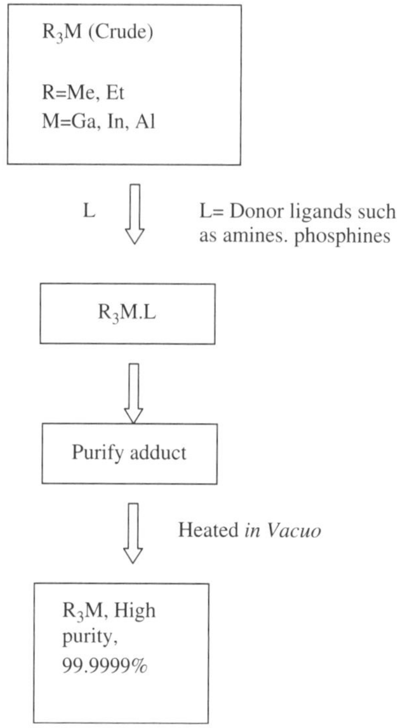

Figure 6.10 Adduct purification of Group III metal-organics.

Figure 6.11 Large-scale trimethylgallium plant (SAFC Hitech, Bromborough UK).

(SPC/SQC) tools have been employed to achieve consistency in process yields and quality from batch-to-batch.

6.5.2 Analysis

The electrical and optical properties of semiconductor materials like lasers, LEDs, transistors and detectors depend upon the purity of precursors and gases. For liquid precursors, inductively coupled plasma (ICP) with either optical emission or mass spectrometry (ICP-OES and ICP-MS) are used to determine the concentration of metallic impurities[61] and Fourier-transform nuclear magnetic resonance (FT-NMR) spectroscopy is employed to measure trace levels of organic and oxygen impurities.[62] Several other techniques based on gas chromatographic separation have been explored for the analysis of organic and oxygen impurities but they were found to be less than ideal due to the highly reactive nature of the metal alkyls. Tables 6.5 and 6.6 show typical examples of the detection limits and purity levels achieved by ICP-OES and ICP-MS, respectively.

FT-NMR spectroscopy was found to be particularly useful in the quantitative measurement of oxygen impurities in TMAl. The oxygen species in TMAl is trimeric dimethylaluminium methoxide, $(Me_2AlOMe)_3$. Starowieyski *et al.*,[63] have shown that a larger fraction of this impurity gets transported in the gas phase than would be predicted by calculations based upon the molar ratio and vapor pressure. This is due to the formation of a mixed species $Me_3Al.MeOAlMe_2$, which is more volatile than Me_2AlOMe itself. Figure 6.12 shows a good correlation of this species in the NMR spectra as a function of concentration, which was further substantiated by the incorporation of oxygen in the AlGaAs and AlAs layers grown from these precursor sources (Table 6.7). Similar analyses have been performed on other metal alkyl sources like TMGa and TMIn.

For Group V hydride gases such as arsine, phosphine and ammonia the analytical techniques employed are atomic pressure ionization mass spectrometry (APIMS), X-ray fluorescence (XRF), and graphite furnace atomic emission spectrometry (GFAES). The impurity sensitivity levels achieved in these gases are at the parts-per-billion (ppb) level. The purification process for these precursor gases have the additional advantage of utilizing various *in situ* getters, purifiers and absorbers at the point-of-use to further improve the performance.

6.5.3 Vapor Phase Transport and Measurements

As noted above, in most cases, the liquid and solid precursors are delivered to the heated substrate in the MOCVD reactor chamber by bubbling the "bubbler gas" through the precursors. The amount of source material transported is calculated by using the source temperature, pressure and the flow rate. For liquids, this method of transport is generally very effective, achieving near 100% saturation of the metal alkyl source in the carrier gas until the contents of the bubbler are depleted. However, for solid metal alkyl precursors, complete carrier gas saturation is difficult to achieve, especially at higher flow rates. For example, TMIn, a moderately volatile solid precursor used for In, suffers from inconsistent pick up throughout the life of the bubbler. Several approaches have been tried to alleviate this problem with a view to increasing the residence time of the bubbler carrier gas through the solid metal alkyl source material:

1. Bubbler design modifications such as reverse-flow configurations, dip leg modifications, dual chambers and multiple chambers.[64]
2. Use of an inert support medium to increase the surface area of the solid metal alkyl source.
3. Solution TMIn systems and solutions of other solid metal alkyls such as Cp_2Mg and CBr_4.[65]

Table 6.5 ICP-OES analysis (using direct injection) of trimethylaluminium.

Element	*Results*	*Detection limit (ppm)*	*Specification (ppm)*
Ag	N/D	<0.4	<0.4
As	N/D	<0.5	<0.5
Au	N/D	<0.5	<0.5
B	N/D	<0.4	<0.4
Ba	N/D	<0.1	<0.1
Be	N/D	<0.02	<0.02
Bi	N/D	<0.5	<0.5
Cd	N/D	<0.02	<0.02
Co	N/D	<0.4	<0.4
Cr	N/D	<0.4	<0.4
Cu	N/D	<0.05	<0.1
Fe	N/D	<0.1	<0.1
Ge	N/D	<0.5	<0.5
Hg	N/D	<0.5	<0.5
I	N/D	<0.5	<1.0
La	N/D	<0.4	<0.4
Li	N/D	<0.4	<0.4
Mg	N/D	<0.02	<0.02
Mn	N/D	<0.03	<0.03
Mo	N/D	<0.5	<0.5
Nb	N/D	<0.5	<0.5
Ni	N/D	<0.5	<0.5
P	N/D	<0.5	<0.5
Pb	N/D	<1.0	<1.0
Pd	N/D	<0.5	<0.5
Pt	N/D	<0.5	<0.5
Rh	N/D	<0.5	<0.5
S	N/D	<1.0	<1.0
Sb	N/D	<1.0	<1.0
Se	N/D	<1.0	<1.0
Si	N/D	<0.03	<0.3
Sn	N/D	<1.0	<1.0
Sr	N/D	<0.1	<0.1
Tb	N/D	<0.5	<0.5
Te	N/D	<0.1	<1.0
Ti	N/D	<0.2	<0.2
V	N/D	<0.5	<0.5
W	N/D	<0.5	<0.5
Y	N/D	<0.02	<0.02
Zn	N/D	<0.1	<0.2
O	N/D	<1.0	<5.0

The last approach (*i.e.*, using TMIn dissolved/suspended as a TMIn.DMDA adduct) continues to gain advantage in the production environment due to the high utilization efficiency of these sources, even at bubbler carrier gas flow rates of over 3 L min^{-1} volumetric flow of the bubbler carrier gas through the bubbler, as well as the concomitant precise compositional control and *in situ* oxygen gettering abilities of the solution-TMIn source itself.

The level of the liquid sources in the stainless steel bubblers is measured using various level-indication systems, employing optical, ultrasonic and capacitance measurements. In another approach, the use of Episensor™ systems based on *in situ* capacitance measurements is very effective with TMAl and TMGa liquid sources – commercial systems are available that can monitor the metal alkyl precursor levels in eight individual bubblers in real time.[66] The signal interface of the Episensor with a

Table 6.6 ICP-MS analysis of trimethylaluminium.

Element	*Results (ppb)*	*Detection limit (ppb)*
Li	N/D	0.04
Mg	0.58	0.01
Ca	2.4	0.51
Cr	N/D	0.75
Mn	1.83	0.92
Fe	1.05	0.02
Ni	4.45	0.34
Cu	2.27	0.12
Zn	2.00	0.04
Cd	N/D	0.04
Sr	N/D	0.03
Be	N/D	0.09
Co	N/D	0.09
Ga	3.64	0.05
As	N/D	0.35
In	0.30	0.08
Ba	N/D	0.06
Tl	N/D	0.01

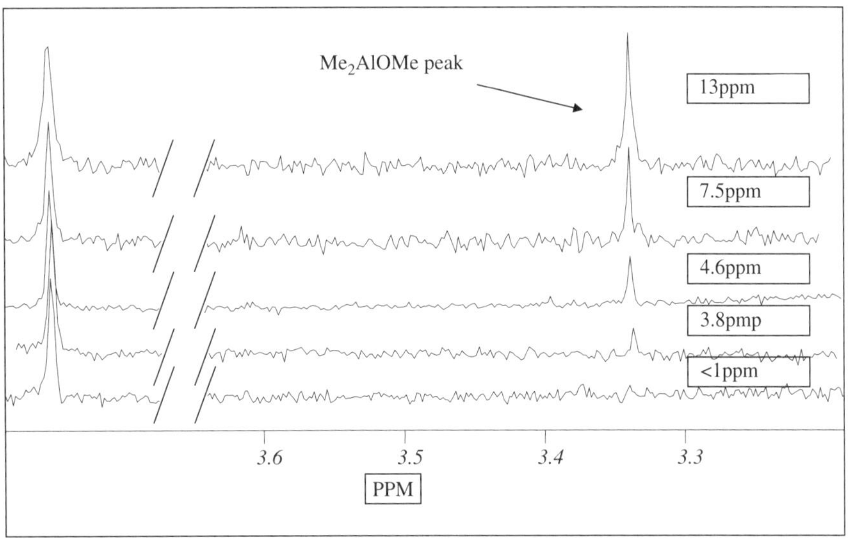

Figure 6.12 ^1H NMR spectra of TMAl samples with varying concentrations of oxygen impurity species.

metal alkyl bulk liquid distribution system has resulted in the development of commercial automatic bubbler-refill systems that operate in real-time without the need to disconnect a bubbler from the tool to replace an empty source bubbler. This system provides enhanced tool "up" time and also reduced possibilities for errors and contamination of the MOCVD reactor system.

6.5.4 Condensed Phase Group V Precursors

While many alternatives to the use of the gaseous Group V precursors arsine and phosphine have been explored, two liquid precursors have emerged as the viable candidates, namely,

Table 6.7 Correlation of oxygen species in TMA *versus* SIMS data.

TMA sample number	*O level in TMA using new NMR technique (ppm)*	*[O] in $Al_xGa_{1-x}As$ by SIMS (cm^{-3})*		*Composition (value of x)*
		p type (C) ($3–5\times10^{18}$ cm^{-3})	*n type (Si) ($1–3\times10^{18}$ cm^{-3})*	
1	>30	1×10^{19}	8×10^{18}	0.9
2	13	1×10^{19}	2×10^{18}	0.9
3	7.5	1.5×10^{18}	6×10^{17}	0.9
4	6.0	6×10^{17}	1.5×10^{17}	0.9
5	4.6	1×10^{17}	1×10^{17}	0.9
6	3.8	2×10^{17}	4×10^{16}	0.9
EpiPure™ TMA	<1 (ND)	$<3\times10^{16}$ (ND)	$<3\times10^{16}$ (ND)	1

Table 6.8 Comparison of toxicity and thermal decomposition behavior of TBP, TBA, phosphine and arsine.

Material	*Toxicity LC_{50} (4 h) (ppm)*	*Toxicity LC_{LO} (ppm)*	*Boiling point (°C)*	*Conc^n after 1 min release into 2000 ft^3 room*	*1 min after release into unscrubbed 20 000 cfm exhaust*	*Temperature at 50% thermal decomposition (°C)*	*Temperature at 100% thermal decomposition (°C)*
AsH_3	5–40	–	–62.5	10 000	1000	680	760
TBA	70–90	46	69	2.5	0.25	300	475
PH_3	11–57	–	–87.7	10 000	1000	700	850
TBP	>1100	>1100	56.1	5	0–1	450	550

tert-butylphosphine (TBP) and *tert*-butylarsine (TBA). Table 6.4 shows their physical properties while Table 6.8 compares their toxicity and thermal decomposition behavior with those of phosphine and arsine. The use of TBP and TBA is primarily limited to certain device structures where a performance advantage for these precursors is realized. In some instances, these liquids are employed where local regulations prohibit the use of arsine and phosphine. Most III-V epitaxial materials manufacturing companies producing As- and P-containing films have adapted arsine and phosphine gas precursors with the associated safety and monitoring systems.

Various metal alkyl antimony precursors have been investigated to grow antimony-based III-V alloys at lower temperatures. Three such sources – trimethylantimony (TMSb), triethylantimony (TESb) and tris(dimethylamino)antimony (TDMSb) – have received considerable attention, while others have been of passing interest, *e.g.*, triisopropylantimony.

The decomposition temperature of metal alkyl antimony sources follows the general trend: TMSb > TESb > TDMSb. The lower decomposition temperature of TDMSb is attributed to the absence of a Sb–C bond and the presence of labile Sb–N bonds. TDMSb decomposes at around 360 °C in the gas phase.[67] This is substantially lowered during heterogeneous decomposition over GaAs surfaces. At above 500 °C, homolysis of N–C bonds commences, complicating the chemistry of deposition. Conceivably, impurities such as nitrogen may be incorporated at higher temperatures. The decomposition of antimony sources was found to influence the growth kinetics of InSb. The increase of total pressure and decrease in carrier gas flow rate enhanced the cracking efficiency of the antimony precursors. TESb had a slight advantage over TMSb due to its lower decomposition temperature and vapor pressure.[68] It is also suggested in the literature that TESb assists in reducing carbon in the films due to β-hydrogen elimination. Wang *et al.*[69] have investigated the growth of GaSb using various antimony precursors TMSb, TESb and TDMSb and concluded that TEGa with TMSb is the preferred source combination due to the mass-transport limited growth.

Liquid precursors of nitrogen are primarily limited to unsym-dimethylhydrazine (DMHz) and to a lesser extent to *tert*-butylhydrazine (TBHz). The main driving force behind the primary use of DMHz is its higher vapor pressure and lower pyrolysis temperature compared to ammonia. The pyrolysis of DMHz occurs *via* N–N bond cleavage and begins at 320 °C; 50% pyrolysis is achieved at 550 °C, with complete decomposition at 800 °C. The decomposition temperature is higher when DMHz is co-pyrolyzed with TEGa, which is presumed to be due to adduct formation between TEGa and DMHz.[70] While GaN can be deposited using DMHz, the most advantageous use of this precursor is in the deposition of dilute nitride materials.[71]

6.6 MOCVD of Specific III-V Materials

In this section we briefly review the actual typical growth parameters that affect the structural, electrical, and optical qualities and properties of epitaxial materials. Understanding the relation between growth parameters and material quality and properties and their control is critical in MOCVD material growth for specific device applications. Several important growth parameters are (1) growth temperature (T_g), which is the actual growth temperature of the growing surface on the substrate; (2) the reactor pressure (P_g), which is a pressure of the growth chamber during the growth; (3) substrate orientation, which is designated by the surface plane index and tilt angle toward a certain direction with error range [*e.g.*, (100) 2° off toward <110> ±0.2°]; (4) the V/III ratio, which is the ratio of molar flow rate of Group V precursor to that of Group III (refer to formulae in Section 6.4.2 for molar flow rate calculation and note that the nominal input V/III ratio can be calculated rather than the actual V/III ratio near the growing surface, which is more relevant to the growth mechanism and is generally expected to be different from the nominal input V/III ratio); and (5) growth rate (R_g), which is the layer deposition thickness in a unit time period ($\mu m\,h^{-1}$, $nm\,min^{-1}$ or $nm\,s^{-1}$, $Å\,s^{-1}$). For most of the procedures for the growth of epitaxial structures for devices, a certain thickness of "buffer layer" composed of the same material as substrate is homoepitaxially grown on the substrate prior to growth of the specific device layers.

6.6.1 GaAs-based Materials

GaAs-based materials are used for optoelectronic devices operating from the near-IR (~980 nm with compressively strained InGaAs) to the visible yellowish-green (~570 nm with InAlGaP) as well as for many high-speed electronic device applications.

6.6.1.1 GaAs and AlGaAs Materials

The GaAs MOCVD epitaxial growth technology is the most mature among the III-V materials. In fact, it was the first III-V material that Manasevit reported growing in 1968.[24] The growth conditions and properties are reasonably well understood and have been applied to large-scale commercial materials growth for various devices. $Al_xGa_{1-x}As$ is more difficult to grow due to the reactivity of aluminium with oxygen and carbon. Especially, incorporated oxygen can (1) form a deep level electronic state inside the bandgap of AlGaAs, creating insulating material; (2) act as a nonradiative recombination center, causing poor radiative recombination efficiency; and (3) induce catastrophic failure of faceted mirrors during the operation of Fabry–Pérot (F-P) lasers. The most popular precursors for $Al_xGa_{1-x}As$ growth include trimethylgallium [$(CH_3)_3Ga$, TMGa], triethylgallium [$(C_2H_5)_3Ga$, TEGa], trimethylaluminium [$(CH_3)_3Al$, TMAl] and arsine (AsH_3). Typical growth temperatures employed for the MOCVD deposition of GaAs and AlGaAs epitaxial films are higher than 600 °C, which is in the diffusion-limited transport growth regime. Since liquid TMAl is a dimer, *i.e.*, it exists in the form of $(CH_3)_6Al_2$, the distribution coefficient of Al in AlGaAs growth using TMAl and TMGa is often higher than 1 and is close to 2. To reduce the incorporation of oxygen, especially for the growth of high-Al composition AlGaAs films, a higher growth temperature (normally higher than 720 °C but depending on Al mole fraction) is required to enhance the volatility of Al sub-oxide which can be adsorbed on the growing surface.[72] Carbon incorporation generally decreases with higher V/III ratio, lower Al mole fraction and higher growth temperature. In addition, materials grown using TEGa and triethylaluminium [$(C_2H_5)_3Al$, TEAl] are reported to have a lower carbon concentration, which is related to the pyrolysis mechanism of triethyl alkyl precursors.[73] Group IV Si atoms preferentially occupy Group III atomic sites in the crystal, acting as electron donors, and hence Si hydride precursors are often employed for the *n*-type doping of III-V epitaxial films. Diluted disilane (Si_2H_6) is preferred over diluted silane (SiH_4) for use as a Si dopant precursor due to the large and temperature-independent distribution coefficient in the typical T_g range that is used for AlGaAs. Diethylzinc [$(C_2H_5)_2Zn$, DEZn] is widely used for *p*-type doping by Group II Zn dopant atoms – but Zn has a high diffusivity in many III-V materials, which may result in a smeared doping profile and in an unwanted junction shift. Group IV carbon (C) atoms can occupy Group V sites in the III-V crystal, thus acting as electron acceptors for *p*-type doping. The advantage of C over Zn *p*-type doping is that the C atoms are not as mobile as Zn. Therefore, in some applications, C is not considered as an unwanted impurity but as an intentional dopant and the incorporation of C is intentionally increased in a controlled way rather than suppressed. For C doping, intentional doping by carbon tetrachloride (CCl_4) and carbon tetrabromide (CBr_4) precursors is commonly used, but achieving *p*-type doping only by using background C impurities due to the decomposition of CH_3 radicals is also possible, especially for AlGaAs with high Al mole fractions grown using TMAl.

6.6.1.2 InAlGaP Materials

For the growth of $In_{\sim 0.47-0.49}(Al_xGa_{1-x})_{\sim 0.53-0.51}P$ materials, the use of trimethyl alkyl precursors is preferred over triethyl alkyl precursors, to minimize parasitic reactions with the hydride As and P sources; therefore, TMIn, TMAl, TMGa and PH_3 are generally employed for the growth of the important InAlGaP III-V ternary and quaternary compounds. The typical T_g is between 600 and 750 °C and the optimum temperature depends on the Al mole fraction (higher T_g with higher Al) and generally a higher V/III ratio is required than is employed for AlGaAs materials. The optical quality of material and surface morphology are reported to be very sensitive to T_g, V/III ratio and substrate orientation. For ternary InGaP and InAlP materials, Cu-Pt type long-range ordering[74] along the $\langle 111 \rangle$ direction has been reported, depending on the growth conditions and the degree of ordering affects the optical quality of the materials. Silicon is an effective dopant for *n*-type doping of InAlGaP. For *p*-type doping of InAlGaP, which has the widest bandgap among the zinc-blende-structure III-V compound semiconductors, magnesium (Mg) may be a more effective dopant than Zn. The metal alkyl compound bis(η^5-cyclopentadienyl)magnesium [$(C_5H_5)_2Mg$, Cp_2Mg] is commonly used as the Mg precursor, despite several difficulties in its use: Cp_2Mg shows memory effects,[75] is prone to oxygen incorporation, and can form Mg-H complexes in the III-V crystal, which cause the Mg acceptor atom to be electrically inactive. This requires Mg dopant activation annealing under N_2 ambient to break the Mg–H bond to make the Mg dopant atoms electrically active.

6.6.2 InP-based Materials

InP-based materials are dominantly used in photonic devices in the current and next-generation of long-haul, high-speed optical communication systems (operating at $\lambda = 1310$ and 1550 nm with InGaAsP or InAlGaAs) and many ultrahigh-speed electronic device applications due to the high mobility of electrons in the conduction band of these materials.

6.6.2.1 InP Material

Similar to the growth of InAlGaP materials, TMIn and PH_3 are typical precursors and a high V/III ratio (generally >200 but this also depends on the details of the reactor design and the T_g) is required to achieve high electron mobility and improved optical quality. The requirement of a high V/III ratio for the growth is related to the limited pyrolysis efficiency of PH_3. The typical T_g is somewhat lower than that of InAlGaP and reported to be 550–650 °C. Silicon doping by using Si_2H_6 and Zn doping by DEZn are used for *n*-type and *p*-type doping, respectively. *p*-Type doping by C is known to be extremely difficult in InP material.

6.6.2.2 InAlGaAs Materials

For the growth of $In_{0.52-0.53}(Al_xGa_{1-x})_{0.48-0.47}As$ materials, TMIn, TMAl, TMGa, TEGa and AsH_3 are generally employed. The typical T_g is between 600 and 700 °C, depending on the desired Al mole fraction (higher T_g with higher Al). Owing to its lower growth temperature, InAlGaAs materials with high Al mole fractions may contain a substantial concentration of oxygen impurities if the reactor conditions and growth parameters are not optimized. Silicon doping by Si_2H_6 and Zn doping by DEZn are generally used for *n*-type and *p*-type doping, respectively. *p*-Type doping by C in InAlGaAs is also possible but the control is more difficult than in AlGaAs materials. For InAlGaAs/InP heterostructures, the interface quality control of the different Group V sublattices is more challenging than for other hetero-interfaces with same Group V elements. Interface switching of heterostructures between materials with same Group V and different Group III elements is generally less challenging; however, interface switching between materials with different Group V may not be abrupt. When the interface is switched from III-P to III-As, the interface is abrupt, while it is generally not abrupt when the interface results from switching from III-As to III-P. This is due to residual AsH_3 (or As on the surface) at the interface during the transition and inefficient PH_3 pyrolysis efficiency for As-terminated surfaces in the initial stage of III-P growth.

6.6.2.3 InGaAsP Materials

In the growth of InGaAsP materials, one can control the lattice parameter and bandgap energy independently by adjusting the mole fraction of Group III and Group V elements independently. The growth of quaternary InGaAsP materials, with the mixed As and P atoms on the Group V sublattice, requires tight control of the growth temperature to achieve both uniformity and run-to-run repeatability. This is because of the temperature sensitivity of the distribution coefficient of P due to the temperature dependence of the PH_3 pyrolysis efficiency in the typical growth temperature range employed for the deposition of InGaAsP epitaxial films (T_g = 500–650 °C).

6.6.3 GaSb-based Materials

GaSb-based materials are important in photonic devices operating in the long-wavelength IR spectral region ($\lambda > 3\,\mu m$). The growth of III-V antimony (Sb) semiconductor materials by MOCVD[76] is different in some important ways from that of III-AsP materials. As noted above, the Sb-containing III-V semiconductor materials, such as GaSb, InSb, InGaSb, GaAsSb, *etc.*, generally employ the metal-organic precursor trimethylantimony [$(CH_3)_3Sb$, TMSb] for Group V source, mainly due to instability of the Sb hydride precursor (SbH_3). In addition, the optimum V/III ratio is much lower than for other III-V semiconductor materials due to the relatively low vapor pressure of Sb. For III-As, III-P and III-N materials, a V/III ratio much higher than unity is required for optimum growth to compensate for the high volatility of Group V elements; otherwise, Group III metallic droplets can form on the surface. Conversely, Sb metallic droplets can be formed if the V/III ratio is too high. The low melting point (T_m) of Sb-containing III-V materials, especially for InSb ($T_m = 525\,°C$), limits the maximum temperature that can be used for MOCVD growth. The typical T_g for $In_xGa_{1-x}Sb$ depends on the In mole fraction and it increases from 475 °C for InSb to 550–600 °C for GaSb. The lower growth temperature limits the pyrolysis efficiency of TMSb, which causes Group V element deficiency, but, at the same time, the V/III ratio cannot be increased too much due to the low vapor pressure of Sb. Control of the V/III ratio is critical in the MOCVD growth of III-Sb semiconductor materials. The low growth temperature limitation also introduces complexities in the growth mechanisms. Growth at lower temperatures can be in the surface-reaction-kinetically limited regime rather than the diffusion-limited regime as observed for the growth of III-AsP materials. Also, III-Sb materials are generally *p*-type when the films are not intentionally doped, in contrast to other III-V compound semiconductors, which generally show *n*-type conductivity for unintentionally doped materials. Silicon may not be effective for *n*-type doping and so alternative donors, Se or Te, are often proposed for use as *n*-type dopant in Sb-based devices.[77]

6.6.4 GaN-based Materials

GaN-based materials are used from visible-green ($\lambda \sim 540$ nm with InGaN) to deep-UV ($\lambda \sim 280$ nm with AlGaN) photonic-device applications and high-speed and high-power electronic device applications.

In contrast to other III-V semiconductors for which the growth generally always begins with a homoepitaxial buffer layer grown on lattice matched bulk substrates, to date, GaN-based materials are generally not grown on bulk GaN substrates. This is because, until recently, "bulk" GaN substrates were not commercially available. The bulk growth of GaN crystals in the same way that is employed in other III-V semiconductors, typically from a liquid melt, is extremely difficult due to the very high vapor pressure of nitrogen at the melting temperature of GaN.[78] Alternative methods of GaN bulk crystal growth have been investigated, such as the ammonothermal crystal growth technique and the Na flux method;[79] however, the quality of bulk substrates does not meet the requirements to be used in actual device applications (as of the time of this writing). Instead, GaN-based materials are grown on a GaN buffer or "template" layer that is grown on a foreign substrate by strained heteroepitaxy. The GaN buffer layer used in many III-N optoelectronic and electronic devices is typically grown on *c*-plane (0001) sapphire or (0001) SiC substrates by MOCVD or HVPE (hydride vapor phase epitaxy). The threading dislocation density of heteroepitaxial GaN layers grown on these non-native substrates decreases with increasing thickness of the GaN layer. Because GaN film growth by HVPE can have a high growth rate, it can be used to grow layers on the order of several mm thick. As an alternative to preparing traditional "bulk" GaN substrates, a thick GaN layer is grown on a foreign substrate, *e.g.*, on (0001 SiC), by HVPE heteroepitaxy followed by removal of the foreign substrate. GaN substrates prepared in such a way are called "free-standing bulk" GaN substrates to distinguish them from GaN substrates prepared by more traditional melt-growth technologies.

6.6.4.1 GaN Material

For heteroepitaxial growth of GaN, the growth of GaN heteroepitaxial layers directly on top of a foreign substrate, *e.g.*, (0001) sapphire, results in the GaN layer having a very high density of crystalline defects and very rough surface morphology that cannot be used for device applications. This is unsurprising considering the huge lattice mismatch ($\sim$12.5%) between the GaN and sapphire bulk in-plane lattice parameters. To prepare a device-quality platform (called a template) for the subsequent heteroepitaxial growth, a two-step growth procedure has been developed.[80] In the two-step heteroepitaxial growth of GaN on non-native substrates, a well-designed heteroepitaxial low-temperature buffer layer is introduced, as opposed to a high-temperature heteroepitaxial buffer in the case of conventional III-V heteroepitaxial growth processes, *e.g.*, the growth of AlGaAs on GaAs. For the GaN growth on a basal-plane (0001) sapphire substrate, for example, a thin ($\sim$10–50 nm) AlN or GaN layer is deposited at a low growth temperature of $T_g = 500$–$800\,^{\circ}C$ (the low-temperature buffer step), subsequently followed by GaN layer growth at high growth temperatures of $T_g = 1000$–$1100\,^{\circ}C$ (the high-temperature buffer step). The low-temperature buffer layer can act as a nucleation layer for III-N materials on sapphire substrates and can also act as a strain-absorbing layer. In the process of two-step heteroepitaxial growth, nucleation, island formation, island coalescence, lateral growth and two-dimensional (layer-by-layer) growth occur sequentially, as the deposition proceeds.[81]

GaN can be effectively doped with various donors but Si is the most common dopant for *n*-type doping and diluted silane (SiH_4) is used for the precursor, since the distribution coefficient of SiH_4 becomes high and temperature-independent at typical GaN growth temperatures. For *p*-type doping of GaN, Mg is known to be the only effective dopant in GaN and doping is accomplished using Cp_2Mg as a precursor. To achieve *p*-type conduction in Mg-doped GaN layers, dopant atom activation annealing under a N_2 ambient is required to make Mg atoms electrically active as acceptors; however, even after activation, the electrically active Mg dopant among incorporated Mg in the crystal is less than 5%. This is partially due to the relatively "deep" location of acceptor levels inside the energy gap of GaN. The activation energy of Mg-doped GaN is reported to be 150–200 meV.[82] This is very large considering that the thermal energy at 300 K, k_BT, is $\sim$25.9 meV.

6.6.4.2 InGaN Materials

III-N ternary InGaN materials are critical elements in high-brightness visible violet, blue and green LEDs and they are generally used in the device active region as the material of which the ultrathin quantum-wells (QWs) are composed. The growth of InGaN is challenging in that the optimum growth conditions of the two terminal binary semiconductors of which InGaN is composed, *e.g.*, InN and GaN, are very different. The In–N bond strength of InN is very weak, which requires low growth temperatures to obtain ternary InGaN layers with a relatively high [In] mole fraction. The In incorporation efficiency is a strong function of the growth temperature and the H_2 partial pressure in the chamber; both these parameters have to be lowered as the desired [In] mole fraction in the solid is increased.[83] Increasing the molar flow rate input of the In precursor (TMIn) in an attempt to compensate for the low In incorporation efficiency is not generally effective and often yields In droplets on the growing surface rather than the incorporation of more In onto lattice, if the TMIn input is excessive.[84] Enhancing the In incorporation by decreasing the growth temperature, however, occurs at the expense of crystalline quality. Therefore, the optimum T_g for InGaN growth needs to be determined considering the trade-off between the requirements for [In] mole fraction and the need for high material quality. InGaN growth also requires very high V/III ratios (often higher than the order of magnitude of 10^3, depending on the reactor chamber geometry and other growth parameters) to compensate for the extremely low pyrolysis efficiency of NH_3 in the T_g range of InGaN and to provide enough atomic N vapor pressure over the growing surface. InGaN materials are also thermodynamically unstable and show phase separation.[85] Spatial fluctuations of the [In] mole fraction of InGaN layers, with high and low [In] mole fraction regions caused by phase separation, creates local variation of the electron potential and this is believed to be beneficial in the operation of blue and violet light-emitting devices despite the existence of a high density of threading dislocations in III-nitride based visible LEDs. In this model, electrons and holes are localized at the potential minima (created by the In fluctuations) before they recombine radiatively, preventing them from diffusing to defects that act as a non-radiative recombination pathway.[86]

6.6.4.3 AlGaN Materials

For the growth of $Al_xGa_{1-x}N$ ternary alloys, parasitic reactions between Group III trimethyl alkyl precursor of Al (TMAl) and NH_3 can create a TMAl:NH_3 adduct that can limit the growth efficiency and the mole fraction of [Al] in the solid phase because the adduct (and its by-products) is stable over a wide temperature range. Minimization of the parasitic reaction requires the separation of Group III and Group V precursor until they arrive at the growth surface by careful precursor injector design, reduced chamber pressure and/or lower V/III ratio.[87] Also, the Al–N bond of AlN is very strong, which requires a higher growth temperature and a lower V/III ratio to improve the crystalline quality of AlGaN materials with increasing [Al] mole fraction.[88] With increasing [Al] mole fraction, the bandgap becomes wider and the material becomes more insulating. For AlGaN with high Al composition, *e.g.* $x > 0.5$, effective *n*-type and especially *p*-type doping is extremely difficult. The activation energy of Mg acceptor states in *p*-type AlGaN:Mg is even higher than that of Mg-doped GaN and the acceptor energy state moves even farther from the valence band with increasing [Al], and *n*-type doping, *e.g.*, with Si donors, also becomes difficult for AlGaN with [Al] >0.5.[89]

6.6.4.4 Lateral Epitaxial Overgrowth

For GaN-based materials, lateral epitaxial overgrowth (LEO) has been developed primarily as an attempt to reduce the threading dislocation density in GaN film grown by mismatched hetero-epitaxy.[90] Lateral epitaxial overgrowth utilizes selective-area growth by employing (mostly stripe) patterned dielectric masks such as SiO_2. This technique of localized growth using masked regions to define the geometry of epitaxial layers has been used for many of the III-V semiconductors. The growth of the semiconductor film is suppressed on the dielectric mask due to the almost zero sticking coefficient of Ga ad-atoms on the dielectric mask, while the growth continues on the exposed semiconductor "substrate area" where there is no dielectric mask (called the *window*). Such selective-area vertical growth on the exposed window regions is followed by both the lateral and vertical growth when the initial vertical growth proceeds beyond the mask thickness.[91] The lateral and vertical growth rates depend on the pressure and temperature of the overgrowth,[92] the orientation of stripe mask openings and the ratio of window and mask area.[93] After the epitaxial layer thickness exceeds that of the dielectric mask layer, lateral overgrowth occurs over the mask area and this region of the semiconductor film over the dielectric mask is called the *wing*. The wing fronts coming from the two neighboring window regions that grow laterally in opposite directions eventually merge. Vertical propagation of the high density of threading dislocations in the GaN layer below the mask that existed before the overgrowth is suppressed in material grown over the mask in the wing region, resulting in a much reduced density of threading dislocation in the wing region by an order magnitude of 10^2–$10^3\,cm^{-2}$, while the threading dislocations extend over window region. As an alternative to LEO, a slightly different approach, called "pendeo epitaxy"™, has been developed for GaN films.[94] "Pendeo" (from the Latin, to hang or to be suspended) epitaxy is different from LEO in that the growth does not initiate on open windows but is forced to initiate on the sidewall of etched GaN mesas that have a dielectric mask on the top of the mesa.

While these specialized lateral growth technologies were developed specifically for reducing the very high threading dislocation density of GaN heteroepitaxial structures grown on sapphire and SiC substrates (typically, $\sim 10^9$–$10^{10}\,cm^{-2}$), the current-generation technology of advanced III-N devices is more focused on the growth on "bulk" GaN substrates as described above. Several companies now provide large-area (2-inch diameter and larger) "free-standing" GaN substrates, which permit the homoepitaxial growth of GaN-based devices with relatively low ($<10^5\,cm^{-2}$) dislocation densities. The growth of high-performance III-N injection lasers, for example, is exclusively carried out on free-standing GaN substrates. Consequently, the lateral-growth technologies described above have become of less practical interest in the past few years.

6.7 MOCVD in the Future

The MOCVD process for the growth of III-V materials initially developed and explored by Manasevit in 1968 for the growth of GaAs has in the last 40 years been further studied, refined and developed by many others to become the dominant technology, exploited world-wide, for the growth of virtually all III-V compound semiconductor epitaxial device structures, including the most advanced lasers, heterojunction bipolar transistors, light-emitting diodes, solar cells and other devices. Clearly, in the near future, this expansion of the technology into other commercial device production applications will continue to dominate the field. This is especially true for the large-scale production of III-V semiconductor solar cells for concentrator applications and for the continued rapid expansion of visible laser and LED production. In fact, as III-N-based LED lighting enters the mainstream applications for large-scale white-light illumination in commercial and industrial buildings, the demand for MOCVD growth systems, precursors and substrates will expand dramatically.

Other applications, including the use of AlGaN/GaN heterostructures for various high-power electronic devices, will enable many new, compact high-performance defense-related systems, including improved millimeter-wave and microwave radars, high-efficiency electronic power systems, compact electric vehicles and lighter-weight communication systems.

While it is difficult to predict the role of MOCVD in the production of III-V semiconductor materials and devices in the next 40 years, at this time it appears that it will continue to dominate the growth of III-V epitaxial structures for device commercial applications for virtually all new products that will be developed.

6.8 Summary and Conclusions

Metal-organic chemical vapor deposition technology has been developed into a highly advanced, flexible and efficient epitaxial growth technology for the production of high-quality III-V epitaxial and heteroepitaxial films. While some fundamental process steps remain to be defined, studied and understood in detail, for most processes over-all views of the chemical reactions involved have been developed to the point where computer-based simulations of chemical species and the resulting growth rates in real reactor systems can be calculated and useful results obtained even for the relatively complex reactor geometries used in current-generation systems.

Advances in the analysis, purification and production of metal-organic (as noted above, these can be alternatively called metal alkyl or organometallic) sources have resulted in the development of high-efficiency chemical precursor synthesis reactors capable of the production of several metric tons of high-purity product in a year. Studies of how the purity and properties of metal-organic sources affect the materials quality and, hence, how specific impurities affect the performance of devices has led to the understanding of how to optimize precursor production process and systems to provide the necessary quality at an economically viable price. Further advances in synthesis efficiencies will lead to better cost–benefit ratios for devices and thus to even more demand for these materials and to the continued expansion of precursor production facilities throughout the developed world.

References

1. (a) H. Welker, *Zeitschrift fur Naturforschung*, 1952, **7a**, 744; (b) H. Welker, *Zeitschrift fur Naturforschung*, 1953, **8a**, 248.
2. (a) J. I. Pankove, *Chapter 6. Radiative Transitions* in *Optical Processes in Semiconductors*, Dover, New York, New York, 1971; (b) P. Bhattacharya, *Chapter 3. Optical Processes in*

Semiconductors in *Semiconductor Optoelectronic Devices*, Prentice Hall, Upper Saddle River, New Jersey, 1997.

3. (a) C. Hammond, *Introduction to Crystallography*, Oxford University Press, Oxford, UK, 1992; (b) C. Kittel, *Chapter 1. Crystal Structure* in *Introduction to Solid State Physics*, John Wiley & Sons, Hoboken, New Jersey, 2005.
4. T. Matsuoka, H. Okamoto, M. Nakao, H. Harima and E. Kurimoto, *Appl. Phys. Lett.*, 2002, **81**, 1246.
5. Zh I. Alferov, *Semiconductors*, 1998, **32**, 1.
6. L. G. Meners and H. H. Wieder, *Materials Science Report*, 1988, **3**, 139.
7. M. Passlack, *J. Vac. Sci. Technol. B*, 2005, **B 23**, 1773.
8. J. M. Dallesasse, N. Holonyak Jr., A. R. Sugg, T. A. Richard and N. El-Zein, *Appl. Phys. Lett.*, 1990, **57**, 2844.
9. (a) H. Landolt and R. Bornstein, Numerical Data and Functional Relationships in Science and Technology, Vol. 17 Semiconductors, ed. O. Madelung, M. Schultz and W. Weiss, Springer-Verlag, Berlin, Germany, 1982; (b) H. Landolt and R. Bornstein, Numerical Data and Functional Relationships in Science and Technology, Vol. 22 *Semiconductors (Supplements and extension to Vol. 17)*, ed. O. Madelung, M. Schultz and W. Weiss, Springer-Verlag, Berlin, 1987; (c) Ioffe Physico-Technical Institute, Electronic archive, New Semiconductor Materials: Characteristics and Properties, http://www.ioffe.rssi.ru/SVA/NSM/(accessed on Aug. 2008); (d) S. Adachi, *Optical Constants of Crystalline and Amorphous Semiconductors: Numerical Data and Graphical Information*, Kluwer Academic Publisher, Norwell, Massachusetts, 1999.
10. M. Mansuripur and G. Sincerbox, *Proc. IEEE*, 1997, **85**, 1780.
11. J. A. Savage, *Materials Science Report*, 1987, **2**, 99.
12. E. F. Schubert, *Light Emitting Diodes*, Cambridge University Press, Cambridge, United Kingdom, 2006.
13. (a) M. G. Craford, *IEEE Circuits and Devices Magazine*, 1992, **8** Sep., 24; (b) D. A. Vanderwater, I.-H. Tan, G. E. Hofler, D. C. Defevere and F. A. Kish, *Proc. IEEE*, 1997, **85**, 1752.
14. (a) S. Nakamura, *Science*, 1998, **281**, 956; (b) T. Mukai, *IEEE J. Select. Topic. Quantum Electron.*, 2002, **8**, 264.
15. (a) N. Holonyak Jr, *Proc. IEEE*, 1997, **85**, 1678; (b) G. P. Agrawal and N. K. Dutta, *Semiconductor Lasers*, 2nd ed., Van Nostrand Reinhold, New York, New York, 1993; (c) L. A. Coldren and S. W. Corzine, *Diode Lasers and Photonic Integrated Circuits*, Wiley-Intersience, New York, New York, 1995.
16. *Vertical-Cavity Surface-Emitting Lasers*, ed. C. Wilmsen, H. Temkin and L. A. Coldren, Cambridge University Press, Cambridge, United Kingdom, 1999.
17. S. R. Forrest, *IEEE Spectrum*, 1986, May, 76.
18. J. J. Loferski, *Progress in Photovoltaics: Research and Applications*, 1993, **1**, 67.
19. W. Liu, *Fundamentals of III-V Devices, HBTs, MESFETs, and HETs/HEMTs*, Wiley-Intersience, New York, New York, 1999.
20. P. M. Asbeck, M. F. Chang, K.-C. Wang, D. L. Miller, G. J. Sullivan, N. H. Sheng, E. Sovero and J. A. Higgins, *IEEE Trans. Microwave Theory Tech.*, 1987, **MTT–35**, 1462.
21. H. Kroemer, *Proc. IRE*, 1957, **45**, 1535.
22. H. Morkoc and P. Solomon, *IEEE Spectrum*, 1984, Feb., 28.
23. R. Dingle, H. L Stormer, A. C. Gossard and W. Wiegmann, *Appl. Phys. Lett.*, 1978, **33**, 665.
24. H. M. Manasevit, *Appl. Phys. Lett.*, 1968, **12**, 156.
25. This company originated from the North American Aviation Aircraft Company (famed for the design and production of the P51 Mustang fighter plane of WWII) and subsequently became Rockwell International Corporation.
26. R. A. Ruehrwein, Production of epitaxial film, *U.S. Pat.* 3312571, 1967.

27. (a) R. Didchenko, J. E. Alix and R. H. Toeniskoetter, *J. Inorg. Nucl. Chem.*, 1960, **14**, 35; (b) B. C. Harris and E. H. Tomkins, *Inorg. Chem.*, 1962, **1**, 951.
28. H. M. Manasevit published many papers on the first use of MOCVD for the heteroepitaxial growth of III-V, II-VI, and IV-IV semiconductors on various insulating substrates, only a few of which are listed in this chapter.
29. (a) H. M. Manasevit and W. I. Simpson, *J. Electrochem. Soc.*, 1969, **116**, 1725; (b) H. M. Manasevit, *J. Crystal Growth*, 1972, **13/14**, 306; (c) H. M. Manasevit, *J. Crystal Growth*, 1981, **55**, 1.
30. H. M. Manasevit and W. I. Simpson, Metalorganic chemical vapor deposition of IVA-IVA compounds and composite, *U.S. Pat.* 4066481, 1978.
31. A description of the MOCVD process was omitted from this paper because the North American Rockwell legal department did not want the process revealed since they had decided to patent it.
32. H. M. Manasevit, F. M. Erdman and W. I. Simpson, *J. Electrochem. Soc.*, 1971, **118**, 1864.
33. (a) R. D. Dupuis, P. D. Dapkus, R. D. Yingling and L. A. Moudy, *Appl. Phys. Lett.*, 1977, **31**, 201; (b) R. D. Dupuis and P. D. Dapkus, *Appl. Phys. Lett.*, 1977, **31**, 466.
34. N. Holonyak Jr, R. M. Kolbas, R. D. Dupuis and P. D. Dapkus, *Appl. Phys. Lett.*, 1978, **33**, 73.
35. In the late 1950s compounds containing metal–carbon bonds were becoming recognized as being interesting for various applications in the chemical field. The dominant term for this class of compounds at the time was "metal-organics." As this field developed, the term metal alkyl was applied to specific compounds containing alkyl groups having the general formula C_nH_{2n+1}. The term "organometallic" is now more commonly used for the set of chemical compounds containing metal–carbon bonds. However, the original name for the MOCVD process used the term "metal-organic," emphasizing the "metal" part of the compound, which is of primary interest for MOCVD, and so we shall retain the original term. There is no need for any other term for this process as it is completely accurate and descriptive to call it "MOCVD" and, in fact, it is the term dominantly used in the literature today despite earlier attempts to "rename" the process.
36. (a) K. F. Jensen, D. I. Fotiadis and T. J. Mountziaris, *J. Crystal Growth*, 1991, **107**, 1, and references therein; (b) C. A. Larsen, N.I. Buchan and G. B. Stringfellow, *Appl. Phys. Lett.*, 1988, **52**, 480, and references therein.
37. K. Tamaru, *J. Phys. Chem.*, 1955, **59**, 777.
38. R. Swalin, *Thermodynamics of Solids*, Wiley-Interscience, Hoboken, New Jersey, 1972.
39. C. R. Barrett, A. S. Tetelman and W. D. Nix, *Chapter 5. Kinetics in The Principles of Engineering Materials*, Prentice Hall, Upper Saddle River, New Jersey, 1973.
40. T. Mountziaris and K. F. Jensen, *J. Electrochem. Soc.*, 1991, **138**, 2426, and references therein.
41. D. W. Shaw, *J. Crystal Growth*, 1975, **31**, 130.
42. D. W. Shaw, *J. Electochem. Soc.*, 1968, **115**, 405.
43. G. Costrini and J. J. Coleman, *J. Appl. Phys.*, 1985, **57**, 2249.
44. (a) W. Breaker and A. L. Mossman, *Matheson Gas Data Book*, Matheson, New Jersey, 1980; (b) G. D. Clayton and F. E. Clayton, *Patty's Industrial Hygiene and Toxicology*, Vol. **2A**, General Principles, 3rd ed., Wiley, New York, New York 1978.
45. (a) K. L. Hess and R. J. Riccio, *J. Crystal Growth*, 1968, **77**, 95; (b) E. Johnson, R. Tsui, D. Convey, N. Mellen and J. Curless, *J. Crystal Growth*, 1984, **68**, 497.
46. 316L stainless steel is a standard type by ASTM (American Society for Testing and Materials) International and (AISI) American Iron and Steel Institute. 316L designates the Austenitic-phase stainless steel containing chromium ($\sim$17%), nickel ($\sim$12%), and molybdenum ($\sim$2%) with minimized carbon content for improved corrosion resistance.

47. J. B. Hunter, Ultrapure hydrogen by diffusion through palladium alloys, *Symposium on the Production of Hydrogen Petroleum Division*, The American Chemical Society Meeting, New York, Fall, 1963.
48. D. R. Gaskell, *Chapter 7. Phase equilibrium of a one-component system*, in *Introduction to the Thermodynamics of Materials*, 3rd ed., Taylor and Francis, Washington DC, 1995.
49. Thomas Swan Scientific Equipment Ltd. is headquartered in Cambridge, U.K; the company was acquired by Aixtron group and is now called Aixtron Ltd.
50. (a) C. Martin, M. Dauelsberg, H. Protzmann, A. R. Boyd, E. J. Thrush, M. Heuken, R. A. Talalaev, E. V. Yakovlev and A. V. Kondratyev, *J. Crystal Growth*, 2007, **303**, 318; (b) E. J. Thrush, M. J. Kapers, P. Dawson, M. E. Vickers, J. Barnard, D. Graham, G. Makaronidis, F. D. G. Rayment, L. Considine and C. J. Humphreys, *J. Crystal Growth*, 2003, **248**, 518.
51. (a) T. Arai, J. Hidaka, H. Tokunaga and K. Matsumoto, *J. Crystal Growth*, 1997, **170**, 88; (b) H. Tokunaga, I. Waki, A. Yamaguchi, N. Akutsu and K. Matsumoto, *J Crystal Growth*, 1998, **189/190**, 519.
52. Taiyo Nippon Sanso Corporation is headquartered in Tokyo, Japan.
53. K. Matsumoto, T. Arai and H. Tokunaga, *Vacuum*, 1998, **51**, 699.
54. (a) H. Jurgensen, *Microelectronics Engineering*, 1992, **18**, 119; (b) D. Brien, M. Dauelsberg, K. Christiansen, J. Hofeldt, M. Deufel and M. Heuken, *J. Crystal Growth*, 2007, **303**, 330.
55. Aixtron AG is headquartered in Aachen, Germany.
56. D. Schmitz, G. Stauch, H. Jurgensen and M. Heyen, *J. Crystal Growth*, 1991, **107**, 188.
57. (a) M. A. McKee, P. E. Norris, R. A. Stall, G. S. Tompa, C. S. Chern, D. Noh, S. S. Kang and T. J. Jasinski, *J. Crystal Growth*, 1991, **107**, 445; (b) G. S. Tompa, P. A. Zawadzki, K. Moy, M. McKee, A. G. Thomson, A. I. Gurary, E. Wolak, P. Esherick, W. G. Breiland, G. H. Evans, N. Bulitka, J. Hennessy and C. J. L. Moore, *J. Crystal Growth*, 1994, **145**, 655; (c) G. S. Tompa, W. G. Breiland, A. Gurary, P. A. Zawadzki, G. H. Evans, P. Esherick, B. Kroll and R. A. Stall, *Microelectronics Journal*, 1994, **25**, 757.
58. Veeco Compound Semiconductor Inc. TurboDisc MOCVD Systems is headquartered in Somerset, NJ, U.S.A.
59. (a) G.E. Coates, M.L.H. Green, K. Wade, *Organometallic Compounds*, 1967, **1**, Main Group Elements, Methune & Company Ltd. 1967; (b) J. J. Eisch, *J. Amer. Chem. Soc*, 1962, **84**, 3605; (c) C.J. Smit, A.J. Van Der Lee, G.J.M. Van Eijden, *US Patent*, 1995, 5.473,090; (d) C.J. Smit, E.W. J. Van Hannik, G.J.M. Van Eijden, US Pat., 1993, 5,248,800.
60. A. C. Jones, *CVD of Compound Semiconductors*, 1997, Chapter 2, 43, VCH, Weinheim, Germany, 1997.
61. (a) I. Bertenyi and R. M. Barnes, *Anal. Chem.*, 1986, **58**, 1734; (b) A. C. Jones, G. Wales, P. J. Wright and P. E. Oliver, *Chemtronics*, 1987, **2**, 83.
62. S. A. Rushworth, L. M. Smith, M. S. Ravetz, K. M. Coward, R. Odedra, R. Kanjolia, S. W. Bland, F. Dimroth and A. W. Bett, *J. Crystal Growth*, 2003, **248**, 86.
63. K. B. Starowieyski, M. Kaczorek and K. Pakula, *J. Organomet. Chem.*, 2000, **601**, 133.
64. (a) L. M. Smith, R. Odedra, A. J. Kingsley, K. M. Coward, S. A. Rushworth, G. Williams, T. A. Leese, A. J. Purdie and R. K. Kanjolia, *J. Crystal Growth*, 2004, **272**, 37; (b) H. Cunning, G. W. Williams, R. Odedra, and R. K. Kanjolia, WO 2007057631, May 24, 2007; M. L.. Timmons, R. Colby, and R. S. Stennick, U.S Patent 6607785; N. H Tran, D. L. Deavenport, T. Ko, and N. El-Zein, *U.S. Pat.* 20070221127.
65. (a) D. M. Frigo, and A. W. Gal, *U.S. Pat.*, 5232860; (b) M. S. Ravetz, L. M. Smith, S. A. Rushworth, A. B. Leese, R. Kanjolia, J. I. Davies and R. T. Blunt, *J. Electron. Mater.*, 2000, **29**, 156; (c) Y. Qi, C. Musante, K. M. Lau, L. Smith, R. Odedra and R. Kanjolia, *J. Electron. Mater.*, 2001, **30**, 1382.
66. R. K. Kanjolia, R. Odedra, A. J. Kingsley, K. M. Coward, S. Travis, S. A. Rushworth and L. M. Smith, *J. Crystal Growth*, 2004, **261**, 236.

67. S. Salim, C. K. Lim and K. F. Jensen, *Chem. Mater.*, 1995, **7**, 507.
68. M. Behet, B. Stoll and K. Heime, *J. Crystal Growth*, 1994, **134**, 434.
69. C. A. Wang, S. Salim, K. F. Jensen and A. C. Jones, *J. Crystal Growth*, 1997, **170**, 55.
70. E. Bourret-Courchesne, Q. Ye, D. W. Peters, J. Arnold, M. Ahmed, S. J. C. Irvine, R. Kanjolia, L. M. Smith and S. A. Rushworth, *J. Crystal Growth*, 2000, **217**, 47.
71. (a) N. Tansu, J. Y. Yeh and L. J. Mawst, *IEEE J Select. Topic. Quantum Electronics*, 2003, **9**, 1220; (b) Q. Gao, H. H. Tan, C. Jagdish, B. Q. Sun, M. Gal, L. Ouyang and J. Zou, *J. Crystal Growth*, 2004, **264**, 92.
72. M. J. Tsai, M. M. Tashima and R. L. Moon, *J. Electron. Mater.*, 1984, **13**, 437.
73. R. Bhat, P. O'Connor, H. Temkin, R. Dingle and V. G. Keramidas, *Inst. Phys. Conf. Ser.*, 1982, **63**, 101.
74. A. Gomyo and T. Suzuki, *Phys. Rev. Lett.*, 1988, **60**, 2645.
75. Y. Ohba and A. Hatano, *J. Crystal Growth*, 1994, **145**, 214.
76. R. M. Biefeld, *Materials Science and Engineering: Report*, 2002, **R36**, 105.
77. C. von Eichel-Streiber, M. Behet, M. Heuken and K. Heine, *J. Crystal Growth*, 1997, **170**, 783.
78. I. Grzegory, *J. Phys: Condens. Matter.*, 2001, **13**, 6875.
79. A. Denis, G. Goglio and G. Demazeau, *Materials Science and Engineering: Report*, 2006, **R50**, 167.
80. H. Amano, N. Sawaki, I. Akasaki and Y. Toyoda, *Appl. Phys. Lett.*, 1986, **48**, 353.
81. K. Hiramatsu, S. Itoh, H. Amano, I. Akasaki, N. Kuwano, T. Shiraishi and K. Oki, *J. Crystal Growth*, 1991, **115**, 628.
82. J. K. Sheu and G. C. Chi, *J. Phys: Condens. Matter.*, 2002, **14**, R657.
83. M. Bosi and R. Fornari, *J. Crystal Growth*, 2004, **265**, 434.
84. O. V. Bord, R. A. Talalaev, S. Yu. Karpov and Yu. N. Makarov, *Phys. Stat. Sol. (a)*, 1999, **176**, 279.
85. A. Koukitu and H. Seki, *J. Crystal Growth*, 1998, **189/190**, 13.
86. S. Nakamura, *Science*, 1998, **281**, 956.
87. C. H. Chen, H. Liu, d. Steigerwald, W. Imler, C. P. Kuo and M. G. Craford, *J. Electron. Mater.*, 1996, **25**, 1004.
88. (a) S. Keller, G. Parish, P. T. Fini, S. Heikman, C.-H. Chen, N. Zhang, S. P. DenBaars and U. K. Mishra, *J. Appl. Phys.*, 1999, **86**, 5850; (b) N. Fujimoto, T. Kitano, G. Narita, N. Okada, K. Balakrishnan, M. Iwaya, S. Kamiyama, H. Amano, I. Akasaki, K. Shimono, T. Noro, T. Takagi and A. Bandoh, *Phys. Stat. Sol.* (c), 2006, **3**, 1617.
89. J. K. Sheu and G. C. Chi, *J. Phys.: Condens. Matter.*, 2002, **14**, R657.
90. O.-H. Nam, M. D. Bremser, T. S. Zheleva and R. F. Davis, *Appl. Phys. Lett.*, 1997, **71**, 2638.
91. T. S. Zheleva, O.-H. Nam, M. D. Bremser and R. F. Davis, *Appl. Phys. Lett.*, 1997, **71**, 2472.
92. K. Hiramatsu, K. Nishiyama, A. Motogaito, H. Miyake, Y. Iyechika and T. Maeda, *Phys. Stat. Sol.* (a), 1999, **176**, 535.
93. J. Park, P. A. Grudowski, C. J. Eiting and R. D. Dupuis, *Appl. Phys. Lett.*, 1998, **73**, 333.
94. K. Linthicum, T. Gehrke, D. Thomson, E. Carson, P. Rajagopa, T. Smith, D. Batcherlor and R. F. Davis, *Appl. Phys. Lett.*, 1999, **75**, 196.

CHAPTER 7

Chemical Vapor Deposition of Metals: W, Al, Cu and Ru

BING LUO AND WAYNE L. GLADFELTER

Department of Chemistry, University of Minnesota, Minneapolis, MN 55455, USA

7.1 Introduction

Metal films are used as diffusion barriers, anticorrosion surfaces, interconnects and resistors in microelectronics, reflective surfaces, antifriction and wear surfaces and decorative applications. They can be deposited by dip and powder coating methods, electrochemical and electroless depositions from liquid solutions and by physical vapor deposition (PVD) and chemical vapor deposition (CVD). Following several general comments, this chapter focuses on the CVD of a select group of metals: tungsten, aluminium, copper and ruthenium. These have been studied thoroughly and have current and/or future industrial significance. Several reviews related to metal CVD have been published.[1–9]

7.1.1 Deposition Chemistry – General Comments

A *useful* metal CVD process requires a *thermodynamically* and *kinetically* accessible route to *reduce* a *volatile* metal complex to the zerovalent state. To emphasize the need for reduction, consider the sample of known precursors for various metals shown in Figure 7.1. Among the common precursors, only the metal carbonyls begin the process in the formal, zerovalent state. While this would seem to make them ideal candidates for CVD, this is true only for $Ni(CO)_4$[10,11] because in most cases the metal surface is active enough to split CO, leaving behind metal films contaminated with carbon and oxygen.

Of the 67 naturally occurring metallic elements, useful CVD processes have been identified for less than half of these. Accessible elements include those bordering the metalloid line among the p-block elements to the vanadium triad in the d-block. None of the elements to the left of this group, including the f-block elements have been deposited by a *useful* CVD process. These represent the most electropositive elements in the periodic table, which are typically found as compounds in their highest possible oxidation state. Such compounds are difficult to reduce.

Chemical Vapour Deposition: Precursors, Processes and Applications
Edited by Anthony C. Jones and Michael L. Hitchman

Published by the Royal Society of Chemistry, www.rsc.org

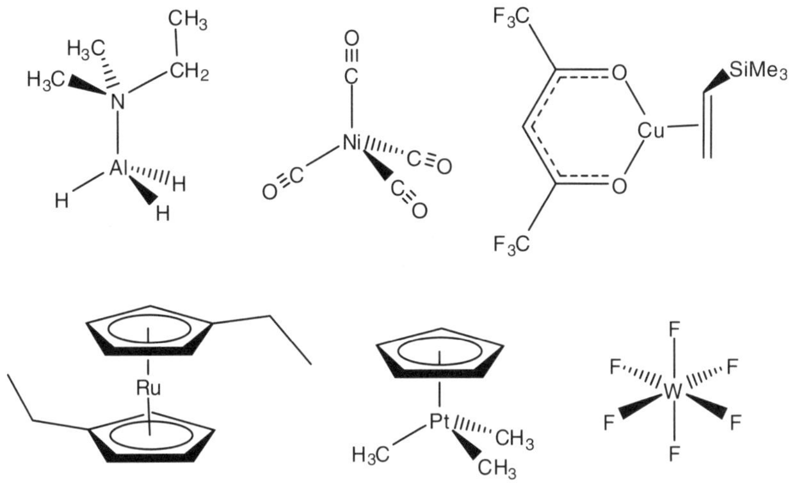

Figure 7.1 Selected metal precursors.

There are three generic reactions that can be used for precursor reduction and deposition:

1. Addition of an external reducing agent such as hydrogen;
2. Reductive elimination of ligands;
3. Disproportionation.

Equation 7.1 illustrates the use of hydrogen to reduce WF_6 to tungsten metal. Hydrogen is used in many depositions even when its role as a stoichiometric reductant would seem unnecessary. In such depositions it serves both as a carrier gas and as an aid to increase the purity of the metal film. In the latter role, it helps to remove carbon, oxygen or other impurities that may be left behind by the ligands on the precursor. Other reducing agents such as silane or alcohols have been reported, but none is as widely used as H_2:

$$WF_6 + 3H_2 \rightarrow W + 6HF \tag{7.1}$$

The use of amine alanes to deposit metallic aluminium provides an excellent example of a deposition in which a reductive elimination transforms the Al^{3+} into Al^0 [Equation (7.2)]:[12]

$$(Me_2EtN)AlH_3 \rightarrow Al + 3/2H_2 + Me_2EtN \tag{7.2}$$

The hydride ligands (formally H^-) provide the electrons when they dissociate as molecular hydrogen from the aluminium surface. As emphasized later, the reductive elimination of hydrogen serves as the electron source in every useful Al CVD process even when no hydrides are present in the precursor (*e.g.*, $Al[CH_2CH(CH_3)_2]_3$).

Disproportionation involves one metal serving both as an electron donor and acceptor. Monovalent copper is one of the more common metal/oxidation state combinations that exhibits disproportionation, and it serves the basis for one of the important copper CVD processes [Equation (7.3)]:[13]

$$2Cu(hfac)(vtms) \rightarrow Cu + Cu(hfac)_2 + 2vtms \tag{7.3}$$

7.1.2 Induction Periods, Selectivity, Microstructure and the Critical Role of Nucleation

Induction periods, *i.e.* initial growth rates that are slower than found during steady state deposition, are frequently observed during the CVD of metals on nonmetallic substrates. This rate dependence on substrate has been used to deposit metal films selectively on metallic surfaces. When CVD is used to deposit metals on nonmetallic substrates, large grain microstructures often lead to rough films. These three important observations can be related to the difference between the rate of nucleation of the metal film and its rate of steady state growth.

The chemical reactions involved in the transformation of a molecular precursor into a metal film will depend on the surface. At the very least one would expect the activation parameters for similar reactions (*e.g.*, ligand dissociation) to differ on one surface *vs.* another; however, it is more probable that the reaction mechanism itself will be dependent on the surface. Metal surfaces are outstanding catalysts for many organic reactions and one can expect that their ability to catalyze reactions will decrease the activation energy of the deposition process. In essence, we should anticipate that conditions leading to selective deposition on metallic surfaces would be the norm. On nonmetallic surfaces, films having large grains are formed because the probability would be greater for a precursor to desorb or diffuse to an existing nucleus rather than react to form a new nucleus. The presence of an induction period would also be expected when depositing on a nonmetallic surface because of the higher activation energies often required for the nucleation events.

7.2 CVD of Tungsten

The thermodynamically stable form of tungsten is α-tungsten, which exhibits a body-centered cubic structure, a melting point of $3410 \pm 20\,°C$[14] and an electrical resistivity of $5.3\,\mu\Omega\,cm$ at 27 °C.[15] The meta-stable β phase exists below 630 °C and is transformed into the α phase above 700 °C.[7] The properties of W have been extensively studied and documented.[15] Tungsten shows relatively low chemical reactivity, for example, bulk W is inert to O_2 and only reacts slowly in concentrated mixtures of HNO_3 and HCl at room temperature.[16]

The extensive use of W metal as the via and plug interconnect material in VLSI commenced in mid-1980s,[17] while W CVD studies for these applications started in early 1970s.[18,19] By the mid-1990s, W CVD using the gaseous WF_6 (mp 2.3 °C and bp 17.1 °C)[20] precursor had been investigated comprehensively. These studies have been reviewed in multiple accounts.[1–5,7] In the early use of W metallization in VLSI, depositions were mainly conducted on Si substrates, with great interest in selective deposition in the presence of SiO_2. Currently, W is almost universally deposited on TiN/Ti barriers and the process involves a seed layer deposition using WF_6 and SiH_4 followed by deposition using WF_6 and H_2.[21] This method takes advantage of the merits and minimizes the problems of both depositions; depositions from WF_6 and SiH_4 afford high growth rates with short induction periods but poor step coverages while depositions from WF_6 and H_2 give conformal films but require longer induction periods.[21] The current research focus is more toward W ALD, which is promising in achieving conformal step coverages in the higher aspect ratio structures expected in future ULSI (Ultra Large Scale Integration) devices.

Other precursors, including $W(CO)_6$ and $CpW(CO)_3CH_3$, where Cp = cyclopentadienyl, were widely studied in an attempt to lower deposition temperatures and overcome problems associated with HF formation. In most cases, however, unacceptable levels of C were present in films, resulting in higher electrical resistivities.[7] Studies on these precursors have been reviewed in detail elsewhere.[7] Since then, progress on new precursor development has been limited.[22–24] In the following sections we begin by examining the reaction of WF_6 with Si and SiO_2 surfaces in the absence of an external reductant. This will set the stage for a review of the WF_6/H_2 and WF_6/SiH_4 CVD processes. Lastly, recent work on the ALD of W is surveyed.

7.2.1 Deposition on Si and SiO_2 from WF_6

The deposition on Si substrates using WF_6 is a self-limiting process. Film deposition was found to be fast initially [*e.g.*, 100 nm min^{-1} (ref. 25)], but growth stopped in the range from *ca.* 10 to 100 nm.[2,25,26] Factors influencing the final film thickness included the characteristics of the surface oxide layers and the amounts of trace H_2O present during the depositions.[2,26,27]

Ultrahigh vacuum studies on single crystal Si surfaces provided much information on the adsorption and reactivity of WF_6. WF_6 physisorbed on clean Si(100) at low temperatures and desorbed at –143 °C without decomposition.[28] Dosing WF_6 at room temperature on Si(100) and Si(111) led to dissociative adsorption, affording W and SiF_x ($x = 1-3$).[3,28–30] This self-limiting deposition stopped after all exposed Si atoms had reacted.[28,30] At elevated temperatures, WF_x desorbed, allowing further WF_6 adsorption and reaction. Mass spectrometric analysis showed that two reactions [Equations (7.4) and (7.5)],were operative with the formation of SiF_4 [Equation (7.4)] predominating below 450 °C:[3]

$$2WF_6 + 3Si \rightarrow 2W + 3SiF_4 \qquad (7.4)$$

$$WF_6 + 3Si \rightarrow W + 3SiF_2 \qquad (7.5)$$

On SiO_2, no deposition occurred from WF_6 or WF_6/H_2 below 440 °C,[31] suggesting the possibility of selective deposition on Si when both Si and SiO_2 were present. This was successful; however, loss of selectivity was a major problem, hindering the practical application of this technology. In addition to deposition temperature, other factors, including deposition rate, precursor partial pressure and deposition time, influence the effectiveness of the selective depositions. Several causes for the selectivity loss were proposed.[5] One proposal was that the partially fluorinated species such as WF_5 and WF_4 desorbed from the W surfaces, and re-adsorbed and disproportionated on the nearby SiO_2 surfaces to form W nuclei.[32,33] The observation that higher nucleation densities were detected on SiO_2 close to initial W deposit was consistent with this hypothesis.[34]

Pretreatment of the substrate with a He plasma followed by exposure to WF_6 and H_2 allowed W deposition on SiO_2.[35] The initial film growth was due to the reaction of WF_6 with elemental Si produced in the plasma treatment. Tungsten deposition on SiO_2 was also achieved by irradiating a WF_6/H_2 mixture with a 193 nm laser.[31] The laser treatment afforded an activation energy of 41 kJ mol^{-1} in the steady-state film growth – much lower than those obtained from depositions without irradiation (see below). Similarly, activating WF_6 with an Ar plasma, perhaps producing F and WF_x ($x = 1–6$) radicals, allowed W nucleation on SiO_2 at temperatures below 400 °C.[36]

7.2.2 WF_6–H_2 System

Tungsten CVD from WF_6 and H_2 proceeds according to Equation (7.1).[1,18,19,31,34,37–46] The formation of HF was detected by *in situ* mass spectrometry.[47] High-quality films with resistivities close to bulk metal were obtained.

Kinetic studies showed that depositions in the mass transport limited regime occurred above 600 °C and/or under total pressures above 150 Torr.[40] In the surface reaction limited regime, the steady-state W film growth rates were one-half order with respect to the H_2 partial pressure and zero [Equation (7.6)] or one-sixth order with respect to the WF_6 partial pressure [Equation (7.7)], where k_1 and k_2 are proportional to $e^{(-Ea/RT)}$; E_a is the activation energy.[1,25,41,42,44,46] Most studies gave an activation energy between 64 and 73 kJ mol^{-1}:

$$\text{Rate} = k_1 P_{H_2}^{1/2} \qquad (7.6)$$

$$\text{Rate} = k_2 P_{\text{H}_2}^{1/2} P_{\text{WF}_6}^{1/6} \tag{7.7}$$

The early proposal[1,25] that H_2 dissociation was the rate-limiting step was inaccurate on the basis of more recent analyses.[42] The UHV studies showed that H_2 dissociatively adsorbed on single crystal W surfaces with low energy barriers.[48–50] The H_2 initial sticking coefficients on W surfaces were high for a wide range of temperatures [*e.g.* 0.4 on W(100) at 300–500 °C].[51] Thus the H_2 dissociation on the as-grown polycrystalline W surfaces in CVD is expected to be a fast process.

More consistent with the observed kinetics, the formation or desorption of HF was proposed to be the rate-limiting step(s).[41,42,44] Formation of HF involved the reactions of the adsorbed H, giving a half-order dependence of the growth rate on the H_2 partial pressure as shown in Equation (7.6). The zero-order dependence on WF_6 was obtained by assuming that the second reactant, *e.g.*, F or WF_x ($x = 1$–6), was the most abundant surface species.[41,42,44] When the HF desorption was proposed as the rate-limiting step, the growth rate in Equation (7.7) was derived. Alternative pathways were also possible, which included reactions between the gas phase WF_6 and adsorbed H or between the gas phase H_2 and adsorbed F or WF_x ($x = 1$–6).[2] In addition, a complicated modeling study involving multiple species chemistry and a multidimensional transport model for the conservation of mass, momentum, energy and chemical species also appeared.[52]

The formation of F and partially fluorinated WF_x was supported in the UHV studies of the adsorption of WF_6 on single crystal and polycrystalline W surfaces.[53,54] A study[54] employing TPD and XPS revealed that following adsorption of WF_6 on W(100) at −173 °C it dissociated to form adsorbed WF_5 and F by −123 °C. The adsorbed WF_5 was proposed to undergo several disproportionation events to form W, F and WF_6. The as-formed WF_6 desorbed at temperatures below *ca.* 170 °C while F was removed from the surface as partially fluorinated WF_x species at higher temperatures. In this study, deuterium fluoride was not detected in the TPD after adsorption of WF_6 on a deuterium-saturated W(100) surface. Instead, the adsorbed deuterium desorbed in the form of D_2 at temperatures below 120 °C. This was due to higher activation energies for reactions between the adsorbed D and adsorbed F or WF_x ($x = 1$–6) compared to the D re-combinative desorption.[54] These UHV results did not conflict with those obtained in CVD studies where HF was the gaseous product because, in CVD, H_2 was constantly supplied and the temperatures were much higher. In fact, the UHV results supported the proposal that the adsorbed H was involved in the rate-limiting step(s).

Notably, the initial film growth adopted different reaction pathways from the steady-state growth. When the depositions were conducted on Si substrates, the initial W depositions occurred *via* the reactions of WF_6 with Si according to Equations (7.4) and (7.5).[25] This can be largely attributed to the lack of dissociative adsorption of H_2 on Si at typical CVD temperatures. UHV studies showed that the initial sticking coefficients of H_2 on Si(100) and Si(111) were of the order of 10^{-8} at 327 °C.[55]

7.2.3 WF_6–SiH_4 System

Extensive studies of W depositions through SiH_4 reduction of WF_6 started in late 1980s.[30,56–60] Earlier results using the WF_6/SiH_4 system were focused on producing tungsten silicide (WSi_x) films.[4] Tungsten and other metal silicides exhibiting lower resistivities than polysilicon had been used in conjunction with polysilicon in the gate electrodes and interconnects in VLSI technology.[4,17] WSi_x continues to be applied in current microelectronic technology and the depositions have focused on the use of WF_6 and SiH_2Cl_2.[4,61–64] Compared to the WF_6/H_2 combination, W deposition from WF_6/SiH_4 gave higher growth rates (up to 1 μm min^{-1}) at relatively low temperatures (250–500 °C) with good selectivity.[4,59] Generally, films containing low levels of Si ($\leq 3\%$) were deposited with SiH_4/WF_6 ratios of ≤ 1.[60,65–67] These films exhibited the α-W crystalline phase, and the resistivities dropped below 10 μΩ cm. Films containing higher percentages of Si and, eventually, WSi_2 were deposited as the SiH_4/WF_6 ratios exceeded one.[60,65–67] The films prepared with moderate SiH_4/WF_6 ratios, *e.g.*, between 1 and 3, typically exhibited mixed phases of α- and β-W. The β-W phase was proposed to be stabilized by the presence of Si.[60,65–67]

The W film growth kinetics using WF_6/SiH_4 were more complex than those found in the WF_6/H_2 system. Hydrogen was typically used as a carrier gas or co-dosed with WF_6 and SiH_4, but no dependence of the growth rates on H_2 partial pressures was found.[60,66–68] In the study using a hot-wall LPCVD reactor, there was no distinctly linear regime found in the plots of the logarithmic growth rates *vs.* the reciprocal deposition temperatures for several experiments with SiH_4/WF_6 ratios of less than 1.[68] The measurements were conducted at temperatures of 145 to 400 °C with a fixed total pressure of 0.5 Torr. The growth rates increased with temperature below 300 °C and decreased above it. This general trend was also observed in other studies.[59,66,69] The rate equations in the *hot-wall reactor* study[68] are expressed in Equations (7.8) and (7.9), where k_3 and k_4 were temperature-dependent rate constants:

$$\text{Rate} = k_3 P_{\text{SiH}_4} \tag{7.8}$$

for $SiH_4/WF_6 \leq 0.3$

$$\text{Rate} = k_4 P^2_{\text{SiH}_4} P^{-1}_{\text{WF}_6} \tag{7.9}$$

for $0.5 \leq SiH_4/WF_6 \leq 1.0$

For most kinetic studies conducted in *cold-wall reactors*, the growth rates abided by Equation (7.8) for SiH_4/WF_6 ratios ≤ 1.[58,66,67] In a later study, where care was taken to ensure the depositions were run in a surface-reaction limited regime,[69] for SiH_4/WF_6 ratios smaller than 0.7, the growth rate was slightly negative in P_{WF6} and nearly quadratic in P_{SiH4}. For $0.7 < SiH_4/WF_6 < 1$, there was no simple equation to correlate the growth rate with P_{SiH4} and P_{WF6}. For SiH_4/WF_6 ratios close to 1, the growth rate was limited only by gas phase diffusion. In a more recent study conducted in a non-flowing reactor, the growth rates were found to depend on the partial pressures of all reactants SiH_4, WF_6 and H_2.[70] Notably, however, the measurements were performed at 95 °C, which is much lower than those employed in typical depositions.

The detailed mechanisms were difficult to obtain because of the complexity of the reaction system.[68] The overall reactions, however, as depicted in Equations (7.10) and (7.11), were identified on the basis of *in situ* IR, AES and MS measurements.[70–72] The reaction in Equation (7.10) preferentially occurred at low SiH_4/WF_6 ratios, whereas Equation (7.11) dominated at high SiH_4/WF_6 ratios.[70,71] These studies clarified the uncertainty in an earlier UHV study in which only the reaction in Equation (7.11) was suggested:[73]

$$WF_6 + 2SiH_4 \rightarrow W + 2SiHF_3 + 3H_2 \tag{7.10}$$

$$2WF_6 + 3SiH_4 \rightarrow 2W + 3SiF_4 + 6H_2 \tag{7.11}$$

Tungsten depositions from WF_6 and SiH_4 have been studied on various substrates.[3] Selective depositions on Si in the presence of SiO_2 have also been the subject of intense research.[25,56,59,60,69,71,72,74,75] The depositions on Si substrates were most likely initiated by the reaction of WF_6 with Si as described above.[3,30,76] On TiN, nucleation proceeded by slow formation of a uniform W monolayer followed by rapid formation of three-dimensional islands.[77–79] Ultra-high vacuum studies suggested that the W monolayer was produced from the direct reaction between WF_6 and TiN[3,80] or, in case of the presence of surface oxides, *via* the reaction of WF_6 with Si that formed on the surface from the reaction of SiH_4.[81]

7.2.4 Tungsten ALD

Tungsten ALD studies have focused primarily on the WF_6/Si_2H_6 system;[82–92] however, several reports using WF_6/SiH_4 have also appeared.[93–95] Depositions using WF_6/Si_2H_6 were conducted at temperatures from 152 to 327 °C. The growth rates, determined on a W seed layer that had been prepared on SiO_2/Si by pre-exposure to several WF_6/Si_2H_6 cycles, ranged from 0.25 to 0.52 nm cycle^{-1}.[82,90] The WF_6 reaction was self-limiting. For Si_2H_6, however, a rapid rate increase at lower exposures was followed by a slower W ALD growth for Si_2H_6 exposures exceeding 4×10^4 L.[90] Based on the studies using various techniques,[83,85,86,88,90] the surface reactions involved in the ALD were proposed in Equations (7.12–7.15), where the subscript (a) indicates surface species.

For the WF_6 half cycle:

$$WSiHFSiH_{3(a)} + 2WF_6 \rightarrow 2W + WF_{4(a)} + 2SiF_4 + 3/2H_2 + HF \tag{7.12}$$

For the Si_2H_6 half cycle:

$$WF_{4(a)} + Si_2H_6 \rightarrow WSiHF_{2(a)} + SiHF_3 + 3/2H_2 \tag{7.13}$$

$$WSiHF_{2(a)} + {}^1\!/_2Si_2H_6 \rightarrow WSiHFSiH_{3(a)} + {}^1\!/_2H_2 \tag{7.14}$$

Combining Equations (7.12–7.14) gives the overall reaction:

$$2WF_6 + 3/2Si_2H_6 \rightarrow 2W + 2SiF_4 + SiHF_3 + 7/2H_2 + HF \tag{7.15}$$

The W films produced from WF_6/Si_2H_6 on SiO_2/Si were amorphous with smooth surfaces.[82] No Si and F were found in the films based on X-ray photoelectron spectroscopy (XPS). Small amounts ($\leq$4%) of C and O present in the films were attributed to contamination of the CO background. Consistent with the amorphous characteristics, a film 32 nm thick exhibited a resistivity of 122 μΩ cm, which is substantially higher than that of the W bulk metal and CVD crystalline films. In the WF_6/SiH_4 ALD system, pretreatment of the SiO_2 and TiN substrates with B_2H_6 induced rapid W nucleation and improved the step coverage in high-respect-ratio structures.[93,94] X-Ray diffraction showed that the films contained both α and β phases and the resistivities were in the range 125–178 μΩ cm.

7.3 CVD of Aluminium

Aluminium is a face-centered cubic metal with a melting point of 660 °C.[96] Its films exhibit high electrical conductivity (2.7 μΩ cm), high reflectivity and low gas permeability. These factors coupled with its low cost and the ability to deposit aluminium using a wide variety of techniques have continued to make aluminium attractive for many applications. Chemical vapor deposition is of interest in microelectronic applications where aluminium serves as an interconnect material. This application places a premium on the low resistivity of aluminium, which falls third in line behind silver and copper among the elements. In many devices copper has become the metal used for interconnects. In addition to its lower resistivity copper is less susceptible to electromigration than aluminium. In recent years, two factors, one practical and the second more fundamental, have caused a renewed interest in aluminium CVD. The practical factor stems from the lower cost of aluminium and the deposition process. The fundamental issue stems from the concern that the resistance of copper lines begins to increase faster than that for aluminium lines as the feature size decreases.[97,98] Several reviews of aluminium CVD have appeared.[5–6,99]

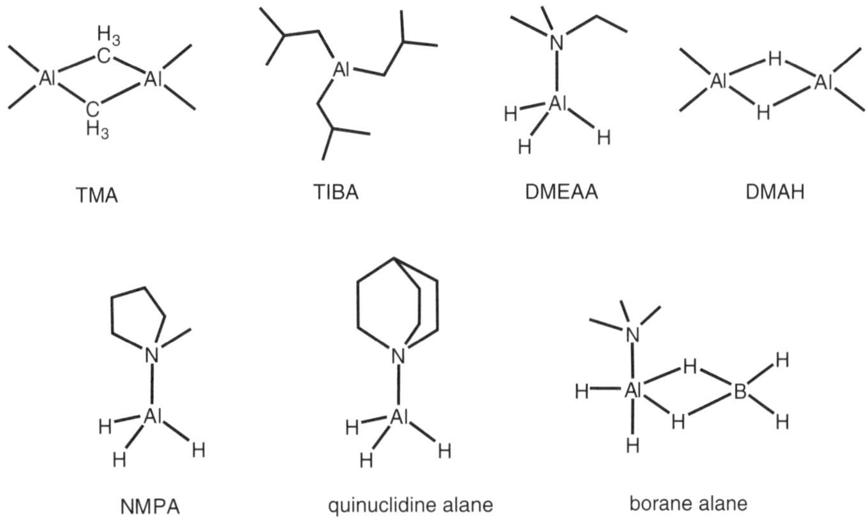

Figure 7.2 Selected Al precursors.

7.3.1 Precursors

The design of precursors for aluminium CVD must address the precursor volatility, deposition mechanism, cost of synthesis and ease of use including safety concerns in handling the precursor. All known aluminium precursors react with oxygen and water, and in some cases this reaction is sufficiently rapid and exothermic to render the compound pyrophoric. Aluminium is trivalent and in all known precursors it binds three, formally anionic ligands, either hydrides or alkyls or a mixture of the two. With a total of only six valence electrons, tricoordinate precursors, such as triisobutylaluminium (TIBA) (Figure 7.2), are stable as monomeric species due to the large steric requirements of the ligands. In complexes bearing smaller ligands such as methyl and hydride, aluminium increases its coordination number by forming bridging ligands through electron deficient, three-center two-electron bonds. While not a useful precursor for metallic aluminium deposition, the prototypical example is trimethylaluminium (TMA), which exists as a dimer in both the liquid and gaseous states.[100,101] In $[Al(CH_3)_3]_2$ and other aluminium organometallics, the Al–C bond is sufficiently strong to render simple bond homolysis ineffective as a mechanism leading to metallic aluminium. In each of the alkyl-containing precursors described below an alternative, low energy mechanism exists for eliminating the Al–C bond efficiently.

Although aluminium hydride, AlH_3, is known it is not a useful precursor. Unlike its lighter congener, boron, alane forms a nonvolatile, three-dimensional network solid state material.[102–104] In trialkylamine derivatives of alane, such as dimethylethylaminealane (DMEAA),[105] both electronic and coordinate saturation are achieved, and a facile route to aluminium metal exists.

7.3.1.1 Triisobutylaluminium (TIBA)

Among the first precursors considered for industrial application,[106–120] $Al[CH_2CH(CH_3)_2]_3$ is a pyrophoric liquid at room temperature that is in equilibrium with the isobutene and diisobutylaluminium hydride, $[Al(CH_2CH(CH_3)_2)_2H]_3$ (DIBAH). The trimeric nature of DIBAH causes it to have a lower vapor pressure at 100 °C (0.3 Torr) than TIBA (16 Torr), which is monomeric. Typical substrate temperatures for aluminium deposition were in the range 200–400 °C.[108–109,112–114,117]

In an elegant study[121] conducted in a UHV deposition system Bent *et al.* established the steady state deposition kinetics and mechanism of TIBA on single-crystal aluminium surfaces. Molecules

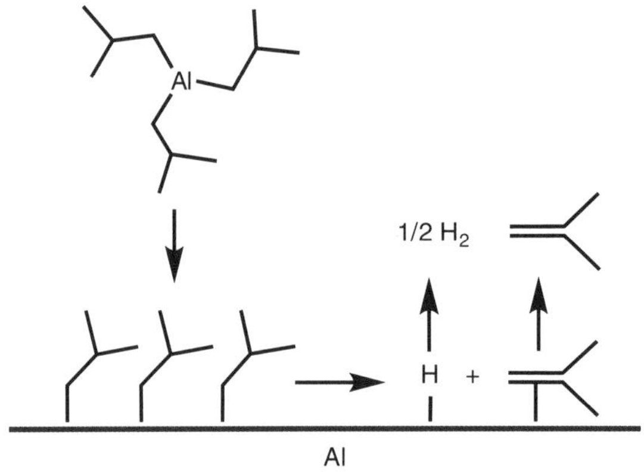

Figure 7.3 Mechanism of Al deposition from TIBA.

of TIBA were found to adsorb irreversibly on a clean aluminium surface. The three isobutyl ligands became equivalent by migrating to adjacent aluminium atoms. The rate-determining step in the deposition involved cleavage of the Al–C bond by a β-hydrogen elimination. Subsequent desorption of isobutene and molecular H_2 was rapid at the deposition temperatures. The formal reduction of Al^{3+} to Al metal results from the desorption of H_2. The deposition kinetics were modeled using Equation (7.16):

$$\text{Rate} = \frac{An_s e^{\frac{-E_a}{RT}}}{\frac{An_s}{\sigma s} n_s e^{\frac{-E_a}{RT}} + 1} \tag{7.16}$$

where A and E_a are the activation parameters, n_s is the number of adsorbed TIBA molecules per unit area at saturation coverage (1.4×10^{14} molecules cm^{-2}), σ is the flux of TIBA, s = sticking probability (1.0), R is the gas constant and T is the substrate temperature.

Equation (7.16) was based on the reactions shown in Figure 7.3. Faster kinetics ($A = 3.8 \times 10^{11}\ s^{-1}$ and $E_a = 115.8\ kJ\ mol^{-1}$) were observed on the Al(111) face compared to the Al(100) face ($A = 1.4 \times 10^{13}\ s^{-1}$ and $E_a = 136.3\ kJ\ mol^{-1}$).

Although the kinetics were measured under high vacuum conditions on single crystal substrates, the deposition rates predicted by Equation (7.16) at typical LPCVD conditions (several Torr) were lower only by a factor of approximately 2–5 than measurements in typical CVD systems.

This study also revealed that a competitive β-methyl elimination can become significant above 330 °C. Propane desorbs from the surface, leaving behind a methyl group that ultimately degrades into methane and hydrogen and a carbon-contaminated film. Although this can be avoided by conducting a slower deposition at lower temperature, the undesirable effects of carbon on the electrical and optical properties of aluminium prompted the search for alternative precursors.

7.3.1.2 Nucleation Promoters and Selective Deposition

Especially at the lower end of the temperature range for Al deposition, it is difficult to produce continuous, reflective films on nonmetallic surfaces, including SiO_2. This phenomenon results from a high barrier to nucleation, and early studies resolved the problem by exposing the substrate surface to a transition metal complex such as $TiCl_4$ prior to its exposure to TIBA.[108–110] Such pretreatment served to promote the formation of nuclei and allowed for the deposition of a uniform, highly reflective aluminium film. Use of a nucleation promoter was valuable for depositing reflective Al films at low temperatures using amine alanes and other precursors.[122,123]

7.3.1.3 Amine Alanes

Interest in amine alanes as possible precursors was motivated by the need to produce carbon-free films. As shown above, compounds such as $(Me_2EtN)AlH_3$ only have a weak Al–N and three Al–H bonds. Tertiary amines do not bind strongly to aluminium surfaces and the hydride ligands undergo facile elimination as H_2 from an aluminium surface. In light of the detailed kinetic and mechanistic work on TIBA that established that the rate-determining step in the deposition was Al–C bond cleavage by β-hydrogen elimination to form isobutene, it was not surprising that the substrate temperature for CVD using TMAA or DMEAA dropped significantly.[105,122,124,125]

High quality, polycrystalline aluminium films were deposited on various substrates at temperatures as low as 100 °C. For nonmetallic substrates, including silicon, silicon oxide, glass and polymers, smooth, highly reflective films required pretreatment of the substrate with a nucleation catalyst such as $TiCl_4$.[122] The fine-grain microstructure for many of these films were textured such that the Al(111) planes were parallel to the substrate surface. The degree of texture decreased as the deposition temperature increased; by 250 °C a random orientation of grains was observed. A second consequence of increasing the deposition temperature was an increase in grain size. These films exhibited a frosty appearance due to the increased surface roughness caused by the large grains. Consistent with the weak Al–N bond, the carbon and nitrogen concentrations in the films were typically near or below the detection thresholds for X-ray photoelectron analyses.

Surface studies by DuBois and co-workers[126] established that Equation (7.16) could also be used to predict the steady state growth rate of aluminium. The activation energy ($74.4\,kJ\,mol^{-1}$) for TMAA was notably lower than that reported for TIBA. Adsorption of $(Me_3N)AlH_3$ on a clean aluminium single crystal leads to cleavage of the Al–N bond, which is followed by desorption of Me_3N and H_2, leaving a clean surface to continue the cycle (Figure 7.4). It was suggested that the rate-determining step was cleavage of the Al–N bond present in the TMAA. In a related study,[127] a similar sequence of mechanistic steps was proposed for $(Et_3N)AlH_3$; however, the Al–N bond dissociates at even lower temperatures and it was proposed that H_2 desorption becomes the rate-limiting step.

Trimethylamine alane is a solid at room temperature, whereas DMEAA is a liquid with mp = 5 °C.[105,128] The latter has a vapor pressure near 1 Torr at room temperature, and it is not pyrophoric like many of the aluminium alkyl precursors, including dimethylaluminium hydride.

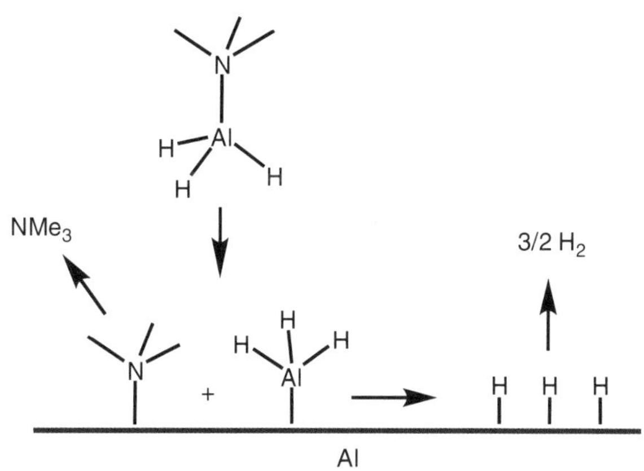

Figure 7.4 Mechanism of Al deposition from TMAA.

The only drawback is related to the long-term stability of the precursor in the precursor vessel.[128] Several precursors have been designed to improve their lifetime while maintaining a relatively high vapor pressure (Figure 7.2). Borane adducts of amine alanes tend to stabilize the hydrides.[123,129] Recent work has focused on (*N*-methylpyrrolidine)alane (NMPA).[130–133] In general, as the cone angle[134] of the amine decreases, the strength of the Al–N bond increases. By incorporating two of the three substituents on the nitrogen in NMP into a five-membered ring, the cone angle is decreased. NMPA is also a liquid at room temperature. Attempts to stabilize the amine alane precursors for long-term storage in a precursor vessel have included addition of excess amine. Stabilization is presumably enhanced by the equilibrium shown in Equation (7.17). For NMe_3, $(Me_3N)_2AlH_3$ is crystalline and has been structurally characterized.[135] The bond to the second amine is weak:

$$(Me_3N)AlH_3 + NMe_3 \rightleftharpoons (Me_3N)_2AlH_3 \tag{7.17}$$

Quinuclidine has an even smaller cone angle and the corresponding complex with alane is the most stable of the known tertiary amine complexes of alane.[136] As a result CVD might occur only at higher temperatures. In addition, H_3Al(quinuclidine) is a crystalline solid with a lower vapor pressure, making it less attractive as a Al CVD precursor.

7.3.1.4 Dimethylaluminium Hydride

Many papers describe the use of dimethylaluminium hydride (DMAH) to deposit high purity aluminium films by CVD.[137–146] For blanket coverage on various substrates deposition temperatures are typically greater than 270 °C. At such high temperatures, there was no need to pretreat the substrate surfaces with $TiCl_4$ or other nucleation promoters. Between 230 and 270 °C selective deposition on silicon, titanium and TiN in the presence of SiO_2 was observed.[138] Above 350 °C, carbon incorporation into the film becomes significant.

The molecularity of DMAH differs in the gaseous and liquid states. In the gas phase DMAH exists as a dimer (Figure 7.2) or trimer,[147–150] and the room temperature vapor pressure is approximately 2 Torr.[147] Bridging hydrides in dimethylaluminium hydride (DMAH) allow the aluminium to achieve a coordination number of four. Higher oligomers and polymers comprise the liquid at room temperature and have the undesirable effect of raising the room temperature viscosity of the liquid to 6400 cP[151] (similar to that of honey). This property, coupled with its pyrophoric behavior, makes DMAH difficult to handle.

As noted above, trimethylaluminium is not useful as a precursor for metallic aluminium because there is no low energy pathway leading to cleavage of the Al–C bonds. The stark contrast in effectiveness for Al CVD between TMA and DMAH is surprising and emphasizes the critical role of the hydride ligand. Earlier work[152] established that the correct stoichiometry of the CVD reaction using DMAH is shown in Equation (7.18):

$$3[(CH_3)_2AlH]_2 \rightarrow 2Al + 3H_2 + 2[Al(CH_3)_3]_2 \tag{7.18}$$

In a formal sense, the ligand redistribution reaction (an exchange of a hydride for a methyl between two aluminiums) provides for an effective route to AlH_3, which can reductively eliminate hydrogen and deposit aluminium. The methyls are removed from the surface by forming TMA.

A combined experimental and computational study has provided mechanistic details for the overall reaction shown in Equation (7.18).[153,154] The reactions are summarized in Equations (7.19–7.23), where the subscripts (g) and (a) denote gas species and absorbed species, respectively. Consistent with observations in the CVD process, the surface studies established that dehydrogenation of surface bound methyls leads to carbon incorporation into the films at high temperatures:

$$[(CH_3)_2AlH]_{2(g)} \rightarrow 2Al(CH_3)_{2(a)} + 2H_{(a)} \quad (7.19)$$

$$2H_{(a)} \rightarrow H_{2(g)} \quad (7.20)$$

$$Al(CH_3)_{2(a)} \rightleftharpoons Al(CH_3)_{(a)} + CH_{3(a)} \quad (7.21)$$

$$Al(CH_3)_{(a)} \rightleftharpoons Al_{(a)} + CH_{3(a)} \quad (7.22)$$

$$Al(CH_3)_{2(a)} + CH_{3(a)} \rightarrow Al(CH_3)_{3(g)} \quad (7.23)$$

Using a scanning tunneling microscope, Ganz and co-workers studied the initial stage of aluminium deposition on Si(001)-(2×1) surfaces.[155] At room temperature DMAH dimers adsorb intact, whereas they react with the surface at temperatures exceeding 80 °C. DMAH, however, would not stick to hydrogen-terminated silicon surfaces. This differential reactivity allowed Ganz and co-workers to implement a nanolithographic procedure based on STM tip-induced hydrogen desorption followed by exposure of the surface to DMAH.[156] Aluminium lines as narrow as 2 nm were formed.

7.4 CVD of Copper

The properties of Cu are well documented.[96] This face-centered cubic metal melts at 1084 °C. The bulk material has a resistivity of 1.678 μΩ cm at 20 °C, and a thermal conductivity of 401 W $m^{-1} K^{-1}$ at 27 °C.

The resistivities of nano-scale Cu wires are greater than that of the bulk material, and the resistivity increase is nonlinear with respect to the size decrease. As the wire dimensions approach the Cu room-temperature electron mean free path (*ca.* 40 nm),[97,98] the contribution from surface and grain boundary scattering becomes significant, adding to the bulk resistivity. Several theoretical models were developed to describe the size effects on metal resistivity.[157–160] The studies on Cu[97,98,160–165] revealed that the film microstructures, which were dependent on the deposition methods and substrates, determined the level of the resistivity increase. Figure 7.5, illustrating a general trend, is adopted from the 2005 international technology roadmap for semiconductors (ITRS).

7.4.1 Deposition from Copper(II) Precursors

7.4.1.1 Copper(II) Bis(β-diketonates)

Copper(II) bis(β-diketonates), especially $Cu(hfac)_2$, where hfac = 1,1,1,5,5,5-hexafluoropentane-2,4-dionato, have been studied extensively. Anhydrous $Cu(hfac)_2$ exists as blue-violet crystals with a mp of 95–98 °C.[166,167] The crystal structure of this monomeric, square-planar compound has been determined (Figure 7.6).[167] Its room-temperature vapor pressure is *ca.* 3 mTorr,[168] and the vapor pressures at elevated temperatures are known.[169,170] Several hydrates of $Cu(hfac)_2$ exist.[167] The monohydrate, $Cu(hfac)_2(H_2O)$, existing in three polymorphs,[167,171,172] is a blue-green crystal with a mp of 133–136 °C.[173] It is volatile and has been used as a CVD precursor,[7] but the anhydrous compound would be preferred if precise control over precursor delivery is needed because the hydrate slowly loses water when sublimed/evaporated in dry carrier gas.[174]

In CVD, the precursor temperatures for $Cu(hfac)_2$ ranged from 60 to 120 °C and the deposition temperatures were typically from 250 to 400 °C.[7,8] The lowest-resistivity films (*e.g. ca.* 2.0 μΩ cm) were obtained with co-introduction of H_2.[175–177] The overall reaction for this process is depicted in

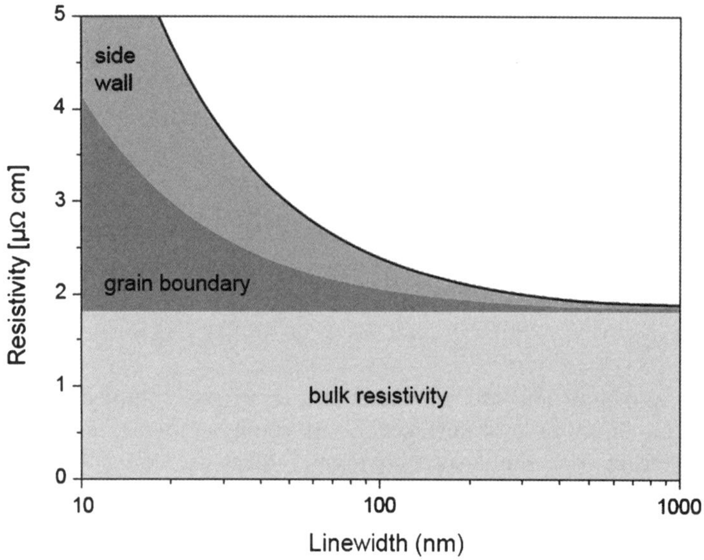

Figure 7.5 Predicted Cu resistivity as a function of line width, where the "grain boundary" and "side wall" indicate the resistivities contributed from grain boundary scattering and interface scattering, respectively.

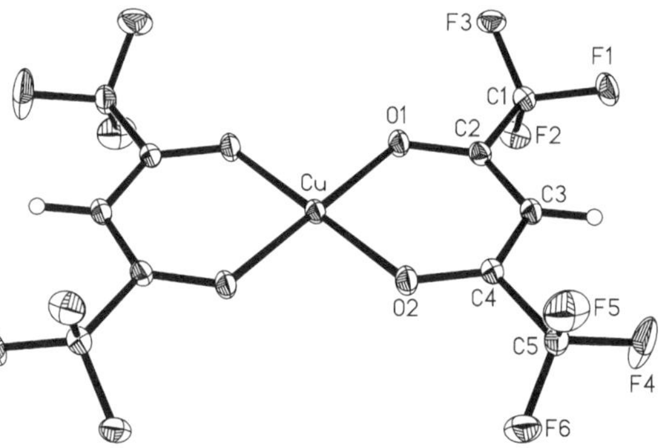

Figure 7.6 ORTEP drawing of $Cu(hfac)_2$ with the displacement ellipsoids at the 50% probability level.

Equation (7.24). Details of the reaction kinetics and mechanisms on various substrates have been studied under typical CVD as well as ultrahigh vacuum (UHV) conditions:[13,175,176,178–181]

$$Cu(hfac)_2 + H_2 \rightarrow Cu + 2(hfac)H \tag{7.24}$$

Without using H_2, pure Cu films with resistivities of 3–7 μΩ cm were obtained in a cold wall reactor at 340–390 °C, but C, O and F impurities were found for films deposited at higher temperatures.[170] Addition of H_2O or alcohols during depositions increased the grow rates and improved the film morphologies by enhancing $Cu(hfac)_2$ surface reaction and Cu nucleation.[174,182–185] The presence of H_2 was also necessary when H_2O was used, otherwise copper oxides formed in the films.[171,186] In addition, plasma assisted CVD (PACVD) has long been a choice in Cu CVD to increase deposition rates at lower temperatures.[7,8] Recently, in the search for novel Cu diffusion barriers, $Cu(hfac)_2$ was used to deposit Cu on a 3-mercaptopropyltrimethoxysilane self-assembled

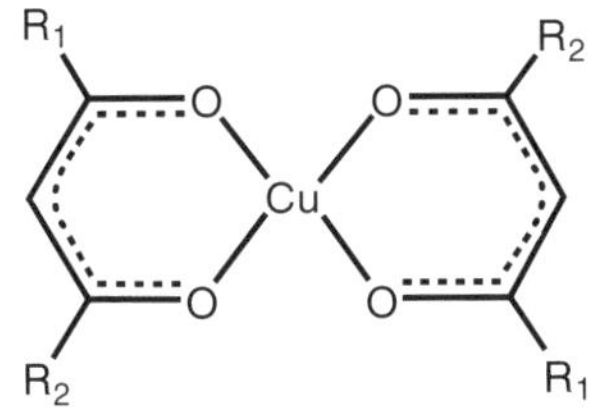

	R_1	R_2
$Cu(acac)_2$:	CH_3	CH_3
$Cu(tmhd)_2$:	$C(CH_3)_3$	$C(CH_3)_3$
$Cu(tfac)_2$:	CH_3	CF_3
$Cu(fhd)_2$:	$C(CH_3)_3$	CF_3
$Cu(ppm)_2$:	$C(CH_3)_3$	C_2F_5
$Cu(fod)_2$:	$C(CH_3)_3$	C_3F_7
$Cu(hfac)_2$:	CF_3	CF_3
$Cu(tdf)_2$:	C_3F_7	C_3F_7

Figure 7.7 Selected Cu(II) bis(β-diketonate) precursors.

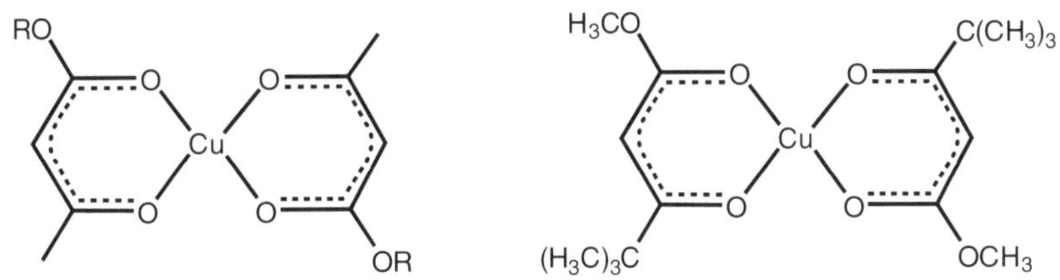

R = CH_3, C_2H_5, $C(CH_3)_3$, $C_2H_5OCH_3$, $CH_2C_6H_5$ $Cu(mpa)_2$

Figure 7.8 Selected Cu(II) bis(β-diketoesterate) precursors.

monolayer (SAM) modified SiO_2.[187,188] In another new method, $Cu(hfac)_2$ was employed for depositions in a supercritical CO_2 ($scCO_2$) environment.[189,190] This process took advantage of the solubility of $Cu(hfac)_2$ in $scCO_2$ and the high diffusivity and zero surface tension of $scCO_2$. Promising results in filling deep and narrow features were obtained.

Figure 7.7 illustrates other common Cu(II) bis(β-diketonate) precursors. These compounds have been described in previous reviews.[7–9] In general, the alkyl-substituted precursors are less volatile than the fluorinated precursors, *e.g.*, the precursor vessel temperatures for $Cu(acac)_2$ used in CVD were 110–190 °C.[191,192] The non-fluorinated precursors might be advantageous if fluorine contamination is a concern on certain substrates that are currently being studied for Cu diffusion barriers.[189]

7.4.1.2 Other Copper(II) Precursors

Other precursors include Cu(II) bis(β-ketoesterates)[193–199] (Figure 7.8) and adducts of Cu(II) bis(β-diketonates) with neutral bidentate ligands (*e.g.*, diamines and amino alcohols) in which the coordination on Cu was five (in dimers) or six (in monomers).[200–202] Most of the precursors and their use in Cu CVD have been reviewed.[7–9] Another group of Cu(II) precursors gaining increasing interest is aminoalkoxides.[203–211] An early study using $Cu[OCH_2CH_2N(CH_3)_2]_2$ in inert atmosphere showed that Cu films were obtained at 200–210 °C, but the resistivities were much higher than the bulk Cu and significant C and O contamination was found for films deposited at higher temperatures.[204] More recent studies using new aminoalkoxide precursors,[206–211] including $Cu[OC(CF_3)_2CH_2NHCH_2CH_2OCH_3]_2$, $Cu[OCH(CH_3)CH_2N(CH_2CH_3)_2]_2$ and $Cu[OC(CF_3)_2CH_2CH(CH_3)NH(CH_3)]_2$ afforded improved results, especially in the attempt to eliminate the use of H_2 as a co-reactant.

7.4.2 Deposition from Copper(I) Precursors

The advantages of Cu(I) precursors over Cu(II) precursors in Cu CVD include high-quality film deposition at lower precursor delivery and deposition temperatures and the reduced need to use H_2 as a co-reactant.[8] The most widely used Cu(I) CVD precursors are adducts of Cu(hfac) with a general formula of Cu(hfac)(L), where L is an alkene, alkyne, diene or phosphine.[7,8] Other precursors, including Cu(I) cyclopentadienyl and *t*-butoxide and their derivatives, are used less extensively.[7,8] The following sub-sections briefly review the most thoroughly studied Cu(I) precursor, Cu(hfac)(vtms), where vtms=vinyltrimethylsilane. Some general concepts in using this precursor have also been applied to others Cu(I) precursors, for which details can be found in reviews[7–9] and original papers.[172,212–229] New Cu(I) compounds for Cu ALD are summarized separately.

7.4.2.1 *Properties and General Description of Cu(hfac)(vtms)*

CVD using liquid Cu(hfac)(vtms) was reported in 1991.[230] Its crystal structure obtained at a low temperature[231] exhibits a nearly planar geometry around the Cu center (Figure 7.1). The vapor pressure at slightly above room temperatures is *ca.* 0.1 Torr.[230,232] Precautions should be taken in handling this precursor because of its high air sensitivity. Slow decomposition has been reported even at room temperature,[216] but this can be suppressed by the presence of excess vtms.[233,234] In CVD, the precursor temperatures are typically 25–40 °C, and deposition temperatures are 100–250 °C. He, N_2, Ar or a mixture with H_2 were used as the carrier gas.

7.4.2.2 *Nucleation in CVD using Cu(hfac)(vtms)*

CVD using Cu(hfac)(vtms) has been conducted on metals, semiconductors, nitrides and oxides. Except for a few metal substrates, Cu nucleation has been a major challenge in obtaining high-quality films. UHV studies showed that Cu, Si, TiN and TiCN all exhibited high reactivities towards dissociation of Cu(hfac)(vtms).[13,235–239] Molecular dynamics simulations predicted that Ta would also be effective at inducing Cu(hfac)(vtms) to react.[240] Nucleation behavior, however, was much different on these substrates.[241–243] Discrete islands formed from Cu(hfac)(vtms) on Ta and TaN_x, whereas a continuous film was deposited on a preformed Cu surface. At the early stages of deposition on Ru, densely packed, small nuclei of metallic copper were formed. The superior result on Ru was attributed to its high Cu wettability, which leads to a smaller activation energy for nucleation.[243,244]

Atomic force microscopy revealed that the number density of Cu nuclei decreased according to the order Pt $>$ Si $>$ SiO_2 with *ca.* two orders of magnitude difference between Pt and SiO_2.[245] Table 7.1 lists the nucleation activation energies on these and other substrates, along with film growth activation energies. Taking advantage of the low deposition rate on SiO_2, Cu was selectively deposited on TiN, W, Ta, Al and PtSi in the presence of SiO_2.[230] A later study,[246] however, showed that deposition occurred on both SiO_2 and W patterned on SiO_2. Infrared studies showed that, from 125 to 175 °C, Cu(hfac)(vtms) dissociated to Cu(hfac) and vtms at OH and other active sites on SiO_2.[247] Passivation of the SiO_2 surface with silyls [*e.g.*, $Si(CH_3)_3$ or $Si(CH_3)_2CH_2CH_2C(CH_3)_3$], which were generated from reactions with organosilicon compounds including hexamethyldisilazane and dimethyl-3,3-dimethylbutyl-dimethylaminosilane, substantially reduced the precursor chemisorption, and, as a result, no Cu deposition took place.[215,248]

Several methods were employed to promote Cu nucleation. Addition of H_2O at the beginning of the depositions on Ta, TiN and TaN enhanced nucleation, leading to dense, smooth Cu films.[172,249–252] Pre-treatment of TiN and SiO_2 with an iodine plasma allowed adsorption of I on the surface and induced higher Cu deposition rates.[253] Interestingly, the initially adsorbed I species segregated onto the newly formed Cu surface in the film growth process, acting as a catalyst to

Table 7.1 Copper nucleation and film growth activation energies.

Substrate	Activation energy ($kJ\,mol^{-1}$) Nucleation	Film growth	LPCVD reactor,[a] carrier gas, total pressure	Deposition temperature (°C)[b]	Ref.
Pt	20		Cold wall, Ar, 2.2 Torr	200–250	245
n-type Si	50.0				
p-type Si	70.6				
SiO_2	110				
TaN_x	163	59	Cold wall, DLI,[c] He, 1 Torr	100–225	251
TaN_x with H_2O	25				
W		42	Hot wall and warm wall, none or N_2, 0.5 Torr	140–170 (200)	246
Polyetherimide, Si, Mo		33–42	Hot wall, N_2 or H_2; 5, 20, 56 Torr	110–150 to180 (300)	262
Cu, Ti		46	Cold wall, DLI,[c] Ar or H_2, 10–40 Torr	140–180 (250) on Cu 100–180 (250) on Ti	259
Ti		103	Warm wall, N_2, 0.5 Torr	140–160 (230)	391
Si, Ti, TiW, Al		17–25	Cold wall, He or H_2, 0.25 Torr	130–180 (220)	232
TiN		70	Hot wall, DLI,[c] H_2, 2–10.5 Torr	100–180 (300)	261
TiN		6464	Cold wall, Ar, 1.5 Torr	100–200 on TiN	255
I-treated TiN		25		50–100 (200) on TiN/I	
Ta		44	Cold wall, DLI,[c] He or H_2; 0.06 and 0.15 Torr	120–180 (240) for 0.15 Torr 120–200 (240) for 0.06 Torr	234
Ta, Ru		70–75	Warm wall, H_2, 5 Torr	90–110 (200)	243
Ta,		81	Cold wall, DLI,[c] Ar/He, 1.5 Torr	150–190 (250)	241
TaN_x ($x < 0.5$)					

[a]All depositions used LPCVD reactors.
[b]The temperature ranges for activation energy measurements were in surface reaction-limited regimes. Values in parentheses were the highest temperatures used in the studies.
[c]DLI=direct liquid injection, for which the precursor is delivered in the liquid phase to a vaporizer where it is vaporized and mixed with a carrier gas.

maintain the high-rate deposition (*ca.* 100 times higher than without I adsorption).[253–255] Pretreating TiN substrates with a hydrogen plasma[256] or with WF_6[257] also facilitated Cu nucleation and yielded high-quality films. Depositing a seed layer of In on TaN significantly increased the Cu nuclei density, allowing a rapid formation of a continuous Cu layer.[258]

7.4.2.3 Deposition Mechanisms and Kinetics

The overall reaction for Cu deposition from Cu(hfac)(vtms) is shown in Equation (7.3). Under various conditions, pure Cu films were deposited, indicating that $Cu(hfac)_2$ and vtms desorbed from the surface without decomposition.

Several models exist to describe the detailed reaction mechanism. The mechanism shown in Equations (7.25–7.28), where the subscripted (g) and (a) stand for gas species and adsorbed species, respectively, was suggested by most authors based on the kinetics and UHV surface studies.[13,232,246,259,260] It was not established whether vtms was released directly to the gas phase as in Equation (7.26a) or through an intermediate surface-bonded state as in Equation (7.26b,c):[232,246]

$$2Cu(hfac)(vtms)_{(g)} \rightarrow 2Cu(hfac)(vtms)_{(a)} \tag{7.25}$$

$$2Cu(hfac)(vtms)_{(a)} \rightarrow 2Cu(hfac)_{(a)} + 2vtms_{(g)} \tag{7.26a}$$

or

$$2Cu(hfac)(vtms)_{(a)} \rightarrow 2Cu(hfac)_{(a)} + 2vtms_{(a)} \tag{7.26b}$$

$$2vtms_{(a)} \rightarrow 2vtms_{(g)} \tag{7.26c}$$

$$2Cu(hfac)_{(a)}Cu_{(a)} + Cu(hfac)_{2(a)} \tag{7.27}$$

$$Cu(hfac)_{2(a)} \rightarrow 2Cu(hfac)_{2(g)} \tag{7.28}$$

A gas-phase reaction model was proposed to explain the kinetics obtained in a hot-wall reactor.[261] In this model, Cu(hfac)(vtms) dissociated in the gas phase to form Cu(hfac) and vtms. Cu(hfac) was then adsorbed onto the substrate and reacted *via* a disproportionation as in the previous mechanism [*e.g.* Equations (7.27) and (7.28)]. In a different model appearing to fit the experimental data,[262,263] several assumptions were made, including that the disproportionation occurred only between a gas phase Cu(hfac)(vtms) molecule and an adsorbed Cu(hfac) species and all elemental steps were irreversible.

A summary of activation energies obtained under various conditions is given in Table 7.1 The apparent film growth activation energies ranged from *ca.* 20 to 100 kJ mol^{-1}.

The relationship between film growth rates and precursor partial pressures was successfully explained using the mechanism in Equations (7.25–7.28).[246,264,265] At low temperatures, the deposition rate increased initially, then leveled off. This was due to saturation of the surface adsorption sizes. Studies on the effects of using H_2 as a co-reactant showed that the degree of growth rate increase depended on the H_2 partial pressure. More importantly, the induction periods on TiN and SiO_2 were shortened, and the impurity contents were reduced.[266–268,259,262]

7.4.3 Copper ALD

ALD produces conformal films with controllable rates, which is particularly attractive for film depositions in high-aspect-ratio structures in ULSI. Cu ALD is a relatively new process. Early attempts using CuCl and H_2 on Ta substrates or CuCl and Zn on Al_2O_3 and Al substrates afforded rough films.[218,269] The film growth was not self-limiting and Zn contamination was found in the Cu films.[218] Later, H_2O was introduced with a dosing sequence of $CuCl/H_2O/H_2$, allowing ALD on fused SiO_2 or single-crystal Al_2O_3 at 375–475 °C, but island growth on the oxide substrates gave poor film morphologies.[270] Using the $Cu(tmhd)_2/H_2$ combination, pure Cu films were deposited on Pt/Pd *via* a self-limiting process at temperatures from 190 to 260 °C, while no depositions occurred on Ta, Fe, Ni, TaN and indium tin oxide.[271] Using a hydrogen plasma dose after the $Cu(tmhd)_2$ dose produced conformal Cu films with 1–3% carbon contamination on Au, SiO_2 and TaN_x at temperatures from 90 to 250 °C.[272] When $Cu(hfac)_2(H_2O)_x$ was used as the Cu source and alcohols or formaldehyde solution as the reductant, ALD was achieved on glass, Ta, TiN and TaN. The best results were obtained *via* a two-step process in which $Cu(hfac)_2$ and H_2O were introduced in the first step and isopropanol or formaldehyde solution was introduced in the second step.[273,274]

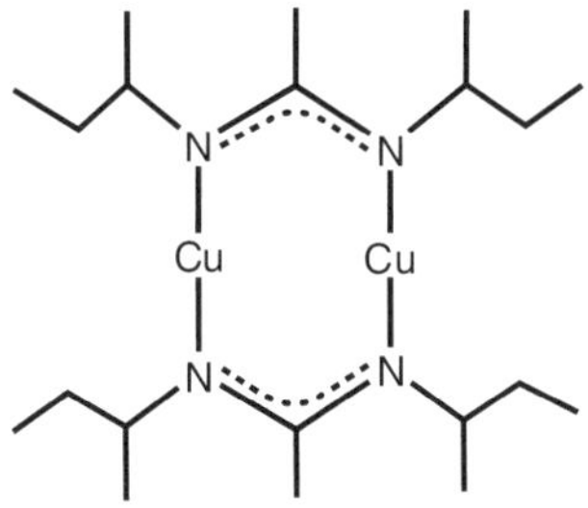

Figure 7.9 Structure of Cu(I) *N,N′*-di-sec-butylacetamidinate.

$Cu(acac)_2$ was also used in Cu ALD on several substrates with the reducing agent in the form of H atom.[275] Growth rates were very low (0.018 nm cycle^{-1}).

Complications in the above ALD process were largely due to the lack of proper reducing agents compatible with the existing precursors. New ALD precursors were investigated and a group of Cu(I) amidinates were synthesized,[276–278] among which the Cu(I) *N,N′*-di-sec-butylacetamidinate (Figure 7.9) was found to be most suitable for Cu ALD.[279] With the use of H_2 as reducing agent, pure Cu ALD films were deposited at low temperatures (150–190 °C) on metals, oxides and nitrides.

7.5 Ruthenium CVD

Interest in Ru CVD has increased in the past decade because of the imminent needs for new electrodes and diffusion barriers in ULSI microelectronics. Ruthenium is a refractory metal (mp = 2273 °C[280]) with a low resistivity (7.16 μΩ cm at 5 °C[281]). It is inert towards mineral acids and slowly reacts with aqueous solutions of Cl_2 and Br_2.[280] RuO_2, the only stable, solid oxide of Ru, is also a metallic conductor with a resistivity of 35.2 μΩ cm at 27 °C.[282] Dry etching of Ru is possible due to the high volatility of RuO_4.[280,283]

Both Ru and RuO_2 are promising for use as the bottom electrodes in gigabyte-scale dynamic random access memories (DRAMs) fabricated using high-κ materials, including Ta_2O_5 and (Sr, Ba)TiO_3. Traditional electrode materials such as W, Al and poly-Si will react under the O_2 and high temperature conditions employed during the CVD of these dielectrics.[284] In addition, Ru and RuO_2 are also attractive candidates for electrodes used in next-generation complementary metal oxide semiconductor (CMOS) transistors. As the gate length of CMOS transistors is reduced to below 50 nm, problems associated with the currently used poly-Si electrode, such as gate depletion and dopant penetration, become significant.[285] To achieve high performance in the transistors, the new metal electrodes should have work functions of *ca.* 4 eV for n-type MOS and *ca.* 5 eV for *p*-type MOS devices.[286,287] Ruthenium, its oxide and alloys with Pt or Ta exhibit work functions in the range of 4.2–5.2 eV and demonstrate excellent compatibilities not only with the traditional SiO_2 gate but also with various new high-κ materials such as ZrO_2 and HfO_2.[288–294] The third potential application of Ru and RuO_2 is for diffusion barrier and seed layer for Cu interconnects.[295–299] It was found that Ru did not form alloys with Cu even at high temperatures,[295] and, among a range of metals, Ru exhibited the best Cu wettability.[300]

7.5.1 Ruthenium CVD from Ru β-Diketonates

Selected Ru β-diketonates are illustrated in Figure 7.10, which is based on single-crystal structures of $Ru(acac)_3$[301,302] (acac = acetylacetonato) and $Ru(hfac)_3$.[303]

The most widely studied precursors in this group are $Ru(tmhd)_3$[304–308] (tmhd = 2,2,6,6-tetramethyl-3,5-heptanedionato) and $Ru(od)_3$ (od = 2,4-octanedionato).[309–313] $Ru(tmhd)_3$ exists as red

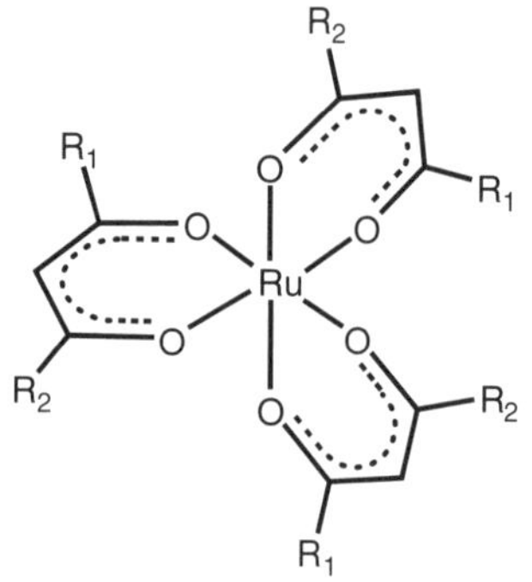

	R_1	R_2
$Ru(tmhd)_3$	$C(CH_3)_3$	$C(CH_3)_3$
$Ru(od)_3$	CH_3	$CH_2CH_2CH_2CH_3$
$Ru(mhd)_3$	CH_3	$CH_2CH(CH_3)_2$
$Ru(acac)_3$	CH_3	CH_3
$Ru(hfac)_3$	CF_3	CF_3
$Ru(tfac)_3$	CH_3	CF_3

Figure 7.10 Selected Ru β-diketonates.

crystals (mp 216–218 °C)[307,314] and sublimes under vacuum (0.01 Torr) at temperatures from 160 to 195 °C.[314] This precursor was mainly used to deposit RuO_2 films under O_2.[304–307] Interestingly, however, films of pure Ru metal were deposited even under high O_2 concentrations when $Ru(tmhd)_3$ was delivered in a THF solution *via* a liquid injection method.[308] It was proposed that oxidation of THF consumed the excess O_2, preventing the formation of RuO_2. $Ru(od)_3$ is a viscous liquid at room temperature. Thermogravimetric analysis (TGA) under Ar revealed that it nearly completely vaporized by 313 °C.[309] Ruthenium metal films were deposited at low temperatures (260–400 °C) using $Ru(od)_3$ and O_2.[310–313] The precursor was delivered either as a gas by maintaining a precursor vessel temperature above 160 °C or by direct liquid injection of a methanol or butyl acetate solution. Control over the O_2 partial pressure was important for obtaining pure Ru films. RuO_2 was obtained at higher O_2 concentrations.[309] Oxygen-free Ru films were also obtained at 400 °C *via* co-introduction of NH_3.[313] Ammonia was proposed to reduce any RuO_2 formed during the deposition.

$Ru(acac)_3$ exhibits a relatively low volatility,[314] and its use in CVD was reported in the early studies.[315,316] For $Ru(mhd)_3$, where mhd = 6-methylheptane-2,4-dionato, a CVD study using a methanol solution as the precursor source recently appeared.[317] $Ru(tfac)_3$ (tfac = trifluoroacetylacetonato) was used to deposit RuO_2 films at low temperatures (350 °C);[318,319] however, no CVD of the metal films was reported. $Ru(hfac)_3$ exhibits a higher volatility than $Ru(tfac)_3$ and $Ru(acac)_3$,[320,321] but little is known about its use in CVD.

7.5.2 $Ru(CO)_5$, $Ru_3(CO)_{12}$ and other Precursors Containing the CO Ligand

Figure 7.11 shows typical Ru precursors containing CO ligands.

$Ru(CO)_5$ (mp −16 °C) is formed from the reaction of $Ru_3(CO)_{12}$ with CO at high pressures and temperatures.[322] A study showed that it deposited Ru metal at 150 °C,[323] but the use of it as a CVD precursor has been limited by its sensitivity to heat and light. $Ru_3(CO)_{12}$ is an orange-red crystalline (mp 155 °C).[324] TGA showed a mass loss from 115–305 °C with the residue accounting for 42.2% of the initial mass (theoretical Ru percentage in $Ru_3(CO)_{12}$, 47.4%).[325] Although the precursor underwent partial decomposition in the precursor vessels, pure Ru films were successfully deposited at 150 °C on Ta and 175 °C on glass.[298,316,326]

$Ru(CO)_4(hfb)$, where hfb = hexafluoro-2-butyne, is a colorless crystalline solid with a vapor pressure of 1.5 Torr at 25 °C.[327–330] CVD studies showed that the Ru films deposited from $Ru(CO)_4(hfb)$ at 500 °C without using H_2 contained a significant amount of C, and at low temperatures a binuclear intermediate, $Ru_2(CO)_6[C_4(CF_3)_4]$ formed.[329,330] Depositions in the presence of H_2 greatly reduced C contamination and improved film morphology and conductivity.[330]

Dinuclear $[Ru(CO)_3(3,3\text{-}(CF_3)_2\text{-pz})]_2$, where $3,3\text{-}(CF_3)_2$-pz is 3,5-bis(trifluoromethyl)pyrazolato, is a yellow crystal that melts at 149–151 °C.[331] TGA showed complete evaporation below 200 °C.

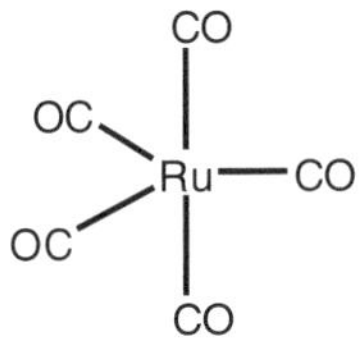

$Ru(CO)_5$

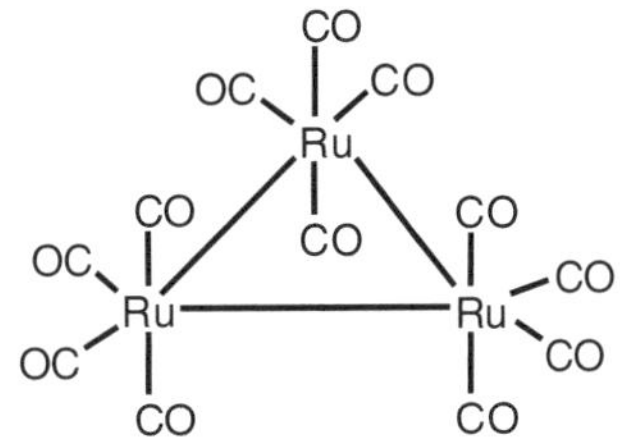

$Ru_3(CO)_{12}$

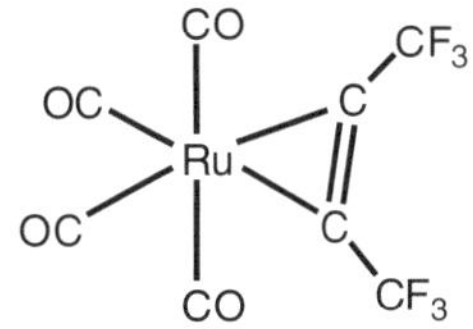

$Ru(CO)_4$(hfb)

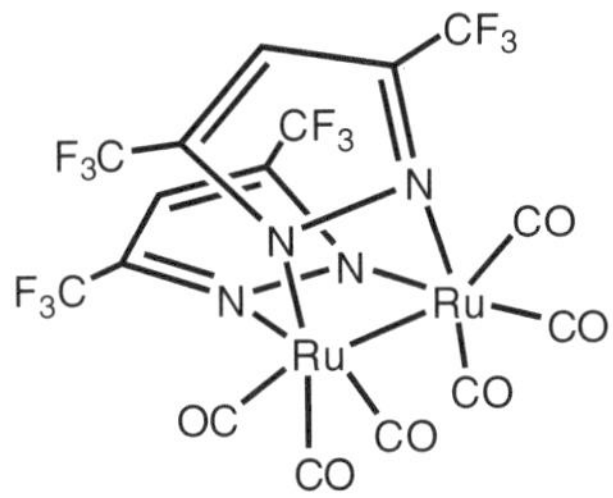

$[Ru(CO)_3(3,3\text{-}(CF_3)_2\text{-pz})]_2$

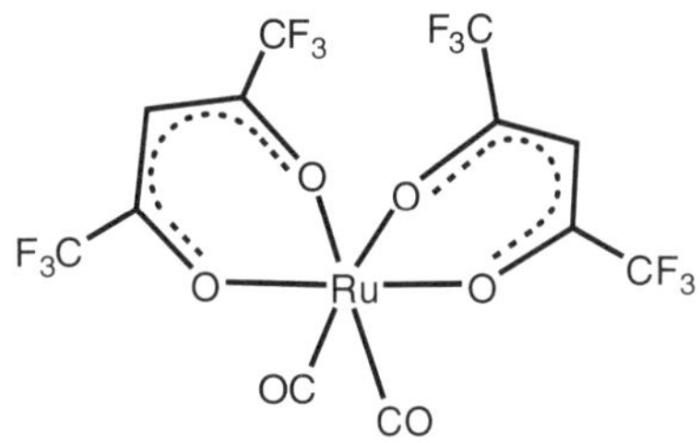

$Ru(CO)_2(hfac)_2$

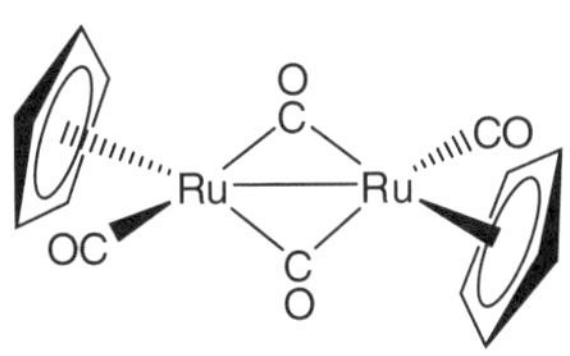

$[CpRu(CO)_2]_2$

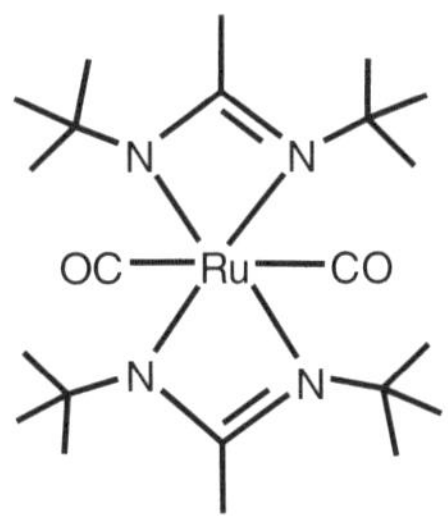

$Ru(^tBu\text{-Me-amd})_2(CO)_2$

Figure 7.11 Selected Ru precursors containing CO ligands.

Depositions were conducted at precursor temperatures of 80 or 110 °C.[331] Pure Ru films with resistivities of 23–40 μΩ cm were obtained on Si under H_2 at 300–450 °C while RuO_2 films were deposited at the same temperatures using O_2.

Among the β-diketonate derivatives, $Ru(CO)_2(L)_2$, where L = acac, hfac, tmhd and tfac,[332] $Ru(CO)_2(hfac)_2$ (mp 64–66 °C) was reported to be the most suitable precursor to Ru films.[333,334] This orange crystalline solid exhibited a high volatility (1.4 Torr) at room temperature. Ruthenium films of comparable qualities were deposited under H_2 at 400–450 °C or under a dilute O_2 atmosphere at 300–400 °C.[332] $Ru(CO)_2(tmhd)_2$ (mp 70–72 °C) has a lower volatility than the hfac compound. Depositions under H_2 or a low concentration of O_2 in Ar also afforded Ru films, but these films contained higher C levels than the films deposited using $Ru(CO)_2(hfac)_2$.[332] High quality RuO_2 films were prepared from $Ru(CO)_2(tmhd)_2$ and O_2.[335]

Dimeric $[CpRu(CO)_2]_2$ is a yellow crystalline solid with *ca.* 0.01 Torr vapor pressure at room temperature.[336] Mass spectrometric analysis showed that substantial sublimation occurred from 120 to 180 °C.[337] Pure Ru films were deposited on patterned Si_3N_4 at 300 °C and flat (Ba, Sr)TiO_3 at

350 °C with the presence of O_2. Low temperatures ($\leq$260 °C) were preferred for the formation of conformal films on the patterned Si_3N_4.[337]

Bis(N,N'-di-*tert*-butylacetamidinato)ruthenium(II) dicarbonyl, Ru(^{t}Bu-Me-amd)$_2$(CO)$_2$, has a vapor pressure of about 0.05 Torr at 130 °C.[338] It was used in pulsed CVD to deposit Ru films on SiO_2 at temperatures above 200 °C. A particularly interesting aspect of this precursor is that neither an oxidant (*e.g.*, O_2) nor a reducing agent (*e.g.*, H_2) was needed for producing Ru films. Formation of Ru was proposed to occur according to Equation (7.29), in which the acetamidinate ligands underwent reductive elimination to form a novel dimer.

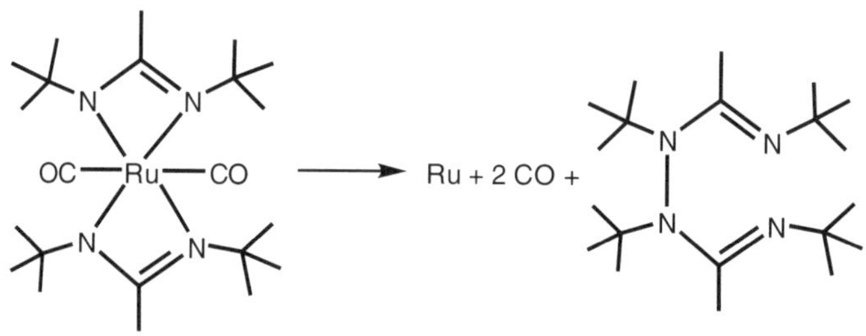

(7.29)

7.5.3 Ruthenium CVD from Sandwich or Half-sandwich Precursors

Figure 7.12 illustrates typical precursors of this group.[322,339–347] Some properties of selected precursors are given in Table 7.2. The CVD results of the most widely studied precursors, Cp_2Ru and $(EtCp)_2Ru$, where EtCp=ethylcyclopentadienyl, are summarized below in Tables 7.3 and 7.4, respectively.

The first CVD study using Cp_2Ru, carried out under 1 atm of H_2 in a hot wall reactor, produced pure Ru films on glass at 590–600 °C.[348] Under low H_2 partial pressures, vacuum or an inert

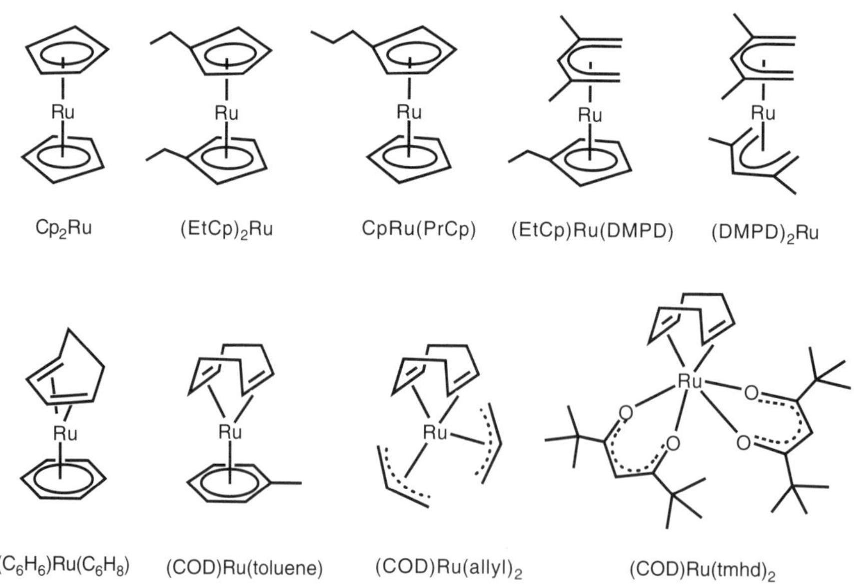

Figure 7.12 Sandwich and half-sandwich Ru precursors.

Table 7.2 Melting points and vapor pressures of selected sandwich Ru precursors.

Precursor	*Mp (°C)*	*Volatility*	*Ref.*
Cp_2Ru	199–200	3 Torr at 130 °C	355
$(EtCp)_2Ru$	6	0.02 Torr at 50 °C, 0.1 Torr at 73 °C	360,362
CpRu(PrCp)	–	0.2 Torr at 85 °C	345
(EtCp)Ru(DMPD)	17	0.05 Torr at 63 °C, 0.1 Torr at 75 °C	372,373
$(DMPD)_2Ru$	89	0.03 Torr at 65 °C,0.1 Torr at 82 °C	373
$(C_6H_6)Ru(C_6H_8)$	114–115	–	340
(1,5-COD)Ru(toluene)	72	Vacuum distillable at 100 °C	377
$(1,5\text{-COD})Ru(allyl)_2$	Decomposed at 130–140	–	341

atmosphere, no deposition was found at temperatures up to 800 °C.[316,349,350] Hence, except for the above study, all depositions using Cp_2Ru[316,349–356] and most of the other precursors in this group, including $(EtCp)_2Ru$,[357–366,299,344,346] CpRu(PrCp), where PrCp = propylcyclopentadienyl,[345,367,368] (EtCp)Ru(DMPD), where DMPD = 2,4-dimethylpentadienyl,[346,369–372] and $Ru(DMPD)_2$,[373] employed O_2 or an oxygen plasma as a co-reactant. The organics in the precursors were removed by reactions with chemisorbed O through a combustion mechanism.[360,374,375]

Using Cp_2Ru, films of Ru, RuO_2 or a mixture of the two were produced from *ca.* 200 to 600 °C (Table 7.3). Precise control over the O_2 to precursor ratio was essential for depositions of pure Ru. The lowest resistivities of the Ru films produced from Cp_2Ru were close to bulk metal.

Deposition using $(EtCp)_2Ru$ has been more extensively studied because in part it is a liquid at room temperature (Table 7.4).[299,344,346,357–366] Similar to Cp_2Ru, low O_2 partial pressures afforded pure Ru films while high O_2 partial pressures led to a mixture of Ru and RuO_x or RuO_2. Typically, the reaction-limited deposition regions were below 300–350 °C.[344,357,358,362,365] The addition of an iodine source (CH_3I or C_2H_5I) at the beginning and/or during the deposition promoted the nucleation on SiO_2/Si forming smoother films.[362,363] Pretreatment of a TiN surface with $PdCl_2$ showed similar effects.[364] The deposition conditions and film characteristics using CpRu(PrCp) were similar to those for $(EtCp)_2Ru$.[345] Ruthenium nucleation on more ionic substrates, *e.g.*, Ta_2O_5 and TiO_2, exhibited lower energy barriers than on substrates with a higher degree of covalency, *e.g.*, Si_3N_4, SiO_2 and TiN.[368] Argon plasma treatment of TiN improved Ru nucleation because of the selective removal of N and subsequent oxidation to form more ionic TiN_xO_y surface species.[368]

Comparative studies showed that the decomposition temperatures of several precursors decreased in the order $(EtCp)_2Ru$, (EtCp)Ru(DMPD) and $Ru(DMPD)_2$.[346,373] The minimum temperatures needed to produce a film and the temperatures for depositing best-quality films (*e.g.*, low resistivity and low surface roughness) also decreased in this order.[346,373] Low-temperature depositions without using O_2 were achieved using compounds bearing diene and benzene or toluene ligands, $(C_6H_6)Ru(C_6H_8)$, where C_6H_8 = 1,3-cyclohexadiene,[376] and (COD)Ru(toluene), where COD=1,5-cyclooctadiene.[377] Films exhibited low C impurity levels when deposited under optimized conditions. When $(COD)Ru(allyl)_2$ was used as the precursor, film depositions were successful at 300 °C on Al_2O_3 and Si under N_2, H_2 or O_2. Depending on the gas used, Ru, RuO_2 or mixed Ru and RuO_2 films were produced.[378] Depositions employing $(COD)Ru(tmhd)_2$ were mainly carried out under O_2 *via* liquid injections of the precursor using THF or alkane solutions.[294,347,379,380] Ruthenium films were also produced at 280–320 °C by alternatively dosing the precursor and O_2, but self-limiting behavior was not observed.[379,380]

7.5.4 Ruthenium CVD from RuO_4

RuO_4 (mp 27 °C)[283] was used to deposit pure RuO_2 at low temperatures.[381–384] It was prepared by oxidation of the metal with strong oxidants at elevated temperatures or by oxidation of Ru(VI) salts

Table 7.3 CVD using Cp_2Ru.

Reactor[a]	Substrate	Carrier gas, flow rate (sccm)	Reactant gas, flow rate (sccm)	Total pressure (Torr)	Precursor temperature (T_p) (°C)	Deposition temperature (T_d) (°C)	Films	Ref.
Hot	Glass	H_2, 100	H_2	760	94	590–600	Ru	348
Hot	Si, SiO_2/Si	O_2	O_2	1, 5	–	575–600	Ru, Ru + RuO_2, RuO_2	316
Hot	Si, SiO_2/Si, quartz	N_2, 5	O_2, 500–1000	2.5–10	140–170	550–575	Ru, Ru + RuO_2, RuO_2	351
Cold	Si	Ar, 100	O_2, 50	0.2	130	230–315	Ru	355
Cold	SiO_2/Si	Ar, 1–20	O_2 plasma, 10–200	0.04	80–120	500	Ru, Ru + RuO_2, RuO_2	349
Cold	SiO_2/Si, treated with O plasma	Ar, 5–20	O_2, 5–200	0.04	95–115	225–500	Ru, Ru + RuO_2, RuO_2	350, 354
Hot	SiO_2/Si	Ar, 50–200	O_2, 50–300	3	65–85	200–350	RuO_2	353
Hot	quartz	O_2,	O_2,	40	–	350–700	RuO_2	352
Cold	$LaAlO_3$ MgO	N_2, 50	O_2, 200	6	85	300–600	RuO_2	356

[a] Hot = hot wall; cold = cold wall.

Table 7.4 CVD using $(EtCp)_2Ru$.

Reactor[a]	Substrate	Carrier gas, flow rate (sccm)	Reactant gas, flow rate (sccm)	Total pressure (Torr)	Precursor temperature (T_p) (°C)	Deposition temperature (T_d) (°C)	Films	Ref.
Cold	SiO_2/Si	Ar, 100	O_2, 50	0.2	Room temp., DLI[b]	243–315	Ru	344
Cold	SiO_2/Si	Ar, 50	O_2, 50	5	Room temp., DLI[b] (THF solution)	300–500	Ru	357
Warm	Si, SiO_2/Si	Ar, 100[c]	O_2, 0–400	3	85	300–400	Ru, Ru + RuO_2	358
Warm	SiO_2/Si	Ar, 15	O_2, 0.1–5	0.5	50	200–400	Ru, Ru + RuO_2	360
Warm	SiO_2/Si	Ar, 15	O_2, 0.2 or 5	0.5	50	220–400	Ru, Ru + RuO_x	359
–	5 nm Ru/SiO_2/Si	N_2, 250	O_2, 50–2000	–	–	355	Ru, Ru + RuO_2	361
Cold	TiN/Ti/Si	Ar, 200	O_2, 50; CH_3I or C_2H_5I, 700	5.7	75	270–400	Ru + RuO_2	362
Cold	TiN/Ti/Si	Ar, 150	O_2, 50; C_2H_5I, 50	3	110	280–450	Ru	363
Cold	SiO_2/Si	Ar, 100, (total, 600)	Not mentioned	10	60	260–500	Ru	346
Warm	SiO_2/Si	None[d]	O_2, 50–300	0.3	115	320–480	Ru, RuO_x, RuO_2	365
Cold	TiN/Ti/Si, $PdCl_2$-pre-treated TiN/Ti/Si	Ar, 150	O_2, 50	3	90	280–400	Ru	364
Warm	Si, SiO_2/Si, TiO_2/SiO_2/Si	N_2, 50	O_2, 50–500	1.2	Room temp., DLI,[b] (toluene solution)	330–460	Ru	366

[a]Cold = cold wall; warm = warm wall.
[b]Direct liquid injection.
[c]Total flow rate is 1000 sccm, including Ar carrier gas, Ar dilute gas and O_2.
[d]MKS Instruments1153A low vapor pressure source delivery system, and 100 sccm Ar was used to clean the backside of the heater.

with Cl_2.[385] It can also be obtained commercially as a dilute aqueous solution and extraction with pentane or CCl_4 afforded the corresponding organic solutions. Pure RuO_4 was obtained by evaporating the organics at low temperatures. It is a powerful oxidant and contact with a reducing agent can produce explosions.[385] The depositions used freshly prepared RuO_4 without a carrier gas[384] or used a carrier gas (H_2, Air or N_2) passing through an aqueous or organic solution.[381,384] When H_2 was used, Ru films were formed.[382,384] A modified method was reported to deposit RuO_2 using *in situ* generated RuO_4 from the reaction of Ru and ozone.[383]

7.5.5 Ruthenium ALD

The Ru sources used in ALD included Cp_2Ru,[374,386,387] $(EtCp)_2Ru$,[299,388] (EtCp)Ru(DMPD),[389] $Ru(tmhd)_3$[390] and $Ru(^tBu\text{-}Me\text{-}amd)_2(CO)_2$.[338] In all cases except for $Ru(^tBu\text{-}Me\text{-}amd)_2(CO)_2$, O_2 was used as the second reactant.

In ALD using Cp_2Ru, the Ru growth rates were reported to be 0.45[386] and 1 Å $cycle^{-1}$.[387] Depositions in both studies required incubation periods. *In situ* Auger electron spectroscopic (AES) analysis revealed a slower nucleation on a H-terminated Si substrate than that on SiO_2 or HfO_2, which was attributed to its lower surface energy.[387] The film growth exhibited self-limiting behavior at a fixed temperature, but growth rates increased as the substrate temperatures increased,[386] indicating larger amounts of Ru precursor adsorbing and reacting in each cycle at higher temperatures. This atypical ALD behavior could be attributed to the increased reactivities of the adsorbed oxygen at higher temperatures. UHV studies revealed three oxygen states existed on a Ru seed layer and the reaction toward $(EtCp)_2Ru$ increased as the temperature was raised from *ca.* 230 to 580 °C.[375]

ALD using $(EtCp)_2Ru$ on TiN afforded a growth rate of *ca.* 1.5 Å $cycle^{-1}$. Unlike the behavior of Cp_2Ru, RuO_2 was observed when insufficient $(EtCp)_2Ru$ was supplied in a pulse.[388] When (EtCp)Ru(DMPD) was used, there was essentially no incubation period for ALD on Si, SiO_2, TiO_2 and TiN.[389] Here, the growth rate increased from 0.36 Å $cycle^{-1}$ at 230 °C to 0.5 Å $cycle^{-1}$ at 280 °C. ALD was also achieved using $Ru(tmhd)_3$, but the impurity levels in the films were higher than those in the films deposited using Cp_2Ru.[390]

As previously described, $Ru(^tBu\text{-}Me\text{-}amd)_2(CO)_2$ afforded Ru films in CVD without using a second reactant.[338] This implied the reaction of this precursor on surfaces was not self-limiting. However, by controlling alternate doses of $Ru(^tBu\text{-}Me\text{-}amd)_2(CO)_2$ and NH_3, ALD was achieved at the same temperatures (200–300 °C) as used in CVD. Conformal Ru films on flat surfaces as well as in high-aspect-ratio (40 : 1) structures were produced.[338] In addition, the deposition rates were higher in ALD than those obtained in pulsed CVD under similar conditions but without using NH_3.

7.6 Concluding Remarks

The deposition of metal films by chemical vapor deposition continues to be important in applications requiring uniform thickness over complex shapes or where selective deposition offers an approach that minimizes additional lithographic steps. For most metals a choice exists among several possible precursors and the decision to use one over another depends on demands of film purity, process conditions and throughput, step coverage capability and precursor handling requirements. As there is currently no viable alternative to WF_6, tungsten is an exception to this rule. Continued research is needed to devise depositions for the more electropositive metals. Also, the importance of nucleation in CVD processes demands fundamental studies aimed at understanding the initial stages of film formation.

References

1. W. A. Bryant, *J. Electrochem. Soc.*, 1978, **125**, 1534–1543.
2. M. L. Hitchman, A. D. Jobson and L. F. T. Kwakman, *Appl. Surf. Sci.*, 1989, **38**, 312–337.
3. M. L. Yu, K. Y. Ahn and R. V. Joshi, *IBM J. Res. Develop.*, 1990, **34**, 875–883.
4. J. E. J. Schmitz *Chemical Vapor Deposition of Tungsten And Tungsten Silicides,* Noyes Publications, Park Ridge, N.J., USA, 1992.
5. W. L. Gladfelter, *Chem. Mater.*, 1993, **5**, 1372–1388.
6. J. R. Creighton and J. E. Parmeter, *Critical Reviews in Solid State and Materials Sciences*, 1993, **18**, 175–238.
7. T. T. Kodas and M. J. Hampden-Smith, Editors, *The Chemistry of Metal CVD*, 1994. VCH, Weinheim and New York.
8. V. N. Vertoprakhov and S. A. Krupoder, *Russian Chem. Rev.*, 2000, **69**, 1057–1082.
9. J. Rickerby and J. H. G. Steinke, *Chem. Rev.*, 2002, **102**, 1525–1549.
10. J. P. Day, R. G. Pearson and F. Basolo, *J. Am. Chem. Soc.*, 1968, **90**, 6933–6938.
11. W. Kraeuter, D. Baeuerle and F. Fimberger, *Appl. Phys. A*, 1983, **A31**, 13–18.
12. A. Ludviksson, D. W. Robinson and J. W. Rogers Jr, *Thin Solid Films*, 1996, **289**, 6–13.
13. G. S. Girolami, P. M. Jeffries and L. H. Dubois, *J. Am. Chem. Soc.*, 1993, **115**, 1015–1024.
14. A. Cezairliyan, *High Temp. Sci.*, 1972, **4**, 248–252.
15. *ASM Handbook online, Vol. 2: Properties and Selection-Nonferrous Alloys and Pure Metals.* Online accessed through the Library of the University of Minnesota at the twin cities. Accessed date: November 2007.
16. F. A. Cotton, G. Wilkinson, M. Bochmann and C. Murillo, *Advanced Inorganic Chemistry, 6th Edition*, Wiley, New York, 1999.
17. R. Liu, C.-S. Pai and E. Martinez, *Solid-State Electron.*, 1999, **43**, 1003–1009.
18. J. M. Shaw and J. A. Amick, *RCA Rev.*, 1970, **31**, 306–316.
19. C. M. Melliar-Smith, A. C. Adams, R. H. Kaiser and R. A. Kushner, *J. Electrochem. Soc.*, 1974, **121**, 298–303.
20. E. J. Barber and G. H. Cady, *J. Phys. Chem.*, 1956, **60**, 505–506.
21. P. J. Ireland, *Thin Solid Films*, 1997, **304**, 1–12.
22. D. A. Buchanan, F. R. McFeely and J. J. Yurkas, *Appl. Phys. Lett.*, 1998, **73**, 1676–1678.
23. K. K. Lai and H. H. Lamb, *Thin Solid Films*, 2000, **370**, 114–121.
24. S. Jeon, K. Yong, S.-G. Park and S.-W. Rhee, *J. Vac. Sci. Tech., B*, 2006, **24**, 1428–1431.
25. E. K. Broadbent and C. L. Ramiller, *J. Electrochem. Soc.*, 1984, **131**, 1427–1433.
26. M. L. Green, Y. S. Ali, T. Boone, B. A. Davidson, L. C. Feldman and S. Nakahara, *J. Electrochem. Soc.*, 1987, **134**, 2285–2292.
27. P. A. C. Groenen, J. G. A. Hoelscher and H. H. Brongersma, *Appl. Surf. Sci.*, 1994, **78**, 123–132.
28. R. B. Jackman and J. S. Foord, *Surf. Sci.*, 1988, **201**, 47–58.
29. J. A. Yarmoff and F. R. McFeely, *J. Appl. Phys.*, 1988, **63**, 5213–5219.
30. M. L. Yu, B. N. Eldridge and R. V. Joshi, *Tungsten Other Refract. Met. VLSI Appl. 4, Proc. Workshop*, Editors: R. S. Blewer and C. M. McConica, Materials Research Society, Pittsburgh, PA, USA. 1989, 221–230.
31. T. F. Deutsch and D. D. Rathman, *Appl. Phys. Lett.*, 1984, **45**, 623–625.
32. J. R. Creighton, *J. Electrochem. Soc.*, 1989, **136**, 271–276.
33. J. R. Creighton, *J. Vac. Sci. Technol. A*, 1989, **7**, 621–624.
34. N. Desatnik and B. E. Thompson, *J. Electrochem. Soc.*, 1994, **141**, 3532–3539.
35. T. Noma, K. S. Seol, M. Fujimaki and Y. Ohki, *J. Appl. Phys.*, 1999, **85**, 8423–8426.
36. Y. Saito and T. Takagi, *J. Electrochem. Soc.*, 1996, **143**, 3670–3674.
37. A. Miller and G. D. Barnett, *J. Electrochem. Soc.*, 1962, **109**, 973–976.

38. J. F. Berkeley, A. Brenner and W. E. Reid Jr, *J. Electrochem. Soc.*, 1967, **114**, 561–568.
39. W. A. Bryant and G. H. Meier, *J. Electrochem. Soc.*, 1973, **120**, 559–565.
40. C. E. Morosanu and V. Soltuz, *Thin Solid Films*, 1978, **52**, 181–194.
41. Y. Pauleau and P. Lami, *J. Electrochem. Soc.*, 1985, **132**, 2779–2784.
42. C. M. McConica and K. Krishnamani, *J. Electrochem. Soc.*, 1986, **133**, 2542–2548.
43. Y. W. Park, C. O. Park and J. S. Chun, *Thin Solid Films*, 1991, **201**, 167–175.
44. T. G. M. Oosterlaken, G. J. Leusink, G. C. A. M. Janssen and S. Radelaar, *J. Electrochem. Soc.*, 1996, **143**, 1668–1675.
45. L. Vazquez, R. C. Salvarezza, E. Albano, A. J. Arvia, A. H. Creus, R. A. Levy and J. M. Albella, *Chem. Vap. Deposition*, 1998, **4**, 89–91.
46. I. S. Chang and M.-H. Hon, *Thin Solid Films*, 1998, **333**, 108–113.
47. T. Gougousi, Y. Xu, J. N. Kidder Jr, G. W. Rubloff and C. R. Tilford, *J. Vac. Sci. Tech. B*, 2000, **18**, 1352–1363.
48. P. W. Tamm and L. D. Schmidt, *J. Chem. Phys.*, 1971, **54**, 4775–4787.
49. P. W. Tamm and L. D. Schmidt, *J. Chem. Phys.*, 1971, **55**, 4253–4259.
50. R. P. H. Gasser, T. N. Morton, J. M. Overton and A. K. Szczepura, *Surf. Sci.*, 1971, **28**, 574–580.
51. H. F. Berger, C. Resch, E. Groesslinger, G. Eilmsteiner, A. Winkler and K. D. Rendulic, *Surf. Sci.*, 1992, **275**, L627–L630.
52. K. J. Kuijlaars, C. R. Kleijn and H. E. A. van den Akker, *Thin Solid Films*, 1995, **270**, 456–461.
53. D. A. Bell, J. L. Falconer and C. M. McConica, *J. Electrochem. Soc.*, 1995, **142**, 2401–2404.
54. W. Chen and J. T. Roberts, *Surf. Sci.*, 1995, **324**, 169–184.
55. M. Duerr and U. Hoefer, *Surf. Sci. Reports*, 2006, **61**, 465–526.
56. K. Y. Ahn, P. M. Fryer, J. M. E. Harper, R. V. Joshi, C. W. Miller and E. G. Colgan, *Tungsten Other Refract. Met. VLSI Appl. 4, Proc. Workshop*, Editors: R. S. Blewer and C. M. McConica, Materials Research Society, Pittsburgh, PA, USA. 1988, 35–53.
57. Y. Kusumoto, M. Takakuwa, H. Hashinokuchi, T. Ikuta and I. Nakayama, *Tungsten Other Refract. Met. VLSI Appl. 3, Proc. Workshop*, Editor: V. A. Wells, Materials Research Society, Pittsburgh, PA, USA. 1988.
58. J. E. J. Schmitz, A. J. M. Van Dijk and M. W. M. Graef, *Proc.-Electrochem. Soc.*, 1987, **87–8**, 625–634.
59. T. Ohba, T. Suzuki, T. Hara, Y. Furumura and K. Wada, *Tungsten Other Refract. Met. VLSI Appl. 4, Proc. Workshop*, Editors: R. S. Blewer and C. M. McConica, Materials Research Society, Pittsburgh, PA, USA. 1989. 17–25.
60. J. E. J. Schmitz, M. J. Buiting and R. C. Ellwanger, *Tungsten Other Refract. Met. VLSI Appl. 4, Proc. Workshop*, Editors: R. S. Blewer and C. M. McConica, Materials Research Society, Pittsburgh, PA, USA. 1989. 27–33.
61. Y. H. Song, J. H. Park, S. E. Lee, J. Y. Lee and J. S. Shin, *Jpn. J. Appl. Phys., Part 2*, 2007, **46**, L120–L122.
62. J. Widiez, M. Vinet, B. Guillaumot, X. Garros, S. Minoret, T. Poiroux, O. Weber, L. Thevenod, P. Holliger, B. Previtali, V. Barral, K. S. A. Cherif, P. Grosgeorges, A. Toffoli, S. Maitrejean, M. Casse, F. Martin, D. Lafond, O. Faynot, M. Mouis and S. Deleonibus, International Symp. on VLSI Techn., *Systems, and Applications (VLSI-TSA)*, Publisher: IEEE, Piscataway, NJ, USA. *Hsinchu, Taiwan, Apr. 24-26, 2006*, 2006, 117–118.
63. T. Saito, Y. Shimogaki, Y. Egashira, K. Sugawara, K. Takahiro, S. Nagata, S. Yamaguchi and H. Komiyama, *Thin Solid Films*, 2006, **513**, 36–42.
64. T. Saito, Y. Shimogaki, Y. Egashira, K. Sugawara, K. Takahiro, S. Nagata, S. Yamaguchi and H. Komiyama, *Microelectron. Eng.*, 2006, **83**, 1994–2000.

65. M. Suzuki, N. Kobayashi, K. Mukai and S. Kondo, *J. Electrochem. Soc.*, 1990, **137**, 3213–3218.
66. R. S. Rosler, J. Mendonca and M. J. Rice Jr, *J. Vac. Sci. Tech. B*, 1988, **6**, 1721–1727.
67. H. L. Park, S. S. Yoon, C. O. Park and J. S. Chun, *Thin Solid Films*, 1989, **181**, 85–93.
68. J. A. M. Ammerlaan, P. J. van der Put and J. Schoonman, *J. Appl. Phys.*, 1993, **73**, 4631–4636.
69. C. A. Van der Jeugd, G. C. A. M. Janssen and S. Redelaar, *J. Appl. Phys.*, 1992, **72**, 1583–1588.
70. D. A. Bell, C. M. McConica, K. L. Baker and E. Kuchta, *J. Electrochem. Soc.*, 1996, **143**, 296–302.
71. N. Kobayashi, H. Goto and M. Suzuki, *J. Appl. Phys.*, 1991, **69**, 1013–1019.
72. N. Kobayashi, Y. Nakamura, H. Goto and Y. Homma, *J. Appl. Phys.*, 1993, **73**, 4637–4643.
73. M. L. Yu and B. N. Eldridge, *J. Vac. Sci. Technol., A*, 1989, **7**, 625–629.
74. N. Kobayashi, Y. Nakamura, H. Goto and Y. Homma, *Proc. -Electrochem. Soc.*, 1993, **93–2**, 250–256.
75. P. Gouy-Pailler, P. Lami and R. Morales, *Thin Solid Films*, 1994, **241**, 374–377.
76. Y. Yamamoto, T. Matsuura and J. Murota, *Surf. Sci.*, 1998, **408**, 190–194.
77. M. S. Marangon, G. Queirolo and C. Savoia, *Appl. Surf. Sci.*, 1995, **91**, 157–161.
78. Y. Kajikawa, T. Tsumura, S. Noda, H. Komiyama and Y. Shimogaki, *Jpn. J. Appl. Phys., Part 1*, 2004, **43**, 3945–3950.
79. S. Noda, T. Tsumura, J. Fukuhara, T. Yoda, H. Komiyama and Y. Shimogaki, *Jpn. J. Appl. Phys., Part 1*, 2004, **43**, 6974–6977.
80. M. L. Yu, K. Y. Ahn and R. V. Joshi, *J. Appl. Phys.*, 1990, **67**, 1055–1061.
81. S. L. Lantz, A. E. Bell, W. K. Ford and D. Danielson, *J. Vac. Sci. Technol., A*, 1994, **12**, 1032–1038.
82. J. W. Klaus, S. J. Ferro and S. M. George, *Thin Solid Films*, 2000, **360**, 145–153.
83. J. W. Klaus, S. J. Ferro and S. M. George, *Appl. Surf. Sci.*, 2000, **162–163**, 479–491.
84. J. W. Klaus, S. J. Ferro and S. M. George, *J. Electrochem. Soc.*, 2000, **147**, 1175–1181.
85. J. W. Elam, C. E. Nelson, R. K. Grubbs and S. M. George, *Surf. Sci.*, 2001, **479**, 121–135.
86. J. W. Elam, C. E. Nelson, R. K. Grubbs and S. M. George, *Thin Solid Films*, 2001, **386**, 41–52.
87. M. Schuisky, J. W. Elam and S. M. George, *Appl. Phys. Lett.*, 2002, **81**, 180–182.
88. R. K. Grubbs, N. J. Steinmetz and S. M. George, *J. Vac. Sci. Tech. B*, 2004, **22**, 1811–1821.
89. R. K. Grubbs, C. E. Nelson, N. J. Steinmetz and S. M. George, *Thin Solid Films*, 2004, **467**, 16–27.
90. F. H. Fabreguette, Z. A. Sechrist, J. W. Elam and S. M. George, *Thin Solid Films*, 2005, **488**, 103–110.
91. F. H. Fabreguette, R. A. Wind and S. M. George, *Appl. Phys. Lett.*, 2006, **88**, 013116/013111-013116/013113.
92. J. W. Elam, J. A. Libera, M. J. Pellin, A. V. Zinovev, J. P. Greene and J. A. Nolen, *Appl. Phys. Lett.*, 2006, **89**, 053124/053121-053124/053123.
93. S.-H. Kim, E.-S. Hwang, B.-M. Kim, J.-W. Lee, H.-J. Sun, T. E. Hong, J.-K. Kim, H. Sohn, J. Kim and T.-S. Yoon, *Electrochem. Solid-State Lett.*, 2005, **8**, C155–C159.
94. S.-H. Kim, N. Kwak, J. Kim and H. Sohn, *J. Electrochem. Soc.*, 2006, **153**, G887–G893.
95. W. Lei, L. Henn-Lecordier, M. Anderle, G. W. Rubloff, M. Barozzi and M. Bersani, *J. Vac. Sci. Tech. B*, 2006, **24**, 780–789.
96. D. R. Lide, *CRC Handbook of Chemistry and Physics, Internet Version, 87th Edition*, Taylor and Francis, Boca Raton, FL, 2007. Online accessed in November, 2007 through the Libraries of the University of Minnesota at the twin city.
97. S. M. Rossnagel and T. S. Kuan, *J. Vac. Sci. Tech., B*, 2004, **22**, 240–247.
98. S. Maitrejean, R. Gers, T. Mourier, A. Toffoli and G. Passemard, *Microelectron. Eng.*, 2006, **83**, 2396–2401.

99. M. G. Simmonds and W. L. Gladfelter, *The Chemistry of Metal CVD*, Editors: T. T. Kodas and M. J. Hampden-Smith, VCH, Weinheim and NewYork, 1994, Page 45–103.
100. L. O. Brockway and N. R. Davidson, *J. Am. Chem. Soc.*, 1941, **63**, 3287–3297.
101. N. Muller and D. E. Pritchard, *J. Am. Chem. Soc.*, 1960, **82**, 248–249.
102. J. W. Turley and H. W. Rinn, *Inorg. Chem.*, 1969, **8**, 18–22.
103. F. M. Brower, N. E. Matzek, P. F. Reigler, H. W. Rinn, C. B. Roberts, D. L. Schmidt, J. A. Snover and K. Terada, *J. Am. Chem. Soc.*, 1976, **98**, 2450–2453.
104. V. A. Yartys, R. V. Denys, J. P. Maehlen, C. Frommen, M. Fichtner, B. M. Bulychev and H. Emerich, *Inorg. Chem.*, 2007, **46**, 1051–1055.
105. M. G. Simmonds, E. C. Phillips, J. W. Hwang and W. L. Gladfelter, *Chemtronics*, 1991, **5**, 155–158.
106. H. O. Pierson, *Thin Solid Films*, 1977, **45**, 257–263.
107. A. Malazgirt and J. W. Evans, *Metall. Trans. B*, 1980, **11**, 225–232.
108. M. J. Cooke, R. A. Heinecke, R. C. Stern and J. W. C. Maes, *Solid State Tech.*, 1982, **25**, 62–65.
109. M. L. Green, R. A. Levy, R. G. Nuzzo and E. Coleman, *Thin Solid Films*, 1984, **114**, 367–377.
110. R. A. Levy, M. L. Green and P. K. Gallagher, *J. Electrochem. Soc.*, 1984, **131**, 2175–2182.
111. J. Y. Tsao and D. J. Ehrlich, *Appl. Phys. Lett.*, 1984, **45**, 617–619.
112. T. Kobayashi, A. Sekiguchi, N. Hosokawa and T. Asamaki, *Jpn. J. Appl. Phys. Part 2*, 1988, **27**, L1775–L1777.
113. A. Sekiguchi, T. Kobayashi, N. Hosokawa and T. Asamaki, *Jpn. J. Appl. Phys. Part 2*, 1988, **27**, L2134–L2136.
114. B. E. Bent, L. H. Dubois and R. G. Nuzzo, *Mater. Res. Soc. Symp. Proc.*, 1989, **131**, 327–338.
115. L. F. T. Kwakman, B. G. Sluijk, H. Piekaar and E. H. A. Granneman, *Tungsten Other Refract. Met. VLSI Appl. 4, Proc. Workshop*, Editors: R. S. Blewer and C. M. McConica, Materials Research Society, Pittsburgh, PA, USA. 1989, 315–321.
116. A. Sekiguchi, T. Kobayashi, N. Hosokawa and T. Asamaki, *J. Vac. Sci. Technol., A*, 1990, **8**, 2976–2979.
117. W. Y. C. Lai, R. Liu, K. P. Cheung, C. Case, L. F. T. Kwakman and D. Huibregtse, *Tungsten Other Adv. Met. ULSI Appl. 1990, Proc. Workshop, 7th*, Editors: G. C. Smith and R. Blumenthal Materials Research Society, Pittsburgh, PA, USA. 1991, 169–175.
118. T. Kobayashi, A. Sekiguchi, N. Akiyama, N. Hosokawa and T. Asamaki, *J. Vac. Sci. Technol., A*, 1992, **10**, 525–538.
119. K. I. Lee, Y. S. Kim and S. K. Joo, *J. Electrochem. Soc.*, 1992, **139**, 3578–3581.
120. T. Suzuki, *J. Mater. Res.*, 1997, **12**, 1866–1871.
121. B. E. Bent, R. G. Nuzzo and L. H. Dubois, *J. Am. Chem. Soc.*, 1989, **111**, 1634–1644.
122. W. L. Gladfelter, D. C. Boyd and K. F. Jensen, *Chem. Mater.*, 1989, **1**, 339–343.
123. J. A. Glass Jr, S. S. Kher and J. T. Spencer, *Chem. Mater.*, 1992, **4**, 530–538.
124. D. B. Beach, S. E. Blum and F. K. LeGoues, *J. Vac. Sci. Technol., A*, 1989, **7**, 3117–3118.
125. M. G. Simmonds, I. Taupin and W. L. Gladfelter, *Chem. Mater.*, 1994, **6**, 935–942.
126. L. H. Dubois, B. R. Zegarski, C. T. Kao and R. G. Nuzzo, *Surf. Sci.*, 1990, **236**, 77–84.
127. L. H. Dubois, B. R. Zegarski, M. E. Gross and R. G. Nuzzo, *Surf. Sci.*, 1991, **244**, 89–95.
128. D. M. Frigo, G. J. M. van Eijden, P. J. Reuvers and C. J. Smit, *Chem. Mater.*, 1994, **6**, 190–195.
129. J. A. Glass Jr, S. Kher and J. T. Spencer, *Thin Solid Films*, 1992, **207**, 15–18.
130. M. Sakamoto, K. Masu, F. Nakamura, T. Yoshihama, Y. Miyaguchi, S. J. Lim, M. Hatanaka and Y. Furumura, *Adv. Met. Conf. 2003, Proc. Conf., Tokyo, Japan, Sept. 29-Oct. 1 and Montreal, QC, Canada, Oct. 21-23, 2003*, Editor: G. W. Ray, Materials Research Society, Warrendale, PA, USA. 2004, 595–599.

131. Y. Liu, L. J. Overzet and M. J. Goeckner, *Thin Solid Films*, 2006, **510**, 48–54.
132. Y. Liu, L. Overzet and M. Goeckner, *Thin Solid Films*, 2007, **515**, 6730–6736.
133. K.-I. Choi, S.-H. Han, S. Yun, D.-Y. Kim, J. W. Hong, S. W. Lee, B. H. Kim, S.-T. Kim, U. I. Chung, J.-T. Moon and B.-I. Ryu, *Jpn. J. Appl. Phys., Part 1*, 2006, **45**, 2987–2991.
134. C. A. Tolman, *Chem. Rev.*, 1977, **77**, 313–348.
135. C. W. Heitsch, C. E. Nordman and R. W. Parry, *Inorg. Chem.*, 1963, **2**, 508–512.
136. J. L. Atwood, K. W. Butz, M. G. Gardiner, C. Jones, G. A. Koutsantonis, C. L. Raston and K. D. Robinson, *Inorg. Chem.*, 1993, **32**, 3482–3487.
137. K. Tsubouchi, K. Masu, N. Shigeeda, T. Matano, Y. Hiura and N. Mikoshiba, *Appl. Phys. Lett.*, 1990, **57**, 1221–1223.
138. K. Tsubouchi and K. Masu, *J. Vac. Sci. Technol., A*, 1992, **10**, 856–862.
139. N. Zhu, T. Cacouris, R. Scarmozzino and R. M. Osgood Jr, *J. Vac. Sci. Tech. B*, 1992, **10**, 1167–1176.
140. K. Tsubouchi and K. Masu, *Kotai Butsuri*, 1994, **29**, 599–605.
141. E. Kondoh and T. Ohta, *J. Vac. Sci. Tech. A*, 1995, **13**, 2863–2871.
142. T. Nitta and M. Hanabusa, *J. Appl. Phys.*, 1996, **69**, 340–342.
143. H. Sakaue, Y. Katsuda, S. Konagata, S. Shingubara and T. Takahagi, *Jpn. J. Appl. Phys., Part 1*, 1996, **35**, 1010–1013.
144. M. Naik, T. Guo, L. Chen, R. Mosely and I. Beinglass, *J. Vac. Sci. Technol., A*, 1998, **16**, 1233–1238.
145. T. Amazawa, *J. Electrochem. Soc.*, 1998, **145**, 4327–4332.
146. M. Sugiyama, T. Iino, H. Itoh, J. Aoyama, H. Komiyama and Y. Shimogaki, *Jpn. J. Appl. Phys., Part 2*, 1999, **38**, L1528–L1531.
147. T. Wartik and H. I. Schlesinger, *J. Am. Chem. Soc.*, 1953, **75**, 835–839.
148. A. Haaland, G. A. Anderson, A. Almenningen and F. R. Forgaard, *J. Chem. Soc., Chem. Commun.*, 1971, 480–481.
149. B. G. Willis and K. F. Jensen, *J. Phys. Chem. A*, 2000, **104**, 7881–7891.
150. G. Vass, G. Tarczay, G. Magyarfalvi, A. Boedi and L. Szepes, *Organometallics*, 2002, **21**, 2751–2757.
151. J.-H. Lee, M.-Y. Park, J.-H. Yun and S.-W. Rhee, *Thin Solid Films*, 1999, **348**, 130–133.
152. D. R. Strongin and P. B. Comita, *J. Phys. Chem.*, 1991, **95**, 1329–1333.
153. B. G. Willis and K. F. Jensen, *Surf. Sci.*, 2001, **488**, 286–302.
154. B. G. Willis and K. F. Jensen, *Surf. Sci.*, 2001, **488**, 303–324.
155. T. Mitsui, E. Hill, R. Curtis and E. Ganz, *Phys. Rev. B*, 1999, **59**, 8123–8127.
156. T. Mitsui, E. Hill and E. Ganz, *J. Appl. Phys.*, 1999, **85**, 522–524.
157. K. Fuchs, *Proc. Cambridge Phil. Soc.*, 1938, **34**, 100–108.
158. E. H. Sondheimer, *Advances in Physics*, 1952, **1**, 1–42.
159. A. F. Mayadas and M. Shatzkes, *Phys. Rev. B*, 1970, **1**, 1382–1389.
160. W. Steinhogl, G. Schindler, G. Steinlesberger and M. Engelhardt, *Phys. Rev. B*, 2002, **66**, 075414/075411-075414/075414.
161. J. Vancea, H. Hoffmann and K. Kastner, *Thin Solid Films*, 1984, **121**, 201–216.
162. W. F. A. Besling, M. Broekaart, V. Arnal and J. Torres, *Microelectron. Eng.*, 2004, **76**, 167–174.
163. W. Wu, S. H. Brongersma, M. Van Hove and K. Maex, *Appl. Phys. Lett.*, 2004, **84**, 2838–2840.
164. W. Zhang, S. H. Brongersma, T. Clarysse, V. Terzieva, E. Rosseel, W. Vandervorst and K. Maex, *J. Vac. Sci. Tech., B*, 2004, **22**, 1830–1833.
165. W. Zhang, S. H. Brongersma, Z. Li, D. Li, O. Richard and K. Maex, *J. Appl. Phys.*, 2007, **101**, 063703/063701-063703/063711.
166. J. A. Bertrand and R. I. Kaplan, *Inorg. Chem.*, 1966, **5**, 489–491.

167. A. W. Maverick, F. R. Fronczek, E. F. Maverick, D. R. Billodeaux, Z. T. Cygan and R. A. Isovitsch, *Inorg. Chem.*, 2002, **41**, 6488–6492.
168. C. R. Moylan, T. H. Baum and C. R. Jones, *Appl. Phys. A*, 1986, **A40**, 1–5.
169. W. R. Wolf, R. E. Sievers and G. H. Brown, *Inorg. Chem.*, 1972, **11**, 1995–2002.
170. D. Temple and A. Reisman, *J. Electrochem. Soc.*, 1989, **136**, 3525–3529.
171. J. Pinkas, J. C. Huffman, D. V. Baxter, M. H. Chisholm and K. G. Caulton, *Chem. Mater.*, 1995, **7**, 1589–1596.
172. A. Jain, T. T. Kodas, T. S. Corbitt and M. J. Hampden-Smith, *Chem. Mater.*, 1996, **8**, 1119–1127.
173. L. L. Funck and T. R. Ortolano, *Inorg. Chem.*, 1968, **7**, 567–573.
174. B. Lecohier, B. Calpini, J. M. Philippoz and H. Van den Bergh, *J. Appl. Phys.*, 1992, **72**, 2022–2026.
175. W. G. Lai, Y. Xie and G. L. Griffin, *J. Electrochem. Soc.*, 1991, **138**, 3499–3504.
176. D.-H. Kim, R. H. Wentorf and W. N. Gill, *J. Electrochem. Soc.*, 1993, **140**, 3267–3272.
177. D.-H. Kim, R. H. Wentorf and W. N. Gill, *J. Electrochem. Soc.*, 1993, **140**, 3273–3279.
178. N. S. Borgharkar and G. L. Griffin, *J. Electrochem. Soc.*, 1998, **145**, 347–352.
179. K.-I. Hanaoka, K. Tachibana and H. Ohnishi, *Thin Solid Films*, 1995, **262**, 209–217.
180. J. Wang, R. B. Little, W. G. Lai and G. L. Griffin, *Thin Solid Films*, 1995, **262**, 31–38.
181. J. Farkas, M. J. Hampden-Smith and T. T. Kodas, *J. Phys. Chem.*, 1994, **98**, 6753–6762.
182. N. Awaya and Y. Arita, *Jpn. J. Appl. Phys., Part 1*, 1993, **32**, 3915–3919.
183. J.-Y. Kim, P. J. Reucroft and D.-K. Park, *Thin Solid Films*, 1996, **289**, 184–191.
184. H.-H. Lee, C. Lee, Y.-L. Kuo and Y.-W. Yen, *Thin Solid Films*, 2006, **498**, 43–49.
185. L. Wang and G. L. Griffin, *J. Electrochem. Soc.*, 2006, **153**, C137–C141.
186. J.-Y. Kim, Y.-K. Lee, H.-S. Park, J.-W. Park, D.-K. Park, J.-H. Joo, W.-H. Lee, Y.-K. Ko, P. J. Reucroft and B.-R. Cho, *Thin Solid Films*, 1998, **330**, 190–195.
187. X. Liu, Q. Wang and S. Wu, *Chem. Vap. Deposition*, 2006, **12**, 679–684.
188. X. Liu, Q. Wang, S. Wu and Z. Liu, *J. Electrochem. Soc.*, 2006, **153**, C142–C145.
189. T. Momose, M. Sugiyama and Y. Shimogaki, *Jpn. J. Appl. Phys., Part 2*, 2005, **44**, L1199–L1202.
190. E. Kondoh and K. Shigama, *Thin Solid Films*, 2005, **491**, 228–234.
191. N. Awaya and Y. Arita, *Jpn. J. Appl. Phys., Part 1*, 1991, **30**, 1813–1817.
192. Y. Pauleau and A. Y. Fasasi, *Chem. Mater.*, 1991, **3**, 45–50.
193. S. Hwang, H. Choi and I. Shim, *Chem. Mater.*, 1996, **8**, 981–983.
194. S. T. Hwang, I. Shim, K. O. Lee, K. S. Kim, J. H. Kim, G. J. Choi, Y. S. Cho and H. Choi, *J. Mater. Res.*, 1996, **11**, 1051–1060.
195. G. Ramaswamy, A. K. Raychaudhuri, J. Goswami and S. A. Shivashankar, *J. Appl. Phys.*, 1997, **82**, 3797–3807.
196. A. Devi and S. A. Shivashankar, *J. Mater. Sci. Lett.*, 1998, **17**, 367–369.
197. S. Mukhopadhyay, K. Shalini, R. Lakshmi, A. Devi and S. A. Shivashankar, *Surf. Coat. Tech.*, 2002, **150**, 205–211.
198. H. Guillon, S. Daniele and L. G. Hubert-Pfalzgraf, *J. de Phys. IV*, 1999, **9**, 861–868.
199. H. Guillon, S. Daniele, L. G. Hubert-Pfalzgraf and C. Bavoux, *Inorg. Chim. Acta*, 2000, **304**, 99–107.
200. J. Pinkas, J. C. Huffman, J. C. Bollinger, W. E. Streib, D. V. Baxter, M. H. Chisholm and K. G. Caulton, *Inorg. Chem.*, 1997, **36**, 2930–2937.
201. S. I. Troyanov, O. Y. Gorbenko and A. A. Bosak, *Polyhedron*, 1997, **16**, 1595–1598.
202. O. Y. Gorbenko, S. I. Troyanov, A. Meetsma and A. A. Bosak, *Polyhedron*, 1997, **16**, 1999–2004.
203. S. C. Goel, K. S. Kramer, M. Y. Chiang and W. E. Buhro, *Polyhedron*, 1990, **9**, 611–613.
204. V. L. Young, D. F. Cox and M. E. Davis, *Chem. Mater.*, 1993, **5**, 1701–1709.

205. R. Becker, J. Weiss, M. Winter, K. Merz and R. A. Fischer, *J. Organomet. Chem.*, 2001, **630**, 253–262.
206. P.-F. Hsu, Y. Chi, T.-W. Lin, C.-S. Liu, A. J. Carty and S.-M. Peng, *Chem. Vap. Deposition*, 2001, **7**, 28–31.
207. R. Becker, J. Weiss, A. Devi and R. A. Fischer, *J. de Phys. IV*, 2001, **11**, Pr3/569-Pr563/575.
208. Y. Chi, P.-F. Hsu, C.-S. Liu, W.-L. Ching, T.-Y. Chou, A. J. Carty, S.-M. Peng, G.-H. Lee and S.-H. Chuang, *J. Mater. Chem.*, 2002, **12**, 3541–3550.
209. R. Becker, A. Devi, J. Weiss, U. Weckenmann, M. Winter, C. Kiener, H.-W. Becker and R. A. Fischer, *Chem. Vap. Deposition*, 2003, **9**, 149–156.
210. J. W. Park, H. S. Jang, M. Kim, K. Sung, S. S. Lee, T.-M. Chung, S. Koo, C. G. Kim and Y. Kim, *Inorg. Chem. Commun.*, 2004, **7**, 463–466.
211. E. Lay, Y.-H. Song, Y.-C. Chiu, Y.-M. Lin, Y. Chi, A. J. Carty, S.-M. Peng and G.-H. Lee, *Inorg. Chem.*, 2005, **44**, 7226–7233.
212. M. E. Gross, *J. Electrochem. Soc.*, 1991, **138**, 2422–2426.
213. S. L. Cohen, M. Liehr and S. Kasi, *J. Vac. Sci. Tech., A*, 1992, **10**, 863–868.
214. R. Kumar, F. R. Fronczek, A. W. Maverick, W. G. Lai and G. L. Griffin, *Chem. Mater.*, 1992, **4**, 577–582.
215. J. Farkas, M. J. Hampden-Smith and T. T. Kodas, *J. Electrochem. Soc.*, 1994, **141**, 3547–3555.
216. E. S. Choi, S. K. Park, H. K. Shin and H. H. Lee, *Appl. Phys. Lett.*, 1996, **68**, 1017–1019.
217. C.-H. Jun, Y. T. Kim, J.-T. Baek, H. J. Yoo and D.-R. Kim, *J. Vac. Sci. Tech., A*, 1996, **14**, 3214–3219.
218. M. Juppo, M. Ritala and M. Leskela, *J. Vac. Sci. Tech., A*, 1997, **15**, 2330–2333.
219. E. C. Plappert, T. Stumm, H. Van den Bergh, R. Hauert and K. H. Dahmen, *Chem. Vap. Deposition*, 1997, **3**, 37–43.
220. H. Choi and S. Hwang, *Chem. Mater.*, 1998, **10**, 2326–2328.
221. S.-W. Kang, S.-H. Han and S.-W. Rhee, *Thin Solid Films*, 1999, **350**, 10–13.
222. M.-Y. Park, J.-H. Son, S.-W. Kang and S.-W. Rhee, *J. Mater. Res.*, 1999, **14**, 975–979.
223. T. Y. Chen, J. Vaissermann and P. Doppelt, *Inorg. Chem.*, 2001, **40**, 6167–6171.
224. T. Y. Chen, J. Vaissermann, E. Ruiz, J. P. Senateur and P. Doppelt, *Chem. Mater.*, 2001, **13**, 3993–4004.
225. M. Zhang, A. Sekiguchi, O. Okada, A. Itsuki and K. Ogi, *Jpn. J. Appl. Phys., Part 1*, 2001, **40**, 4825–4828.
226. B.-M. Kim and H. H. Lee, *Chem. Vap. Deposition*, 2001, **7**, 242–245.
227. K.-K. Choi and S.-W. Rhee, *Thin Solid Films*, 2002, **409**, 147–152.
228. C. Cavallotti, V. Gupta, C. Sieber and K. F. Jensen, *Phys. Chem. Chem. Phys.*, 2003, **5**, 2818–2827.
229. M. Joulaud, L. Omnes, T. Mourier, D. Mayer and P. Doppelt, *Proc.-Electrochem. Soc.*, 2003, **2003–8**, 1268–1274.
230. J. A. T. Norman, B. A. Muratore, P. N. Dyer, D. A. Roberts and A. K. Hochberg, *J. de Phys. IV*, 1991, **1**, C2/271-C272/278.
231. J. A. T. Norman, B. A. Muratore, P. N. Dyer, D. A. Robers, A. K. Hochberg and L. H. Dubois, *Mater. Sci. Eng., B*, 1993, **B17**, 87–92.
232. M. B. Naik, S. K. Lakshmanan, R. H. Wentorf, R. R. Reeves and W. N. Gill, *J Cryst. Growth*, 1998, **193**, 133–147.
233. T. Nguyen, L. J. Charneski and S. T. Hsu, *J. Electrochem. Soc.*, 1997, **144**, 2829–2833.
234. C.-L. Lin, P.-S. Chen, Y.-C. Lin, B.-Y. Tsui and M.-C. Chen, *J. Electrochem. Soc.*, 2003, **150**, C451–C456.
235. V. M. Donnelly and M. E. Gross, *J. Vac. Sci. Tech., A*, 1993, **11**, 66–77.

236. K. V. Guinn, V. M. Donnelly, M. E. Gross, F. A. Baiocchi, I. Petrov and J. E. Greene, *Surf. Sci.*, 1993, **295**, 219–229.
237. Y. S. Chung, H. S. Lee, Y. S. Lee and S. Kim, *Surf. Sci.*, 2001, **482–485**, 312–317.
238. L. Pirolli and A. V. Teplyakov, *Surf. Sci.*, 2007, **601**, 155–164.
239. L. Pirolli and A. V. Teplyakov, *Surf. Sci.*, 2006, **600**, 3313–3320.
240. E. Machado, M. Kaczmarski, P. Ordejon, D. Garg, J. Norman and H. Cheng, *Langmuir*, 2005, **21**, 7608–7614.
241. R. Kroger, M. Eizenberg, D. Cong, N. Yoshida, L. Y. Chen, S. Ramaswami and D. Carl, *J. Electrochem. Soc.*, 1999, **146**, 3248–3254.
242. D.-K. Kwak, H.-B. Lee, J.-W. Han and S.-W. Kang, *Electrochem. Solid-State Lett.*, 2006, **9**, C171–C173.
243. H. Kim and Y. Shimogaki, *J. Electrochem. Soc.*, 2007, **154**, G13–G17.
244. H. Kim, T. Koseki, T. Ohba, T. Ohta, Y. Kojima, H. Sato, S. Hosaka and Y. Shimogaki, *Appl. Surf. Sci.*, 2006, **252**, 3938–3942.
245. L.-S. Hong and M.-Z. Lin, *Jpn. J. Appl. Phys., Part 2*, 1997, **36**, L711–L713.
246. A. Jain, K. M. Chi, T. T. Kodas and M. J. Hampden-Smith, *J. Electrochem. Soc.*, 1993, **140**, 1434–1439.
247. J. Farkas, M. J. Hampden-Smith and T. T. Kodas, *J. Electrochem. Soc.*, 1994, **141**, 3539–3546.
248. A. Jain, J. Farkas, T. T. Kodas, M. J. Hampden-Smith, A. V. Gelatos, R. Marsh and C. J. Mogab, *Mater. Res. Soc. Symp. Proc.*, 1993, **315**, 105–110.
249. A. V. Gelatos, R. Marsh, M. Kottke and C. J. Mogab, *Appl. Phys. Lett.*, 1993, **63**, 2842–2844.
250. M.-J. Mouche, J.-L. Mermet, F. Pires, E. Richard, J. Torres, J. Palleau and F. Braud, *Appl. Surf. Sci.*, 1995, **91**, 129–133.
251. D. Yang, J. Hong, D. F. Richards and T. S. Cale, *J. Vac. Sci. Tech. B*, 2002, **20**, 495–506.
252. T. Momose and Y. Shimogaki, *Jpn. J. Appl. Phys., Part 1*, 2006, **45**, 8618–8623.
253. O.-K. Kwon, J.-H. Kim, H.-S. Park and S.-W. Kang, *Electrochem. Solid-State Lett.*, 2003, **6**, C109–C111.
254. E. S. Hwang and J. Lee, *Electrochem. Solid-State Lett.*, 2000, **3**, 138–140.
255. E. S. Hwang and J. Lee, *Chem. Mater.*, 2000, **12**, 2076–2081.
256. G. Braeckelmann, D. Manger, A. Burke, G. G. Peterson, A. E. Kaloyeros, C. Reidsema, T. R. Omstead, J. F. Loan and J. J. Sullivan, *J. Vac. Sci. Tech., B*, 1996, **14**, 1828–1836.
257. D. H. Kim, Y. J. Lee, C. O. Park, J. W. Park and J. J. Kim, *Chem. Eng. Commun.*, 1996, **152–153**, 307–317.
258. T.-Y. Chang, J.-J. Tze and D.-S. Tsai, *Appl. Surf. Sci.*, 2004, **236**, 165–174.
259. N. Awaya and Y. Arita, *Thin Solid Films*, 1995, **262**, 12–19.
260. J. A. T. Norman, D. A. Roberts, A. K. Hochberg, P. Smith, G. A. Petersen, J. E. Parmeter, C. A. Apblett and T. R. Omstead, *Thin Solid Films*, 1995, **262**, 46–51.
261. Y. K. Chae, Y. Shimogaki and H. Komiyama, *J. Electrochem. Soc.*, 1998, **145**, 4226–4233.
262. M. L. H. Ter Heerdt, J. J. Overdijk, P. J. Van der Put and J. Schoonman, *Chem. Mater.*, 1999, **11**, 3470–3475.
263. M. L. H. ter Heerdt, P. J. van der Put and J. Schoonman, *Chem. Vap. Deposition*, 2001, **7**, 199–203.
264. S.-Y. Lee, S.-K. Rha, W.-J. Lee, D.-W. Kim, J.-S. Hwang and C.-O. Park, *Jpn. J. Appl. Phys., Part 1*, 1997, **36**, 5249–5252.
265. W.-J. Lee, S.-K. Rha, S.-Y. Lee, D.-W. Kim and C.-O. Park, *Thin Solid Films*, 1997, **305**, 254–258.
266. P.-J. Lin and M.-C. Chen, *J. Electronic Mater.*, 1999, **28**, 567–571.
267. P.-J. Lin and M.-C. Chen, *Jpn. J. Appl. Phys., Part 1*, 1999, **38**, 4863–4867.
268. S. Kim, J.-M. Park and D.-J. Choi, *Thin Solid Films*, 1998, **315**, 229–237.

269. P. Maartensson and J. O. Carlsson, *Chem. Vap. Deposition*, 1997, **3**, 45–50.
270. T. Toerndahl, M. Ottosson and J.-O. Carlsson, *Thin Solid Films*, 2004, **458**, 129–136.
271. P. Martensson and J.-O. Carlsson, *J. Electrochem. Soc.*, 1998, **145**, 2926–2931.
272. C. Jezewski, W. A. Lanford, C. J. Wiegand, J. P. Singh, P.-I. Wang, J. J. Senkevich and T.-M. Lu, *J. Electrochem. Soc.*, 2005, **152**, C60–C64.
273. R. Solanki and B. Pathangey, *Electrochem. and Solid-State Lett.*, 2000, **3**, 479–480.
274. J. Huo, R. Solanki and J. McAndrew, *J. Mater. Res.*, 2002, **17**, 2394–2398.
275. A. Niskanen, A. Rahtu, T. Sajavaara, K. Arstila, M. Ritala and M. Leskela, *J. Electrochem. Soc.*, 2005, **152**, G25–G28.
276. B. S. Lim, A. Rahtu and R. G. Gordon, *Nature Mater.*, 2003, **2**, 749–754.
277. B. S. Lim, A. Rahtu, J.-S. Park and R. G. Gordon, *Inorg. Chem.*, 2003, **42**, 7951–7958.
278. Z. Li, S. T. Barry and R. G. Gordon, *Inorg. Chem.*, 2005, **44**, 1728–1735.
279. Z. Li, A. Rahtu and R. G. Gordon, *J. Electrochem. Soc.*, 2006, **153**, C787–C794.
280. J. A. Rard, *Chem. Rev.*, 1985, **85**, 1–39.
281. F. Justi, *Z. Naturforschung*, 1949, **4a**, 472–474.
282. W. D. Ryden, A. W. Lawson and C. C. Sartain, *Phys. Rev. B*, 1970, **1**, 1494–1500.
283. Y. Koda, *J. Chem. Soci., Chem. Commun.*, 1986, 1347–1348.
284. D.-S. Yoon, J. S. Roh, H. K. Baik and S.-M. Lee, *Cri. Rev. Solid State and Mater. Sci.*, 2002, **27**, 143–226.
285. Y.-C. Yeo, *Thin Solid Films*, 2004, **462–463**, 34–41.
286. I. De, D. Johri, A. Srivastava and C. M. Osburn, *Solid-State Electron.*, 2000, **44**, 1077–1080.
287. I. Polishchuk, P. Ranade, T.-J. King and C. Hu, *IEEE Electron Device Lett.*, 2001, **22**, 444–446.
288. H. Zhong, G. Heuss, V. Misra, H. Luan, C.-H. Lee and D.-L. Kwong, *Appl. Phys. Lett.*, 2001, **78**, 1134–1136.
289. V. Misra, H. Zhong and H. Lazar, *IEEE Electron Device Lett.*, 2002, **23**, 354–356.
290. K. Froehlich, K. Husekova, D. Machajdik, J. C. Hooker, N. Perez, M. Fanciulli, S. Ferrari, C. Wiemer, A. Dimoulas, G. Vellianitis and F. Roozeboom, *Mater. Sci. Eng., B*, 2004, **B109**, 117–121.
291. Z. Zhang, S. C. Song, C. Huffman, M. M. Hussain, J. Barnett, N. Moumen, H. N. Alshareef, P. Majhi, J. H. Sim, S. H. Bae and B. H. Lee, *Electrochem. Solid-State Lett.*, 2005, **8**, G271–G274.
292. R. M. Todi, A. P. Warren, K. B. Sundaram, K. Barmak and K. R. Coffey, *IEEE Electron Device Lett.*, 2006, **27**, 542–545.
293. M. Tapajna, K. Husekova, D. Machajdik, A. P. Kobzev, T. Schram, R. Luptak, L. Harmatha and K. Froehlich, *Microelectron. Eng.*, 2006, **83**, 2412–2416.
294. K. Frohlich, R. Luptak, K. Husekova, K. Cico, M. Tapajna, U. Weber, P. K. Baumann, J. Lindner and J. P. Espinos, *J. Electrochem. Soc.*, 2006, **153**, F176–F179.
295. O. Chyan, T. N. Arunagiri and T. Ponnuswamy, *J. Electrochem. Soc.*, 2003, **150**, C347–C350.
296. D. Josell, D. Wheeler, C. Witt and T. P. Moffat, *Electrochem. Solid-State Lett.*, 2003, **6**, C143–C145.
297. R. Chan, T. N. Arunagiri, Y. Zhang, O. Chyan, R. M. Wallace, M. J. Kim and T. Q. Hurd, *Electrochem. Solid-State Lett.*, 2004, **7**, G154–G157.
298. Q. Wang, J. G. Ekerdt, D. Gay, Y. M. Sun and J. M. White, *Appl. Phys. Lett.*, 2004, **84**, 1380–1382.
299. S.-H. Kwon, O.-K. Kwon, J.-S. Min and S.-W. Kang, *J. Electrochem. Soc.*, 2006, **153**, G578–G581.
300. H. Kim, Y. Naito, T. Koseki, T. Ohba, T. Ohta, Y. Kojima, H. Sato and Y. Shimogaki, *Jpn. J. Appl. Phys., Part 1*, 2006, **45**, 2497–2501.
301. G. K. J. Chao, R. L. Sime and R. J. Sime, *Acta Cryst. B*, 1973, **29**, 2845–2849.

302. T. S. Knowles, M. E. Howells, B. J. Howlin, G. W. Smith and C. A. Amodio, *Polyhedron*, 1994, **13**, 2197–2203.
303. I. R. Baird, S. J. Rettig, B. R. James and K. A. Skov, *Can. J. Chem.*, 1999, **77**, 1821–1833.
304. T. Takagi, I. Oizuki, I. Kobayashi and M. Okada, *Jpn. J. Appl. Phys., Part 1*, 1995, **34**, 4104–4107.
305. G. R. Bai, A. Wang, C. M. Foster and J. Vetrone, *Thin Solid Films*, 1997, **310**, 75–80.
306. J. Vetrone, C. M. Foster, G. R. Bai, A. Wang, J. Patel and X. Wu, *J. Mater. Res.*, 1998, **13**, 2281–2290.
307. D.-J. Lee, S.-W. Kang and S.-W. Rhee, *Thin Solid Films*, 2002, **413**, 237–242.
308. M. Kadoshima, T. Nabatame, M. Hiratani, Y. Nakamura, I. Asano and T. Suzuki, *Jpn. J. Appl. Phys. Part 2*, 2002, **41**, L347–L350.
309. J.-H. Lee, J.-Y. Kim and S.-W. Rhee, *Electrochem. Solid-State Lett.*, 1999, **2**, 622–623.
310. J.-H. Lee, J.-Y. Kim, S.-W. Rhee, D. Yang, D.-H. Kim, C.-H. Yang, Y.-K. Han and C.-J. Hwang, *J. Vac. Sci. Tech., A*, 2000, **18**, 2400–2403.
311. J. W. Lee, K.-M. Kim, H.-S. Song, K.-C. Jeong, J. M. Lee and J. S. Roh, *Jpn. J. Appl. Phys., Part 1*, 2001, **40**, 5201–5205.
312. H.-J. Sun, Y. Kim, H.-S. Song, J.-M. Lee, J.-S. Roh and H.-c. Sohn, *Jpn. J. Appl. Phys., Part 1*, 2004, **43**, 1566–1570.
313. H.-J. Sun, Y. Kim, S.-E. Park, K. Hong, J.-S. Roh and H.-c. Sohn, *Jpn. J. Appl. Phys., Part 1*, 2004, **43**, 5482–5486.
314. M. Lashdaf, T. Hatanpaa and M. Tiitta, *J. Therm. Analysis Cal.*, 2001, **64**, 1171–1182.
315. J. C. Viguie and J. Spitz, *J. Electrochem. Soc.*, 1975, **122**, 585–588.
316. M. L. Green, M. E. Gross, L. E. Papa, K. J. Schnoes and D. Brasen, *J. Electrochem. Soc.*, 1985, **132**, 2677–2685.
317. K.-W. Kim, N.-S. Kim, Y.-S. Kim, I.-S. Choi, H.-J. Kim, J.-C. Park and S.-Y. Lee, *Jpn. J. Appl. Phys., Part 1*, 2002, **41**, 820–825.
318. P. Hones, T. Gerfin and M. Graetzel, *Appl. Phys. Lett.*, 1995, **67**, 3078–3080.
319. P. Hones, F. Levy, T. Gerfin and M. Gratzel, *Chem. Vap. Deposition*, 2000, **6**, 193–198.
320. A. F. Bykov, N. B. Morozova, I. K. Igumenov and S. V. Sysoev, *J. Therm. Anal.*, 1996, **46**, 1551–1565.
321. M. A. V. Ribeiro da Silva, M. J. S. Monte and E. Giera, *J. Chem. Thermodyn.*, 2001, **33**, 369–376.
322. P. Rushman, G. N. Van Buuren, M. Shiralian and R. K. Pomeroy, *Organometallics*, 1983, **2**, 693–694.
323. A. D. Berry, D. J. Brown, R. Kaplan and E. J. Cukauskas, *J. Vac. Sci. Technol. A*, 1986, **4**, 215–218.
324. P. Banditelli, A. Cuccuru and F. Sodi, *Thermochimica Acta*, 1976, **16**, 89–93.
325. L. M. Fillman and S. C. Tang, *Thermochimica Acta*, 1984, **75**, 71–84.
326. E. P. Boyd, D. R. Ketchum, H. Deng and S. G. Shore, *Chem. Mater.*, 1997, **9**, 1154–1158.
327. M. R. Gagne and J. Takats, *Organometallics*, 1988, **7**, 561–563.
328. G. Marinelli, W. E. Streib, J. C. Huffman, K. G. Caulton, M. R. Gagne, J. Takats, M. Dartiguenave, C. Chardon, S. A. Jackson and O. Eisenstein, *Polyhedron*, 1990, **9**, 1867–1881.
329. Y. Senzaki, F. B. McCormick and W. L. Gladfelter, *Chem. Mater.*, 1992, **4**, 747–749.
330. Y. Senzaki, W. L. Gladfelter and F. B. McCormick, *Chem. Mater.*, 1993, **5**, 1715–1721.
331. Y.-H. Song, Y.-L. Chen, Y. Chi, C.-S. Liu, W.-L. Ching, J.-J. Kai, R.-S. Chen, Y.-S. Huang and A. J. Carty, *Chem. Vap. Deposition*, 2003, **9**, 162–169.
332. Y.-H. Lai, Y.-L. Chen, Y. Chi, C.-S. Liu, A. J. Carty, S.-M. Peng and G.-H. Lee, *J. Mater. Chem.*, 2003, **13**, 1999–2006.

333. F.-J. Lee, Y. Chi, C.-S. Liu, P.-F. Hsu, T.-Y. Chou, S.-M. Peng and G.-H. Lee, *Chem. Vap. Deposition*, 2001, **7**, 99–101.
334. W.-Y. Cheng, L.-S. Hong, J.-C. Jiang, Y. Chi and C.-C. Lin, *Thin Solid Films*, 2005, **483**, 31–37.
335. R.-s. Chen, Y.-s. Huang, Y.-l. Chen and Y. Chi, *Thin Solid Films*, 2002, **413**, 85–91.
336. A. P. Humphries and S. A. R. Knox, *J. Chem. Soc., Dalton Trans.*, 1975, 1710–1714.
337. K. C. Smith, Y. M. Sun, N. R. Mettlach, R. L. Hance and J. M. White, *Thin Solid Films*, 2000, **376**, 73–81.
338. H. Li, D. B. Farmer, R. G. Gordon, Y. Lin and J. Vlassakb, *J. Electrochem. Soc.*, 2007, **154**, D642–D647.
339. G. Wilkinson, *J. Am. Chem. Soc.*, 1952, **74**, 6146.
340. D. Jones, L. Pratt and G. Wilkinson, *J. Chem. Soc.*, 1962, 4458–4463.
341. J. Powell and B. L. Shaw, *J. Chem. Soc. A*, 1968, 159–161.
342. G. B. Shul'pin, *Zhurnal Obshchei Khimii*, 1981, **51**, 2152–2153.
343. P. Pertici, G. Vitulli, R. Lazzaroni, P. Salvadori and P. L. Barili, *J. Chem. Soc., Dalton Trans.*, 1982, 1019–1022.
344. T. Aoyama and K. Eguchi, *Jpn. J. Appl. Phys. Part 2*, 1999, **38**, L1134–L1136.
345. S. Y. Kang, H. J. Lim, C. S. Hwang and H. J. Kim, *J. Electrochem. Soc.*, 2002, **149**, C317–C323.
346. T. Shibutami, K. Kawano, N. Oshima, S. Yokoyama and H. Funakubo, *Electrochem. Solid-State Lett.*, 2003, **6**, C117–C119.
347. S. K. Dey, J. Goswami, A. Das, W. Cao, M. Floyd and R. Carpenter, *J. Appl. Phys.*, 2003, **94**, 774–777.
348. D. E. Trent, B. Paris and H. H. Krause, *Inorg. Chem.*, 1964, **3**, 1057–1058.
349. S.-E. Park, H.-M. Kim, K.-B. Kim and S.-H. Min, *Thin Solid Films*, 1999, **341**, 52–54.
350. S.-E. Park, H.-M. Kim, K.-B. Kim and S.-H. Min, *J. Electrochem. Soc.*, 2000, **147**, 203–209.
351. J. Si and S. B. Desu, *J. Mater. Res.*, 1993, **8**, 2644–2648.
352. S. Y. Mar, Y. S. Huang and K. K. Tiong, *Thin Solid Films*, 1995, **258**, 104–109.
353. W.-C. Shin and S.-G. Yoon, *J. Electrochem. Soc.*, 1997, **144**, 1055–1060.
354. S.-E. Park, H.-M. Kim, K.-B. Kim and S.-H. Min, *Electrochem. Solid-State Lett.*, 1998, **1**, 262–264.
355. T. Aoyama, M. Kiyotoshi, S. Yamazaki and K. Eguchi, *Jpn. J. Appl. Phys., Part 1*, 1999, **38**, 2194–2199.
356. P. Lu, S. He, F. X. Li and Q. X. Ji, *Thin Solid Films*, 1999, **340**, 140–144.
357. T. Nabatame, M. Hiratani, M. Kadoshima, Y. Shimamoto, Y. Matsui, Y. Ohji, I. Asano, T. Fujiwara and T. Suzuki, *Jpn. J. Appl. Phys. Part 2*, 2000, **39**, L1188–L1190.
358. S. Y. Kang, K. H. Choi, S. K. Lee, C. S. Hwang and H. J. Kim, *J. Electrochem. Soc.*, 2000, **147**, 1161–1167.
359. Y. Matsui, M. Hiratani, T. Nabatame, Y. Shimamoto and S. Kimura, *Electrochem. Solid-State Lett.*, 2002, **5**, C18–C21.
360. Y. Matsui, M. Hiratani, T. Nabatame, Y. Shimamoto and S. Kimura, *Electrochem. Solid-State Lett.*, 2001, **4**, C9–C12.
361. P. G. Ganesan, M. Eizenberg and C. Dornfest, *J. Electrochem. Soc.*, 2002, **149**, G510–G516.
362. J. J. Kim, D. H. Jung, M. S. Kim, S. H. Kim and D. Y. Yoon, *Thin Solid Films*, 2002, **409**, 28–32.
363. J. J. Kim, M. S. Kim and D. Y. Yoon, *Chem. Vap. Deposition*, 2003, **9**, 105–109.
364. H. Han, J. J. Kim and D. Y. Yoon, *J. Vac. Sci. Tech., A*, 2004, **22**, 1120–1123.
365. F. Papadatos, S. Consiglio, S. Skordas, E. T. Eisenbraun, A. E. Kaloyeros, J. Peck, D. Thompson and C. Hoover, *J. Mater. Res.*, 2004, **19**, 2947–2955.
366. T. Hur'yeva, M. Lisker and E. P. Burte, *Chem. Vap. Deposition*, 2006, **12**, 429–434.

367. H. J. Lim, S. Y. Kang, C. S. Hwang and H. J. Kim, *J. Vac. Sci. Tech., A*, 2003, **21**, 1381–1385.
368. S. Y. Kang, C. S. Hwang and H. J. Kim, *J. Electrochem. Soc.*, 2005, **152**, C15–C19.
369. G. Asano, H. Morioka, H. Funakubo, T. Shibutami and N. Oshima, *Appl. Phys. Lett.*, 2003, **83**, 5506–5508.
370. O. Auciello, *Journal of Applied Physics*, 2006, **100**, 051614/051611-051614/051615.
371. K. Kawano, H. Kosuge, N. Oshima and H. Funakubo, *Electrochem. Solid-State Lett.*, 2006, **9**, C175–C177.
372. K. Kawano, A. Nagai, H. Kosuge, T. Shibutami, N. Oshima and H. Funakubo, *Electrochem. Solid-State Lett.*, 2006, **9**, C107–C109.
373. K. Kawano, H. Kosuge, N. Oshima and H. Funakubo, *Electrochem. Solid-State Lett.*, 2007, **10**, D60–D62.
374. T. Aaltonen, A. Rahtu, M. Ritala and M. Leskela, *Electrochem. Solid-State Lett.*, 2003, **6**, C130–C133.
375. B. Luo, Q. Wang and J. M. White, *Chem. Vap. Deposition*, 2004, **10**, 311–317.
376. J. Choi, Y. Choi, J. Hong, H. Tian, J.-S. Roh, Y. Kim, T.-M. Chung, Y. W. Oh, Y. Kim, C. G. Kim and K. No, *Jpn. J. Appl. Phys., Part 1*, 2002, **41**, 6852–6856.
377. A. Schneider, N. Popovska, F. Holzmann, H. Gerhard, C. Topf and U. Zenneck, *Chem. Vapor Dep.*, 2005, **11**, 99–105.
378. D. Barreca, A. Buchberger, S. Daolio, L. E. Depero, M. Fabrizio, F. Morandini, G. A. Rizzi, L. Sangaletti and E. Tondello, *Langmuir*, 1999, **15**, 4537–4543.
379. S. K. Dey, J. Goswami, S. Bhaskar, W. Cao and W. C. Noh, *J. Vac. Sci. Tech., B*, 2004, **22**, L32–L34.
380. J. Goswami, W. Noh and S. K. Dey, *Chem. Vap. Deposition*, 2005, **11**, 94–98.
381. Z. Yuan, R. J. Puddephatt and M. Sayer, *Chem. Mater.*, 1993, **5**, 908–910.
382. J. Gatineau, K. Yanagita and C. Dussarrat, *Microelectronic Eng.*, 2006, **83**, 2248–2252.
383. R. Hashaikeh, I. S. Butler and J. A. Kozinski, *Thin Solid Films*, 2006, **515**, 1918–1921.
384. J. Sankar, T. K. Sham and R. J. Puddephatt, *J. Mater. Chem.*, 1999, **9**, 2439–2444.
385. R. N. Goldberg and L. G. Hepler, *Chem. Rev.*, 1968, **68**, 229–252.
386. T. Aaltonen, P. Alen, M. Ritala and M. Leskela, *Chem. Vap. Deposition*, 2003, **9**, 45–49.
387. K. J. Park, D. B. Terry, S. M. Stewart and G. N. Parsons, *Langmuir*, 2007, **23**, 6106–6112.
388. O.-K. Kwon, J.-H. Kim, H.-S. Park and S.-W. Kang, *J. Electrochem. Soc.*, 2004, **151**, G109–G112.
389. S. K. Kim, S. Y. Lee, S. W. Lee, G. W. Hwang, C. S. Hwang, J. W. Lee and J. Jeong, *J. Electrochem. Soc.*, 2007, **154**, D95–D101.
390. T. Aaltonen, M. Ritala, K. Arstila, J. Keinonen and M. Leskelae, *Chem. Vap. Deposition*, 2004, **10**, 215–219.
391. B. H. W. Toh, D. W. McNeill and H. S. Gamble, *J. Mater. Sci.: Mater. Electron.*, 2005, **16**, 437–443.

CHAPTER 8

Chemical Vapour Deposition of Metal Oxides for Microelectronics Applications

ANTHONY C. JONES,[a] HELEN C. ASPINALL[a] AND PAUL R. CHALKER[b]

[a] Department of Chemistry, University of Liverpool, Crown Street, Liverpool, L69 7ZD, UK;
[b] Department of Engineering, University of Liverpool, Liverpool, L69 3BX, UK

8.1 Introduction

Thin films of metal oxides are finding an ever increasing application in the fields of microelectronics and telecommunications, such as next-generation computer memories, microprocessors, infrared detectors, optical wave-guides and electro-optic storage[1] (Table 8.1).

Several processes have been developed for the deposition of metal oxide films, including physical vapour deposition (radiofrequency and magnetron sputtering, ion beam sputtering, molecular beam epitaxy, laser ablation);[2–5] solution deposition (sol–gel, metal-organic decomposition);[6,7] and chemical vapour deposition (metalorganic chemical vapour deposition, MOCVD)[8–14] atomic layer deposition, ALD.[15] Of these techniques, MOCVD and ALD are particularly well suited to modern manufacturing methods for microelectronics, as they offer the potential for large area growth, and have the advantages of good composition control, high film uniformity, good doping control and, most importantly, they give excellent conformal step coverage on non-planar device geometries.

Metal oxide precursors (usually alkoxides, β-diketonates or alkylamides) generally have very low vapour pressures ($\ll$1 Torr at room temperature). Therefore, early studies into oxide MOCVD used extensively modified reactors in which the precursor container and inlet pipe-work was heated at high temperature (e.g. 150–250 °C) throughout the MOCVD process, to evaporate and efficiently transport the precursor in the gas-phase (Chapter 1, Section 1.5 and Figure 1.14). Although this approach has been used for the growth of several complex ferroelectric oxides such as $PbTiO_3$, $Pb(Zr,Ti)O_3$ (ref. 11) and high-T_c superconducting oxides,[13] it was found that most oxide precursors had insufficient thermal stability to withstand heating for long periods, leading to precursor decomposition in the bubbler or in the inlet pipe-work, and to poor oxide layer uniformity and reactor blockages. This has led to the increased use of liquid injection MOCVD,[16–18] in which

Chemical Vapour Deposition: Precursors, Processes and Applications
Edited by Anthony C. Jones and Michael L. Hitchman

Published by the Royal Society of Chemistry, www.rsc.org

Table 8.1 Electronics applications of metal oxides.

Oxide class	*Material*	*Applications*
Dielectric	TiO_2, Al_2O_3, Ta_2O_5, Nb_2O_5 ZrO_2, $ZrSi_xO_y$, HfO_2, $HfSi_xO_y$, $(HfO_2)_x(Al_2O_3)_{1-x}$ Ln_2O_3, $LnSi_xO_y$ (Ln = La, Pr, Gd, Nd, etc.) $LnAlO_3$, $LnScO_3$ $La_xHf_{1-x}O_{2-\delta}$, $La_xZr_{1-x}O_{2-\delta}$ $(Zr,Sn)TiO_4$	High-κ gate dielectric layers in CMOS technology. Capacitor layers in DRAMs
Ferroelectric	$SrTiO_3$, $(Ba,Sr)TiO_3$, $Pb(Zr,Ti)O_3$ $Pb(Sc,Ta)O_3$ $Pb(Mg,Nb)O_3$ $SrBi_2(Ta_xNb_{1-x})_2O_9$ $Bi_4Ti_3O_{12}$	DRAMs, NVFERAM computer memories. Infrared detectors, microelectromechanical devices, transducers, ceramic capacitors
Conducting	$(La,Sr)CoO_3$, RuO_2, $SrRuO_3$, $LaNiO_3$	Ferroelectric capacitor electrodes
Semiconducting	ZnO	Microsensors, micromachined actuators, surface acoustic wave devices
Magnetoelectric	$BiFeO_3$	Non-volatile memories
Magnetic	$(Ni,Zn)Fe_2O_4$ $(Mn,Zn)Fe_2O_4$	Recording media, high frequency read heads
Superconducting	$YBa_2Cu_2O_{7-x}$ Bi-Sr-Ca-Cu-O Tl-Bi-Ca-CuO	Josephson junctions, bolometers, SQUIDS

thermal decomposition of the precursor is prevented by using a precursor solution that is held at room temperature until required in the MOCVD process (see Chapter 1, Section 1.5 and Chapter 2, Section 2.5.1).

The surface controlled and saturative nature of the ALD process allows precise control of film deposition at the monolayer level and gives better step coverage (or conformality) than MOCVD on non-planar surfaces containing trenches or vias,[15] (see Chapter 4 for a detailed description of ALD, and also Chapter 1, Section 1.3.2 and Chapter 2, Section 2.6.1). Therefore, ALD has become the chosen technique for the deposition of very thin oxide films required in CMOS gate dielectrics, and for oxide deposition in high aspect ratio ($\sim$100 : 1) DRAM structures. There has also been a growing use of liquid injection techniques in ALD to minimize decomposition of thermally-sensitive precursors.

In this chapter, we focus on the precursors used for the MOCVD and ALD of various dielectric, ferroelectric and superconducting oxides used in electronics applications. The rapid growth in the number and complexity of these metal oxides is making it increasingly necessary to "tailor" the physical properties of the CVD precursor to optimize its physical properties and deposition characteristics. Some general principles of precursor chemistry and molecular design are described in the next section.

8.2 Precursor Chemistry

This section discusses the general chemistry and classes of precursors used in the MOCVD and ALD of electroceramic oxides. A crucial requirement for precursors used in both techniques is that they have adequate volatility for efficient evaporation and transport in the vapour phase without decomposition during the evaporation step. However, the introduction of liquid injection methods means that the volatility requirements are not quite as demanding as they once were in conventional MOCVD. The precursors should also decompose cleanly and have a good shelf-life (Chapter 1, Section 1.4.3). In the most general sense, volatility is achieved by reduction of intermolecular forces, the strongest of which are those that lead to dimer, oligomer or polymer formation. Precursors should ideally be monomeric so that for a neutral complex $[ML_n]$, n mono-anionic ligands completely fill the coordination sphere of M^{n+}, preventing the formation of bridged oligomers or polymers. This is relatively easy to achieve for small metals such as Al (six-coordinate radius of $Al^{3+} = 67.5$ pm[19]), but for large metals with a lower charge (*e.g.* six-coordinate radius of $Ba^{2+} = 149$ pm[19]) this is a very significant challenge.

The steric demand of a ligand can be increased by increasing the bulk of the substituents on the donor atoms or by increasing the number of donor atoms, or by a combination of these approaches. The cone angle θ of a ligand, as defined in Figure 8.1(a), is a useful concept in considering the steric demands of a monodentate ligand. Notably, an alkoxide ligand Figure 8.1(b), where the anionic O-donor carries only one substituent, cannot be as sterically demanding as a dialkylamide ligand, Figure 8.1(c), where the anionic N-donor carries two substituents.

It is frequently impossible to prepare monomeric complexes with monodentate ligands, and therefore chelating ligands such as those shown in Figure 8.2 are required.

Group 2 metals such as Ba and Sr, which have large radii and can accommodate only two monoanionic ligands, pose exceptional problems, and for these metals it is often necessary to add neutral donor ligands such as triglyme ($CH_3O(CH_2CH_2O)_3CH_3$) and tetraglyme ($CH_3O(CH_2CH_2O)_4CH_3$) (see schematic structures in the footnote to Table 8.7 below), to fill the coordination sphere.

Once a monomeric complex has been obtained, there is still scope for improving volatility by reduction of attractive intermolecular forces. One way to do this is by fluorination of ligands, which has been investigated both for β-diketonates[20] and for alkoxides;[21] in both cases there is a significant increase in volatility for the fluorinated complexes. The effect of fluorination on volatility has been rationalized in two ways: firstly, the increased number of lone pairs on the periphery of the complex results in increased intermolecular repulsion and, secondly, the replacement H atoms with less polarizable F atoms results in a decrease in intermolecular attractions. Unfortunately, this

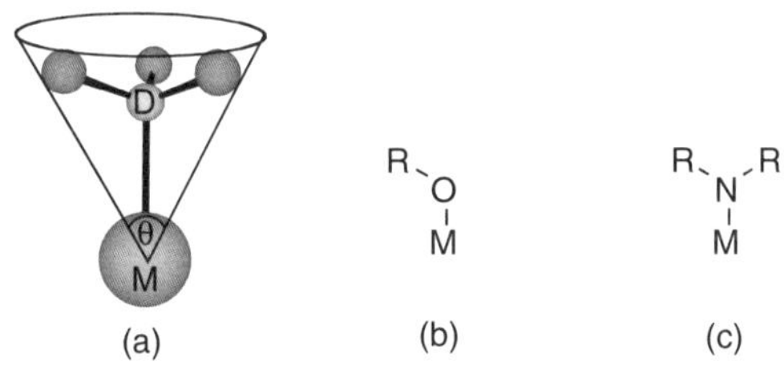

Figure 8.1 Steric effects of various ligands found in oxide precursors.

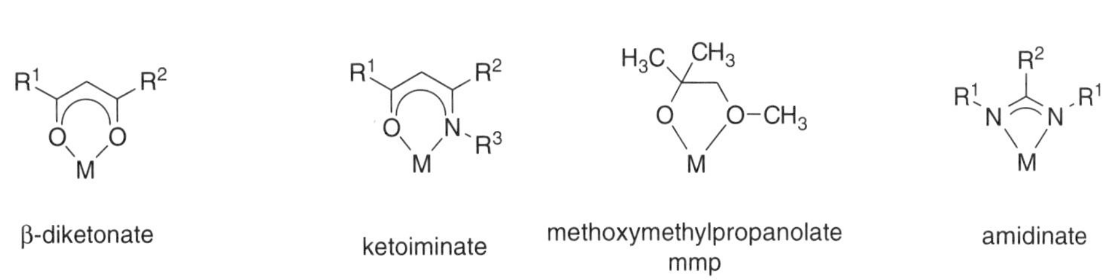

Figure 8.2 Chelating ligands commonly found in oxide MOCVD precursors.

increase in volatility comes at a high cost: the use of fluorinated β-diketonates often leads to fluorine contamination in the resulting oxide films, which can only be reduced by the addition of H_2O during film growth.[13] The presence of fluorine in the growth system can also lead to degradation of the Si substrate, which is highly undesirable in Si-based microelectronics applications.

The ideal chemical properties of precursors for MOCVD and ALD are very different and in some ways mutually exclusive. Precursors for MOCVD must decompose cleanly at temperatures low enough to be compatible with device processing but high enough to allow clean evaporation without premature thermal decomposition whereas ALD precursors must show facile reaction with the oxygen source (usually water) and, to achieve self-limiting growth, must have sufficient thermal stability to prevent premature thermal decomposition on the substrate. For complexes where the metal–ligand bonding is highly electrostatic, consideration of relative pK_as of water and protonated ligand will determine whether the ALD process with water is thermodynamically viable, but will not give any indication of the kinetics of the process. In Section 8.2.1 we will see that precursor volatility is often achieved by increasing the steric bulk at the metal to prevent oligomer formation. This steric hindrance makes reaction with water much less facile, and successful design of precursors for ALD requires a careful balance of volatility with reactivity. One further consideration is that facile reaction with water is an excellent characteristic for the ALD process but makes a precursor much more difficult to handle and store without decomposition. The next sections discuss individual classes of ligand with reference to how they fulfil the requirements of volatility and reactivity.

8.2.1 β-Diketonates and Related Ligands

β-Diketonate ligands can be modified to tune the steric and electronic properties, and complexes of most metals with these ligands have been known for over a century. Table 8.2 shows a range of β-diketonate and related ligands that have been utilized in precursors for the MOCVD and ALD of metal oxides.

The importance of steric factors in determining the volatility of β-diketonate complexes is well-illustrated by a series of Pb complexes in which volatility increases in the order [Pb(acac)$_2$] < [Pb(dhd)$_2$] < [Pb(thd)$_2$],[22] due to the increased shielding of the positively charged metal centre and reduction of intermolecular oxygen–metal interactions in the solid state. More dramatic effects are seen for the rare earths where [Ln(acac)$_3$] are polymeric and involatile for all but the smallest elements Yb, Lu and Sc. The more sterically demanding thd ligand gives dimeric [Ln(thd)$_3$]$_2$ for La-Dy and monomeric [Ln(thd)$_3$] for Ho-Lu. The dimeric complexes dissociate in the vapour phase and sublime as monomers.

Another method of forming monomeric [M(diket)$_n$] complexes is to add neutral donor ligands to increase the coordination number of M. This is particularly important in [M(diket)$_2$] complexes of the large Group 2 metals (*e.g.* Sr^{2+} six-coordinate radius = 132 pm; Ba^{2+} six-coordinate radius = 149 pm), which exist as low-volatility oxygen-bridged polynuclear clusters. For example, [Sr(thd)$_2$] is trimeric (Figure 8.3)[23] and [Ba(thd)$_2$] is tetrameric.[24]

Much research effort has been aimed at increasing the vapour pressure of these complexes for MOCVD applications, especially in the field of high-T_c superconducting oxides.[13,20,21,25,26] Addition of polyethers or polyamines (see Table 8.6 below) leads to a range of monomeric complexes, including [M(thd)$_2$(polyether)][27] (Figure 8.4), [M(hfac)$_2$(polyether)] (M = Sr, Ba; polyether = triglyme or tetraglyme)[28] and [M(thd)$_2$(pmdeta)] (M=Sr, Ba; pmdeta=pentamethyldiethylenetriamine, $Me_2NCH_2CH_2N\{Me\}CH_2CH_2NMe_2$).[29]

The presence of the adducting ligands is undoubtedly useful in protecting the [Ba(thd)$_2$] and [Sr(thd)$_2$] complexes from hydrolysis and stabilizing the complex in solution during liquid injection MOCVD.[29] However, in the gas-phase the non-fluorinated-thd complexes dissociate. Complexes

Table 8.2 Some β-diketonate and related ligands used in metal oxide MOCVD and ALD precursor complexes.

Structure	*Name*	*Abbreviation*
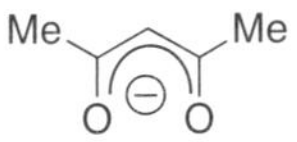	Pentane-2,4-dionate (acetylacetonate)	acac
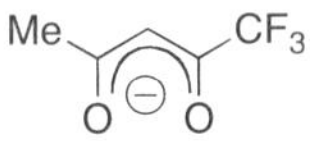	1,1,1-Trifluoropentane-2,4-dionate (trifluoroacetylacetonate)	tfac
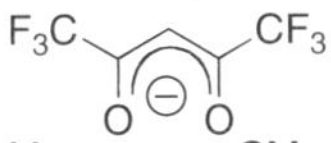	1,1,1,5,5,5-Hexafluoropentane-2,4-dionate (hexafluoroacetylacetonate)	hfac
	1,1-Dimethylhexane-3,5-dionate	dhd
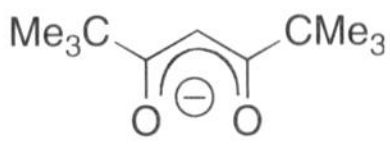	2,2,6,6-Tetramethylheptane-3,5-dionate	thd
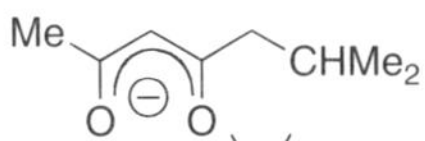	6-Methylheptane-2,4-dionate	mhd
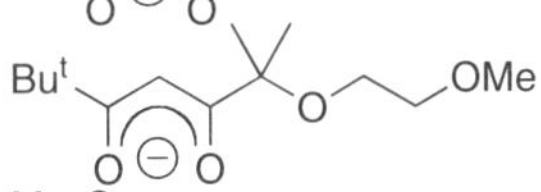	2-(2-Methoxyethoxy)-2,6,6-trimethylheptane-3,5-dionate	methd
	2,2,7-Trimethyloctane-3,5-dionate	tmod
	Methyl 4,4-dimethyl-3-oxopentanoate	mdop
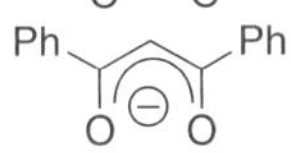	1,3-Diphenylpropane-1,3-dionate (dibenzoylmethanate)	dbm
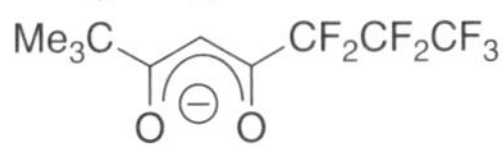	6,6,7,7,8,8,8-Heptafluoro-2,2-dimethyloctane-3,5-dionate	fod
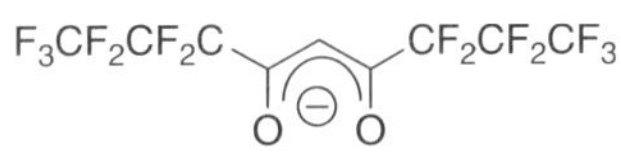	1,1,1,2,2,3,3,7,7,8,8,9,9,9-Tetradecafluorononane-4,6-dionate	tdf
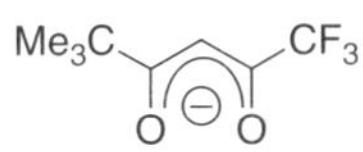	1,1,1-Trifluoro-5,5-dimethylhexane-2,4-dionate	pta
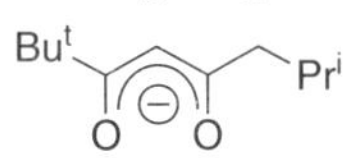	2,2,7-Trimethyl-3,5-octanedionate	tod
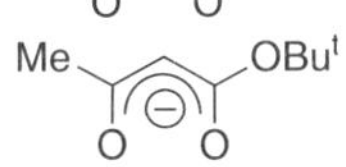	*t*-Butyl acetoacetate	tboac
	*N*,*N*-Diethylacetoacetamide	deacam
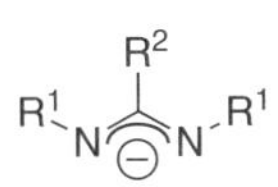	Amidinate	AMD
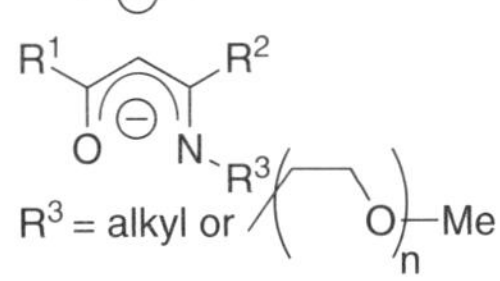	β-Ketoiminate	ketoim
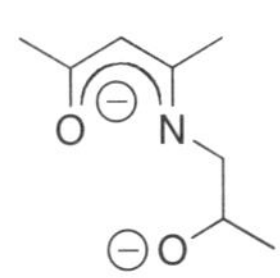	(*E*)-4-(2-Oxidopropylimino)pentan-2-olate	meip

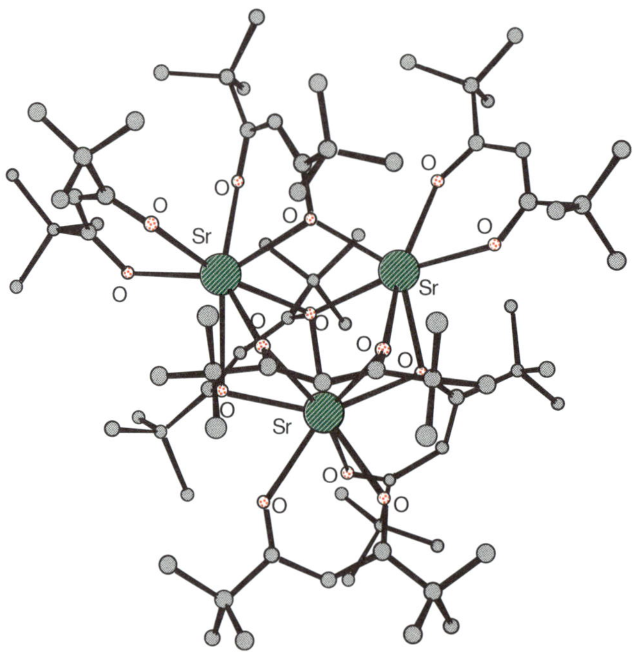

Figure 8.3 Structure of the $[Sr(thd)_2]_3$ trimer.

with the more electron-withdrawing fluorinated ligands are more Lewis acidic, and so adducts with neutral Lewis bases do not dissociate in the gas phase; however, these complexes lead to fluorine contamination in MOCVD-grown oxide films. Addition of the neutral N-donors bipyridyl and phenanthroline to polymeric $[Ln(acac)_3]$ gives volatile monomers that have been used as precursors for MOCVD of rare earth oxides.[30,31]

A systematic investigation of the effects of fluorination on the volatility of $[M(diket)_n]$ complexes[20] shows that fluorination of the ligand leads to an increase in volatility. For a series of [Sc(β-diket)$_3$] complexes the vapour pressures at 155 °C are as follows: $[Sc(acac)_3]$, 3.22 Torr; $[Sc(thd)_3]$, 3.31 Torr; and $[Sc(hfac)_3]$, 28.89 Torr. The difference in steric bulk between acac and thd has little effect on volatility in these monomeric complexes whereas the effect of fluorination in the hfac complex is dramatic.

Thermal decomposition of β-diketonate complexes is generally a high-energy process. Mechanistic studies indicate that, for complexes with thd (which lack a β-hydrogen in the ligand), successive stripping of Bu^t groups by C–C homolysis is the first step, followed by dissociation of the C_3O_2 fragment as shown in Figure 8.5(a). In the absence of added O_2, elemental metal is deposited.[32] If the ligand has a β-hydrogen, then there is a more facile decomposition pathway, affording a metal-oxy species, as shown for Ba in Figure 8.5(b).[33]

Complexes containing fluorinated β-diketonate ligands almost invariably decompose to give either metal fluorides or oxides that are heavily contaminated with fluoride, even in the presence of O_2, as found for $[Sr(hfac)_2(tetraglyme)]$.[34] This means that, despite their enhanced volatility, complexes with such ligands are generally unsuitable as precursors for MOCVD of oxides.

β-Diketonate complexes have many attractive features: they are straightforward to make, easy to handle, have the potential for modification to fine tune their properties, and generally have good volatility. However, their high thermal stability means that C contamination is often a problem in MOCVD. The relatively high acidity of most β-diketones (*e.g.* pK_a for Hacac = 8.2) means that their complexes are not protonated by water to any significant extent, and so ALD of oxides using these precursors requires O_3 as the oxygen source.

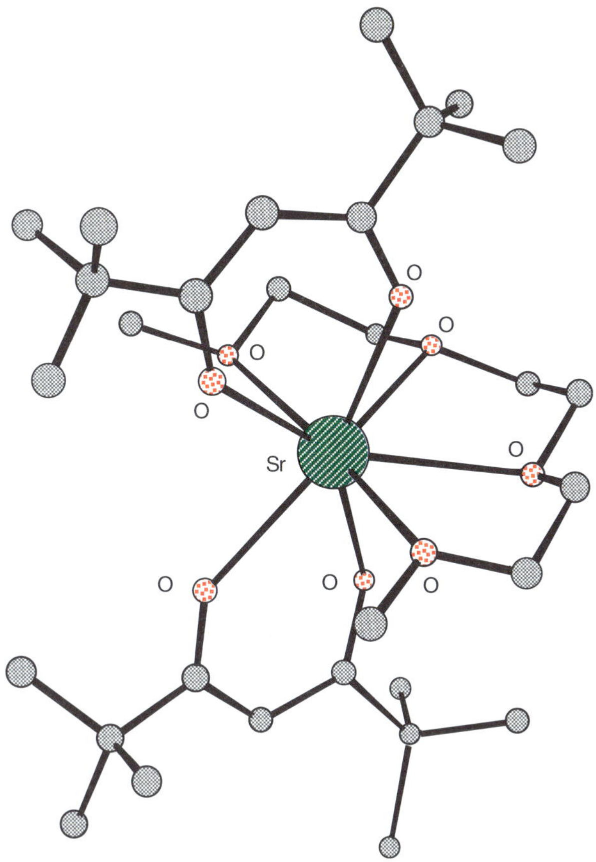

Figure 8.4 Crystal structure of the [Sr(thd)$_2$(triglyme)] complex.

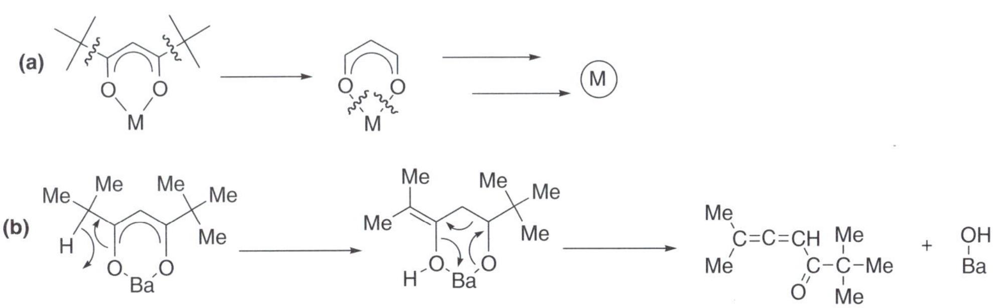

Figure 8.5 Decomposition pathways for β-diketonate ligands.

8.2.2 β-Ketoiminates

In β-ketoiminate ligands (Table 8.2) the N donor atom can carry either a simple alkyl group, resulting just in an increase in steric demand of the ligand, or it can carry a donor functionalized substituent, such as a polyether, to give a multidentate ligand.

Polyether functionalized β-ketoimines have been developed, and these multidentate ligands have allowed the preparation of volatile monomeric complexes with Group 2 metals[35–38] and rare earths.[39,40] The volatility of the rare earth complexes compares favourably with that of analogous [Ln(thd)$_3$] complexes; however, the Ba complexes leave up to 20–40% residue on sublimation, and there is some dissociation of free ligand on heating.

8.2.3 Alkoxides

Metal alkoxides [M(OR)$_x$] have been widely used as precursors for MOCVD of oxides,[41] and the factors that influence their volatility were established in the 1950s.[42] A simple monodentate alkoxide [RO$^-$] can act as a terminal, doubly or triply bridging ligand (Figure 8.6), and so oligomeric species [M(OR)$_x$]$_n$ can readily be formed to achieve coordinative saturation at the metal.

The degree of association, n, has a major influence on the volatility of the complex. It depends on the metal radius and the steric demands of R, and has been determined for the Group 4 alkoxides [M(OR)$_4$] as shown in Table 8.3, where the effects on volatility are clearly evident.[41,42] The non-integral values of n are a result of the presence of a mixture of species, and in solution the composition of the mixture is concentration dependent.

The incorporation of an additional neutral donor function into alkoxide ligands is another way to reduce the degree of association. Anwander and Herrmann have developed a series of ether and amine functionalized alcohols (Table 8.4), and pioneered their use in synthesis of volatile metal complexes.[43,44]

Substitution of alkyl groups on the carbon atom adjacent to the oxygen (α-position) in the donor functionalized alkoxide ligand favours chelating modes over bridging or assembling modes. Therefore, Hmmp (HOCMe$_2$CH$_2$OMe) has been an important ligand for the synthesis of MOCVD and ALD precursors, as it strongly promotes the formation of monomeric complexes. The complexes [Hf(mmp)$_4$] (Figure 8.7a) and [Zr(mmp)$_4$] have been characterized by X-ray diffraction, and have been shown to be monomeric octahedral six-coordinate complexes. The Hf and Zr complexes have two bidentate and two "dangling" mmp ligands whereas in the Bi complex all of the ligands are bidentate. The six-coordinate heteroleptic complexes [M(mmp)$_2$(OBu^t)$_2$] (M = Zr, Hf) (Figure 8.7b) have also been prepared and characterized crystallographically, and used in MOCVD.[45,46]

The coordination chemistry of mmp with the rare earths is complex and only two homoleptic [Ln(mmp)$_3$]$_n$ complexes have been characterized crystallographically. The smallest lanthanide Lu gives dimeric [Lu(mmp)$_3$]$_2$ in which one Lu atom is six-coordinate and one is seven coordinate, and two of the [mmp] ligands are "dangling."[47] The largest lanthanide La gives a trimeric complex with two six-coordinate and one seven-coordinate La atoms, and four "dangling" mmp ligands (Figure 8.8).[48]

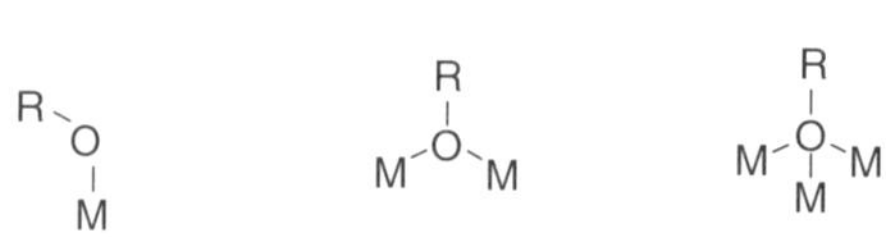

Figure 8.6 Coordination modes of an alkoxide ligand.

Table 8.3 Volatility of simple Ti and Zr alkoxide complexes as a function of molecular association.

R	[Ti(OR)$_4$]$_n$		[Zr(OR)$_4$]$_n$	
	n	bp (°C)/pressure (Torr)	n	bp (°C) /pressure (Torr)
Et	2.4	103/0.1	3.6	180/0.1
Pr^i	1.4	49/0.1	3.0	160/0.1
Bu^t	1.0	64.5/0.4	1.0	55/0.2

Table 8.4 Some donor functionalized alkoxide ligands.

Formula	*Name*	*Abbreviation*
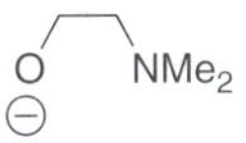	2-Dimethylaminoethanolate	dmae
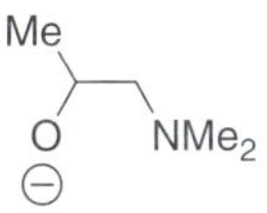	1-Dimethylaminopropan-2-olate	dmap
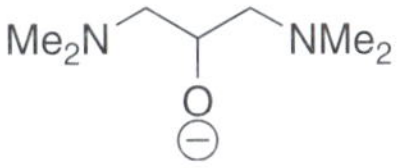	1,3-Bis(dimethylamino)propan-2-olate	bis-dmap
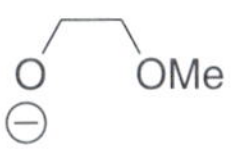	2-Methoxyethanolate	mee
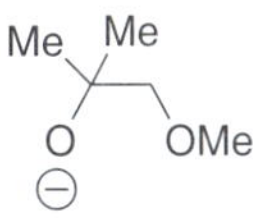	1-Methoxy-2-methyl-2-propanolate	mmp
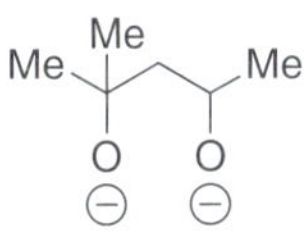	2-Methylpentane-2,4-bis(olate)	mpd
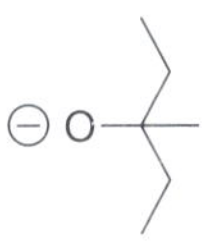	3-Methyl-3-pentoxide	mp
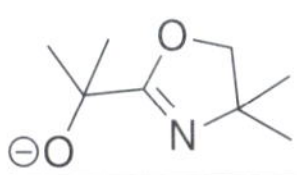	2-(4,4-Dimethyl-4,5-dihydrooxazol-2-yl)propan-2-olate	dmop

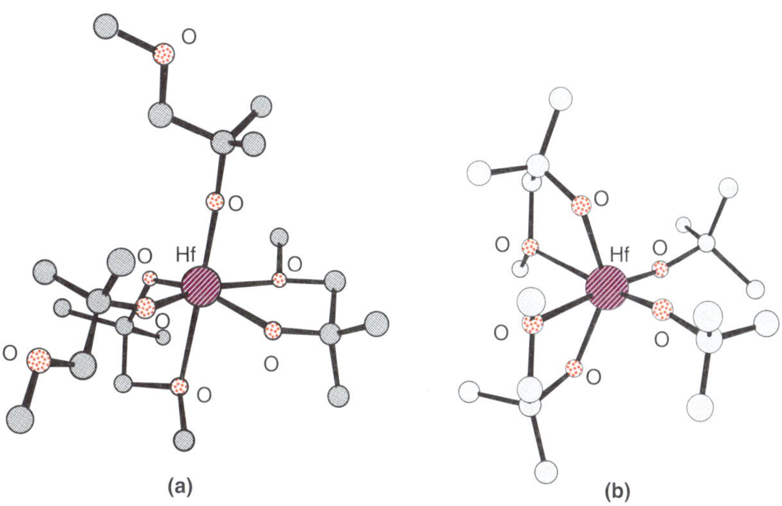

Figure 8.7 Molecular structure of (*a*) [Hf(mmp)$_4$] and (*b*) [Hf(OBu^t)$_2$(mmp)$_2$].

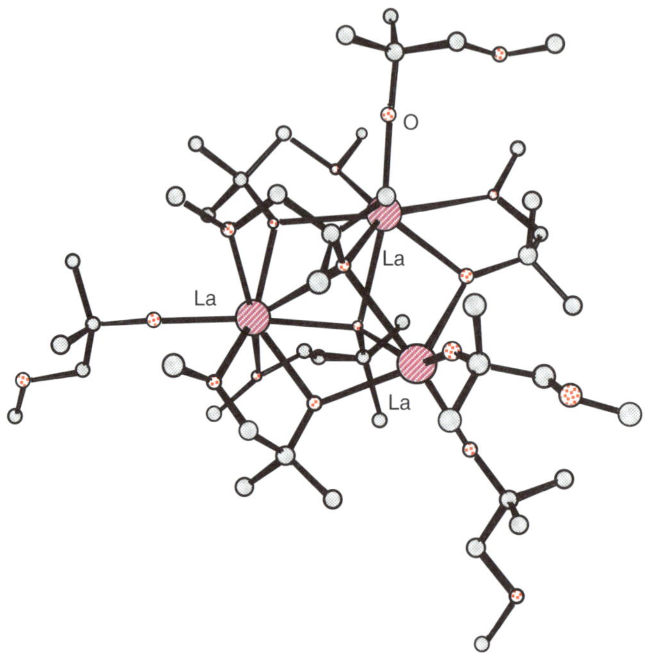

Figure 8.8 Molecular structure of $[La(mmp)_3]_3$.

Generally, alkoxide ligands $[RO^-]$ are much stronger donors than neutral ether functions, and, despite the chelate effect, the OMe group of the mmp ligand is clearly a rather poor donor. A further illustration of the importance of α-Me group in reducing the degree of association in mmp complexes is that the less sterically hindered analogue $HOCH_2CH_2OMe$ gives a cyclic decamer with rare earths.[49]

Amino-functionalized alkoxides have been somewhat less studied, but both homoleptic and heteroleptic complexes of Zr and Hf with the ligands dmae, dmap and bis-dmap (see Table 8.4) have been used in MOCVD and ALD. However, the lack of alkyl substituents on the α-carbon results in complexes such as $[Zr(OPr^i)_3(dmap)]$,[50] $[Zr(OPr^i)_3(bis\text{-}dmap)]$[50] and $[Zr(OBu^t)_2(dmae)_2]$[51] are asymmetric dimers with relatively low volatilities.

A recent development in donor-functionalized alkoxide ligands is the use of the oxazolinyl substituent as a neutral donor as in the oxazolinyl alcohol (Hdmop) and oxazolinylphenol (Hdmomph) (Table 8.4).

The imine nitrogen of dmop is expected to be a stronger donor than the OMe group of mmp, and, together with the added steric bulk of the Me substituents on the oxazoline ring, this should favour the formation of monomeric complexes with larger metal ions. From a reactivity point of view, oxazolinylphenol ligands such as Hdmomph are particularly interesting as their lack of readily eliminated β-hydrogens blocks one facile decomposition route (see discussion below), giving them enhanced thermal stability that is important for ALD. Some Zr and Hf complexes with dmop have been characterized by X-ray diffraction,[52,53] and the structures of $[Hf(dmop)_4]$ (a) and $[Hf(OBu^t)_2(dmop)_2]$ (b) are shown in Figure 8.9.

The combination of the Lewis acidity of many simple metal alkoxides and the propensity of the alkoxide O donor to act as a bridging ligand means that many heterometallic alkoxides are available, and these complexes frequently have the good volatility required for MOCVD and ALD. In favourable cases the heterometallic alkoxide has the metals in the correct ratio to deposit a film of the desired stoichiometry. For example, $[LaAl(OPr^i)_6(Pr^iOH)]_2$ (Figure 8.10) has been used as a single source precursor for MOCVD of $LaAlO_3$.[54]

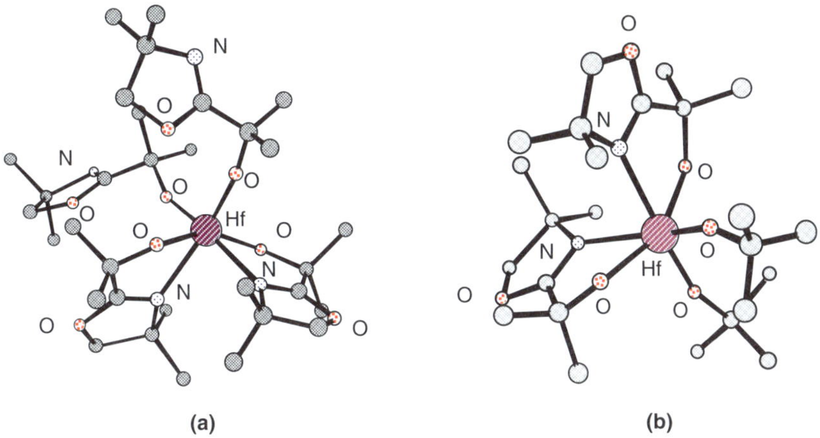

Figure 8.9 Molecular structure of (*a*) [Hf(dmop)$_4$] and (*b*) [Hf(OBu^t)$_2$(dmop)$_2$].

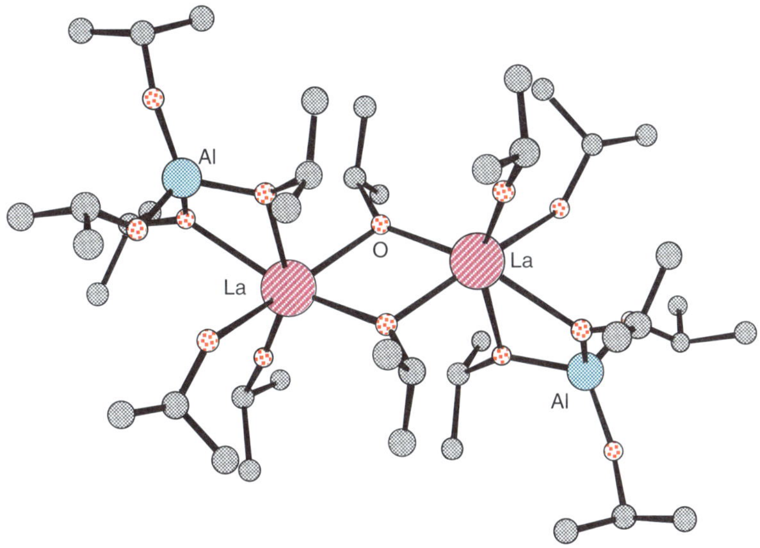

Figure 8.10 Molecular structure of [LaAl(OPr^i)$_6$(Pr^{i}OH)]$_2$.

Metal alkoxides generally have a significantly lower thermal stability than the corresponding β-diketonates, and so oxide films can be deposited at lower temperatures in MOCVD processes using alkoxide precursors. A crucial factor in determining the thermal stability of metal alkoxides is the presence of a hydrogen on the second carbon from the coordinated oxygen atom, commonly called a β-hydrogen. This β-hydrogen can often be readily eliminated to form an M–OH group and liberate an alkene (Figure 8.11).

This β-hydrogen elimination reaction gives a facile decomposition route to metal oxide species even in the absence of added oxygen. High purity oxides can thus be grown at relatively low temperatures, although in most MOCVD processes using alkoxide precursors, oxygen is added to minimize the presence of C impurities in the film. Based on GC-MS studies a mechanism has been proposed for the decomposition of [Hf(mmp)$_4$] in the absence of oxygen (Scheme 8.1).[55]

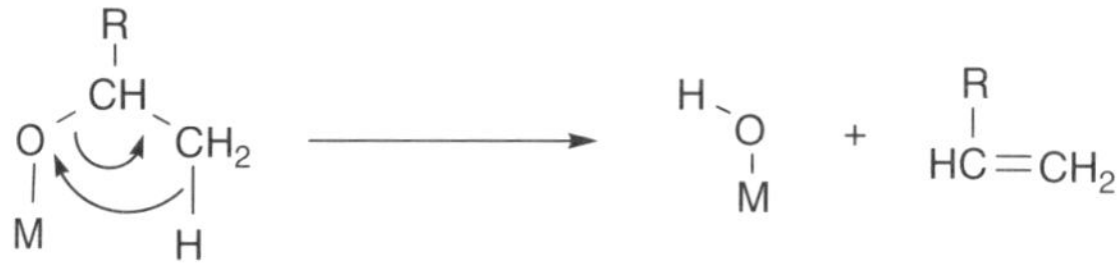

Figure 8.11 Elimination of a β-hydrogen atom from an alkoxide ligand.

$$[Hf(mmp)_4] \xrightarrow{\beta\text{-H elimination}} [Hf(OH)_4] + 4Me_2C{=}CH(OMe)$$

$$[Hf(OH)_2(mmp)_2] \xrightarrow[\text{intramolecular H-transfer}]{\beta\text{-H elimination}} HfO_2 + 2Hmmp + 2Me_2C{=}CH(OMe)$$

Scheme 8.1 Mechanism proposed for the thermal decomposition of $[Hf(mmp)_4]$ in the absence of oxygen.

$$Zr(OBu^t)_4 + H_2O \longrightarrow ZrO(OBu^t)_2 + 2Bu^tOH$$

$$2Bu^tOH \longrightarrow 2H_2C{=}CMe_2 + 2H_2O$$

$$\text{overall} \quad Zr(OBu^t)_4 \longrightarrow ZrO_2 + 4H_2C{=}CMe_2 + 2H_2O$$

Scheme 8.2 Elimination of the OBu^t ligand by autocatalytic hydrolysis.

The main features of the mechanism are the elimination of a β-hydride from a coordinated mmp to form $[Hf(OH)_4]$ and liberate the alkene $(CH_3)_2C{=}CHOCH_3$, and the transfer of a hydrogen atom from a Hf-OH to an adjacent mmp ligand to liberate Hmmp, both of which facilitate the clean elimination of the mmp ligand.

In the presence of small amounts of H_2O (*e.g.* residual surface adsorbed water) a related autocatalytic hydrolysis reaction can occur for OBu^t (Scheme 8.2), as observed for Zr alkoxides by Bradley and Faktor.[41]

In this scheme the rate of the autocatalytic hydrolytic decomposition is clearly directly linked to the rate of decomposition of the liberated alcohol.

The marked difference in thermal stability between metal alkoxides and β-diketonates can sometime be used to "tune" the thermal stability of a precursor, so that it is more compatible with its co-precursors This is especially important in liquid injection MOCVD, where there is usually only a single evaporator. For instance, $[Zr_2(OPr^i)_6(thd)_2]$ has a thermal stability between that of $[Zr(thd)_4]$ and $[Zr(OPr^i)_4]$ and is therefore a good thermal match to $[Pb(thd)_2]$ and $[Ti(OPr^i)_2(thd)_2]$. This allows the liquid injection MOCVD of $Pb(Zr,Ti)O_3$ with improved film uniformity (Section 8.4.1 and Figure 8.25 below).

The main requirement for an ALD precursor is facile reaction with H_2O. For complexes where the metal–ligand bonding is highly electrostatic, consideration of relative pK_as of water and protonated ligand will determine whether the ALD process with water is thermodynamically viable. We have shown above that precursor volatility is often achieved by using sterically demanding alkoxide ligands; however, the large steric hindrance around the metal centre may hinder the access of H_2O, and lead to reduced ALD growth rates. In addition, the presence of β-hydrogen atoms in the ligand may result in low thermal stability, preventing self-limiting growth.

Three factors have been identified as important in controlling ALD growth from metal alkoxide precursors:[56] Firstly, ease of β-hydride elimination from the alkoxide ligand to form reactive [M-OH] sites will be a major factor controlling precursor decomposition at low substrate temperatures (225–300 °C). Secondly, steric shielding by alkoxide ligands of the metal centre in $[M(OR)_x]$ precursors will have an influence on the surface reaction between [OH] and the metal centre and will thus affect growth rates. Thirdly, the thermal stability of the [M–OR] bonds will be a significant factor at higher substrate temperatures (*e.g.* >350 °C).

8.2.4 Alkylamides

Simple dialkylamides R_2N^- are more sterically demanding than the corresponding alkoxides RO^-, and therefore monomeric metal dialkylamides are more readily available than their alkoxide counterparts. There are several examples of volatile transition metal dialkylamides $[M(NR_2)_x]$. These complexes are widely used for the MOCVD of metal nitrides (*e.g.* TiN),[57] and are now being investigated for the MOCVD of oxide films. For example, ZrO_2 and HfO_2 films have been deposited from $[Zr(NEt_2)_4]$[58] and $[Hf(NEt_2)_4]$,[59,60] respectively. Although a metal-oxygen bond is not present in the precursor, the alkylamide ligand is readily displaced in the presence of excess oxygen, which allows the growth of high purity oxide films at relatively low substrate temperatures.

Simple volatile dialkylamides are not available for the rare earths; the monomeric complexes that are available, *e.g.* $[Ln(NPr^i_2)_3(THF)]$, have poor thermal stability and cannot be sublimed.

8.2.5 Amidinates

The bidentate N-donor amidinate ligands (Table 8.2) can be considered as N-donor analogues of the allyl group and their complexes with transition metals, main group and f-elements are well known.[61] Steric tuning can be achieved by variation of the R^1 and R^2 substituents, and the amidinate ligand is sometimes considered as a steric equivalent of cyclopentadienyl. $[M(amidinate)_n]$ adopt various structures, depending on the oxidation state of M and the steric bulk of the amidinate ligand. All $[M(amidinate)_3]$ are six-coordinate monomers; $[M(amidinate)_2]$ exist as four-coordinate monomers or dimers, and [M(amidinate)] are either two-coordinate dimers or cyclic tetramers. Volatile complexes have been prepared and characterized for transition elements in oxidation states +1, +2 and +3,[62] for lanthanum[62] and for Mg.[63] Figure 8.12 shows the crystal structure of a La complex.

Amidinate complexes generally show significantly higher thermal stability than dialkylamide complexes, in part due to the chelate effect. This enhanced thermal stability, in combination with

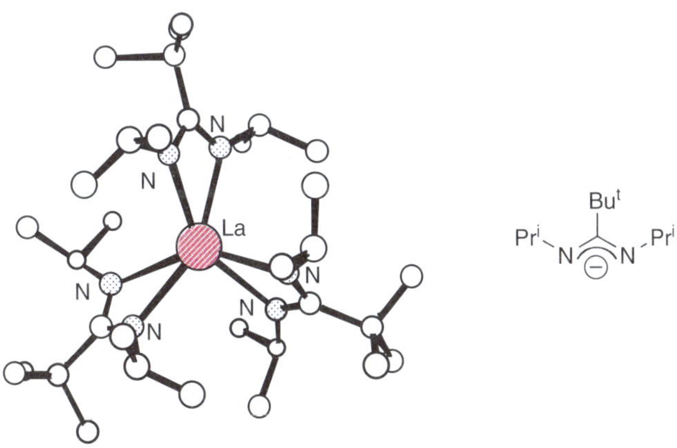

Figure 8.12 Molecular structure of $[La\{CBu^t(NPr^i)_2\}_3]$.

the Brønsted basicity of the amidinate group, makes these complexes attractive as potential precursors for ALD. They have been used as precursors for ALD of first row transition metals,[64,65] Y_2O_3[66] and rare earth aluminates.[67,68]

8.2.6 Organometallic Precursors

Simple alkyls [MR_4] of the Group 4 metals are coordinatively unsaturated and, except when R is sterically demanding and lacking β-hydrogens (*e.g.* CH_2CMe_3), these compounds have very poor thermal stability. For example, [$ZrMe_4$] decomposes above −15 °C. However, the "bent metallocene" complexes [Cp_2MR_2] (see structure of [Cp_2HfMe_2] in Figure 8.13) are more sterically and electronically saturated and are therefore significantly more robust.

[Cp_2ZrMe_2] and [Cp_2HfMe_2] can be sublimed at *ca.* 70–110 °C *in vacuo* without decomposition.[69] These 16-electron complexes react cleanly with H_2O, readily eliminating CpH and CH_4 and are good ZrO_2 and HfO_2 ALD precursors. However, they are less suitable for MOCVD, giving layers that are often heavily contaminated with carbon.

The first organolanthanides to be prepared were the tris-cyclopentadienyls [$LnCp_3$] in 1956.[70] These complexes adopt a range of structures depending on the Ln^{3+} radius, and are only monomeric in the solid state for Ln around the middle of the series such as Sm. Because the bonding is essentially ionic these complexes, unlike classical transition metal cyclopentadienyls such as ferrocene, are highly reactive with H_2O and O_2. They are, however, thermally robust and can be sublimed without decomposition at 200–250 °C and 10^{-4} Torr. The volatility of the simple [$LnCp_3$] complexes increases from La to Yb, *i.e.* with decreasing Ln^{3+} radius. Substituents can be introduced into the Cp ring to increase volatility, as shown for complexes of Nd in Figure 8.14.[71]

Simple neutral alkyl complexes of the lanthanides are only known with extremely bulky alkyl groups such as [$(Me_3Si)_2CH^-$]. Heterometallic complexes [$Ln(AlMe_4)_3$] have been synthesized and, although they are pyrophoric when exposed to air, they have sufficient thermal stability to be sublimed intact at 90 °C and 10^{-3} Torr.[72,73] Applications of organolanthanides (compounds that involve a direct bond between C and Ln) in materials science have been reviewed.[74]

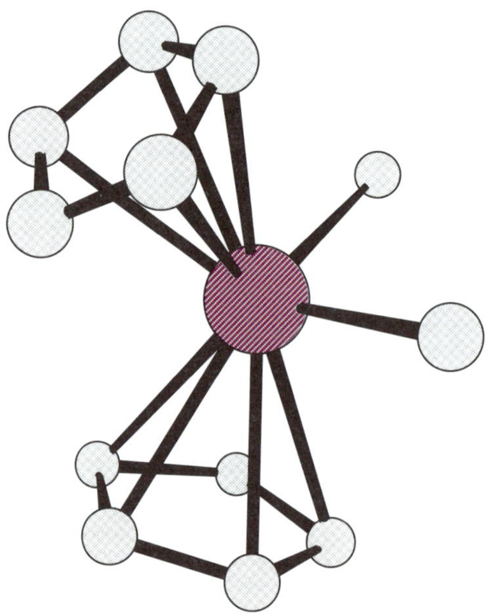

Figure 8.13 Molecular structure of [Cp_2HfMe_2].

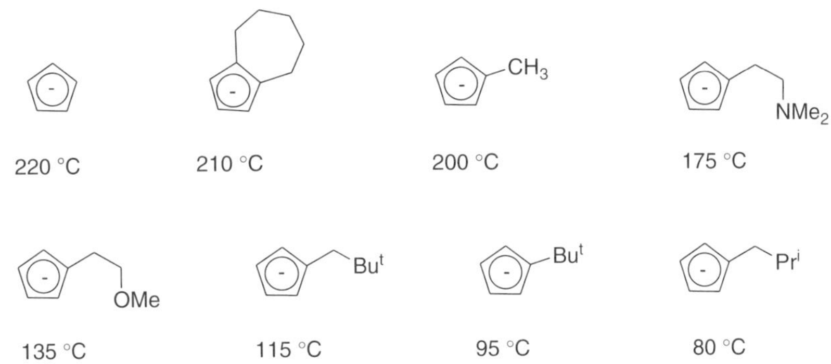

Figure 8.14 Effect of alkyl substitution on the Cp ring on the bp of $[NdCp_3]$ complexes.

Table 8.5 Vapour pressure data for various metal oxide precursors.[a] [Vapour pressure equations are courtesy of SAFC Hitech (Bromborough, Wirral, Merseyside, UK)]

Chemical precursor	*Vapour pressure equation*	
	A	*B*
$[TiCl_4]$	7.640	1947.0
$[Ti(NMe_2)_4]$	8.600	2850.0
$[Zr(NEtMe)_4]$	7.3218	2759.9
$[(MeCp)_2ZrMe_2]$	7.1955	2755.8
$[(MeCp)_2ZrMe(OMe)]$	7.2474	2586.4
$HfCl_4$	11.50	5103.0
$[Hf(NMe_2)_4]$	9.9457	3270.8
$[Hf(NEt_2)_4]$	7.3066	2896.1
$[Hf(NEtMe)_4]$	9.447	3432.3
$[Cp_2HfMe_2]$	8.062	3142.9
$[(MeCp)_2HfMe_2]$	8.0218	3072.1
$[(MeCp)_2HfMe(OMe)]$	6.4697	2555.9
$[Hf(OBu^t)_4]$	8.2277	2710.6
$[Hf(OBu^t)_2(mmp)_2]$	8.2865	3374.2
$[Hf(mmp)_4]$	3.1171	1464.0
$[Ta(NMe_2)_5]$	11.265	4124.9
$[Ta(OEt)_5]$	9.966	4288.0
$[Bi(thd)_3]$	10.27	5199.0
$[Bi(mmp)_3]$	8.251	2257.0
$[La(thd)_3]$	2.4732	1903.0
$[Ru_3CO_{12}]$	13.00	5490.0
$[Cp_2Ru]$	10.881	3221.6
$[(2,4\text{-dimethylpentadienyl})_2Ru]$	8.619	3502.0
$[Si(NMe_2)_4]$	7.037	1969.5
$[AlMe_3]$	8.224	2134.9
$[ZnMe_2]$	7.800	1560.0

[a]Vapour pressure is given by $\log_{10}P(\text{mmHg}) = A - B/T(\text{K})$.

As shown in Chapter 1 (Section 1.4.2) the accurate determination of vapour pressure is a key element for the successful exploitation of a chemical precursor. The vapour pressures of various precursors are given in Tables 8.5 and 8.6, which illustrate the trends in vapour pressure as a function of molecular structure described above.

8.3 CVD of Dielectric Oxides

The increased computing power of microprocessors used in products such as computers and mobile phones can be largely attributed to advances in silicon integrated circuit (IC) technology, which has undergone a continual reduction of the dimensions of the feature sizes in the building block metal-oxide-semiconductor field effect transistor (MOSFET).[75] Over three decades, amorphous SiO_2 has been used as the insulating dielectric layer in these devices due to its stable high quality Si-SiO_2 interface, and excellent electrical isolation properties. To maintain the capacitance in the shrinking MOSFET without breaking down, the use of materials with a substantially higher dielectric constant than SiO_2 ($\kappa \sim 3.9$) is required. Consequently, there has been much research into alternative high-κ dielectric oxides or silicates,[76] such as TiO_2,[76] Ta_2O_5,[76] Al_2O_3,[77] ZrO_2,[78] HfO_2,[79] Zr- and Hf-silicate,[80,81] $(HfO_2)_x(Al_2O_3)_{1-x}$,[82,83] La_2O_3,[84] Pr_2O_3,[85] Y_2O_3,[86] Gd_2O_3,[86,87] and Nd_2O_3.[88]

The dynamic random access memory (DRAM) is another key device component in the Si microelectronics industry as the primary information storage medium, and there is presently a huge drive to increase the effective capacitor area without increasing its footprint in the cell. The replacement of SiO_2 capacitor layers by alternative high-κ dielectric oxides is key to progress in this area, and HfO_2/Al_2O_3 nanolaminates are currently under intensive investigation.[89]

To replace SiO_2, candidate high-κ materials must meet several exacting requirements. The dielectric must have a κ value sufficiently high enough to meet the scaling needs of one or more generations of device feature sizes. The material must be thermodynamically stable with respect to the silicon substrate and sufficiently kinetically stable to withstand short times at high back-end process temperatures (>950 °C). The dielectric should have large band offsets with respect to silicon to avoid carrier injection and the material should form a good electrical interface with silicon containing a low density of interface states.[90]

8.3.1 CVD of ZrO_2 and HfO_2

8.3.1.1 MOCVD of ZrO_2 and HfO_2

Zirconium dioxide (ZrO_2) and hafnium dioxide (HfO_2) have high permittivities ($\kappa \sim 25$) compared SiO_2 and are relatively thermodynamically stable in contact with silicon.[76] A range of issues have been encountered with the development of Zr and Hf oxide precursors, for instance [$ZrCl_4$][91] and [$HfCl_4$][92] are low volatility solids, which need substrate deposition temperatures (≥ 800 °C) and there is the possibility of chloride contamination in the films. The β-diketonates [Zr(acac)$_4$] and [Hf(acac)$_4$] (acac = pentane-2,4-dionate) (Figure 8.15) require high evaporation temperatures (~200 °C) for sufficient vaporization,[93] and the oxide films are often heavily contaminated with carbon.[94] The β-diketonate [Zr(thd)$_4$] (thd = 2,2,6,6-tetramethylheptane-3,5-dionate) has been successfully used for the MOCVD of high purity ZrO_2,[94,95] although it is also relatively involatile, requiring high substrate temperatures (>600 °C) for oxide growth. The fluorinated complexes [Zr(tfac)$_4$] and [Hf(tfac)$_4$] (tfac = 1,1,1-trifluoropentane-2,4-dionate) are more volatile,[96] but the presence of fluorine, which can react adversely with the Si substrate, is highly undesirable. Recently, two new β-diketonate complexes, [Zr(tod)$_4$] and [Hf(tod)$_4$] (tod = 2,7,7-trimethyl-3,5-octanedionate) have been used to deposit ZrO_2 and HfO_2 by liquid injection MOCVD. The complexes were found to give higher oxide growth rates at lower deposition temperatures than the corresponding thd complexes.[97]

The metal nitrate complexes [$Zr(NO_3)_4$] and [$Hf(NO_3)_4$] are volatile eight coordinate complexes, isostructural with [$Ti(NO_3)_4$] (Figure 8.16), and they have been used for the MOCVD of high purity ZrO_2 and HfO_2 films at growth temperatures as low as 300 °C.[93,98] The films were found to contain 10–20% excess oxygen and nitrogen concentrations were below XPS detection limits (~0.2 at.%).

The organometallic Zr complexes [Cp_2ZrMe_2], [$(MeCp)_2ZrMe_2$] and [$(MeCp)_2Zr(BH_4)_2$] have been used for the MOCVD of ZrO_2 in the temperature range 400–550 °C.[99] The oxide films grown

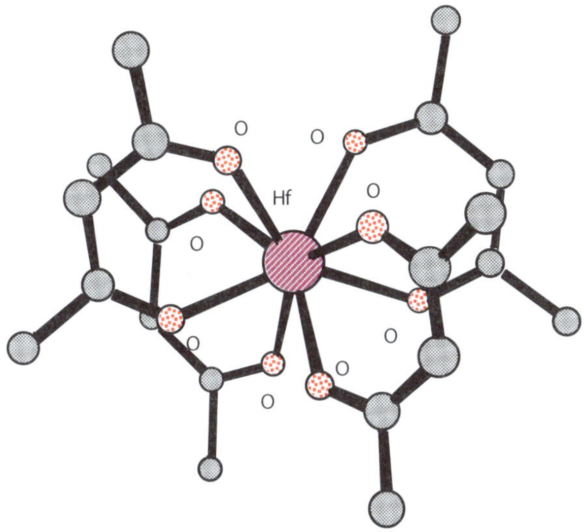

Figure 8.15 Molecular structure of $[Hf(acac)_4]$.

from $[(MeCp)_2ZrMe_2]$ and $[(MeCp)_2Zr(BH_4)_2]$ were of poor quality, containing up to 30 at.% of carbon, although the presence of alkoxide ligands in the molecule, as in $[(MeCp)_2ZrMe(OR)]$ (OR = OMe, OBu^t), significantly reduces carbon contamination in the ZrO_2 to ~7 at.%,[100] probably due to inhibition of a $[Cp\text{-}CH_2\text{-}Zr\text{-}CH_2\text{-}Cp]$ intermediate (Figure 8.17), a probable precursor[99] to carbon incorporation.

Metal alkoxides are attractive MOCVD precursors as they allow lower deposition temperatures, and under optimum growth conditions allow the deposition of carbon-free films.[101] However, most $[Zr(OR)_4]$ and $[Hf(OR)_4]$ complexes are dimeric or polymeric with limited volatility, as discussed in Section 8.2.3. The volatile monomeric complexes $[Zr(OBu^t)_4]$ and $[Hf(OBu^t)_4]$ have been successfully used for the MOCVD of ZrO_2[102–104] and HfO_2,[105] respectively. One strategy used to limit oligomerization in Zr and Hf-alkoxides is to satisfy the coordination sphere of the central metal atom by introducing bidentate chelating β-diketonates such as thd, β-ketoesters (*tert*-butyl acetoacetate, "tbaoac") or β-ketoamides (*N*,*N*-diethylacetoacetamide, "deacam") (Table 8.2).

These heteroleptic complexes potentially combine the advantages of good ambient stability associated with metal β-diketonates with the lower thermal stability and higher volatility of metal alkoxides. The complex $[Zr_2(OPr^i)_6(thd)_2]$ is a centrosymmetric dimer containing terminal thd groups and two bridging OPr^i ligands and has been used for the liquid injection MOCVD of ZrO_2 over the temperature range 300–600 °C.[106] The similar complexes $[Zr(OBu^t)_3(acac)]$, $[Zr(OBu^t)_2\text{-}(acac)_2]$ and $[Zr(OBu^t)(acac)_3]$ have been used for the MOCVD of ZrO_2 employing ultrasonic nebulization of the precursors. The insertion of the acac group into $[Zr(OBu^t)_4]$ made the oxide deposition rate higher and less sensitive to substrate temperature.[107]

The complexes $[Zr(OPr^i)_2(tbaoac)_2]$ and $[Zr(OPr^i)_2(deacam)_2]$ are monomeric, and the former has been used for the MOCVD of ZrO_2 in the absence of an added oxidant.[108]

The donor-functionalized mmp ligand (discussed in Section 8.2.3) gives a series of volatile monomeric complexes, $[Zr(OBu^t)_2(mmp)_2]$, $[Hf(OBu^t)_2(mmp)_2]$, $[Zr(mmp)_4]$ and $[Hf(mmp)_4]$ (Figure 8.7),[45,46] and these have been used for the liquid injection MOCVD of ZrO_2 (κ ~ 24) and HfO_2 (κ ~ 26) thin films over the temperature range 350–650 °C.[46] High-frequency capacitance–voltage (*C-V*) measurements obtained at 100 kHz on $[Al/MO_2/p\text{-}Si]$ metal oxide semiconductor (MOS) capacitors showed clear accumulation, depletion and inversion conditions. MOS capacitors incorporating films deposited from each of these precursors all exhibited flat band

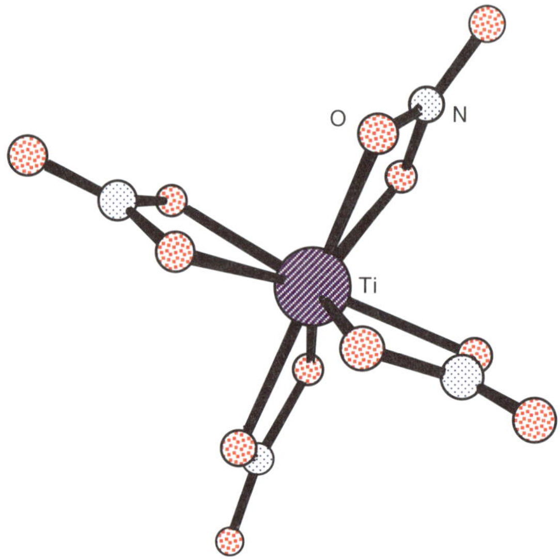

Figure 8.16 Molecular structure of $[Ti(NO_3)_4]$.

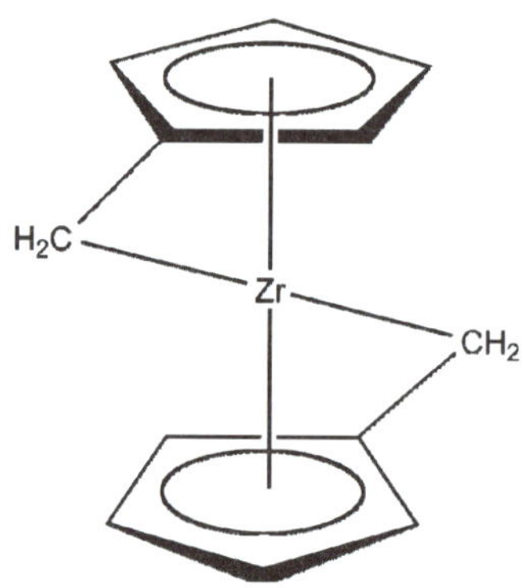

Figure 8.17 Probable intermediate leading to carbon deposition during the MOCVD of ZrO_2 using $[(MeCp)_2ZrMe_2]$ and $[(MeCp)_2Zr(BH_4)_2]$.

voltage shifts (ΔV_{FB}) of less than -1.5 V and only -0.5 V in the case of $[Hf(mmp)_4]$. In all cases the oxide trapped charge density (N_{OT}) was less than $5 \times 10^{11}\ cm^{-2}$, which is directly comparable with an unannealed MOS capacitor based on thermal SiO_2 gate dielectric. The capacitors based on films grown from the hafnium precursors showed clockwise hysteresis due to the presence of a small amount of mobile positively charged ions in the films.[109] A proposed mechanism[55] for the decomposition of $[Hf(mmp)_4]$ in the absence of oxygen is shown in Scheme 8.1 (Section 8.2.3). $[Hf(OBu^t)_2(mmp)_2]$ has subsequently been widely used by other workers for the MOCVD of HfO_2.[110,111]

Another Hf alkoxide complex, $[Hf(mp)_4]$ (mp = $OCMeEt_2$), has been used to deposit high purity HfO_2 (carbon not detected) with no additional oxygen source. The thermolysis products were analysed by GC-MS and ^1H NMR and it was proposed that HfO_2 forms by the β-hydrogen elimination from the [Me] group on the α-carbon and β-hydrogen elimination from the CH_2 moiety in the ethyl group.[112]

The complexes $[M(OBu^t)_2(dmop)_2]$ (M = Hf, Zr) have been used for the liquid injection MOCVD of HfO_2 and ZrO_2 (carbon ~1.9–8 at.%).[53] The homoleptic alkoxide $[Hf(dmop)_4]$ is a six-coordinate monomeric complex, (Figure 8.9a), and this complex has also been used for the deposition of HfO_2 films (C: 2.1–11.6 at.%; N: 0.4–1.9 at.%) by liquid injection MOCVD.[52] Films deposited below 400 °C were amorphous, becoming monoclinic above 450 °C, and they contained traces of carbon (2.1–11.6 at.%) and nitrogen (0.4–1.9 at.%). High frequency *C-V* data for MOS $[Al/HfO_2/n\text{-}Si]$ capacitor structures grown from $[M(OBu^t)_2(dmop)_2]$ (M = Zr, Hf) and $Hf(dmop)_4$ showed hysteresis due to the presence of mobile ions in the films, and shifts in the flatband voltages indicated the presence of trapped charge at the metal oxide–silicon interface.[52,53]

Metal alkylamide complexes have considerable potential as MOCVD precursors for the deposition of both metal nitride and metal oxide thin films. It was shown some time ago that $[Zr(NEt_2)_4]$ can be used for the MOCVD of high quality ZrO_2 films in the temperature range 500–580 °C.[58] More recently, the liquid alkylamide $[Hf(NEt_2)_4]$ has been used to deposit HfO_2 by conventional (*i.e.* "bubbler-based") MOCVD.[59,60] The levels of residual carbon and nitrogen were found to decrease with increasing oxygen flow rates and increasing substrate temperatures, and, at 450 °C, stoichiometric polycrystalline HfO_2 containing little residual impurities was deposited. It was proposed on the basis of NMR data[60] that $[Hf(NEt_2)_4]$ is a monomeric molecule, but the structure of the analogous complex $[Hf(NMe_2)_4]$ has been shown to be dimeric (Figure 8.18), containing two five-coordinate Hf atoms and bridging alkylamide groups;[113] the complex has been used for the liquid injection MOCVD of HfO_2.[113]

Although reported many years ago,[114] the hydroxylamide complexes $[Zr(ONR_2)_4]$ and $[Hf(ONR_2)_4]$ have received little attention as MOCVD precursors. These complexes have been shown to be quasi-eight coordinate monomers, which should make them less reactive and easier to handle in liquid injection MOCVD than coordinatively unsaturated $[M(NR_2)_4]$ complexes. The $[Hf(ONEt_2)_4]$ complex has been used to deposit HfO_2 by liquid injection MOCVD.[113]

Theoretically, HfO_2 exhibits a higher permittivity in the cubic phase (κ ~ 29) or the tetragonal phase (κ ~ 70) than in the monoclinic phase (κ ~ 16–18).[115] Normally, the cubic and tetragonal phases can only be obtained at high temperature (2700 and 1700 °C, respectively), but it is possible to stabilize these phases by the addition of yttrium.[116] Thin films of Y-doped HfO_2 have been deposited by liquid injection MOCVD using $[Hf(OBu^t)_2(mmp)_2]$ (Figure 8.7b) and $[Y(thd)_3]$,[117] and it was found that a pure cubic phase of HfO_2 was stabilized at an Y content of ~6.5 at.%. A maximum permittivity of ~22 was obtained for Y contents of 6.5–10 at.%, and the films exhibited low leakage current densities (5×10^{-7} A cm^{-2} at 1 V for a 6.4 nm film).

8.3.1.2 ALD of ZrO_2 and HfO_2

ZrO_2 and HfO_2 films have been grown by ALD using the halides $[ZrCl_4]$,[118] $[ZrI_4]$,[119] $[HfCl_4]$ and $[HfI_4]$.[120–123] In contrast to MOCVD, the high thermal stability of metal halides is an advantage in ALD, as exchange reactions between reactive $HfCl_x$ moieties and surface [OH] species predominate over thermal decomposition reactions, and so halide precursors show good self-limiting growth behaviour. Although Zr and Hf halides have given pure oxide films by ALD, they are high melting solid sources, and so there is the risk of particle transport to the substrate. Halide contamination of the film is also a serious potential problem in microelectronics applications.

Although the use of $[Cp_2ZrMe_2]$ in MOCVD can lead to carbon-contaminated ZrO_2 films,[99] both $[Cp_2ZrMe_2]$[124] and $[Cp_2HfMe_2]$[125] deposit high purity ZrO_2 and HfO_2 (C and H < 0.1 at.%) by ALD, and give self-limiting growth at the optimum growth temperature of 350 °C. The use of O_3 in

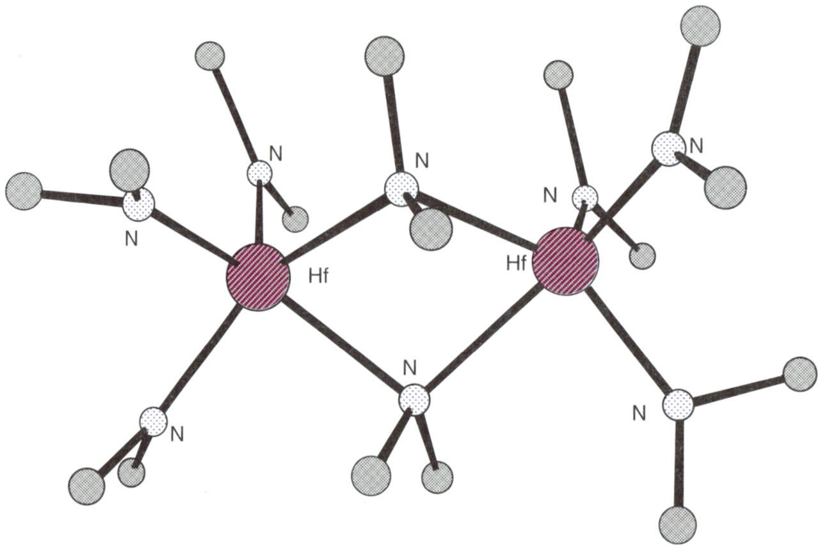

Figure 8.18 Crystal structure of $[Hf(NMe_2)_4]_2$.

place of H_2O as oxygen source led to a significant increase in growth rate (~0.53 compared to 0.43 Å cycle^{-1}).

Metal alkylamides $[M(NR_2)_4]$ (M = Zr, Hf) also give self-limiting ALD growth,[126] and high purity HfO_2 (C, N and H contents = 0.3–0.6, 0.1–0.2 and 2–3 at.%, respectively) has been deposited by ALD using $[Hf(NEtMe)_4]$.[127] Self-limiting growth was demonstrated over the temperature range 200–250 °C, but above 250 °C thermal decomposition of the precursor led to a breakdown of self-limiting growth, as shown by the variation in HfO_2 growth rate with substrate temperature (Figure 8.19). *C-V* measurements showed that the HfO_2 permittivities varied between 11 and 14.[127]

Cp-based Zr and Hf complexes have a significantly higher thermal stability than $[Hf(NR_2)_4]$ complexes, as illustrated by the variation of HfO_2 growth rate with substrate temperature for $[Cp_2HfMe_2]$ and $[(MeCp)_2HfMe(OMe)]$ (Figure 8.20). This facilitates self-limiting ALD of HfO_2 using $[(MeCp)_2HfMe(OMe)]$ at substrate temperatures as high as 500 °C.

Both Zr- and Hf-alkoxides have also been widely investigated as ALD precursors. Zirconium tetra-*tert*-butoxide, $[Zr(OBu^t)_4]$ has been used for the ALD of high purity ZrO_2 using H_2O as the oxygen source.[128] The films had permittivities, as high as 32, that compare favourably with the permittivities of 18–24 obtained using halide precursors. However, $[Zr(OBu^t)_4]$ failed to give truly self-limiting growth due to decomposition of the precursor by β-hydride elimination (Scheme 8.2, Section 8.2.3).

Several zirconium- and hafnium alkoxide ALD precursors containing alternative alkoxide groups to $[OBu^t]$ have therefore been examined in an effort to stabilize the precursor to surface decomposition and to modify its reactivity towards surface [OH] species. The deposition of ZrO_2 has been achieved using $[Zr(OBu^t)_2(dmae)_2]$, $[Zr(OPr^i)_2(dmae)_2]$ and $[Zr(dmae)_4]$,[129] but none of the complexes displayed self-limiting growth, due to the decomposition of the alkoxide groups on growth surface. Mechanistic studies[130] using in situ quadrupole mass spectroscopy (QMS) and a quartz crystal microbalance (QCM) indicated that this was due to the elimination of a β-hydride from $[Zr\text{-}OPr^i]$ or [Zr-dmae] groups to liberate an alkene and generate surface-bound [Zr-OH] species.

HfO_2 films have been deposited by ALD using $[Hf(mmp)_4]$ or $[Hf(OBu^t)_2(mmp)_2]$ with H_2O in the temperature range 275–425 °C,[131,132] although film growth was not self-limiting due to the β-hydride elimination of the mmp and OBu^t groups to generate reactive Hf-OH surface sites (Scheme 8.2 above). Oxide growth rates from $[Hf(mmp)_4]$ and $[Hf(OBu^t)_2(mmp)_2]$ were very low (~0.02–0.03 and ~0.04–0.06 nm $cycle^{-1}$, respectively) and were much lower than from $[Hf(NEtMe)_4]$ (~0.09 nm $cycle^{-1}$). This is probably due to steric inhibition of the surface exchange reaction between [OH] species and $[Hf(mmp)_x]$ precursor moieties.

HfO_2 films have also been deposited by liquid injection ALD using a heptane solution of $[Hf(mmp)_4]$ and H_2O.[56] The purity of the ALD-grown films (C = 0.8–2.5 at.%) was equal to the films deposited by MOCVD. The purity also compared favourably to films grown by conventional ALD using $[Hf(mmp)_4]$ (C = 2.2–6 at.%, determined by TOF-ERDA),[132] showing that the heptane solvent does not contribute carbon to the films.

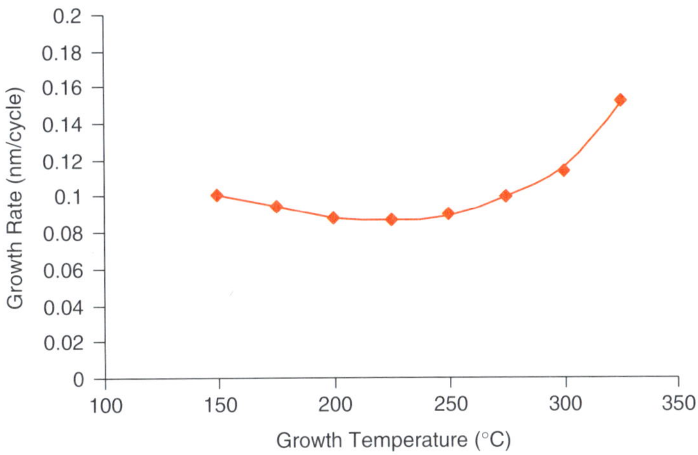

Figure 8.19 Variation of HfO_2 ALD growth rate with deposition temperature for $[Hf(NEtMe)_4]$.

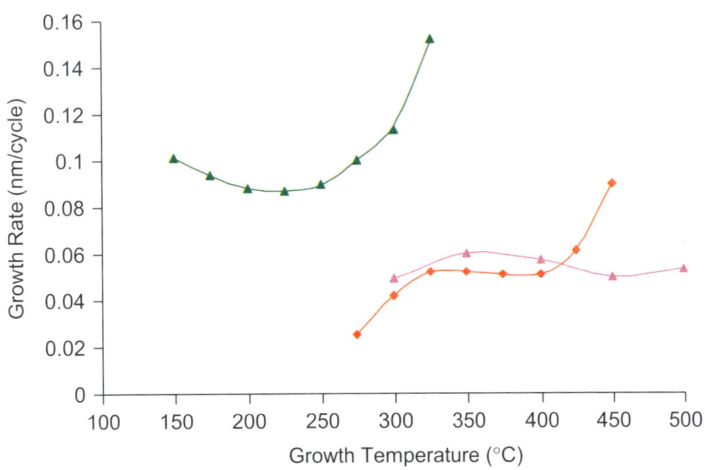

Figure 8.20 Variation in HfO_2 ALD growth rate with substrate temperature for $[Hf(NMeEt)_4]$ (green), $[Cp_2HfMe_2]$ (red) and $[(MeCp)_2HfMe(OMe)]$ (purple).

8.3.2 CVD of Zr- and Hf-silicate

Neither ZrO_2 nor HfO_2 are completely chemically or thermally stable on silicon. Both ZrO_2 and HfO_2 crystallize at relatively low temperatures (~450–500 °C), as shown by the X-ray diffraction data in Figure 8.21 for HfO_2 films grown using $[Hf(OBu^t)_2(mmpl)_2]$.[53] This leads to current leakage along the polycrystalline grain boundaries and promotes the formation of a low-κ SiO_2 layer during the high temperatures (~900 °C) involved in CMOS processing.

A possible solution to this problem is to deposit a metal silicate, such as $ZrSi_xO_y$ or $HfSi_xO_y$, directly onto the silicon substrate.[76] These form films that remain amorphous up to high temperature (~800–900 °C), leading to reduced leakage currents and they are thermodynamically stable on silicon. The permittivities of these alloys are lower than the pure metal oxides, but this is may be considered to be an acceptable trade-off for the greatly improved interface stability. For instance, $ZrSi_xO_y$ has an intermediate dielectric constant of approx. 12–25, but this is still significantly higher than that of SiO_2 ($\kappa = 3.9$).[76]

Various precursor combinations have been investigated; for instance, $ZrSi_xO_y$ and $HfSi_xO_y$ have been deposited by liquid injection MOCVD using $[Zr(NEt_2)_4]$[133] or $[Hf(NEt_2)_4]$[134] in combination with $[Si(NMe_2)_4]$. $[Hf(NEt_2)_4]$ has also been used with several different Si precursors, $[HSi(NEt_2)_3]$, $[Si(NCO)_4]$ and $[Si(OEt)_4]$, to deposit $HfSi_xO_y$ by conventional MOCVD.[135] The level of carbon in the film depends strongly on the nature of the Si precursor and substrate temperature. For $[HSi(NEt_2)_3]$ and $[Si(NCO)_4]$ carbon levels increase significantly (up to 20–30 at.%) with increasing substrate temperature. This was in marked contrast to $[Si(OEt)_4]$, which showed an inverse trend in carbon incorporation, giving layers with no detectable carbon at a substrate temperature of ~560 °C.

Liquid injection MOCVD of $HfSi_xO_y$ has been demonstrated using $[Hf(OBu^t)_2(mmp)_2]$ and $[Si(mmp)_4]$.[136] $ZrSi_xO_y$ has been deposited from $[Zr(acac)_2(OSiMe_2Bu^t)_2]$ in the absence of an added Si source.[137] The combination of metal alkylamide $[M(NR_2)_4]$ (M = Zr, Hf) and alkoxy silanols, $[(RO)_3SiOH]$, has been proposed as a general method for the CVD of $ZrSi_xO_y$ and $HfSi_xO_y$, and the ALD of $HfSi_xO_y$ has been achieved using alternate pulses of $[Hf(NMe_2)_4]$ and $[(Bu^tO)_3SiOH]$.[138] The novel "single-source" precursor $[Hf(OSiBu^tMe_2)_4(Et_2NH)]$ (Figure 8.22), has been used for the growth of $HfSi_xO_y$ films by liquid injection MOCVD,[139] but a disadvantage of using single-source precursors is that the fixed Hf : Si ratio makes it difficult to alter

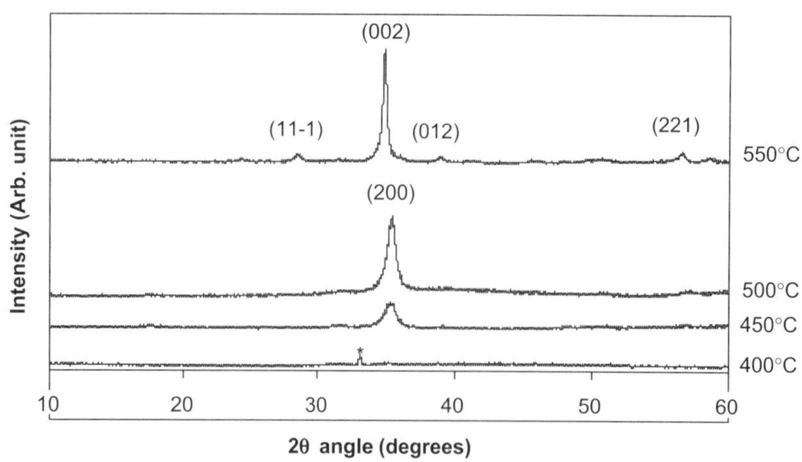

Figure 8.21 XRD data illustrating the crystallization of HfO_2 as the deposition temperature is increased from 400 to 550 °C.

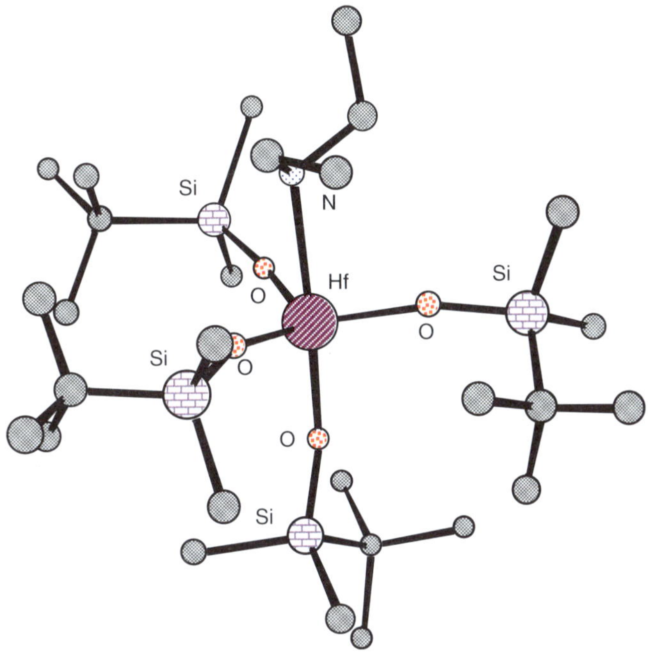

Figure 8.22 Crystal structure of [Hf(OSiBu^{t}Me$_2$)$_4$(Et$_2$NH)].

the metal : silicon concentration ratio in the film and a Si level of ~10–15 at.% was obtained that showed little variation over a range of growth temperatures and oxygen partial pressures.

The use of two separate sources, [Hf(NMe$_2$)$_4$] and [Bu^{t}Me$_2$SiOH], allowed fine-tuning of the silicon concentration in the HfSi$_x$O$_y$ film up to maximum level of 12–15%.[139] It was proposed that [Hf(OSiMe$_2$Bu^t)$_4$], or a related species, was formed *in situ* as the reactive intermediate in the gas phase.

8.3.3 CVD of Hf-aluminate

Another approach to stabilizing the Si–HfO$_2$ interface and reducing leakage currents is to alloy HfO$_2$ with Al$_2$O$_3$ to form hafnium-aluminate, $(HfO_2)_x(Al_2O_3)_{1-x}$, which is amorphous up to high processing temperatures (*e.g.* 900 °C for Al concentrations of 31%), and has been reported to retain a relatively high dielectric constant of $\kappa \sim 15$.[76]

$(HfO_2)_x(Al_2O_3)_{1-x}$ (Al = 8.4–38.5 at.%) has been grown on Si(100) by liquid injection MOCVD using the metal alkoxide precursors [Hf(mmp)$_4$] and [Al(OPr^i)$_3$].[83] The films became amorphous when only very small fractions of Al (~2 at.%) were present. Medium energy ion scattering (MEIS) detected a thin layer of native SiO$_2$ between the $(HfO_2)_x(Al_2O_3)_{1-x}$ film and the Si(100) substrate,[140]and the effect of annealing $(HfO_2)_x(Al_2O_3)_{1-x}$ films containing varying amounts of Al on the thickness of the SiO$_2$ interfacial layer was investigated. At low Al concentrations (10 at.%), the $(HfO_2)_x(Al_2O_3)_{1-x}$ films crystallized on annealing at 1000 °C, leading to a 125-fold increase in the SiO$_2$ layer thickness. However, in high Al content $(HfO_2)_x(Al_2O_3)_{1-x}$ films (Al=38 at.%) crystallization is suppressed, so that there was no appreciable increase in the thickness of the SiO$_2$ interfacial layer after annealing.

Hf-aluminate nanolaminate films have been deposited by ALD using alternating oxide deposition cycles and the precursor combinations [HfCl$_4$]/[AlMe$_3$]/H$_2$O[82,141] or [Hf(NMeEt)$_4$]/[AlMe$_3$]/H$_2$O.[142] A major potential application of Hf-aluminate is in very high aspect ration DRAM stack or trench structures, and so ALD with its superior step coverage to MOCVD is the preferred deposition technique.

8.3.4 CVD of Lanthanide Oxides

Although ZrO_2, HfO_2 and their related silicates have been the most intensively investigated materials for CMOS gate dielectric applications,[76] there has also been much interest in the rare earth oxides.[76,143] The rare-earth oxides M_2O_3 (M = La, Pr, Gd and Nd) are good insulators due to their large band-gaps (3.9 eV for Pr_2O_3, 5.6 eV for Gd_2O_3), they have high dielectric constants (Gd_2O_3 $\kappa = 16$, La_2O_3 $\kappa = 30$ and Pr_2O_3 $\kappa = 26$–30) and good thermodynamic stability on silicon, making them attractive materials for high-κ dielectric applications.[76] Another attractive feature of some rare earth oxides such as Pr_2O_3 and Gd_2O_3 is their relatively close lattice match to silicon (*e.g.* $\alpha(Gd_2O_3) = 10.812$ Å; $2a(Si) = 10.862$ Å), which offers the possibility of epitaxial growth, eliminating problems related to current leakage along grain boundaries in polycrystalline films. It has also been shown that Gd_2O_3 can be grown epitaxially on GaAs(100), to give an insulating and passivating oxide layer,[144] which has potential applications in a wide range of III-V photonic and electronic devices.

8.3.4.1 MOCVD of Lanthanide Oxides

The large radii of Ln^{3+} ions (*e.g.* six-coordinate radii: $La^{3+} = 117.2$, $Pr^{3+} = 113$, $Gd^{3+} = 107.8$ pm) presents a real challenge for the synthesis of volatile precursors for MOCVD: coordinative saturation requires either high coordination numbers (up to 12) or very sterically demanding ligands. High coordination numbers are often achieved by the formation of oligomeric species with bridging ligands, resulting in extremely low volatility.

The oldest known class of volatile lanthanide complexes, known since the early years of the 20th century, is the tris-β-diketonates. They are easily prepared and have excellent ambient stability. $[Ln(thd)_3]$ complexes have been investigated for MOCVD of oxides of La,[145] Pr,[146] Gd[147] and Nd.[148] However, these precursors are thermally quite robust and require high growth temperatures that are generally incompatible with microelectronics applications, where high growth temperatures can lead to problems such as increased dopant diffusion, and the oxide films produced are often contaminated with trace carbon.

A range of Pr-oxides (PrO_2, Pr_6O_{11}, Pr_2O_3) have been grown recently by conventional MOCVD using $[Pr(thd)_3]$ (thd = 2,2,6,6,-tetramethylheptane-3,5-dionate).[146,149,150] The deposited phase of Pr-oxide was critically dependent on oxygen partial pressure, with PrO_2 deposited at an oxygen partial pressure of 2 Torr, and Pr_2O_3, Pr_6O_{11} at $P_{O2} = 0.7$ Torr, whilst the required Pr_2O_3 phase was obtained at $P_{O2} = 10^{-3}$ Torr and a substrate temperature of 750 °C. The Pr_2O_3 films were polycrystalline with good dielectric properties. The fluorinated β-diketonate $[Pr(hfac)_3(diglyme)]$ was also investigated,[150] but led only to the unwanted oxyfluoride phase, PrOF.

The β-diketonate precursor $[Gd(thd)_3(tetraglyme)]$ was used for the MOCVD of Gd_2O_3 but only poor quality films were obtained.[147] Neodymium oxide with a cubic structure has been deposited by aerosol pyrolysis using $[Nd(F_3C_5H_4O_2)_3]$ dissolved in n-butanol,[151] whilst cubic Nd_2O_3 for use in corrosion protection was deposited by MOCVD using $[Nd(thd)_3]$.[148] La_2O_3 films have been deposited by MOCVD using $[La(thd)_3]$,[145,152] $[La(thd)_3(tetraglyme)]$[153] and $[La(acac)_3]$.[154]

A new class of volatile fluorine-free lanthanide β-ketoiminate complexes has been developed for the MOCVD of lanthanide oxides (Table 8.2). The complexes $[Ln(ketoim)_3]$ with R = Me, Et or Bu^t are six-coordinate monomers and $[Ce(ketoim)_3]$ was used to deposit CeO_2 thin films by pulsed liquid injection MOCVD.[40]

The lanthanide tris-silylamides [Ln{N(SiMe$_3$)$_2$}$_3$] were the first example of monomeric three-coordinate lanthanide complexes,[155] and they show reasonable volatility. MOCVD using these precursors resulted in variable incorporation of silicate into the oxide film.[156] Simple Si-free alkylamides of the lanthanides (*e.g.* [Ln(NPr$^i_2$)$_3$(THF)]) are not sufficiently volatile to be used in MOCVD and so there are as yet no suitable amide precursors for MOCVD of lanthanide oxides.

As discussed in Section 8.2.3, the donor-functionalized alkoxide mmp is not able to stabilize monomeric complexes with the large rare earth ions; "[Lu(mmp)$_3$]" is a dimer in the solid state[47] and "[La(mmp)$_3$]" is a trimer (Figure 8.8).[48] However, these complexes are rendered volatile by the addition of 3 mol. equivalent tetraglyme/[Ln(mmp)$_3$] or 1 mol. equivalent of Hmmp, which also stabilizes the precursor in solution, probably by minimizing intermolecular oligomerization. Toluene solutions of [Ln(mmp)$_3$]/tetraglyme] (Ln = La,[157] Pr,[158,159] Nd[160] and Gd[161]) have been successfully used for the liquid injection MOCVD of Ln-oxides over a wide range of substrate temperatures (250–600 °C).

XRD data showed that the Pr-oxide films were deposited in the Pr_6O_{11} phase. At low deposition temperatures the Gd_2O_3 and Nd_2O_3 deposited on Si(100) were amorphous, but above 450 °C crystallization occurs to form c-type Gd_2O_3 [preferred (111) orientation] and c-type Nd_2O_3 [preferred (100) orientation]. The X-ray diffraction data of a La_2O_3 film deposited at 450 °C were consistent with the random powder diffraction pattern of the hexagonal La_2O_3 phase. X-Ray diffraction data of a Gd_2O_3 film deposited on GaAs(100) at 450 °C were dominated by the (222) reflection, indicating a strong preferred orientation or even a heteroepitaxial relation with the underlying GaAs.

8.3.4.2 *ALD of Lanthanide Oxides*

There are relatively few volatile precursors available for the ALD of rare earth oxides. The simple lanthanide alkylamides [Ln(NR$_2$)$_3$] are unstable and involatile, and so the volatile lanthanide β-diketonates have been extensively investigated. However, lanthanide β-diketonates do not readily react with water, and require a strong oxidizing agent, such as ozone to provide oxide films with low carbon and hydrogen content, but ALD growth rates are still very low (0.1–0.4 Å cycle^{-1}). The use of ozone can also lead to the formation of a low permittivity interfacial SiO_x layer between the oxide and the silicon substrate. This increases the apparent equivalent oxide thickness [EOT = $(3.9/\kappa) \times$ thickness of the high-κ material] of the dielectric layer rather than decreasing it, which is desirable for smaller device feature sizes. Materials grown using [Ln(thd)$_3$] and O_3 include Sc_2O_3,[162] Y_2O_3,[163,164] La_2O_3[165] and CeO_2,[166,167] as well as other Ln_2O_3 films (Ln = Nd, Sm, Eu, Gd, Dy, Ho, Tm).[168–170] Y_2O_3 grown using [Y(thd)$_3$] have been investigated as a high-κ dielectric.[171] Owing to the strongly oxidizing conditions, the silicon became oxidized to form an 11–14 Å interfacial layer of SiO_2. In addition, during high temperature annealing, the Y_2O_3 film reacted with the SiO_2 layer to form an yttrium-silicate. Therefore, ozone-based ALD processes are unlikely to be suitable for the growth of very thin oxide films with equivalent oxide thicknesses of 10 Å or below.

To overcome the problems associated with ozone, water-based ALD processes have been developed for the deposition of Ln_2O_3 using cyclopentadienyl-based lanthanide precursors, such as [Cp$_3$Ln] and [(MeCp)$_3$Ln] (Ln = Sc,[162] Y,[172] Er[173] and Gd[174]) (Section 8.2.6) and Lu_2O_3 has been grown by ALD using a Lu source with silylated-Cp groups.[175] These Cp-based complexes are volatile and highly reactive to water, leading to increased ALD growth rates (0.75–1.5 Å cycle^{-1}). Self-limiting growth was demonstrated by [Cp$_3$Sc]/H_2O in the temperature range 250–350 °C,[162] and in the range 250–300 °C for [(MeCp)$_3$Er]/H_2O. However, ALD using [(MeCp)$_3$Gd] and H_2O was not self-limiting due to partial decomposition of the [(MeCp)$_3$Gd] precursor.[174] The use of O_3

as oxidant generally leads to an increase in growth rates.[162,173] Ln_2O_3 films deposited from [Cp_3Ln] and [$(MeCp)_3Ln$] complexes are generally of high purity with low carbon and hydrogen contents (often <1 at.%). A recent patent describes the use of *tris*-cyclopentadienyl lanthanide complexes for the deposition of a large range of lanthanide oxides by ALD.[176]

Rare earth amidinates (Section 8.2.5) have been designed specifically as ALD precursors, and the ALD of Y_2O_3 films has been demonstrated using [$Y(Pr^iNC(Me)NPr^i)_3$] and H_2O.[66] Self-limiting growth was demonstrated over the temperature range 150–280 °C, with growth rates of 0.8 Å cycle^{-1}, and the films were high purity (C,N <0.5 at.%). The Y_2O_3 films had cubic polycrystalline structure with a refractive index of 1.8. At growth temperatures above 280 °C, thermal decomposition of the precursor was observed.

Recent ALD studies[177] showed that [$Pr\{N(SiMe_3\}_3$] produced a Pr-silicate, $PrSi_xO_y$, rather than a pure oxide film, and so several studies have been carried out using the volatile Si-free alkoxides [$Ln(mmp)_3$] (Ln = Gd, Pr).[56] Gd_2O_3 thin films with no detectable carbon have been deposited by liquid injection ALD between 200 and 300 °C using a solution of [$Gd(mmp)_3$] in toluene with 3 mol. equivalent of added tetraglyme. Liquid injection ALD has also been used for the growth of PrO_x films (carbon = non-detectable to 3 at.%) between 200 and 300 °C using a solution of [$Pr(mmp)_3$] in toluene with 3 mol equiv. of added tetraglyme.[56] In both cases, film growth was not self-limiting due to decomposition of the precursor, probably *via* a β-hydride elimination mechanism analogous to that shown in Scheme 8.1 (Section 8.2.3).

8.3.5 CVD of Multi-component Lanthanide Oxides

8.3.5.1 CVD of Lanthanide Silicates

Analogous to the deposition of $HfSi_xO_y$, the deposition of lanthanide silicates $LnSi_xO_y$ has been investigated, as these films remain amorphous up to high temperatures ($\sim$850 °C), reducing current leakage and stabilizing the interface between the silicon substrate and the high-κ metal oxide.[76] In contrast to simple [$Ln(NR_2)_3$] complexes, lanthanide tris-silylamides, [$Ln[N(SiMe_3)_2]_3$], are volatile and relatively stable and these have been investigated as precursors for the MOCVD of lanthanide silicates.

The [$Pr\{N(SiMe_3)_2\}_3$] complex allowed the MOCVD growth of $PrSi_xO_y$ films over the temperature range 350–550 °C, without an added silicon source.[159] AES analysis showed that the films were praseodymium silicate with average silicon concentrations ranging from approx. 16 to 22 at.%. The silicon concentration varied with depth in some of the films, but showed no obvious dependence on substrate temperature.

$PrSi_xO_y$ has also been deposited by ALD using [$Pr\{N(SiMe_3)_2\}_3$]. Despite the lower deposition temperatures used in ALD, some decomposition of the [$N(SiMe_3)_2$] ligand occurs, leading to significant Si incorporation (4–12 at.%), which increased with increasing deposition temperature (200–300 °C).[177]

The lack of ambient stability of La_2O_3 films and their tendency to react with the Si substrate to form a La-silicate interfacial layer[178] has led to efforts focussing on the deposition and properties of lanthanum silicate ($LaSi_xO_y$) films.[179] Although $LaSi_xO_y$ films have a lower permittivity than La_2O_3, they are more stable and are less likely to contain oxygen deficiencies. There have been few studies on the MOCVD of $LaSi_xO_y$, but it has been shown that $LaSi_xO_y$ films could be deposited by liquid injection MOCVD using the "single-source" precursor [$La\{N(SiMe_3)_2\}_3$] in the presence of oxygen,[156] but without an added Si source. AES measurements indicated that the silicon content of the films varied between 8.5 and 15.2 at.% across the range of deposition temperatures investigated.

8.3.5.2 CVD of Lanthanide Aluminates

Lanthanum aluminate, $LaAlO_3$, is a potential substrate, or buffer layer for various perovskite films (*e.g.* superconductors, ferroelectrics, metal conducting oxides).[180] It has also attracted attention as an alternative to SiO_2 as the gate dielectric layer in sub-0.1 μm CMOS technology.[181] $LaAlO_3$ has a high dielectric constant ($\kappa \sim$ 22-27), a large band gap of 6.2 eV and high band offsets on Si (1.8 eV for electrons and 3.2 eV for holes). $LaAlO_3$ also has a high chemical and thermal stability on Si and remains amorphous up to high temperatures (~860 °C),[182] inhibiting the formation of a low-permittivity SiO_2 interfacial layer during CMOS processing and reducing leakage currents.

The β-diketonates $[La(thd)_3]$ and $[Al(thd)_3]$ have been used for the MOCVD of $LaAlO_3$.[183,184] The "single-source" alkoxide precursor $[LaAl(OPr^i)_6(Pr^iOH)]_2$ (Figure 8.10) has been used to deposit $LaAlO_3$ by liquid injection MOCVD.[54] Films were deposited over a wide range of substrate temperatures (350–550 °C). Energy dispersive X-ray data showed that the films had variable La : Al ratios (0.48–1.03) with the optimum deposition temperature range for the required 1 : 1 ratio being 450–500 °C. At higher substrate temperatures (500–550 °C) the films became La deficient. This may be due to decomposition of the precursor, leading to partial separation of the La and Al components and to pre-deposition of involatile La-oxide. Thin films of $PrAlO_x$ and $NdAlO_x$ have also been deposited by liquid injection MOCVD using $[LnAl(OPr^i)_6(Pr^iOH)]_2$ precursors, and the optimum temperatures for deposition of near-stoichiometric $LnAlO_3$ films were found to be 500 and 450 °C for $PrAlO_x$ and $NdAlO_x$, respectively.[185]

ALD of $LaAlO_3$ has been achieved using $[La(thd)_3]$,[165] $[La[N(SiMe_3)_2]_3]$,[186] $[La(Pr^iAMD)_3]$ (Pr^iAMD = *N,N′*-diisopropylacetamidinate) and $[AlMe_3]$.[68] The growth of multicomponent oxides such as $LaAlO_3$ by ALD can be challenging. For instance, the use of separate La and Al precursors requires careful control of the alternating La, Al and H_2O precursor pulses to deposit stoichiometric $LaAlO_3$. In addition, the most commonly used aluminium precursor in ALD is trimethylaluminium, $[AlMe_3]$, which is a hazardous pyrophoric chemical requiring specialist handling techniques and stringent safety precautions. The use of a [La-Al] single source precursor is a potentially simpler approach, and high purity $LaAlO_3$ films (carbon non-detectable by AES) have been deposited by liquid injection ALD using a toluene solution of $[LaAl(OPr^i)_6(Pr^iOH)]_2$ and H_2O.[187] The films were La deficient (La : Al ratios 0.50–0.61) and remained amorphous during annealing up to 850 °C. *C-V* data for an Al/$LaAlO_x$/n-Si MOS structure showed that the $LaAlO_x$ films had good dielectric properties with a κ of ~13.

Praseodymium aluminate ($PrAlO_x$) ($\kappa \sim$ 25.0) and neodymium aluminate ($NdAlO_x$) ($\kappa \sim$ 22.5) are also potential high-κ materials for CMOS and DRAM devices. Films with the stoichiometry $Pr_{1.15}Al_{0.85}O_3$ have been deposited by conventional ALD using tris(*N,N′*,-diisopropylacetamidinato)praseodymium, $[Pr(Pr^iAMD)_3]$, $[AlMe_3]$ and H_2O.[66] The films were high purity (carbon not detected by RBS) and annealed films demonstrated excellent dielectric properties (low hysteresis, minimal ΔV_{FB}, very low leakage currents and $\kappa = 18$). Thin films of $PrAlO_x$ and $NdAlO_x$ have also been deposited by liquid injection ALD using the [Ln-Al] single-source precursor $[LnAl(OPr^i)_6(Pr^iOH)]_2$.[185] The films were carbon-free, and were also deficient in the lanthanide element (Pr : Al ~ 0.54–0.71; Nd : Al ~ 0.30–0.42) at all ALD growth temperatures. This resulted in less than optimum permittivities for the $PrAlO_x$ (Pr : Al = 0.71) and $NdAlO_x$(Nd : Al = 0.42) films of 14 and 7, respectively.

8.3.5.3 CVD of $La_xHf_{1-x}O_{2-\delta}$ and $La_xZr_{1-x}O_{2-\delta}$

Lanthanum hafnium oxide, $La_2Hf_2O_7$, has also attracted interest in gate dielectric applications as it remains amorphous to high temperatures (>950 °C), and has a relatively high κ value (~18).[188] A lanthanum hafnium oxide film of stoichiometry $La_{1.3}HfO_{4.1}$ has been deposited by liquid injection

MOCVD using [La(mmp)$_3$]/tetraglyme and [Hf(OBu^t)$_2$(mmp)$_2$].[189] The films remained amorphous to 900 °C and had good dielectric properties (low hysteresis, low density of interface states and $\kappa \sim 20$).

Similarly, $La_xZr_{1-x}O_{2-\delta}$ has been deposited by liquid injection MOCVD and ALD using the Cp-based precursors [(^{i}PrCp)$_3$La] and [(MeCp)$_2$ZrMe(OMe)].[190] $La_xZr_{1-x}O_{2-\delta}$ with low La concentrations ($x = 0.09$, 0.22) crystallized in the cubic phase on annealing at 700 °C, whereas at higher La concentrations ($x = 0.35$, 0.63), the films remained amorphous after annealing. The films deposited by ALD showed good dielectric properties with low hysteresis voltages and negligible flatband voltage shifts. The permittivity of the films ranged from 11 to 14, and leakage current densities were in the range 2.6×10^{-6} to 5.3×10^{-7} A cm^{-2}.

8.3.5.4 CVD of $LnScO_3$

Ternary lanthanide oxides such as $LaScO_3$, $GdScO_3$, $DyScO_3$ and $YScO_3$ have attracted great interest as potential high-κ gate oxides in MOSFETs.[191] $GdScO_3$ and $DyScO_3$ remain amorphous even after prolonged thermal annealing at high temperatures.[192] In the amorphous state $LnScO_3$ materials possess excellent dielectric properties such as high-permittivity, low leakage current and small hysteresis in the *C-V* curve. They also have a high band offset with silicon and they are thermally stable.

Amorphous $YScO_3$ films have been deposited by ALD using two precursor systems; [M(thd)$_3$] (M = Sc, Y) with O_3 or [(MeCp)$_3$Y] and [Cp$_3$Sc] with H_2O.[193] The deposition rate on Si using the [M(thd)$_3$] precursors at 335 and 350 °C was found to be very low (0.18 Å cycle^{-1}), while the Cp-based precursors gave much higher growth rates of 1.07 Å cycle^{-1} at substrate temperatures of 300 °C. Stoichiometric $YScO_3$ films of high purity (C < 0.2 at.%, H < 1 at.%) were obtained from both precursor systems.

8.3.6 CVD of TiO_2

Thin films of titanium oxide (TiO_2) have several potential applications in microelectronics. TiO_2 has a very large dielectric constant ($\kappa = 86$–170 for the rutile phase) and good thermodynamic stability with respect to silicon, making it a potential alternative to SiO_2 as the gate dielectric material in metal oxide semiconductor field effect transistors (MOSFETS).[76,194] Titanium oxide is also a component of several ferroelectric oxides such as $SrTiO_3$, (Ba,Sr)TiO_3 and Pb(Zr,Ti)O_3 used in infrared detectors and non-volatile computer memories.[1,12]

8.3.6.1 MOCVD of TiO_2

Titanium tetrachloride, [TiCl$_4$], has been widely employed as a precursor to TiO_2;[195–197] however, its use can lead to severe chloride contamination in MOCVD-grown films.[195] Titanium isopropoxide, [Ti(OPr^i)$_4$], has thus often been used for the MOCVD of high purity TiO_2 at relatively low substrate temperatures (< 500 °C).[198–202] However, [Ti(OPr^i)$_4$] contains an unsaturated four-coordinate Ti(IV) centre, which is highly reactive to air and moisture and can lead to problems in storage and handling, especially in solution-based liquid injection MOCVD applications.

Chelating β-diketonate groups have been inserted to increase the coordinative saturation of the Ti(IV) centre. [Ti(OPr^i)$_2$(acac)$_2$] has been used for the MOCVD of TiO_2,[96] and the related complex [Ti(OPr^i)$_2$(thd)$_2$] was used in preference to [Ti(OPr^i)$_4$] for the liquid injection MOCVD of Pb(Zr,Ti)O_3.[16] The structure of [Ti(OPr^i)$_2$(thd)$_2$] has not been determined, but NMR studies on similar Ti alkoxy-β-diketonates[203] indicate that it is probably a six-coordinate monomer in solution. The presence of the β-diketonate group gives [Ti(OPr^i)$_2$(thd)$_2$] a significantly higher thermal stability than [Ti(OPr^i)$_4$], and is a good match to the thermal stabilities of Ba, Sr and Zr β-diketonate co-precursors commonly used in multi-component oxide growth. Therefore, [Ti(OPr^i)$_2$(thd)$_2$] is now widely employed for the liquid injection MOCVD of Pb(Zr,Ti)O_3,[29,204,205] Pb(La,Ti)O_3,[206] $SrTiO_3$[204] and (Ba,Sr)TiO_3.[29]

Bidentate donor functionalized alkoxide ligands, such as dmae (Table 8.4), also increase the coordinative saturation at the metal centre, and [Ti(OPr^i)$_3$(dmae)] and [Ti(OPr^i)$_2$(dmae)$_2$] have been used to deposit TiO_2 at 300–450 °C by liquid injection MOCVD.[207] A liquid mixture of [Ta(OEt)$_4$(dmae)] (90%)/[Ti(OEt)$_2$(dmae)$_2$](10%) was used to deposit a Ta_2O_5/TiO_2 mixed oxide,[208] and [Ti(dmae)$_4$] has been used in combination with [Ba(thd)$_2$(pmdeta)] and [Sr(thd)$_2$(pmdeta)] to deposit (Ba,Sr)TiO_3 by liquid injection MOCVD.[91] [Ti(dmae)$_4$] was a more efficient low temperature source of Ti over the temperature range 420–480 °C than [Ti(OPr^i)$_2$(thd)$_2$], leading to (Ba,Sr)TiO_3 with improved uniformity.[209]

Diolate alkoxide ligands have also been used to increase the saturation on the Ti(IV) centre in Ti-oxide MOCVD precursors: [Ti(mpd)(thd)$_2$] (mpd = 2-methylpentane-2,4-diolate) (Table 8.4) has been used for the liquid injection MOCVD of (Ba,Sr)TiO_3[210] and the dimeric alkoxy diolate [Ti(OCMe$_2$CMe$_2$O)(dmae)$_2$]$_2$ has been used for the liquid injection MOCVD of TiO_2 over the temperature range 325–450 °C.[211] A similar complex, [Ti(mpd)(dmae)$_2$], has also been used for the liquid injection MOCVD of TiO_2,[212] and was found to extend the available temperature window for oxide growth compared with the conventional source [Ti(OPr^i)$_4$].[212] The methoxy bridged dimer [Ti(mpd)(mdop)(μ-OMe)]$_2$ (mdop = methyl 4,4-dimethyl-3-oxopentanoate) (Table 8.2) has also been used with [Ba(methd)$_2$] and [Sr(methd)$_2$] [methd = 2-(2-methoxyethoxy)-2,6,6-trimethylheptane-3,5-dionate] (Table 8.2) for the MOCVD of Ba(Sr,Ti)O_3, and it was reported that it leads to more efficient incorporation of the Ti component in the oxide.[210]

It has been claimed that the monomeric β-ketoiminate complex [Ti(2meip)$_2$] (Table 8.2), incorporates Ti more efficiently than [Ti(OPr^i)$_2$(thd)$_2$] at <500 °C, and is therefore a superior precursor for the MOCVD of (Ba,Sr)TiO_3.[213]

The anhydrous metal nitrate precursor [Ti(NO$_3$)$_4$] (Figure 8.16) is also an efficient source for the deposition of high purity TiO_2 at low substrate temperatures (230 °C).[93] The absence of carbon in the molecule was seen as an advantage for optimizing layer purity. [Ti(NMe$_2$)$_4$] has also been shown to be a good precursor for the growth of TiO_2 by conventional MOCVD. Films grown in the presence of oxygen were free from TiN and TiO_xN_y impurities. Films annealed at 600 °C in air formed the anatase phase and had dielectric constants of ~ 50.[214]

(Zr,Sn)TiO_4 has a dielectric constant of 20–27, and so has potential as a high-κ dielectric material in CMOS and DRAM devices.[215] There have been very few MOCVD studies on (Zr,Sn)TiO_4, but it has been deposited by liquid injection MOCVD using a "cocktail" of the liquid precursors [Zr(OBu^t)$_4$], [Sn(OBu^t)$_4$] and [Ti(OBu^t)$_4$].[216] A range of "single source" [Ti-Sn] precursors, [Ti(OPr^i)$_3${N(SnMe$_3$)$_2$}], [Ti(OBu^t)$_3${N(SnMe$_3$)$_2$}] and [Ti(OPr^i)$_3${N(SnMe$_3$)(SiMe$_3$)}] have also been investigated, using [Zr(OBu^t)$_4$] as the Zr precursor.[217]

8.3.6.2 ALD of TiO_2

The most commonly used precursor for the ALD of TiO_2 is $TiCl_4$,[218–221] usually with H_2O as the oxygen precursor, although hydrogen peroxide has been used. Problems associated with this process are the corrosive nature of $TiCl_4$, the liberation of HCl as a reaction by-product, and the possibility of chloride residues in the film. Although TiI_4 has been used,[222,223] and is not as corrosive as $TiCl_4$, similar purity concerns remain.

Titanium alkoxides are halide-free and are not corrosive, and so [Ti(OPr^i)$_4$],[219,224–226] [Ti(OEt)$_4$],[227,228] [Ti(OMe)$_4$][229] and [Ti(OBu^t)$_4$] have been investigated using H_2O as the oxygen source. It was established that the precursor decomposition onset temperature varied in the order

[Ti(OBu^t)$_4$]<[Ti(OPr^i)$_4$]<[Ti(OEt)$_4$]<[Ti(OMe)$_4$], and the relatively high stability of [Ti(OMe)$_4$] allowed ALD over the widest temperature range. The resulting films deposited at 250°C and above were polycrystalline anatase, whilst amorphous films were deposited at 200 °C.[229]

8.3.7 CVD of Al_2O_3

Thin films of aluminium oxide, Al_2O_3, have applications in electronics as passivating layers and as an alternative gate dielectric oxide.[192] Al_2O_3 has been investigated as an alternative gate dielectric due to several favourable properties.[76] These include, high band gap (8.7 eV), good thermodynamic stability on Si and it remains amorphous under most conditions encountered in microelectronics processing. Although its permittivity is relatively low ($\kappa \sim 8$–10) Al_2O_3 films have shown a reduction in leakage current of ~ 100 times compared with SiO_2 of the same equivalent thickness; Al_2O_3 has proved to be a good short-term alternative to SiO_2, especially in DRAM devices.[76]

8.3.7.1 MOCVD of Al_2O_3

Early studies into the MOCVD of Al_2O_3 involved the hydrolysis or oxidation of aluminium trihalides, such as [$AlCl_3$].[230,231] However, the high deposition temperatures required (typically ~700–950 °C) are unsuitable for microelectronics applications. Lower deposition temperatures (400–700 °C) have been obtained by the use of [$AlMe_3$] in the presence of either O_2[232,233] or N_2O.[234,235] [$AlEt_3$] has also been used for the MOCVD of Al_2O_3 at deposition temperatures below 400 °C.[77]

The high reactivity of [AlR_3] compounds to O_2 can lead to problems of premature reaction and to particulate contamination of the films. Incomplete oxidation of the [AlR_3] precursor can also lead to the incorporation of residual carbon in the films. The use of aluminium isopropoxide, [Al(OPr^i)$_3$], with O_2[236,237] or N_2O[238] avoids the problem of pre-reaction and allows the MOCVD of Al_2O_3 at 200–750 °C. However, [Al(OPr^i)$_3$] is a trimeric solid (mp 118 °C) and tends to oligomerize on prolonged storage, leading to changes in vapour pressure that have a significant effect on the deposition rate of the Al_2O_3 film. A solution to this problem has been to form [Al(OPr^i)$_3$] in situ in the gas-phase by mixing vapours of [$AlMe_3$] and Pr^{i}OH upstream from the reactor cell, and this allowed the growth of high purity Al_2O_3 (C<0.5 at.%) by conventional MOCVD over the temperature range 400–600 °C.[239] [Al(acac)$_3$] has also been investigated. Films grown in the absence of oxygen,[240] or in the presence of air,[241] showed heavy carbon contamination, although those grown using [Al(acac)$_3$] with O_2 or H_2O were of higher quality.[242,243]

8.3.7.2 ALD of Al_2O_3

Early studies used various Al precursors for the ALD of Al_2O_3, including [$AlCl_3$], [Al(OEt)$_3$], [Al(OPr^n)$_3$] and [$AlMe_3$] in combination with H_2O, O_3, H_2O_2 or O_2.[244–247] More recently [$AlMe_3$] has generally been used,[248] because its high reactivity to H_2O coupled with its relatively high thermal stability makes it an almost ideal ALD precursor. However, [$AlMe_3$] is less well suited to liquid injection ALD and so the much less reactive [Al(mmp)$_3$] has been used with H_2O.[249] Self-limiting growth was observed at 250 °C, but trace carbon (2.4 at.%) was observed in the Al_2O_3 films.

8.3.8 CVD of Ta_2O_5

8.3.8.1 MOCVD of Ta_2O_5

Owing to its high dielectric constant (κ ~ 25), high thermodynamic stability, low leakage currents and high breakdown voltage, thin films of Ta_2O_5 have been intensively investigated as an

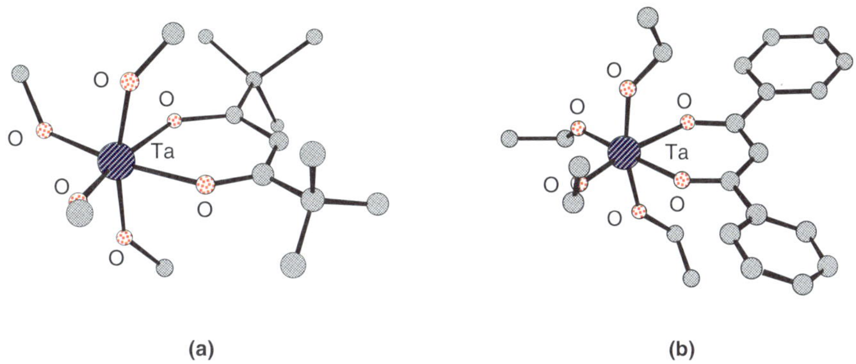

Figure 8.23 Crystal structures of (*a*) [Ta(OMe)$_4$(thd)] and (*b*) [Ta(OEt)$_4$(dbm)].

alternative to SiO_2 in 64 and 256 Mbit DRAM applications and CMOS gate dielectric applications.[250] Although [$TaCl_5$],[251,252] [Ta(thd)$_4$Cl],[253] [TaF_5][254] and [Ta(OMe)$_5$][255,256] have been investigated, the most commonly used precursor for the MOCVD of Ta_2O_5 is tantalum ethoxide, [Ta(OEt)$_5$], in the presence of oxygen.[250,257–261] Films of Ta_2O_5 have been deposited by low pressure MOCVD at 340–400 °C.[257]

[Ta(OEt)$_5$] is an ethoxide bridged dimeric molecule with a relatively low vapour pressure ($\sim$0.075 Torr at 108 °C),[8] but the addition of the chelating dmae ligand ($OCH_2CH_2NMe_2$) yields the monomeric [Ta(OEt)$_4$(dmae)] complex (as shown by ^{1}H NMR), which has a significantly higher vapour pressure than [Ta(OEt)$_5$].[8]

[Ta(OEt)$_4$(dmae)] has also been used with [Ti(OEt)$_2$(dmae)$_2$] to deposit high-permittivity (TiO_2)-(Ta_2O_5) films by liquid injection MOCVD.[208] The mixed alkoxide-β-diketonate complexes [Ta(OR)$_4$(thd)] (R = Me, Et, Pr^i) (Figure 8.23(a)) have a higher thermal stability than the parent [Ta(OR)$_5$] complexes, and are a better "thermal match" to [Sc(thd)$_3$] (Section 8.4.2).

Ta_2O_5 films, with similar carbon content and leakage currents to those grown using [Ta(OEt)$_5$] have been grown by MOCVD using the alkylamide precursor [Ta(NMe$_2$)$_5$] in the presence of a large excess of oxygen.[262]

8.3.8.2 *ALD of Ta_2O_5*

There have been very few reports on the ALD of Ta_2O_5, but [Ta(OEt)$_5$] has been investigated in the temperature range 150–450 °C, using H_2O as the oxygen precursor.[263] Because of the thermal decomposition of [Ta(OEt)$_5$], self-limiting ALD growth was achieved only below 350 °C. All the grown films were amorphous and were stoichiometric Ta_2O_5 with carbon and hydrogen impurity levels of up to 4 and 3 at.%, respectively. Permittivities of films grown at 250 and 325 °C were 21 and 25, respectively.

8.4 CVD of Ferroelectric Metal Oxides

Ferroelectric oxides are an important class of materials. The arrangement of cations and anions within a ferroelectric gives rise to polar axes along which a spontaneous polarization can occur below the material's Curie temperature. Applying a sufficiently large electric field can switch the polarization direction and the resulting polarization can be measured *via* the material's surface charge. Consequently, these oxides have potential applications in non-volatile ferroelectric random access memory (NV-FERAM) devices,[1] which exploit this capacity of the material to be polarized in two opposite directions, used as the 0 and 1 binary states in the oxide memory device.

8.4.1 CVD of $Pb(Zr,Ti)O_3$

The lead-based perovskite $Pb(Zr,Ti)O_3$ has an exceptionally high remnant polarization, and has been extensively investigated for NV-FERAM applications.[264]

There have been many studies into the MOCVD of $Pb(Zr,Ti)O_3$. The monomeric $[Pb(thd)_2]$ complex (Figure 8.24) is the preferred lead precursor,[8] due to its reduced toxic risk compared to the more volatile alkyls, $[PbEt_4]$ or $[Et_3Pb(OCH_2C(CH_3)_3)]$, which present a severe risk of toxic inhalation.

The most widely used precursor combination has been $[Pb(thd)_2]$, $[Zr(thd)_4]$ and $[Ti(OPr^i)_2(thd)_2]$.[16,204] However, the vapour-phase transport and MOCVD process characteristics of these precursors are not well matched. $[Zr(thd)_4]$ is much less volatile than $[Pb(thd)_2]$ and has a significantly higher thermal stability, depositing oxide at higher substrate temperatures.[95] This leads to compositional uniformity problems in the resulting $Pb(Zr,Ti)O_3$ film, and to blockages in the injector and inlet pipe-work of the MOCVD reactor. Consequently, efforts have been made to "tailor" the Zr precursor to improve its volatility and make its thermal decomposition behaviour more compatible with $[Pb(thd)_2]$.

The heteroleptic Zr complex $[Zr(OPr^i)_2(thd)_2]$ was proposed as an alternative to $[Zr(thd)_4]$ for the liquid injection MOCVD of ZrO_2 and $Pb(Zr,Ti)O_3$, with the aim of combining the high ambient stability of Zr β-diketonates with the lower thermal stability and higher volatility of Zr alkoxides.[18] It was shown that $[Zr(OPr^i)_2(thd)_2]$ had a significantly lower thermal stability than $[Zr(thd)_4]$, with maximum oxide growth rates occurring at $\sim 500\,^{\circ}C$ compared with $\sim 600\,^{\circ}C$ for $[Zr(thd)_4]$.[18] This leads to a marked improvement in the quality of $Pb(Zr,Ti)O_3$ films deposited by liquid injection MOCVD.[29] Two sets of precursors were investigated, $[Pb(thd)_2(pmdeta)]/[Zr(thd)_4]/[Ti(OPr^i)_2(thd)_2]$ and $[Pb(thd)_2(pmdeta)]/[Zr(OPr^i)_2(thd)_2]/[Ti(OPr^i)_2(thd)_2]$ (pmdeta=pentamethyl-diethylenetriamine), and it was shown that Zr incorporation efficiency in $Pb(Zr,Ti)O_3$ was significantly higher in films grown using $[Zr(OPr^i)_2(thd)_2]$ than in films grown using $[Zr(thd)_4]$, and was much closer to the Pb and Ti incorporation efficiencies from $[Pb(thd)_2(pmdeta)]$ and $[Ti(OPr^i)_2(thd)_2]$. This led to improved control of the $Pb(Zr,Ti)O_3$ film composition with excellent run-to-run reproducibility, and the $Pb(Zr,Ti)O_3$ films grown from the new $[Zr(OPr^i)_2(thd)_2]$ source also demonstrated good ferroelectric properties.[29]

Variable-temperature ^1H NMR studies indicated that $[Zr(OPr^i)_2(thd)_2]$ in toluene solution exists as a dynamic mixture of three species, $[Zr(OPr^i)_4]$, $[Zr(thd)_4]$ and $[Zr(OPr^i)_3(thd)]$. Recrystallization

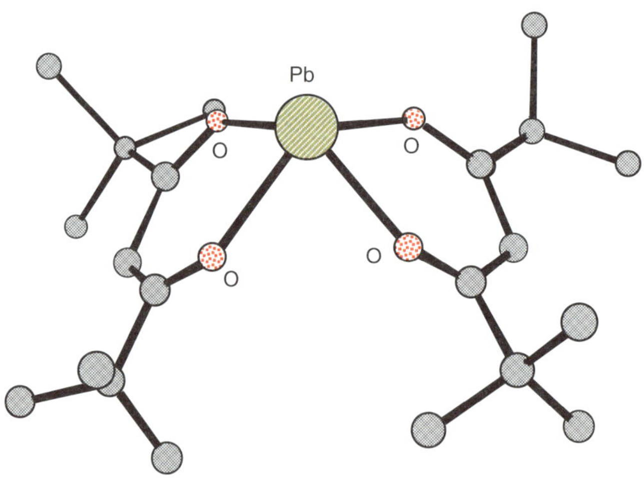

Figure 8.24 Crystal structure of $[Pb(thd)_2]$.

of the solid products from this solution yielded the symmetric dimer $[Zr_2(OPr^i)_6(thd)_2]$ (Figure 8.25) as the most stable complex.[265]

$[Zr_2(OPr^i)_6(thd)_2]$ has significant advantages over $[Zr(thd)_4]$. For instance, it is more volatile (see TGA data in Figure 8.26), and it deposits oxide films over a wider temperature range.

The physical properties (*e.g.* volatility, thermal stability) of $[Zr_2(OPr^i)_6(thd)_2]$ are well-matched to those of $[Pb(thd)_2]$ with the optimum temperature range for oxide deposition from $[Zr_2(OPr^i)_6(thd)_2]$ (300–600 °C) overlapping with the oxide deposition temperature zones of $[Pb(thd)_2]$ and $[Ti(OPr^i)_2(thd)_2]$.[14] $[Zr_2(OPr^i)_6(thd)_2]$ proved to be a good precursor for the deposition of ZrO_2 by liquid injection MOCVD,[107] and $[Zr_2(OPr^i)_6(thd)_2]$ has been used with $[Pb(thd)_2]$ and $[Ti(OPr^i)_2(thd)_2]$ for the liquid injection MOCVD of $Pb(Zr,Ti)O_3$. The $Pb_{1.1}(Zr_{0.75}Ti_{0.25})_{0.9}O_3$ films demonstrated high compositional uniformity and excellent surface morphology.[266] Polarization–voltage hysteresis data for the films showed high remnant polarization values and low coercive field strengths, $-E_c$ and E_c, of the order of –23 and +20 kV cm^{-1}, respectively, characteristic of high quality ferroelectric films.[1,29]

A fluorinated mixed alkoxide/β-diketonate, $[Zr(OPr^i)_2(hfip)_2]$ (hfip = $OCH(CF_3)_2$) which is probably a monomeric six-coordinate complex, has also been proposed as a possible precursor for ZrO_2 and $Pb(Zr,Ti)O_3$,[267] but trace fluorine contamination was observed under certain growth conditions in ZrO_2 films deposited by conventional MOCVD, which is a serious concern in microelectronics applications.

8.4.2 CVD of $Pb(Sc_{0.5}Ta_{0.5})O_3$

The pyroelectric material lead scandium tantalate, $Pb(Sc_{0.5}Ta_{0.5})O_3$, has exceptionally good pyroelectric figures of merit, with a significant increase in detectivity compared with a standard pyroelectric material (*e.g.* modified lead zirconate). Therefore $Pb(Sc_{0.5}Ta_{0.5})O_3$ has a range of applications in un-cooled thermal imaging such as night-sight technologies and fire detection.

Conventional precursors for the MOCVD of $Pb(Sc_{0.5}Ta_{0.5})O_3$ include $[Pb(thd)_2]$, $[Sc(thd)_3]$ and $[Ta(OEt)_5]$.[268] However, these are generally incompatible for liquid injection MOCVD applications. For instance, $[Sc(thd)_3]$ (Figure 8.27a) deposits oxide films at much higher temperatures

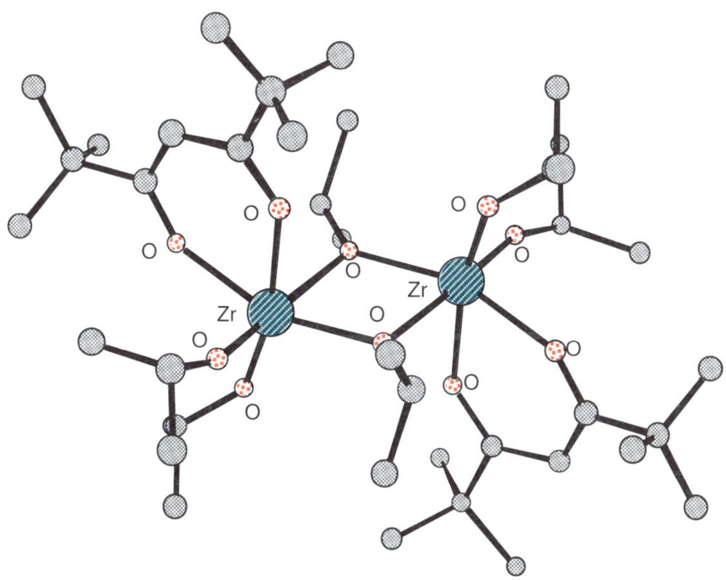

Figure 8.25 Crystal structure of $[Zr_2(OPr^i)_6(thd)_2]$.

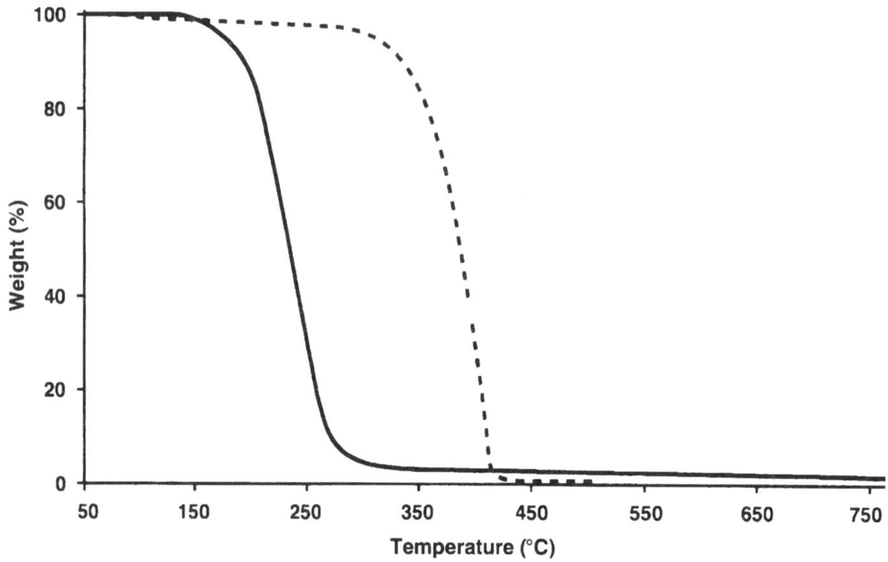

Figure 8.26 TGA data for $[Zr_2(OPr^i)_6(thd)_2]$ (——) and $[Zr(thd)_4]$(----).

(550–600 °C) than the less thermally stable $[Ta(OEt)_5]$ precursor (oxide deposition at 400–450 °C), leading to $Pb(Sc_{0.5}Ta_{0.5})O_3$ with poor layer uniformity. $[Sc(thd)_3]$ is also much less volatile than $[Ta(OEt)_5]$ and requires an evaporator temperature of >250 °C to give acceptable oxide growth rates, which can lead to the decomposition of $[Ta(OEt)_5]$. Consequently, a more thermally stable Ta source has been sought for the reproducible liquid injection MOCVD of $Pb(Sc_{0.5}Ta_{0.5})O_3$.

The addition of a β-diketonate group to $[Ta(OR)_5]$ results in more thermally stable complexes of the type $[Ta(OR)_4(acac)]$ and $[Ta(OR)_4(thd)]$.[269,270] Accordingly, the optimum oxide deposition temperature from $[Ta(OMe)_4(acac)]$ and $[Ta(OEt)_4(thd)]$ is higher than from the parent $[Ta(OR)_5]$ complexes, and increases in the order $[Ta(OMe)_4(acac)] < [Ta(OPr^i)_4(thd)] < [Ta(OEt)_4(thd)] < [Ta(OMe)_4(thd)]$. These $[Ta(OR)_4(\beta\text{-diketonate})]$ precursors are monomeric six-coordinate octahedral complexes and are generally more volatile and less air-sensitive than the parent alkoxides.[269] These precursors are also a better "thermal match" to $[Sc(thd)_3]$ than the parent $[Ta(OR)_5]$ precursors, and undergo much less decomposition in the evaporator during liquid injection MOCVD. This leads to major process advantages and $Pb(Sc_{0.5}Ta_{0.5})O_3$ has been deposited at 500–600 °C using the optimum precursor combination $[Pb(thd)_2]$, $[Sc(thd)_3]$ and $[Ta(OMe)_4(thd)]$ in a single THF solution.[271]

$[Ta(OEt)_4(dbm)]$ (Figure 8.23b) has an even higher thermal stability than other $[Ta(OR)_4(\beta\text{-diketonate})]$ complexes, and is an even better "thermal match" to $[Sc(thd)_3]$, having a similar optimum temperature for oxide deposition (550–600 °C).[272] The remarkable thermal stability of the $[Ta(OEt)_4(dbm)]$ complex was attributed to resonance stabilization of the [C–O] bonds in the β-diketonate group by delocalization of negative charge into an extended conjugated electron system involving the π-orbitals of the phenyl groups in the dibenzoyl-methanate ligand. The analogous Nb complex $[Nb(OEt)_4(dbm)]$ showed a similar increase in thermal stability relative to $[Nb(OEt)_5]$.[272]

The physical properties of the scandium β-ketonate precursor have been improved by modification of the β-diketonate ligand. Substitution of an iso-butyl group for a *tert*-butyl group in $[Sc(thd)_3]$ leads to the complex $[Sc(tmod)_3]$ (see ligand nomenclature in Table 8.2). The complex exists as a mixture of *fac* and *mer* isomers (Figure 8.27b and c). TGA data (Figure 8.28) has shown that $[Sc(tmod)_3]$ is more volatile than $[Sc(thd)_3]$ and $[Sc(mhd)_3]$ (Table 8.2), possibly due to a disruption of the molecular packing of the scandium β-diketonate complex in the solid state.[273]

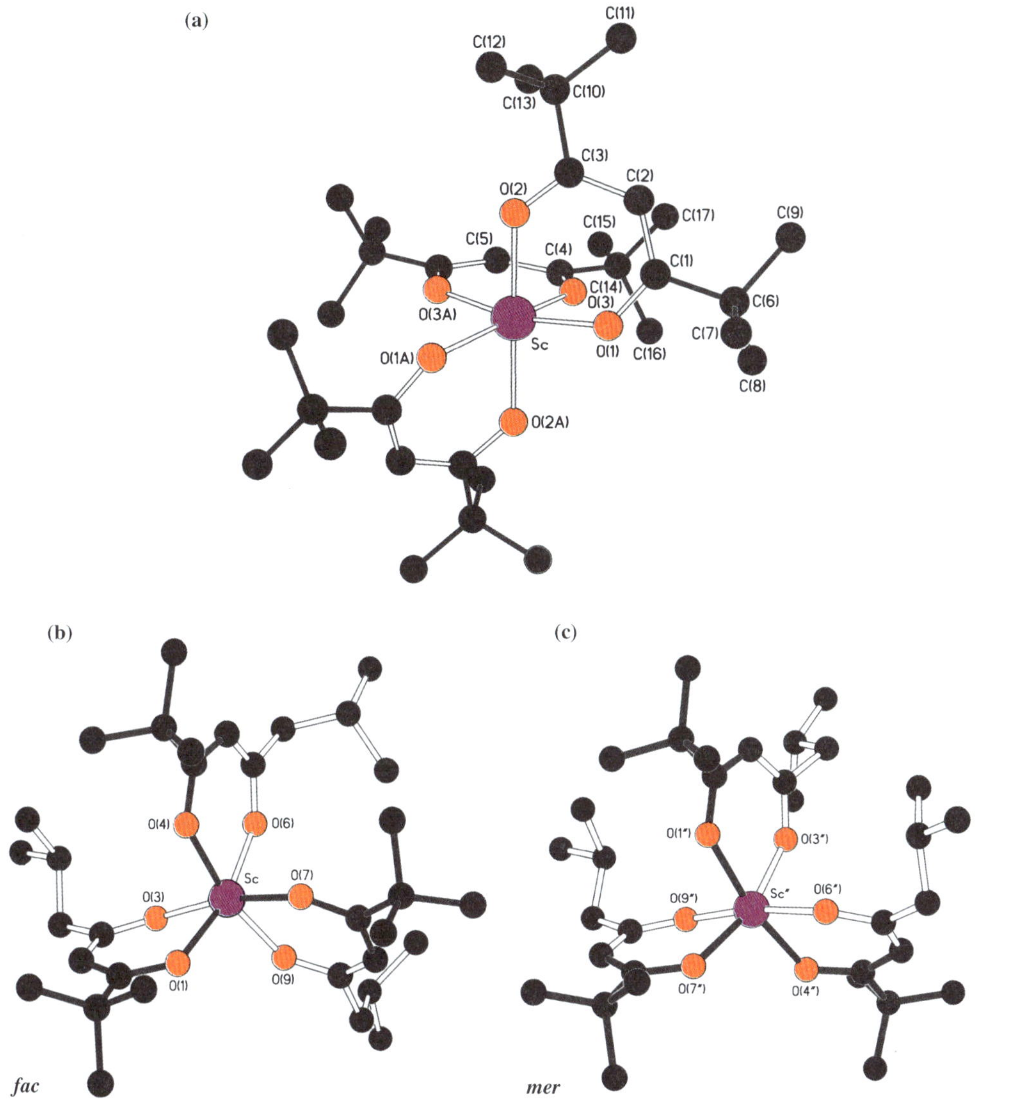

Figure 8.27 Molecular structures of (*a*) [Sc(thd)$_3$]; (*b*) [Sc(tmod)$_3$] (*fac* isomer) and (*c*) [Sc(tmod)$_3$] (*mer* isomer)

8.4.3 CVD of Pb(Mg$_{0.33}$Nb$_{0.66}$)O$_3$

The perovskite phase of lead magnesium niobate, Pb(Mg$_{0.33}$Nb$_{0.66}$)O$_3$ (PMN), has a high dielectric constant, low thermal expansion, broadened dielectric maxima and a high electrostrictive strain coefficient.[274] This makes it attractive for various micro-electromechanical (MEM) and electronic applications, such as in transducers, ceramic capacitors and DRAMs.[275] There have been relatively few MOCVD studies into Pb(Mg$_{0.33}$Nb$_{0.66}$)O$_3$, and most have employed conventional precursors such as [PbEt$_4$], [Mg(acac)$_2$], [Mg(thd)$_2$] and [Nb(OEt)$_5$].[276,277] To solve problems of incompatibility between the Nb-alkoxide precursor and the Mg-β-diketonates (the alkoxide is more volatile and decomposes at much lower temperatures), a dimeric "single-source" precursor, [Mg{Nb(OEt)$_6$}$_2$]-(EtOH)$_2$] (Figure 8.29), containing Mg and Nb in the required 1 : 2 ratio, has been investigated using

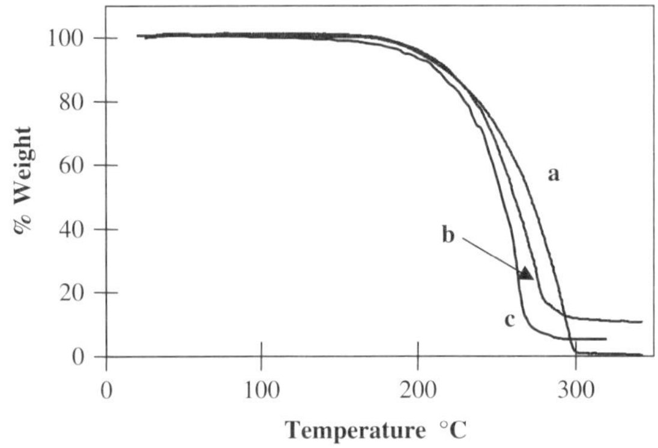

Figure 8.28 TGA data for (a) [Sc(thd)$_3$], (b) [Sc(mhd)$_3$] and (c) [Sc(tmod)$_3$].

liquid injection MOCVD.[278] At low temperature the films were Mg-deficient, but PMN films containing the required 1 : 2 Mg : Nb ratio were obtained at high substrate temperatures (500–600 °C).

8.4.4 CVD of $SrBi_2(Ta_xNb_{1-x})_2O_9$

Thin films of the layered perovskites strontium bismuth tantalate, $SrBi_2Ta_2O_9$, strontium bismuth niobate, $SrBi_2Nb_2O_9$, and strontium bismuth tantalate niobate, $SrBi_2(Ta_xNb_{1-x})_2O_9$, have important potential applications in non-volatile ferroelectric (FERAM) computer memories,[279,280] owing to their low operating voltage and high resistance to polarization fatigue. The application of liquid injection MOCVD techniques to the deposition of $SrBi_2Ta_2O_9$ and $SrBi_2(Ta_xNb_{1-x})_2O_9$ has been restricted by a lack of stable and compatible precursors. Conventional precursors include [Sr(thd)$_2$], [Bi(C_6H_5)$_3$], [BiMe$_3$], [Bi($OC_5H_{11}{}^t$)$_3$] and [Ta(OEt)$_5$] or [Ta(OPr^i)$_4$(thd)],[280,281] but these are generally incompatible, having very different physical properties and decomposition characteristics that can lead to poor layer uniformity. A potential solution to this problem is the use of "single-source" precursors that contain elements required in the oxide film in a single molecule; accordingly, the Sr-Ta and Sr-Nb heterometal alkoxides [Sr{Ta(OEt)$_6$}$_2$], [Sr{Ta(OPr^i)$_6$}$_2$] and [Sr{Nb(OEt)$_6$}$_2$][280,282] have been used for the MOCVD of $SrBi_2Ta_2O_9$ and $SrBi_2(Ta_xNb_{1-x})_2O_9$. Advantageously, the metal : metal ratio in the precursor matches that in the deposited oxide film, but the vastly different volatilities of the [Sr(OR)$_2$] and [Ta(OR)$_5$] fragments may partition the Sr and Ta components, leading to an excess of the more volatile Ta component in the gas phase. Also, the [Sr{Ta(OR)$_6$}$_2$] precursors contain an unsaturated Sr centre, making them susceptible to attack by moisture and reducing their shelf-life. Therefore donor-functionalized alkoxide ligands, such as 2-dimethylaminoethoxide (dmae, $OCH_2CH_2NMe_2$) have been inserted into the heterometal alkoxide to increase the ligand coordination around the central Sr atom, and render the complexes less susceptible to decomposition.

The complexes [Sr{Ta(OEt)$_5$(dmae)}$_2$] and [Sr{Ta(OEt)$_5$(bis-dmap)}$_2$] (Figure 8.30) have been used in liquid injection MOCVD to deposit stoichiometric strontium tantalate, $SrTa_2O_6$.[283] The use of [Sr{Ta(OEt)$_5$(dmae)}$_2$] leads to improved stoichiometry control in $SrBi_2Ta_2O_9$ grown by liquid injection MOCVD,[284] and $SrBi_2Ta_2O_9$ films with good ferroelectric properties have been grown by liquid injection MOCVD using [Sr{Ta(OEt)$_5$(dmae)}$_2$] and [Bi(C_6H_5)$_3$].[284] Other donor functionalized alcohols such as 2-methoxyethanol (mee, OCH_2CH_2OMe) are also effective in stabilizing Sr-Ta and Sr-Nb heterometal alkoxides, and it has been shown that, in contrast to [Sr{Ta(OPr^i)$_6$}$_2$], the Sr : Ta mole ratio remains constant during the distillation of [Sr{Ta(OEt)$_5$(mee)}$_2$].[285] The analogous Sr-Nb complex [Sr{Nb(OEt)$_5$(dmae)}$_2$] has also been synthesized,[286] although its use for the MOCVD of $SrBi_2(Ta_xNb_{1-x})_2O_9$ has not yet been demonstrated.

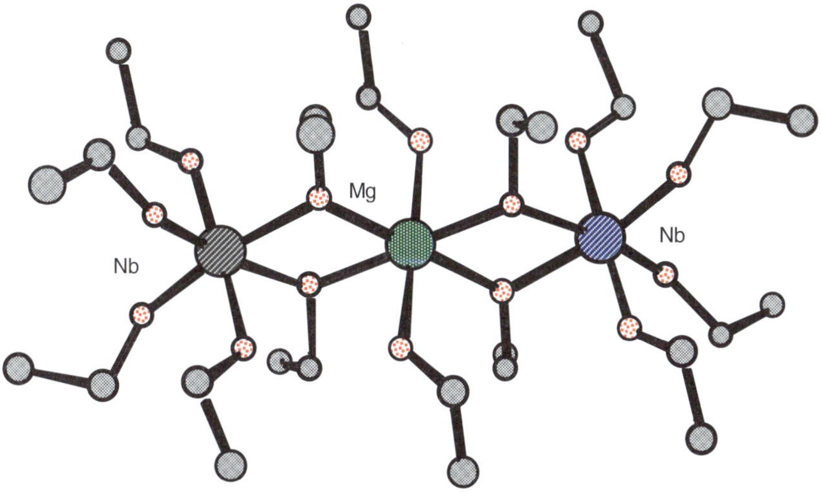

Figure 8.29 Molecular structure of $[Mg\{Nb(OEt)_6\}_2(EtOH)_2]$.

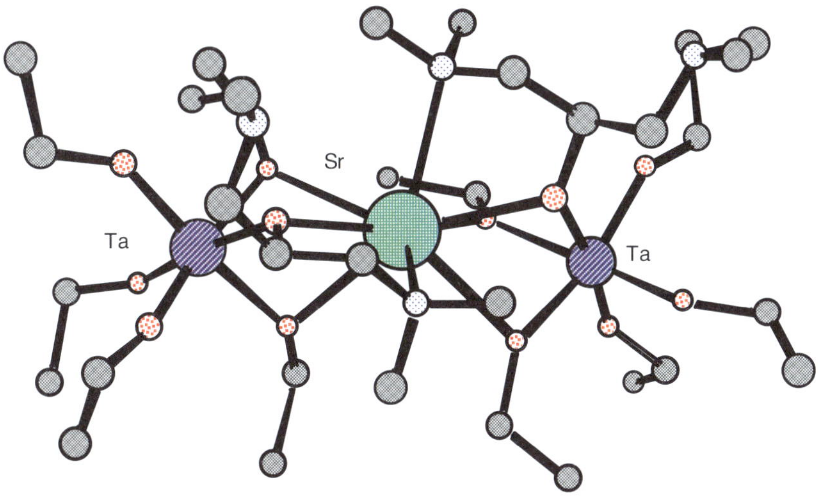

Figure 8.30 Molecular structure of $[Sr\{Ta(OEt)_5(\text{bis-dmap})\}_2]$.

The progress in heterometal alkoxide chemistry described above led to a search for a compatible Bi source. The conventional sources $[Bi(C_6H_5)_3]$ and $[Bi(thd)_3]$ are not particularly compatible with Sr-Ta and Sr-Nb alkoxides, having significantly higher thermal stabilities and different evaporation characteristics. Bismuth alkoxides are potentially more suitable co-precursors, as they are likely to have similar volatility and decomposition characteristics to Sr-Ta and Sr-Nb heterometal alkoxides. However, simple bismuth alkoxides, $[Bi(OR)_3]$ ($R = Me$, Et, Pr^i) are poorly characterized and have limited stability and volatility, subliming in only very low yields under high vacuum.[287] This is due to the high tendency of the Bi(III) centre to expand its coordination sphere to six, which favours the formation of polymeric $[Bi(OR)_3]_n$ aggregates. To inhibit oligomerization, sterically demanding alkoxide ligands have again been employed and $[Bi(OBu^t)_3]$ has been shown to be a volatile monomeric species,[288] but is highly air sensitive and has only a limited stability in solution. $[Bi(OC_5H_{11}{}^t)_3]$ is also volatile, and has been used with $[Sr\{Ta(OEt)_6\}_2]$ for the liquid injection MOCVD of

$SrBi_2Ta_2O_9$.[282] The sterically hindered ligand mmp inhibits metal alkoxide polymerization, and has been used to form the volatile monomeric complex $[Bi(mmp)_3]$[289,290] (Figure 8.31).

$[Bi(mmp)_3]$ evaporates in a similar temperature range to $[Sr\{Ta(OEt)_5(dmae)\}_2]$, and was shown to deposit oxide at similar substrate temperatures.[290] $[Sr\{Ta(OEt)_5(dmae)\}_2]$ and $[Bi(mmp)_3]$ have been used to deposit $SrBi_2Ta_2O_9$ at 500 °C by liquid injection MOCVD.[291] The simultaneous evaporation of $[Sr\{Ta(OEt)_5(dmae)\}_2]$ and $[Bi(mmp)_3]$ precursor solutions led to a competition between the incorporation of Sr and Bi in the deposited oxide. To overcome this, a series of $SrTaO_x/Bi_2O_3$ superlattices were deposited, followed by post-deposition annealing.[291]

There have been few attempts to deposit $SrBi_2Ta_2O_9$ by ALD, presumably due to the difficulty in controlling three separate deposition cycles of Sr precursor/H_2O, Bi precursor/H_2O and Ta precursor/H_2O. However, the use of the [Sr-Ta] "single-source" precursor offers a simpler route. Stoichiometric $SrTa_2O_6$ was successfully deposited by ALD using $[Sr\{Ta(OEt)_5(dmae)\}_2]$ and H_2O,[292] and the same precursor was used with $[Bi\{N(SiMe_3)_2\}_3]$ to deposit Sr-Bi-Ta-O at 190–200 °C by ALD.[293] The as-deposited films were amorphous, but post-deposition annealing at 800 °C in oxygen gave the layered perovskite phase of $SrBi_2Ta_2O_9$.

8.4.5 CVD of Bismuth Titanate ($Bi_4Ti_3O_{12}$)

Bismuth titanate, $Bi_4Ti_3O_{12}$, is a layered perovskite exhibiting several useful ferroelectric and electro-optic properties, and has potential applications in non-volatile memory devices[279,294] and ferroelectric-effect transistors. Various precursor combinations have been used for its MOCVD. These include $BiPh_3$ in combination with $[Ti(OEt)_4]$,[295] $[Ti(OBu^n)_4]$[296,297] and $[Ti(OPr^i)_4]$[298,299] and $BiMe_3$ in combination with $[Ti(OPr^i)_4]$.[300] Similar to other Bi-containing oxides, such as $SrBi_2(Ta_xNb_{1-x})_2O_9$, the Bi precursor has several disadvantages that have hampered progress in the development of an MOCVD process for $Bi_4Ti_3O_{12}$ growth. For instance, $[BiPh_3]$ has a significantly higher thermal stability than Ti alkoxides, leading to a marked increase in the Bi : Ti ratio with increasing substrate temperature,[299] which causes problems in controlling compositional

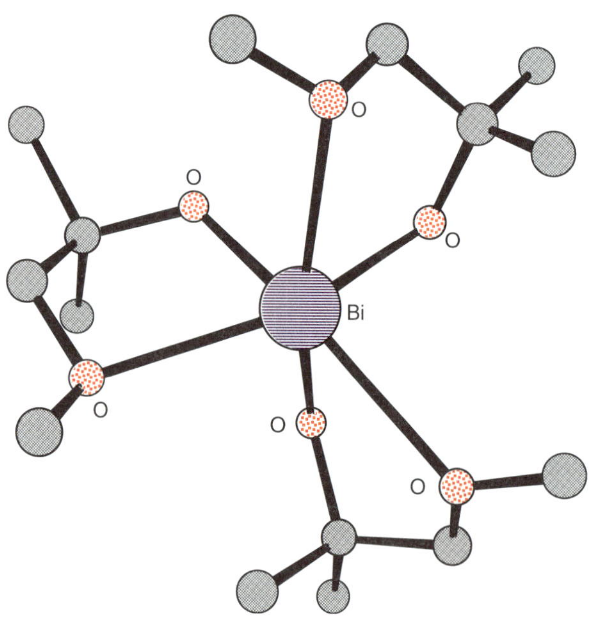

Figure 8.31 Molecular structure of $[Bi(mmp)_3]$.

uniformity across the substrate. Although the less thermally stable complex [$BiMe_3$] is a better match to the [$Ti(OR)_4$] precursor, it is highly reactive and potentially explosive, which inevitably limits its widespread application. However, [$Bi(mmp)_3$] is a good match to [$Ti(OR)_4$] precursors and it has been used with [$Ti(OPr^i)_2(mmp)_2$] for the liquid injection MOCVD of Bi-titanate.

There have been few studies on the ALD of $Bi_4Ti_3O_{12}$. Although attempts to deposit Bi_2O_3 using [$BiPh_3$] and H_2O failed, when [$BiPh_3$]/H_2O cycles were mixed with [$Ti(OPr^i)_4$]/H_2O cycles Bi-Ti-O films were deposited.[301] However, Bi was initially incorporated into the film in metallic form and post-deposition was required to obtain the oxide. In addition, Bi incorporation was limited to a Bi : Ti ratio of only 0.6. A liquid injected mixture of [$Bi(mmp)_3$] and [$Ti(mmp)_4$] with an O_2/O_3 mixture has been used for the ALD of $Bi_4Ti_3O_{12}$.[302] Growth rates of 0.5–1 Å cycle^{-1} were reported at substrate temperatures of 250–350 °C, but no proof of self-limiting growth was given.

8.4.6 CVD of $SrTiO_3$ and (Ba,Sr)TiO_3

Owing to their high dielectric constant, good thermal stability and good high frequency characteristics, $SrTiO_3$ and (Ba,Sr)TiO_3 have attracted particular attention as capacitor materials for giga-bit scale dynamic random-access memories.[303] In this application, the dielectric layer must be crystalline, to obtain the very high permittivities typically observed ($\kappa > 300$). $SrTiO_3$ and (Ba,Sr)TiO_3 have also received attention as alternative gate dielectric materials in CMOS transistors,[76] but the thermodynamic instability of $SrTiO_3$ on Si makes it necessary to deposit epitaxial layers, which requires precise control of deposition at the sub-monolayer level. Nevertheless, transistors showing encouraging results have been fabricated using $SrTiO_3$ gate dielectrics.[304,305]

8.4.6.1 MOCVD of $SrTiO_3$ and (Ba,Sr)TiO_3

Various precursors have been used for the MOCVD of $SrTiO_3$, $BaTiO_3$ and (Ba,Sr)TiO_3. For instance, $SrTiO_3$ has been grown by conventional (bubbler-based) MOCVD using [$Sr(thd)_2$] and [$Ti(OPr^i)_4$].[306–308] Relative dielectric constants as high as 160 were obtained for a $SrTiO_3$ thickness above 20 nm. A SiO_2 equivalent thickness of 0.48 nm was obtained from the thicker films.[306] Similarly, a 110 nm $SrTiO_3$ film had a dielectric constant of 210, and a SiO_2 equivalent thickness of 1.1 nm was demonstrated by a 40 nm thick $SrTiO_3$ film.[307] A disadvantage of [$Sr(thd)_2$] is that it has a relatively low volatility and requires source temperatures in excess of 200 °C. Moreover, there is a tendency towards thermally-induced oligomerization during film growth, resulting in involatile products and instability in transport properties. To avoid this the fluorinated precursor [$Sr(hfac)_2$(tetraglyme)] has been used, in combination with [$Ti(OPr^i)_4$].[309] Unlike the non-fluorinated Sr- and Ba-β-diketonates, stable adducts are formed with polyethers such as tetraglyme and diglyme, and [$Sr(hfac)_2$(tetraglyme)] could be evaporated at a source temperature approx. 100 °C lower than that typically used for [$Sr(thd)_2$], and there was no degradation of the precursor during extended use. Epitaxial $SrTiO_3$ films were deposited by conventional MOCVD on (100)$LaAlO_3$ substrates at 810 °C, and there was no evidence of fluorine contamination (XRD, AES measurements).[309] An alternative solution to the problem of the thermal decomposition of [$Sr(thd)_2$] is to use liquid injection MOCVD, and $SrTiO_3$ has been deposited by this technique using the precursor combination [$Sr(thd)_2$] and [$Ti(OPr^i)_2(thd)_2$] in hexane.[310]

An alternative approach to the problems associated with the low volatility and lack of stability of the [$Sr(thd)_2$] source is to use a "single-source" precursor containing both Sr and Ti in the same molecule. The heterometallic precursor [$Sr_2Ti_2(OPr^i)_8(thd)_4$] contains the Sr and Ti in the 1 : 1 ratio required for $SrTiO_3$, and the precursor has been used successfully in liquid injection MOCVD to deposit $SrTiO_3$ in the perovskite phase at 500 °C, which is lower than required by the conventional precursor mixture $[Sr(thd)_2]/[Ti(OPr^i)_2(thd)_2]$. The films were free from carbonate contamination and had good dielectric properties, with SiO_2 EOT as low as 1.36 nm.[311]

Similar precursor combinations have been employed for the conventional MOCVD of $BaTiO_3$ and $(Ba,Sr)TiO_3$. For instance $BaTiO_3$ has been deposited using $[Ba(thd)_2]$ and $[Ti(OPr^i)_4]$.[312,313] The dielectric constant of the as-deposited $BaTiO_3$ films was ~250. To avoid thermal decomposition of $[Ba(thd)_2]$ and $[Sr(thd)_2]$, there has been a growing use of liquid injection MOCVD for the deposition of $(Ba,Sr)TiO_3$, and the precursors $[Ba(thd)_2]$, $[Sr(thd)_2]$ with $[TiO(thd)_2]$ have been utilized in this technique. To reduce intermolecular and internuclear interactions in precursor solutions, the use of Lewis-base stabilized group II precursors has been proposed.[29] The adduct precursor combinations $[Ba(thd)_2(tetraglyme)]/[Sr(thd)_2(tetraglyme)]/[Ti(OPr^i)_2(thd)_2]$ and $[Ba(thd)_2(pmdeta)]/[Sr(thd)_2(pmdeta)]/[Ti(OPr^i)_2(thd)_2]$ have been used for the liquid injection MOCVD of $(Ba,Sr)TiO_3$. All the films showed good uniformity and excellent run-to-run reproducibility. Films grown from the tetraglyme adducts had a permittivity of 280 (electrical leakage $= 1 \times 10^{-9}$ at 22.5 V μm^{-1}) and those grown from the pmdeta complexes had a permittivity of 240 (electrical leakage $= 0.9 \times 10^{-9}$ at 24.0 V μm^{-1}).

There have been very few studies on the ALD of $SrTiO_3$ and $BaSrTiO_3$, but the ALD of $SrTiO_3$ and $BaTiO_3$ has been demonstrated using Sr- and Ba-cyclopentadienyl compounds, $[Ti(OPr^i)_4]$ and H_2O as precursors. After annealing, a permittivity of 180 was measured for $SrTiO_3$ and 165 for $BaTiO_3$.[225,226] $SrTiO_3$ films have also been deposited by plasma-enhanced ALD using $[Sr(thd)_2]$[314] or $[Sr(methd)_2]$[315] and oxygen.

8.5 CVD of Conducting, Semiconducting and Magnetic Oxides

8.5.1 MOCVD of RuO_2, $LaNiO_3$ and $LaSrCoO_3$

Ruthenium dioxide, RuO_2, a transition metal oxide with a rutile structure, exhibits metallic conductivity and high thermal and chemical stability. These properties give it potential applications in electronics as a metallization material in very large-scale integrated circuits,[316,317] and as electrodes for ferroelectric thin films.[318] Platinum is the conventional electrode material in ferroelectric devices, but it exhibits poor fatigue resistance due to lattice mismatch and chemical incompatibility with the ferroelectric oxide layer. The use of a conducting oxide electrode improves the stability and morphology of the device structure.

There have been many studies on the MOCVD of RuO_2, and these have mostly employed Ru-carbonyls, β-diketonates or Ru-cyclopentadienyl complexes, such as $[Ru_3(CO)_{12}]$,[319] $[Ru(acac)_3]$,[319] $[Ru(thd)_3]$,[320,321] $[Ru(tfac)_3]$,[322] $[Cp_2Ru]$ and $[(EtCp)_2Ru]$.[323] In a comparative study RuO_2 films grown from $[Cp_2Ru]$ were found to be similar to those grown from $[Ru(acac)_3]$, with an electrical resistivity of 89.9 μΩ cm.[319] In another study, RuO_2 films deposited using $[Ru(thd)_3]$ had resistivities of 45–60 μΩ cm, comparable to that of bulk RuO_2, and superior to the higher resistivity films obtained using $[Ru_3(CO)_{12}]$.[321] A lower deposition temperature (250 °C) and the addition of a small amount of oxygen (2.5 % in the gas-phase) improved the step coverage of the RuO_2 films.[321] In addition to the above precursors, several heteroleptic Ru complexes have been investigated, such as $[Ru(CO)_3(C_6H_8)]$,[324] $[Ru(CO)_4(hfb)]$ (hfb=hexafluoro-2-butyne])[325] and $[Ru(C_3H_5)_2(COD)]$ (COD=1,4-cyclooctadiene). Strontium ruthenate, $SrRuO_3$, has a pseudo-perovskite structure and is also electrically conductive; it has been deposited by hot wall MOCVD using bis(2,4-dimethylpentadienyl)ruthenium and $[Sr(thd)_2]$.[326]

Other conducting oxides with applications as electrodes for ferroelectric capacitor layers include $LaNiO_3$ and $LaSrCoO_3$. $LaNiO_3$ has been deposited at 630 °C by liquid injection MOCVD using $[La(thd)_3]$ and $[Ni(thd)_2]$, and a maximum resistivity of 300 μΩ cm was measured. Epitaxial $La_{1-x}Sr_xCoO_3$ and $PbTiO_3/La_{1-x}Sr_xCoO_3$ heterostructures were deposited by aerosol MOCVD and powder flash evaporation MOCVD using [La(thd)$_3$], [Sr(thd)$_2$], [Co(thd)$_2$], [Pb(thd)$_2$] and [Ti(OPr^i)$_2$(thd)$_2$].

8.5.2 CVD of ZnO

Zinc oxide, ZnO, is an intrinsic p-type semiconductor and shows piezoelectric properties, making it useful for microsensor devices, micromachined actuators,[327,328] solar cells[329] and surface acoustic wave devices.[330] Traditional methods for the growth of ZnO by MOCVD have involved the pyrolysis of diethylzinc (Et$_2$Zn)[331–333] or dimethylzinc (Me$_2$Zn)[334] in the presence of oxygen and/or H$_2$O. However, a severe premature reaction of the highly oxygen-sensitive [R$_2$Zn] precursor leads to particulate deposition upstream of the substrate. This can be avoided by the use of less reactive oxidants, such as CO$_2$,[335] N$_2$O[335,336] and NO$_2$,[336] or by the use of the alcohols methanol, ethanol or *tert*-butyl alcohol in combination with [Et$_2$Zn][337] or the adduct [Me$_2$Zn(THF)].[338] It was proposed that reaction between R′OH and [R$_2$Zn] probably leads to alcohol adducts or the zinc alkoxides [RZn(OR′)] as gas-phase intermediates;[337,338] support for this proposal comes from the successful use of the tetrameric "single-source" precursors [MeZn(OPr^i)] and [MeZn(OBu^t)] for the growth of ZnO in the absence of an added oxygen source.[339] ZnO grown from [MeZn(OPr^i)] and [MeZn(OBu^t)] exhibited high purity (C not detected by AES), and a pyrolysis mechanism was proposed involving the abstraction of a β-hydrogen from [OPr^i] or [OBu^t], with the elimination of methane, an alkene and formation of ZnO.[339]

8.5.3 CVD of Magnetic and Magnetoelectric Oxides

Thin films of the spinel-type ferrites MFe$_2$O$_4$ (M = Ni, Zn) have applications in magnetic cores of coils, transformers, recording magnetic heads and memory elements. To obtain films with large magnetic domain structures it is generally necessary to carry out sol–gel or PVD deposition processes at 600–900 °C. However, NiFe$_2$O$_4$ and ZnFe$_2$O$_4$ films with good magnetic properties have been deposited at lower temperature (450 °C) by plasma-enhanced MOCVD using the β-diketonate precursors [Ni(acac)$_2$] or [Zn(acac)$_2$] in combination with [Fe(acac)$_3$].[340] The precursor combination [Ni(thd)$_2$], [Zn(thd)$_2$] and [Fe(thd)$_3$] has been used for the growth of NiFe$_2$O$_4$ and ZnFe$_2$O$_4$ and nickel zinc ferrite at 500 °C by liquid injection MOCVD. The layers were shown by X-ray diffraction to be crystalline with no preferred orientation.[341]

BiFeO$_3$ is a magnetoelectric material that can present an electric and a magnetic order simultaneously, offering promise for reduced operating power and non-volatile memory applications.[342] The growth of BiFeO$_3$ has been investigated by liquid injection MOCVD using [Bi(thd)$_3$] or [Bi(mmp)$_3$] with [Fe(thd)$_3$].[343] It was found that [Bi(mmp)$_3$] was a more efficient source of Bi than [Bi(thd)$_3$] and was a better "thermal-match" to [Fe(thd)$_3$], with both precursors depositing oxide in a similar temperature regime (550–600 °C).

8.6 CVD of High-T_c Superconducting Oxides

The discovery of high-T_c superconducting oxides in the mid-1980s caused an enormous stir in the scientific community, and led to intense efforts to prepare these materials as thin films. The first materials system investigated was Ba-La-Cu-O, which had a T_c of the order of 30 K.[344] Subsequently, a range of copper oxide-based high-T_c superconducting oxides have been announced, including Y-Ba-Cu-O (YBCO) (T_c = 93 K),[345] Bi-Sr-Ca-Cu-O (BSCCO) (T_c = 105 K)[346] and Tl-Ba-Ca-Cu-O (TBCCO) (T_c = 120 K).[347] These materials were deposited in a triaxial orientation on monocrystalline substrates, and unlike common forms of bulk oxide superconductors they displayed physical properties dominated by the intrinsic behaviour of the superconducting phase, and not grain boundary effects. They also support relatively large critical currents,[348] which has led to potential applications in SQUID magnetometers[349] and optically switchable band reject filters.

High-T_c superconducting oxides were initially deposited by PVD techniques,[350] but MOCVD techniques have been intensively investigated. Several excellent and comprehensive reviews have summarized developments in precursor chemistry and MOCVD processes,[13,351–353] and so only a brief overview of high-T_c superconductor precursor chemistry is given here.

Metal β-diketonates are by far the most widely used precursors for the MOCVD of high-T_c oxide superconductors (Table 8.2 and Section 8.2.1). Table 8.6 gives physical data[13] for several commonly used YBCO precursors. These precursors are generally low-volatility and so need to be vapourized at between 100 and 250 °C.

Most precursor development has been carried out on the alkaline metal sources used for YBCO, BSCCO and TBCCO growth. The β-diketonate complexes of these large highly positively charged metal ions (e.g. Sr, atomic radius ~1.91 Å, and Ba, atomic radius ~ 1.96 Å) tend to form low volatility oxygen-bridged polynuclear clusters to fulfil the saturation requirement of the central metal atom, e.g. $[Sr(thd)_2]$ is trimeric (Figure 8.4)[23] and $[Ba(thd)_2]$ is tetrameric.[24] Consequently, much research effort has been aimed at increasing the vapour pressure of these complexes for high-T_c MOCVD applications.[13,25,26] Furthermore, $[Ba(thd)_2]$ exhibits a variable molecular composition as a function of the synthetic route employed. For instance, $[Ba(thd)_2]$ prepared in aqueous media retains chemically-bound water to form oligomeric complexes such as $[Ba_5(thd)_5(OH)(H_2O)_3]$[354] and $[Ba_6(thd)_{12}(H_2O)_{13}]$,[355] and hydrated $[Ba(thd)_2]$ can decompose during heating in the MOCVD source container to eliminate thdH, which causes severe problems in the reproducibility of gas-phase transport. However, this decomposition can be suppressed by the addition of thdH to the carrier gas stream,[351,354,356] and addition of Lewis bases such as the free ligand, tetrahydrofuran, ammonia or amines to the gas phase has been commonly used to improve the vapour pressure and transport characteristics of $[Ba(thd)_2]$ and $[Sr(thd)_2]$.[357–364]

In an attempt to inhibit oligomerization, neutral donor ligands have been added to $[Ba(thd)_2]$ and $[Sr(thd)_2]$. The addition of polyethers or polyamines leads to a range of monomeric complexes, including $[M(thd)_2(polyether)]$[24] (Figure 8.4), with M = Sr, Ba; polyether = triglyme, $CH_3O(CH_2CH_2O)_3CH_3$ or tetraglyme, $CH_3O(CH_2CH_2O)_3CH_3$),[28] and $[M(thd)_2(pmdeta)]$, with M = Sr, Ba; pmdeta = pentamethyldiethylene-triamine, $Me_2NCH_2CH_2N\{Me\}CH_2CH_2NMe_2$.[29] The presence of the adducting ligands is undoubtedly useful in protecting the $[Ba(thd)_2]$ and $[Sr(thd)_2]$ complexes from hydrolysis and stabilizing the complex in solution during liquid injection MOCVD,[29] but in the gas phase the non-fluorinated-thd complexes dissociate.

Table 8.6 Physical properties of selected precursors used for the MOCVD of YBCO (From ref. 13 and references therein).

Precursor formula	*Melting point (°C)*	*Vapour pressure equation*	*Vaporization temperature range*
$[Cu(thd)_2]$	195(dec)	$Log_{10}P(mmHg) = 14.9 - 6510/T^b$	80–150
$[Cu(acac)_2]$	284(dec)	$Log_{10}P = 12.1 - 5720/T^b$	110–145
$[Y(thd)_3]$	173	$Log_{10}P = 4.0 - 6110/T^b$	90–170
$[Ba(thd)_2]^a$	<200	$Log_{10}P = 8.4 - 4710/T^c$	180–280
$[Ba(hfac)_2(tet)]$	155	$Log_{10}P = 11.6 - 5320/T^c$	105–175
$[Ba(hfac)_2(18\text{-}C\text{-}6)]$	260(dec)	$Log_{10}P = 10.7 - 5480/T^b$	175–189
$[Ba(tdf)_2(tet)]$	70		96–146

[a]Properties vary according to preparation method and sample history.
[b]Valid for solid compound.
[c]Valid for molten compound. tet = tetraglyme.
18-c-6=18-crown-6-polyether.

In contrast, fluorinated Ba and Sr β-diketonates form stable adducts, such as such as [M(hfac)$_2$.L] (M = Ca, Sr, Ba; L = tetraglyme, triglyme),that do not dissociate in the gas phase and exhibit increased and stable vapour pressures.[28,365,366] This exceptional stability can be attributed to the electron-withdrawing ability of the fluorine atoms.[354] The superior vapour pressure characteristics and thermal stability of these "second-generation" complexes have been widely exploited in the MOCVD growth of thin films of YBCO,[367–369] BSCCO[370] and TBCCO.[366,371,372]

A complication in using [M(hfac)$_2$.L] and related fluorinated β-diketonates is that the oxide films frequently require an additional H_2O/O_2 annealing treatment to eliminate fluoride contamination. This has led to efforts to develop fluorine-free sources, such as [Ba(thd)$_2$(phen)$_2$], which has been used for the MOCVD of YBCO,[373,374] although other studies have indicated that the complex dissociates when sublimed.[375]

Another approach used to increase the saturation around the Ba or Sr centre in ML$_2$ complexes, and thereby increase the vapour pressure, is the use of elaborate multidentate ligands. These combine a mono-anionic "head" with an uncharged "tail" containing a donor functional group capable of completing the coordination sites round the central M^{2+} ion. Anionic ligand heads that have been used include cyclopentadienyl, alkoxy, β-diketonate[356] and β-ketoiminate groups.[38]

There has been much less development work on the non-alkaline earth precursors for the MOCVD of YBCO, BSCCO and TBCCO. [Cu(thd)$_2$] and [Y(thd)$_3$] are the most widely used Cu and Y precursors for YBCO deposition, although the less volatile [Cu(acac)$_2$] complex has been used quite frequently.[351,352,356] Both [Cu(thd)$_2$] and [Y(thd)$_3$] vapourize well below their melting points and retain excellent chemical stability at source operating temperatures. There has been a limited amount of work directed at obtaining lower melting Cu and sources. The fluorinated

Table 8.7 Some metalorganic precursors used for the MOCVD of high-T$_c$ superconducting oxides (YBCO, BSCCO and TBCCO).

Metal	*Precursor*
Yttrium	[Y(thd)$_3$], [Y(fod)$_3$], [Y(thd)$_3$(PyNO)]
Barium	[Ba(thd)$_2$], [Ba(fod)$_2$], [Ba(hfac)$_2$(tet)], [Ba(tdf)$_2$(tet)], [Ba(thd)$_2$(tet)], [Ba(thd)$_2$(phen)], [Ba(thd)$_2$(pmdeta)]
Copper	[Cu(thd)$_2$], [Cu(acac)$_2$], [Cu(hfac)$_2$], [Cu(fod)$_2$], [Cu(acim)$_2$]
Bismuth	[BiPh$_3$], [Bi(OEt)$_3$]
Strontium	[Sr(thd)$_2$], [Sr(hfac)$_2$(tet)], [Sr(thd)$_2$(pmdeta)]
Calcium	[Ca(thd)$_2$], [Ca(fod)$_2$], [Ca(hfac)$_2$(tet)], [Ca(hfac)$_2$(tri)]
Thallium	[Tl(thd)$_3$], [TlCp]

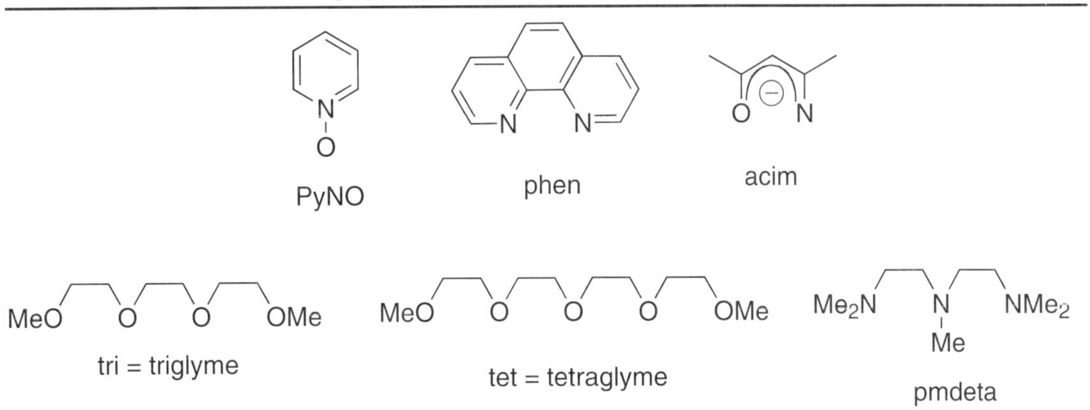

β-diketonate [Cu(tdf)$_2$] melts at 68 °C,[376] and the 4-*tert*-butyl pyridine-N-oxide adduct of [Y(thd)$_3$] melts at 97 °C,[377] compared with a melting point of 173 °C for [Y(thd)$_3$].

Early studies into the MOCVD of BSCCO thin films utilized triphenylbismuth, [BiPh$_3$], with [Sr(thd)$_2$], [Ca(thd)$_2$] and [Cu(thd)$_2$].[378] Subsequently, fluorinated precursor systems have been employed, such as [Sr(hfac)$_2$], [Cu(hfac)$_2$], [Ca(hfac)$_2$] with BiPh$_3$,[379] or [Sr(fod)$_2$], [Ca(fod)$_2$] and [Cu(fod)$_2$] with BiPh$_3$.[380] "Second-generation" fluorinated alkaline earth precursors, such as [Sr(hfac)$_2$(tetraglyme)] and [Ca(hfac)$_2$(triglyme)], have also been used with [Cu(acac)$_2$] and [BiPh$_3$] for the MOCVD of BSCCO, although it is necessary to add water for the in situ MOCVD of BSCCO, owing to the thermodynamic stability of the alkaline earth fluorides relative to the oxides. Similarly, "second-generation" Ba and Ca fluorinated β-diketonate precursors have been investigated for the MOCVD of TBCCO.[371,372] Table 8.7 gives a selection of precursors used for the MOCVD of high-T_c superconducting oxides.

8.7 Conclusions

The methodologies of MOCVD and ALD can be tailored to the manufacture of a host of different metal oxide thin film materials. Specifically, both MOCVD and ALD require the availability of precursors with appropriate physical properties and decomposition characteristics, such as adequate volatility for efficient vapour phase transport and low residual contamination. An adequate temperature window must exist between precursor evaporation and decomposition, and for many electronics applications the oxide deposition temperature is often restricted to the region of 500 °C and below to prevent degradation of the device structure. The state of the art in precursor development has reached the point where design and synthesis is capable of subtly adjusting the thermal and transport properties of molecular precursors. These developments have, in turn, led to the development of the highly sophisticated processes used in the manufacture of twenty-first century electronics.

Acknowledgement

All molecular structures shown in this chapter are courtesy of the Cambridge Structural Database (see "The Cambridge Structural Database; a quarter of a million crystal structures and rising", F. H. Allen, *Acta Crystallogr.*, 2002, **B58**, 380).

References

1. O. Auciello and R. Ramesh, eds., *Electroceramic Thin Films*, Parts I and II, *MRS Bulletin*, 1996, 21(6/7).
2. S. B. Krupanidhi, N. Maffei, M. Sayer and K. El-Assal, *J. Appl. Phys.*, 1983, **54**, 6601.
3. K. Sreenivas, M. Sayer, D. J. Baar and M. Nishioka, *Appl. Phys. Lett.*, 1988, **52**, 709.
4. K. Saenger, R. Roy, K. Etzold and J. Cuomo, *Mater. Res. Soc. Symp. Proc.*, 1991, **200**, 115.
5. R. Ramesh, A. Inam, W. K. Chan, F. Tillerot, B. Wilkens, C. C. Chang, T. Sands, J. M. Tarascon and V. G. Keramidas, *Appl. Phys. Lett.*, 1991, **59**, 3542.
6. K. D. Budd, S. K. Dey and D. A. Payne, *British Ceramic Society Proceedings*, 1985, **36**, 107.
7. R. W. Vest and J. Xu, *Ferroelectrics*, 1989, **93**, 21.
8. A. C. Jones, *Chem. Vap. Deposition*, 1998, **4**, 169.
9. L. A. Wills, W. A. Feil, B. W. Wessels, L. M. Tonge and T. J. Marks, *J. Cryst. Growth*, 1991, **107**, 712.
10. T. Kawahara, M. Yamamuka, T. Makita, J. Naka, A. Yuuki, N. Mikami and K. Ono, *Jpn. J. Appl. Phys. Part 1*, 1994, **33**, 5129.

11. M. de Keijser and G. J. M. Dormans, *MRS Bull.*, 1996, **21**, 37.
12. M. de Keijser, G. J. M. Dormans, J. F. M. Cillessen, D. M. Deleeuw and H. W. Zandbergen, *Appl. Phys. Lett.*, 1991, **58**, 2636.
13. I. M. Watson, *Chem. Vap. Deposition*, 1997, **3**, 9.
14. A. C. Jones, *J. Mater. Chem.*, 2002, **12**, 2576.
15. M. Leskelä and M. Ritala, in *Handbook of Thin Films*, ed. H. S. Nalwa, Academic Press, 2002.
16. R. A. Gardiner, P. C. van Buskirk and P. S. Kirlin, *Mater. Res. Soc. Symp. Proc.*, 1994, **335**, 221.
17. N. Bourhila, F. Felten, J. P. Senateur, F. Schuster, R. Madar and A. Abrutius, *Electrochem. Soc. Proc.*, 1997, **31**, 201.
18. A. C. Jones, T. J. Leedham, P. J. Wright, M. J. Crosbie, P. A. Lane, D. J. Williams, K. A. Fleeting, D. J. Otway and P. O'Brien, *Chem. Vap. Deposition*, 1998, **4**, 46.
19. R. D. Shannon, *Acta Crystallographica Section A*, 1976, **32**, 751.
20. B. D. Fahlman and A. R. Barron, *Adv. Mater. Opt. Electron.*, 2000, **10**, 223.
21. D. C. Bradley, H. Chudzynska, M. E. Hammond, M. B. Hursthouse, M. Motevalli and W. Ruowen, *Polyhedron*, 1992, **11**, 375.
22. G. K. Schweitzer, B. P. Pullen and Y. H. Fang, *Anal. Chim. Acta*, 1968, **43**, 332.
23. J. Brooks, H. O. Davies, T. J. Leedham, A. C. Jones and A. Steiner, *Chem. Vap. Deposition*, 2000, **6**, 66.
24. S. R. Drake, M. B. Hursthouse, K. M. A. Malik and D. J. Otway, *J. Chem. Soc., Dalton Trans.*, 1993, 2883.
25. D. J. Otway and W. S. Rees, *Coord. Chem. Rev.*, 2000, **210**, 279.
26. W. A. Wojtczak, P. F. Fleig and M. J. Hampden-Smith, *Adv. Organomet. Chem.*, 1996, **40**, 215.
27. S. R. Drake, M. B. Hursthouse, K. M. A. Malik and S. A. S. Miller, *J. Chem. Soc., Chem. Commun.*, 1993, 478.
28. K. Timmer, K. Spee, A. Mackor, H. A. Meinema, A. L. Spek and P. Van der Sluis, *Inorg. Chim. Acta*, 1991, **190**, 109.
29. J. F. Roeder, T. H. Baum, S. M. Bilodeau, G. T. Stauf, C. Ragaglia, M. W. Russell and P. C. Van Buskirk, *Adv. Mater. Opt. Electron.*, 2000, **10**, 145.
30. M. P. Singh and S. A. Shivashankar, *J. Cryst. Growth*, 2005, **276**, 148.
31. M. P. Singh, C. S. Thakur, K. Shalini, S. Banerjee, N. Bhat and S. A. Shivashankar, *J. Appl. Phys.*, 2004, **96**, 5631.
32. H. A. Luten, W. S. Rees and V. L. Goedken, *Chem. Vap. Deposition*, 1996, **2**, 149.
33. R. G. Gordon, S. Barry, R. N. R. Broomhall-Dillard and D. J. Teff, *Adv. Mater. Opt. Electron.*, 2000, **10**, 201.
34. G. G. Condorelli, A. Baeri and I. L. Fragalà, *Chem. Mater.*, 2002, **14**, 4307.
35. D. L. Schulz, B. J. Hinds, D. A. Neumayer, C. L. Stern and T. J. Marks, *Chem. Mater.*, 1993, **5**, 1605.
36. D. L. Schulz, B. J. Hinds, C. L. Stern and T. J. Marks, *Inorg. Chem.*, 1993, **32**, 249.
37. D. A. Neumayer, J. A. Belot, R. L. Feezel, C. Reedy, C. L. Stern and T. J. Marks, *Inorg. Chem.*, 1998, **37**, 5625.
38. T. J. Marks, *Pure Appl. Chem.*, 1995, **67**, 313.
39. J. A. Belot, A. C. Wang, R. J. McNeely, L. Liable-Sands, A. L. Rheingold and T. J. Marks, *Chem. Vap. Deposition*, 1999, **5**, 65.
40. N. L. Edleman, A. C. Wang, J. A. Belot, A. W. Metz, J. R. Babcock, A. M. Kawaoka, J. Ni, M. V. Metz, C. J. Flaschenriem, C. L. Stern, L. M. Liable-Sands, A. L. Rheingold, P. R. Markworth, R. P. H. Chang, M. P. Chudzik, C. R. Kannewurf and T. J. Marks, *Inorg. Chem.*, 2002, **41**, 5005.
41. D. C. Bradley, *Chem. Rev.*, 1989, **89**, 1317.

42. D. C. Bradley, R. C. Mehrotra and D. P. Gaur, *Metal Alkoxides*, Academic Press, New York, 1978.
43. W. A. Herrmann, R. Anwander and M. Denk, *Chemische Berichte-Recueil*, 1992, **125**, 2399.
44. W. A. Herrmann, N. W. Huber and O. Runte, *Angew. Chem., Int. Ed. Engl.*, 1995, **34**, 2187.
45. P. A. Williams, J. L. Roberts, A. C. Jones, P. R. Chalker, J. F. Bickley, A. Steiner, H. O. Davies and T. J. Leedham, *J. Mater. Chem.*, 2002, **12**, 165.
46. P. A. Williams, J. L. Roberts, A. C. Jones, P. R. Chalker, N. L. Tobin, J. F. Bickley, H. O. Davies, L. M. Smith and T. J. Leedham, *Chem. Vap. Deposition*, 2002, **8**, 163.
47. R. Anwander, F. C. Munck, T. Priermeier, W. Scherer, O. Runte and W. A. Herrmann, *Inorg. Chem.*, 1997, **36**, 3545.
48. H. C. Aspinall, J. F. Bickley, J. M. Gaskell, A. C. Jones, G. Labat, P. R. Chalker and P. A. Williams, *Inorg. Chem.*, 2007, **46**, p. 5852.
49. O. Poncelet, L. G. Hubert-Pfalzgraf, J. C. Daran and R. Astier, *J. Chem. Soc., Chem. Commun.*, 1989, 1846.
50. K. A. Fleeting, P. O'Brien, A. C. Jones, D. J. Otway, A. J. P. Andrew and D. J. Williams, *J. Chem. Soc. Dalton Trans.*, 1999, 2853.
51. R. Matero, M. Ritala, M. Leskelä, A. C. Jones, P. A. Williams, J. F. Bickley, A. Steiner, T. J. Leedham and H. O. Davies, *J. Non-Cryst. Solids*, 2002, **303**, 24.
52. Y. F. Loo, R. O'Kane, A. C. Jones, H. C. Aspinall, R. J. Potter, P. R. Chalker, J. F. Bickley, S. Taylor and L. M. Smith, *Chem. Vap. Deposition*, 2005, **11**, 299.
53. Y. F. Loo, R. O'Kane, A. C. Jones, H. C. Aspinall, R. J. Potter, P. R. Chalker, J. F. Bickley, S. Taylor and L. M. Smith, *J. Mater. Chem.*, 2005, **15**, 1896.
54. T. D. Manning, Y. F. Loo, A. C. Jones, H. C. Aspinall, P. R. Chalker, J. F. Bickley, L. M. Smith and G. W. Critchlow, *J. Mater. Chem.*, 2005, **15**, 3384.
55. S. Horii, K. Yamamoto, M. Asai, H. Miya and M. Niwa, *Jpn. J. Appl. Phys. Part 1*, 2003, **42**, 5176.
56. R. J. Potter, P. R. Chalker, T. D. Manning, H. C. Aspinall, Y. F. Loo, A. C. Jones, L. M. Smith, G. W. Critchlow and M. Schumacher, *Chem. Vap. Deposition*, 2005, **11**, 159.
57. R. M. Fix, R. G. Gordon and D. M. Hoffman, *Chem. Mater.*, 1990, **2**, 235.
58. A. Bastianini, G. A. Battiston, R. Gerbasi, M. Porchia and S. Daolio, *J. Phys. IV*, 1995, **5**, 525.
59. Y. Ohshita, A. Ogura, A. Hoshino, S. Hiiro and H. Machida, *J. Cryst. Growth*, 2001, **233**, 292.
60. Y. Ohshita, A. Ogura, A. Hoshino, S. Hiiro, T. Suzuki and H. Machida, *Thin Solid Films*, 2002, **406**, 215.
61. F. T. Edelmann, *Coord. Chem. Rev.*, 1994, **137**, 403.
62. B. S. Lim, A. Rahtu, J. S. Park and R. G. Gordon, *Inorg. Chem.*, 2003, **42**, 7951.
63. A. R. Sadique, M. J. Heeg and C. H. Winter, *Inorg. Chem.*, 2001, **40**, 6349.
64. B. S. Lim, A. Rahtu and R. G. Gordon, *Nature Materials*, 2003, **2**, 749.
65. Z. W. Li, S. T. Barry and R. G. Gordon, *Inorg. Chem.*, 2005, **44**, 1728.
66. P. de Rouffignac, J. S. Park and R. G. Gordon, *Chem. Mater.*, 2005, **17**, 4808.
67. P. de Rouffignac and R. G. Gordon, *Chem. Vap. Deposition*, 2006, **12**, 152.
68. B. S. Lim, A. Rahtu, P. de Rouffignac and R. G. Gordon, *Appl. Phys. Lett.*, 2004, **84**, 3957.
69. E. Samuel and M. D. Rausch, *J. Am. Chem. Soc.*, 1973, **95**, 6263.
70. J. M. Birmingham and G. Wilkinson, *J. Am. Chem. Soc.*, 1956, **78**, 42.
71. W. A. Herrmann, R. Anwander, F. C. Munck and W. Scherer, *Chemische Berichte-Recueil*, 1993, **126**, 331.
72. W. J. Evans, R. Anwander and J. W. Ziller, *Organometallics*, 1995, **14**, 1107.
73. H. M. Dietrich, G. Raudaschl-Sieber and R. Anwander, *Angew. Chem., Int. Ed.*, 2005, **44**, 5303.

74. Y. K. Gunko and F. T. Edelmann, *Comments on Inorganic Chemistry*, 1997, **19**, 153.
75. P. A. Packan, *Science*, 1999, **285**, 2079.
76. G. D. Wilk, R. M. Wallace and J. M. Anthony, *J. Appl. Phys.*, 2001, **89**, 5243.
77. T. M. Klein, D. Niu, W. S. Epling, W. Li, D. M. Maher, C. C. Hobbs, R. I. Hegde, I. J. R. Baumvol and G. N. Parsons, *Appl. Phys. Lett.*, 1999, **75**, 4001.
78. M. Copel, M. Gribelyuk and E. Gusev, *Appl. Phys. Lett.*, 2000, **76**, 436.
79. B. H. Lee, L. G. Kang, R. Nieh, W. J. Qi and J. C. Lee, *Appl. Phys. Lett.*, 2000, **76**, 1926.
80. G. D. Wilk and R. M. Wallace, *Appl. Phys. Lett.*, 1999, **74**, 2854.
81. G. D. Wilk and R. M. Wallace, *Appl. Phys. Lett.*, 2000, **76**, 112.
82. M. Y. Ho, H. Gong, G. D. Wilk, B. W. Busch, M. L. Green, W. H. Lin, A. See, S. K. Lahiri, M. E. Loomans, P. I. Raisanen and T. Gustafsson, *Appl. Phys. Lett.*, 2002, **81**, 4218.
83. P. A. Marshall, R. J. Potter, A. C. Jones, P. R. Chalker, S. Taylor, G. W. Critchlow and S. A. Rushworth, *Chem. Vap. Deposition*, 2004, **10**, 275.
84. Y. H. Wu, M. Y. Yang, A. Chin, W. J. Chen and C. M. Kwei, *IEEE Electron Device Lett.*, 2000, **21**, 341.
85. H. J. Osten, J. P. Liu and H. J. Mussig, *Appl. Phys. Lett.*, 2002, **80**, 297.
86. J. Kwo, M. Hong, A. R. Kortan, K. T. Queeney, Y. J. Chabal, J. P. Mannaerts, T. Boone, J. J. Krajewski, A. M. Sergent and J. M. Rosamilia, *Appl. Phys. Lett.*, 2000, **77**, 130.
87. J. A. Gupta, D. Landheer, J. P. McCaffrey and G. I. Sproule, *Appl. Phys. Lett.*, 2001, **78**, 1718.
88. M. D. Kannan, S. K. Narayandass, C. Balasubramanian and D. Mangalaraj, *Physica Status Solidi a-Applied Research*, 1991, **128**, 427.
89. M. H. Cho, Y. S. Roh, C. N. Whang, K. Jeong, H. J. Choi, S. W. Nam, D. H. Ko, J. H. Lee, N. I. Lee and K. Fujihara, *Appl. Phys. Lett.*, 2002, **81**, 1071.
90. J. Robertson, *Reports on Progress in Physics*, 2006, **69**, 327.
91. R. N. Tauber, A. C. Dumbri and R. E. Caffrey, *J. Electrochem. Soc.*, 1971, **118**, 747.
92. C. F. Powell, in *Chemically Deposited Non-Metals*, eds. C. F. Powell, J. H. Oxley and J. M. Blocher, John Wiley & Sons Inc., New York, 1966, pp. 343.
93. R. C. Smith, T. Z. Ma, N. Hoilien, L. Y. Tsung, M. J. Bevan, L. Colombo, J. Roberts, S. A. Campbell and W. L. Gladfelter, *Adv. Mater. Opt. Electron.*, 2000, **10**, 105.
94. M. Pulver and G. Wahl, *Electrochem. Soc. Proc.*, 1977, **97–25**, 960.
95. A. C. Jones, T. J. Leedham, P. J. Wright, M. J. Crosbie, D. J. Wiliams, P. A. Lane and P. O'Brien, *Mater. Res. Soc. Proc.*, 1998, **495**, 11.
96. M. Balog, M. Schieber, M. Michman and S. Patai, *J. Cryst. Growth*, 1972, **17**, 298.
97. S. V. Pasko, L. G. Hubert-Pfalzgraf, A. Abrutis, P. Richard, A. Bartasyte and V. Kazlauskiene, *J. Mater. Chem.*, 2004, **14**, 1245.
98. D. G. Colombo, D. C. Gilmer, V. G. Young, S. A. Campbell and W. L. Gladfelter, *Chem. Vap. Deposition*, 1998, **4**, 220.
99. S. Codato, G. Carta, G. Rossetto, G. A. Rizzi, P. Zanella, P. Scardi and M. Leoni, *Chem. Vap. Deposition*, 1999, **5**, 159.
100. J. M. Gaskell, A. C. Jones, K. Black, P. R. Chalker, T. Leese, A. Kingsley, R. Odedra and P. N. Heyes, *Surf. Coat. Technol.*, 2007, **201**, 9095.
101. J. J. Gallegos, T. L. Ward, T. J. Boyle, M. A. Rodriguez and L. P. Francisco, *Chem. Vap. Deposition*, 2000, **6**, 21.
102. H. J. Frenck, E. Oesterschulze, R. Beckmann, W. Kulisch and R. Kassing, *Materials Science and Engineering a-Structural Materials Properties Microstructure and Processing*, 1991, **139**, 394.
103. B. J. Gould, I. M. Povey, M. E. Pemble and W. R. Flavell, *J. Mater. Chem.*, 1994, **4**, 1815.
104. Y. Takahashi, T. Kawae and M. Nasu, *J. Cryst. Growth*, 1986, **74**, 409.

105. S. Pakswer and P. Skoug, in *Thin Dielectric Oxide Films Made by Oxygen Assisted Pyrolysis of Alkoxides*, eds. J. M. Blocher and J. C. Withers, The Electrochemical Society, Los Angeles, 1970, p. 619.
106. A. C. Jones, T. J. Leedham, P. J. Wright, M. J. Crosbie, D. J. Wiliams, K. A. Fleeting, H. O. Davies, D. J. Otway and P. O'Brien, *Chem. Vap. Deposition*, 1998, **4**, 197.
107. D.-Y. Kim, C.-H. Lee and S. J. Park, *J. Mater. Res.*, 1996, **11**, 2583.
108. U. Patil, M. Winter, H. W. Becker and A. Devi, *J. Mater. Chem.*, 2003, **13**, 2177.
109. S. Taylor, P. A. Williams, J. L. Roberts, A. C. Jones and P. R. Chalker, *Electron. Lett.*, 2002, **38**, 1285.
110. C. Dubourdieu, E. Rauwel, C. Millon, P. Chaudouet, F. Ducroquet, N. Rochat, S. Rushworth and V. Cosnier, *Chem. Vap. Deposition*, 2006, **12**, 187.
111. S. Van Elshocht, U. Weber, T. Conard, V. Kaushik, M. Houssa, S. Hyun, B. Seitzinger, P. Lehnen, M. Schumacher, J. Lindner, M. Caymax, S. De Gendt and M. Heyns, *J. Electrochem. Soc.*, 2005, **152**, F185.
112. T. S. Yang, K. S. An, E. J. Lee, W. Cho, H. S. Jang, S. K. Park, Y. K. Lee, T. M. Chung, C. G. Kim, S. Kim, J. H. Hwang, C. Lee, N. S. Lee and Y. Kim, *Chem. Mater.*, 2005, **17**, 6713.
113. P. A. Williams, A. C. Jones, N. L. Tobin, P. R. Chalker, S. Taylor, P. A. Marshall, J. E. Bickley, L. M. Smith, H. O. Davies and G. W. Critchlow, *Chem. Vap. Deposition*, 2003, **9**, 309.
114. A. Singh, C. K. Sharma, A. K. Rai, V. D. Gupta and R. C. Mehrotra, *J. Chem. Soc. A*, 1971, **1**, 2440.
115. X. Y. Zhao and D. Vanderbilt, *Phys. Rev. B*, 2002, **65**.
116. J. Y. Dai, P. F. Lee, K. H. Wong, H. L. W. Chan and C. L. Choy, *J. Appl. Phys.*, 2003, **94**, 912.
117. E. Rauwel, C. Dubourdieu, B. Holländer, N. Rochat, F. Ducroquet, M. D. Rossell, G. Van Tendeloo and B. Pelissier, *Appl. Phys. Lett.*, 2006, **89**.
118. M. Ritala and M. Leskelä, *Appl. Surf. Sci.*, 1994, **75**, 333.
119. K. Kukli, K. Forsgren, J. Aarik, T. Uustare, A. Aidla, A. Niskanen, M. Ritala, M. Leskelä and A. Harsta, *J. Cryst. Growth*, 2001, **231**, 262.
120. M. Ritala, M. Leskelä, L. Niinistö, T. Prohaska, G. Friedbacher and M. Grasserbauer, *Thin Solid Films*, 1994, **250**, 72.
121. J. Aarik, A. Aidla, A. A. Kiisler, T. Uustare and V. Sammelselg, *Thin Solid Films*, 1999, **340**, 110.
122. K. Forsgren, A. Harsta, J. Aarik, A. Aidla, J. Westlinder and J. Olsson, *J. Electrochem. Soc.*, 2002, **149**, F139.
123. M. J. Cho, J. Park, H. B. Park, C. S. Hwang, J. Jeong and K. S. Hyun, *Appl. Phys. Lett.*, 2002, **81**, 334.
124. M. Putkonen, J. Niinistö, K. Kukli, T. Sajavaara, M. Karppinen, H. Yamauchi and L. Niinistö, *Chem. Vap. Deposition*, 2003, **9**, 207.
125. J. Niinistö, M. Putkonen, L. Niinistö, S. L. Stoll, K. Kukli, T. Sajavaara, M. Ritala and M. Leskelä, *J. Mater. Chem.*, 2005, **15**, 2271.
126. D. M. Hausmann, E. Kim, J. Becker and R. G. Gordon, *Chem. Mater.*, 2002, **14**, 4350.
127. K. Kukli, M. Ritala, T. Sajavaara, J. Keinonen and M. Leskelä, *Chem. Vap. Deposition*, 2002, **8**, 199.
128. K. Kukli, M. Ritala and M. Leskelä, *Chem. Vap. Deposition*, 2000, **6**, 297.
129. A. C. Jones, P. A. Williams, J. L. Roberts, T. J. Leedham, H. O. Davies, R. Matero, M. Ritala and M. Leskelä, *Mater. Res. Soc. Symp. Proc.*, 2002, **716**, 145.
130. R. Matero, M. Ritala, M. Leskelä, T. Sajavaara, A. C. Jones and J. L. Roberts, *Chem. Mater.*, 2004, **16**, 5630.

131. K. Kukli, M. Ritala, M. Leskelä, T. Sajavaara, J. Keinonen, A. C. Jones and J. L. Roberts, *Chem. Vap. Deposition*, 2003, **9**, 315.
132. K. Kukli, M. Ritala, M. Leskelä, T. Sajavaara, J. Keinonen, A. C. Jones and J. L. Roberts, *Chem. Mater.*, 2003, **15**, 1722.
133. Y. Senzaki, A. K. Hochberg and J. A. T. Norman, *Adv. Mater. Opt. Electron.*, 2000, **10**, 93.
134. B. C. Hendrix, A. S. Borovik, C. Xu, J. F. Roeder, T. H. Baum, M. J. Bevan, M. R. Visokay, J. J. Chambers, A. L. P. Rotondaro, H. Bu and L. Colombo, *Appl. Phys. Lett.*, 2002, **80**, 2362.
135. Y. Ohshita, A. Ogura, M. Ishikawa, T. Kada, A. Hoshino, T. Suzuki, H. Machida and K. Soai, *Chem. Vap. Deposition*, 2006, **12**, 130.
136. M. Schumacher, P. K. Baumann and T. Seidel, *Chem. Vap. Deposition*, 2006, **12**, 99.
137. S. Zürcher, M. Morstein, N. D. Spencer, M. Lemberger and A. Bauer, *Chem. Vap. Deposition*, 2002, **8**, 171.
138. R. G. Gordon, J. Becker, D. Hausmann and S. Suh, *Chem. Mater.*, 2001, **13**, 2463.
139. J. L. Roberts, P. A. Marshall, A. C. Jones, P. R. Chalker, J. F. Bickley, P. A. Williams, S. Taylor, L. M. Smith, G. W. Critchlow, M. Schumacher and J. Lindner, *J. Mater. Chem.*, 2004, **14**, 391.
140. R. J. Potter, P. A. Marshall, P. R. Chalker, S. Taylor, A. C. Jones, T. C. Q. Noakes and P. Bailey, *Appl. Phys. Lett.*, 2004, **84**, 4119.
141. R. B. Clark-Phelps, A. Srivasta, L. Cleveland, T. E. Seidel and O. Sneh, *Mater. Res. Soc. Symp. Proc.*, 2001, **670**, K2–21.
142. S. H. Kim and S. W. Rhee, *Chem. Vap. Deposition*, 2006, **12**, 125.
143. M. Leskelä and M. Ritala, *J. Solid State Chem.*, 2003, **171**, 170.
144. M. Hong, J. Kwo, A. R. Kortan, J. P. Mannaerts and A. M. Sergent, *Science*, 1999, **283**, 1897.
145. A. Weber and H. Suhr, *Mod. Phys. Lett. B*, 1989, **3**, 1001.
146. R. Lo Nigro, R. G. Toro, G. Malandrino, V. Raineri and I. L. Fragalà, *Adv. Mater.*, 2003, **15**, 1071.
147. J. McAleese, J. C. Plakatouras and B. C. H. Steele, *Thin Solid Films*, 1996, **286**, 64.
148. S. Chevalier, G. Bonnet and J. P. Larpin, *Appl. Surf. Sci.*, 2000, **167**, 125.
149. R. Lo Nigro, V. Raineri, C. Bongiorno, R. Toro, G. Malandrino and I. L. Fragalà, *Appl. Phys. Lett.*, 2003, **83**, 129.
150. R. Lo Nigro, R. G. Toro, G. Malandrino, I. L. Fragalà, P. Rossi and P. Dapporto, *J. Electrochem. Soc.*, 2004, **151**, F206.
151. M. Langlet and R. D. Shannon, *Thin Solid Films*, 1990, **186**, L1.
152. J. B. Cheng, A. D. Li, Q. Y. Shao, H. Q. Ling, D. Wu, Y. Wang, Y. J. Bao, M. Wang, Z. G. Liu and N. B. Ming, *Appl. Surf. Sci.*, 2004, **233**, 91.
153. J. Jun, H. Jun and D. J. Choi, *J. Mater. Sci. Lett.*, 2002, **21**, 1847.
154. M. V. Cabanas, C. V. Ragel, F. Conde, J. M. Gonzalez-Calbet and M. Vallet-Regi, *Solid State Ionics*, 1997, **101**, 191.
155. D. C. Bradley, J. S. Ghotra and F. A. Hart, *J. Chem. Soc., Dalton Trans.*, 1973, **1**, 1021.
156. H. C. Aspinall, P. A. Williams, J. Gaskell, A. C. Jones, J. L. Roberts, L. M. Smith, P. R. Chalker and G. W. Critchlow, *Chem. Vap. Deposition*, 2003, **9**, 7.
157. H. C. Aspinall, J. Gaskell, P. A. Williams, A. C. Jones, P. R. Chalker, P. A. Marshall, L. M. Smith and G. W. Critchlow, *Chem. Vap. Deposition*, 2004, **10**, 13.
158. H. C. Aspinall, J. Gaskell, P. A. Williams, A. C. Jones, P. R. Chalker, P. A. Marshall, J. F. Bickley, L. M. Smith and G. W. Critchlow, *Chem. Vap. Deposition*, 2003, **9**, 235.
159. H. C. Aspinall, J. Gaskell, P. A. Williams, A. C. Jones, P. R. Chalker, P. A. Marshall, L. M. Smith and G. W. Critchlow, *Chem. Vap. Deposition*, 2004, **10**, 83.

160. Y. F. Loo, R. L. Potter, A. C. Jones, H. C. Aspinall, J. M. Gaskell, P. R. Chalker, L. M. Smith and G. W. Critchlow, *Chem. Vap. Deposition*, 2004, **10**, 306.
161. H. C. Aspinall, J. M. Gaskell, Y. F. Loo, A. C. Jones, P. R. Chalker, R. J. Potter, L. M. Smith and G. W. Critchlow, *Chem. Vap. Deposition*, 2004, **10**, 301.
162. M. Putkonen, M. Nieminen, J. Niinistö and L. Niinistö, *Chem. Mater.*, 2001, **13**, 4701.
163. H. Mölsä, L. Niinistö and M. Utriainen, *Adv. Mater. Opt. Electron.*, 2004, **4**, 389.
164. M. Putkonen, T. Sajavaara, L. S. Johansson and L. Niinistö, *Chem. Vap. Deposition*, 2001, **7**, 44.
165. M. Nieminen, M. Putkonen and L. Niinistö, *Appl. Surf. Sci.*, 2001, **174**, 155.
166. H. Mölsä and L. Niinistö, *Mater. Res. Soc. Symp. Proc.*, 1994, **335**, 341.
167. J. Päiväsaari, M. Putkonen and L. Niinistö, *J. Mater. Chem.*, 2002, **12**, 1828.
168. L. Niinistö, J. Päiväsaari, J. Niinisto, M. Putkonen and M. Nieminen, *Physica Status Solidi a-Applied Research*, 2004, **201**, 1443.
169. J. Päiväsaari, M. Putkonen and L. Niinistö, *Thin Solid Films*, 2005, **472**, 275.
170. A. Kosola, J. Päiväsaari, M. Putkonen and L. Niinistö, *Thin Solid Films*, 2005, **479**, 152.
171. E. P. Gusev, E. Cartier, D. A. Buchanan, M. Gribelyuk, M. Copel, H. Okorn-Schmidt and C. D'Emic, *Microelectron. Eng.*, 2001, **59**, 341.
172. J. Niinisto, M. Putkonen and L. Niinistö, *Chem. Mater.*, 2004, **16**, 2953.
173. J. Päiväsaari, J. Niinistö, K. Arstila, K. Kukli, M. Putkonen and L. Niinistö, *Chem. Vap. Deposition*, 2005, **11**, 415.
174. J. Niinistö, N. Petrova, M. Putkonen, L. Niinistö, K. Arstila and T. Sajavaara, *J. Cryst. Growth*, 2005, **285**, 191.
175. G. Scarel, E. Bonera, C. Wiemer, G. Tallarida, S. Spiga, M. Fanciulli, I. L. Fedushkin, H. Schumann, Y. Lebedinskii and A. Zenkevich, *Appl. Phys. Lett.*, 2004, **85**, 630.
176. J. Niinistö, M. Putkonen, M. Ritala, P. I. Raisanen, A. Niskanen and M. Leskelä, US Patent No. 6, 358, 546 (2005).
177. K. Kukli, M. Ritala, T. Pilvi, T. Sajavaara, M. Leskelä, A. C. Jones, H. C. Aspinall, D. C. Gilmer and P. J. Tobin, *Chem. Mater.*, 2004, **16**, 5162.
178. S. Guha, E. Cartier, M. A. Gribelyuk, N. A. Bojarczuk and M. C. Copel, *Appl. Phys. Lett.*, 2000, **77**, 2710.
179. M. Copel, E. Cartier and F. M. Ross, *Appl. Phys. Lett.*, 2001, **78**, 1607.
180. C. M. Carlson, J. C. Price, P. A. Parilla, D. S. Ginley, D. Niles, R. D. Blaugher, A. Goyal, M. Paranthaman, D. M. Kroeger and D. K. Christen, *Phys. C*, 1998, **304**, 82.
181. J. H. Jun, D. J. Choi, K. H. Kim, K. Y. Oh and C. J. Hwang, *Jpn. J. Appl. Phys. Part 1*, 2003, **42**, 3519.
182. B. E. Park and H. Ishiwara, *Appl. Phys. Lett.*, 2001, **79**, 806.
183. A. A. Molodyk, I. E. Korsakov, M. A. Novojilov, I. E. Graboy, A. R. Kaul and G. Wahl, *Chem. Vap. Deposition*, 2000, **6**, 133.
184. S. W. Kang and S. W. Rhee, *J. Electrochem. Soc.*, 2002, **149**, C345.
185. J. M. Gaskell, S. Przybylak, A. C. Jones, H. C. Aspinall, P. R. Chalker, K. Black, R. J. Potter, P. Taechakumput and S. Taylor, *Chem. Mater.*, 2007, **19**, 4796.
186. K. Kukli, M. Ritala, V. Pore, M. Leskelä, D. Sajavaara, R. I. Hegde, D. C. Gilmer, P. J. Tobin, A. C. Jones and H. C. Aspinall, *Chem. Vap. Deposition*, 2006, **12**, 158.
187. J. M. Gaskell, A. C. Jones, H. C. Aspinall, S. Przybylak, P. R. Chalker, K. Black, H. O. Davies, P. Taechakumput, S. Taylor and G. W. Critchlow, *J. Mater. Chem.*, 2006, **16**, 3854.
188. G. Apostopoulos, G. Vellianitis, A. Dimoulas, J. C. Hooker and T. Conard, *Appl. Phys. Lett.*, 2004, **84**, 260.
189. Y. F. Loo, S. Taylor, R. T. Murray, A. C. Jones and P. R. Chalker, *J. Appl. Phys.*, 2006, **99**.
190. J. M. Gaskell, A. C. Jones, P. R. Chalker, M. Werner, H. C. Aspinall, S. Taylor, P. Taechakumput, and P. N. Heys, *Chem. Vap. Deposition*, 2007, **13**, 671.

191. T. Heeg, M. Wagner, J. Schubert, C. Buchal, M. Boese, M. Luysberg, E. Cicerrella and J. L. Freeouf, *Microelectron. Eng.*, 2005, **80**, 150.
192. C. Zhao, T. Witters, B. Brijs, H. Bender, O. Richard, M. Caymax, T. Heeg, J. Schubert, V. V. Afanas'ev, A. Stesmans and D. G. Schlom, *Appl. Phys. Lett.*, 2005, **86**.
193. P. Myllymäki, M. Nieminen, J. Niinistö, M. Putkonen, K. Kukli and L. Niinistö, *J. Mater. Chem.*, 2006, **16**, 563.
194. S. A. Campbell, D. C. Gilmer, X. C. Wang, M. T. Hsieh, H. S. Kim, W. L. Gladfelter and J. H. Yan, *IEEE Trans. Electron Devices*, 1997, **44**, 104.
195. R. N. Goshtagore, *J. Electrochem. Soc.*, 1970, **117**, 529.
196. S. Hayashi and T. Hirai, *J. Cryst. Growth*, 1976, **36**, 157.
197. L. M. Williams and D. W. Hess, *Journal of Vacuum Science & Technology a-Vacuum Surfaces and Films*, 1983, **1**, 1810.
198. E. T. Fitzgibbons, K. J. Sladek and W. H. Hartwig, *J. Electrochem. Soc.*, 1972, **119**, 735.
199. M. Yokozawa, H. Iwasa and I. Teramoto, *Jpn. J. Appl. Phys.*, 1968, **7**, 96.
200. C. P. Fictorie, J. F. Evans and W. L. Gladfelter, *Journal of Vacuum Science & Technology a-Vacuum Surfaces and Films*, 1994, **12**, 1108.
201. J. P. Lu, J. D. Wang and R. Raj, *Thin Solid Films*, 1991, **204**, L13.
202. N. Rausch and E. P. Burte, *J. Electrochem. Soc.*, 1993, **140**, 145.
203. P. Comba, H. Jakob, B. Nuber and B. K. Keppler, *Inorg. Chem.*, 1994, **33**, 3396.
204. P. Schäfer and R. Waser, *Adv. Mater. Opt. Electron.*, 2000, **10**, 169.
205. P. C. Van Buskirk, S. M. Bilodeau, J. F. Roeder and P. S. Kirlin, *Jpn. J. Appl. Phys. Part 1*, 1996, **35**, 2520.
206. D. B. Beach and C. E. Vallet, *Mater. Res. Soc. Symp. Proc.*, 1996, **415**, 225.
207. A. C. Jones, T. J. Leedham, P. J. Wright, M. J. Crosbie, K. A. Fleeting, D. J. Otway, P. O'Brien and M. E. Pemble, *J. Mater. Chem.*, 1998, **8**, 1773.
208. C. Jimenez, M. Paillous, R. Madar, J. P. Senateur and A. C. Jones, *J. Phys. IV*, 1999, **9**, 569.
209. J. H. Lee, J. Y. Kim, J. Y. Shim and S. W. Rhee, *Journal of Vacuum Science & Technology a-Vacuum Surfaces and Films*, 1999, **17**, 3033.
210. J. S. Lee, S. H. Hong, K. J. Woo and W. I. Lee, *Chem. Vap. Deposition*, 2004, **10**, 67.
211. A. C. Jones, P. A. Williams, J. F. Bickley, A. Steiner, H. O. Davies, T. J. Leedham, A. Awaluddin, M. E. Pemble and G. W. Critchlow, *J. Mater. Chem.*, 2001, **11**, 1428.
212. A. Awaluddin, M. E. Pemble, A. C. Jones and P. A. Williams, *J. Phys. IV*, 2001, **11**, 531.
213. Y. S. Min, Y. J. Cho, D. Kim, J. H. Lee, B. M. Kim, S. K. Lim, I. M. Lee and W. I. Lee, *Chem. Vap. Deposition*, 2001, **7**, 146.
214. J. B. Woods, D. B. Beach, C. L. Nygren and Z. L. Xue, *Chem. Vap. Deposition*, 2005, **11**, 289.
215. R. B. van Dover, L. D. Schneemeyer and R. M. Fleming, *Nature*, 1998, **392**, 162.
216. E. L. Mays, D. W. Hess and W. S. Rees, *J. Cryst. Growth*, 2004, **261**, 309.
217. J. F. Eichler, O. Just and W. S. Rees, *J. Mater. Chem.*, 2004, **14**, 3139.
218. M. Ritala, M. Leskelä, E. Nykanen, P. Soininen and L. Niinistö, *Thin Solid Films*, 1993, **225**, 288.
219. M. Ritala, M. Leskelä, L. Niinistö and P. Haussalo, *Chem. Mater.*, 1993, **5**, 1174.
220. J. Aarik, A. Aidla, V. Sammelselg, H. Siimon and T. Uustare, *J. Cryst. Growth*, 1996, **169**, 496.
221. J. Aarik, A. Aidla, T. Uustare and V. Sammelselg, *J. Cryst. Growth*, 1995, **148**, 268.
222. K. Kukli, M. Ritala, M. Schuisky, M. Leskelä, T. Sajavaara, J. Keinonen, T. Uustare and A. Harsta, *Chem. Vap. Deposition*, 2000, **6**, 303.
223. J. Aarik, A. Aidla, T. Uustare, K. Kukli, V. Sammelselg, M. Ritala and M. Leskelä, *Appl. Surf. Sci.*, 2002, **193**, 277.
224. H. Doring, K. Hashimoto and A. Fujishima, *Berichte Der Bunsen-Gesellschaft-Physical Chemistry Chemical Physics*, 1992, **96**, 620.

225. M. Vehkamäki, T. Hatanpaa, T. Hanninen, M. Ritala and M. Leskelä, *Electrochemical and Solid State Letters*, 1999, **2**, 504.
226. M. Vehkamäki, T. Hanninen, M. Ritala, M. Leskelä, T. Sajavaara, E. Rauhala and J. Keinonen, *Chem. Vap. Deposition*, 2001, **7**, 75.
227. M. Ritala, M. Leskelä and E. Rauhala, *Chem. Mater.*, 1994, **6**, 556.
228. J. Aarik, J. Karlis, H. Mandar, T. Uustare and V. Sammelselg, *Appl. Surf. Sci.*, 2001, **181**, 339.
229. V. Pore, A. Rahtu, M. Leskelä, M. Ritala, D. Sajavaara and J. Keinonen, *Chem. Vap. Deposition*, 2004, **10**, 143.
230. M. Kamoshida, I. V. Mitchell and J. W. Mayer, *Appl. Phys. Lett.*, 1971, **18**, 292.
231. D. A. Mehta, S. R. Butler and F. J. Feigl, *J. Electrochem. Soc.*, 1973, **120**, 1707.
232. T. H. Hua and M. Armgarth, *J. Electron. Mater.*, 1987, **16**, 27.
233. R. S. Ehle, B. J. Baliga and W. Katz, *J. Electron. Mater.*, 1983, **12**, 587.
234. M. Ishida, I. Katakabe, T. Nakamura and N. Ohtake, *Appl. Phys. Lett.*, 1988, **52**, 1326.
235. L. H. Hall and W. C. Robinett, *J. Electrochem. Soc.*, 1971, **118**, 1624.
236. J. A. Aboaf, *J. Electrochem. Soc.*, 1967, **114**, 948.
237. J. Saraie, J. Kown and Y. Yodogawa, *J. Electrochem. Soc.*, 1985, **132**, 890.
238. S. S. Yom, W. N. Kang, Y. S. Yoon, J. I. Lee, D. J. Choi, T. W. Kim, K. Y. Seo, P. H. Hur and C. Y. Kim, *Thin Solid Films*, 1992, **213**, 72.
239. A. C. Jones, D. J. Houlton, S. A. Rushworth and G. W. Critchlow, *J. Phys. IV*, 1995, **5**, 557.
240. L. A. Ryabova and Y. S. Savitska, *Thin Solid Films*, 1968, **2**, 141.
241. O. B. Ajayi, M. S. Akanni, J. N. Lambi, C. Jeynes and J. F. Watts, *Thin Solid Films*, 1990, **185**, 123.
242. T. Maruyama and S. Arai, *Appl. Phys. Lett.*, 1992, **60**, 322.
243. J. S. Kim, H. A. Marzouk, P. J. Reucroft, J. D. Robertson and C. E. Hamrin, *Thin Solid Films*, 1993, **230**, 156.
244. T. Suntola, J. Antson, A. Pakkala and S. Lindfors, *SID Digest*, 1980, **11**, 108.
245. L. Hiltunen, H. Kattelus, M. Leskelä, M. Makela, L. Niinistö, E. Nykanen, P. Soininen and M. Tiitta, *Mater. Chem. Phys.*, 1991, **28**, 379.
246. G. S. Higashi and C. G. Fleming, *Appl. Phys. Lett.*, 1989, **55**, 1963.
247. J. F. Fan, K. Sugioka and K. Toyoda, *Jpn. J. Appl. Phys. Part 2*, 1991, **30**, L1139.
248. M. Juppo, A. Rahtu, M. Ritala and M. Leskelä, *Langmuir*, 2000, **16**, 4034.
249. Y. S. Min, Y. J. Cho and C. S. Hwang, *Chem. Mater.*, 2005, **17**, 626.
250. H. Treichel, A. Mitwalsky, G. Tempel, G. Zorn, W. Kern, N. Sandler and P. A. Lane, *Mater. Res. Soc. Symp. Proc.*, 1993, **282**, 557.
251. M. Matsui, S. Oka, K. Yamagishi, K. Kuroiwa and Y. Tarui, *Jpn. J. Appl. Phys. Part 1*, 1988, **27**, 506.
252. M. Matsui, H. Nagayoshi, G. Muto, S. Tanimoto, K. Kuroiwa and Y. Tarui, *Jpn. J. Appl. Phys. Part 1*, 1990, **29**, 62.
253. K. Tominaga, R. Muhammet, I. Kobayashi and M. Okada, *Jpn. J. Appl. Phys. Part 2*, 1992, **31**, L585.
254. R. A. B. Devine, L. Vallier, J. L. Autran, P. Paillet and J. L. Leray, *Appl. Phys. Lett.*, 1996, **68**, 1775.
255. K. Yamagishi and Y. Tarui, *Jpn. J. Appl. Phys. Part 2*, 1986, **25**, L306.
256. C. H. An and K. Sugimoto, *J. Electrochem. Soc.*, 1992, **139**, 1956.
257. S. Zaima, T. Furuta, Y. Yasuda and M. Iida, *J. Electrochem. Soc.*, 1990, **137**, 1297.
258. W. R. Hitchens, W. C. Krusell and D. M. Dobkin, *J. Electrochem. Soc.*, 1993, **140**, 2615.
259. E. P. Burte and N. Rausch, *J. Non-Cryst. Solids*, 1995, **187**, 425.
260. W. Kern, A. Chen and N. Sandler, *Journal De Physique Ii*, 1991, **1**, 311.

261. S. Kamiyama, P. Y. Lesaicherre, H. Suzuki, A. Sakai, I. Nishiyama and A. Ishitani, *J. Electrochem. Soc.*, 1993, **140**, 1617.
262. K. A. Son, A. Y. Mao, Y. M. Sun, B. Y. Kim, F. Liu, A. Kamath, J. M. White, D. L. Kwong, D. A. Roberts and R. N. Vrtis, *Appl. Phys. Lett.*, 1998, **72**, 1187.
263. K. Kukli, M. Ritala and M. Leskelä, *J. Electrochem. Soc.*, 1995, **142**, 1670.
264. A. Mansingh, *Ferroelectrics*, 1990, **102**, 69.
265. K. A. Fleeting, P. O'Brien, D. J. Otway, A. J. P. White, D. J. Williams and A. C. Jones, *Inorg. Chem.*, 1999, **38**, 1432.
266. A. C. Jones, T. J. Leedham, P. J. Wright, D. J. Williams, M. J. Crosbie, H. O. Davies, K. A. Fleeting and P. O'Brien, *J. Eur. Ceram. Soc.*, 1999, **19**, 1431.
267. M. Morstein, I. Pozsgai and N. D. Spencer, *Chem. Vap. Deposition*, 1999, **5**, 151.
268. D. H. Liu and H. Chen, *Mater. Lett.*, 1996, **28**, 17.
269. M. J. Crosbie, P. J. Wright, D. J. Williams, P. A. Lane, J. Jones, A. C. Jones, T. J. Leedham, P. O'Brien and H. O. Davies, *J. Phys. IV*, 1999, **9**, 935.
270. A. C. Jones, H. O. Davies, T. J. Leedham, M. J. Crosbie, P. J. Wright, P. O'Brien and K. A. Fleeting, *Mater. Res. Soc. Symp. Proc.*, 2000, **606**, 51.
271. A. C. Jones, H. O. Davies, T. J. Leedham, P. J. Wright, P. A. Lane, M. J. Crosbie, D. J. Williams, J. C. Jones and C. L. Reeves, *Integrated Ferroelectrics*, 2000, **30**, 19.
272. P. A. Williams, A. C. Jones, P. J. Wright, M. J. Crosbie, J. F. Bickley, A. Steiner, H. O. Davies and T. J. Leedham, *Chem. Vap. Deposition*, 2002, **8**, 110.
273. K. A. Fleeting, H. O. Davies, A. C. Jones, P. O'Brien, T. J. Leedham, M. J. Crosbie, P. J. Wright and D. J. Williams, *Chem. Vap. Deposition*, 1999, **5**, 261.
274. J. P. Maria, W. Hackenberger and S. Trolier-McKinstry, *J. Appl. Phys.*, 1998, **84**, 5147.
275. K. R. Udayakumar, J. Chen, P. J. Schuele, L. E. Cross, V. Kumar and S. B. Krupanidhi, *Appl. Phys. Lett.*, 1992, **60**, 1187.
276. S. Stemmer, G. R. Bai, N. D. Browning and S. K. Streiffer, *J. Appl. Phys.*, 2000, **87**, 3526.
277. Y. Takeshima, K. Shiratsuyu, H. Takagi and K. Tomono, *Jpn. J. Appl. Phys. Part 1*, 1995, **34**, 5083.
278. H. O. Davies, A. C. Jones, T. J. Leedham, P. J. Wright, M. J. Crosbie, P. A. Lane, A. Steiner and J. F. Bickley, *Adv. Mater. Optics Electron.*, 2000, **10**, 177.
279. C. A. Paz de Araujo, J. D. Cuchiaro, L. D. McMillan, M. C. Scott and J. F. Scott, *Nature*, 1995, **374**, 627.
280. H. Funakubo, K. Ishikawa, T. Watanabe, M. Mitsuya and N. Nukaga, *Advan. Mater. Optics Electron.*, 2000, **10**, 193.
281. J. F. Roeder, B. C. Hendrix, F. Hintermaier, D. A. Desrochers, T. H. Baum, G. Bhandari, M. Chappuis, P. C. Van Buskirk, C. Dehm, E. Fritsch, N. Nagel, H. Wendt, H. Cerva, W. Honlein and C. Mazure, *J. Eur. Ceram. Soc.*, 1999, **19**, 1463.
282. C. Isobe, K. Hironaka and S. Hishikawa, *Adv. Mater. Optics Electron*, 2000, **10**, 183.
283. M. J. Crosbie, P. J. Wright, H. O. Davies, A. C. Jones, T. J. Leedham, P. O'Brien and G. W. Critchlow, *Chem. Vap. Deposition*, 1999, **5**, 9.
284. W. C. Shin, K. J. Choi, E. S. Choi, C. M. Park and S. G. Yoon, *Integrated Ferroelectrics*, 2000, **30**, 27.
285. N. Kadokura, Y. Okuhara, M. Mitsuya and H. Funakubo, *Chem. Vap. Deposition*, 2000, **6**, 225.
286. A. C. Jones, N. L. Tobin, P. A. Marshall, R. J. Potter, P. R. Chalker, J. F. Bickley, H. O. Davies, L. M. Smith and G. W. Critchlow, *J. Mater, Chem*, 2004, **14**, 887.
287. R. C. Mehrotra and A. K. Rai, *Indian Journal of Chemistry*, 1966, **4**, 537.
288. A. Haaland, H. P. Verne, H. V. Volden, R. Papiernik and L. G. Hubert-Pfalzgraf, *Acta Chem. Scand.*, 1993, **47**, 1043.

289. W. A. Herrmann, N. W. Huber, R. Anwander and T. Priermeier, *Chemische Berichte-Recueil*, 1993, **126**, 1127.
290. P. A. Williams, A. C. Jones, M. J. Crosbie, P. J. Wright, J. F. Bickley, A. Steiner, H. O. Davies, T. J. Leedham and G. W. Critchlow, *Chem. Vap. Deposition*, 2001, **7**, 205.
291. P. R. Chalker, R. J. Potter, J. L. Roberts, A. C. Jones, L. M. Smith and M. Schumacher, *J. Cryst. Growth*, 2004, **272**, 778.
292. M. Vehkamäki, M. Ritala, M. Leskelä, A. C. Jones, H. O. Davies, T. Sajavaara and E. Rauhala, *J. Electrochem. Soc.*, 2004, **151**, F69.
293. M. Vehkamäki, T. Hatanpaa, M. Ritala and M. Leskelä, *J. Mater. Chem.*, 2004, **14**, 3191.
294. J. F. Scott and C. A. Paz de Araujo, *Science*, 1989, **246**, 1400.
295. J. Si and S. B. Desu, *J. Appl. Phys.*, 1993, **73**, 7910.
296. H. Wang, L. W. Fu and S. X. Shang, *J. Appl. Phys.*, 1993, **73**, 7963.
297. B. Jiang, J. L. Peng and L. A. Bursill, *Mod. Phys. Lett. B*, 1999, **13**, 933.
298. T. Nakamura, R. Muhammet, M. Shimizu and T. Shiosaki, *Jpn. J. Appl. Phys. Part 1*, 1993, **32**, 4086.
299. D. A. Neumayer, P. R. Duncombe, R. B. Laibowitz, T. Shaw, R. Purtell and A. Grill, *Integrated Ferroelectrics*, 1998, **21**, 331.
300. T. Watanabe, K. Saito and H. Funakubo, *J. Cryst. Growth*, 2002, **235**, 389.
301. M. Schuisky, K. Kukli, M. Ritala, A. Harsta and M. Leskelä, *Chem. Vap. Deposition*, 2000, **6**, 139.
302. Y. J. Cho, Y. S. Min, J. H. Lee, B. S. Seo, J. K. Lee, Y. S. Park and J. H. Choi, *Integrated Ferroelectrics*, 2003, **59**, 1483.
303. P. Y. Lesaicherre, H. Yamaguchi, Y. Miyasaka, H. Watanabe, H. Ono and M. Yoshida, *Integrated Ferroelectrics*, 1995, **8**, 201.
304. R. A. McKee, F. J. Walker and M. F. Chisholm, *Mater. Res. Soc. Symp. Proc.*, 1999, **567**, 415.
305. K. Eisenbeiser, J. M. Finder, Z. Yu, J. Ramdani, J. A. Curless, J. A. Hallmark, R. Droopad, W. J. Ooms, L. Salem, S. Bradshaw and C. D. Overgaard, *Appl. Phys. Lett.*, 2000, **76**, 1324.
306. M. Kiyotoshi and K. Eguchi, *Appl. Phys. Lett.*, 1995, **67**, 2468.
307. H. Yamaguchi, P. Y. Lesaicherre, T. Sakuma, Y. Miyasaka, A. Ishitani and M. Yoshida, *Jpn. J. Appl. Phys. Part 1*, 1993, **32**, 4069.
308. T. T. Arai, Y. Inaishi, Y. Sawado, I. Kobayashi and J. Hidaka, *Jpn. J. Appl. Phys. Part 1*, 1996, **35**, 4875.
309. S. R. Gilbert, B. W. Wessels, D. B. Studebaker and T. J. Marks, *Appl. Phys. Lett.*, 1995, **66**, 3298.
310. J. P. Senateur, C. Dubourdieu, F. Weiss, M. Rosina and A. Abrutis, *Adv. Mater. Optics Electron.*, 2000, **10**, 155.
311. C. Dubourdieu, H. Roussel, C. Jimenez, M. Audier, J. P. Senateur, S. Lhostis, L. Auvray, F. Ducroquet, B. J. O'Sullivan, P. K. Hurley, S. Rushworth and L. Hubert-Pfalzgraf, *Materials Science and Engineering B-Solid State Materials for Advanced Technology*, 2005, **118**, 105.
312. D. L. Kaiser, M. D. Vaudin, L. D. Rotter, Z. L. Wang, C. S. Cline, C. S. Hwang, R. B. Marinenko and J. G. Gillen, *Mater. Res. Soc. Symp. Proc.*, 1995, **361**, 355.
313. B. S. Kwak, K. Zhang, E. P. Boyd, A. Erbil and B. J. Wilkens, *J. Appl. Phys.*, 1991, **69**, 767.
314. J. H. Lee, Y. J. Cho, Y. S. Min, D. Kim and S. W. Rhee, *Journal of Vacuum Science & Technology a-Vacuum Surfaces and Films*, 2002, **20**, 1828.
315. D. S. Kil, J. M. Lee and J. S. Roh, *Chem. Vap. Deposition*, 2002, **8**, 195.
316. K. Sakiyama, S. Onishi, K. Ishihara, K. Orita, T. Kajiyama, N. Hosoda and T. Hara, *J. Electrochem. Soc.*, 1993, **140**, 834.
317. L. Krusin-Elbaum, M. Wittmer and D. S. Yee, *Appl. Phys. Lett.*, 1987, **50**, 1879.

318. H. Maiwa, N. Ichinose and K. Okazaki, *Jpn. J. Appl. Phys. Part 1*, 1994, **33**, 5223.
319. M. L. Green, M. E. Gross, L. E. Papa, K. J. Schnoes and D. Brasen, *J. Electrochem. Soc.*, 1985, **132**, 2677.
320. J. H. Lee, J. Y. Kim, S. W. Rhee and W. Rhee, *Electrochemical and Solid State Letters*, 1999, **2**, 622.
321. J. M. Lee, J. C. Shin, C. S. Hwang, H. J. Kim and C. G. Suk, *J. Vac. Sci. Technol. A*, 1998, **16**, 2768.
322. P. Hones, T. Gerfin and M. Gratzel, *Appl. Phys. Lett.*, 1995, **67**, 3078.
323. T. Aoyama and K. Eguchi, *Jpn. J. Appl. Phys. Part 2*, 1999, **38**, L1134.
324. S. Uhlenbrock and B. A. Vaartstra (Micron Technology, Inc., USA). US Patent No. 6, 114, 557 (1999).
325. Y. Senzaki, D. Colombo, W. L. Gladfelter and F. B. McCormick, *Proc.-Electrochem. Soc.*, 1997, **97–25**, 933.
326. R. Breitkopf, L. J. Meda, T. Haas and R. U. Kirss, *Mater. Res. Soc. Symp. Proc.*, 1998, **495**, 51.
327. D. L. Polla, R. S. Muller and R. M. White, *IEEE Electron Device Lett.*, 1986, **7**, 254.
328. B. A. Martin, S. W. Wenzel and R. M. White, *Sensors and Actuators a-Physical*, 1990, **22**, 704.
329. J. A. Aranovich, D. Golmayo, A. L. Fahrenbruch and R. H. Bube, *J. Appl. Phys.*, 1980, **51**, 4260.
330. G. S. Kino and R. S. Wagers, *J. Appl. Phys.*, 1973, **44**, 1480.
331. A. P. Roth and D. F. Williams, *J. Appl. Phys.*, 1981, **52**, 6685.
332. W. Kern and R. Heim, *J. Electrochem. Soc.*, 1970, **117**, 562.
333. S. K. Ghandhi, R. J. Field and J. R. Shealy, *Appl. Phys. Lett.*, 1980, **37**, 449.
334. F. T. J. Smith, *Appl. Phys. Lett.*, 1983, **43**, 1108.
335. C. K. Lau, S. K. Tiku and K. M. Lakin, *J. Electrochem. Soc.*, 1980, **127**, 1843.
336. R. Solanki and G. J. Collins, *Appl. Phys. Lett.*, 1983, **42**, 662.
337. S. Oda, H. Tokunaga, N. Kitajima, J. Hanna, I. Shimizu and H. Kokado, *Jpn. J. Appl. Phys. Part 1*, 1985, **24**, 1607.
338. T. Kaufmann, G. Fuchs, M. Webert, S. Frieske and M. Gackle, *Cryst. Res. Technol.*, 1989, **24**, 269.
339. J. Auld, D. J. Houlton, A. C. Jones, S. A. Rushworth, M. A. Malik, P. O'Brien and G. W. Critchlow, *J. Mater. Chem.*, 1994, **4**, 1249.
340. V. P. Ovsiannikov, G. V. Lashkarev, Y. A. Mazurenko and M. E. Bugaeva, *Proc.-Electrochem. Soc.*, 1997, **97–25**, 1020.
341. P. A. Lane, P. J. Wright, M. J. Crosbie, A. D. Pitt, C. L. Reeves, B. Cockayne, A. C. Jones and T. J. Leedham, *J. Cryst. Growth*, 1998, **192**, 423.
342. M. Fiebig, *Journal of Physics D-Applied Physics*, 2005, **38**, R123.
343. J. Thery, C. Dobourdieu, T. Baron, C. Ternon, H. Roussel and F. Pierre, *Chem. Vap. Deposition*, 2006, **13**, 232.
344. J. G. Bednorz and K. A. Muller, *Zeitschrift Fur Physik B-Condensed Matter*, 1986, **64**, 189.
345. M. K. Wu, J. R. Ashburn, C. J. Torng, P. H. Hor, R. L. Meng, L. Gao, Z. J. Huang, Y. Q. Wang and C. W. Chu, *Phys. Rev. Lett.*, 1987, **58**, 908.
346. H. Maeda, Y. Tanaka, M. Fukutomi and T. Asano, *Jpn. J. Appl. Phys. Part 2*, 1988, **27**, L209.
347. Z. Z. Sheng and A. M. Hermann, *Nature*, 1988, **332**, 138.
348. T. L. Hylton and M. R. Beasley, *Phys. Rev. B*, 1990, **41**, 11669.
349. L. P. Lee, K. Char, M. S. Colclough and G. Zaharchuk, *Appl. Phys. Lett.*, 1991, **59**, 3051.
350. M. Schieber, *J. Cryst. Growth*, 1991, **109**, 401.
351. M. Leskelä, H. Mölsä and L. Niinistö, *Supercond. Sci. Technol.*, 1993, **6**, 627.
352. K. H. Dahmen and T. Gerfin, *Progress in Crystal Growth and Characterization of Materials*, 1993, **27**, 117.

353. C. Gómez-Aleixandre, O. Sanchez, J. M. Albella, J. Santiso and A. Figueras, *Adv. Mater.*, 1995, **7**, 111.
354. S. B. Turnipseed, R. M. Barkley and R. E. Sievers, *Inorg. Chem.*, 1991, **30**, 1164.
355. A. Drozdov and S. Troyanov, *J. Phys. IV*, 1995, **5**, 373.
356. A. R. Barron and W. S. Rees, *Adv. Mater. Opt. Electron.*, 1993, **2**, 271.
357. J. M. Zhang, F. Dimeo, B. W. Wessels, D. L. Schulz, T. J. Marks, J. L. Schindler and C. R. Kannewurf, *J. Appl. Phys.*, 1992, **71**, 2769.
358. P. H. Dickinson, T. H. Geballe, A. Sanjurjo, D. Hildenbrand, G. Craig, M. Zisk, J. Collman, S. A. Banning and R. E. Sievers, *J. Appl. Phys.*, 1989, **66**, 444.
359. S. Matsuno, F. Uchikawa and K. Yoshizaki, *Jpn. J. Appl. Phys. Part 2*, 1990, **29**, L947.
360. J. Zhao, D. W. Noh, C. Chern, Y. Q. Li, P. Norris, B. Gallois and B. Kear, *Appl. Phys. Lett.*, 1990, **56**, 2342.
361. C. S. Chern, J. Zhao, Y. Q. Li, P. Norris, B. Kear and B. Gallois, *Appl. Phys. Lett.*, 1990, **57**, 721.
362. R. Singh, S. Sinha, N. J. Hsu, J. T. C. Ng, P. Chou, R. P. S. Thakur and J. Narayan, *J. Appl. Phys.*, 1991, **69**, 2418.
363. R. Singh, J. T. C. Ng, R. K. Singh, F. Qian, N. J. Hsu and D. J. Krueger, *IEEE Transactions on Applied Superconductivity*, 1993, **3**, 1.
364. J. M. Buriak, L. K. Cheatham, J. J. Graham, R. G. Gordon and A. R. Barron, *Mater. Res. Soc. Symp. Proc.*, 1991, **204**, 545.
365. R. Gardiner, D. W. Brown, P. S. Kirlin and A. L. Rheingold, *Chem. Mater.*, 1991, **3**, 1053.
366. G. Malandrino, D. S. Richeson, T. J. Marks, D. C. Degroot, J. L. Schindler and C. R. Kannewurf, *Appl. Phys. Lett.*, 1991, **58**, 182.
367. S. J. Duray, D. B. Buchholz, S. N. Song, D. S. Richeson, J. B. Ketterson, T. J. Marks and R. P. H. Chang, *Appl. Phys. Lett.*, 1991, **59**, 1503.
368. S. J. Duray, D. B. Buchholz, H. Zhang, S. N. Song, D. L. Schulz, V. P. Dravid, T. J. Marks, J. B. Ketterson and R. P. H. Chang, *Journal of Vacuum Science & Technology a-Vacuum Surfaces and Films*, 1993, **11**, 1346.
369. C. I. M. A. Spee, E. A. Van der Zouwen-Assink, K. Timmer, A. Mackor and H. A. Meinema, *J. Phys. IV*, 1991, **1**, 295.
370. J. M. Zhang, B. W. Wessels, D. S. Richeson, T. J. Marks, D. C. Degroot and C. R. Kannewurf, *J. Appl. Phys.*, 1991, **69**, 2743.
371. D. L. Schulz, D. S. Richeson, G. Malandrino, D. Neumayer, T. J. Marks, D. C. Degroot, J. L. Schindler, T. Hogan and C. R. Kannewurf, *Thin Solid Films*, 1992, **216**, 45.
372. D. A. Neumayer, D. L. Schulz, D. S. Richeson, T. J. Marks, D. C. Degroot, J. L. Schindler and C. R. Kannewurf, *Thin Solid Films*, 1992, **216**, 41.
373. H. Zama, K. Sakai and S. Oda, *Jpn. J. Appl. Phys. Part 2*, 1992, **31**, L1243.
374. S. Oda, H. Zama and S. Yamamoto, *J. Cryst. Growth*, 1994, **145**, 232.
375. W. S. Rees, H. A. Luten, M. W. Carris, C. R. Caballero, W. Hesse and V. L. Goedken, *Mater. Res. Soc. Symp. Proc.*, 1993, **310**, 375.
376. B. C. Richards, S. L. Cook, D. L. Pinch and G. W. Andrews, *J. Phys. IV*, 1995, **5**, 407.
377. H. A. Meinema, K. Timmer, J. L. Linden and C. I. M. A. Spee, *Mater. Res. Soc. Symp. Proc.*, 1994, **335**, 193.
378. K. Endo, S. Hayashida, J. Ishiai, Y. Matsuki, Y. Ikedo, S. Misawa and S. Yoshida, *Jpn. J. Appl. Phys. Part 2*, 1990, **29**, L294.
379. M. Nemoto and M. Yamanaka, *J. Mater. Res.*, 1990, **5**, 1.
380. N. Hamaguchi, J. Vigil, R. Gardiner and P. S. Kirlin, *Jpn. J. Appl. Phys. Part 2*, 1990, **29**, L596.

CHAPTER 9

Metal-organic Chemical Vapour Deposition of Refractory Transition Metal Nitrides

ROLAND A. FISCHER AND HARISH PARALA

Lehrstuhl für Anorganische Chemie II, Organometallics and Materials, Ruhr-Universität Bochum, Universitätsstr, 150, 44801 Bochum, Germany

9.1 Introduction and Overview

Thin films and coatings of the binary nitrides MN of the refractory, "early" transition metals (M = Ti, Zr, Hf, V, Nb, Ta, Cr, Mo, W) have many applications because of their rather unique combination of chemical and physical properties. Transition metal nitrides exhibit impressive properties that include, high melting points, superior hardness, high corrosion resistance, good chemical resistance, and electrical conductivity and metallic appearance. These important characteristics have led to several useful thin film applications such as hard coatings, decorative coatings, and corrosion and abrasion resistant layers.[1,2] In addition, transition metal nitrides have been extensively investigated as diffusion barrier materials due to their good thermal stability and electrical conductivity in Cu metallization schemes in ultra-large scale integration (ULSI) devices. Besides diffusion barrier layers, transition metal nitrides and metal-silicon nitrides are promising gate electrodes for complementary metal oxide semiconductor (CMOS) devices.[3,4]

Titanium nitride, which is the technologically most important nitride of this materials family, is a better electrical conductor than titanium metal, and its hardness exceeds all elemental metals, being even close to diamond. Its melting point is around 3000 °C and its gold-like optical appearance makes it very suitable for decorative coatings.[5,6] Owing to their outstanding mechanical properties and high-temperature chemical resistance, the nitrides in general, and in particular TiN, are widely used as wear-resistant and friction-reducing protective coatings on machine tools.[7,8] Their optical properties also make them attractive as wavelength selective transparent films, in particular as solar control coatings. A heat-mirror must be a chemically and physically robust, thin conformal coating with good visible light transmission.[9] Their low electrical resistivity combined with chemical

Chemical Vapour Deposition: Precursors, Processes and Applications
Edited by Anthony C. Jones and Michael L. Hitchman

Published by the Royal Society of Chemistry, www.rsc.org

inertness allows applications of the metal nitrides as diffusion barriers and gate electrodes being compatible with novel low- and high-*k* materials.[10,11] In this respect, HfN and TaN in particular play an important role within the context of the development of advanced interconnect metallization schemes (aluminium, copper, silver) in micro-electronics.[12,13] The high-T_c superconductive properties of some of metal nitrides, *e.g.* HfN ($T_c = 9.8$ K) and NbN ($T_c = 17.3$ K), are attractive for several advanced applications in superconductive microelectronics, including high speed electronics, submillimeter heterodyne receivers, digital devices for Petaflops-scale computers, and devices for functional and topographic brain investigation using multichannel SQUIDs systems.[14–16] Notably, a new series of superconductors based on the layer structured, semiconducting nitrides of the type MNX (M = Zr, Hf; X = halide) have been studied more recently.[17] By electrons doping the nitride layers during intercalation of alkali metals the compounds become superconductors with transition temperatures as high as 13 and 25.5 K for β-ZrNCl and β-HfNCl systems, respectively.[18] However, fabrication of thin films of these materials, in particular by CVD methods, remains a challenge. We emphasize some applications of nitrides in heterogeneous catalysis.[19] Interestingly, the carbides and nitrides of Mo and W can adsorb and activate hydrogen, a property that is important for their use in hydrogenation, hydrodesulfurization and hydrodenitrogenation reactions.[20,21] CVD preparation of high-surface area metal nitride supported catalysts[22] may be advantageous over classical aqueous co-precipitation and temperature programmed calcinations/nitridation procedures.[23]

For diffusion barriers, adhesion and seed layers as well as for gate electrode applications, physical vapour deposition (PVD) techniques have been used in modern semiconductor device fabrications (>100 nm nodes). For sub-100 nm, high aspect ratio device structures, PVD-based technology has inherent limitations due to its poor conformal coverage of films, especially at the side walls and corners of the device feature.[24] Chemical vapour deposition (CVD) and its variants such as metal-organic chemical vapour deposition (MOCVD) and atomic layer deposition (ALD) techniques will be required to ensure the uniform thin film growth of nitrides.[25,26] Moreover, the growing interest in nanotechnology also needs atomically controlled nano-layer deposition techniques like ALD.[27]

The main focus of this chapter is related to two important applications of transition metal nitrides: (a) as barrier layers and (b) as gate electrodes; the following paragraphs provide a short introduction to these applications. For additional and more detailed information, the reader is directed to the more specialized books,[28,29] and reviews[3,30,39] referred to throughout this chapter.

9.2 Applications of Transition Metal Nitrides

9.2.1 Diffusion Barrier Layers

In the new generation of microelectronic devices, Cu has already been partly replaced by Al and Al alloys in ultra-large scale integrated metallization below 0.18 μm due to its higher conductivity and better resistance to electro-migration than Al and Al alloys.[31] However, the most serious problems with Cu are its fast diffusion in silicon and the formation of highly resistive Cu_3Si compound, leading to a decrease in device performance. For the successful integration of Cu metallization into integrated circuits (ICs), a better diffusion barrier layer to prevent the inter-diffusion or the reaction between the Cu and the adjacent layer is necessary. The important properties of a barrier layers include high conductivity, high thermal stability, good adhesion, chemical inertness with copper and silicon and no grain boundaries. Figure 9.1 gives a schematic view of the diffusion barrier layer. Barrier material must be stable during both manufacturing and operation of the device. Since the diffusion takes place primarily along the grain boundaries, amorphous or nanocrystalline microstructures are preferred as barrier layer.[32,33] Even very ultrathin barrier layers must perform

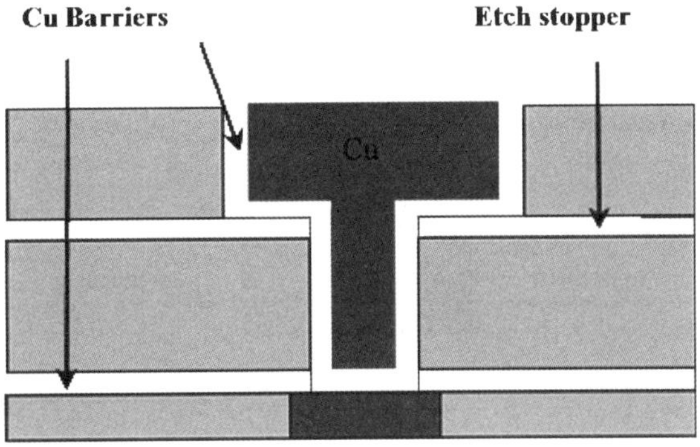

Figure 9.1 Schematic of a diffusion barrier layer.

effectively, as the barrier thickness is expected to decrease to 2.5 nm in the 22 nm mode by the year 2016.[32]

Transition metal nitrides, metal silicides and metal-silicon nitrides are widely used and extensively investigated as diffusion barrier layers. Titanium nitride is one of the most studied materials and, recently, many other nitride materials (HfN, ZrN, TaN, NbN) have attracted much interest as barrier layers.[34–38]

9.2.2 Gate Electrode Applications

In integrated circuits (ICs) such as microprocessors and semiconductor memories a metal oxide semiconductor field effect transistor (MOSFET) is one of the main components. The main parts of a MOSFET are source, drain, gate, gate oxide and channel.

A CMOS device consists of an n-channel and p-channel MOSFET integrated on the same chip. A p-channel MOS (PMOS) can be formed by changing the substrate to an n-type semiconductor, and the source and drain to p^+ regions. One PMOS and one NMOS device form a complementary metal oxide semiconductor (CMOS) structure. Figure 9.2 shows a schematic illustration of the cross-section of a modern CMOS transistor.

As the dimensions of devices are continuously scaled down, traditional materials can no longer cope. The scaling requirements for future CMOS technologies are generally guided by the International Technology Road map for Semiconductors (ITRS). The ITRS identifies the technological challenges and needs facing the semiconductor industry.[32] Due to aggressive scaling of the CMOS devices, several problems have been encountered lately, such as ploy-Si depletion, high resistivity, boron penetration and Fermi level pinning.[39] In the 45 nm technology node and beyond, metal gate electrodes in conjunction with thin high-k gate dielectric films are required to overcome problems associated with the silicon dioxide based materials and the use of polycrystalline silicon (poly-Si) as gate electrode. So far, silicon dioxide based materials have been used as gate oxides, but when the gate oxide is less than 1.2 nm thick the tunnelling currents will be too high across the gate oxide and the reliability and performance of the device will be reduced. A dielectric layer with a high dielectric constant is needed to maintain the capacitance between the gate electrode and the Si channel without decreasing the SiO_2 thickness.[40] The most widely studied high-k materials to replace SiO_2-based materials are aluminium oxide, zirconium oxide and hafnium oxide, all of which have much higher permittivity than SiO_2.[41–43] As the gate oxide thickness decreases, the capacitance associated with the depletion layer in the polycrystalline silicon gate electrode at the interface with the gate oxide becomes significant. It is expected, therefore, that metal gates will replace polycrystalline silicon, as this will eliminate the gate depletion. However, the requirements for

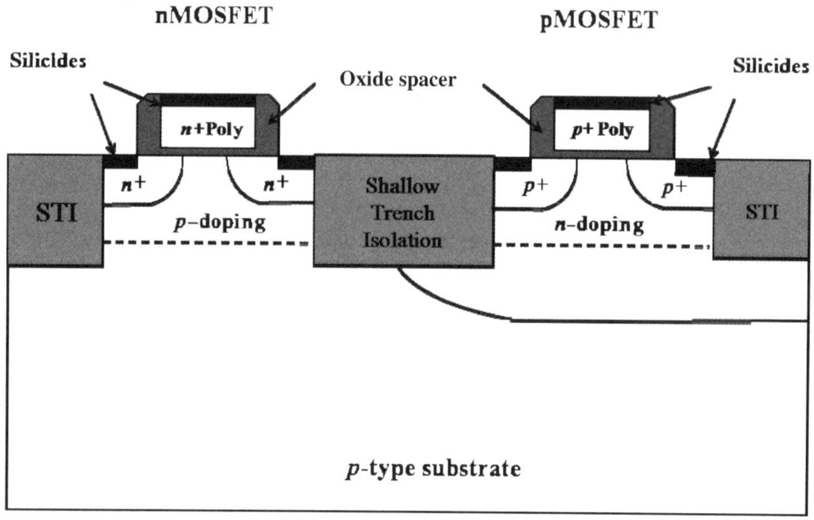

Figure 9.2 Schematic cross-section of a CMOS device.

the metal gate are complex.[3] The material has to be stable during manufacturing and operation of the device and it must not react with materials under or above, the impurity content should be low, and, most importantly, the material must have an appropriate work function.[40] Transition metal nitrides and transition-silicon-metal nitrides are considered promising gate electrode materials to replace polycrystalline silicon in CMOS devices as they have high thermal stability, comparable work functions and are inert to reaction with other materials. Among the transition metal nitrides, TaN and HfN are emerging as strong candidates.[44,45] Optimal work function for NMOS devices is 0.2 eV below the conduction band edge of silicon, while for PMOS devices it is 0.2 eV above the valence band edge of silicon.[46,47] This means that the work functions for the n-type and p-type gates should be approximately 4 and 5 eV. The work function can be measured by various methods[48] and for the gate stack in MOS-devices, the *C-V* and *I-V* methods are used.[49,50] Table 9.1 summarizes the some of the different stacks as well as extracted work functions of transition metal nitrides and extraction methods. An interval in work function is given that is due to variation in composition (nitrogen content) or variation in work function after several thermal treatments.

9.3 Crystal Chemistry and Materials Properties

9.3.1 Titanium Nitride, Zirconium Nitride and Hafnium Nitride

Owing to the relativistic effects, there is a tendency toward an increase in the stability of the highest oxidation state for the heavier transition metals.[62,63] In group IVB compounds, this tendency is well demonstrated by the higher stability of ZrO_2 and HfO_2 against reduction as compared to TiO_2, which can be readily reduced to Ti_2O_3.[64] Owing to this difference in stability, ZrO_2 and HfO_2 are better candidates as dielectric layers than TiO_2. Similarly, the higher stability of the +4 oxidation state leads to the coexistence of the metallic MN (+3 oxidation state) and insulating M_3N_4 (+4 oxidation state) phases of Zr and Hf nitrides. They adopt the rock-salt NaCl structure type and are highly conductive owing to a finite density of states at the Fermi level from the residual electron contributed by each atom in the metal d band. TiN is the only stable Ti nitride phase under normal conditions and has useful properties such as golden colour, good electrical and thermal conductivity, high melting point and hardness, good chemical inertness and adhesion to other metals,

Table 9.1 Work function of important transition metal nitride compounds.

Metal nitride	*Dielectric material*	*Work function (eV)*	*Method*	*Ref.*
TiN	SiO_2	4.15–5.10	MOS capacitor V_{fb} and FN tunnelling	51–53
TiN	Al_2O_3	4.85–5.2	MOS capacitor V_{fb}	52
TiAlN	SiO_2	4.36–5.1	MOS capacitor V_{fb}	54
ZrN	SiO_2	4.0–4.9	MOS capacitor V_{fb} and FN tunnelling	55
HfN	SiO_2	4.65–4.70	MOS capacitor V_{fb}	56
HfN	HfO_2	4.75–4.80	MOS capacitor V_{fb}	56
TaN	HfO_2	4.4–4.65	MOS capacitor V_{fb}	57
TaN	SiO_2	4.4–4.7	MOS capacitor V_{fb}	58
TaN	HfO_2	4.34–4.41	MOS capacitor V_{fb}	58
TaN	HfO_2	4.5–4.8	MOS capacitor V_{fb}	59
TaSiN	HfO_2	4.4–4.5	MOS capacitor V_{fb}	57
MoN	SiO_2	4.7–5.33	MOS capacitor V_{fb}	60
NbN	SiO_2	4.7	MOS capacitor V_{fb}	61
NbN	HfO_2	4.9	MOS capacitor V_{fb}	61
NbSiN	SiO_2	4.7	MOS capacitor V_{fb}	61
NbSiN	HfO_2	4.8	MOS capacitor V_{fb}	61

semiconductors and insulators.[65] Thin films consisting of Group IVB element mono-nitrides deposited using various chemical and physical vapour deposition techniques are widely used in many fields of applications, ranging from cutting tools to diffusion barrier layers. It is generally difficult to precisely control and vary the concentration of nitrogen in the nitride thin films and metastable higher nitrides with orthorhombic M_3N_4 (M = Zr, Hf) structures have been reported.[66,67] Nitrogen-rich films are transparent and less conductive than mononitrides. Recently, hard cubic Zr_3N_4 thin films have been grown using the modified filtered cathodic arc (FCA) method and their use in wear and oxidation resistant applications was demonstrated.[66] The preparation and characterization of hard multilayers containing TiN and/or ZrN materials have been reviewed by Ziebert *et al.*[68] Table 9.2 summarizes some of the phases of Ti, Zr and Hf nitrides.

9.3.2 Niobium Nitride and Tantalum Nitride

Tantalum and niobium nitride materials exist in various stable and metastable phases and several crystalline phases exist in the phase diagram of Ta-N and Nb-N.[69] When compared to Ta-N phases, there are no reports on the dielectric phase of Nb-N system and it can be expected that all phases of Nb-N are conductive and show metallic behavior.[70–72] Figure 9.3 displays all reported and identified Nb-N phases in order of increasing nitrogen content.

Most of applications of NbN are related to its superconducting properties.[73–76] The superconductivity of NbN was reported long ago, in 1941 by Justi *et al.*, and the highest T_c measured for NbN is approximately 17.3 K.[77] Alloys of Nb-Ti-N also showed superconducting behavior.[78] Application of NbN in low temperature superconducting electronics such as tunnel junctions as well as in single photon detectors has been reported.[79,80] NbN has also been investigated as corrosion resistant coating,[81] wear protective layer[82] and a possible cathode material in vacuum microelectronic devices.[83,84] The mechanical properties of NbN have been enhanced by depositing TiN/NbN, W/NbN, TaN/NbN and Si_3N_4/NbN superlattice films.[85–90] In addition, NbN is also an interesting material for diffusion barrier application due to its low resistivity, high melting point and chemical inertness.[91,92]

Table 9.2 Crystal structures and corresponding JCPDS data files for Ti-N, Zr-N and Hf-N phases.

M-N phase	*Structure*	*Space group*	*Lattice constants*	*JCPDS file*
Ti_2N	Tetragonal	P42/mnm	$a=4.9452$, $c=3.0342$	017-0386
Ti_2N	Tetragonal (BC)	141/amd	$a=4.14$, $c=8.805$	023-1455
TiN	Cubic	Fm3m	$a=4.24173$	038-1420
ζ-Ti_4N_{3-x}	Rhombohedral	R-3m	$a=2.9783$, $c=28.9438$	039-015
η-Ti_3N_{2-x}	Rhombohedral	R-3m	$a=2.9809$, $c=21.6642$	040-0958
α-$TiN_{0.30}$	Hexagonal	P63/mmc	$a=2.9737$, $c=4.7917$	041-1352
$TiN_{0.26}$	Hexagonal	0	$a=2.956$, $c=4.765$	044-1095
ZrN	Cubic	Fm3m	$A=4.56$	002-0956
ZrN	Cubic	Fm3m	$A=4.574$	031-1493
ZrN	Cubic	Fm3m	$A=4.57756$	035-0753
α-$ZrN_{0.28}$	Hexagonal	P63/mmc	$a=3.222$, $c=5.1755$	040-1275
α-Zr_2N	Orthorhombic	Pnnm	$a=5.634$, $b=5.23$, $c=3.253$	046-1204
Hf_3N_2	Rhombohedral	3m	$a=3.206$, $c=23.26$	23-1097
Hf_4N_3	Rhombohedral	3m	$a=3.214$, $c=31.16$	24-0465
HfN	Cubic	Fm3m	$A=4.5253$	33-0592
α-HfN	Hexagonal	P63/mmc	$A=3.221$, $c=5.157$	40-1277

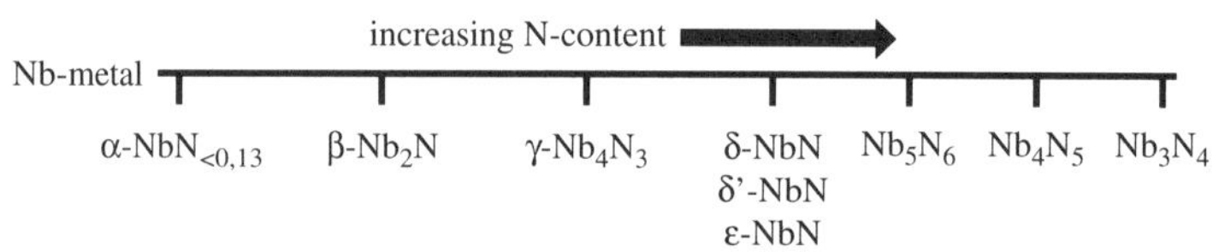

Figure 9.3 Niobium nitride phases (increasing nitrogen content from left to right).

Tantalum forms compounds with nitrogen in oxidation states from nearly zero up to +v.[93,94] Table 9.3 shows the crystal structures and corresponding JCPDS date files for Ta-N and Nb-N phases. At the highest oxidation state (+v) tantalum forms a dielectric Ta_3N_5 phase with nitrogen. In the lower oxidation state, it forms two conductive phases of TaN and Ta_2N and these phases have high melting points and they are extremely hard material.[95] In addition, due to the good conductivity of these phases, they are very interesting for many applications. Tantalum nitrides have been investigated extensively in microelectronic applications, including gate electrodes,[96–100] passivation layers against copper oxidation[101,102] and diffusion barrier layers between metals and silicon.[103–113] TaN has also been used as protective and hard coatings in many sensors and actuators,[114,115] and as a non-magnetic interlayer in NiFeCo/TaN/NiFeCo non-volatile random access memory applications (MRAM).[116] It is also used as a potential biomedical material in developing a artificial heart valve application.[117]

9.3.3 Molybdenum Nitride and Tungsten Nitride

Molybdenum nitride (1750 °C) is thermodynamically relatively stable. Mo_2N has been observed to decompose at 790 °C (Table 9.4).[118] Molybdenum nitride films have very low resistivities (~100 μΩ cm)[119] and the hexagonal MoN phase has also been reported to show superconductivity, with T_c up to 12 K.[120] Molybdenum nitrides have attracted considerable attention because they show excellent catalytic properties, resembling those of noble metals in many hydroprocessing reactions.[121,122] They have been used in hydrodesulfurization,[123,124] hydro-denitrogenation,[125]

Table 9.3 Crystal structures and corresponding JCPDS data files for Nb-N and Ta-N, phases.

Nb/Ta-N phase	*Structure*	*Space group*	*Lattice constants*	*JCPDS file*
$Nb_{4.62}N_{2.14}$	Hexagonal	–	–	30-0869
NNb_2	Hexagonal	–	–	40-1274
Nb_4N_3	Tetragonal	–	$a=4.382$, $c=4.316$	20-0803
$Nb_4N_{3.92}$	Cubic	–	–	34-0337
$NbN_{0.95}$	Hexagonal	P63/mmc	$a=2.968$, $c=5.535$	25-1361
NbN	Hexagonal	P63/mmc	$a=2.96$, $b=2.96$, $c=11.27$	20-0801
δ-$NbN_{1.000}$	Cubic	Fm3m	$a=4.3927$	38-1155
NbN	Cubic	–	–	43-1420
NbN	Cubic	–	–	43-1421
NbN	Hexagonal	–	–	14-0547
β-$TaN_{0.04}$	Cubic	1	$a=10.09$	14-0471
Ta_3N_5	Orthorhombic	Cmcm	$a=3.893$, $b=10.264$, $c=10.264$	32-1283
$TaN_{0.1}$	Cubic	Im3m	$a=3.369$	25-1278
$TaN_{0.8}$	Hexagonal	P6m2	$a=2.931$, $c=2.879$	25-1279
Ta_2N	Hexagonal	P63/mmc	$a=3.0445$, $c=4.9141$	26-0985
$Ta_6N_{2.57}$	Hexagonal	P-3lm	$a=5.285$, $c=4.919$	3-1370
Ta_4N	Orthorhombic	C	$a=5.16$, $b=3.11$, $c=9.94$	32-1282
TaN	Cubic	Fm3m	$a=4.33$	32-1283
TaN	Hexagonal	P6/mmc	$a=5.1918$, $c=2.9081$	39-1485

Table 9.4 Crystal structures and corresponding JCPDS data files for Mo-N and W-N phases.

Mo/W-N phase	*Structure*	*Space group*	*Lattice constants*	*JCPDS file*
$Mo_{16}N_7$	Tetragonal	–	$a=8.41$, $c=8.05$	23-1256
Mo_2N	Tetragonal	I41/amd		24-0768
Mo_2N	Tetragonal	I41/amd	$a=4.21$, $c=8.06$	25-1368
γ-Mo_2N	Cubic	Pm3m		25-1366
MoN	Hexagonal	P63/mmc	$a=4.188$, $c=8.048$	25-1367
δ-WN	Hexagonal	P6m2	$a=4.163$	25-1256
β-W_2N	Cubic	Pm3m	$a=5.725$, $c=5.608$	25-1257
			$a=2.893$, $c=2.826$	
			$a=4.126$	

hydrogenation of alkadienes,[126] ammonia synthesis[127] and catalytic decomposition of ammonia[128] and hydrazine.[129] Both MoN and γ-Mo_2N phases exhibit a high bulk modulus,[130,131] and both have been studied as hard coatings.[132,133] Good mechanical properties were observed for the ternary phases of Mo-Ti-N[134,135] and $MoSi_2N_x$.[136] Molybdenum nitride films have also been investigated as diffusion barrier layers.[137–139]

The first crystallographic analysis of tungsten nitride was made in 1930 by Hägg,[140] who identified β-tungsten nitride, which has the composition W_2N (Figure 9.4) and was synthesized by the treatment of metal powder with ammonia. This β-phase was later confirmed by electron diffraction.[141] Kiesling and Liu postulated a new γ-phase in 1951.[142] The first well characterized γ-tungsten nitride was synthesized by Neugebauer *et al.* using β-tungsten powder with dry ammonia to obtain a tetragonal phase of WN.[143] A cubic γ-phase of stoichiometry W_3N_4 was synthesized by heating α-W in an atmosphere of hydrogen and ammonia.[144] A hexagonal δ-phase of tungsten nitride and the rhombohedral WN_2 have been described by Khitrova and Pinsker.[145–150]

Tungsten nitride is also a promising barrier material for copper metallization because of its refractory nature and excellent chemical, thermal and mechanical properties.[118] WN has the lowest electrical resistivity (single crystal) of all transition metal nitrides. There are no known compounds

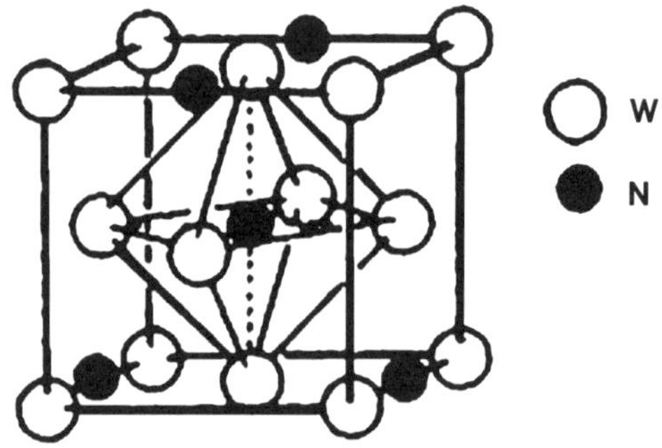

Figure 9.4 Crystal structure of β-W_2N.

that tungsten nitride forms with copper, indicating the stability of WN with copper. It can also be deposited in amorphous form without grain boundaries and there is no fast diffusion into copper. Multi-layered films of CrWN, CrN/WN[151,152] and TiN/WN[153] have been deposited to enhance mechanical properties.

9.4 Thin Film Deposition of Transition Metal Nitrides

The typical method for metal nitride thin films deposition for fundamental studies on physical materials properties and as well for device fabrication in the microelectronic industry is sputtering using metal targets and argon ion bombardment in combination with plasma or hot filament activation of dinitrogen to generate the required nitrogen radicals.[154] There are numerous versions of this general principle of the delivery the individual atomic species of the metal nitride film to the substrate.[155] However, with shrinking dimensions of the microelectronic devices, reaching the lower nanometre range, and the rising three-dimensional complexity of substrates, physical vapour deposition techniques are approaching their limits. The general strengths of chemical vapour deposition, in contrast, are high throughput and large-scale deposition, which is particularly relevant for metal nitride as protective coatings for machine tools. In addition, CVD offers superior step coverage in the case of high aspect ratio substrates, as demanded for ultra-large scale integration in microelectronics. Obviously, the line of sight PVD methods are not suited for coating of internal surfaces of porous materials of any kind, which is required for catalyst preparation in a fluidized bed reactor, for example.[156] CVD, however, is heavily dependent on the availability of molecular precursors with the desired physical properties (volatility) and chemical reactivity (selective decomposition). Both PVD and CVD of nitride materials share the problem of an efficient nitrogen source. Thermal activation of dinitrogen or ammonia, by far the most important N-sources, requires conditions above 1000 °C, which is the typical regime for thermodynamically controlled classical CVD of refractory nitrides, employing metal halides MX_n (X = F, Cl, Br, I), *e.g.* TiX_4, TaX_5, WX_6, as precursors.[157–159] Plasma-enhanced CVD processes have been developed to overcome the associated problems most prominently in the context of the nitrides as hard protective coatings. These type of CVD processes, thermal and plasma activated, using the metal halides as precursors have been covered extensively in earlier reviews and monographs on our subject and are not treated further here.[160,161]

9.4.1 Precursor Chemistry for MOCVD and ALD of Nitrides

The first reports of metal-organic chemical vapour deposition (MOCVD) of metal nitrides appeared in 1991. Tetrakis(dimethylamido)titanium, $[Ti(NMe_2)_4]$ (TDMAT), and its Zr and Hf congeners were used as precursors either as single sources or in combination with ammonia, rather than the metal halides MX_4.[161] One key aspect of MOCVD in comparison to CVD is the possibility to operate the deposition process at moderate temperatures of 200–500 °C in the kinetically controlled regime because the metal-organic precursors are chemically much more labile than the metal halides. The idea was to use certain metal-organic compounds that exhibit the N-source directly attached to the metal atom in the precursor molecule, which is the case for the homoleptic alkylamides of the type $[M(NR_2)_n]$ (M = Ti, Zr, Hf; $n = 4$; and M = Nb, Ta, $n = 5$) or for the heteroleptic alkylamides/alkyl imides of the type $[M(NR')_a(NR_2'')_b]$ (M = Nb, Ta; $a = 1$, $b = 3$; and M = Mo, W; $a = 2$, $b = 2$). These type of precursors have also been studied as so-called single source precursors (SSPs), aiming at the deposition of nitrides even in the absence of additional nitrogen sources.[162] However, the MOCVD of pure nitrides turned out to be difficult, if not impossible, using such precursors as single sources. A fundamental limitation of metal-organic precursors can be the more or less preferred carbon incorporation into the deposited films. The refractory metals are particularly known for their thermodynamically very stable binary carbides, *e.g.* TiC and WC.[163] The nitrides MN form solid solutions, MC_xN_y ($x + y = 1$), which facilitates carbon incorporation.[164] Thus, using $[Ti(NMe_2)_4]$ or $[Nb(NtBu)(NMe_2)_3]$ as representative examples for metal-organic nitride precursors typically gave the ternary thin film materials TiN_xC_y and NbC_xN_y in MOCVD studies. While these carbonitrides are interesting materials in their own right, with related technological importance to the pure nitrides and carbides, they are not covered in more detail here.[165] Using oxygen, nitrous oxide or ozone as oxidants to reduce the intrinsic carbon incorporation as for MOCVD of metal oxides (Chapter 8) is not feasible for MOCVD of metal nitrides, of course. In the presence of such oxygen sources, oxynitrides MN_xO_y will be deposited or even the respective pure binary oxide phases. However, notably, the metal-organic M–N bonded precursors together with water and/or oxidants are in fact very attractive precursors for MOCVD and ALD of metal oxide materials.[166] In addition, the selective deposition on N-doped metal oxides using M–N bonded precursors is of interest.[167] Carbon-free TiN was deposited from $[Ti(NR_2)_4]$ at moderate thermal conditions in the presence of an additional nitrogen source, ammonia in the first place but also alkyl amines, hydrazines, hydrazoic acid or nitrogen radicals provided by plasma sources (Scheme 9.1).[168–172] The same holds in general for similar precursors for the other nitrides. The underlying principle is the kinetically preferred transamination reactions that take place if ammonia reacts with $[Ti(NR_2)_4]$ and related precursors.[173]

In fact this reaction scheme is similar to the deposition of nitrides from the metal halides MX_n and ammonia, which involves an ammonolysis of the metal halide bonds under CVD conditions. However, radical reactions are more important in CVD than in MOCVD.[174] During the interaction of metal-organic precursor with ammonia and subsequent further decomposition steps to the final nitride material the metal component must be chemically reduced from Ti(IV) in the precursor to Ti(III) in the nitride film.

A detailed understanding of the interaction of ammonia (and other nitrogen sources) with the metal-organic precursors in the gas phase and also at the surface, and the dependence on process conditions, is thus critical for a full mechanistic description on the molecular level that will lead, finally, to reliable modelling of the over-all growth process. The chemistry of CVD and MOCVD is characterized by the coupling of mass-transport, gas-phase reactions and surface reactions. In contrast, the ALD growth mode for ultrathin layers or related CVD processes using pulsed dosing of reactants is designed to rule out mass-transport effects and gas-phase reactions. Quite similar to ALD of other materials (compound semiconductors, oxides and metals), ALD of the refractory

$$[Ti(NMe_2)_4] + 2NH_3 \longrightarrow TiN + 4\,HNMe_2 + H_2 + 1/2N_2$$

Scheme 9.1

metal nitrides involves the surface ammonolysis of M–N bonds of the adsorbed precursor. Particularly for ALD, but also for MOCVD, the issue of hydrogen inside the deposited nitride films is very relevant. In contrast to PVD and thermodynamically controlled classical high temperature CVD, kinetically controlled low-temperature deposition processes may typically lead to more or less amorphous, structurally ill-defined thin film materials containing also imido NH and amido NH_2 species rather than the crystalline MN phase. In fact, the inevitable additional nitrogen source ammonia plays an ambiguous role. It is necessary to avoid carbon incorporation and is indispensable as co-reactant for ALD. However, other metal nitrogen phases than the desired ones may be formed, such as not reduced, nitrogen-rich nitride materials, *e.g.* Hf_3N_4 and Ta_3N_5, which are not conducting, rather than the desired reduced HfN or TaN.[175]

9.4.2 Titanium Nitride Deposition

The model case of metal nitride MOCVD and ALD is TiN, using the above-mentioned homoleptic alkylamido precursors $[Ti(NMe_2)_4]$ (DMAT), $[Ti(NEt_2)_4]$ (DEAT) and the Me/Et mixed one $[Ti(NMeEt)_4]$ (TEMAT), which all are commercially available.[173] Numerous related all-nitrogen coordinated titanium precursors have been screened, especially in the early 1990s but without superior results. The relevant literature up to about 1994 was nicely covered by T. Gerfin and K.-H. Dahmen in *CVD of Nonmetals* edited by W. S. Rees.[160] We refer to the discussion given therein.

9.4.2.1 Gas-phase Pyrolysis of $[Ti(NR_2)_4]$ and Carbon Incorporation

Early studies on the gas-phase pyrolysis of $[Ti(NMe_2)_4]$ in the absence of other reactants (*e.g.* ammonia) using transmission FTIR showed that β-hydrogen abstraction from the methyl takes place, leading to dimethylamine, $HNMe_2$, split-off at temperatures above 550 K.[176] The remaining titanium coordinated methylimino species $MeN{=}CH_2$, revealing a Ti–C bond, is believed to be the key intermediate leading to carbon incorporation. Extended investigations of this kind, again using FTIR and *in situ* molecular beam mass spectroscopy, generally substantiate this interpretation but also point to the involvement of more complex gas phase mechanisms and coupling to surface reactions.[177,178] Quite similarly, $[Ti(NMeEt)_4]$ (TEMAT) decomposes into the intermediate forms of imine compounds in an Ar and H_2 atmosphere. Some additional nitrile formation was observed only in a H_2 atmosphere, however. The decomposition rate of TEMAT under H_2 was slower than that in an Ar atmosphere, which resulted in extension of the surface reaction controlled regime of the MOCVD process.[179]

9.4.2.2 Adsorption and Surface Reactions of $[Ti(NMe_2)_4]$

The adsorption and dissociation of $[Ti(NMe_2)_4]$ on a Si (100) 2×1 surface has been studied by a combination of quantum chemical calculations and infrared spectroscopy. The DFT calculations and FTIR data suggest initial interaction through the nucleophilic attack of a surface silicon atom by the lone pair of an amido nitrogen atom. Dissociation of TDMAT is permitted kinetically if it occurs through the scission of the N–Ti bond, while scission of an N–C bond is kinetically hindered despite the thermodynamic stability of the structure produced.[180]

Very few studies have been performed looking at the surface chemistry of TDMAT (or related) on TiN substrates. The adsorption and desorption kinetics for TDMAT and $HNMe_2$ (DMA), which is a decomposition product of TDMAT (Scheme 9.1), were measured on a sputter-deposited TiN surface using laser induced thermal desorption (LITD) techniques. The LITD measurements revealed that DMA has a higher sticking coefficient than TDMAT. The sticking coefficients for both TDMAT and DMA were also dependent on surface coverage. These findings are consistent with DMA re-adsorption and site blocking as a major contributor to non-conformal TiN growth in trench structures using TDMAT.[181]

9.4.2.3 *Gas-phase Reactions of $[Ti(NR_2)_4]$ with Ammonia and Alkylamines*

The presence of excess ammonia as additional nitrogen source in thermal MOCVD of TiN using the alkylamide precursors drastically reduces C-incorporation.[161,182] Mechanistic studies with $^{15}NH_3$ and ND_3 and unlabelled $[Ti(NMe_2)_4]$ gave evidence that the N-component in the films comes from the ammonia and not from the dimethylamido ligand. A level of 94% of ^{15}N was found by RBS in TiN films and $DNMe_2$ was formed exclusively. These findings together with knowledge from preparative chemistry on the reactivity of $[Ti(NR_2)_4]$ with alkylamines[168,183] and ammonia[184,185] substantiate the idea of stepwise ammonolysis of the Ti-NMe_2 moieties by the attack of ammonia and also suggest the formation of dimeric or oligomeric amido/imido species $[(R_2N)_2TiNH]_n$.[170,186] The growth kinetics of the $[Ti(NMe_2)_4]/NH_3$ system were studied and a kinetic model was derived based on a Langmuir–Hinshelwood mechanism.[187] In extension of these studies, the gas-phase kinetics of the reaction of TDMAT with NH_3 was measured using a flow tube reactor and FTIR spectrometer.[188,189] Even at 24 °C, TDMAT reacts quantitatively within seconds with NH_3 in a transamination reaction to form $HNMe_2$ as a direct product (Scheme 9.2). A primary kinetic isotope effect of $k(H)/k(D) = 2.6 \pm 0.4$ was observed with ND_3, indicating that cleavage of an N–H bond is the rate-limiting step. The bimolecular rate constant of $k = 1.2 \pm 0.2 \times 10^{-16}\,cm^3\,molecules^{-1}\,s^{-1}$ is assigned to the elimination of the first dimethylamido ligand. From the kinetic data the activation energy of $8.1 \pm 0.1\,kcal\,mol^{-1}$ was extracted. Clearly, under the CVD process conditions of high NH_3 partial pressures and temperatures, extensive gas reactions occur and TDMAT will be converted into a new species before it reaches the surface of the growing film. Very interestingly, the reaction rate is strongly suppressed in the presence of excess $HNMe_2$. Thus, a reversible initial transamination reaction followed by an elimination step of a second $HNMe_2$ was proposed (Scheme 9.3).

Several computational studies on the complex gas-phase reactions in case of $TiCl_4/NH_3$ and $[Ti(NMe_2)_4]/NH_3$ on the DFT level of theory (B3LYP/6-311G(d)) have been performed, with $[Ti(NH_2)_4]$ being a suitable model for TDMAT.[190–192] The computed barrier height of 8.4 kcal mol^{-1} for the $[Ti(NH_2)_4] + NH_3$ ligand exchange reaction is in very good agreement with the experimental activation energy of 8 kcal mol^{-1} for $[Ti(NMe_2)_4] + NH_3$ ligand exchange. The barrier for formation of $[Ti(NH_2)_2NH]$ by elimination from $[Ti(NH_2)_4]$ is 33.5 kcal mol^{-1}. Further elimination of NH_3 from $[Ti(NH_2)_2NH]$ can yield a diimido product, $[Ti(NH)_2]$, or a nitride/amido product, $[Ti(NH_2)N]$, but the barriers and heats of reaction are sufficiently high to make these reactions unlikely in the gas phase during CVD. Thus, the dimerization of $[Ti(NH_2)_2NH]$ was used as a model for oligomerization of the initial monomeric gas-phase intermediates $[Ti(NR_2)_2NH]$. It turned out that these oligomer amido/imido complexes are very stable. Subsequent elimination of N_2H_4 from these is rather endothermic relative to the dimer/oligomer complex, suggesting that further decomposition and, in particular, reduction of Ti(IV) to Ti(III) may occur on the surface rather than in the gas phase.[193] The interaction of the alkylamide precursors with ammonia and

$$[Ti(NMe_2)]_4 + NH_3 \longrightarrow [(Me_2N)_3Ti\text{-}NH_2] + HNMe_2$$

Scheme 9.2

$$[(Me_2N)_3Ti\text{-}NH_2] \longrightarrow [(Me_2N)_2Ti{=}NH] + HNMe_2$$

Scheme 9.3

alkylamines may not only be relevant for the purity of the films. It has been reported that $[Ti(NEt_2)_4]$ and $[Ti(NMeEt)_4]$ gives higher quality TiN than $[Ti(NMe_2)_4]$ in terms of purity and step coverage as well.[194–197] Apparently, the rate of reaction of $[Ti(NMe_2)_4]$ with NH_3 is too fast and leads to the formation of intermediates with high sticking coefficients or low surface mobility, including some oligomeric species.[198,199] The reaction of $[Ti(NEt_2)_4]$ or $[Ti(NMeEt)_4]$ with NH_3 is much slower and intermediates with smaller sticking coefficients could be formed, which might be a likely reason for the improved film properties.[200]

9.4.2.4 Atomic Layer Deposition of TiN

ALD of TiN based materials has increasing importance for device applications in microelectronics. Generally, the properties of ALD TiN thin films in terms of impurities, density and resistivity in combination with deposition temperatures below 400 °C and the excellent step coverage are superior to those of CVD and MOCVD. Some representative results of ALD of TiN are compiled in Table 9.5.

However, the ALD growth conditions, including the choice of precursors, have a very strong impact on the properties. Ideal ALD processes rule out gas-phase pre-reactions of any kind. Tuning ALD processes requires a detailed understanding of the surface chemistry. TiN ALD from $TiCl_4$ and NH_3 is one of the best ALD processes studied to date,[201–210] but very few surface studies like the following one were performed, so far. ALD of TiN using $TiCl_4$ and ammonia at 400 °C was investigated by *in situ* X-ray photoelectron spectroscopy.[211] During the $TiCl_4$ adsorption step, both a partial loss of chlorine atoms and the reduction of the metal take place. The remaining chlorine loss nitride formation occur within the subsequent ammonia dosing step. Both half-reactions were proven self-limited, stopping after the deposition of sub monolayer quantities of the materials. Multilayer films display a Ti_3N_4 layer on top of the expected TiN. Negligible chlorine impurities

Table 9.5 Overview of some important ALD of TiN.

Precursors	*Substrate temperature (°C)*	*Thin film properties*	*Ref.*
$TiCl_4$, NH_3	350–500	250 μΩ cm; Cl < 5 at.% (500 °C)	201, 202, 395
$TiCl_4$, NH_3	450	70 μΩ cm (450 °C)	204, 396
$TiCl_4$, NH_3, Zn	400–500	50 μΩ cm	202
$TiCl_4$, DMHy	250–500	Cl 2 at.%, C 13 at.%	119
$TiCl_4$, $AllylNH_2$; $tBuNH_2$, NH_3	400–500	310 μΩ cm, Cl 1.4 at.%	397
TiI_4, NH_3,	400–500	I < 0.5%; 70 μΩ cm (500 °C)	398
TiI_4, $tBuNH_2$, NH_3	400–500	150 μΩ cm	397
TiCl4, NH_3	300–400	200 μΩ cm; Cl < 1 at.%	210
TDMAT, NH_3	180	C, O < 10 at.%; 5000 μΩ cm	254, 399, 400
TEMAT, NH_3	150–230	C 4–10%; 230–8000 μΩ cm	401, 402
TDMAT, H_2, N_2, H_2/N_2 plasma	150–350	300–500 μΩ cm (N_2 plasma); C, O < 3% (250 °C, N_2 plasma)	
TDMAT, H_2, N_2, Ar plasma	150–350	Cl 0.5% 80 μΩ cm (350 °C)	403, 404
$TiCl_4$, H_2, N_2, Plasma (a)	150–350	Cl 2.1%; 209 μΩ cm (100 °C); Cl 0.07%; 71 μΩ cm (400 °C)	405, 406
$TiCl_4$, NH_3	100–400	200 μΩ cm (400 °C)	407
$TiCl_4$ H_2, N_2, plasma (b)	350–400	142 μΩ cm (320 °C); Cl 0.6%	
TDMAT, H_2, N_2, H_2/N_2 plasma (c)	270–370		
	150–350	300 μΩ cm; 95% step coverage	408

were deposited on the surface, and originate from readsorption of the HCl by-product. Similarly, the surface chemistry and film growth were examined during ALD of TiN using TDMAT and NH_3.[212] According to *in situ* FTIR the TDMAT reacts with surface NH_x species on the TiN surface to deposit new $[Ti(NMe_2)_2]_x$ species. Subsequent NH_3 exposure consumes the dimethylamino species and regenerates the NH_x species. These findings are consistent with transamination exchange reactions during the alternating TDMAT and NH_3 exposures, similar to the above discussed gas-phase chemistry. The self-limiting characteristics of the TiN ALD process were confirmed by monitoring the weight gain using an *in situ* quartz-crystal microbalance. However, the resistivities of the TiN ALD films were greater than or equal to $10^4\,\mu\Omega\,cm$ and the densities were less than or equal to $3\,g\,cm^{-3}$, corresponding to a porosity of $\sim$40%. The high porosity allows facile oxidation of the TiN films, lowers the film resistivity and limits application as diffusion barrier. Compared to this, ALD of TiN from $TiCl_4$ and ammonia gave superior results.

9.4.2.5 *MOCVD and ALD of TiN on Self-assembled Organic Monolayers (SAMs)*

Another challenge much related to the above ALD surface chemistry is the goal of depositing TiN and similar nitrides on top of organic, rather than inorganic, surfaces or substrates. So-called self-assembled organic monolayers (SAMs), being typically based on the chemisorption of alkanethiols on gold substrates or of alkanesilanes on silicon and/or silicon oxide surfaces, serve as models of organic surfaces in general[213–215] or may even be more directly interesting as functional components for molecular electronics or other molecular thin film devices.[216–219] Very little work has been carried out on the MOCVD and ALD of inorganic materials on top of organic polymers and in particular on SAMs, so far.[220–225] For example, ALD of aluminium oxide on polymers such as polyethylene, poly(methyl methacrylate), polystyrene, poly-propylene and polyvinylchloride has been investigated.[226] A related DFT study on Al_2O_3 ALD on SAMs was performed that emphasized the initial stages of the organic/inorganic interface formation as a function of the surface exposed end groups (*e.g.* $-CH_3$, $-OH$, $-NH_2$).[227] With a similar motivation, the reaction of TDMAT with SAMs of alkyltrichlorosilanes exhibiting different functional end groups was characterized experimentally[228] and was analyzed by computational methods as well.[229] The reaction probability was found to follow the order $-OH > -NH_2 > -CH_3$ in the range -50 to 110 °C. However, only in case of NH_2 termination were the results obtained from angle resolved XPS measurements consistent with TDMAT reaction only at the terminal ($-NH_2$) groups without any penetration through the SAM to the underlying Si/SiO_2 substrate. However, so far, there have been no further studies on the ALD of TiN or other metal nitride layers on SAMs.

9.4.2.6 *Plasma-enhanced MOCVD and ALD of TiN*

With the aim of reducing deposition temperatures significantly below 500 °C, different kind of non-thermal activation methods have been investigated, including, most prominently, plasma enhanced chemical vapour deposition (PECVD). However, unlike the similar $TiCl_4/NH_3$ system, PECVD of TiN using TDMAT as the typical precursor usually leads to TiC_xN_y films due to non-selective decomposition and co-deposition of polymeric hydrocarbon material arising from the alkylamido ligands.[230–234] Electron cyclotron resonance (ECR) plasma offers a higher degree of excitation than RF or DC plasmas, and of thus raising the density of activated nitrogen species in case of dinitrogen and/or ammonia as plasma gases. By introducing TDMAT into the downstream region of such an ECR plasma very pure highly conductive TiN films (45–150 $\mu\Omega$ cm) have been deposited at temperatures as low as 100 °C. ^{15}N-labeling studies confirmed the plasma gas being the source of the N-component in the deposited films.[235,236] Accordingly, very pure (C < 3 at.%) and compositionally highly uniform TiN films with an excellent step coverage above 95% at a very high aspect ratio structure (contact holes) were deposited from TDMAT by remote plasma enhanced ALD (PEALD). Plasmas of hydrogen, nitrogen and ammonia were used. However, electrical characterization data have not been published.[237] A related and very detailed study on thermal ALD, MOALD and PEALD of TiN showed that by using $TiCl_4$ and a remote H_2/N_2 plasma superior results are obtained in terms of electrical properties. Lowest resistivities were 71 μΩ cm at 400 °C and 209 μΩ cm at 100 °C. Typical values for thermal TiN ALD from $TiCl_4$ or from TDMAT range above 400–600 μΩ cm.

9.4.2.7 *MOCVD and ALD of Ternary Ti-Si-N and Ti-Al-N Thin Films*

The properties of MOCVD and in particular of ALD TiN thin layers grown from DTMAT (without ammonia) as diffusion barriers can be improved in terms of purity, resistivity and density by a post deposition N_2/H_2 plasma treatment.[238,239] However, even plasma annealed TiN in general is not considered as a sufficiently robust diffusion barrier for copper metallization, particularly for high aspect ratio structures. The sidewalls of the vias/trenches between Cu and the dielectric represent a critical location for barrier requirement.[240–242] However, the incorporation of Si into the TiN films leads to amorphous Ti-Si-N phases with no or few grain boundaries, which significantly reduces the likelihood of diffusion between Cu and Si layers, and for Ti-Si-N ternary materials superior diffusion barrier properties were demonstrated.[195,243–248] Typical Ti-Si-N films have been prepared by MOCVD of TiN using TDMAT at 350 °C. After N_2/H_2 plasma annealing the films was *in situ* exposed to SiH_4. According to XPS studies Si–N and Ti–N bonded sites were present but Ti–Si sites were absent. Such low temperature MOCVD grown TiN films are composed of TiN nanocrystals embedded in an amorphous hydrocarbon matrix, $C_xN_yH_z$.[249,250] The SiH_4 reacts with this amorphous matrix to form Si_3N_4.[251,252] Along these lines of improved barrier properties of TiN-type layers, quaternary Ti-Si-N-O films have been investigated using metal-oxide-semiconductor (MOS) capacitors. For films with a relatively high Si/Ti ratio of 0.91 the flatband voltage shifts by 0.75 V after 90 min of bias temperature stressing at 150 °C and 0.5 MV cm^{-1}. This behaviour was attributed to the interface states at the Ti-Si-N-O/oxide interface that were generated during the plasma process and could not be fully healed after 250 °C annealing. By adjusting the silane gas flow rate to lower Si incorporation, *e.g.* Si : Ti ratio of 0.48, an electrically stable Ti-Si-N-O film was obtained with fewer interface states.[253] Certainly, ALD is the method of choice for the precisely composition controlled deposition of ternary nitrides such as Ti-Si-N as well as Ti-Al-N using SiH_4 and Me_2AlH-EPP (dimethylaluminium hydride–ethylpiperidine, DMAH-EPP) as the third precursors. $TiSi_xN_y$ ALD using TDMAT with NH_3 and SiH_4 was achieved at 180 °C but the resistivities were high (30000 μΩ cm).[254] Better resistivities of 500–1000 μΩ cm were obtained with PEALD using the sequence of $TiCl_4$ pulse, SiH_4 pulse, $N_2/H_2/Ar$ plasma. Growth rates of 0.73 Å cycle^{-1} and 8.5 at.% of Si with very low Cl $\leq$0.5 at.% were measured.[255,256] Within this context of Ti-Si-N MOCVD and (PE)ALD the reactivity of various silanes with TDMAT was investigated.[257] Liquid $[Ti(NMe_2)_4]$ with SiH_4 at 23 °C gave H_2, $HSi(NMe_2)_3$ and a black solid. $HNMe_2$ was not detected in this reaction. *Ab initio* quantum chemical calculations of the reactions of $[Ti(NR_2)_4]$ (R = Me, H) with SiH_4 indicated that the formation of aminosilanes and $[HTi(NR_2)_3]$ was favoured. However, it is unclear whether similar hydride/alkylamide exchange reactions take place during ALD of Ti-Si-N. Quite similar to Ti-Si-N, the Al-containing ternary nitride Ti-Al-N is interesting as an improved TiN based diffusion barrier and was deposited by ALD techniques using TDAMT or $TiCl_4$ and NH_3 together with TMA[258] or with DMAH-EPP.[259] The obtained films revealed acceptable resistivities of 400 μΩ cm in the case of $TiCl_4$ and 1500 μΩ cm with TDMAT (also about 25 at.% C).

9.4.3 Zirconium and Hafnium Nitride Deposition

Quite similar to TiN, the binary rock-salt type cubic phases ZrN and HfN are interesting as chemically inert refractory materials for machine tools.[260] However, there has been far fewer reports on CVD and MOCVD of ZrN and HfN in comparison to TiN. For example, Hf(C,N) thin films on Si(100) and STD11 steel substrates were deposited by pulsed dc plasma assisted metal-organic chemical vapour deposition (PA-MOCVD) in the temperature range 200–300 °C. Tetrakis(diethylamido)hafnium, $[Hf(NEt_2)]_4$ (TDEAH), was used as the hafnium precursor. The Vickers hardness of the films ranged between 1000 to 2500 HV, depending on the deposition conditions.[261] Very similar results were obtained for Zr(C,N) films deposited by PA-MOCVD using $[Zr(NR_2)_4]$ (R = Me, Et) as precursors. The Vickers hardness of the coatings deposited at 300 °C at maximum growth rate reached 2000 HV.[234,262,263] Like the TiN case, the deposition of the pure, *i.e.* carbon-free, binary MN materials from metal-organic precursors is not straightforward. For device application in microelectronics (diffusion barriers and gate electrodes), ZrN and HfN have interesting properties such as high thermal stability and the low resistivities in the range 10–200 μΩ cm (bulk resistivity of ZrN ~ 13.6 μΩ cm; HfN ~ 33 μΩ cm). For example, HfN is a very promising candidate as gate electrode for HfO_2 based high-*k* dielectrics, and some investigations on PVD deposited HfN gate electrodes have been carried out.[56] In addition, HfN has potential as a buffer layer for epitaxial GaN materials.[264] However, reduction from the M(IV) state in the precursors $[MX_4]$ or $[M(NR_2)_4]$ is more difficult than with TiN. The formation of other phases, in particular the insulating M_3N_4, occurs.[265–267] This problem is particularly relevant in the case of MOCVD and ALD using $[M(NR_2)_4]$ and ammonia as reactants.[161] Highly uniform, smooth and conformal coatings of some amorphous, nitrogen-rich materials with a formal composition of Hf_3N_4 and Zr_3N_4 were produced by atomic layer deposition from the homoleptic $[M(NR_2)_4]$ type precursors and ammonia at low substrate temperatures (150–250 °C). The step coverage was 100% in vias with an aspect ratio of 40 : 1.[268] HfN_x films with low levels of carbon (< 1 at.%) and oxygen (< 1–2 at.%) impurities were formed by MOCVD using $[Hf(NEt_2)_4]$ and NH_3 but the sheet resistance of the film was very high due to the formation of insulating Hf_3N_4 phase.[269] Using methyl and dimethylhydrazine ($MeHNNH_2$, Me_2NNH_2) instead of ammonia together with $[Hf(NEt_2)_4]$ it was possible to achieve MOCVD of the desired cubic HfN phase at comparably high temperatures (800 °C).[270] HfN_xC_y films have been deposited by MOCVD and their use as gate electrodes on SiO_2 and HfO_2 dielectrics investigated.[271] The obtained film revealed a fair resistivity of 7400 μΩ cm. PEALD of HfN using $[Hf(NMe_2)_4]$ and a pulsed N_2/H_2 RF plasma gave resistivities of 4.0 to 1500×10^3 μΩ cm, depending on the conditions.[272] Hafnium nitride films were prepared on the Si substrates by MOCVD using $[Hf(NEt_2)_4]$. The HfN samples were then oxidized in air, followed by rapid-thermal annealing to produce HfO_xN_y thin films.[273] ZrN films deposited using $[Zr(NEt_2)_4]$ and N_2 plasma by PE-ALD exhibited a resistivity of 400 μΩ cm and carbon impurities of 6 at.%.[274]

9.4.4 Niobium and Tantalum Nitride Deposition

As with Group 4 nitrides, Group 5 nitrides have been studied as wear resistance materials (*e.g.* for cutting or drilling tools). Sputtering techniques are typically used for these deposition processes.[275] Multilayered films of different metal nitrides, *e.g.* TaN/NbN and TiN/NbN, have been found to increase the hardness of the films with respect to single phase coatings up to the range of 50 GPa (based on a Knoop hardness test for thin films).[276] CVD of Group 5 nitrides has been investigated using the metal halides, *i.e.* $NbCl_5$ and $TaCl_5$ and NH_3/H_2 at 600–1100 °C.[277–281] The thermally less stable $TaBr_5$ was tested in CVD processes with the co-reactants NH_3/H_2 and with a H_2/N_2-plasma.[110,282] Using the latter technique, films were deposited that consisted of highly conductive TaN in the face-centred cubic (fcc) rocksalt-type phase. The resistivity of these PECVD TaN films (150 μΩ cm) is the lowest value reported for TaN thin films made by CVD-techniques. It is close to the literature value for bulk TaN (128 μΩ cm). Substitution of ammonia as N-source by 1,1,1,3,3,3-hexamethyldisilazane (HMDS) allows the APCVD of cubic TaN and NbN at as low as 500–600 °C.[283] Some related CVD work has been performed on the VN and the V-N-O system, aiming at solar control coatings.[284] VCl_5 and $VOCl_3$ were used as precursors in combination with HMDS at APCVD conditions (300–500 °C). X-ray photoelectron spectroscopy showed that the V/N films were almost stoichiometric. The V/N/O films consist of mixed phases of vanadium oxynitride and V_2O_3.[285] MOCVD of VN_xC_y materials for protective coatings has been studied using precursors such as $[(C_6H_6)_2V]$,[286,287] $[Cp_2VMe_2]$[288] and $[V(NEt_2)_4]$.[289] However, similar to the previous TiN case study, we focus mainly on the MOCVD and ALD of TaN and NbN materials within the context of application in microelectronics emphasized in Section 9.1. Figure 9.5 shows some known precursors for NbN.

9.4.4.1 Tantalum and Niobium Nitride as Gate Electrodes in MOSFETs

Cubic TaN and NbN phases are candidates as gate electrodes for MOSFETs, in particular on new high-*k* dielectric oxides such as HfO_2 or related materials. In addition, they may serve as diffusion barriers for copper metallization schemes, as discussed for TiN materials. According to simulations, the work function of these materials should be around 4.2–4.3 or 3.8–3.9 eV in the case of n-type MOS-devices and 4.8–5.0 eV for an application in p-type-MOS devices (dual gate technology). For these values, the drive currents through the conducting channel are supposed to run through a maximum.[41,57,290] Many studies underline the superior electrical properties of TaN[291–294] and TaSiN[99] thin films for the substitution of a highly doped polycrystalline silicon gate electrode. Again, not only an electrical fit of the materials plays a role but also stability issues with respect to chemical inertness and resistance to thermal stress. Other materials, *e.g.* TiN_x, $TaSi_x$ and WN_x, have work function that are out of range for an application as advanced gate electrode materials.[291] Based on a recent literature search NbN was not mentioned so far in this respect. However, considering its work function of approximately 4.7–4.8 eV[295] and its relatively high thermal stability it is likely that NbN will be also tested. The phase selectivity is less problematic for NbN than for TaN. Deposition of TaN in the presence of excess N-source may lead to the insulating Ta_3N_5 material, as with Hf_3N_4 discussed earlier. There are no insulating niobium nitride phases known; however, nitrogen-rich materials such as Nb_3N_4 are semiconducting.[70–72] An interesting, new approach is the use of tertiary materials consisting of two refractory metals and nitrogen. These tertiary systems (*e.g.* $TaTb_xN$ or Hf_xTa_yN) exhibit work functions that can easily be tuned between 4.2 and 4.6 eV by varying the content of the incorporated second metal (in this case Tb or Hf).[296,297]

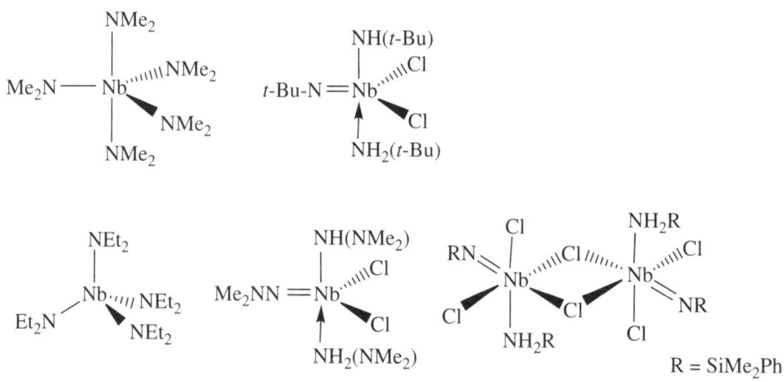

Figure 9.5 Some known precursors for niobium nitride.

9.4.4.2 MOCVD of Tantalum and Niobium Nitride

A review by Winter[298] describes the CVD of metal nitride films, especially TaN, Ta-Si-N, NbN, and related nitride materials for applications as advanced barrier layers in microelectronic devices. Pentakis(dimethylamido)tantalum, $[Ta(NMe_2)_5]$ (PDMAT),[299] which is the analogous alkylamide-type precursor to $[Ti(NMe_2)_4]$ (TDMAT), was used in MOCVD processes, together with ammonia as a reactive gas (Table 9.6).[65,300] Although depositions at temperatures between 200 and 450 °C were carried out successfully, insulating Ta_3N_5 material was deposited instead of cubic conductive TaN. PA-MOCVD with methane/hydrogen[301] ammonia- or hydrogen-plasma[302] was studied to facilitate the reduction from tantalum +v to +iii. Conducting films were obtained that consisted of tantalum nitride and tantalum carbide. Discrimination between TaN and TaC and the solid solution Ta(N,C) by XRD is particularly difficult due to very similar lattice constants. For significant C-levels, the obtained film consist of a mixture of amorphous Ta_3N_5 and nanocrystalline TaC, while the tantalum carbide phase is responsible for the low resistivity and the observed peaks in the XRD-spectra. The precursor PDEAT [pentakis(diethylamido)tantalum] is a topic on its own.[303] In contrast to PDMAT, PDEAT cannot be evaporated without decomposition. At higher temperatures it decomposes to a mixture of an amido/imido compound and the complex $[Ta\{\eta^2\text{-}(EtN{=}CHCH_3)\}(NEt_2)_3]$ containing a η^2 imino-ligand (Figure 9.6).[304]

Table 9.6 Precursors, deposition temperatures and thin film properties of Ta-N deposited by vapour deposition methods.

Precursors	*Substrate temperature (°C)*	*Structure/ phases*	*Thin film properties*	*Ref.*
$(Et_2N)_3Ta{=}Net$	500–650	TaN	Carbon-rich films	409
PDMAT-NH_3	200–400	Ta_3N_5	N : Ta ~ 1.7, > 1×10^6 μΩ cm	300
$[TaCl_2(N^tBu)(NH^tBu)\text{-}(NH_2{}^tBu)]_2$	500–600	Ta_3N_5	Brown/Yellow films	316
TBTDET	450–650	TaN	Good step coverage; 13000 μΩ cm (450 °C) 900 μΩ cm (650 °C)	290, 410
PDMAT & H_2 Plasma	200–350	Tantalum Nitride (Amorph)	C and N rich films 2000 μΩ cm;	411
PDMAT & NH_3 Plasma	200–350	TaN (≥ 300 °C)	4000 μΩ cm $^{-1}$ Ω cm or >4000 μΩ cm N : Ta ~ 0.9, C : Ta ~ 1.5 (200 °C); N : Ta ~ 1.1, C : Ta ~ 0.51 (300 °C)	302
$TaBr_5$, H_2, N_2 and Plasma	350–450	TaN	150 μΩ cm	282
$TaBr_5$, NH_3, and H_2	350–500	TaN, Ta_3N_5	2500 μΩ cm (425 °C)	110, 282, 412
PDEAT	300–375	TaN	Carbon rich films (30% C) good conformality	306
PDEAT & NH_3	300–375	TaN	No C, non uniformity	306
PDEAT & Ar-ions, N-ions	275–400	–	C 20–30% 600 μΩ cm (350 °C)	413
$TaCl_5$ & N_2 Plasma	600–650	Ta_2N_3	–	94
$Ta(NEt_2)_2(NCy_2)_2$-NH_3	340	Amorph	Density: 6.6–7.0 g cm^3 2.5 ± 0.1 × 10^5 μΩ cm	310
$TaCl_5$-$NH(SiMe_3)_2$	400–580	<550 °C Amorph; ≥ 550 °C TaN	500 μΩ cm (550 °C) No chlorine	283

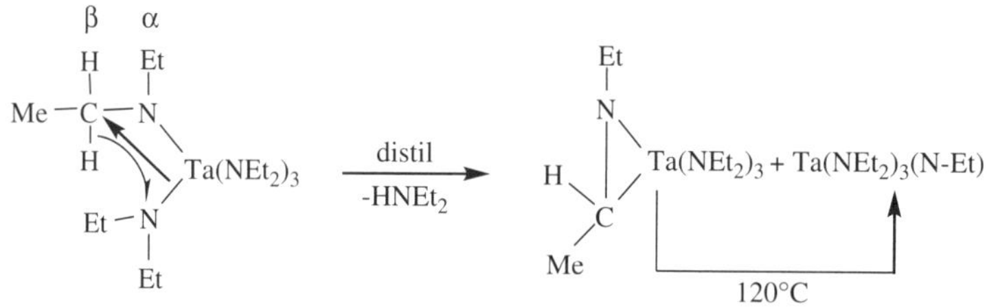

Figure 9.6 Thermal decomposition of PDEAT during distillation.

The mechanism is supposed to involve an activation of a β-CH bond of a NEt_2 ligand and the subsequent cleavage of diethylamine. This compound in turn decomposes at higher temperatures to $[Ta(NEt)(NEt_2)_3]$ and ethylene. This chemistry is very much related to the gas-phase decomposition of $[Ti(NMe_2)]$ and $[Ti(NEt_2)_4]$ discussed earlier.

The use of PDEAT as a single source precursor is not suitable for depositing carbon-free films. All films show carbon contents of above 30%. This issue appears to be not sufficiently recognized in some older literature.[305,306] As with the related titanium alkylamide precursors, introduction of ammonia leads to transamination reactions during growth and the release of $HNEt_2$ and low carbon levels (<5%).[306] However, the resistivities of these films were rather high (~12000 μΩ cm). Plasma assisted MOCVD depositions (H_2-plasma) or the bombardment with argon/nitrogen ions during deposition did not help to reduce the carbon content of the films significantly.

The most established precursors for TaN and NbN, however, are the mixed ligand alkylamido complexes of the type $[M(NR_2)_3(NR')]$ (with R = Me, Et; and R′ = t-Bu, amyl). The best known example is TBTDET (*tert*-butylimidotris(diethylamido)tantalum, $[Ta(NEt_2)_3(N\text{-}t\text{-}Bu)]$).[307]

TBTDET has been successfully used as a single source precursor for MOCVD. Together with ammonia as co-reactant, the resultant low resistivity films, deposited at about 600–650 °C, contained carbon below 10% and showed reflections for cubic Ta(N,C) in the XRD-spectra.[290,308] So far these experiments are the only examples of good quality TaN films deposited by a plasma free deposition process that show a resistivity below 1000 μΩ cm. In general, similar results were obtained using the related precursor TAIMATA (*tert*-amylimidotris(dimethylamido)tantalum, $[Ta(NEt_2)_3(NCMe_2Et)]$).[291,309] Some related tantalum(IV) complexes tested by aerosol assisted MOCVD are the homoleptic complex $[Ta(NEt_2)_4]$, the chemical identity of which is still questionable, and the thermally less stable mixed amido complex $[Ta(NEt_2)_2(Ncy_2)_2]$.[310] Thin, amorphous films with composition $TaN_{1.5}H_{0.3}$ were deposited at 340 °C. However, the resistivity of the films was rather high (*ca.* 250 000μΩ cm). For low-valent Ta(III,IV) nitrides a much better conductivity is expected.[311,312] The complex $[Ta(\eta^2\text{-}Me_2NNSiMe_3)(NtBu)(NMe_2)_2]$ was the first example of the introduction of hydrazido-groups into all-nitrogen coordinated Ta(V) SSPs aimed at facilitating reduction to Ta(III). Carbon-free but insulating Ta(N,Si) films were obtained.[313] Some related volatile bishydrazido complexes $[M(NtBu)(\eta^2\text{-}Me_2NNSiMe_3)_2X]$ (M = Ta, Nb; X = Cl, NMe_2, NH^tBu) were characterized as potential precursors for TaN and NbN films.[314] Other precursors being considered were mixed ligand halide containing amido/imido complexes and hydrazido derivatives, including $[TaCl_2(N\text{-}t\text{-}Bu)(HN\text{-}t\text{-}Bu)(H_2N\text{-}t\text{-}Bu)]_2$,[315,316] $[TaCl_2(NNMe_2)(HNNMe_2)(H_2NNMe_2)]_n$[317] and $[TaCl_3(NSiMe_3)(NC_5H_3Me_2\text{-}3,5)_2]$.[318] However, the results were not superior to TBTDET or TAIMATA. Other related compounds have been claimed as TaN precursors but not tested, yet.[319–321] Precursors that have been used for MOCVD of NbN are $[Nb(NEt_2)_4]$ and $[Nb(NMe_2)_5]$,[300,322] $[NbCl_2(N\text{-}t\text{-}Bu)(HN\text{-}t\text{-}Bu)(H_2N\text{-}t\text{-}Bu)]_2$, $[NbCl_2(NNMe_2)(HNNMe_2)(H_2NNMe_2)]_n$[315] and $[NbCl_3(NSiMe_2Ph)(NH_2SiMe_2Ph]_2$.[318] The homoleptic alkylamido complexes $[Nb(NR_2)_n]$ (n = 4, 5; R = Me, Et) together with ammonia gave the nitrogen-rich, but conducting phase Nb_3N_4. In analogy to the above tantalum case, the mixed alkylamide/alkylimido precursors $[Nb(NR_2)_3(NR')]$ (with R = Me, Et; and R′ = t-Bu, amyl) appear to be the most convenient and promising candidates for high quality NbN in thermally activated or plasma enhanced MOCVD in the presence of additional ammonia, hydrazine or other N-sources.[323,324] A systematic study of the application of commercially available $[Nb(NEt_2)_3(NtBu)]$ (TBTDEN) with ammonia using a commercial AIXTRON 200 RF tool was performed.[325] The best films, 20–50 nm thick, were grown at 500–650 °C at 5 mbar total pressure (NH_3 and H_2), revealing resistivities of 590–800 μΩ cm and a work function of 4.72 eV, which matches the best values of NbN films grown by ALD.[326,327] The densities of the NbN films were 7–8 g cm^{-3}, the N : N ratio was very close to 1 : 1 (±5%) and carbon and oxygen levels were below 3–5 at.% as determined by depth profiling using secondary neutral mass spectrometry (SNMS). If, however, $[Nb(NEt_2)_3(NEt)]$ is used as SSP, NbN_xC_y-type films will be deposited, which have been characterized as diffusion barriers for copper.[328,329] To date, the only examples of suitable SSPs for in fact carbon-free TaN and NbN thin films appear to be the bis-guanidinato complexes $[M(NR_2)\{\eta^2\text{-(N-}i\text{-Pr)}_2C(NR_2)\}_2(\text{N-}t\text{-Bu})]$ (R = Me, Et), which can be synthesized directly from commercial $[M(NR_2)_3(NR')]$ by selective insertion of carbodiimides R″N = C = NR″ (R″ = i-Pr, cyclohexyl, *etc.*) into the metal alkylamide bond.[330–334] The precursors have been tested for standard low pressure MOCVD and as well for liquid injection MOCVD using an AIXTRON 200 FE tool without additional ammonia or other nitrogen sources. SNMS analyses of the *ca.* 100 nm thick films reveal surprisingly low carbon levels of below 1 at.%. Some oxygen impurities (5–10 at.%) were present (possibly due to post deposition oxidation of the films and/or oxygen diffusion from the SiO_2/Si substrate). Resistivities of 7000 μΩ cm (TaN) and 2230 μΩ cm (NbN) were obtained.[325,335] A possible reason for the unusual properties of these compositionally rather complex precursors may be that their coordinative saturated structure prevents C–H activation reactions that the need vacant coordination sites available at the tetra- and penta-coordinated standard precursors $[M(NR_2)_n]$, $[M(NR_2)_3(NR)]$ (M = group 4, 5 metal).

9.4.4.3 ALD of Tantalum and Niobium Nitride

Early work on ALD of TaN using $TaCl_5/NH_3$ dates back to 1988;[201] however, little analytical data was given and full studies were not presented until 1999.[336] Similar to MOCVD the use of the metal chlorides as precursors requires rather high surface temperatures (500 °C) (Figure 9.7). Low-temperature ALD of TaN and NbN needs more labile metal organic precursors such as $[M(NR_2)_2(NR')]$ and/or plasma enhanced ALD techniques.

The reduction of M(V) to the desired M(III) oxidation state is very critical in ALD, especially at low temperatures. The use of DMHy instead of ammonia together with the halides $TaCl_5$ (or $TaBr_5$) also produced high resistivity Ta_3N_5 films (Table 9.7).[119] Other reducing agents such as zinc, trimethylaluminium (TMA) or alkylamines ($tBuNH_2$) were employed and some improvement was obtained.[337–339] Cubic TaN with a comparably low electrical resistivity of 960 mV cm was deposited from $TaCl_5/NH_3$ when zinc was the reductant (desorption of $ZnCl_2$ as by-product). The use of gaseous zinc requires very high temperatures (min. 400 °C) for the whole transfer line system, including the valves. Similar to the related ALD of TiN, co-dosing of TMA led to Al incorporation and resulted in a $TaAl_xN_y$ ternary phase with C contamination. Plasma-enhanced ALD (PEALD)

$$\textbf{(a)}\quad TaCl_{5(g)} + 5NH_{3(g)} \xrightarrow{200\text{-}500°C} Ta_3N_{5(s)} + 15HCl_{(g)}$$

$$\textbf{(b)}\quad TaCl_{5(g)} + Zn_{(g)} + NH_{3(g)} \xrightarrow{400\text{-}500°C} TaN_{(s)} + ZnCl_{2(g)} + 3HCl_{(g)}$$

Figure 9.7 Proposed reactions for the ALD of TaN_x using $TaCl_5$ and (*a*) NH_3 or (*b*) NH_3 and gaseous Zn.

Table 9.7 ALD of tantalum nitride and niobium nitride thin films using metal halides.

Precursors	*Substrate temperature (°C)*	*Structure/ phases*	*Thin film properties*	*Ref.*
$TaCl_5$, NH_3	200–500	Orthorhombic Ta_3N_5 (above 400 °C)	Insulating; below 300 °C high chlorine contamination of the films	336
$TaCl_5$, NH_3 and Zn	400–500	Cubic TaN	Conducting: 900 μΩ cm, low level of contamination	336
$TaCl_5$, NH_3 and $AlMe_3$	250–400	Cubic TaN, or cubic TaC of low crystallinity	1300–7500 μΩ cm, high contamination with C, Cl, Al (total: 50–60%)	337
$TaBr_5$, NH_3 and zinc	400–500	Comparable $TaCl_5$ (ref. 106)		337
$3TaBr_5$, NH_3 and $AlMe_3$	250–400	Low crystalline TaN-films	6600–64 000 μΩ cm, high contamination with Al, C, Cl	337
$TaCl_5$, NH_3 and $SiH(NMe_2)_3$	300–500	Cubic TaN and insulating Si_3N_4 (based on XPS)	36 000–220 000 μΩ cm, low contamination with chlorine or carbon	414
$TaCl_5$, 1,1-DMHy	300–400	Amorphous	Insulating, tantalum nitride (Ta_3N_5)	119
$TaCl_5$, N_2/H_2-plasma	100–400	Cubic TaN	350 μΩ cm (300 °C), film contains Cl	415, 416
$NbCl_5$, NH_3 and Zn	500	Crystalline δ-NbN or Nb_4N_5	200 μΩ cm, $O_2 < 1\%$, Nb : N = 2 : 3	92
$NbCl_5$, NH_3	500	Crystalline δ-NbN or Nb_4N_5	500 μΩ cm, $O_2 < 1\%$, Nb : N = 2 : 3	92
$NbCl_5$, NH_3	250–500	Most probably Nb_4N_5	16 000 μΩ cm (250 °C)	326
$NbCl_5$, DMHy	400	Low crystallinity	600 μΩ cm (500 °C), $O_2 <$ μΩ 1%; Cl 8% (250 °C), <1% (500 °C) 2900 μΩ cm, 5% of Cl	119

of $TaCl_5$ and a mixed hydrogen/nitrogen plasma showed the best results (350 μΩ cm). Highly conductive, cubic TaN was deposited even at 300 °C. In all experiments, ALD-growth (self-limitation) was observed and the films were highly conformal. The oxygen content of the films plays an important role for the electrical properties of the films.[340] All films contained 5–8% of oxygen. The main reason given by the authors is post oxidation of the film, once it was removed from the reactor. ALD films are generally not as dense as bulk TaN or MOCVD deposited TaN, thus displaying a certain porosity, rendering the films susceptible to the diffusion of oxygen. The work function of the PEALD TaN film on SiO_2 has been estimated by *C–V* measurements as 4.4 eV, which indicates possible application as an *n*-FET gate metal electrode.[341]

For metal-organic ALD (MOALD) of TaN the two mixed amido/imido compounds TBTDET ($[Ta(NEt_2)_3(N\text{-}t\text{-}Bu)]$) and TAIMATA ($[Ta(NMe_2)_3(NCMe_2Et)]$) were tested with the co-reactants ammonia as well as hydrogen plasma. The results regarding the use of ammonia are contradictory. Two groups (TAIMATA[342] and TBTDET[343]) reported that amorphous, almost insulating TaN_x films were deposited (Table 9.8). These films had very low densities (~3.6 g cm^{-3}; *cf.* density of bulk TaN of 13.7 g cm^{-3}) and were highly susceptible towards post oxidation in air. These results

Table 9.8 ALD of tantalum nitride thin films using metal organic compounds.

Precursors	*Substrate temperature (°C)*	*Structure/ phases*	*Thin film properties*	*Ref.*
TAIMATA, NH_3	250	Amorphous Ta_3N_5	Insulating, oxidized (20% O_2)	342
TAIMATA, H_2-plasma	250	Cubic TaN	366 μΩ cm, 17% C	342
TBTDET, NH_3	250	Amorphous TaN	500–1000 μΩ cm, 7% C, 7% O_2	417, 418
TBTDET, NH_3	260	Amorphous TaN_x	1.4×10^6 μΩ cm	343
TBTDET, H_2-plasma	260	Cubic TaN	400 μΩ cm, 20% C	343, 419

agree with the general assumption that ammonia is not suitable as a reducing agent at low temperatures, but only as a proton donator for transamination reactions or the protonation of metal chlorides (Figure 9.7). In contrast, van der Straten *et al.* reported the formation of completely amorphous, but very conductive TaN films using TBTDET and ammonia at substrate temperatures of 250 °C (refs 417, 418 below). As no other reports exist in which completely amorphous TaN_x films are conductive, these results have to be considered with care. The results for films, deposited from TBTDET or TAIMATA, using hydrogen plasma are more consistent. Cubic, nicely conducting TaN films with a relative high density (~7.9 g cm^{-3}) were deposited. It seems the hydrogen-plasma does not have the ability to inhibit the incorporation of carbon into the films (17–20 at.%), and so the films may rather be a Ta(N,C) solid solution or may contain graphitic C. It can be concluded that no significant protonation of the amido groups and consequent cleavage of amines occur. Owing to the higher density of the films, no aging effects in air (post oxidation) were observed.

ALD of NbN is less developed and only $NbCl_5$ has been used as precursor for the deposition of NbN_x thin films. Because no insulating NbN phases are known, all depositions led to conducting films. Experiments using ammonia as the co-reactant resulted in NbN_x films that contained significant amounts of chlorine (depending on the substrate temperature). The crystalline phase could not be explicitly assigned (δ-NbN or Nb_4N_5). Time-of-flight elastic recoil detection analysis (TOF-ERDA) clearly showed that the films were nitrogen rich. This indicates the formation of Nb_4N_5. Oxidation of the films does not play an important role. The oxygen content always remained below 1%. The use of zinc as an additional reducing agent did not show fundamental improvement of the quality of the films. Likewise, no positive effect was achieved when dimethylhydrazine was used as the source of nitrogen in the ALD process.

9.4.5 Molybdenum and Tungsten Nitride Deposition

Most CVD work on thin films of Group 6 nitride and carbonitride materials deals with tungsten nitride, using WF_6 and WCl_6 together with ammonia and hydrogen (as reducing agent) as precursors. Similarly, MoF_6 and $MoCl_5$ were used as metal sources. However, in contrast to the Groups 4 and 5 cases, homoleptic, volatile metal carbonyls [$M(CO)_6$] are available and have been employed as metal sources with ammonia as well.

9.4.5.1 CVD and ALD of the Nitrides using the Metal Halides as Precursors

The CVD of β-W_2N from WF_6 and NH_3/H_2 represents the prototypical case, here, and early work dates back to 1987 (Scheme 9.4).[344] Layers with resistivities down to 100 μΩ cm were deposited.[345,346]

$$2WF_{6(g)} + 3NH_3 + 3/2H_2 \xrightarrow{\Delta} N_{2(g)} + 12HF_{(g)} + W_2N_{(s)}$$

Scheme 9.4

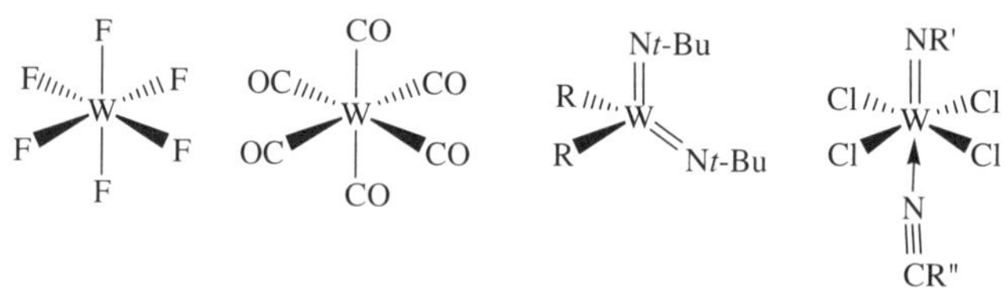

R = NME_2, NEt_2, NH*t*-Bu, pyralozes
R' = *i*-Pr, Ph, Allyl
R'' = Me, Ph

Figure 9.8 Some known precursors for tungsten nitride.

While hydrogen is the most common reducing agent,[345–348] silanes[349,350] or boranes[351] can also be used. The reducing agent has to be dosed very carefully; otherwise, WF_6 will be reduced to elemental tungsten. A serious problem of this process is the occurrence of HF, which forms adducts like NH_4F and NH_4HF_2 as solid by-products, which might lead to fluorine incorporation into the films. Fluorine diffuses rapidly into copper, which damages the metallization. Additionally, WF_6 can damage silicon or silicon oxide surface by etching.[345,352] Nevertheless WF_6/NH_3 has been studied as precursor combination for ALD[351] of tungsten nitride as well as of tungsten carbonitride (Figure 9.8).[353–359] For ALD of tungsten carbonitride, triethylboron was used both as a reducing agent as well as a carbon source. Films obtained by this method were down to 10 nm thick and had conductivities of $\sim 350\,\mu\Omega\,cm$.[351] The films had very low densities[356,373] ($\sim 4.6\,g\,cm^{-3}$) when films less than 5 nm thick were deposited but this increased to $13.1\,g\,cm^{-3}$ at a film thickness of 40 nm. Resistivities are reported to be $350\,\mu\Omega\,cm$ at 24 nm layer thickness.[353] Plasma enhanced ALD of W_2N on silicon using WF_6/NH_3 without additional hydrogen or other reducing agents was reported.[360] The reduction was induced by exposing the ammonia to a plasma pulse, producing reactive N–H species that are easily absorbed to the silicon substrate and react sequentially with the tungsten hexafluoride. This surface nitridation suppresses silicon reduction by WF_6. An interesting result was reported by using the combination of ethylene and silane as reducing agents, pulsed $WF_6/NH_3/C_2H_4/SiH_4/NH_3$.[361] The obtained film had the stoichiometry of W_3N_2 and was phase mixtures of β-W_2N and γ-WN. Both silicon and carbon content were below the detection limit of XPS.

Compared to tungsten nitride, studies on CVD and ALD on molybdenum nitride phases are rare. Figure 9.9 shows some precursors of molybdenum nitride. However, the obtained general results are quite similar to those discussed above. Cubic γ-Mo_2N can be deposited from $MoF_6/NH_3/H_2$.[362] Above 800 °C hexagonal δ-MoN is formed as well as metallic molybdenum. Because these films were produced for hard coating purposes, nothing about the electrical properties was reported. $MoCl_5/NH_3$ was used to deposit Mo_xN materials in the absence of further reducing agents. At 400 °C, the Mo : N ratio was 2 : 1, indicating the formation of cubic γ-Mo_2N, while above 700 °C more nitrogen-rich Mo_xN films containing δ-MoN and γ-Mo_2N together with elemental molybdenum were formed.[363,364] In the temperature range 350–500 °C $MoCl_5$ is also suitable as a precursor for ALD purposes.[362] Similar to the CVD experiments the nitrogen content is dependent on the deposition temperature. XRD patterns showed the formation of γ-Mo_2N, while the film

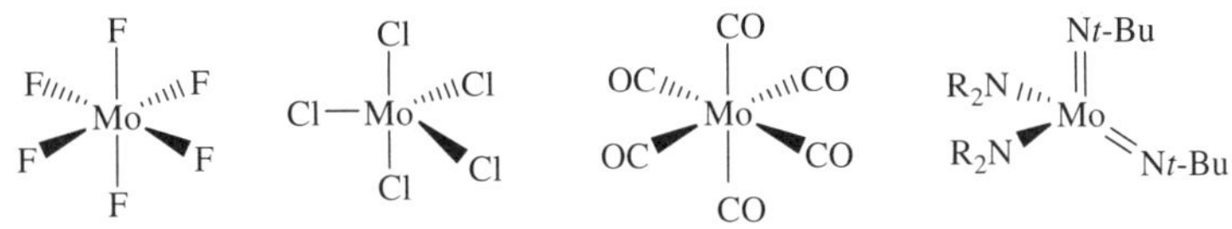

Figure 9.9 Some known precursors for molybdenum nitride.

composition was observed as $MoN_{1.23}$ at 350 °C and $MoN_{1.1}$ an 500 °C, indicating the presence of a nitrogen-rich amorphous phase. At higher temperatures the formation of δ-MoN was again observed.

9.4.5.2 MOCVD and ALD of the Nitrides using Hexacarbonyl Precursors

Tungsten nitride deposition from the low valent W(0) source $[W(CO)_6]$ and ammonia is possible without additional reducing agents at temperatures as low as 200 °C.[365–368] Resistivities as low as 123 μΩ cm for 50 nm thick films[25] and 590 μΩ cm for 15 nm thick films[366] were obtained. The dissociation energy of W–CO bonds is relatively low, with recombination being easily interrupted by the presence of nitrogen-bearing reactants, such as ammonia, leading to the formation of tungsten nitride. Depending on the process conditions the obtained films incorporated 2–16 at.% of carbon, and about 4 at.% of oxygen.[367] Similar to the WF_6 case the film stoichiometry can be easily controlled by adjusting the flow rates of $W(CO)_6$ and ammonia. Application of $[Mo(CO)_6]$ and ammonia lead to mixtures of γ-Mo_2N and δ-MoN.[20] It was possible to obtain a pure δ-MoN phase at 700 °C. Additionally, the formation of pure molybdenum less favoured compared to $MoCl_5$ because of the absence of hydrogen. Not only film deposition on flat substrates could be carried out by this precursor but also Mo_xN deposition inside zeolite structures was reported, which is important for heterogeneous catalysis.[369] Similarly, other low valent organo Group 6 precursors exhibiting metal–carbon bonds, including bis(arene)chromium compounds such as $[(\eta^6\text{-}C_6H_6)_2Cr]$ as substitute for the hexacarbonyl $[Cr(CO)_6]$, have been employed to deposit metal carbonitride films as protective coatings.[370,371]

9.4.5.3 MOCVD and ALD using Amide-type Precursors

A similar strategy as discussed for the Groups 4 and 5 nitrides has been used for the growth of Group 6 nitrides and carbonitride by MOCVD and ALD. The chemical complexity of precursors range from simple amine adduct at WCl_6 over the well-established bis(imido-amido) systems $[M(NR_2)_2(NR')_2]$[372,373] to more sophisticated compounds containing an even more complex mixed ligand sphere with chelating groups such as guanidinates, as discussed in the case of Ta and Nb. The motivation for these investigations is, again, reducing the deposition temperatures and/or shifting the ALD process window to appropriate positions matching the processing requirements for integration. Bis(*tert*-butylimido)-bis(*tert*-butylamido)tungsten(VI) was used as single source precursor in the absence of ammonia and led to heavily carbon contaminated films at 450–650 °C, with 66 at.% carbon and only 25 at.% nitrogen, as well as 3 at.% oxygen.[373] The carbon-containing fragments were produced by fragmentation of the *tert*-butyl-imido and amido ligands. *tert*-Butylamine, isobutene and acetonitrile were the main decomposition products.[374,375] The decomposition reaction involves γ-hydride and β-methyl activation, leading to carbon contamination. Thus, ammonia is essential for removing the carbon containing fragments from the surface.[374] Precursors containing residual NH bonds turned out to be rather non-volatile, possibly due to intermolecular hydrogen bridging. Thus, the related all alkyl substituted compounds such as $[W(NtBu)_2(NR_2)_2]$ (R = Me, Et) are liquids and quite volatile. Relatively pure β-W_2N was deposited on Si(100) by MOCVD.[375] In ALD experiments, together with ammonia as an additional reactive gas an amorphous, a carbon-free material of stoichiometry WN was obtained.[376–378] Films 20 nm thick showed resistivities of 480 μΩ cm. The advantage of ALD was impressively documented by coating objects with an extremely high aspect ratio. A 2 cm long fused silica capillary tube with an inner diameter of 20 μm was homogeneously covered with this tungsten nitride material.[379] Ammonia was essential for the growth of the nitride material, but it was not incorporated into the films, as indicated by labelling studies with $^{15}NH_3$. This finding is in surprising contrast to the mechanisms discussed for TiN MOCVD and ALD from $[Ti(NR_2)_4]$. When ammonia was replaced by hydrogen plasma, the obtained films possessed only very low contents of nitrogen, but consisted mainly of tungsten carbide.[380] Details of the decomposition mechanism of the precursor, however, in the gas-phase and on the surface are not yet elucidated to a satisfying extent. Similarly to the Group 4 and 5 cases, several alternative precursors with a more complex ligand sphere around the metal centre were introduced by keeping the fragment $W(NtBu)_2$ and substituting one or both alkylamido groups NR_2 with pyrazolates[380,381] or guanidinates.[382,383] For example, the precursor $[W(NtBu)_2(NMe_2)\{(NiPr)_2CNMe_2\}]$ gave amorphous films of a W : N composition of 2 : 1 (by SNMS) at 500 °C in the presence of ammonia, revealing excellent resistivity of 94 μΩ cm for a 24 nm thick film. Remarkably, the very closely related derivative $[W(NtBu)_2(H)\{(NiPr)_2CNMe_2\}]$ with a hydride group replacing the NMe_2 ligand yielded films with a different W : N composition of 1 : 1 but again with good resistivity of 600 μΩ cm. In both cases, the levels of C and O impurities were well below 5–6 at.% and in some cases even below the detection limit if the deposition experiments were carried out using excellent equipment and ultrapure ammonia (*e.g.* AIXTRON 200 RF system). Depositions at higher temperatures gave crystalline β-W_2N material with larger grain sizes but higher resistivities of 1240 and 1340 μΩ cm respectively. A series of halide-containing alkylimido precursors of the general formula $[(RN)WCl_4(NCR')]$ (R = *i*-Pr, Ph, allyl; R′ = Me, Ph) have been investigated for tungsten nitride materials in some detail (Scheme 9.5).[384–387] These compounds are not volatile but can be used for liquid injection MOCVD or aerosol assisted MOCVD. The hydrogen carrier gas serves as a reducing agent and removes the chlorine atoms from the reaction as HCl.[388] In the absence of ammonia, the films were contaminated with up to 50 at.% carbon,[386,387] mainly originating from the solvent. The type and presence of a nitrile ligand affected the nitrogen content, too.[385] Despite the high C and O contamination the resistivity of the films deposited below 500 °C were as low as 750 μΩ cm, but increased up to 5500 μΩ cm at higher deposition temperatures.[387] When the precursors are used in the presence of ammonia, the nitrogen content of the films could be significantly increased. But the level of carbon contamination could still not be reduced below 10 at.%.[404] From such complexes, some guanidinato complexes have been synthesized by substitution of one chloride.[389]

While most precursor chemistry deals with W(VI) compounds, one noteworthy report looks at using $[W_2(NMe_2)_6]$ and ammonia.[390] This W(III) compound shows ALD growth behaviour in the temperature window 180–220 °C, which is the lowest deposition temperature of W_xN films reported. The films showed tungsten contents of only 39 at.% and high contamination levels of carbon, oxygen

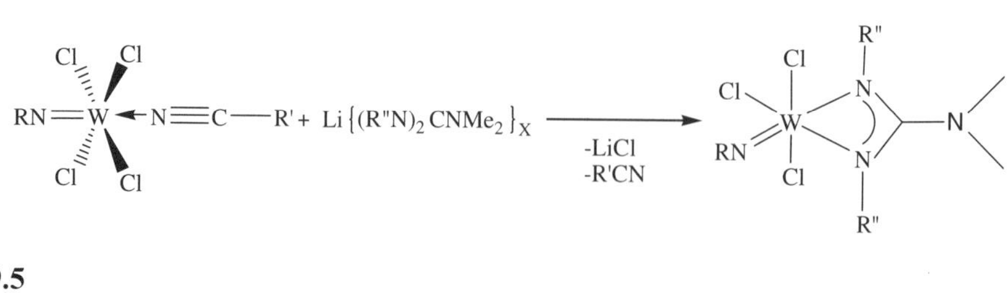

Scheme 9.5

and even hydrogen of 5–13 at.%. The W : N ratios in these films were 1.2–1.35, which is clearly nitrogen-rich towards W_2N. Resistivities of the as-deposited, amorphous films were about 800 μΩ cm. Few metal-organic precursors for Mo/N deposition bearing an all-nitrogen coordination sphere have been reported in the literature. [$Mo(NMe_2)_4$] was reported by Fix *et al.*[391,392] MOCVD-experiment in the presence of ammonia, yielded a gold, amorphous phase of stoichiometry Mo_2N_3, at as low as 200 °C. The obtained films are conductive and free of carbon contamination. Similarly to the tungsten case above, bis(diethylamido-)-bis(*tert*-butylimido)molybdenum gave γ-Mo_2N at 450–600 °C even in the absence of ammonia, but the films were highly contaminated with carbon.[393] The dimethyl amido analogue of this compound, [$Mo(N\textit{t}Bu)_2(NMe_2)_2$], was reported as a precursor for ALD purposes.[409] Highly conformal, conductive molybdenum nitride films with thicknesses down to 44 nm could be deposited by this method. The step coverage of these films was very high. XRDs of these films showed the formation of the tetragonal β-$Mo_{16}N_7$ phase, but the measured Mo : N composition was close to 1 : 1. This indicated the presence of an amorphous, nitrogen-rich phase in the films. Some mixed ligand imido/pyrazolato compounds analogous to the tungsten case above were studied, but without MOCVD or ALD results.[396,397] Molybdenum guanidanato complexes such as [$Mo(NtBu)_2(X)\{(NiPr)_2CNMe_2\}$] (X = NMe_2, Cl, N_3), similar to the tungsten congeners, were studied. However, the MOCVD results were not encouraging.[394]

9.5 Conclusions and Prospects

The MOCVD and ALD of transition metal nitrides (except TiN) are more or less in the early stage of development, and there are many challenges ahead in terms of precursor development and process integration, especially for ALD. There are many important issues to be addressed for the precursor chemistry of transition metal nitrides. Some of the challenges are:

1. Identification of new ligand systems, which will promote the reduction of transition metals and which in turn lead to high quality films.
2. Precursor synthesis optimization (efficient and easy synthesis for scale up, high yield, high purity and low cost, *etc.*).
3. Systematic studies on surface reactions of precursors on various surfaces, focussing on metal nitride MOCVD and ALD processes.

Since the ALD process proceeds through surface reactions, a detailed understanding of initial surface conditions as well as interaction of the precursor on various surfaces are important for the optimization of ALD process. Very promising results have been achieved for MOCVD and ALD of transition metal nitrides. However, a few questions still remain regarding the properties of metal nitride thin films and a detailed investigation is required with respect to device fabrication, integration and performance related problems such as electro-migration resistance, and time-dependent electrical properties, *etc.* Extensive understanding of transition metal nitride MOCVD and ALD processes as well as thin film properties are essential for the successful implementation of transition metal nitride films in future semiconductor device fabrication.

Acknowledgements

Financial support from the German Science Foundation (DFG: CVD SPP-1119) and H. C. Starck is gratefully acknowledged.

References

1. L. E. Toth, In *Transition Metal Carbides and Nitrides*, J. L. Margrave, ed., Refractory Materials. Academic Press, New York, 1971, **7**.
2. H. O. Pierson, *Handbook of Refractory Carbides and Nitrides: Properties, Characteristics and Applications*, Noyes Publications, Westwood, NJ, 1996.
3. A. E. Kaloyeros and E. Eisenbraun, *Annu. Rev. Mater. Sci.*, 2000, **30**, 363.
4. M. Wittmer, *J. Vac. Sci. Technol. A*, 1985, **3**, 1797.
5. S. R. Kurtz and R. G. Gordon, *Thin Solid Films*, 1986, **140**, 277.
6. M. K. Wahlstrom, E. Johasson, E. Veszelei, P. Bennich, M. Olsson and S. Hogmark, *Thin Solid Films*, 1992, **220**, 315.
7. M. Veszelei and E. Veszelei, *Thin Solid Films*, 1993, **236**, 46.
8. H. E. Hintermann, *Thin Solid Films*, 1981, **84**, 215.
9. S. Adachi and M. Takahashi, *J. Appl. Phys.*, 2000, **87**, 1264.
10. Y. Omura, H. Inokawa and K. Izumi, *J. Mater. Res.*, 1991, **6**, 1238.
11. T. Hara, A. Yamanoue, H. Iio, K. Inoue, G. Washidzu and S. Nakamura, *Jpn. J. Appl. Phys., Part 1*, 1991, **30**, 1447.
12. R. C. Glass, L. M. Spellman, S. Tanaka and R. F. Davis, *J. Vac. Sci. Technol. A*, 1992, **10**, 1625.
13. P. Patsalas and S. Logothedies, *J. Appl. Phys.*, 2001, **90**, 4725.
14. X.-J. Chen, V. V. Struzhkin, Z. Wu, M. Somayazulu, J. Qian, S. Kung, A. N. Christensen, Y. Zhao, R. E. Cohen, H.-K. Mao and R. J. Hemley, *Proc. Natl. Acad. Sci. U.S.A.*, 2005, **102**, 3198.
15. J. R. Gao, J. N. Hovenier, Z. Q. Yang, J. J. A. Baselmans, A. Baryshev, M. Hajenius, T. M. Klapwijk, A. J. L. Adam, T. O. Klaassen, B. S. Williams, S. Kumar, Q. Hu and J. L. Reno, *Appl. Phys. Lett.*, 2005, **86**, 244104.
16. E. Baggetta, B. Ebert, N. Hadacek, J. C. Villegier and M. Maignan, *IEEE Transactions on Applied Superconductivity*, 2005, **15**, 453.
17. J. Oro-Sole, C. Frontera, B. Martinez, D. Beltran-Porter, M. R. Palacin and A. Fuertes, *Chem. Commun.*, 2005, 3352.
18. L. P. Zhu and S. Yamanaka, *Chem. Mater.*, 2003, **15**, 1897.
19. M. Nagai, *Applied Catalysis A-General*, 2007, **322**, 178.
20. S. T. Oyama, *Catalysis Today*, 1992, **15**, 179.
21. E. Furimsky, *Appl. Catalysis*, 2003, **240**, 1.
22. P. Serp, P. Kalck and R. Feurer, *Chem. Rev.*, 2002, **102**, 3085.
23. M. Nagai, *Appl. Catalysis A-General*, 2007, **322**, 178.
24. H. Kim, *Surf. Coat. Technol.*, 2006, **200**, 3104.
25. M. Schumacher, P. K. Baumann and T. Seidel, *Chem. Vap. Deposition*, 2006, **12**, 99.
26. H. Kim, *J. Vac. Sci. Technol. A.*, 2003, **21**, 2231.
27. M. Ritala and M. Leskelä, *Nanotechnology*, 1999, **10**, 19.
28. S. M. Sze, In *Physics of Semiconductor Devices*, 2nd ed., John Wiley & Sons Inc., New York, 1981.
29. S. M. Sze, "VLSI Technology", 2nd ed, McGrow-Hill International Edn, 1988.
30. T. Kauerauf, B. Govoreanu, R. Degraeve, G. Groeseneken and H. Maes, *Solid State Electronics*, 2005, **49**, 695.
31. M. Rosenberg, D. C. Edelstein, C.-K. Hu and K. P. Rodbell, *Annu. Rev. Mater. Sci.*, 2000, **30**, 229.
32. *International Technology Roadmap for Semiconductors*, Semiconductor Industry Association, San Jose, CA, 2003.

33. S. Haukka, K.-E. Elers and M. Tuominen, *Mat. Res. Soc. Symp. Proc.*, 2001, **612**, D6.4/1-D6.4/6.
34. J. F. Kang, H. Y. Yu, C. Ren, N. Sa, H. Yang, M.-F. Li, D. S. H. Chan, X. Y. Liu, R. Q. Han and D.-L. Kwong, *J. Electrochem. Soc.*, 2007, **154**, H927.
35. J. Lu, Y. Kuo, S. Chatterjee and J. Y. Tewg, *J. Vac. Sci. Technol. B*, 2006, **24**, 349.
36. B. H. Lee, J. Oh, H. H. Tseng, R. Jammy and H. Huff, *Materials Today*, 2006, **9**, 32.
37. M. Lemberger, S. Thiemann, A. Baunemann, H. Parala, R. A. Fischer, J. Hinz, A. J. Bauer and H. Ryssel, *Surf. Coat. Technol.*, 2007, **201**, 9154.
38. S.-H. Kim, J.-K. Kim, J. H. Lee, N. Kwak, J. Kim, S.-H. Jung, M.-R. Hong, S. H. Lee, J. Collins and H. Sohne, *J. Electrochem. Soc.*, 2007, **154**, D435.
39. P. M. Soloman, *Annu. Rev. Mater. Sci.*, 2000, **30**, 681.
40. E. W. A. Young, S. Mantl and P. B. Griffin, in Silicon MOSFETs-Novel Materials and Alternative Concepts. ed., R. Waser, *Nanoelectronics and Information Technology*. Wiley-VCH, Weinheim, 2005.
41. G. D. Wilk, R. M. Wallace and J. M. Anthony, *J. Appl. Phys.*, 2001, **89**, 5243.
42. R. M. Wallace and G. D. Wilk, *Critical Reviews in Solid State and Materials Sciences*, 2003, **28**, 231.
43. J. Robertson, *European Physical Journal-Appl. Phys.*, 2004, **28**, 265.
44. C. Ren, B. B. Faizhal, D. S. H. Chan, M. F. Li, Y. C. Yeo, A. D. Trigg, N. Balasubramanian and D. L. Kwong, *Thin Solid Films*, 2006, **504**, 174.
45. Y. Sugimoto, M. Kajiwara, K. Yamamoto, Y. Suehiro and D. Wang, *Appl. Phys. Lett.*, 2007, **91**, 112105.
46. T. Kauerauf, B. Govoreanu, R. Degraeve, G. Groeseneken and H. Maes, *Solid-State Electronics*, 2005, **49**, 695.
47. I. De, D. Johri, A. Srivastava and C. M. Osburn, *Solid-State Electronics*, 2000, **44**, 1077.
48. Y. Gotoh, H. Tsuji and J. Ishikawa, *J. Vac. Sci. Technol. B.*, 2003, **21**, 1607.
49. H.-C. Wen, R. Choi, G. A. Brown, T. Böscke, K. Matthews, H. R. Harris, K. Choi, H. N. Alshareef, H. Luan, G. Bersuker, P. Majhi, D.-L. Kwong and B. H. Lee, *IEEE Electronic Device Letters*, 2006, **27**, 598.
50. R. Jha, J. Gurganos, Y. H. Kim, R. Choi, J. Lee and V. Misra, *IEEE Electron Device Lett.*, 2004, **25**, 420.
51. G. Sjoblom, J. Westlinder and J. Olsson, *IEEE Transactions on Electronic Devices*, 2005, **52**, 2349.
52. J. Westlinder, T. Schram, L. Pantisano, E. Cartier, A. Kerber, L. S. Lujan, J. Olsson and G. Groesenken, *IEEE electron Device, Letters*, 2003, **24**, 550.
53. J. Westlinder, G. Sjoblom and J. Olsson, *Microelectronic Engineering*, 2004, **75**, 389.
54. T.-H. Cha, D.-G. Park, T.-K. Kim, S.-A. Jang, I.-S. Roh and J. W. Park, *Appl. Phys. Lett.*, 2002, **81**, 4192.
55. J. Westlinder, J. Malmstrom, G. Sjoblom and J. Olsson, *Solid State Electronics*, 2005, **49**, 1410.
56. H. Y. Yu, M.-F. Li and D.-L Kwong, *IEEE Transaction on electron devices*, 2004, **51**, 609.
57. J. K. Scheffer, S. Samavedam, D. Gilmer, V. Dhandapani, P. Tobin, J. Mogab, B.-Y. Nguyen, B. White, S. Dakshina-Murthy, R. Rai, Z.-X. Jiang, R. Martin, M. V. Raymond, M. Zavala, L. La, J. A. Smith, R. Garcia, D. Raon, M. Kottke and R. B. Gregory, *J. Vac. Sci. Technol. B.*, 2003, **21**, 11.
58. C. Ren, H. Y. Yu, J. F. Kang, Y. T. Hou, M.-F. Li, W. D. Wang, S. H. D. Chan and D.-L. Kwon, *IEEE electron Device Letters*, 2004, **25**, 123.
59. K. Choi, H. N. Alshareef, H. C. Wen, H. Harris, H. Luan, Y. Senzaki, P. Lysaght, P. Majhi and B. H. Lee, *Appl. Phys. Lett.*, 2006, **89**, 032113.
60. H. Matsuhashi and S. Nishikawa, *OKI Technical Review*, 1994, **60**, 65.

61. N. Van Hoornick, H. De Witte, T. Witters, C. Zhao, C. Conardd, H. Huotari, J. Swerts, T. Schram, J. W. Maes, S. De. Gendt and M. Heyns, *J. Electrochem. Soc.*, 2006, **153**, G437.
62. P. Pyykko, *Chem. Rev*, 1988, **88**, 563.
63. J. E. Huheey, E. A. Keiter and R. L. Keiter, *Inorganic Chemistry Principles of Structure and Reactivity*, 4th edn, HarperCollins, New York, 1993.
64. A. F. Wells, *"Structural Inorganic chemistry"*, Clarendon, Oxford, 1986.
65. D. M. Hoffman, *Polyhedron*, 1994, **13**, 1169.
66. M. Chhowalla and H. E. Unalan, *Nat. Mater.*, 2005, **4**, 317.
67. A. Zerr, G. Miehe and R. Riedel, *Nat. Mater.*, 2003, **2**, 185.
68. C. Ziebert and S. Ulrich, *J. Vac. Sci. Technol. A*, 2006, **24**, 554.
69. *Encyclopedia of Inorganic Chemistry*, ed. R. B. King, John Wiley & Sons Ltd., Chichester, 1994.
70. G. Brauer and W. Kern, *Z. Anorg. Allg. Chem.*, 1983, **507**, 127.
71. W. Lengauer, M. Bohn, B. Wollein and K. Lisak, *Acta Mater.*, 2000, **48**, 2633.
72. R. E. Treece, J. S. Horwitz, S. B. Qadri, E. F. Skelton, E. P. Donovan and D. B. Chrisey, *J. Solid State Chem.*, 1995, **117**, 294.
73. Y. M. Shy, L. E. Toth and R. Somasundaram, *J. Appl. Phys.*, 1973, **44**, 5539.
74. T. Ishiguro, K. Matsushima and K. Hamasaki, *J. Appl. Phys.*, 1993, **73**, 1151.
75. R. E. Treece, J. S. Horwitz, J. H. Claassen and D. B. Chrisey, *Appl. Phys. Lett.*, 1994, **65**, 2860.
76. A. B. Kaul, T. D. Sands and T. Van Duzer, *J. Mater. Res.*, 2001, **16**, 1223.
77. G.-I. Oya and Y. Onodera, *J. Appl. Phys.*, 1974, **45**, 1389.
78. W. Mayr, W. Lengauer, V. Buscaglia, J. Bauer, M. Bohn and M. Fialin, *J. Alloys Compd.*, 1997, **262–263**, 521.
79. Z. Wang, H. Terai, A. Kawakami and Y. Uzawa, *Appl. Phys. Lett.*, 1999, **75**, 701.
80. J. Zhang, N. Boiadjieva, G. Chulkova, H. Deslandes, G. N. Goltsman, A. Korneev, P. Kouminov, M. Leibowitz, W. Lo, R. Malinsky, O. Okunev, A. Pearlman, W. Slysz, K. Smirnov, C. Tsao, A. Verevkin, B. Voronov, K. Wilsher and R. Sobolewski, *Electron. Lett.*, 2003, **39**, 1086.
81. M. Fenker, M. Balzer, R. V. Büchi, H. A. Jehn, H. Kappl and J.-J. Lee, *Surf. Coat. Technol.*, 2003, **163–164**, 169.
82. M. Benkahoul, E. Martinez, A. Karimi, R. Sanjinés and F. Lévy, *Surf. Coat. Technol.*, 2004, **180–181**, 178.
83. Y. Gotoh, M. Nagao, T. Ura, H. Tsuji and J. Ishikawa, *Nucl. Instrum. Methods Phys. Res. Sect. B*, 1999, **148**, 925.
84. Y. Gotoh, H. Tsuji and J. Ishikawa, *Rew. Sci. Instrum.*, 2000, **71**, 1002.
85. V. N. Zhitomirsky, I. Grimberg, L. Rapoport, N. A. Travitzky, R. L. Boxman, S. Goldsmith and B. Z. Weiss, *Surf. Coat. Technol.*, 1999, **120–121**, 219.
86. L. Hultman, C. Engström and M. Odén, *Surf. Coat. Technol.*, 2000, **133–134**, 227.
87. Y. Long, R. J. Stearn, Z. H. Barber, S. J. Lloyd and W. J. Clegg, *Mat. Res. Soc. Symp. Proc.*, 2004, **791**, Q5.18.1.
88. J. J. Jeong and C. M. Lee, *Appl. Surf. Sci.*, 2003, **213**, 11.
89. A. Madan, S. A. Barnett, A. Misra, H. Kung and M. Nastasi, *J. Vac. Sci. Technol. A*, 2001, **19**, 952.
90. X. Yu, Q. Lai, G. Li, J. Xu and M. Gu, *J. Mat. Sci. Lett.*, 2002, **21**, 1671.
91. K.-E. Elers, M. Ritala, M. Leskelä and E. Rauhala, *Appl. Surf. Sci. 82/83*, 1994, **82–83**, 468.
92. M. Ritala, T. Asikainen, M. Leskelä, J. Jokinen, R. Lappalainen, M. Utriainen, L. Niinistö and E. Ristolainen, *Appl. Surf. Sci.*, 1997, **120**, 199.
93. C.-S. Shin, Y.-W. Kim, D. Gall, J. E. Greene and I. Petrov, *Thin Solid Films*, 2002, **402**, 172.
94. A. Y. Ganin, L. Kienle and G. V. Vajenine, *Eur. J. Inorg. Chem.*, 2004, **1**, 3233.

95. *CRC Handbook of Chemistry and Physics*, 82nd edn., Ed. D.R. Lide, CRC Press LLC, 2001–2002, p. 4.
96. Y.-S. Suh, G. Heuss, H. Z. Zhong, S.-N. Hong and V. Misra, *VLSI Tech. Dig.*, 2001, 47.
97. S. Gopalan, K. Onishi, R. Nieh, C. S. Kang, R. Choi, H.-J. Cho, S. Krishnan and J. C. Lee, *Appl. Phys. Lett.*, 2002, **80**, 4416.
98. Y.-S. Suh, G. P. Heuss and V. Misra, *Appl. Phys. Lett.*, 2002, **80**, 1403.
99. Y.-S. Suh, G. P. Heuss, V. Misra, D.-G. Park and K.-Y. Lim, *J. Electrochem. Soc.*, 2003, **150**, F79.
100. C. Chaneliere, J. L. Autran, R. A. B. Devine and B. Balland, *Mater. Sci. Eng. Rep.*, 1998, **R22**, 269.
101. J.-C. Chuang and M.-C. Chen, *J. Electrochem. Soc.*, 1998, **145**, 3170.
102. C. Cabral Jr., K. L. Saenger, D. E. Kotecki and J. M. E. Harper, *J. Mater. Res.*, 2000, **15**, 194.
103. T. Laurila, K. Zeng, J. K. Kivilahti, J. Molarius, T. Riekkinen and I. Suni, *Microelectronic Eng.*, 2002, **60**, 71.
104. C.-L. Lin, S.-R. Ku and M.-C. Chen, *Jpn. J. Appl. Phys.*, 2001, **40**, 4181.
105. D. Fischer, T. Scherg, J. G. Bauer, H.-J. Schulze and C. Wenzel, *Microelectronic Eng.*, 2000, **50**, 459.
106. K.-H. Min, K.-C. Chun and K.-B. Kim, *J. Vac. Sci. Technol. B*, 1996, **14**, 3263.
107. Y.-J. Lee, B.-S. Suh, S.-K. Rha and C.-O. Park, *Thin Solid Films*, 1998, **320**, 141.
108. Z.-C. Wu, C.-C. Wang, R.-G. Wu, Y.-L. Liu, P.-S. Chen, Z.-M. Zhu, M.-C. Chen, J.-F. Chen, C.-I. Chang and L.-J. Chen, *J. Electrochem. Soc.*, 1999, **146**, 4290.
109. H.-J. Bae, Y.-H. Shin and C. Lee, *J. Kor. Phys. Soc.*, 1999, **34**, 504.
110. A. E. Kaloyeros, X. Chen, T. Stark, K. Kumar, S.-C. Seo, G. G. Peterson, H. L. Frisch, B. Arkles and J. Sullivan, *J. Electrochem. Soc.*, 1999, **146**, 170.
111. S. J. Im, S.-H. Kim, K.-C. Park, S.-L. Cho and K.-B. Kim, *Mat. Res. Soc. Symp. Proc.*, 2000, **612**, D6.7.1.
112. C. Lee and Y.-H. Shin, *Mat. Chem. Phys.*, 1998, **57**, 17.
113. S.-L. Cho, K.-B. Kim, S.-H. Min, H.-K. Shin and S.-D. Kim, *J. Electrochem. Soc.*, 1999, **146**, 3724.
114. C. Linder, A. Dommann, G. Staufert and M.-A. Nicolet, *Sens. Actuators A*, 1997, **61**, 387.
115. I. Ayerdi, E. Castaño, A. García-Alonso and J. Gracia, *Sens. Actuators A*, 1997, **60**, 72.
116. T. Yeh, D. Swanson, L. Berg and P. Karn, *IEEE Trans. Magn.*, 1997, **33**, 3631.
117. Y. X. Leng, H. Sun, P. Yang, J. Y. Chen, J. Wang, G. J. Wan, N. Huang, X. B. Tian, L. P. Wang and P. K. Chu, *Thin Solid Films*, 2001, **398–399**, 471.
118. *CRC Handbook of Chemistry and Physics*, 82nd edn., Ed. D.R. Lide, CRC Press LLC, 2001–2002, p. 4.
119. M. Juppo, M. Ritala and M. Leskelä, *J. Electrochem. Soc.*, 2000, **147**, 3377.
120. C. L. Bull, P. F. McMillan, E. Soignard and K. Leinenweber, *J. Solid State Chem.*, 2004, **177**, 1488.
121. E. Furimsky, *Appl. Catal. A: Gen.*, 2003, **240**, 1.
122. P. Liu and J. A. Rodriguez, *Catal. Lett.*, 2003, **91**, 247.
123. T. Kadono, T. Kubota and Y. Okamoto, *Catal. Today*, 2003, **87**, 107.
124. S. Gong, H. Chen, W. Li and B. Li, *Catal. Commun.*, 2004, **5**, 621.
125. M. Nagai, Y. Yamamoto and R. Aono, *Colloids Surf. A: Physicochem. Eng. Aspects*, 2004, **241**, 257.
126. Y. Li, Y. Fan, J. He, B. Xu, H. Yang, J. Miao and Y. Chen, *Chem. Eng. J.*, 2004, **99**, 213.
127. R. Kojima and K. Aika, *Appl. Catal. A: Gen.*, 2001, **219**, 141.
128. H. J. Lee, J.-G. Choi, C. W. Colling, M. S. Mudholkar and L. T. Thompson, *Appl. Surf. Sci.*, 1995, **89**, 121.
129. X. Chen, T. Zhang, M. Zheng, Z. Wu, W. Wu and C. Li, *J. Catal.*, 2004, **224**, 473.

130. E. Soignard, P. F. McMillan, T. D. Chaplin, S. M. Farag, C. L. Bull, M. S. Somayazulu and K. Leinenweber, *Phys. Rew. B*, 2003, **68**, 132101-1.
131. J. E. Lowther, *J. Alloys Compd.*, 2004, **364**, 13.
132. R. Sanjinés, P. Hones and F. Lévy, *Thin Solid Films*, 1998, **332**, 225.
133. P. Hones, N. Martin, M. Regula and F. Lévy, *J. Phys. D: Appl. Phys.*, 2003, **36**, 1023.
134. C. Wiemer, R. Sanjinés and F. Lévy, *Surf. Coat. Technol.*, 1996, **86–87**, 372.
135. J. Kozlowski, J. Markowski, A. Prajzner and J. Zdanowski, *Surf. Coat. Technol.*, 1998, **98**, 1440.
136. H. Kung, T. R. Jervis, J.-P. Hirvonen, T. E. Mitchell and M. Nastasi, *Nanostruct. Mater.*, 1996, **7**, 81.
137. J.-Y. Lee and J.-W. Park, *Jpn. J. Appl. Phys.*, 1996, **35**, 4280.
138. J.-C. Chuang, S.-L. Tu and M.-C. Chen, *Thin Solid Films*, 1999, **346**, 299.
139. Y. He and J. Y. Feng, *J. Cryst. Growth*, 2004, **263**, 203.
140. G. Hägg, *Z. Physik. Chem.*, 1930, **7**, 339.
141. Z. G. Pinsker and S. V. Kaverin, *Kristallografiya*, 1957, **2**, 386.
142. R. Kiesling and Y. H. Liu, *J. Metals*, 1951, **3**, 639.
143. J. Neugebauer, A. J. Hegedus and T. Millner, *Z. Anorg.Allgem. Chem.*, 1959, **302**, 50.
144. F. A. Gunther and H. G. Schneider, *Kristallografiya*, 1966, **11**, 683.
145. V. I. Khitrova and Z. G. Pinsker, *Kristallografiya*, 1958, **3**, 545.
146. V. I. Khitrova and Z. G. Pinsker, *Kristallografiya*, 1960, **5**, 711.
147. V. I. Khitrova and Z. G. Pinsker, *Kristallografiya*, 1961, **6**, 882.
148. V. I. Khitrova, *Kristallografiya*, 1963, **8**, 873.
149. V. I. Khitrova, *Kristallografiya*, 1961, **6**, 549.
150. V. I. Khitrova, *Kristallografiya*, 1962, **7**, 374.
151. F. B. Wu, S. K. Tien and J. G. Duh, *Surf. Coat. Technol.*, 2005, **200**, 1514.
152. P. Hones, R. Consiglio, N. Randall and F. Lvy, *Surf. Coat. Technol.*, 2000, **125**, 179.
153. W.-D. Münz, T. Hurkmans, T. Trinh, D. B. Lewis and J. S. Brooks, *Surf. Coat. Technol.*, 2000, **125**, 179.
154. J. L. Vossen and W. Kern, *Thin Film Processes II*, Academic, New York, 1991.
155. S. M. Rossnagel, *J. Vac. Sci. Technol. B*, 1998, **16**, 2585.
156. C. Vahlas, B. Caussat, P. Serp and G. N. Angelopoulos, *Materials Science and Engineering R-Reports*, 2006, **53**, 1.
157. J. P. Dekker, P. J. Vanderput, H. J. Veringa and J. Schoonman, *J. Electrochem. Soc.*, 1994, **141**, 787.
158. K. Hieber, *Thin Solid Films*, 1974, **24**, 157.
159. T. Nakajima, K. Watanabe and N. Watanabe, *J. Electrochem. Soc.*, 1987, **134**, 3175.
160. T. Gerfin, K. H. Dahmen, CVD of conducting materials, in *CVD of Nonmetals*, Ed, W. S. Rees, Jr, VCH, Weinheim, 1996.
161. R. Fix, R. G. Gordon and D. M. Hoffman, *Chem. Mater.*, 1991, **3**, 1138.
162. H. T. Chiu, S. H. Chuang, C. E. Tsai, G. H. Lee and S. M. Peng, *Polyhedron*, 1998, **17**, 2187.
163. A. Krajewski, L. D'Alessio and G. De Maria, *Cryst. Research Growth*, 1998, **33**, 341.
164. E. R. Engbrecht, Y. M. Sun, S. Smith, K. Pfiefer, J. Bennett, J. M. White and J. G. Ekerdt, *Thin Solid Films*, 2002, **418**, 145.
165. S.-H. Jhi, J. Ihm, S. G. Louie and M. L. Cohen, *Nature*, 1999, **399**, 132.
166. D. M. Hausmann, E. Kim, J. Becker and R. G. Gordon, *Chem. Mater.*, 2002, **14**, 4350.
167. W. J. Maeng, S. J. Park and H. Kim, *J. Vac. Sci. Technol. B*, 2006, **24**, 2276.
168. J. P. Endle, Y. M. Sun, J. M. White and J. G. Ekerdt, *J. Vac. Sci. Technol. A*, 1998, **16**, 1262.
169. C. Amato-Wierda and D. A. Wierda, *J. Mater. Research*, 2000, **15**, 2414.
170. A. Intemann, H. Koerner and F. Koch, *J. Electrochem. Soc.*, 1993, **140**, 3215.

171. D. H. Kim, G. T. Lim, S. K. Kim, J. W. Park and J. G. Lee, *J. Vac. Sci. Technol. B*, 1999, **17**, 2197.
172. H. Tsuda and Y. Numasawa, *Japn. J. Appl. Phys. Lett. Part 2*, 1995, **34**, L 691.
173. S. T. Wang and Z. Zhang, *Prog. Chem.*, 2003, **15**, 374.
174. M. L. Hitchman and K. F. Jensen, *Handbook of Chemical Vapor Deposition: Principles and Applications*, Academic, San Diego, CA, 1993.
175. W. W. Wang, T. Nabatame and Y. Shimogaki, *Thin Solid Films*, 2006, **498**, 75.
176. L. H. Dubois, B. R. Zegarski and G. S. Girolami, *J. Electrochem. Soc.*, 1992, **139**, 3603.
177. J. Driessen, J. Schoonman and K. F. Jensen, *J. Electrochem. Soc.*, 2001, **148**, G178.
178. E. T. Norton and C. Amato-Wierda, *Chem. Mat.*, 2001, **13**, 4655.
179. S. J. Kim, B. H. Kim, H. G. Woo, S. K. Kim and D. H. Kim, *Bull. Korean Chem. Soc.*, 2006, **27**, 219.
180. J. C. F. Rodriguez-Reyes and A. V. Teplyakov, *J. Phys. Chem. C*, 2007, **111**, 4800.
181. L. A. Okada and S. M. George, *Appl. Surf. Sci.*, 1999, **137**, 113.
182. J. N. Musher and R. G. Gordon, *J. Electrochem. Soc.*, 1996, **143**, 736.
183. D. C. Bradley and I. M. Thomas, *Can. J. Chem.*, 1962, **40**, 1355.
184. A. W. Jackson, O. Shebanova, A. L. Hector and P. F. McMillan, *J. Solid State Chem.*, 2006, **179**, 1383.
185. G. M. Brown and L. Maya, *J. Am. Ceram. Soc.*, 1988, **71**, 78.
186. L. H. Dubois, *Polyhedron*, 1994, **13**, 1329.
187. T. S. Cale, M. B. Chaara, G. B. Raupp and I. J. Raaijmakers, *Thin Solid Films*, 1993, **236**, 294.
188. B. H. Weiller and B. V. Partido, *Chem. Mat.*, 1994, **6**, 260.
189. B. H. Weiller, *J. Am. Chem. Soc.*, 1996, **118**, 4975.
190. J. B. Cross and H. B. Schlegel, *Chem. Mater.*, 2000, **12**, 2466.
191. J. B. Cross, S. M. Smith and H. B. Schlegel, *Chem. Mater.*, 2001, **13**, 1095.
192. M. D. Allendorf and C. F. Melius, *Surf. Coat. Technol.*, 1998, **109**, 191.
193. T. T. Huang, K. Tan, M. H. Lin and Q. E. Zhang, *Chin. J. Chem.*, 2007, **25**, 910.
194. I. J. Raaijmakers, R. N. Vrtis, J. Yang, S. Ramaswami, A. Lagendijk, D. A. Roberts and E. K. Broadbent, *Mater. Res. Soc. Symp. Proc.*, 1992, **260**, 99.
195. I. J. Raijmakers, *Thin Solid Films*, 1994, **247**, 85.
196. I. J. Raaijmakers and J. Yang, *Appl. Surf. Sci.*, 1993, **73**, 31.
197. S. Panda, J. Kim, B. H. Weiller, D. J. Economou and D. M. Hoffman, *Thin Solid Films*, 1999, **357**, 125.
198. J. J. Hsieh, *J. Vac. Sci. Technol. A*, 1993, **11**, 78.
199. J. C. Rey, L.-Y. Cheng, J. P. McVettie and K. C. Saraswat, *J. Vac. Sci. Tech.*, 1991, **A9**, 1083.
200. B. H. Weiller and S. D. Adamson, *J. Electrochem. Soc.*, 1997, **144**, L40.
201. L. Hiltunen, M. Leskela, M. Makela, L. Niinisto, E. Nykanen and P. Soininen, *Thin Solid Films*, 1988, **166**, 149.
202. M. Ritala, M. Leskela, E. Rauhala and P. Haussalo, *J. Electrochem. Soc.*, 1995, **142**, 2731.
203. J. W. Lim, H. S. Park and S. W. Kang, *J. Appl. Phys.*, 2000, **88**, 6327.
204. C. H. Ahn, S. G. Cho, H. J. Lee, K. H. Park and S. H. Jeong, *Metals Mater. Int.*, 2001, **7**, 621.
205. J. Uhm and H. Jeon, *Jpn. J. Appl. Phys., Part 1*, 2001, **40**, 4657.
206. K. E. Elers, V. Saanila, P. J. Soininen, W. M. Li, J. T. Kostamo, S. Haukka, J. Juhanoja and W. F. A. Besling, *Chem. Vapor Deposition*, 2002, **8**, 149.
207. A. Satta, J. Schuhmacher, C. M. Whelan, W. Vandervorst, S. H. Brongersma, G. P. Beyer, K. Maex, A. Vantomme, M. M. Viitanen, H. H. Brongersma and W. F. A. Besling, *J. Appl. Phys.*, 2002, **92**, 7641.
208. S. Smith, W.-M. Li, K. E. Elers and K. Pfeifer, *Microelectron. Eng.*, 2002, **64**, 247.
209. J. Gelatos, L. Chen, H. Chung, R. Thakur and A. Sinha, *Solid State Technol.*, 2003, **46**, 44.
210. J. Kim, H. Hong, K. Oh and C. Lee, *Appl. Surf. Sci.*, 2003, **210**, 231.

211. H. Tiznado and F. Zaera, *J. Phys. Chem. B*, 2006, **110**, 13491.
212. J. W. Elam, M. Schuisky, J. D. Ferguson and S. M. George, *Thin Solid Films*, 2003, **436**, 145.
213. A. Ulman ed., *An Introduction to Ultrathion Organic Films, From Langmuir-Blodgett to Self-Assembly*, Academic Press, MA, 1991.
214. L. H. Dubois and R. G. Nuzzo, *Ann. Rev. Phys. Chem.*, 1992, **43**, 437.
215. A. Ulman, *Chem. Rev.*, 1996, **96**, 1533.
216. W. Salaneck, R. Stafstrom and J.-L. Bredas (eds.), *Conjugated Polymer Surfaces and Interfaces, Electronic and Chemical Structure of Interfaces for Polymer Light Emitting Devices*, Cambridge University Press, Cambridge, 1996.
217. A. Aviram and M. Ratner, (eds.), *Molecular Electronics: Science and Technology*, NY Academy of Science, New York, 1998, Vol. **852**.
218. H. Ishii, K. Sugiyama, E. Ito and K. Seki, *Adv. Mater.*, 1999, **11**, 605.
219. J. Chen, M. A. Reed, S. M. Dirk, D. W. Price, A. M. Rawlett, J. M. Tour, D. S. Grubisha and D. W. Bennett, *NATO Sci. Ser. II: Math., Phys. Chem.*, 2003, **96**, 59.
220. U. Weckenmann, S. Mittler, S. Kramer, A. K. A. Aliganga and R. A. Fischer, *Chem. Mat.*, 2004, **16**, 621.
221. C. Winter, U. Weckenmann, R. A. Fischer, J. Kashammer, V. Scheumann and S. Mittler, *Chem. Vapor Deposition.*, 2000, **6**, 199.
222. R. A. Fischer, U. Weckenmann, C. Winter, J. Kashammer, V. Scheumann and S. Mittler, *J. Phys. IV: Proc.*, 2001, **11**, Pr3/1183–Pr3/1190.
223. P. Wohlfart, J. Weiss, J. Kashammer, M. Kreiter, C. Winter, R. A. Fischer and S. Mittler-Neher, *Chem. Vap. Deposition*, 1999, **5**, 165.
224. P. Wohlfart, J. Weiss, J. Kashammer, C. Winter, V. Scheumann, R. A. Fischer and S. Mittler-Neher, *Thin Solid Films*, 1999, **340**, 274.
225. J. Weiss, H. J. Himmel, R. A. Fischer and C. Wöll, *Chem. Vap. Deposition*, 1998, **4**, 17.
226. C. A. Wilson, R. K. Grubbs and S. M. George, *Chem. Mater.*, 2005, **17**, 5625.
227. Ye Xu and C. B. Musgrave, *Chem. Mater.*, 2004, **16**, 646.
228. A. S. Killampalli, P. F. Ma and J. R. Engstrom, *J. Am. Chem. Soc.*, 2005, **127**, 6300.
229. M. Haran, J. R. Engstrom and P. Clancy, *J. Am. Chem. Soc.*, 2006, **128**, 836.
230. C. L. Liang, G. A. Cheng, R. T. Zheng, H. P. Liu, J. C. Li, H. F. Zhang, G. J. Ma and Y. L. Jiang, *Surf. Coat. Technol.*, 2007, **201**, 5537.
231. C. K. Wang, L. M. Liu, M. Liao, H. C. Cheng and M. S. Lin, *Jpn. J. Appl. Phys. Part 1-Regul. Pap. Short Notes Rev. Pap.*, 1996, **35**, 4274.
232. J. Driessen, A. D. Kuypers and J. Schoonman, *J. Phys. IV*, 1999, **9**, 605.
233. G. Ruhl, R. Rehmet, M. Knizova, R. Merica and S. Veprek, *Chem. Mat.*, 1996, **8**, 2712.
234. C. Taschner, K. Bartsch and A. Leonhardt, *Surf. Coat. Technol.*, 1993, **61**, 158.
235. A. Weber, R. Nukulski, C. P. Klages, M. E. Gross, W. L. Brown, E. Dons and R. M. Charatan, *J. Electrochem. Soc.*, 1994, **141**, 849.
236. A. Weber, C. P. Klages, M. E. Gross, R. M. Charatan and W. L. Brown, *J. Electrochem. Soc.*, 1995, **142**, L79.
237. J. Y. Kim, D. Y. Kim, H. O. Park and H. T. Jeon, *J. Electrochem. Soc.*, 2005, **152**, G29.
238. M. Danek, M. Liao, J. Tseng, K. Littau, D. Saigal, H. Zhang, R. Mosely and M. Eizenberg, *Appl. Phys. Lett.*, 1996, **68**, 1015.
239. M. McCarrol, M. Xi, M. Chang, Z. Lando, S. Wang, H. Luo, V. Hornback, D. Allman and N. Chiesel, in *Conference Proceedings ULSI XIV*, p. 363, Materials Research Society, Pittsburgh, PA, 1999.
240. K. C. Park and K. B. Kim, *J. Electrochem. Soc.*, 1995, **142**, 3109.
241. D. H. Kim, S. L. Cho, K. B. Kim, J. J. Kim, J. W. Park and J. J. Kim, *Appl. Phys. Lett.*, 1996, **69**, 4182.

242. S. H. Kim, D. S. Chung, K. C. Park, K. B. Kim and S. H. Min, *J. Electrochem. Soc.*, 1999, **146**, 1455.
243. J. S. Reid, E. Kolawa, C. M. Garland, M.-A. Nicolet, F. Cardone, D. Gupta and R. P. Ruiz, *J. Appl. Phys.*, 1996, **79**, 1109.
244. P. M. Smith and J. S. Custer, *Appl. Phys. Lett.*, 1997, **70**, 3116.
245. J.-S. Park, D. K. Sohn, B. H. Lee, J.-U. Bae, J. S. Byun and J. W. Park, *J. Electrochem. Soc.*, 1999, **146**, 1579.
246. T. Harada, S. Hirao, S. Fuji, S. Hashimoto and M. Shishino, in *Conference Proceedings ULSI XIV*, p. 363, Materials Research Society, Warrendale, PA, 1999.
247. E. Eisenbraun, A. Upham, R. Dash, W. X. Zeng, J. Hoefnagels, S. Lane, D. Anjum, K. Dovidenko, A. Kaloyeros, B. Arkles and J. J. Sullivan, *J. Vac. Sci. Technol. B*, 2000, **18**, 2011.
248. C. Marcadal, M. Eizenberg, A. Yoon and L. Chen, *J. Electrochem. Soc.*, 2002, **149**, C52.
249. M. Eizenberg, K. Littau, S. Ghanayem, A. Mak, Y. Maeda, M. Chang and A. K. Sinha, *Appl. Phys. Lett.*, 1994, **65**, 2416.
250. M. Eizenberg, K. Littau, S. Ghanayem, M. Liao, R. Mosely and A. K. Sinha, *J. Vac. Sci. Technol. A*, 1995, **13**, 590.
251. Y. C. Ee, Z. Chen, L. Chan, K. H. See, S. B. Law, S. Xu, Z. L. Tsakadze, P. P. Rutkevych, K. Y. Zeng and L. Shen, *J. Vac. Sci. Technol. B*, 2005, **23**, 2444.
252. J. Bonitz, R. Ecke, S. E. Schulz and T. Gessner, *Microelectron. Eng.*, 2005, **82**, 618.
253. Y. C. Ee, J. S. Juneja, P. I. Wang, T. M. Lu, H. Bakhru, L. Chan, S. B. Law, C. Yong, Z. Chen and S. Xu, *J. Electrochem. Soc.*, 2006, **153**, G470.
254. J. S. Min, J. S. Park, H. S. Park and S. W. Kang, *J. Electrochem. Soc.*, 2000, **147**, 3868.
255. J. S. Park and S. W. Kang, *Electrochem. Solid State Lett.*, 2004, **7**, C87.
256. J. S. Park, S. W. Kang and H. Kim, *J. Vac. Sci. Technol. B*, 2006, **24**, 1327.
257. X. Z. Liu, Z. Z. Wu, H. Cai, Y. H. Yang, T. N. Chen, C. E. Vallet, R. A. Zuhr, D. B. Beach, Z. H. Peng, Y. D. Wu, T. E. Concolino, A. L. Rheingold and Z. L. Xue, *J. Am. Chem. Soc.*, 2001, **123**, 8011.
258. M. Juppo, P. Alen, M. Ritala and M. Leskela, *Chem. Vapor Depoition*, 2001, **7**, 211.
259. J. Koo, J.-W. Lee, T. Doh, Y. Kim, Y.-D. Kim and H. Jeon, *J. Vac. Sci. Technol. A*, 2001, **19**, 2831.
260. M. H. Staia, D. G. Bhat, E. S. Puchi-Cabrera and J. Bost, *Wear*, 2006, **261**, 540.
261. J. G. Han, Y. K. Cho, J. S. Yoon, K. T. Rie, W. C. Roh, D. Jung, S. W. Lee and J. H. Boo, *Surf. Coat. Technol.*, 2000, **131**, 73.
262. H. Berndt, A. Q. Zeng, H. R. Stock and P. Mayr, *Surf. Coat. Technol.*, 1995, **74**, 369.
263. H. Wendel and H. Suhr, *Appl. Phys. A-Mater. Sci. Process*, 1992, **54**, 389.
264. R. Armitage, Q. Yang, H. Feick, J. Gebauer, E. R. Weber, S. Shinkai and K. Sasaki, *Appl. Phys. Lett.*, 2002, **81**, 1450.
265. P. Kroll, *J. Phys.-Condens. Matter.*, 2004, **16**, S1235.
266. J. W. Li, D. Dzivenko, A. Zerr, C. Fasel, Y. P. Zhou and R. Riedel, *Z. Anorg. Allg. Chem.*, 2005, **631**, 1449.
267. Z. G. Zhang, T. W. Liu, J. Xu, X. L. Deng and C. Dong, *Surf. Coat. Technol.*, 2006, **200**, 4918.
268. J. S. Becker, E. Kim and R. G. Gordon, *Chem. Mat.*, 2004, **16**, 3497.
269. W. W. Wang, T. Nabatame and Y. Shimogaki, *Jpn. J. Appl. Phys. Part 2–Lett. Express Lett.*, 2004, **43**, L1445.
270. Y. Kim, A. Baunemann, H. Parala, A. Devi and R. A. Fischer, *Chem. Vapor Deposition*, 2005, **11**, 294.
271. W. Wang, T. Nabatame and Y. Shimogaki, *J. Electrochem. Soc.*, 2007, **154**, G25.
272. E. J. Kim and D. H. Kim, *Electrochem. Solid State Lett.*, 2006, **9**, C123.

273. Y.-H. Chou, H.-T. Chiu, T.-F. Kuo, C.-C. Chi and S.-H. Chuang, *Appl. Phys. Lett.*, 2006, **89**, 252901.
274. S. Cho, K. Lee, P. Song, H. Jeon and Y. Kim, *Jpn. J. Appl. Phys.*, 2007, **46**, 4085.
275. M. S. Wong, W. D. Sproul, X. Chu and S. A. Barnett, *J. Vac. Sci. Technol. A*, 1993, **11**, 1528.
276. X. Yu, Q. Lai, G. Li, J. Xu and M. Gu, *J. Mater. Sci. Lett.*, 2002, **21**, 1671.
277. G.-I. Oya and Y. Onodera, *J. Appl. Phys.*, 1974, **45**, 1389.
278. K. Hieber, *Thin Solid Films*, 1974, **24**, 157.
279. T. Takahashi, H. Itoh and T. Yamaguchi, *J. Cryst. Growth*, 1979, **46**, 69.
280. G. M. Demyashev, V. R. Tregulov and R. K. Chuzhko, *J. Cryst. Growth*, 1983, **63**, 135.
281. G. M. Demyashev, V. R. Tregulov and A. A. Gavrich, *J. Cryst. Growth*, 1988, **87**, 33.
282. X. Chen, H. L. Frisch and A. E. Kaloyeros, *J. Vac. Sci. Technol. B*, 1999, **17**, 182.
283. A. C. Newport, J. E. Bleau, C. J. Carmalt, I. P. Parkin and S. A. O'Neill, *J. Mater. Chem.*, 2004, **14**, 3333.
284. G. S. Elwin and I. P. Parkin, *Chem. Vapor Depos.*, 2000, **6**, 59.
285. A. Newport, C. J. Carmalt, I. P. Parkin and S. A. O'Neill, *Eur. J. Inorg. Chem.*, 2004, 4286.
286. S. Abisset and F. Maury, *Surf. Coat. Technol.*, 1998, **109**, 200.
287. F. Maury, F. Teyssandier and L. Valade, *Ann. Chim.-Sci. Mat.*, 1998, **23**, 637.
288. L. Valade, C. Danjoy, B. Chansou, E. Riviere, J. L. Pellegatta, R. Choukroun and P. Cassoux, *Appl. Organomet. Chem.*, 1998, **12**, 173.
289. P. Bonnefond, R. Feurer, A. Reynes, F. Maury, B. Chansou, R. Choukroun and P. Cassoux, *J. Mater. Chem.*, 1996, **6**, 1501.
290. M. H. Tsai, S. C. Sun, H. T. Chiu, C. E. Tsai and S. H. Chuang, *Appl. Phys. Lett.*, 1995, **67**, 1128.
291. S. G. Park, Y. K. Lee, S. B. Kang, H. S. Jung, S. J. Doh, J.-H. Lee, J. H. Choi, G. H. Kim, G. H. Choi, U. I. Chung, and J. T. Moon, *Electron Device Meeting 2003, Technical Digest, IEEE Intern.* 2003, 13.6.1.
292. W. P. Bai, N. Lu, A. Ritenour, M. L. Lee, D. A. Antoniadis and D.-L. Kwong, *IEEE Electr. Dev. Lett.*, 2006, **27**.
293. M. Kadoshima, K. Akiyama, K. Yamamoto, H. Fujiwara, T. Yasuda and A. Toriumi, *J. Vac. Sci. Technol. B*, 2005, **23**, 42.
294. K.-L. Fang and B.-Y. Tsui, *J. Appl. Phys.*, 2003, **93**, 5546.
295. Y. Gotoh, H. Tsuji and J. Ishikawa, *J. Vac. Sci. Technol. B*, 2003, **21**, 1607.
296. C. Ren, H. Y. Yu, X. P. Wang, H. H. H. Ma, D. S. H. Chan, M.-F. Li, Y.-C. Yeo, C. H. Tung, N. Balasubramanian, A. C. H. Huan, J. S. Pan and D.-L. Kwong, *IEEE Electr. Dev. Lett.*, 2005, **26**, 75.
297. C.-L. Cheng, K.-S. Chang-Liao, T.-C. Wang, T.-K. Wang and H. C.-H. Wang, *IEEE Electr. Dev. Lett.*, 2006, **27**, 148.
298. C. H. Winter, *Aldrichimica ACTA*, 2000, **33**, 3.
299. D. C. Bradley and M. H. Gitlitz, *Journal of the Chemical Society, A*, 1969, 980.
300. R. Fix, R. G. Gordon and D. M. Hoffman, *Chem. Mater.*, 1993, **5**, 614.
301. E. R. Engbrecht, Y.-M. Sun, S. Smith, K. Pfiefer, J. Bennett, J. M. White and J. G. Ekerdt, *Thin Solid Films*, 2002, **418**, 145.
302. K.-N. Cho, C.-H. Han, K.-B. Noh, J.-E. Oh, P. Su-Hyoun, C.-S. Park, S.-I. Lee, M. Y. Lee and J. G. Lee, *Jpn. J. Appl. Phys.*, 1998, **37**, 6502.
303. D. C. Bradley and I. M. Thomas, *Can. J. Chem.*, 1962, **40**, 1355.
304. Y. Takahashi, N. Onoyama, Y. Ishikawa and S. Motojima, *Chem. Letters*, 1978, **525**.
305. G.-C. Jun, S.-L. Cho, K.-B. Kim, H.-K. Shin and D.-H. Kim, *Jpn. J. Appl. Phys.*, 1998, **37**, L30.
306. S.-I. Cho, K.-B. Kim and S.-H. Min, *J. Electrochem. Soc.*, 1999, **146**, 3724.
307. X. Zhao, N. P. Magtoto and J. A. Kelber, *Thin Solid Films*, 2005, **478**, 188.

308. Y. Senzaki, A. K. Hochber and J. A. T. Norman, *Adv. Mater. Opt. Electron.*, 2000, **10**, 93.
309. M. Kadoshima, K. Akiyama, K. Yamamoto, H. Fujiwara, T. Yasuda and A. Toriumi, *J. Vac. Sci. Technol. B*, 2005, **23**, 42.
310. J.-S. M. Lehn, P. Van der Heide, Y. Wang, S. Suh and D. M. Hoffman, *J. Mater. Chem.*, 2004, **14**, 3239.
311. H. Machida, A. Hoshino, T. Suzuki, A. Ogura and Y. Ohshita, *J. Cryst. Growth*, 2002, **237–239**, 586.
312. Y. Oshita, A. Ogura, A. Hoshino, S. Hiiro and H. Machida, *J. Cryst. Growth*, 2000, **220**, 604.
313. A. Baunemann, Y. Kim, M. Winter and R. A. Fischer, *Dalton Trans.*, 2006, 121.
314. D. Gaess, K. Harms, M. Pokoj, W. G. Stolz and J. Sundermeyer, *Inorg. Chem.*, 2007, **46**, 6688.
315. K. C. Jayaratne, G. P. A. Yap, B. S. Haggerty, A. L. Rheingold and C. H. Winter, *Inorg. Chem.*, 1996, **35**, 4910.
316. C. H. Winter, K. C. Jayaratne and J. W. Proscia, *Mat. Res. Soc. Symp. Proc.*, 1994, **327**, 103.
317. C. H. Winter, T. S. Lewkebandara, and K. C. Jayaratne, US Pat 5591483, 1997.
318. J. E. Bleau, C. J. Carmalt, S. A. O'Neill, I. P. Parkin, A. J. P. White and D. S. Williams, *Polyhedron*, 2005, **24**, 463.
319. W. A. Herrmann, N. W. Huber, P. Härter, M. Denk and F. Dyckhoff, *Chem. Ber.*, 1992, **125**, 117.
320. J. M. Decams, S. Daniele, L. G. Hubert-Pfalzgraf, J. Vaissermann and S. Lecocq, *Polyhedron*, 2001, **20**, 2405.
321. S. Suh and D. M. Hoffman, *Inorg. Chem.*, 1996, **35**, 5015.
322. X. Liu, J. R. Babcock, M. A. Lane, J. A. Belot, A. W. Ott, M. Metz, C. R. Kannewurf, R. P. H. Chang and T. J. Marks, *Chem. Vapor Depos.*, 2001, **7**, 25.
323. H. T. Chiu, J.-C. Lin, S.-H. Chuang, G.-H. Lee and S.-M. Peng, *J. Chin. Chem. Soc.*, 1998, **45**, 355.
324. M. Yasuhara and H. Kadokura, *Jpn. Kokai. Tokkyo Koho*, 2006, 131606.
325. T. Thiede, Master Thesis, Ruhr-University Bochum, Germany, 2007.
326. P. Alen, M. Ritala, K. Arstila, J. Keinonen and M. Leskela, *Thin Solid Films*, 2005, **491**, 235.
327. N. Van Hoornick, H. De Witte, T. Witters, C. Zhao, T. Conard, H. Huotari, J. Swerts, T. Schram, J. W. Maes, S. De Gendt and M. Heyns, *J. Electrochem. Soc.*, 2006, **153**, G437.
328. C. W. Wu, W. C. Gau, J. C. Hu, T. C. Chang, C. H. Chen, C. J. Chu and L. J. Chen, *Surf. Coat. Technol.*, 2003, **163**, 214.
329. W. C. Gau, C. W. Wu, T. C. Chang, P. T. Liu, C. J. Chu, C. H. Chen and L. J. Chen, *Thin Solid Films*, 2002, **420**, 548.
330. A. Baunemann, D. Rische, A. Milanov, Y. Kim, M. Winter, C. Gemel and R. A. Fischer, *Dalton Trans.*, 2005, 3051.
331. A. Baunemann, M. Winter, K. Csapek, C. Gemel and R. A. Fischer, *Eur. J. Inorg. Chem.*, 2006, 4665.
332. A. Baunemann, M. Lemberger, A. J. Bauer, H. Parala and R. A. Fischer, *Chem. Vapor Depos.*, 2007, **13**, 77.
333. J. M. Decams, L. G. Hubert-Pfalzgraf and J. Vaissermann, *Polyhedron*, 1999, **18**, 2885.
334. L. G. Hubert-Pfalzgraf, J. M. Decams and S. Daniele, *J. Phys. IV*, 1999, **9**, 953.
335. A. Baunemann, Dissertation, Ruhr University Bochum, 2006.
336. M. Ritala, P. Kalsi, D. Riihelä, K. Kukli, M. Leskela and J. Jokinen, *Chem. Mat.*, 1999, **11**, 1712.
337. P. Alén, M. Juppo, M. Ritala, T. Sajavaara, J. Keinonen and M. Leskela, *J. Electrochem. Soc.*, 2001, **148**, G566.
338. P. Alén, M. Juppo, M. Ritala, M. Leskelä, T. Sajavaara and J. Keinonen, *J. Mater. Res.*, 2002, **17**, 107.

339. M. Ritala, M. Leskela, E. Rauhala and P. Haussalo, *J. Electrochem. Soc.*, 1995, **142**, 2731.
340. J. Imahori, T. Oku and M. Murakami, *Thin Solid Films*, 1997, **301**, 142.
341. H. Kim and S. M. Rossnagel, in Proceedings of the 49th International American Vacuum Society Conference, Denver, CO, 2002.
342. J. W. Hong, K. I. Choi, Y. K. Lee, S. G. Park, S. W. Lee, J. M. Lee, S. B. Kang, G. H. Choi, S. T. Kim, U.-I. Chung and J. T. Moon, *Proc. IEEE International Interconnect Technology Conference*, 2004, **7–9**, 9.
343. J.-S. Park, *Electrochem. Solid St.*, 2001, **4**, C17.
344. T. Nakajima, K. Watanabe and N. Watanabe, *J. Electrochem. Soc.*, 1987, **134**, 3175.
345. Y. T. Kim and S. K. Min, *Appl. Phys. Lett.*, 1991, **59**, 929.
346. Y. T. Kim, C. W. Lee and S. K. Min, *Appl. Phys. Lett.*, 1992, **61**, 537.
347. C. W. Lee, Y. T. Kim and J. Y. Lee, *Appl. Phys. Lett.*, 1994, **64**, 619.
348. C. W. Lee and Y. T. Kim, *Solid-State Electronics*, 1995, **38**, 679.
349. K.-M. Chang, T.-H. Yeh, I. C. Deng and C.-W. Shih, *J. Appl. Phys.*, 1997, **82**, 1469.
350. J. W. Klaus, S. J. Ferro and S. M. George, *Appl. Surf. Sci.*, 2000, **162–163**, 479.
351. S.-H. Kim, J.-K. Kim, N. Kwak, H. Sohn, J. Kim, S.-H. Jung, M.-R. Hong, S. H. Lee and J. Collins, *Electrochem. Solid-State Lett.*, 2006, **9**, C54.
352. S. Ganguli, L. Chen, T. Levine, B. Zheng and M. Chang, *J. Vav. Sci. Technol. B*, 2000, **18**, 237.
353. S.-H. Kim, S. S. Oh, K.-B. Kim, D.-H. Kang, W.-M. Li, S. Haukka and M. Tuominen, *Appl. Phys. Lett.*, 2003, **82**, 4486.
354. A. M. Hoyas, J. Schuhmacher, D. Shamiryan, J. Waeterloos, W. Besling, J. P. Celis and K. Maex, *J. Appl. Phys.*, 2004, **95**, 381.
355. S.-H. Kim, S. S. Oh, H.-M. Kim, D.-H. Kang, K.-B. Kim, W.-M. Li, S. Haukka and M. Tuominen, *J. Electrochem Soc.*, 2004, **151**, C272.
356. A. M. Hoyas, J. Schuhmacher, C. M. Whelan, T. Fernandez Landaluce, D. Vanhaeren, K. Maex and J. P. Celis, *J. Appl. Phys.*, 2006, **100**, 114903.
357. A. M. Hoyas, Y. Travaly, J. Schuhmacher, T. Sajavaara, C. M. Whelan, B. Eyckens, O. Richard, S. Giangrandi, B. Brijs, W. Vandervorst, K. Maex, J. P. Celis, A. M. Jonas and A. Vantomme, *J. Appl. Phys.*, 2006, **99**, 063515.
358. A. M. Hoyas, C. M. Whelan, J. Schuhmacher, J. P. Celis and K. Maex, *Electrochem. Solid-State Lett.*, 2006, **9**, F64.
359. K.-S. Kim, M.-S. Lee, S.-S. Yim, H.-M. Kim, K.-B. Kim, H.-S. Park, W. Koh, W.-M. Li, M. Stokhof and H. Sprey, *Appl. Phys. Lett.*, 2006, **89**, 081913.
360. C. W. Lee and Y. T. Kim, *J. Vac. Sci. Technol., B*, 2006, **24**, 1432.
361. J. E. Kelsey, C. Goldberg, G. Nuesca, G. Peterson, A. E. Kaloyeros and B. Arkles, *J. Vav. Sci. Technol. B*, 1999, **17**, 1101.
362. P. Alen, M. Ritala, K. Arstila, J. Keinonen and M. Leskelae, *J. Electrochem. Soc.*, 2005, **152**, G361.
363. S. L. Roberson, D. Finello and R. F. Davis, *Thin Solid Films*, 1998, **324**, 30.
364. S. L. Roberson, D. Finello and R. F. Davis, *Surf. Coat. Technol.*, 1998, **102**, 256.
365. S.-C. Sun, and H.-T. Chiu (Taiwan Semiconductor Manufacturing Company, Taiwan). Application: US 6114242, 2000.
366. Y. M. Sun, E. R. Engbrecht, T. Bolom, C. Cilino, J. H. Sim, J. M. White, J. G. Ekerdt and K. Pfeifer, *Thin Solid Films*, 2004, **458**, 251.
367. W. A. Nugent and R. L. Harlow, *Inorg. Chem.*, 1980, **19**, 777.
368. B. H. Lee and K. Yong, *J. Electrochem. Soc.*, 2004, **151**, C594.
369. T. Kadono, T. Kubota and Y. Okamoto, *Catal. Today*, 2003, **87**, 107.
370. F. Maury, *Appl. Organomet. Chem.*, 1992, **6**, 619.
371. F. Maury, *J. Phys. IV*, 1995, **5**, 449.

372. M. H. Tsai, S. C. Sun, H. T. Chiu and S. H. Chuang, *Appl. Phys. Lett.*, 1996, **68**, 1412.
373. E. L. Crane, H.-T. Chiu and R. G. Nuzzo, *J. Phys. Chem. B*, 2001, **105**, 3549.
374. Y.-W. Yang, J.-B. Wu, J. Wang, Y.-F. Lin and H.-T. Chiu, *Surf. Sci.*, 2006, **600**, 743.
375. J.-B. Wu, Y.-W. Yang, Y.-F. Lin and H.-T. Chiu, *J. Vav. Sci. Technol. A*, 2003, **21**, 162.
376. J. S. Becker and R. G. Gordon, *Appl. Phys. Lett.*, 2003, **82**, 2239.
377. J. S. Becker, S. Suh, S. Wang and R. G. Gordon, *Chem. Mater.*, 2003, **15**, 2969.
378. D.-H. Kim, Y. J. Kim, Y. S. Song, B.-T. Lee, J. H. Kim, S. Suh and R. Gordon, *J. Electrochem. Soc.*, 2003, **150**, C740.
379. A. Rugge, J. S. Becker, R. G. Gordon and S. H. Tolbert, *Nano Lett.*, 2003, **3**, 1293.
380. C. L. Dezelah IV, O. M. El-Kadri, M. J. Heeg and C. H. Winter, *J. Mater. Chem.*, 2004, **14**, 3167.
381. O. M. El-Kadri, M. J. Heeg and C. H. Winter, *Dalton Trans.*, 2006, 1943.
382. D. Rische, A. Baunemann, M. Winter and R. A. Fischer, *Inorg. Chem.*, 2006, **45**, 269.
383. D. Rische, H. Parala, E. Gemel, M. Winter and R. A. Fischer, *Chem. Mater.*, 2006, **18**, 6075.
384. O. J. Bchir, K. M. Green, M. S. Hlad, T. J. Anderson, B. C. Brooks and L. McElwee-White, *J. Cryst. Growth*, 2004, **261**, 280.
385. O. J. Bchir, K. M. Green, M. S. Hlad, T. J. Anderson, B. C. Brooks, C. B. Wilder, D. H. Powell and L. McElwee-White, *J. Organomet. Chem.*, 2003, **684**, 338.
386. O. J. Bchir, S. W. Johnston, A. C. Cuadra, T. J. Anderson, C. G. Ortiz, B. C. Brooks, D. H. Powell and L. McElwee-White, *J. Cryst. Growth*, 2003, **249**, 262.
387. S. W. Yong, S. K. Young, T. J. Anderson, L. R. Laurel, I. Ghiviriga and L. McElwee-White, *J. Am. Chem. Soc.*, 2006, **128**, 13781.
388. O. J. Bchir, K. C. Kim, T. J. Anderson, V. Craciun, B. C. Brooks and L. McElwee-White, *J. Electrochem. Soc.*, 2004, **151**, G697.
389. C. B. Wilder, L. L. Reitfort, K. A. Abboud and L. McElwee-White, *Inorg. Chem.*, 2006, **45**, 263.
390. C. L. Dezelah IV, O. M. El-Kadri, K. Kukli, K. Arstila, R. J. Baird, J. Lu, L. Niinistoe and C. H. Winter, *J. Mater. Chem.*, 2007, **17**, 1109.
391. R. M. Fix, R. G. Gordon and D. M. Hoffman, *J. Am. Chem. Soc.*, 1990, **112**, 7833.
392. R. Fix, R. G. Gordon and D. M. Hoffman, *Thin Solid Films*, 1996, **288**, 116.
393. V. Miikkulainen, M. Suvanto and T. A. Pakkanen, *Chem. Mater.*, 2007, **19**, 263.
394. D. Rische, PhD Thesis, Ruhr-University Bochum, 2007.
395. A. Satta, G. Beyer, K. Maex, K. Elers, S. Haukka and A. Vantomme, *Mater. Res. Soc. Symp. Proc.*, 2000, **612**, D6.5.1.
396. H. Jeon, J.-H. Koo, J.-W. Lee, Y.-S. Kim, K. M. Kang, Y. D. Kim and Y. D. Kim, *Mater. Res. Soc. Symp. Proc.*, 2000, **616**, 211.
397. M. Juppo, P. Alen, M. Ritala, T. Sajavaara, J. Keinonen and M. Leskela, *Electrochem. Solid-State Lett.*, 2002, **5**, C4.
398. M. Ritala, M. Leskela, E. Rouhala and J. Jokinen, *J. Electrochem. Soc.*, 1998, **145**, 2914.
399. J.-W. Lim, J.-S. Park and S.-W. Kang, *J. Appl. Phys.*, 2000, **87**, 4632.
400. J-H. Yun, E.-S. Choi, C.-M. Jang and C.-S. Lee, *Jpn. J. Appl. Phys., Part 2*, 2002, **41**, L418.
401. J.-S. Min, Y.-W. Son, W.-G. Kang, S.-S. Chun and S.-W. Kang, *Jpn. J. Appl. Phys. Part 1*, 1998, **37**, 4999.
402. D.-J. Kim, Y.-B. Jung, M.-B. Lee, Y.-H. Lee and J.-H. Lee, *Thin Solid Films*, 2000, **372**, 276.
403. J.-S. Park and S.-W. Kang, in Proceedings of the American Vacuum Society Topical Conference on Atomic Layer Deposition, Seoul, Korea, 2002.
404. J.-D. Kwon, J.-S. Park, and S.-W. Kang, in Proceedings of the American Vacuum Society Topical Conference on Atomic Layer Deposition, Seoul, Korea, 2002.
405. S. B. S. Heil, E. Langereis, F. Roozeboom, M. C. M. V. de Sanden and W. M. M. Kessels, *J. Electro. Chem.*, 2006, **153**, G956.

406. S. B. S. Heil, J. L. van Hemmen, C. J. Hodson, N. Singh, J. H. Klootwijk, F. Roozeboom, M. C. M. V. de Sanden and W. M. M. Kessels, *J. Vac. Sci. Technol., A*, 2007, **25**, 1357.
407. K. E. Elers, J. Winkler, K. Weeks and S. Marcus, *J. Electrochem. Soc.*, 2005, **152**, G589.
408. J. Y. Kim, S. Seo, D. Y. Kim, H. Jeon and Y. Kim, *J. Vac. Sci. Technol., A*, 2004, **22**, 8.
409. H.-T. Chiu and W.-P. Chang, *J. Mater. Sci. Lett.*, 1992, **11**, 96.
410. M. H. Tsai, S. C. Sun, C. P. Lee, H. T. Chiu, C. E. Tsai, S. H. Chuang and S. C. Wu, *Thin Solid Films*, 1995, **270**, 531.
411. C.-H. Han, K.-N. Cho, J.-E. Oh, S.-H. Paek, C.-S. Park, S.-I. Lee, M. Y. Lee and J. G. Lee, *Jpn. J. Appl. Phys.*, 1998, **37**, 2646.
412. X. C. Chen, G. G. Peterson, C. Goldberg, G. Nuesca, H. L. Frisch, A. E. Kaloyeros, B. Arkles and J. Sullivan, *J. Mater. Res.*, 1999, **14**, 2043.
413. S. J. Im, S.-H. Kim, K.-C. Park, S.-L. Cho and K.-B. Kim, *Mat. Res. Soc. Symp. Proc.*, 2000, **612**, D6.7.1–D6.7.6.
414. P. Alen, T. Aaltonen, M. Ritala, M. Leskelae, T. Sajavaara, J. Keinonen, J. C. Hooker and J. W. Maes, *J. Electrochem. Soc.*, 2004, **151**, G523.
415. H. Kim, C. Lavoie, M. Copel, V. Narayanan, D.-G. Park and S. M. Rossnagel, *J. Appl. Phys.*, 2004, **95**, 5848.
416. H. Kim, A. J. Kellock and S. M. Rossnagel, *J. Appl. Phys.*, 2002, **92**, 7080.
417. O. Van der Straten, Y. Zhu, K. Dunn, E. T. Eisenbraun and A. E. Kaloyeros, *J. Mater. Res.*, 2004, **19**, 447.
418. O. Van der Straten, Y. Zhu, E. Eisenbraun and A. E. Kaloyeros, *Mat. Res. Soc. Symp. Proc.*, 2002, **716**, B11.3.1.
419. J.-S. Park, H.-S. Park and S.-W. Kang, *J. Electrochem. Soc.*, 2002, **149**, C28.

CHAPTER 10

CVD of Functional Coatings on Glass

IVAN P. PARKIN AND ROBERT G. PALGRAVE

Department of Chemistry, University College London, 20 Gordon Street, London, WC1H OAJ, UK

10.1 Introduction

Glass is one of the most important materials in the modern world, used in a wide range of applications, from architectural glazing and high precision optics to food containers. It is well known that the bulk properties of glass can be modified by the addition of various minerals during manufacture. Properties such as colour and thermal expansion can be altered in this way. Surface modification of glass through the application of thin films is a more recent innovation, which has become an important industrial technique. The attraction of glass coating is the ability to alter surface chemical and physical properties, while preserving bulk properties. CVD is an important technique in glass coating due to the sheer range of thin films that can be deposited and the successful integration of CVD coaters into existing glass manufacturing technology. In terms of area deposited, glass coating is currently the largest application of CVD, and the volume of primary research published concerning CVD of thin films on glass suggests that the sector is likely to grow significantly in the future.

As the industrial application of glass coatings become increasingly sophisticated, the material properties of thin films and the chemistry of their deposition must be better understood.[1,2] Many applications of glass take advantage of its high visible light transmittance and clarity. In most cases coatings applied to the glass must also have high visible transmittance and low scattering. For example, window glass requires greater than 80% transmission throughout the visible region, and less than 0.2% haze or scattering. Other applications have similarly stringent requirements. Successful coatings will therefore be transparent, of uniform thickness and contain few impurities. They may also be required to be durable and resistant to a wide range of environmental conditions, depending on their specific use; for example, a window coating on an exterior pane must be as durable as the glass itself to survive a lifetime of use. Finally, they must, of course, carry out whatever function they were designed for. Despite these restrictions, many different functional coatings have been commercialized, and many more are in development. Table 10.1 illustrates some of the functions of coatings that have been applied to glass.

Chemical Vapour Deposition: Precursors, Processes and Applications
Edited by Anthony C. Jones and Michael L. Hitchman

Published by the Royal Society of Chemistry, www.rsc.org

Table 10.1 Overview of functional glass coatings.

Glass application	*Coating functions*
Architectural	Low-emissivity, solar control, anti-reflective, reflective, self-cleaning, electrochromic, thermochromic, conducting, barrier layers
Automotive/aerospace	Anti-fogging, solar control, electrochromic, conducting
Containers	Low friction, toughening
Glass fibre	Lubrication
Displays	Electrochromic, conducting
Optics	Anti-reflection, anti-scratch, photochromic

10.1.1 Architectural Glazing

Architectural glazing is perhaps the most active area of coating innovation and also the fastest growing commercially. One target is the reduction of energy used to either heat or cool a building by changing the thermal properties of window glass. Low-emissivity (Low-E) coatings are established commercial products that are designed to improve the thermal insulating properties of glass. Untreated glass is a poor insulator, and windows are a major source of heat loss from a building, especially in modern buildings where a larger proportion of the exterior wall is constructed from glass. Double and triple glazing improves insulation but uses proportionally more material at a correspondingly greater cost. A low-E coating is reflective in the infrared region, and works to reflect heat back into a room, thus improving insulation and saving on heating costs in cold climates. Fluorine-doped SnO_2 is the most widely used low-E coating. In hot climates, solar control coatings are used to limit the solar energy passing through a window, reducing the cost of cooling a building. Commercially, these products are relatively mature, although further research is driven by European and US legislation requiring new buildings to be increasingly energy efficient.

A more recently introduced commercial product is self-cleaning coatings, which chemically break down dirt using sunlight. Future products may make use of thermochromic, electrochromic and photochromic coatings, which can change their transmission and reflective properties in response to a change in temperature, electric current or light, respectively.

10.1.2 Automotive/Aerospace Glazing

Existing automotive and aerospace glass coatings include conducting layers for resistive heating of windows to prevent frost. More recently, self dimming rear view mirrors using electrochromic coatings, and anti-fogging coatings that are highly hydrophilic, have been introduced.

10.1.3 Container Coating

The scale of the container industry is vast and is probably the largest single user of chemical vapour deposition. Between 75 and 95% of all glass containers are coated by CVD with an excess of 180 billion units per annum. The coatings are applied for several reasons. The primary one is the speed at which containers can be moved around the plant. By creating containers that have lower friction, faster processing speeds are achieved and, indeed, incredibly high process speeds can be achieved; containers can now be produced at a rate of 700 min^{-1}. The coatings also increase the mechanical strength of the glass and lead to lower breakages, they reduce container weight and increase burst resistance.

10.1.4 Industrial Glass Manufacture and Coating

The two large-scale industrial glass manufacturing methods in use today are the float glass process, which produces a continuous ribbon of highly uniform, flat glass and the forming method, which uses moulding to produce individual glass objects such as containers. Float glass is used in any application where flat glass is required; an excellent example is window glass, which is simply cut to the required size from the float glass ribbon. In addition, flat glass can be shaped in a separate process to form the curved windows used in the automotive, aerospace and other industries. Formed glass is extensively used in the manufacture of containers.

Industrial glass coating can be divided into two categories: on-line coating, which takes place as the glass is manufactured (by either method described above), and off-line coating, which is a separate process that takes place after manufacture. On-line coating is more constrained in that the temperature, pressure, deposition time and chemicals used must all be compatible with the manufacturing process. Growth rates must usually be high, as the time available for deposition is short, and coatings must naturally be high quality. Only APCVD offers these characteristics and, as such, APCVD holds a virtual monopoly for on-line glass coating. Off-line coating, in contrast, is much more flexible in that the deposition parameters are not tied to the manufacturing conditions. Consequently, a wider range of CVD techniques can be used, and there is also competition from other coating techniques (*e.g.* physical deposition techniques, sol–gel, *etc.*)

10.1.5 On-line Coating Using the Float Glass Process

The float-glass process was introduced by Pilkington Glass, and today is used to manufacture over 90% of the world's flat glass (Figure 10.1).[3] It produces highly standardized and optically transparent glass. The raw materials are mixed in the correct quantities before passing into the furnace at 1500 °C. The molten glass is then fed onto a bath of liquid tin at a constant rate; the continuous band of glass then floats along the surface of the tin. The temperature is maintained at 1000 °C so that irregularities and defects melt as the glass proceeds along the tin. After a sufficient time, the glass is cooled to about 600 °C and on leaving the bath it is sufficiently hard to resist damage from the rollers that pass the glass into the annealing lehr. The annealing removes some of the internal stresses within the glass, so that it can be cut and shaped without incurring damage. The process is fully automated, with production lines about 0.5 km long, which can operate continuously for 10–15 years, producing about 6000 km of glass per year. The use of a molten metal bath allows extremely regular and flat glass, typically with less than a 1 nm variation across a surface, to be consistently produced. By varying the temperature and speed at which the glass passed along the bath a range of thicknesses from 4 to 20 mm can be achieved. To prevent oxidation of the tin, the float bath is contained in a protective inert nitrogen atmosphere. This atmosphere must be maintained constantly, since the molten glass itself often contains oxygen and sulfur compounds that will be evolved at the glass surface. Consequently, there is often about 10% hydrogen added to the nitrogen atmosphere, which helps to cleanse the atmosphere of the more damaging impurities through reduction.

It is in this environment that on-line CVD takes place. As the glass moves at 900 m h^{-1} at 4 m widths, the coating chemistry must be very fast yet also very uniform; with high fixed costs, it is uneconomical to slow the glass speed. Typical film thicknesses achievable on-line range from 100–5000 Å, meaning that film growth rates must be up to 2000 Å s^{-1}.[4] Several regions have been used on a manufacturing line for deposition: the float bath itself, the area between the float and the annealing lehr and the annealing lehr itself after the annealing zone. These zones have very specific temperature ranges of 750–600, 600–570 and less than 500 °C. The higher temperature zones are the ones that are typically chosen for most on-line CVD driven by the necessity of high activation energies for the chemical vapour deposition and more particularly the high growth rate

Figure 10.1 The float glass process. Top left: gas burners heating the reactant glass mixture. Top middle: glass floating on a bed of molten tin under a H_2/N_2 atmosphere; CVD can take place in this environment or after it leaves the lehr furnace. Top right: glass passes on rollers down the plant after CVD coating and cooling. Bottom left: glass being packaged and stored. Bottom middle: glass is cut on line and packaged. Bottom right: a fully functional glass manufacturing plant – typically 1 km long.

requirements. This coupled with the very short transit time (a few seconds) of the glass within the coating zone mean that only atmospheric pressure chemical vapour deposition fits the growth requirements.

10.1.6 On-line Coating of Glass Containers

Modern manufacture of glass containers is achieved by the moulding of molten glass and the application of compressed air to form the cavity. In most instances two forms of coating are applied to glass containers: cold end coatings (CES) and hot-end coating (HEC). The CES is usually either partly oxidized polyethylene that is spray coated from a dilute aqueous solution or a solid state coating that is deposited as a vapour. In either case the coating is *ca.* 50 nm thick and is applied to the glass at 135–165 °C. This temperature is critical above 165 °C the film formed is anhydrous and this limits adhesion of labels. The HEC protects the glass from damage and also provides an adhesive surface to which the CEC coating can adhere. The HEC is normally a hard ceramic coating *ca.* 10 nm thick. The two main materials used for this purpose are titanium dioxide and tin oxide. The HEC layer is applied immediately after glass forming and is deposited at 450–600 °C. Several precursors are used, including dimethyltin, tin tetrachloride, monobutyltin trichloride and titanium tetrachloride. The HEC deposition conditions have several requirements: it must be fast in that only 0.1–0.5 s are available for deposition, be of low-cost, provide a uniform thickness, be easy to maintain, safe and insensitive to ambient conditions. Many coating technologies have been tried to compete with the HEC/CEC process but to date all have failed. Efforts have been made to avoid the use of the HEC CVD treatment and have a single CEC step; however, generally, the product has inferior mechanical and handling properties, as adhesion of the CEC film on an uncoated glass surface is poor. This coating technology area is very mature and sales of coated containers have fallen steadily due to competition from plastics. Despite this, by area and weight container coatings are the largest application of CVD.

10.1.7 Glass as a Substrate

The substrate naturally has a large effect on the process of CVD, and knowledge of its physical and chemical properties, especially of the surface, are important for understanding deposition on glass as a whole.[5] Glass is an amorphous, or disordered, solid, kinetically trapped in a non-equilibrium state. It has the viscosity of a solid, but the long-range order of a liquid. It has no melting point; the viscosity decreases continuously with increasing temperature, the rate determined by the glass composition. The glass used in almost all practical applications (windows, containers, *etc.*) is a multicomponent system, based on SiO_2 but also containing oxides of many other elements, including B, Ca, Mg, Na, K, Fe, Al or S. The bulk concentrations of these elements are carefully controlled to give the desired mechanical properties. The composition of the surface region is different to that of the bulk, with metal cations tending to be in higher concentration near the surface. The surface itself is hydroxyl terminated; Si–OH bonds are formed through reaction of the bare SiO_2 surface with atmospheric water. Surface OH groups promote the wetting of the surface by water or organic liquids that can hydrogen bond. Alkaline earth ions near the surface can also react with water to form hydroxides or with CO_2 to form carbonates. Migration of ions from the glass substrate into a CVD coating can occur, and this can have detrimental effects on the functionality of the coating. To prevent this, a barrier layer of SiO_2 is often applied to the surface of the glass before application of further CVD coatings.

Float glass absorbs light strongly in the UV portion of the spectrum below approximately 350 nm. In the visible region, the absorption, scattering and reflection are all very low. Absorption and reflection in the infrared are also low, and therefore glass is a poor thermal insulator in architectural terms.

10.1.8 Influence of Precursor Chemistry on Glass

The requirements for a precursor for CVD on glass are largely the same as for CVD in general. The precursor must be compatible with the desired CVD method (volatile in the case of APCVD, soluble in the case of AACVD, *etc.*) and must react to give the desired material, preferably with only gaseous by-products. The scale of the CVD that is undertaken on glass can dictate additional requirements. The area covered is very large compared to other applications of CVD (such as semiconductor device fabrication) so the volume of precursors used is high, and precursor cost becomes an important consideration. As discussed above, on-line APCVD is more cost-efficient than off-line CVD; thus the ideal precursors for coating glass are cheap, volatile compounds suitable for APCVD.

Single source precursors must necessarily contain all the elements present in the target film. A precursor for the deposition of a transition metal compound MX is commonly a complex containing M–X bonds, *e.g.* metal phosphine complexes are used to deposit metal phosphides. It is possible to engineer the precursor to achieve a fast, clean reaction, *e.g.* by using good leaving groups or chelating ligands to strengthen M–X bonds. The mechanism of decomposition can often be deduced from thermo-gravimetric analysis or consideration of the leaving group chemistry. However, single source precursors, due to their complexity, tend to be an expensive option. Dual source precursors are generally chemically simpler than their single source counterparts. MX can be deposited by the use of a precursor containing M, often a reactive halide complex, and a precursor

Table 10.2 Materials used in various functional coatings, and precursors used in their deposition.

Functionality	*Material(s) used*	*Typical precursors*	*Ref.*
Low-emissivity	F-doped SnO_2	$SnCl_4$	6
Conductive	Sn-doped In_2O_3	$InCl_3$, $SnCl_4$	7,8
Self-cleaning	TiO_2	$TiCl_4$, $Ti(O^iPr)_4$	9
Electrochromic/photochromic	WO_3	WCl_6, $W(CO)_6$	10
	MoO_3	$Mo(CO)_6$	11
Reflective	TiN	$TiCl_4/NH_3$	12
Thermochromic	VO_2	VCl_4, $VOCl_3$	13–15

containing X, usually an organic compound. The mechanism is usually very difficult to deduce, as it is rarely clear when the M–X bond forms, and whether either or both of the precursors have already decomposed beforehand. Because of lower cost there is a tendency towards dual source precursors for CVD on glass. Metal halides are popular choices, and many have been found with acceptable vapour pressures and decomposition temperatures. They are used with an oxygen source (such as O_2, H_2O or organic oxygen-containing molecules) to deposit metal oxides, or a nitrogen source (usually NH_3) to deposit nitrides.

10.1.9 Overview of Functional Coatings Applied to Glass

Table 10.1 illustrates the wide variety of functional glass coatings available commercially or in development. The materials used for each type of coating and the precursors typically used to deposit them are shown in Table 10.2. Leading references for each type of coating are also given. Figure 10.2 shows the relationship of these functional properties to the solar spectrum, which is important in coatings used in glazing.

10.2 CVD of Transparent Conducting Coatings on Glass

It is not an overstatement to say that thin films of conducting materials form the cornerstone of modern technology; the electronics and computing industry rely on them. Transparent conducting films make possible a separate range of devices that require optical transparency. For example, they can act as electrodes to induce the colouration of electrochromic materials, discussed later in this chapter. Thin films of metals are conducting, but highly absorbing and reflecting in some or the entire visible spectrum, making them unsuitable for many optical applications. Many semiconductors are transparent; they can absorb light at energies greater than the band gap energy, so semiconductors with band gaps greater than the highest energy visible light (approximately 3.0 eV) will be transparent. However, semiconductors generally have much lower conductivity than metals: the conductivity of silver is almost a factor 10^8 greater than that of silicon.[16] Doping with a higher valency element (n-type doping) or creation of oxide vacancies within the lattice (in the case of oxide semiconductors) can increase the conductivity of semiconductors without costing visible transparency. Metal oxide semiconductors that are transparent in the visible region are known as transparent conducting oxides (TCOs). High conductivity in TCOs leads to increased infrared reflection, which has led to TCO layers being used as low-E coatings.

10.2.1 Indium Tin Oxide

Indium oxide is a wide band gap semiconductor ($E_g = 3.75$ eV). Doping with higher valence metal ions results in an n-type semiconductor; doping with tin gives particularly good results. Indium tin

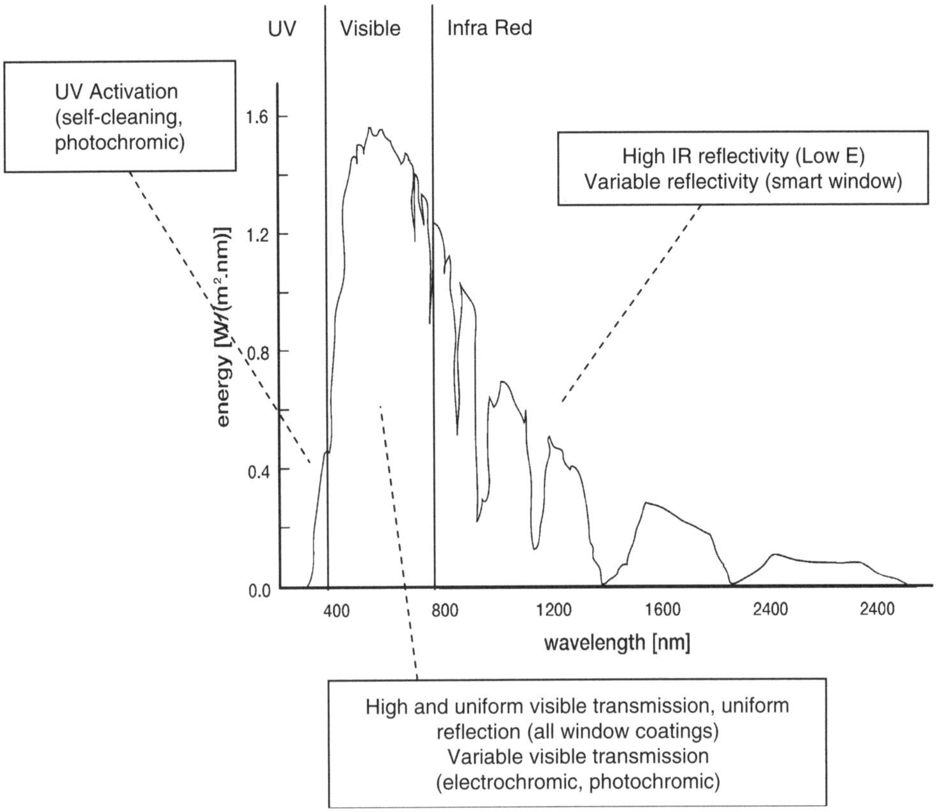

Figure 10.2 Spectral properties of various window coatings, and their relationship to the solar emission spectrum at ground level.

oxide (ITO) exhibits high visible transmittance (>90%) and good resistivity in thin film form (around 200 μΩ cm). The resistivity of ITO is related to the carrier concentration and mobility, which are in turn related to the level of dopant, the level of oxygen vacancies, the crystallinity, crystal orientation, film thickness and impurity and defect levels.[7,16–18] Introducing the Sn dopant to In_2O_3 initially increases conductivity, by providing spare electrons, thus increasing the carrier concentration. However, high levels of Sn doping cause scattering of the charge carriers, reducing carrier mobility and reducing conductivity. The highest conductivities are seen in films with 5–10 at.% Sn.[8] High levels of tin also increase the visible absorption. Table 10.3 summarizes the precursors, deposition parameters and film resistivities for the CVD of ITO films.

Identifying precursors for ITO with suitable volatility is a problem. $SnCl_4$ has good volatility, although it is reactive and difficult to handle, but InF_3 and $InCl_3$ are involatile. Indium β-diketonate complexes have been synthesized in the search for a volatile indium containing precursor. To date, no single source precursors to ITO have been reported, so the amount of Sn doping must be determined by variation of experimental parameters. Indium and tin acetylacetonate complexes proved successful precursors for APCVD giving films with controllable stoichiometry and high conductivities.[21,22] In these depositions, the crystallinity of the films, and hence the conductivity, was strongly affected by the deposition temperature. Films deposited around 450 °C gave the highest electrical conductivity. In(2-ethylhexanoate) and $In(DPM)_3$ have also been used as APCVD precursors to ITO. AACVD using $InCl_3$ and $SnCl_4$ has been reported.[24] As found in other forms of CVD, the deposition temperature strongly affected the conductivity.

Table 10.3 CVD precursors and conditions used to produce ITO thin films on glass.

Precursor 1	*Precursor 2*	*CVD method*	*Substrate temperature range (°C)*	*Film resistivity (μΩ cm)*	*Ref.*
$In(DPM)_3$ [a]	$Bu_2Sn(ac)_2$	AP	400–600	200	19
$In(acac)_3$	$Sn(acac)_2Br_2$	AA	600	100–200	20
$In(acac)_3$	$Sn(acac)_2$	AP	350–500	200	21,22
In(2-ethylhexanoate)	$SnCl_4$	AP	400	290	23
$InCl_3$	$SnCl_4$	AA	500	150	8,24

[a]DPM = 2,2,6,6-tetramethyl-3,5-heptane-dionato.

Table 10.4 Summary of CVD routes to doped SnO_2 thin films on glass.

Precursor 1	*Precursor 2, 3*	*Dopant*	*CVD method*	*Substrate temperature range (°C)*	*Film resistivity (μΩ cm)*	*Ref.*
$SnCl_4$	O_2	Cl	AP	300–500	1000	27
$Sn(Bu)_2(ac)_2$	O_2, $SbCl_3$	Sb	AP	– [a]	2200	28
$Sn(Bu)_2(ac)_2$	O_2, CCl_3CF_3	F	AP	– [a]	1400	28
$Sn(CF_3CO_2)_2$	–	F	AP	200–500	600	6
$Sn(Me)_4$	O_2, CF_3Br	F	AP	350–600	400–800	29,30
$SnCl_4$	O_2, NH_4F	F	AP	600	– [a]	31,32
$SnCl_4$	O_2, F_2	F	AP	400–550	500–1100	33
SnR_4 [b]	–	F	AP	550	460–2000	26

[a]Not reported.
[b]Various perfluoroalkyl ligands were used.

10.2.2 F, Cl, Sb Doped Tin Oxide

Like indium oxide, tin oxide (SnO_2) can be doped to increase its conductivity while retaining its visible light transparency. Chlorine, fluorine and antimony are commonly used as dopants, although fluorine doped tin oxide receives most attention due to higher conductivity and better IR reflective properties. Fluorine doped tin oxide films show slightly lower conductivity than ITO films, but are used widely as low-E coatings to reduce a window's thermal emissivity. Fluorine doping increases the conductivity and IR reflectance. The highest conductivity is seen between 1 and 3 at.% F. Thicker films show greater conductivity and lower emissivity, but at the expense of lower visible transmission; the conductivity can decrease sharply in films under 1-μm thick.[25,26] Several approaches to the deposition of doped tin oxide coatings have been taken. Volatile tin precursors are available, such as $SnCl_4$ and $Sn(Me)_4$, but these need to be combined with an oxygen source and a source of the required dopant. In the case of F doping, the most convenient precursors are chlorofluorocarbons, bromofluorocarbons, HF and F_2, which each have excellent volatility but significant handling and environmental problems. This has led to efforts to produce a volatile precursor containing Sn and F which can be used with an oxygen source, or alternatively a single source precursor containing Sn, F and O which would need no additional oxygen source. Table 10.4 outlines the precursors and conditions used for the deposition of doped SnO_2 films.

Deposition using $SnCl_4$ in the presence of O_2 leads to Cl doped SnO_2 films.[27] Addition of a fluorine source (F_2, NH_4F) leads to deposition of F doped SnO_2 films with higher conductivity.[31–33] Tetramethyltin (TMT) has also been used as a tin precursor, and gave films with low resistivities in combination with CF_3Br and O_2.[29,30] Single source precursors such as $Sn(CF_3CO_2)_2$ have been investigated with some success.[6] These avoid use of dangerous F-containing precursors.

10.3 CVD of Reflective Coatings on Glass

10.3.1 Titanium Nitride Thin Films

Titanium nitride (TiN) is a hard, metallic material, which is used in several applications, chiefly reflective or protective coating. It can be used as a conducting layer when transparency is not required (although for glass based applications this is rarely the case). TiN displays stoichiometries ranging from $TiN_{0.42}$ to $TiN_{1.2}$, as the N atoms occupy interstitial sites in the close packed Ti lattice. The bulk resistivity of 22 μΩ cm in $TiN_{1.0}$ is lower than in titanium metal.[12,34] Titanium nitride films have been prepared using both dual source APCVD and single source LPCVD. Table 10.5 summarizes the deposition parameters and film properties discussed in this section.

The APCVD reaction of $TiCl_4$ and $HN(SiMe_3)_2$ produced gold coloured, adherent coatings at substrate temperatures of 350–550 °C.[34] Films deposited at 400 °C and above were crystalline by XRD. All films were highly reflective to red and near-IR light, as expected of TiN. The reflection spectrum of a film deposited at 550 °C has a maximum at 990 nm, where 55% of incident light is reflected. The position of the reflection maximum was dependent on the temperature of deposition, and is related to the degree of crystallinity of the films. The composition varied with deposition temperature. Stoichiometric $TiN_{1.00}$ was only deposited at 350 °C; at higher substrate temperatures, the composition fluctuated between $TiN_{0.85}$ and $TiN_{0.95}$, with no obvious trend. Some oxygen was present in the bulk of the film, although silicon, carbon nor chlorine could be detected, demonstrating the clean decomposition of the precursors to TiN. The mechanism probably proceeds by the elimination of Me_3SiCl, although this species has not been directly detected.

The LPCVD reaction of titanium imido complexes of the form $[TiCl_2(NR)(L)_x]$ gave TiN films at a substrate temperature of 600 °C.[35] A wide range of R groups and mono- and tri-dentate L ligands were used. In summary, the quality of the films depended on the nature of the R groups and the lability of the L ligands. It appeared advantageous to form a strong Ti–NR bond with electron donating rather than withdrawing R groups. The R group should be a good leaving group, as should the L ligands, therefore monodentate ligands are superior to chelating ligands. Stoichiometric, crystalline $TiN_{1.00}$ films were produced from $[TiCl_2(N^tBu)(py)_3]$, whereas inferior films were produced from $[TiCl_2(N^tBu)(Me_3[9]aneN_3)]$ ($Me_3[9]aneN_3$ stands for tris(N-methyltriamine)-1,4,7-triazacyclononane), which contains a tridentate ligand rather than three monodentate ligands, and worse films still were obtained from $[TiCl_2(NC_6F_5)(Me_3[9]aneN_3)]$, with an electron-withdrawing R group. Notably, since the decomposition of the precursor proceeds through elimination of the L ligands, weaker Ti–amine bonds and stronger Ti–imido promote TiN film formation. It should therefore be expected that the imido nitrogen forms the TiN film whilst the amine leaves as a waste product. This work indicated that the choice of precursor was of key importance in determining the pathway of the single source deposition and that the relatively easy to prepare $[TiCl_2(N^tBu)(py)_3]$ was the best precursor evaluated.

Table 10.5 Summary of CVD routes to TiN thin films on glass.

Precursor 1	*Precursor 2*	*CVD method*	*Substrate temperature range (°C)*	*Maximum growth rate (nm min^{-1})*	*TiE$_x$ composition range (E = N, P)*	*Film resistivity (μΩ cm)*	*Ref.*
$TiCl_4$	$HN(SiMe_3)_2$	AP	350–550	300	0.85–1.00	–[a]	34
$[TiCl_2(N^tBu)(py)_3]$	–	LP	600	15	1.00	200	35
$TiCl_4$	N_2	PE	530	50	–[a]	–[a]	36
$Ti(NMe_2)_4$	NH_3	LP	200–450	–[a]	1.2	2500	37
$Ti(NEtMe)_4$	N_2	PE	550	–[a]	–[a]	–[a]	38
$Ti(NEtMe)_4$	NH_3	AP	350	60	0.9	2000	39

[a]Not measured.

Alternative routes to TiN have been investigated by several research groups, examples of which are given in Table 10.5. Plasma enhanced CVD using $TiCl_4$ and N_2 has been used to deposit TiN coatings.[36] An electrical discharge causes activation of N_2, which reacts to form TiN and probably Cl_2. Tetrakis(dialkylamino)titanium $[Ti(NR_2)_4]$ ($R_2 = Me_2$, MeEt, Et_2) is a popular family of precursors that can be used either in APCVD or plasma enhanced CVD, and as single or dual source precursors.[37–39] In APCVD studies, ammonia is used as the second precursor, although $[Ti(NMe_2)_4]$ (TDMAT) can deposit TiN films as a single source precursor, reaction with NH_3 reduces carbon and oxygen contamination and increases the density of the films.[39] Mass spectroscopic studies of the decomposition of TDMAT and $[Ti(NEt_2)_4]$ (TDEAT) at elevated temperatures suggested a radical chain reaction initiated either by homolytic Ti–N bond cleavage or hydrogen radical attack, showing a possible route to single source deposition.[38] Presumably, the higher levels of carbon contamination in the TiN films produced from the single source route is due to incomplete decomposition of the amino ligands. Correspondingly, the use of ammonia reduced carbon content, since the reaction of titanium precursor and ammonia is expected to be rapid at the reactor temperatures.[36]

10.4 CVD of Electrochromic and Photochromic Coatings

10.4.1 Introduction to Metal Oxide Based Electrochromic and Photochromic Devices

Electrochromic materials reversibly change their colour during electrochemical redox reactions. Such materials might be used in a wide variety of devices that rely on rapid, reversible changes in optical properties. An electrochromic material that switches between a transparent state and a coloured state, when coated onto glass, might be used to make windows that can be tinted – or "switched off" – at the touch of a button.[40] This could be used as a solar control coating, as the window could be tinted or bleached (either automatically or manually) in response to external environmental conditions. In hot weather, the tinted coating would reflect incoming solar heat, reducing the need for air conditioning. In cool weather, the clear window would allow solar radiation into the building. Other applications are also possible, such as electrochromic displays, or the commercially available electrochromic rear view mirrors. Electrochromic devices are the most complex of the functional coatings described in this chapter, both in terms of their structure and the difficulty of their construction. They are electrochemical cells in which the electrochromic material is housed, usually as a working electrode, although solution phase electrochromic electrolytes have been developed. From our perspective we only need consider solid state electrochromic materials, although of course the conducting CVD coatings previously described could be used in conjunction with a liquid phase electrochromic material. Solid-state electrochromic glass coatings, therefore, typically require a transparent conducting layer, invariably indium doped tin oxide (ITO), upon which an electrochromic layer is deposited, becoming the working electrode. An electrolyte and counter electrode are also required. The electrolyte can be solid or liquid, and the counter electrode is ideally a second electrochromic material with opposite colouration (*i.e.* cathodic colouration as opposed to anodic, or *vice versa*), so that the coloured states reinforce each other. The whole assembly is then topped with another layer of ITO coated glass, or alternative conducting substrate, creating a sandwich structure (Figure 10.3).[11,41,42]

Integrating such an assembly into architectural glazing is by no means easy. In addition to the difficulty in producing the required layers, a power source and control system is necessary. The voltages required to induce and maintain colouration are low in modern optimized electrochromic devices, such that it has been suggested that such a device might be powered photovoltaically, without need for an external power source.[43]

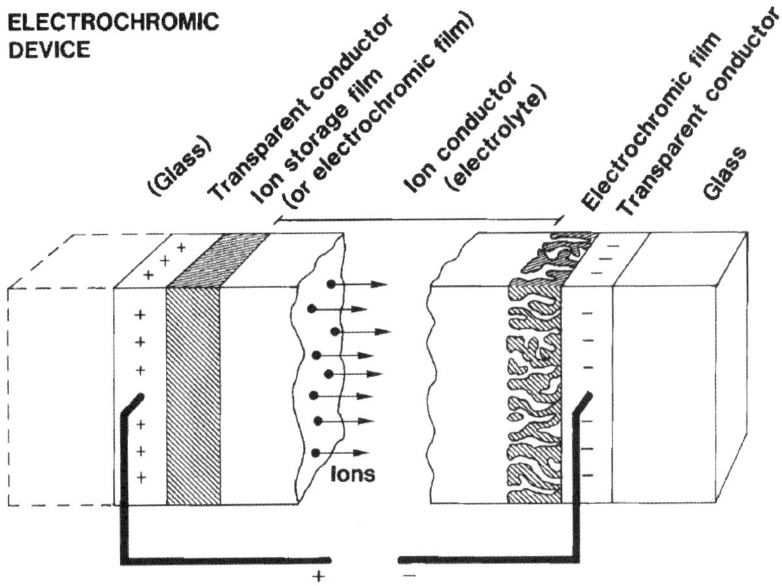

Figure 10.3 Structure of an electrochromic device deposited on glass. (Figure reproduced by permission from The Royal Society of Chemistry from ref. 11.)

The properties of an electrochromic device as a whole are largely dictated by the electrochromic layer. To create a successful device, the electrochromic colour change must be reversible over a large number of cycles, without degradation or loss of performance. How many cycles are necessary depends on the application; a pane of window glass would be expected to last over 20 years, and survive wide variations in temperature and humidity. A mirror or handheld display device might be expected to endure rather less. The rapidity of switching also depends upon application. While it may be acceptable for a window to take several minutes to change colour, this would be useless in most dynamic displays. The requirements for the intensity of the coloured state and the clarity of the bleached state will similarly be application specific. Various measures of the electrochromic performance of a device have been used. The colouration efficiency (η) of an electrochromic material is the constant of proportionality between injected charge density (Q) and the change in optical absorption of the film at a specific wavelength (ΔOD):

$$\Delta \mathrm{OD} = \eta Q \tag{10.1}$$

It measures the rate of change of film absorption with applied potential, which is linear within the range of stability of the electrode. Since devices consist of two electrodes, both of which may be electrochromic with colouration efficiencies η_1 and η_2, the absorption of a complete device (A) at a given wavelength and at charge density Q is given by:

$$A = A_{\mathrm{bl}} + Q(\eta_1 + \eta_2) \tag{10.2}$$

where A_{bl} is the absorption of the bleached state ($Q = 0$).[42] Note that cathodically colouring materials have positive colouration efficiencies and anodic colouring materials have negative colouration efficiencies. Other measures of electrochromic performance include switching time (also known as response time), cycle life and minimum and maximum absorbance. Switching time is the time taken for the device to change from maximum to minimum absorption or *vice versa*. There will typically be different colouring and bleaching switching times. Cycle life is the number of colouration/bleaching cycles a device can withstand before noticeable loss of performance occurs. As noted, the relative importance of these performance measures will depend on the specific application, and a distinction must be made between properties of an electrochromic material and properties of the device as a whole. Many reports are made on the properties of a single electrochromic thin film, while proportionally fewer are made on a complete electrochromic device, such as a window or mirror. Furthermore, most reports in both categories, and especially the second, concern thin films deposited by sputtering, sol–gel or vacuum evaporation. CVD is currently less popular as a deposition technique for electrochromic films. Reasons such as poor morphological control or greater purity of evaporated films may be cited. However, as metal oxide electrochromic devices move toward viable commercial products, no doubt the need for cheap, on-line coatings will become more pressing.

10.4.2 Mechanism of Electrochromism

Many materials fall into the category of "electrochromic," as defined above. However, a much smaller proportion is likely to be useful as functional coatings. Tungsten oxide, vanadium(V) oxide and molybdenum oxide are the most thoroughly investigated inorganic electrochromic materials. In each case, the colour change is achieved through partial reduction of the metal ions, accompanied by intercalation of small, positively charged ions, such as H^+ or Li^+, which balance the charge. Tungsten oxide (WO_3) is the leading electrochromic material, for two main reasons. Firstly, in thin film form and in the bleached state, it is almost totally transparent to visible light, making it ideal for many glass based applications. Secondly, there is a large difference in transmission between the bleached and coloured states, making the change easily detectable by eye. In the case of tungsten oxide the intercalation of ions that causes the electrochromic transition can be written as shown in Scheme 10.1.

The resulting material, M_xWO_3, contains partially reduced tungsten ions. Visible light absorption arises from electron transitions between neighbouring, differently charged, metal ions, known as polaron transitions.[44] A very large amount of research has been directed at tungsten oxide and the construction of electrochromic devices, and a large amount of information is available on the desirable materials properties of an electrochromic thin film.[45–50] Detailed reviews exist covering this research over the past 20 years,[11,51–53] and therefore only the pertinent conclusions are discussed here. WO_3 is a pale yellow material that becomes dark blue under cathodic electrochromic colouration, *i.e.*, insertion of H^+, Li^+ or Na^+ ions into the lattice with accompanying reduction of tungsten ions. Thin films of WO_3 are unstable in acid or base solution under electrochemical cycling conditions; therefore, if a protic electrolyte is to be used, the WO_3 electrode needs to be protected by an overlayer. The overlayer must be ion permeable to allow intercalation; organic polymers or hydrated Ta_2O_5 have been successfully used.[42] The mobility of ions into and out of the WO_3 structure is of prime importance, as it determines the switching time and the maximum colouration of the film. Since high surface area, *i.e.* high porosity, increases ion mobility it is very desirable in electrochromic films. High crystallinity reduces the mobility of ions within the material, and so reduces the electrochromic effect. For this reason amorphous tungsten oxide (a-WO_3) films are favoured. A complication arises as sub-stoichiometric tungsten oxides, of the formula WO_{3-x} ($0<x<0.5$) can easily be formed by CVD and other deposition methods. Owing to the presence of reduced tungsten states, the material is dark blue, showing the same polaron bands as observed in

$$WO_3 + xM^+ + xe^- \leftrightarrow M_xWO_3$$

Scheme 10.1

M_xWO_3. Sub-stoichiometric tungsten oxide is unsuitable as an electrochromic material as its "bleached" state remains highly coloured even in the absence of intercalated ions. WO_{3-x} is produced by CVD reactions in oxygen-deficient environments, but can be oxidized to WO_3 by subsequent heating in air, although this may also have the unwanted side effect of increasing the crystallinity and reducing the electrochromic efficiency of the material. A wide range of electrolytes have been investigated in conjunction with tungsten oxide, such as protic polymers and solid state inorganic Li based electrolytes (*e.g.* lithium phosphorous oxynitride, $LiAlF_4$, $LiNbO_3$).[52,54] Working electrochromic windows of over 1000 cm^2 have been produced that show a change in visible light transmission ($\Delta T\%$) of over 60% between the bleached and the coloured state. Further scale up of such devices is hindered by factors such as the sheet resistance of the conducting layers and difficulty in batch producing and assembling large devices within necessary quality standards. It is fair to say that research into non-WO_3 electrochromic materials is rather less advanced, and is mainly orientated towards the search for an appropriate anodic colouring material to complement WO_3. Investigations into CVD deposited electrochromic films of V_2O_5 and MoO_3 is discussed later in this chapter.

A related phenomenon also observed in WO_3 is photochromism – the change in colour due to irradiation. Several materials show photochromism, which operates through several different mechanisms, although again WO_3 has received most attention. The mechanism of photochromism in WO_3 is complicated by the fact that there are two possible causes of the blue colouration induced by UV irradiation: sub-stoichiometric WO_{3-x} or the tungsten bronze, M_xWO_3. These two materials are difficult to distinguish, as their absorption properties are the same, since they arise from the same polaron transitions. Initial theories on the photochromic mechanism focused on the intercalation of protons, assuming the mechanism to be similar to that of electrochromism, which was and is far more widely studied.[55] However, in the mid-1990s, Bechinger *et al.* published work challenging this theory.[56–59] They found that the photochromic process was only totally reversible in an oxygen-containing atmosphere; in an atmosphere of nitrogen or argon, the blue colour induced by irradiation did not completely fade even after extended periods in the dark. The reduction in absorption during the bleaching process can be modelled accurately by the sum of two exponential decays, suggesting that there are two components to the process. It is thought that one of these components involves adsorption of atmospheric oxygen. Bechinger proposed that one of these components is the formation of a tungsten bronze, the source of the H^+ ions being the photo-destruction of incorporated water molecules. Gavrilyuk later showed that photochromism was caused by the photo-destruction of organic adsorbates, providing H^+ ions that intercalate into the film.[60] It was shown that the destruction of deuterated methanol on the surface of the film caused colouration and led to an increased deuterium content of the film; this deuterium increase was proportional to the increase in optical absorption. This mechanism, known as the photo-injection of hydrogen (PIH), is currently the most widely accepted mechanism for photochromism, although it does not explain the irreversibility of the colour change in an inert atmosphere.

Photochromic coatings might be applicable in the same types of devices as electrochromic coatings, namely solar control coatings, display and optical devices. There are, however, important differences between the two types of coating. Electrochromic coatings can be more easily controlled; an electrochromic device could be connected to a system of electronics to operate on any cue (touch of a button, increase in temperature, increase in visible light level, *etc.*). Photochromic coatings can only be activated by irradiation of an appropriate wavelength. In general, electrochromic coatings switch faster, especially the change from coloured to bleached state, which can be very slow in photochromic coatings. However, photochromic coatings are far simpler in construction (and therefore cheaper) being a single layer of metal oxide. For some applications, this simplicity may make photochromic coatings more desirable. As is the case for electrochromic coatings, amorphous WO_3 shows superior photochromism, due to better ion intercalation. For the same reason, higher surface area films are also more effective. It is the case, therefore, that the materials requirements for electrochromic and photochromic WO_3 are very similar; an effective electrochromic coating will likely be an effective photochromic one.

10.4.3 CVD of Electrochromic and Photochromic Tungsten Oxide Coatings

Stable oxides of tungsten exist for the +4 and +6 oxidation states. There are also many sub-stoichiometric oxides containing mixed oxidation states of tungsten. These range from oxides with a well-defined structure (*e.g.* $W_{20}O_{58}$, $W_{18}O_{49}$), which result from certain adjustments to the WO_3 lattice, to less ordered oxides that can display any stoichiometry from $WO_{2.5}$ to $WO_{3.0}$.[61] CVD can be used to deposit anything from tungsten metal to WO_3. Control over the stoichiometry of the deposited material is important for good functional properties. What follows is a brief overview of the varied precursors that have been used to deposit tungsten oxide, followed by more detailed sections covering the characteristics of each type of deposition.

Tungsten oxide can be deposited onto glass using a range of precursors. Tungsten(VI) halides are popular choices as they are volatile and readily hydrolysed. WF_6, a colourless liquid (bp 17 °C), is the most volatile and reactive of the tungsten hexahalides, but causes problems due to HF formation on decomposition.[62] WCl_6, a blue solid (bp 347 °C), has been used as a cleaner alternative with some success, although it is less volatile.[63] WBr_6 is too involatile for CVD use. Tungsten(VI) oxo halides have been explored as alternatives to the homoleptic halides. $WOCl_4$, a red solid (mp 211 °C), is more readily hydrolysed in the vapour phase than WCl_6, yet less so in the solid phase, due to formation of an extended network of W–O bonds, making it a useful alternative where ease of precursor handling is important.[61] Aside from the halide based complexes, tungsten hexacarbonyl ($W(CO)_6$) is the most well studied WO_3 precursor.[62,64–68] It is one of the less toxic transition metal carbonyls, and is a volatile subliming solid that decomposes at around 400 °C. As it contains tungsten(0), it is capable of depositing metallic tungsten films and sub-stoichiometric tungsten oxides with ease. An additional oxygen source must be used to deposit fully oxidized WO_3. Mono-substituted tungsten carbonyls [*i.e.* $LW(CO)_5$] have been developed to adjust volatility and reactivity. Several tungsten alkoxides, oxo alkoxides and aryl oxides have been investigated, *e.g.* $W(OEt)_6$, $WO(OR)_4$ (R = Me, Et, ^{i}Pr, Bu) and $W(OPh)_6$ and in some cases their fluorinated derivatives.[69,70] The advantage is that these are single source precursors; no second oxygen precursor is necessary, although it may be desirable to ensure the deposition of fully stoichiometric WO_3. Low volatility is a problem in most of these compounds despite fluorination, and low pressure or aerosol-assisted techniques are generally used to deposit coatings from them. Tetra (allyl)tungsten, $W(\eta^3\text{-}C_3H_5)_4$, polyoxometallates (POMs) are an unusual class of precursor that have been used to deposit WO_3 by AACVD. POMs are clusters of tungsten and oxygen, of which a large number of stoichiometries are known. Examples include $[W_6O_{19}]^{2-}$ and $[W_{10}O_{32}]^{4-}$. These are rather unconventional CVD precursors, due to their large size and the fact that they are anionic.

10.4.4 Atmospheric Pressure CVD of WO_3

10.4.4.1 Tungsten Hexachloride

Tungsten hexachloride has only recently been investigated as a CVD precursor. Work on this precursor, although formative, is encouraging, as good quality films have been deposited with the important advantage that HF is not produced as a by-product. The APCVD reaction of WCl_6 and various oxygen-containing precursors deposited tungsten oxide films of varying quality at temperatures of 450–625 °C.[63] Very high flow rates were needed to deposit films from WCl_6, which is perhaps the reason it had not previously been identified as a suitable precursor for tungsten oxide films. Furthermore, the material has a very narrow usable temperature range when employed as precursor, with the melting and decomposition temperatures being separated by 20 °C. The different oxygen precursors investigated (ethanoic anhydride, ethanol, methanol, water and ethanoic acid) produced tungsten oxide with different stoichiometries at varying growth rates. Growth rates at 625 °C ranged from 200 to 1000 nm min^{-1}, with ethanoic anhydride giving the fastest growth and ethanoic acid the slowest. The fastest growth rates were not due to more porous, less dense films, but genuinely due to a greater deposition of material. However, notably, in the case of ethanoic anhydride, the faster growth rate could be due to a greater molar flow of the precursor. Other precursors investigated that failed to deposit a film included ethyl ethanoate, 2-propanol and 2-methyl-2-propanol.

The stoichiometries of the films, and correspondingly their colours, also varied with precursor. The exact stoichiometries could not be determined in some cases, due to the incorporation of a large amount of water in the films; however, the colour of the film could be used as a guide. Yellow, fully stoichiometric WO_3 films were deposited at lower temperatures. Some precursors – ethanol and methanol – deposited blue films at 625 °C, while others – ethanoic anhydride, water and ethanoic acid – continued to deposit yellow films even at 625 °C. Variation in stoichiometry and colour with temperature has been observed in other tungsten oxide depositions, and is usually explained by the more rapid reaction at higher temperatures. Indeed, the growth rates, and hence rates of reaction, of the blue films deposited from ethanol and methanol were greater than those of the yellow films deposited from ethanoic acid and ethyl ethanoate. Ethanoic anhydride is an anomaly in the trend, giving a yellow film at 625 °C despite having the fastest growth rate. This might be explained by the greater molar flow of ethanoic anhydride used, and the larger proportion of oxygen in the precursor, which results in an oxygen-rich atmosphere. Films deposited from different precursors had had very different morphologies. Ethanol and methanol caused deposition a dense mat of thin micron scale needles, while water deposited round individual micron scale platelets, and ethanoic acid deposited fused platelets. Despite these differences, all films gave an XRD pattern consistent with monoclinic WO_3. The mechanism of the reaction is unknown, but some carbon contamination in the films points to extensive decomposition of the oxygen precursors. Elimination of HCl is likely, and correspondingly no chlorine contamination was observed in the films. The need for very high flow rates shows that the reaction is very fast, or the reactive species that deposits the film is very short lived.

None of the films deposited from WCl_6 showed great photochromism, possibly due to high crystallinity. The greatest effect was seen in films deposited from ethanoic anhydride, which showed a change in optical density at 633 nm of 0.30 after 1 hour of ultraviolet irradiation. The photochromic performance of this film may be attributed to a high surface area, although other factors may also be important.

A second APCVD route to tungsten oxide is through the reaction of $WOCl_4$ with ethyl ethanoate or ethanol.[71] The two oxygen precursors gave very similar results. Films were deposited from 350 to 650 °C in both cases, with yellow, fully stoichiometric films produced at lower temperatures (350–450 °C) and blue, sub-stoichiometric films produced at higher temperatures. Growth rates of 200 nm min^{-1} were achieved by both precursors. Interestingly, $WOCl_4$ was able to deposit tungsten oxide films with ethyl ethanoate while WCl_6 was not, showing the greater reactivity of gas phase $WOCl_4$.

10.4.4.2 Tungsten Hexacarbonyl

Tungsten hexacarbonyl can deposit films on glass at substrate temperatures of 200 °C and above. Carbon monoxide is a good leaving group and, in the absence of oxygen, metallic tungsten coatings are formed.[62,64] Introducing oxygen to the reaction results in deposition of WO_{3-x}, with the value of x depending on the partial pressure of O_2.[62] This affects both the appearance and functional performance of the coatings as discussed previously. Ratios of $W(CO)_6 : O_2$ of 1 : 6 or lower are necessary to deposit transparent WO_3 films; a lower oxygen flow results in blue colouration characteristic of partially reduced tungsten oxide.[10] Correspondingly, XPS data indicated the presence of W^{4+} and W^{5+} ions in films grown with low oxygen flow rates.[72] Alternatively, films grown in air without an additional oxygen source also appear to be fully oxidized WO_3.[73] The deposition temperature strongly affects the growth rates of films, but not the stoichiometry, which appears to be largely determined by the O_2 flow rate. Growth rates of up to 380 nm min^{-1} have been recorded for films grown at 400 °C, while at lower temperatures growth rates fall off significantly, to less than 3 nm min^{-1} at 200 °C.[10] High growth rates are sometimes associated with porous, low density films, and so may be desirable for electrochromic applications. Calculations using the refractive indices and densities of the films showed a large increase in porosity in films grown at higher temperatures. However, a high deposition temperature also increases crystallinity, which hampers electrochromism, and so a balance must be struck. With glass substrates, deposition temperatures of 300 °C and below produced X-ray amorphous films, while at 400 °C monoclinic WO_3 was deposited. As expected, the polycrystalline films deposited at higher temperatures showed a lower colouration efficiency (25.5–41 $cm^2 C^{-1}$), while those deposited at lower temperatures performed better, showing colouration efficiencies of 205–230 $cm^2 C^{-1}$.[10,64,73]

Derivatives of $W(CO)_6$ have also been used to deposit tungsten oxide. Tungsten(0) hexacarbonyl isonitriles of formula [$RNCW(CO)_5$], where R = *n*-pentyl or methylbutyl, are volatile liquids. They have been used to deposit tungsten oxide on glass at 275–300 °C. As is the case with $W(CO)_6$, oxygen was required for the deposition of tungsten oxide films.

10.4.5 Aerosol-assisted CVD of WO_3

AACVD has been used to deposit WO_3 from various precursors, although in most cases electrochromic or photochromic characterization has been lacking. The ability to use involatile precursors has been exploited in depositions using unusual precursors such as polyoxometallates (POM) (anionic metal oxide clusters).[74]

The AACVD reaction of $W(OAr)_6$ (Ar = C_6H_5, C_6H_4F, $C_6H_3F_2$) in acetone and toluene resulted in tungsten oxide films at substrate temperatures of 300–500 °C.[69] AACVD depositions at 300 °C resulted in amorphous films in all cases, while at higher temperatures XRD indicated that crystalline films were produced. Yellow films were not deposited, even at the lower temperatures. Both toluene and acetone were used as solvents with $W(OC_6H_5)_6$, while only toluene was used with the fluorinated precursors. For $W(OC_6H_5)$, the use of toluene or acetone made little difference to the growth rate or composition of the film. This suggests that the solvent does not play an important role in the deposition, and further that acetone does not act as an oxygen source in this case. The precursors containing fluorinated aryl groups gave similar growth rates to the unfluorinated precursor, showing that the reaction is not sensitive to small changes in the ligand electronic properties. Cursory studies of the photochromism of WO_3 and doped WO_3 deposited from $W(OPh)_6$ showed a reversible switching under 254 nm UV irradiation.[70]

Tungsten oxo-salicylate complexes were investigated as AACVD precursors to WO_3.[75] Depositions carried out at 600 °C using the precursors dissolved in dichloromethane resulted in brown films with a high carbon content. Post deposition annealing reduced the carbon content and yielded dense, yellow polycrystalline WO_3 films.

Polyoxometallates of formula $[n\text{-}Bu_4N]_2[W_6O_{19}]$ and $[n\text{-}Bu_4N]_4H_3[PW_{11}O_{39}]$ were used to deposit WO_3 and P-doped WO_3, respectively, at 410–480 °C.[74] Acetone was used as a solvent in each case. The resulting films were blue, with Raman patterns indicative of a WO_{3-x} stoichiometry, and had a low carbon content. In the case of the P-containing precursor, the level of P doping in the resulting film was very similar to the proportion of P in the precursor.

10.5 Vanadium Dioxide Thermochromic Coatings

10.5.1 Introduction to Thermochromic VO_2

A thermochromic material changes its optical properties with changing temperature. A thin film of thermochromic material might be used to create an "intelligent window": one that changes its transmittance and reflectance properties in response to environmental changes to regulate the temperature within a room. A suitable material would become more reflective (and therefore less transmitting) in the infrared region at higher temperature. Hence, at low temperatures, in the transmitting state, solar energy is allowed into the room to cause warming. At high temperatures, in the reflecting state, the coating reflects most of the infrared solar energy whilst ideally transmitting as much visible light as possible. These windows would be suited to geographical regions where the temperature may change dramatically during a day or over the year, such as most of Europe, Asia and North America. A thermochromic film is therefore a type of advanced solar control coating that actively responds to a user's needs. Investigative studies into this type of window coating are only just beginning.

By far the most promising candidate material for coating a thermochromic window is vanadium dioxide. VO_2 undergoes a metal–semiconductor transition (MST) transition at a critical temperature (T_c), changing from a transmitting state typical of a thin semiconductor film to a reflecting state typical of a metallic film.[76] This change is a phase transition between rutile and monoclinic polymorphs, which is reversible in thin films of the material.[77] Figure 10.4 shows the transition between monoclinic VO_2 at low temperature and rutile VO_2 at high temperature using powder X-ray diffraction.

However, several problems have so far hindered commercialization of thermochromic windows. Firstly, the T_c of pure VO_2 is 68 °C, which is clearly too high to be useful in regulating room temperature. This problem can be overcome by doping with tungsten oxide or other metal ions,

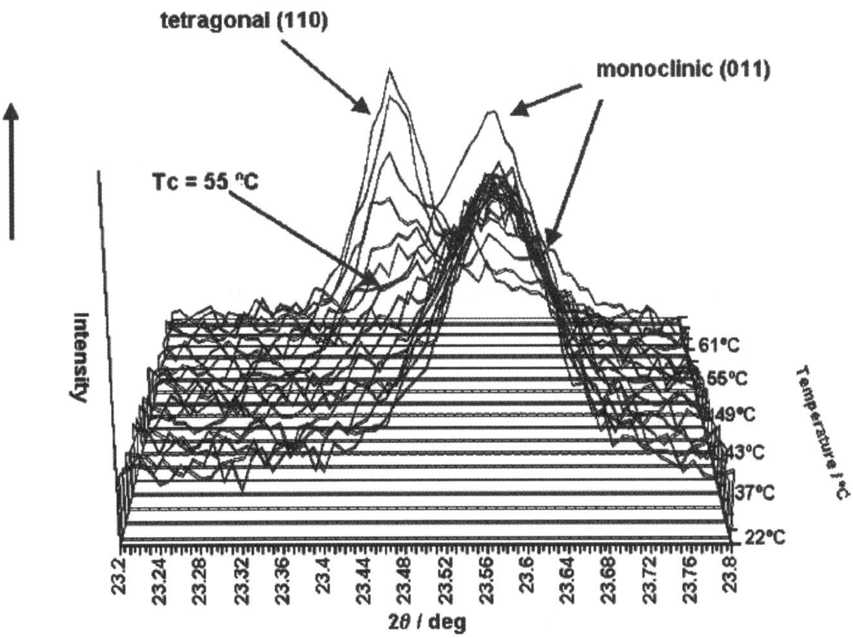

Figure 10.4 High-resolution temperature-dependent XRD of a W-doped VO_2 film showing the transition at $T_c = 55\,^{\circ}C$. (Figure reproduced by permission from The Royal Society of Chemistry from ref. 13.)

which reduces T_c significantly. Secondly, thin films of VO_2 appear yellow-brown and unattractive as window coatings. This is perhaps a more serious a problem, as cosmetic appearance is vital in commercial glazing applications. There are also several important issues of concern to the CVD practitioner. Stable oxides exist for the +3, +4 and +5 oxidation states of vanadium: V_2O_3, VO_2, V_2O_5 and also mixed oxidation states, *e.g.* V_6O_{13} and V_3O_7.[78] Of these, only VO_2 displays thermochromism with T_c low enough for useful application in glazing (V_2O_5 also undergoes a thermochromic transition with a T_c of 257 °C[79]). VO_2 itself has several polymorphs, including the rutile and monoclinic phases which interconvert at the critical temperature. In addition, there are two metastable VO_2 polymorphs that do not undergo metal–semiconductor transition, $VO_2(A)$ and $VO_2(B)$. Therefore, control of stoichiometry and crystal phase is essential for the deposition of thermochromic films.

The thermochromic transition in VO_2 can be detected in several ways; those commonly used are infrared spectroscopy and Raman spectroscopy. Infrared spectroscopy can be used to measure transmittance at a given wavelength, usually 2.5 µm (4000 cm^{-1}) above and below T_c. Recording a complete heating–cooling cycle reveals a hysteresis in the switching temperature. The width of the hysteresis loop (ΔT_h) is dependent on the crystallinity and microstructure of the film, and can range from 2 to 20 °C.[80] Strongly orientated polycrystalline epitaxial films show the smallest ΔT_h, while randomly orientated films show the largest values.[81] Whether a large or small ΔT_h is desirable for thermochromic windows is debatable. Raman spectroscopy can also be used to characterize the metal–semiconductor transition. The monoclinic polymorph of VO_2, present below T_c, shows sharp Raman bands at 192, 222, 307, 388 and 614 cm^{-1}. As T_c is approached during a heating cycle, the intensity of the Raman bands decreases significantly. No peaks are observed at T_c, and further heating causes the emergence of a single broad band at 555 cm^{-1}, which is characteristic of tetragonal VO_2.[14]

10.5.2 Atmospheric Pressure CVD of VO_2

Vanadium oxides have been deposited onto glass by APCVD using VCl_4 and $VOCl_3$.[14,82] Both precursors can deposit thermochromic VO_2 under the correct conditions, but it was also found that both can also deposit V_2O_5 and other oxides. All the depositions discussed here were carried out in a cold wall horizontal bed static reactor. Both precursors required an oxidant, and water was effective in each case. The phase of vanadium oxide deposited depends upon the ratio of precursor to oxidant, the temperature and the flow rate. Both VO_2 and V_2O_5 can be deposited in this way by changing these parameters. Other phases are sometimes present in small quantities. Scheme 10.2 shows idealized deposition reactions.

Both precursors behave very similarly in their pattern of reactivity. At high temperatures, VO_2 is more thermodynamically stable than V_2O_5, therefore high temperatures, slow flow rates and high oxidant to precursor ratios favour VO_2 formation, while the opposites favour V_2O_5 (Table 10.6). Notably, a higher concentration of oxidant favours vanadium(IV) oxide over vanadium(V) oxide, when the more oxidized product might be expected. In depositions using VCl_4, VO_2 is favoured

$$2VOCl_3 + 3H_2O \rightarrow V_2O_5 + 6HCl$$

$$VOCl_3 + H_2O \rightarrow VO_2 + \tfrac{1}{2}Cl_2$$

$$2VCl_4 + 5H_2O \rightarrow V_2O_5 + 8HCl + H_2$$

$$VCl_4 + 2H_2O \rightarrow VO_2 + 4HCl$$

Scheme 10.2

Table 10.6 Deposition conditions and switching temperatures of thermochromic vanadium dioxide thin films on glass.

Precursors	*CVD type*	*Oxidant*	*Precursor : oxidant ratio*	*Deposition temperature (°C)*	*Composition (dopant)*	*Switching temperature (°C)*	*Ref.*
VCl_4	AP	H_2O	1 : 10	450	V_2O_5	–	14
VCl_4	AP	H_2O	1 : 10	550	VO_2	68	14
VCl_4	AP	H_2O	[a]	450–600	V_2O_5	–	82
$VOCl_3$	AP	H_2O	1 : 5	600–650	V_2O_5 + VO_2	68	84
$VOCl_3$	AP	H_2O	2.5 : 1	350–650	V_2O_5	–	84
$VOCl_3$	AP	H_2O	[a]	350–600	V_2O_5	–	82
$VO(acac)_2$	LP (20 mbar)	–	–	370–520	VO_2	68	80
VCl_4 + $W(OEt)_6$	AP	H_2O	1 : 10	500–600	VO_2 (W: 0.3–1.2%)	42–52	15
$VOCl_3$ + WCl_6	AP	H_2O	[a]	650	VO_2 (W: 0.6–3.1%)	5–45	13
$VOCl_3$ + $MoCl_5$	AP	H_2O	[a]	650	VO_2 (Mo: 0.1–1.0%)	47–58	83
$VOCl_3$ + $NbCl_5$	AP	H_2O	–	650	VO_2 (Nb: 0.4%)	55	83

[a]Not reported.

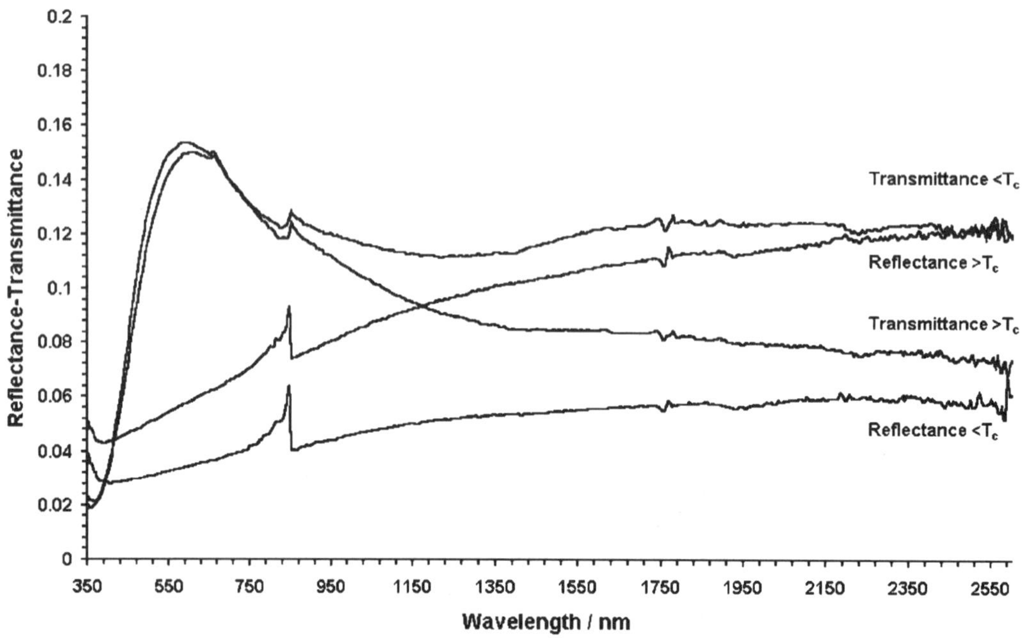

Figure 10.5 Change in reflectance and transmission of a VO_2 film above and below the transition temperature. (Figure reproduced with permission from The Royal Society of Chemistry from ref. 13.)

over V_2O_5 at oxidant : precursor ratios as high as 57 : 1.[14] The microstructure of these VO_2 films was not studied in depth, but where reported the films were composed of rod-shaped particles approximately 50 nm in diameter. VO_2 films from both precursors appear yellow or brown. Those deposited below around 600 °C were poorly adherent and easily removed from the glass by abrasion, although above this temperature most passed the Scotch tape test and other durability tests. All VO_2 films showed T_c at the expected 68 °C. Figure 10.5 shows a typical set of reflectance/transmittance data for a VO_2 film on glass.[13] Note the large increase in reflectance when the film is heated above T_c, especially in the near-IR region.

Transition metal doped VO_2 films show altered T_c depending on the nature and concentration of the dopant. Relatively high dopant levels are needed (more than 1 at.%) and the size of the ion compared to V^{4+} is important. Low valent metal ions, such as Cr^{3+}, increase T_c, whereas high valent ions decrease T_c. To make a functioning thermochromic window, the critical temperature of 68 °C clearly must be reduced to around room temperature. The most successful dopant so far tested is tungsten, which when doped into VO_2 reduces the switching temperature by around 20–25 °C at.%$^{-1}$ (Figure 10.6). Thus dopant levels of around 2–2.5 at.% are required to reduce T_c to around room temperature. Dopants can be introduced using a suitable precursor. Tungsten has been doped into VO_2 using WCl_6 in conjunction with $VOCl_3$, and $W(OEt)_6$ in conjunction with VCl_4, using essentially the same conditions as used to deposit undoped VO_2 (Table 10.6).[13,15] It is important for reduction of T_c that the tungsten is doped within the VO_2 lattice, and does not form a separate tungsten oxide phase. Phase separated tungsten oxide would be detectable by X-ray diffraction or Raman spectroscopy; the films discussed here showed no signs of a secondary phase. Tungsten was present as W^{4+} (determined by XPS), suggesting substitutional doping into the VO_2 lattice. The tungsten content of the films could be varied by changing the flow rate through the tungsten precursor bubbler. Using $VOCl_3$ and WCl_6, tungsten concentrations up to 3.1 at.% can be achieved in this way without the formation of a secondary tungsten oxide phase. These precursors are, therefore, able to deposit films with room temperature T_c. $W(OEt)_6$ suffers from a lower

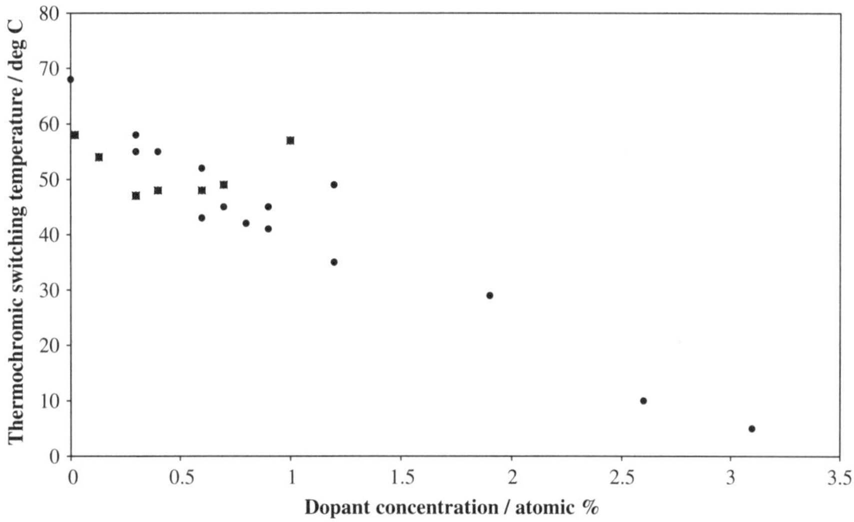

Figure 10.6 Dependence of switching temperature of VO_2 thin films on dopant level; (●) tungsten doping, (■) molybdenum doping.

vapour pressure, and was only able to dope up to 1 at.% tungsten into the VO_2 films. Whichever precursor is used, the T_c appears to have a linear relationship with the tungsten concentration (Figure 10.6).

Other dopants have also been investigated. Molybdenum and niobium have been doped into VO_2 using $MoCl_5$ and $NbCl_5$ precursors, respectively.[83] However, as was the case with $W(OEt)_6$, the volatility and decomposition temperature of these precursors was not ideal, and a maximum of around 1 at.% Mo or Nb was able to be doped into the films. As expected, doping reduced the T_c of VO_2, and, as shown in Figure 10.6, from the data available the rate of decrease caused by Mo doping appears very similar to that caused by W doping. Because only one sample of Nb doped VO_2 was reported, it is not possible to comment on changes in T_c with Nb concentration. Molybdenum was present as Mo^{4+} but in contrast Nb was observed as Nb^{5+}, perhaps indicating a different doping mechanism. As observed in undoped films, both Mo and Nb doped VO_2 films showed crystallographic preferred orientation along the (011) plane, and no secondary oxide phase was observed at any dopant concentration.

10.5.3 Low Pressure CVD of VO_2

Low pressure CVD of VO_2 has been reported using $VO(acac)_2$ as the sole precursor.[80] Considerable control over the phase composition was possible through the use of a hot-wall static reactor and varying the deposition temperature. The thermochromically active monoclinic VO_2 phase was deposited 475 °C. At lower temperatures, the non-active $VO_2(B)$ polymorph was deposited. No other oxides of vanadium were observed. The different phases display different morphologies. Monoclinic VO_2 films were composed of flat micrometre platelets, while $VO_2(B)$ films were made up of needles. The $VO_2(B)$ phase is metastable and can be converted into monoclinic VO_2 by heating in the absence of oxygen. Inactive $VO_2(B)$ films could be converted into active monoclinic VO_2 films in this way, whilst retaining their needle-like microstructure, although the crystal size within such samples was markedly reduced. The films showed a thermochromic transition at the expected T_c of 68 °C, although the hysteresis width was much smaller than observed in APCVD films, probably due to the smaller crystal sizes present.

10.6 Self-cleaning Coatings on Glass

Self-cleaning coatings are desirable for glazing applications to create self-cleaning windows. These have been commercialized successfully by several companies. Currently, all commercial self-cleaning windows are coated with a thin transparent layer of titanium dioxide, a coating that acts to clean the window in sunlight through two distinct properties: photocatalysis causes the coating to chemically break down organic dirt adsorbed onto the window while hydrophilicity causes water to form "sheets" rather than droplets – contact angles are reduced to very low values in sunlight (the coating becomes "super-hydrophilic"), and dirt is washed away. Mills and LeHunte have produced a detailed review of the physical basis of the self-cleaning properties of titania,[85] the intricate details of which are beyond the scope of this chapter. It can be summarized that both self-cleaning properties of TiO_2 are governed by the absorption of ultraviolet light and the generation of electron–hole pairs. The band gap of bulk anatase TiO_2 is 3.2 eV, corresponding to light of wavelength 390 nm, near-ultraviolet (UV) light. A small proportion of the sunlight reaching earth exceeds 3.2 eV, so titania coatings can be solar activated. Titania has become the material of choice for self-cleaning windows, and hydrophilic self-cleaning surfaces in general, because of its favourable physical and chemical properties. Not only is titania highly efficient at photocatalysing dirt in sunlight and reaching the super-hydrophilic state, it is also non-toxic, chemically inert in the absence of light, inexpensive and relatively easy to deposit into thin films.

Several polymorphs of TiO_2 are known, the most significant of which are rutile and anatase. The position of the conduction and valence bands relative to key redox potentials cause pure anatase to be very photoactive (photocatalytic and super-hydrophilic) while pure rutile is less so. Anatase TiO_2 is formed at temperatures exceeding 300 °C, but converts into rutile around 800 °C. This is higher than the temperature usually found in a CVD process, although this temperature is exceeded on a float glass line. Below 300 °C, amorphous titania is usually formed when using glass substrates. Crystalline substrates can give rise to rutile TiO_2 at low temperatures. Titania films deposited by CVD tend to have small crystal sizes, with some commercial TiO_2 coatings having crystal sizes of 7–13 nm.[86] Band gap widening occurs in semiconductors with very small crystal sizes; nanoparticulate titania should be more photoactive, due to a wider band gap leading to greater oxidation power of excited holes and greater reducing power of excited electrons. However, the wider band gap also means that higher energy photons are required to activate the semiconductor, meaning that a smaller proportion of sunlight is useable. The photocatalytic activity of a film is tested by measuring the destruction of an organic compound on the surface of the film. Sunlight or artificial UV light is used to activate the coating, and IR or UV spectroscopy is used to quantify the destruction of the compound. TiO_2 has shown itself capable of breaking down a wide variety of organic compounds, including long-chain carboxylic acids, chlorinated and fluorinated compounds, and aromatic compounds.[87–89]

Titania has a high refractive index and bulk TiO_2 is highly reflective. To achieve the high transparency required for windows, TiO_2 coatings must be under 50 nm thick. In such thin films, the absorption of UV light is also very low. The quantum yields for commercial samples of self-cleaning window coatings were found to be as low as 0.04%,[90,91] meaning that the vast majority of the already small proportion of usable sunlight passes straight through the coating. Thicker films show significantly higher efficiencies, and may be suitable for non-glazing applications. Table 10.7 summarizes CVD routes to TiO_2.

$TiCl_4$ is a widely used volatile precursor that deposits chlorine-free TiO_2 when combined with a suitable oxygen source. In the absence of an oxygen source, no film is deposited. Various organic compounds have been investigated as oxygen sources for these reactions.[92–94] Alcohols were found to be effective; depositions using various compounds, ROH (R = Me, Et, Pr), and EtOAc gave very similar films with good transparency and adherence. The use of water as an oxygen source led to films, with larger crystallite sizes, that were hazier. At 500 °C deposition temperature and above,

Table 10.7 Summary of CVD routes to TiO_2 thin films on glass.

Precursor 1	*Precursor 2*	*CVD method*	*Substrate temperature range (°C)*	*Maximum growth rate (nm min^{-1})*	*Ref.*
$TiCl_4$	ROH	AP	500–600	300	92
$TiCl_4$	EtOAc	AP	400–600	20	93
$Ti(O^iPr)_4$	–	LP	550	–[a]	94
$Ti(dpm)_2(O^iPr)_2$[b]	–	LP	400–520	24	95
$Ti(O^tBu)_4$	–	PE	200	0.4	96
$Ti(O^iPr)_4$	O_2	PA[c]	20	–[a]	97

[a]Not reported.
[b]dpm = 2,2,6,6-tetramethylheptan-3,5-diol.
[c]PA = photo-assisted.

anatase titania was deposited, and all films showed photocatalytic activity when irradiated with $\lambda = 254$ nm light.

Single source precursors to TiO_2 rely on LP, PE or PA (photo-assisted) CVD techniques, as no suitably volatile single source precursors have been produced. $Ti(O^iPr)_4$ has been deposited using low pressure, giving photoactive anatase films with low contamination.[95] $Ti(dpm)_2(O^iPr)_2$ (dpm = 2,2,6,6-tetramethylheptan-3,5-diol) was used to deposit TiO_2 on various substrates, also at low pressure. The precursor decomposes around 400 °C, and the photocatalytic activity of the films was low due to inclusion of rutile phase TiO_2.

References

1. M. D. Allendorf, *Thin Solid Films*, 2001, **392**, 155–163.
2. A. Arnaud, *Journal of Non-Crystalline Solids*, 1997, **218**, 12–18.
3. Pilkington, *Proceedings of the Royal Society of London Series A-Mathematical and Physical Sciences*, 1969, **314**, 1.
4. R. J. McCurdy, *Thin Solid Films*, 1999, **351**, 66–72.
5. P. H. Gaskell, *Journal of Non-Crystalline Solids*, 1997, **222**, 1–12.
6. T. Maruyama and K. Tabata, *Journal of Applied Physics*, 1990, **68**, 4282–4285.
7. R. B. H. Tahar, T. Ban, Y. Ohya and Y. Takahashi, *Journal of Applied Physics*, 1998, **83**, 2631–2645.
8. G. Frank and H. Kostlin, *Applied Physics A-Materials Science & Processing*, 1982, **27**, 197–206.
9. A. Mills, N. Elliott, I. P. Parkin, S. A. O'Neill and R. J. Clark, *Journal of Photochemistry and Photobiology A-Chemistry*, 2002, **151**, 171–179.
10. D. Gogova, A. Iossifova, T. Ivanova, Z. Dimitrova and K. Gesheva, *Journal of Crystal Growth*, 1999, **199**, 1230–1234.
11. G. A. Niklasson and C. G. Granqvist, *Journal of Materials Chemistry*, 2007, **17**, 127–156.
12. S. T. Wang and Z. Zhang, *Progress in Chemistry*, 2003, **15**, 374–378.
13. T. D. Manning and I. P. Parkin, *Journal of Materials Chemistry*, 2004, **14**, 2554–2559.
14. T. D. Manning, I. P. Parkin, R. J. H. Clark, D. Sheel, M. E. Pemble and D. Vernadou, *Journal of Materials Chemistry*, 2002, **12**, 2936–2939.
15. T. D. Manning, I. P. Parkin, M. E. Pemble, D. Sheel and D. Vernardou, *Chemistry of Materials*, 2004, **16**, 744–749.
16. P. P. Edwards, A. Porch, M. O. Jones, D. V. Morgan and R. M. Perks, *Dalton Transactions*, 2004, 2995–3002.
17. I. Hamberg and C. G. Granqvist, *Journal of Applied Physics*, 1986, **60**, R123–R159.
18. M. Kamei, H. Enomoto and I. Yasui, *Thin Solid Films*, 2001, **392**, 265–268.

19. J. Kane, H. P. Schweizer and W. Kern, *Thin Solid Films*, 1975, **29**, 155–163.
20. K. Maki, N. Komiya and A. Suzuki, *Thin Solid Films*, 2003, **445**, 224–228.
21. T. Maruyama and K. Fukui, *Journal of Applied Physics*, 1991, **70**, 3848–3851.
22. L. A. Ryabova, V. S. Salun and I. A. Serbinov, *Thin Solid Films*, 1982, **92**, 327–332.
23. T. Maruyama and K. Fukui, *Thin Solid Films*, 1991, **203**, 297–302.
24. Z. B. Zhou, R. Q. Cui, Q. J. Pang, Y. D. Wang, F. Y. Meng, T. T. Sun, Z. M. Ding and X. B. Yu, *Applied Surface Science*, 2001, **172**, 245–252.
25. C. W. Kwon, G. Campet, J. Portier, A. Poquet, L. Fournes, C. Labrugere, B. Jousseaume, T. Toupance, J. H. Choy and M. A. Subramanian, *International Journal of Inorganic Materials*, 2001, **3**, 211–214.
26. J. E. Stanley, A. C. Swain, K. C. Molloy, D. W. H. Rankin, H. E. Robertson and B. F. Johnston, *Applied Organometallic Chemistry*, 2005, **19**, 644–657.
27. M. Kojima, H. Kato, A. Imai and A. Yoshida, *Journal of Applied Physics*, 1988, **64**, 1902–1905.
28. D. Belanger, J. P. Dodelet, B. A. Lombos and J. I. Dickson, *Journal of the Electrochemical Society*, 1985, **132**, 1398–1405.
29. C. G. Borman and R. G. Gordon, *Journal of the Electrochemical Society*, 1989, **136**, 3820–3828.
30. S. K. Ghandhi, R. Siviy and J. M. Borrego, *Applied Physics Letters*, 1979, **34**, 833–835.
31. T. H. Fang and W. J. Chang, *Applied Surface Science*, 2003, **220**, 175–180.
32. T. H. Fang and W. J. Chang, *Applied Surface Science*, 2005, **252**, 1863–1869.
33. H. L. Ma, D. H. Zhang, S. Z. Win, S. Y. Li and Y. P. Chen, *Solar Energy Materials and Solar Cells*, 1996, **40**, 371–380.
34. A. Newport, C. J. Carmalt, I. P. Parkin and S. A. O'Neill, *Journal of Materials Chemistry*, 2002, **12**, 1906–1909.
35. C. J. Carmalt, A. Newport, I. P. Parkin, P. Mountford, A. J. Sealey and S. R. Dubberley, *Journal of Materials Chemistry*, 2003, **13**, 84–87.
36. K. T. Rie, A. Gebauer, J. Wohle, H. K. Tonshoff and C. Blawit, *Surface & Coatings Technology*, 1995, **74–5**, 375–381.
37. G. S. Sandhu, S. G. Meikle and T. T. Doan, *Applied Physics Letters*, 1993, **62**, 240–242.
38. J. Driessen, A. D. Kuypers and J. Schoonman, *Journal of Vacuum Science & Technology A-Vacuum Surfaces and Films*, 2000, **18**, 1971–1976.
39. D. H. Kim, G. T. Lim, S. K. Kim, J. W. Park and J. G. Lee, *Journal of Vacuum Science & Technology B*, 1999, **17**, 2197–2203.
40. C. G. Granqvist, J. Luther, B. Orel and V. Wittwer, *Solar Energy*, 2000, **69**, 421.
41. S. Papaefthimiou, E. Syrrakou and P. Yianoulis, *Thin Solid Films*, 2006, **502**, 257–264.
42. R. D. Rauh, *Electrochimica Acta*, 1999, **44**, 3165–3176.
43. C. Bechinger, J. N. Bullock, J. G. Zhang, C. E. Tracy, D. K. Benson, S. K. Deb and H. M. Branz, *Journal of Applied Physics*, 1996, **80**, 1226–1232.
44. U. Tritthart, W. Gey and A. Gavrilyuk, *Electrochimica Acta*, 1999, **44**, 3039–3049.
45. A. Henningsson, A. Stashans, A. Sandell, H. Rensmo, S. Sodergren, H. Lindstrom, L. Vayssieres, A. Hagfeldt, S. Lunell and H. Siegbahn, *in Advances in Quantum Chemistry*, 2004, **47**, 23–36.
46. P. V. Ashrit, *Thin Solid Films*, 2001, **385**, 81–88.
47. A. Antonaia, T. Polichetti, M. L. Addonizio, S. Aprea, C. Minarini and A. Rubino, *Thin Solid Films*, 1999, **354**, 73–81.
48. O. Bohnke and B. Vuillemin, *Materials Science and Engineering B-Solid State Materials for Advanced Technology*, 1992, **13**, 243–246.
49. C. Bohnke and M. Rezrazi, *Materials Science and Engineering B-Solid State Materials for Advanced Technology*, 1991, **10**, 313–320.
50. O. Bohnke and C. Bohnke, *Displays*, 1988, **9**, 199–206.

51. C. G. Granqvist, *Handbook of Inorganic Electrochromic Materials*, 1995, Elsevier Science and Technology, Amsterdam.
52. C. G. Granqvist, *Solar Energy Materials and Solar Cells*, 2000, **60**, 201–262.
53. A. Donnadieu, *Materials Science and Engineering B-Solid State Materials for Advanced Technology*, 1989, **3**, 185–195.
54. X. H. Yu, J. B. Bates, G. E. Jellison and F. X. Hart, *Journal of the Electrochemical Society*, 1997, **144**, 524–532.
55. P. Gerard, A. Deneuville, G. Hollinger and T. M. Duc, *Journal of Applied Physics*, 1977, **48**, 4252–4255.
56. C. Bechinger, M. S. Burdis and J. G. Zhang, *Solid State Communications*, 1997, **101**, 753–756.
57. C. Bechinger, D. Ebner, S. Herminghaus and P. Leiderer, *Solid State Communications*, 1994, **89**, 205–207.
58. C. Bechinger, S. Herminghaus and P. Leiderer, *Thin Solid Films*, 1994, **239**, 156–160.
59. C. Bechinger, G. Oefinger, S. Herminghaus and P. Leiderer, *Journal of Applied Physics*, 1993, **74**, 4527–4533.
60. A. I. Gavrilyuk, *Electrochimica Acta*, 1999, **44**, 3027–3037.
61. F. A. Cotton and G. Wilkinson, 1980.
62. R. U. Kirss and L. Meda, *Applied Organometallic Chemistry*, 1998, **12**, 155–160.
63. C. S. Blackman and I. P. Parkin, *Chemistry of Materials*, 2005, **17**, 1583–1590.
64. O. Bohnke, C. Bohnke, A. Donnadieu and D. Davazoglou, *Journal of Applied Electrochemistry*, 1988, **18**, 447–453.
65. D. Davazoglou, G. Leveque and A. Donnadieu, *Solar Energy Materials*, 1988, **17**, 379–390.
66. D. Davazoglou and A. Donnadieu, *Thin Solid Films*, 1987, **147**, 131–142.
67. B. Yous, S. Robin, A. Donnadieu, G. Dufour, C. Maillot, H. Roulet and C. Senemaud, *Materials Research Bulletin*, 1984, **19**, 1349–1354.
68. K. A. Gesheva, T. Ivanova, G. Popkirov and F. Hamelmann, *Journal of Optoelectronics and Advanced Materials*, 2005, **7**, 169–175.
69. W. B. Cross, I. P. Parkin, S. A. O'Neill, P. A. Williams, M. F. Mahon and K. C. Molloy, *Chemistry of Materials*, 2003, **15**, 2786–2796.
70. R. G. Palgrave and I. P. Parkin, *Journal of Materials Chemistry*, 2004, **14**, 2864–2867.
71. S. O'Neill, I. P. Parkin, R. J. H. Clark, A. Mills and N. Elliott, *Chemical Vapor Deposition*, 2004, **10**, 136-+.
72. D. Gogova, K. Gesheva, A. Szekeres and M. Sendova-Vassileva, *Physica Status Solidi A-Applied Research*, 1999, **176**, 969–984.
73. T. Maruyama and S. Arai, *Journal of the Electrochemical Society*, 1994, **141**, 1021–1024.
74. W. B. Cross and I. P. Parkin, *Chemical Communications*, 2003, 1696–1697.
75. W. B. Cross, I. P. Parkin, A. J. P. White and D. J. Williams, *Dalton Transactions*, 2005, 1287–1293.
76. F. J. Morin, *Physical Review Letters*, 1959, **3**, 34–36.
77. C. G. Granqvist, *Thin Solid Films*, 1990, **193**, 730–741.
78. A. Zylbersztejn and N. F. Mott, *Physical Review B*, 1975, **11**, 4383–4395.
79. G. S. Nadkarni and V. S. Shirodkar, *Thin Solid Films*, 1983, **105**, 115–129.
80. M. B. Sahana, G. N. Subbanna and S. A. Shivashankar, *Journal of Applied Physics*, 2002, **92**, 6495–6504.
81. J. F. Denatale, P. J. Hood and A. B. Harker, *Journal of Applied Physics*, 1989, **66**, 5844–5850.
82. M. N. Field and I. P. Parkin, *Journal of Materials Chemistry*, 2000, **10**, 1863–1866.
83. T. D. Manning, I. P. Parkin, C. Blackman and U. Qureshi, *Journal of Materials Chemistry*, 2005, **15**, 4560–4566.
84. T. D. Manning and I. P. Parkin, *Polyhedron*, 2004, **23**, 3087–3095.

85. A. Mills and S. LeHunte, *Journal of Photochemistry and Photobiology A-Chemistry*, 1997, **108**, 1–35.
86. N. P. Mellott, C. Durucan, C. G. Pantano and M. Guglielmi, *Thin Solid Films*, 2006, **502**, 112–120.
87. A. Mills, S. Hodgen and S. K. Lee, *Research on Chemical Intermediates*, 2005, **31**, 295–308.
88. A. Mills and J. S. Wang, *Journal of Photochemistry and Photobiology A-Chemistry*, 1998, **118**, 53–63.
89. A. Mills and J. S. Wang, *Journal of Photochemistry and Photobiology A-Chemistry*, 1999, **127**, 123–134.
90. A. Mills, G. Hill, M. Crow and S. Hodgen, *Journal of Applied Electrochemistry*, 2005, **35**, 641–653.
91. A. Mills, G. Hill, S. Bhopal, I. P. Parkin and S. A. O'Neill, *Journal of Photochemistry and Photobiology A-Chemistry*, 2003, **160**, 185–194.
92. S. A. O'Neill, I. P. Parkin, R. J. H. Clark, A. Mills and N. Elliott, *Journal of Materials Chemistry*, 2003, **13**, 56–60.
93. S. A. O'Neill, R. J. H. Clark, I. P. Parkin, N. Elliott and A. Mills, *Chemistry of Materials*, 2003, **15**, 46–50.
94. S. Mathur, V. Sivakovs, H. Shen, S. Barth, C. Cavelius, A. Nilsson and P. Kuhn, *Thin Solid Films*, 2006, **502**, 88–93.
95. V. G. Besserguenev, R. J. F. Pereira, M. C. Mateus, I. V. Khmelinskii, R. C. Nicula and E. Burkel, *International Journal of Photoenergy*, 2003, **5**, 99–105.
96. M. Karches, M. Morstein, P. von Rohr, R. L. Pozzo, J. L. Giombi and M. A. Baltanas, *Catalysis Today*, 2002, **72**, 267–279.
97. E. Halary, G. Benvenuti, F. Wagner and P. Hoffmann, *Applied Surface Science*, 2000, **154**, 146–151.

CHAPTER 11

Photo-assisted CVD

STUART J. C. IRVINE AND DAN LAMB

Centre for Solar Energy Research, OpTIC Technium/NEWI, Ffordd William Morgan, SE Asaph Business Park, SE Asaph LL17 OJD, North Wales

11.1 Introduction

There are different ways of enhancing the growth rate of CVD or modifying the film properties using various forms of energy assistance. CVD relies on thermal chemistry to bring about a reaction of the precursors either in the gas phase or on the film surface. In some cases it is necessary to achieve thin film deposition at a temperature that is below an efficient pyrolysis temperature and this process can normally be represented by the Arrhenius relationship shown in Equation (11.1):

$$\frac{dz}{dt} = A_0 e^{E_a/RT} \tag{11.1}$$

This is an empirical relationship that may depend on a complex sequential reaction both in the vapour and on the surface. The rate-determining step, for low temperature growth, is normally reaction on the surface, often referred to as the "kinetic growth regime". Energy assistance in the vapour can enhance growth rate through delivery of more reactive species on the surface. Energy assistance on the surface can either be thermal in nature or surface photocatalytic. This chapter considers different photo-assisted processes as a sub-set of energy assisted CVD. The purpose is to create conditions where acceptable deposition rates can be achieved at relatively low substrate temperature. The thermal processes are still important and photo-assistance can be best viewed as a way of modifying the deposition kinetics. It is for this reason that the properties of the film are also modified, in some cases producing higher quality films or enabling formation of a metastable phase. The basic principles of photo-assisted deposition will be considered and examples given of how these are applied in different thin film systems.

11.2 Principles of Photo-assisted CVD

This section covers each of the main mechanisms by which photo-assistance can stimulate the CVD process. As indicated in the introduction, the key to modifying growth rates is the Arrhenius

Chemical Vapour Deposition: Precursors, Processes and Applications
Edited by Anthony C. Jones and Michael L. Hitchman

Published by the Royal Society of Chemistry, www.rsc.org

expression and it will be used as the basis to explain the kinetic process involved for each process. There are extensive texts that can provide useful reference and further background reading on each of these topics, *e.g.* ref. 1.

11.2.1 Photothermal Processes

The principle of photothermal processes is to create local heating on a substrate using a focussed laser beam. This is often pulsed or scanned across the surface to minimize the average heat loading on the substrate while creating localized growth of a thin film. The origins of this technology were in depositing metal strips down to a few microns width onto a semiconductor or insulating substrate.[1,2] More recent work has explored the potential for photothermal processes in the formation of rods and other micro-structures. The emphasis on photothermal processes is on local area deposition while keeping the background temperature low. The kinetics are essentially the same as for a conventional thermal CVD process and this can be illustrated using the Arrhenius plot in Figure 11.1.

One exception to this simplistic view is that gas phase transport can be more rapid as the consumption of precursors on the surface is very low due to the localized nature of the deposition. Therefore, the normal boundary layer depletion and diffusion mechanisms do not apply but can be modelled by a more rapid 3D diffusion process.

A typical photothermal CVD experimental set-up is shown schematically in Figure 11.2. The key components of the system are the laser with focus optics and the deposition chamber, which is on a micro-position stage. The laser beam is scanned across the surface in a suitable pattern by x-y translation of the reaction chamber. The laser can be either cw or pulsed. In cw mode and rapid pulse mode the photothermal process will deposit a line. In slower pulse mode the deposit will be a series of dots that could be grown further into rods. Thus, this is a very powerful method for patterning a deposit across a surface.

The choice of the laser source depends on whether the requirement is for a pulse or cw beam. In both cases, the power density needs to be high enough to give a transient temperature rise of a few hundred degrees Celsius. The most versatile pulsed source is the Nd:YAG laser, which has a primary emission wavelength at 1.06 μm but can be readily doubled or tripled to emit in the visible or the UV. The most commonly used cw source is the argon ion laser, which has a primary green line at 514 nm but also has useful lines extending to shorter wavelengths. The choice of wavelength is determined by the absorption characteristics of the substrate and the growing film.

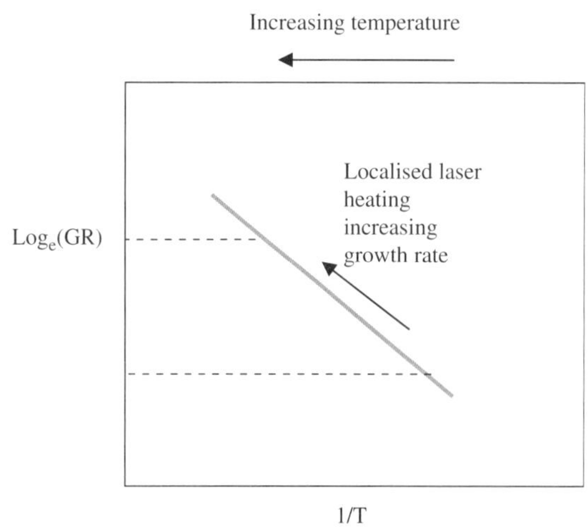

Figure 11.1 Representation of photothermal CVD using an Arrhenius plot.

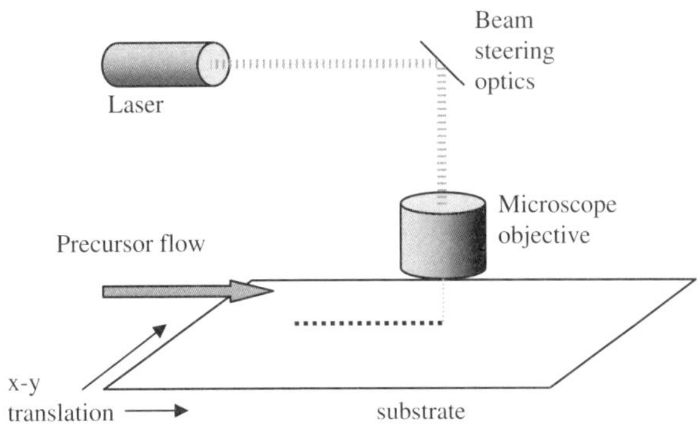

Figure 11.2 Schematic of a photothermal CVD experimental setup.

One problem with metal films is the high reflectivity, which will correspondingly reduce the absorbed power. If the substrate is a semiconductor then the chosen wavelength has to be shorter than the bandgap maximum absorption wavelength. The most efficient temperature rise will be for the laser energy to be absorbed close to the surface – absorption at greater depths in the substrate will have a decreasing effect on increasing the surface temperature. Another important property of the substrate is the thermal diffusion coefficient, where a high thermal diffusion coefficient will require more laser power to produce the required surface temperature rise. These factors can best be understood by inspecting the equation for temperature rise at the centre (coordinates 0,0) of a Gaussian profile laser beam, shown in Equation (11.2):

$$T(0,0,t) = \frac{P(1-R)}{\pi^{3/2}\omega K}\arctan\left[\frac{4Dt}{\omega^2}\right]^{1/2} \tag{11.2}$$

The temperature is represented by T, the laser power by P, reflectivity of the surface R, thermal diffusion coefficient D, time of laser exposure t, $1/e$ beam width ω, and K is the thermal conductivity of the substrate. The reduction in the beam width ω will have a large effect on the temperature rise, so a good focussing system is important. Similarly, as the thermal conductivity of the substrate (K) is reduced the temperature will rise. The extent of temperature rise will depend on the growth kinetics represented in Figure 11.1. The slope of the Arrhenius relationship in Figure 11.1 depends on the activation energy, such that a large activation energy will require a smaller change in temperature to bring about selective CVD. The time taken to gain the critical temperature rise will determine the rate at which a line can be written by photothermal CVD. Clearly, a higher power laser with minimum beam width will give the maximum line writing speed.

Precursors used in photothermal CVD are essentially the same as would be used in conventional CVD processes; for metal deposition, either fluorides or carbonyls would be used. For germanium and silicon, either chlorides or hydrides can be used and similarly for their oxides. When depositing a compound it is important that the combination of precursors do not cause a room temperature reaction as this would lead to poor localization of the deposit. The use of conventional precursors is one advantage of photothermal CVD.

11.2.2 Photolysis

Low temperature thin-film deposition, below the pyrolysis threshold temperature, can also be achieved by photolytic reaction of the precursors. This is a vapour phase reaction to give partial or complete bond homolysis of the precursors, yielding a more reactive species on the surface. In

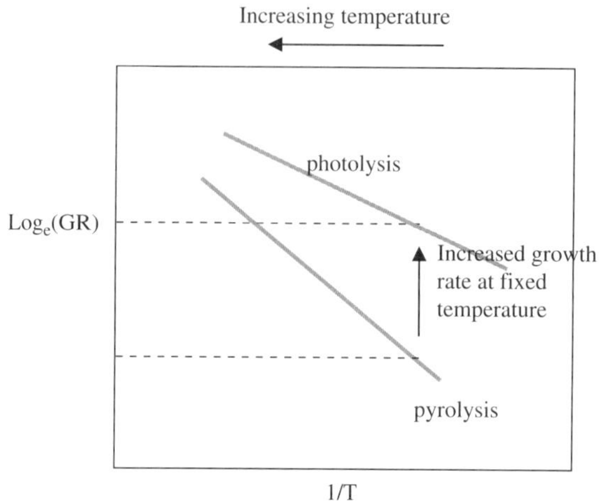

Figure 11.3 Representation of photolysis using an Arrhenius plot.

principal, this can be achieved at room temperature as the cleavage of ligands to release the metal atom can occur without the input of heat energy. In general this is not dependant on the bond energies, provided the photon energy is sufficient to break the bond. In contrast with photothermal processes, photolysis will give rise to a different Arrhenius line with a reduced activation energy. The temperature of the thermal reaction does not change (Figure 11.3).

Photolysis efficiency depends on the absorption cross section of the precursor to create an electronically excited state. The absorption cross section is critically dependant on the excitation wavelength and the molecular absorption bands of the precursor.[3] For many of the simple alkyl organometallics, such as Me_2Cd and Me_2Zn, this absorption occurs in the UV between 200 and 300 nm. Organometallics such as Me_2Zn and Et_3Ga have absorption bands between 200 and 250 nm with peak cross sections in excess of $1 \times 10^{-17}\,\text{cm}^{-2}$. These absorption bands are broad, with vibrational/rotational bands superimposed on the electronic transitions, so any wavelength within the band is likely to be effective in photolysis.[4] Hydrides have peak absorption wavelengths below 200 nm, *e.g.* silane peaks at 115 nm and arsine peaks at 180 nm. A detailed survey of absorption properties is given by M. Rothschild in Chapter 3 of ref. 5. Table 11.1 shows absorption cross sections and suitable radiation sources for excitation of these precursors. In general, a range of lamp and laser sources exist for the 200–250 nm band but are less available below 190 nm, mainly because of the poor transmission through silica at these wavelengths. As can be seen later, windowless lamps can be used to overcome these difficulties but a simpler arrangement is more desirable. Most practical photochemical CVD reactors would require strong absorption cross sections ($>5 \times 10^{18}\,\text{cm}^2$) above 190 nm. The photon absorption can be simply calculated from the absorption cross section (σ_a), partial pressure (p) and optical path length (L) to yield a reaction rate R_a, in the vapour, [Equation (11.3)]:

$$R_a = (I_O/h\nu)F\exp(-\sigma_a pL/kT) \quad (11.3)$$

where I_o is the incident UV intensity, $h\nu$ is the photon energy, F is the photochemical quantum efficiency, k is Boltzmann's constant and T is the vapour temperature in kelvin. This is a modified Beer's law relationship for absorption in the gas that takes into account the proportion of excited molecules undergoing bond homolysis. In other words, this gives the practical photolysis yield in a photolysis reactor. Figure 11.4 shows this process schematically.

Table 11.1 Absorption cross sections and suitable radiation sources for a selection of common precursors used in CVD.

Precursor	*Absorption wavelength (nm)*	*Peak cross section (cm^{-2})*	
SiH_4	115	1×10^{-16}	Only 2-photon at 193 nm
SiH_2Cl_2	151	1.8×10^{-17}	
$SiHCl_3$	110–140		
$Si_2(CH_3)_6$	138, 163, 191		
Si_2H_6	115–135, 165	1.5×10^{-16}, 5×10^{-17}	Weak 1-photon at 193 nm
GeH_4	120		
PH_3	183	4.0×10^{-17}@193 nm	
$P(CH_3)_3$	202	3.4×10^{-17}@193 nm	
$P(C_2H_5)_3$	210	8.5×10^{-18}@193 nm	
AsH_3	180		
$As(CH_3)_3$	200	4×10^{-17}@193 nm	
$As(C_2H_5)_3$	210		
SbH_3	197		
H_2S	193	6×10^{-18}	
H_2Se	197		
$(CH_3)_2Te$	200, 249	7.4×10^{-17}	
$(C_2H_5)_2Te$	198, 220, 247	1.1×10^{-16}@247 nm	
$(CH_3)_2Zn$	200	4×10^{-17}	
$(C_2H_5)_2Zn$	220		
$(CH_3)_2Cd$	215	4×10^{-17}	
$(CH_3)_2Hg$	200	4×10^{-17}	
$(CH_3)_3Al$	190	2×10^{-17}@193 nm	
$(C_2H_5)_3Al$	190–210	4×10^{-18}	
$(CH_3)_3Ga$	200	2×10^{-17}@193 nm	
$(CH_3)_3In$	210	1×10^{-17}@193 nm	
$(C_2H_5)_3In$	230		
$(CH_3)_4Sn$	186		

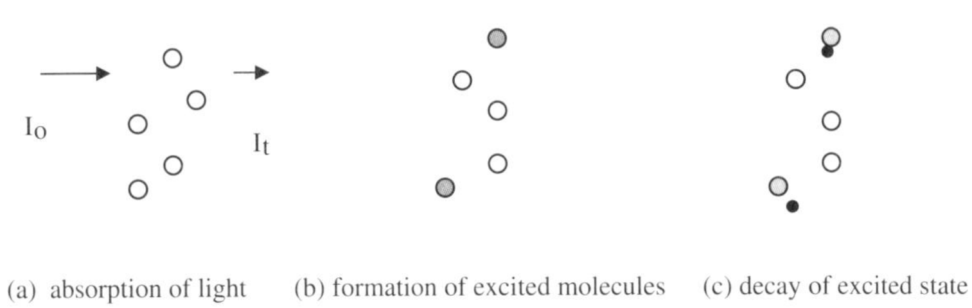

Figure 11.4 Schematic of vapour photolysis of an organometallic.

The simplest photolysis reaction for P-CVD is the one-photon photoabsorption–dissociation reaction. An example is the photolysis of DMCd:

$$(CH_3)_2Cd + h\nu \rightarrow CH_3 + CH_3Cd \rightarrow 2CH_3 + Cd \tag{11.4}$$

The first step is the excitation of the organometallic into an excited electronic state that is either antibonding or has sufficient vibrational energy to cleave a radical.[6] The bond energy for the second radical is 21 kcal mol^{-1}, compared with 46 kcal mol^{-1} for the first step and is sufficiently weak that it will cleave thermally, to yield Cd atoms that can diffuse to the surface.[7] The photodissociation process is efficient (η ∼ 1) and the kinetics are first order. The photofragments do not always yield the metal atoms so readily in other examples such as Et_3Ga, which loses two methyl radicals in the primary photoabsorption but requires a secondary photoabsorption to cleave the remaining radical. This differs from the DMCd example in that the most stable radical is the last one to be removed and is therefore the limiting reaction step:[8]

$$Ga(CH_3)_3 + h\nu \rightarrow GaCH_3 + 2CH_3 \tag{11.5}$$

$$GaCH_3 + h\nu \rightarrow Ga + CH_3 \tag{11.6}$$

More complex multi-step photodissociation is encountered with metal carbonyls, which are a common source of precursors for PCVD of transition metals. The decomposition path also depends on the photon energy absorbed. For example, chromium hexacarbonyl can lose one or two CO ligands according to the photon energy. At 351 nm only one CO ligand is released, Equation (11.7):

$$Cr(CO)_6 + h\nu \rightarrow Cr(CO)_5 + CO \tag{11.7}$$

At 248 nm two CO ligands are released *via* an excited photo-fragment, Equations (11.8) and (11.9):

$$Cr(CO)_6 + h\nu \rightarrow Cr(CO)_5^* + CO \tag{11.8}$$

$$Cr(CO)_5^* \rightarrow Cr(CO)_4 + CO \tag{11.9}$$

Precursors that have very short wavelength absorption bands can be excited *via* two-photon absorption at longer wavelengths, as is the case for SiH_4, which can dissociate *via* two-photon absorption from a 193 nm excimer laser.

The photolysis reactions considered in this section are all vapour phase reactions that are employed as a means to yield metal atoms that will diffuse to the surface. These processes are suitable for low temperature growth but not for localized deposition which requires surface photo-stimulated reactions. The photon absorption in the vapour can be very high, with most of the photons absorbed, but a monolayer coverage on the surface will only give a few percent efficiency with absorption cross section $\sim 1 \times 10^{-17}\,cm^{-2}$. This has a consequence of reducing the growth rate when compared with vapour photolysis and would be too slow for laser writing applications. Selected area deposition can be achieved by projection imaging of a pattern with UV radiation onto the substrate. The substrate is kept at a low temperature, where the surface coverage will be high, for good surface selectivity. The growth rate increases for lower substrate temperatures, going through a maximum around one monolayer coverage. This approach has given the smallest feature sizes, with sub-micron resolution demonstrated by Ehrlich and Tsao,[9] and is ultimately constrained by the diffraction limited spot size for the light wavelength. The feature sizes can be smaller than for photothermal CVD but deposition rates are much lower.

Reaction chambers for photolysis-assisted CVD are similar to more conventional CVD chambers with either a horizontal or vertical chamber. The main difference is the optical window, which needs to be kept clear of reaction products. This can be very challenging for atmospheric pressure systems but practical arrangements based on flow separation and window purging have been used (see Stafford *et al.*[10]).

11.2.3 Photocatalysis

The efficiency of light absorption in the adsorbate is poor compared with vapour absorption, so photolysis is largely a vapour phase reaction. Other methods of energy assistance have to be found if either the available radiation absorption is poor in the precursors or the energy assistance needs to be surface selective. Poor absorption may occur due to the electronic absorption band being too short a wavelength for the readily available visible and UV radiation sources. In these cases, semiconductor substrates can be used to absorb the incident radiation by using a photon energy that is greater than the bandgap (short wavelength), normally in the blue or the near UV. The absorption of light in a semiconductor creates an electron–hole pair that will eventually recombine, either by the emission of light (fluorescence) or by the release of energy through heat. However, if the minority carrier, *i.e.* electrons in a p-type semiconductor, can diffuse to the surface before recombining then it will change the charge state of the surface. This can assist in the catalytic action of the surface by enhancing the desorption rate of organic products or by promoting reaction between the organic ligands to create a stable organic leaving group. As with the previous examples, all that is required is a modification to the low temperature kinetics of film deposition to have some dramatic effects on growth rate.

A test for photocatalytic enhancement of growth rate is to measure the growth rate as a function of the incident light wavelength; an enhancement will be observed for shorter wavelengths than that corresponding to the bandgap wavelength:

$$\lambda = hc/E_g \tag{11.10}$$

Where λ is the wavelength of the incident radiation, h is Planck's constant, c is the velocity of light and E_g is the bandgap energy of the semiconductor substrate. This method was used by Fujita *et al.*[11] for the growth of ZnSe. For a bandgap of 2.6 eV a visible blue/green laser is sufficient to generate the electron–hole pairs. The excess charge transfer to the surface helps the desorption of organic products, which in turn enables the growth to occur at lower temperatures than the normal thermal processes would allow. Using the example of ZnSe, the growth rate normally decreases with temperature below 600 °C when using the precursors Me_2Zn and Me_2Se. Fujita *et al.*[11] used a 500 W xenon arc lamp, with filters to avoid heating of the substrate, to realise growth rates above 1 μm h^{-1} down to 300 °C. Figure 11.5 shows the wavelength dependence of growth rate at 400 °C, clearly illustrating the significance of the bandgap energy. The 458 nm line of the argon ion laser is also a suitable light source for enhancing the growth of ZnSe.[12] Other materials that have been grown using light induced charge transfer are GaAs[13], ZnTe[14] and CdTe.[8]

The mechanisms for photocatalysis can be subtle and are not well understood in terms of the observed enhancement in growth rate. The best way of looking at these mechanisms is to start with the thermal reaction mechanisms and see where the photocatalysis can make a difference in terms of enhancing the film deposition rate. For ZnSe it is desirable to grow at low temperature (around 300 °C) to avoid the formation of point defects. The original work on MOVPE of ZnSe[15] was carried out using Me_2Zn and H_2Se in a stream of hydrogen carrier gas. The hydride reacts with the organometallic in the vapour at room temperature *via* the formation of an adduct. Although this was successful in achieving a low growth temperature, the strong vapour phase reaction was difficult to control and alternatives were sought to avoid this. One route was to use an organoselenium that was less likely to form an adduct bond with the zinc precursor. However, the thermal stability of such selenium precursors was blamed for an increase in practical growth temperatures to in excess of 400 °C. Looking at the alkane series, starting with Me_2Se, the C–Se bond becomes weaker with heavier alkyl ligands in the series:

$$Me_2Se > Et_2Se > iPr_2Se > tBu_2Se$$

Each of the above selenium precursors have been used in combination with a zinc precursor, the adduct $Me_2Zn.NEt_3$, to determine growth rate characteristics.[16–18] With the exception of tBu_2Se, the other precursors do not yield significantly different growth rate curves, as shown in Figure 11.6 for Me_2Se and Et_2Se, despite the Se–C bond energies being 66 and 62 kcal mol^{-1} respectively.

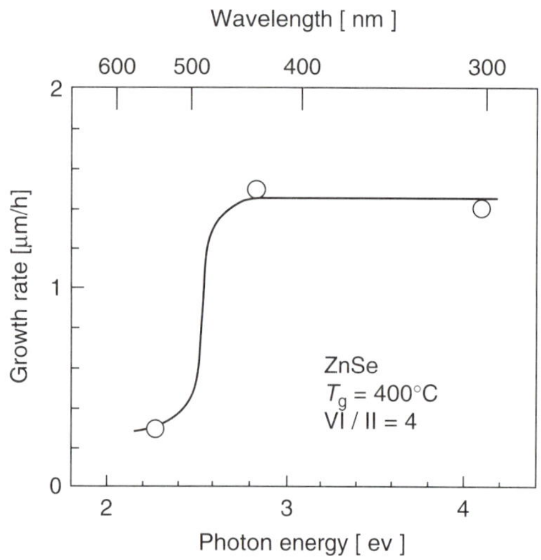

Figure 11.5 Wavelength and energy dependence of growth rate in photo-assisted MOVPE of ZnSe. (After Fujita *et al.*[11])

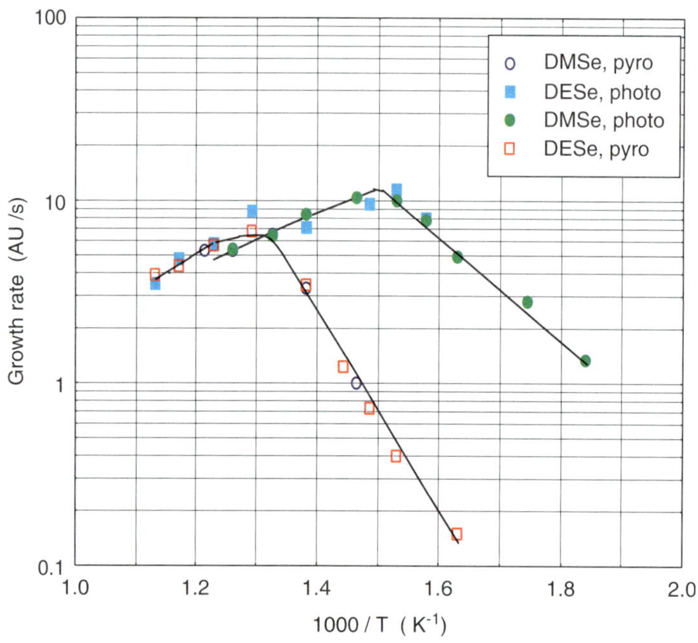

Figure 11.6 Plot of growth rate *versus* the inverse of the substrate temperature for ZnSe growth. (After Ahmed *et al.*[20])

The different behaviour of tBu_2Se can be attributed to a decomposition path *via* β-hydrogen elimination. The similarity of the pyrolytic (*i.e.* without bandgap illumination) curves illustrates that the carbon–Se bond homolysis is not the critical kinetic step. There are two parts to these curves, high and low temperature parts, which are distinguished by different slopes. The low temperature part (high $1000/T$) is indicative of a kinetic step in the growth of the ZnSe film and the high temperature part is associated with desorption of unreacted precursors and has a positive activation energy.[19]

Plots for the photo-catalytic growth show that there is a large enhancement for the low temperature, kinetically limited, growth regime. Note that this is quite different to a photothermal process, which would move the kinetics up the same Arrhenius line, but is similar to photolysis, creating a different Arrhenius plot with different activation energy. The illumination, in this example, was from the 468 nm line of an argon ion laser with intensity of 50–100 mW cm^{-2}. Again there appears to be very little difference in the growth rate for the two precursors. The activation energies are much smaller than the bond homolysis activation energies (Table 11.2), therefore this is not the critical step in the photo-catalytic growth of ZnSe.

Ahmed *et al.*[20] have proposed a model for the photo-assisted growth of ZnSe that assumes that the initiation step is triggered by the absorption of light in the semiconductor substrate and sets off a chain reaction that results in the decomposition of both the Zn and Se precursor. Bearing in mind that the pyrolysis efficiency of the precursors is low at the typical photo-catalytic temperature of 300 °C, the Me_2Zn precursor adsorbed onto the surface will decompose slowly by thermal processes alone. The charge transfer, induced by the photo-absorption in the substrate, will assist in removing Me radicals from the adsorbed Me_2Zn. This in turn will allow chain reactions based on further reaction of the released Me radical. Much of these processes are likely to be the same as for the normal thermal reaction that occurs at higher temperature.

A defining experiment was carried out by Ahmed and Irvine[21] to see whether there was a hydrogen radical step for low temperature growth of ZnSe. The hydrogen radicals are formed from the reaction of methyl radicals with the hydrogen carrier gas. The experiment was to replace the hydrogen carrier gas with helium after growing an initial nucleation layer in the presence of hydrogen. This ensured that the surface was the same for both experiments (with and without hydrogen) and, thereby, that the catalytic effect of the surface was the same. *In situ* laser interferometry was used to gain a real time indication of whether growth was occurring and, if it was, to determine the growth rate.

The initial ZnSe nucleation layer was grown thermally at 480 °C. This ensured a uniform ZnSe buffer layer for photo-absorption of the 458 nm radiation for subsequent photo-assisted growth. The carrier gas was then switched to He and the layer cooled to 340 °C, a typical temperature for the photo-catalytic growth of ZnSe.[21] The laser intensity was increased from 122 to 158 mW cm^{-2} and at each intensity the precursors were left flowing over the surface for 5 min. During this experiment there was no significant change in the interferometer signal, indicating no growth of the ZnSe film, even under conditions of higher laser intensity, approximately three times the intensity normally used for photo-assisted growth with these precursors. This experiment was repeated several times at different temperatures from 340 to 420 °C and a final proof where the carrier gas

Table 11.2 Comparison of activation energies derived from growth rate plots for ZnSe with the bond homolysis energies for the Se precursors (after Ahmed *et al.*[17]).

Conds	*Temperature range (°C)*	E_{act} *for growth (kcal mol^{-1})*	E_{act} *for alkyl–Se bond (kcal mol^{-1})*
DESe Pyro	340–500	−23.4	62
	500–610	+7.11	
DESe Photo	340–380	−12.71	
	380–610	+4.96	
DMSe Pyro	410–480	−23.4	66
DMSe Photo	270–380	−12.71	
	380–500	+4.98	

was switched from hydrogen to helium during growth at 380 °C. In every case, there was no measurable growth with He carrier gas. These results show conclusively that hydrogen carrier gas is essential for the photo-assisted growth of ZnSe and supports the hydrogen radical mechanism.[21] This example illustrates that photocatalysis is likely to enhance growth rate *via* the normal thermochemical reactions. The growth rate enhancement was confirmed by Yu,[22] who also was able to show that the crystalline quality was unchanged by the photocatalysis for an equivalent growth rate to thermal growth, consistent with stimulation of the thermal kinetics.

11.2.4 Photosensitization

Many commonly used precursors, such as the hydrides, have very short wavelength UV absorption bands that are not readily compatible with the readily available UV sources. In these cases, an alternative to primary photolysis is desirable and one of these alternative approaches is to use a photosensitizer to transfer the photon energy to the precursor. This is particularly useful where the precursor does not have a strong absorption band close to a laser or lamp source, such as for silane and disilane. Tarui *et al.*[23] used mercury photosensitization to grow amorphous silicon films from disilane and a mercury vapour mixture. The 185 nm line from a mercury arc lamp was used to excite the mercury atoms in the vapour into the $Hg(^1P_1)$ state. The energy available for transfer to a disilane molecule is 6.7 eV, which is sufficient for dissociation. Photosensitization promotes the disilane molecule into an excited state at a photon wavelength that is very weakly absorbed in the disilane, where the nearest absorption band is at 165 nm. Figure 11.7 shows this process schematically.

Mercury is a commonly used photosensitizer because the mercury arc lamp is a convenient and cheap source of UV radiation. Other applications include the Silox process for the growth of SiO_2 films *via* the photosensitized decomposition of N_2O:[24]

$$Hg + h\nu \rightarrow Hg^* \tag{11.11}$$

$$Hg^* + N_2O \rightarrow N_2 + O(^3P) + Hg \tag{11.12}$$

$$2O + SiH_4 \rightarrow SiO_2 + 2H_2 \tag{11.13}$$

There are issues with this approach of incorporation of Hg impurity into the growing films. Mercury photosensitization has also been used for the growth of mercury-containing compounds where this is not an issue. Photo-assisted growth, using a high-pressure Hg arc lamp, in an ambient of Hg of approximately 10 mbar has been used to stimulate the growth of HgTe films at 200 °C, which is up to 200 °C below the normal pyrolysis temperature.[25]

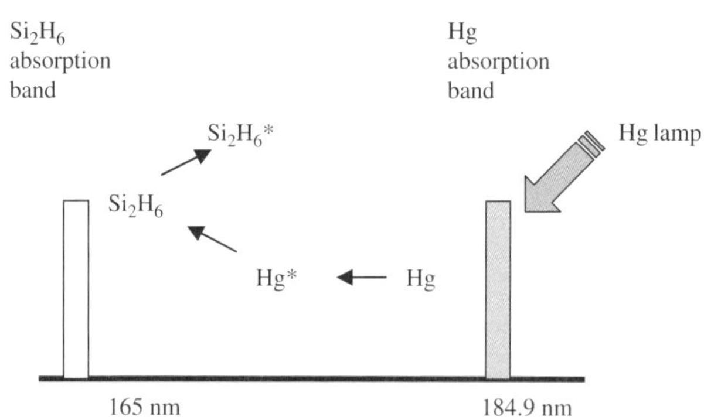

Figure 11.7 Schematic of the Hg photosensitization process for disilane.

A more recent example is the low temperature growth of SiC films reported by Myong *et al.*[26] The precursors were SiH_4, C_2H_4 and B_2H_6 for boron doping. None of these precursors would normally photo-dissociate with the 184.9 and 253.7 nm lines of a low pressure Hg arc lamp. Deposition was carried out in the temperature range 80–350 °C, hundreds of degrees below pyrolysis temperatures. Nano-crystalline films were obtained at the lower temperature end of this range and good electrical conductivity at just 120 °C. These authors used the photosensitized SiC film as a window layer in a thin film a-Si solar cell. The very low temperature deposition would enable the formation of thin film solar cells onto plastic substrates.

11.3 Lamps and Lasers for Photolysis

Photolysis within CVD depends upon the interaction of a photon, of sufficient energy, to cause electronic transitions with one or more of the precursors, resulting in dissociation of the precursor molecule. The possibility of this process occurring depends upon matching the precursors absorption cross-section and the photon energy. Most CVD precursors have absorption cross-sections in the UV-C (100–280 nm) and as a result the energy required to produce photolysis is usually obtained from a laser source or equivalent discharge lamp. Table 11.3 shows the most commonly employed sources and their respective wavelengths.

In addition to laser sources, lamps such as mercury, deuterium and xenon provide a broad spectral range with photolysis inducing intensities in the UV. Of these, the mercury lamp is more commonly used due to its intense peak at 254 nm. Figure 11.8 shows the spectral irradiance for three different Hg power lamps.

When moving to far-UV absorbing precursors the means of transmitting the light source to the deposition zone become more challenging. Synthetically fused quartz such as Suprasil® can transmit wavelengths as short as 150 nm and will enable the use of excimer lamps such as Xe_2*. For shorter wavelength sources such as F_2, Kr_2* and Ar_2* metal-fluoride window materials of the type MgF_2, CaF_2 and LiF can be employed. Depending on experimental design, an alternative for the shorter wavelength sources may be to use a windowless system employing an internal plasma discharge (Figure 11.9). When an electric field is generated between the two electrodes the carrier gas, passing through the field, becomes ionized and subsequently loses energy in the form of photons.

Table 11.3 Excimer sources and their associated peak energy wavelengths.

Excimer	*Wavelength (nm)*
He_2*	74
Ne_2*	83
Ar_2*	126
Kr_2*	146
F_2	157
Xe_2*	172 & 175
ArF	193
CaF_2	193
KrCl	222
KrF	248
Cl_2	259
XeBr	282
XeCl	308
N_2	337
XeF	351

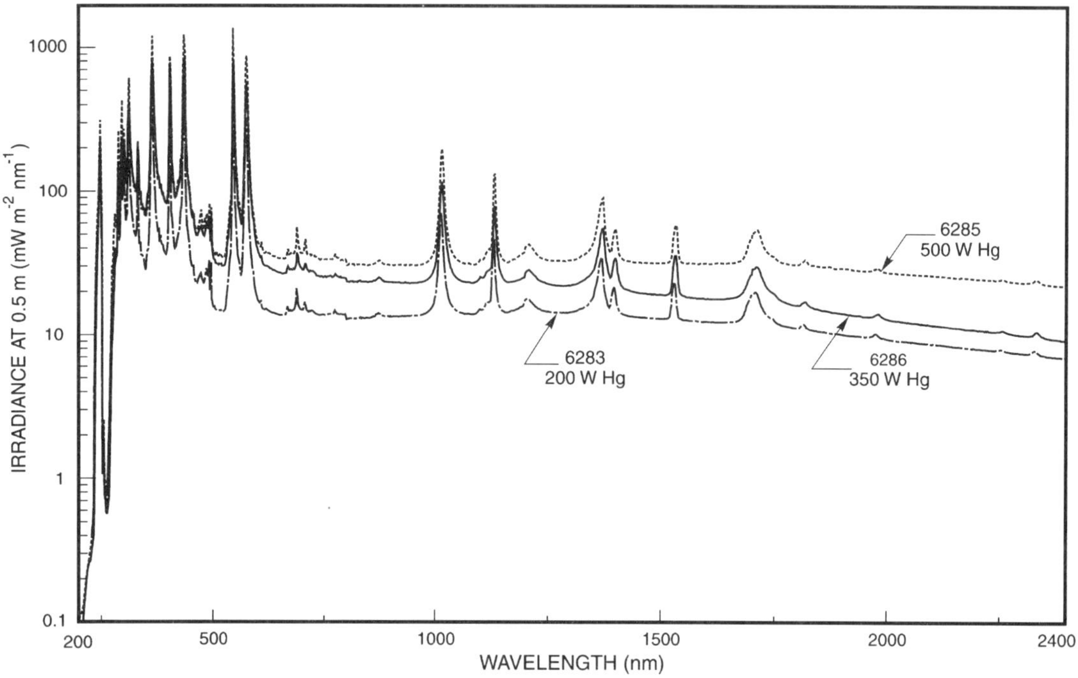

Figure 11.8 Spectral irradiance of 200, 350 and 500 W Hg lamps. (Newport Spectra-Physics)

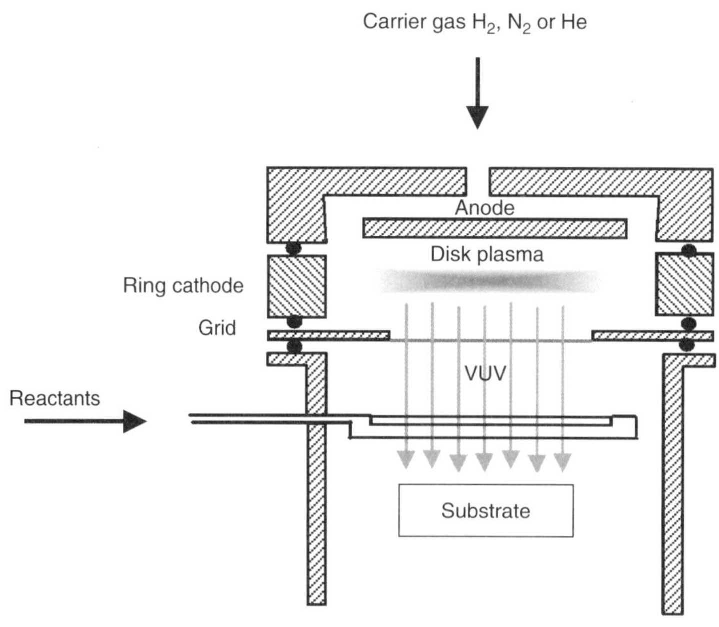

Figure 11.9 Schematic of a windowless lamp for very short wavelength UV photolysis.

11.4 Further Examples of Photolysis in Photo-assisted CVD

Zhang *et al.*[27] have produced thin tantalum pentoxide films by photo-induced CVD employing a 222 nm KrCl* excimer to generate dissociation of the precursors tetraethoxydimethylaminoethoxide and nitrous oxide. The use of the excimer brings about a decrease in activation energy from 2.2 eV (pyrolytically) to 0.08 eV (photolytically).

Figure 11.10 demonstrates the effectiveness of photo-CVD in reducing the thermal activation energy needed for film growth. In comparison to thermal-CVD the photo-CVD growth rate does not extensively change over a wide temperature range (25–400 °C). The negative activation energy between 25 and 100 °C is explained by a condensation reaction whereby the tantalum precursor ($Ta(OC_2H_5)_5$), which is maintained at 125 °C in a bubbler, condenses onto the substrate surface at temperatures below this.

Low temperature CVD is often desirable whether it be to maintain the integrity of the substrate or material to be deposited on or from a growth view where dopant redistribution, defect generation and atom diffusion are all minimized.

Room temperature photo-assisted CVD of germanium sulfide was reported by Tomovska *et al.*[28] Using an ArF laser the co-photolysis of gaseous c-C_2H_4S-TMG mixtures was studied. The ArF laser operated at 193 nm with a pulse energy of 60 mJ and frequency 10 Hz. The laser beam incident on the substrate had fluences of 150 mJ cm^{-2} that avoided dielectric breakdown but produced effective photolysis of both precursors. This was the first example of the reaction between elemental germanium and sulfur in the gas phase at room temperatures producing nano-sized germanium sulfide.

Dai *et al.*[29] have studied the laser deposition of tin oxide thin films using a 193 nm ArF excimer laser at energy 50 mJ cm^2 pulse^{-1} and frequency 10 Hz. The precursors $SnCl_4$ and O_2 were used.

The mechanism proposed was that subsequent photochemical reactions may remove further chloride atoms from the trichloride radical. Also, the O_2 can be photodissociated to form

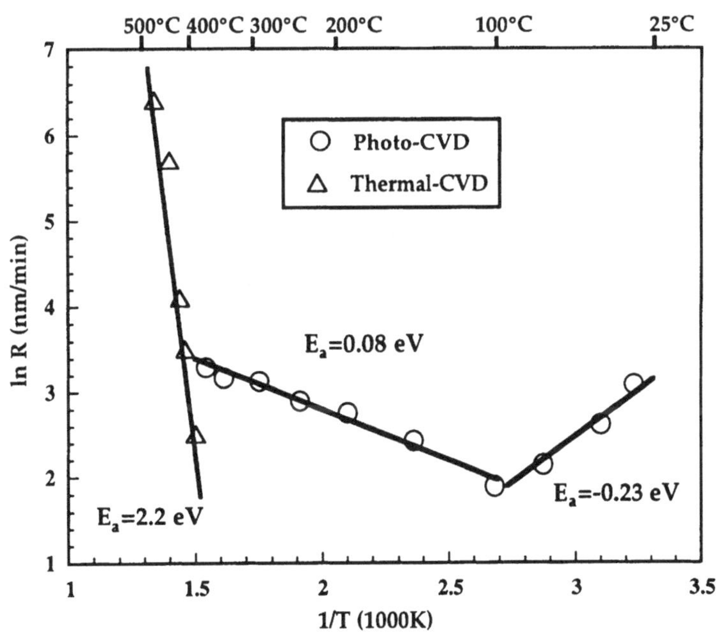

Figure 11.10 Arrhenius plot of the deposition rate of pyrolytically and photolytically enhanced growth of Ta_2O_5 films. (After Zang *et al.*[27])

monoatomic oxygen, which then goes on to react with the $SnCl_3$ and $SnCl_2$. This photochemical reaction resulted in two solid products, $SnOCl_2$ and SnO_2, which were confirmed by X-ray photoelectron spectroscopy. From the work by Dai *et al.*[29] the balanced reaction steps are shown in Equations (11.14–11.17):

$$SnCl_4 + h\nu(6.42\ \text{eV}) \rightarrow SnCl_3 + Cl \tag{11.14}$$

$$O_2 + h\nu(6.42\ \text{eV}) \tag{11.15}$$

$$\rightarrow 2SnCl_3 + 2O \rightarrow 2SnOCl_2 + Cl_2 \tag{11.16}$$

$$SnOCl_2 + O \rightarrow SnO_2 + Cl_2 \tag{11.17}$$

Figure 11.11 demonstrates that, for the above reaction mechanism, increasing the laser energy density increases the SnO_2 deposition rate up to the point where ablation of the film becomes dominant and the deposition rate decreases.

The experimental setup employed by Dai *et al.*[29] also demonstrated selective area growth. The ArF laser was normal incidence to the substrate surface and, with the use of stencil masks, growth was controlled to the area of illumination (Figure 11.12).

Fujita[30] made a study of II-VI organometallic compounds and their application to *in situ* measurements of the photolysis in a metalorganic CVD reactor.

For the matching of photo-assisted light sources, to specific precursors, knowledge of the said precursor's vapour phase UV absorption is essential. This information can be gained by containing the gases in a suitable quartz cell (capable of transmitting 190 nm, *i.e.* Suprasil® quartz). Using a spectrometer scanning between 190 and 400 nm and with knowledge of the path length, precursor pressure and temperature the photoabsorption cross-sections can be calculated using the relationship described by Equation (11.18):

$$\sigma(\lambda) = (kT/PL)\ln\{I_o(\lambda)/I(\lambda)\} \tag{11.18}$$

Where $I_o(\lambda)$ is the incident light and $I(\lambda)$ is the transmitted light at a particular wavelength; P is the pressure of the precursor vapour and L is the path length of the sample cell. Table 11.4 gives the absorption cross-sections for some commonly used metalorganic CVD precursors over a range of laser source wavelengths.

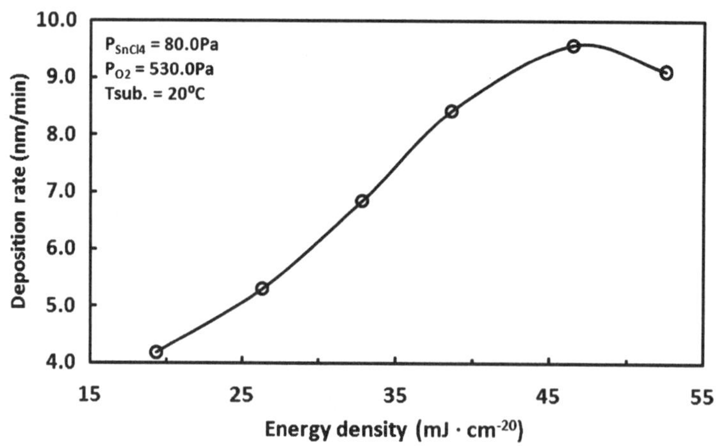

Figure 11.11 Deposition rate of SnO_2 as a function of ArF laser energy density. (After Dai *et al.*[29])

Table 11.4 Absorption cross-section (×10–17 cm^2 $molecule^{-1}$) of II-VI organometallic compounds at the wavelengths of excimer lasers (after Fujita *et al.*[30]).

Species	*Cross section* *ArF (193 nm)*	*KrCl (222 nm)*	*KrF (248 nm)*	*XeCl (308 nm)*	*XeF (350 nm)*
$Cd(CH_3)_2$	0.9	3.0	0.4	0	0
$Te(CH_3)_2$	2.8	0.3	3.6	0.05	<0.03
$Te(C_2H_5)_2$	4.8	1.2	3.6	0.05	<0.03
$Zn(CH_3)_2$	1.8	0.9	0	0	0
$S(CH_3)_2$	0.6	0.17	<0.01	0	0
$S(C_2H_5)_2$	1.1	0.8	<0.05	0	0
$Se(CH_3)_2$	1.1	0.9	0.15	0.02	0
$Se(C_2H_5)_2$	1.9	1.3	0.2	0.03	0

Figure 11.12 Optical micrograph of the selective area deposition of SnO_2 films. (After Dai *et al.*[29])

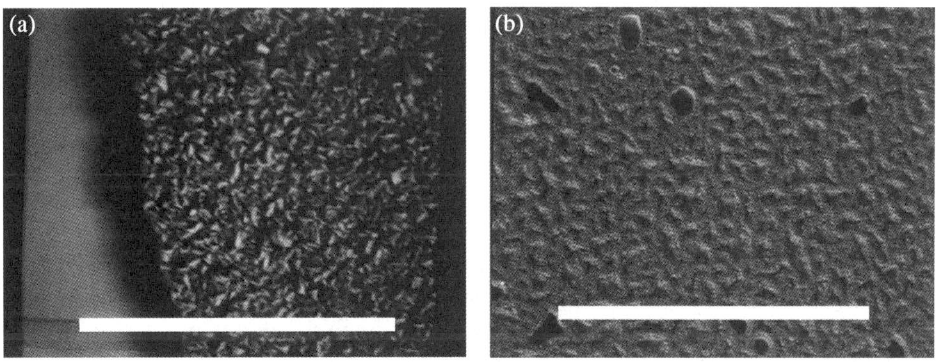

Figure 11.13 SEM micrographs of GaN on sapphire at 650 °C (*a*) with UV irradiation of substrate and (*b*) without. Scale bar: 100 μm. (After Tomar *et al.*[31])

In efforts to deposit GaN at temperatures below the usual 1000 °C Tomar *et al.*[31] employed a mercury grid lamp with 20 W intensity at 254 nm over 5 × 2 in^2 to provide UV photo-assisted deposition of GaN onto sapphire substrates by MOCVD. The photo-assisted growth produced oriented films at temperatures as low as 650 °C (where the thermal growth did not). Figure 11.13(a)

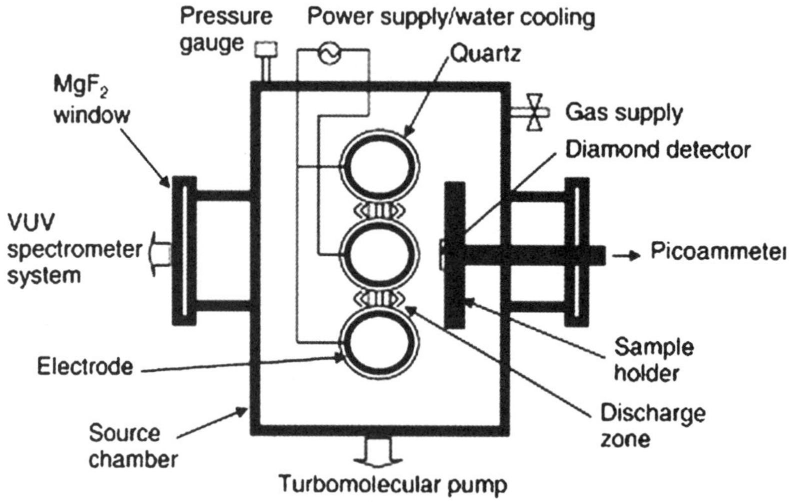

Figure 11.14 Experimental setup for generating VUV excimer reactions. When using the 126 nm Ar_2* excimer the quartz outer lamp envelope was dispensed with. (After Yu *et al.*[32])

and (b) shows SEM micrographs of the photo-assisted and thermally deposited GaN films, respectively. The modification to the surface morphology of the photo-assisted film denotes a change in crystalline structure to a more oriented growth.

Yu *et al.*[32] have tested the potential of large area diamond films, grown by CVD, for use as detectors for the vacuum ultraviolet (VUV) excimer lamps operating at 222, 172 and 126 nm. This experiment, when applied to the 126 nm excimer source, demonstrates the use of the windowless system. In the experiment, the CVD produced diamond films were placed on an unheated substrate holder and the reaction chamber evacuated. Figure 11.14 shows the experimental set up for producing these excimer reactions. When using the Ar_2* (126 nm) excimer source the quartz outer lamp envelope was dispensed with and the excimer reactions took place freely between two neighbouring electrodes, effectively utilizing the chamber wall to host the excimer reactions.

11.5 Conclusions

This chapter has outlined the fundamental mechanisms in photo-assisted CVD either through using lamp sources or lasers. The drive has been to carry out CVD at temperatures below that allowed by thermal kinetics at the ambient temperature of the reaction chamber and substrate. This allows a greater flexibility in determining film properties, substrate type and deposition area. A wider range of materials can be deposited by photo-assisted CVD through developments in reaction chambers and precursors. These processes are likely to become increasingly important with greater demand for structured surfaces and deposition of a wide range of materials onto plastic substrates.

References

1. *Laser Microfabrication: Thin Film Processes and Lithography*, Eds. D. J. Ehrlich and J. Y. Tsao, Academic Press, Inc. (1989).
2. C. J. Chen and R. M. Osgood, *J. Chem. Phys.*, 1984, **81**, 327.
3. D. Baurle, *Laser Processing and Chemistry*, Springer, Berlin, 2000.

4. S. J. C. Irvine, J. B. Mullin, D. J. Robbins and J. L. Glasper, *J. Electrochem. Soc.*, 1985, **132**, 968.
5. R. L. Jackson, T. H. Baum, T. T. Kodas, D. J. Ehrlich, G. W. Tyndall and P. B. Comita, in *Laser Microfabrication*, ed. D. J. Ehrlich and J. Y. Tsao, Academic Press, Inc., 1989, ch.7.
6. C. J. Chen and R. M. Osgood, *J. Chem. Phys.*, 1984, **81**, 327.
7. B. Liu, R. F. Hicks and J. J. Zinck, *J Crystal Growth*, 1992, **123**, 500.
8. S. J. C. Irvine, H. Hill, J. E. Hails, J. B. Mullin, S. J. Barnett, G. W. Blackmore and O. D. Dosser, *J. Vac. Sci. Technol.*, 1990, **A8**, 1059.
9. D. J. Ehrlich and J. Y. Tsao, *Appl. Phys. Lett.*, 1984, **44**, 270.
10. A. Stafford, S. J. C. Irvine and C. L. Griffiths, *J. Crystal Growth*, 1997, **170**, 182–187.
11. S. Fujita, A. Tanabe, T. Sakamoto and M. Isemura, *J. Crystal Growth*, 1988, **93**, 259.
12. A. Yoshikawa, T. Okamoto and T. Fujimoto, *J. Crystal Growth*, 1991, **107**, 653.
13. T. Yamada, R. Iga and H. Sugiura, *Appl. Phys. Lett.*, 1991, **59**, 958.
14. M. Nishio, H. Ogawa and A. Yoshida, *J. Crystal Growth*, 1991, **115**, 284.
15. W. Stutius, *J. Crystal Growth*, 1982, **59**, 1.
16. H. Stanzl, K. Wolf, B. Hahn and W. Gebhardt, *J. Crystal Growth*, 1994, **145**, 918.
17. W. Taudt, S. Lampe, F. Sauerlander, J. Sollner, H. Hamadeh, M. Heuken, A. C. Jones, S. Rushworth, P. O'Brien and M. A. Malik, *J. Crystal Growth*, 1996, **169**, 243.
18. M. Heuken, J. Sollner, W. Taudt, S. Lampe and H. Hamadeh, *J. Crystal Growth*, 1997, **170**, 30.
19. S. J. C. Irvine and J. Bajaj, *J. Crystal Growth*, 1994, **145**, 74.
20. M. U. Ahmed, S. J. C. Irvine and A. Stafford, *J. Materials Science: Materials in Electronics*, 1999, **10**, 595.
21. M. U. Ahmed and S. J. C. Irvine, *J. Electronic Materials*, 2000, **29**, 169.
22. G. Yu, X. W. Fan, J. Y. Zhang, B. J. Yang, X. Zhao, D. Shen and Y. M. Lu, *J. Cryst. Growth*, 1999, **221**, 77–82.
23. Y. Tarui, K. Aota, T. Sugiura and T. Saitoh, *Mat. Res. Soc. Symp. Proc.*, 1984, **l29**, 109.
24. W. Peters, *Technical Digest: International Electron Devices Meeting*, 1981, 240.
25. S. J. C. Irvine, J. B. Mullin and J. Tunnicliffe, *J. Crystal Growth*, 1984, **68**, 188.
26. S. Y. Myong, T. H. Kim, K. S. Lim, K. H. Kim, B. T. Ahn, S. Miyajima and M. Konagai, *Solar Energy Materials and Solar Cells*, 2004, **81**, 485.
27. J. Zang, B. Hopp, Z. Geretovszky and I. Boyd, *Appl. Surf. Sci.*, 2000, **168**, 307–311.
28. R. Tomovska, V. Vorlicek, J. Bohacek, J. Subrt and J. Pola, *J. Photochemistry and Photobiology*, 2006, **182**, 107–111.
29. G. Dai, X. Jiang and Y. Zhang, *Thin Solid Films*, 1988, **320**, 216–219.
30. Y. Fujita, S. Fujii and T. Luchi, *J. Vac. Sci. Tecnol.*, 1989, **A7**, 2276–280.
31. M. S. Tomar, R. Rutherford, C. New and K. A. Kuenhold, *Sol. Energy Mat. & Sol. Cells*, 2000, **63**, 437–443.
32. J. J. Yu and I. W. Boyd, *Diamond and Related Mat.*, 2007, **16**, 494–497.

CHAPTER 12

Plasma Enhanced Chemical Vapour Deposition Processes

SERGEI E. ALEXANDROV[a] AND MICHAEL L. HITCHMAN[b]

[a] Department of Electronic Materials Technology, St Petersburg State Polytechnical University, Polytechnical Str 29, 195 251 St Petersburg, Russia; [b] Thin Film Innovations Ltd., Block 7 Kelvin Campus, West of Scotland Science Park, Glasgow G20 0TH, UK

12.1 Introduction

As has been mentioned elsewhere, most chemical reactions in CVD are thermodynamically endothermic and/or have an apparent kinetic energy of activation associated with them. Generally, this is an advantage since the reactions can be controlled by regulating the energy input. As is apparent, though, from the extensive discussion in other chapters, this means energy has to be supplied to the gaseous precursor system to initiate the chemical reactions leading to deposition.

Traditionally, CVD processes have been initiated and controlled using heat as the source of energy. However, for the CVD of high quality layers with acceptable values of growth rate, often high deposition temperatures are required. Because, though, of the effect of such temperatures on thermally sensitive substrates and the generation of defects in both the substrate and the depositing layer alternative methods of energy supply to reacting systems have been developed. In particular, the energy of plasmas and photons has been widely used for activation of CVD processes. While the use of such sources of energy does not mean that no substrate heating is required, they do allow depositions under much milder thermal conditions. Chapter 11 considers photo-induced processes, while plasma processes are the subject of this chapter.

The promotion of chemical reactions by plasmas is based on two basic functions. First, the formation of chemically active species used for layer formation as a result of inelastic collisions of precursor molecules with energetic particles (mainly electrons) formed in the plasma. Second, delivery of energy to the substrate surface for enhancement of surface processes such as nucleation, particle migration and heterogeneous kinetics. The plasma parameters determine the composition, structure and the properties of the deposited films, as will be illustrated later.

Chemical Vapour Deposition: Precursors, Processes and Applications
Edited by Anthony C. Jones and Michael L. Hitchman

Published by the Royal Society of Chemistry, www.rsc.org

Many methods have been developed to generate plasmas, but electrical discharges in gaseous media are extensively used.[1,2] As has already been mentioned, one of the main purposes of using plasmas as sources of energy for CVD processes instead of heat is to overcome high deposition temperatures. Therefore, electrical discharges sustained in a gaseous atmosphere that form non-equilibrium, non-thermal plasmas have been widely used. An example of the industrial use of this type of plasma for CVD is in microelectronics processing. Plasma enhanced CVD (PECVD) has been employed since the mid-1960s for the formation of dielectric films at relatively low temperatures (up to 300 °C). Other types of industrial PECVD equipment and processes have been developed with most of them being based on the use of non-equilibrium, non-thermal plasmas sustained by radiofrequency (RF) or microwave (MW) sources.

However, although the use of PECVD processes for commercial applications has been growing, the technological basis of that growth has been largely empirical. PECVD processes are extremely complex and the mechanisms of reaction are often poorly understood. So, for example, despite hundreds of papers on the study of the PECVD of Si from SiH_4, which is one of the most extensively used systems, the detailed mechanism of the deposition processes is still not clear. Having said that, several excellent reviews and texts summarize what is known about the chemistry of PECVD processes[3–5] as well as the physics of plasmas.[1,2] In addition, a special recent issue of the journal *Chemical Vapor Deposition* has been dedicated to recent developments in more traditional PECVD.[6] Consequently, because of the vast literature on more conventional PECVD, we have chosen to concentrate on two less conventional types of PECVD processes based on non-thermal plasmas that have been of particular interest in recent years. The first is remote PECVD, which will be illustrated using the deposition of silicon nitride films. The second relates to PECVD processes with non-thermal plasmas generated at atmospheric pressure.

12.2 Remote Plasma Enhanced CVD (RPECVD) Processes

12.2.1 Introduction

Most PECVD systems are based on the use of an RF glow discharge sustained in an appropriate gas mixture, and the substrates are usually placed on one of the electrodes in the discharge generation region. This conventional PECVD approach has several disadvantages, *e.g.*:

- Electrically active defects can be generated near the surface of the substrate because of bombardment by energetic particles.
- Plasma particle bombardment of the growing film may cause generation of pinholes and charges trapped in the film.
- Complex reactions can result from the simultaneous production of many different reacting species and this can cause unpredictable compositions of the deposited films.
- The main process parameters, such as the total pressure, RF-power, gas composition and gas flow rate are all interdependent and their individual influences can be difficult to control.

Much of the research in the field of PECVD of various materials, and in particular silicon nitride and oxide, has been directed towards overcoming some of these problems.[4] One of the most intensively studied techniques, during the last two decades or so, has been that of remote PECVD (RPECVD).[7,8] In this technique the plasma generation region and the area of the reactor used for substrate location are spatially separated and only some of the input reactants are excited by the glow discharge and then transported to the deposition zone. Other reactants, usually the less stable ones, are introduced downstream from the discharge generation region. Active species formed in the discharge generation region and in the afterglow zone are delivered to the region of the

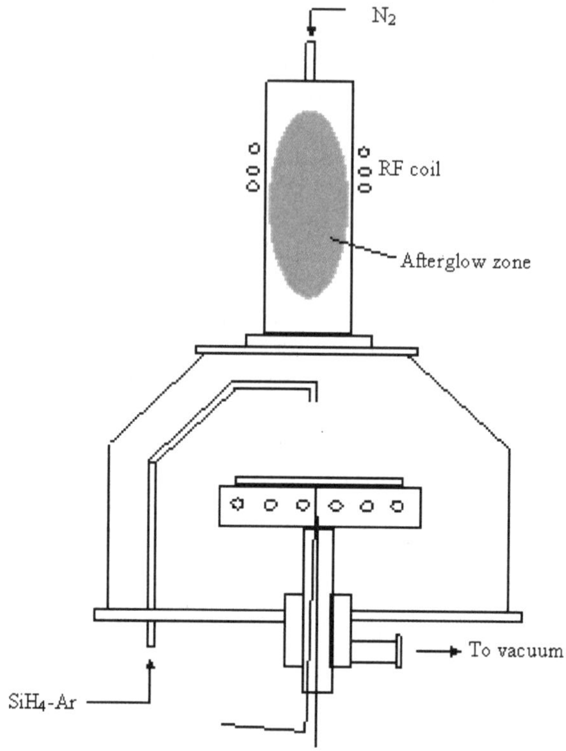

Figure 12.1 Schematic diagram of an inductively coupled RPECVD reactor for silicon nitride deposition. (Reproduced with permission from ref. 8.)

substrate, where they can react homogeneously with non-excited reactants and/or heterogeneously on the substrate surface to form a deposited film.

Two approaches for RPECVD process are based on the method of RF power coupling into the reaction chamber. In most remote processes,[7,8] RF power is coupled into the system inductively for reactant excitation (Figure 12.1). Figure 12.2 shows a schematic diagram of a capacitively coupled RF powered RPECVD system developed about ten years ago for SiN_xH_y deposition.[9]

12.2.2 Advantages and Disadvantages of RPECVD

As indicated above, RPECVD offers a potential method for minimizing the density of electrically active defects generated near a substrate surface by bombarding energetic particles during deposition. To investigate this, various dielectric materials have been deposited on the surface of special test gateless GaAs structures (Figure 12.3) using different deposition techniques.[8]

By comparing the values of saturation current (the maximum possible current in the top layer) obtained from the static *I-V* curves of these structures before and after deposition, it is possible to evaluate the influence of plasma bombardment on the surface properties of the semiconductor. To prevent the possible influence of electric charge, which can be accumulated in the growing dielectric films during deposition, the deposited layers were chemically etched before measurements. Generation of defects near a semiconductor surface will cause a decrease of saturation current because of a decrease of effective thickness in the layer above. Table 12.1 shows the results of these

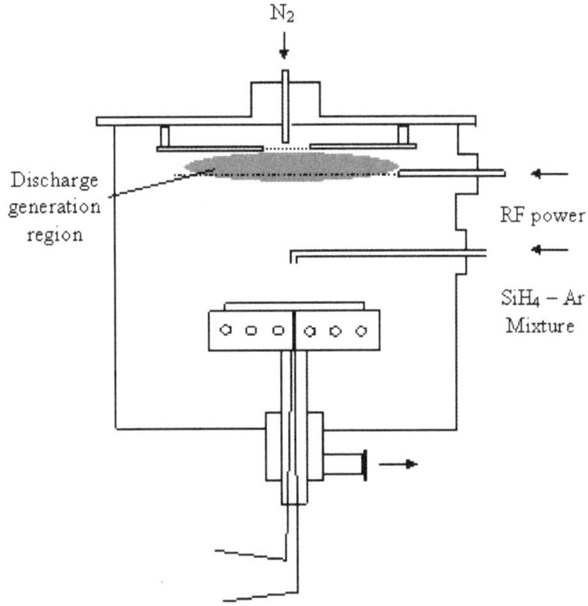

Figure 12.2 Schematic diagram of a RPECVD reactor with capacitively coupled RF power for silicon nitride deposition. (Reproduced with permission from ref. 9.)

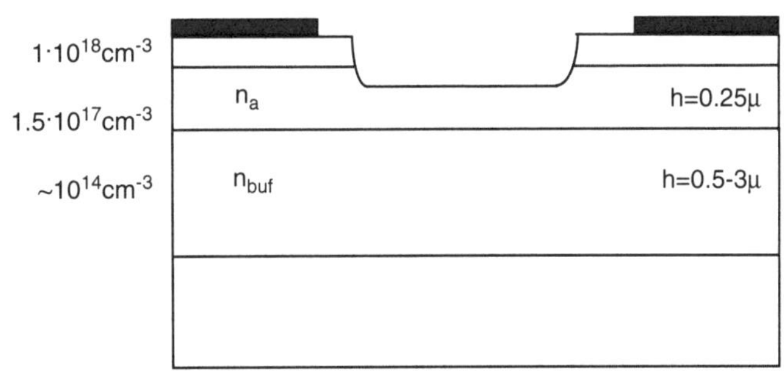

Figure 12.3 Test gateless GaAs structure. (Reproduced with permission from ref. 8.)

experiments. In all cases the dielectric deposition processes did not influence the general appearance of the *I-V* curves, but it can be seen from the data contained in Table 12.1 that the saturation current of the structures decreased 30–40% after deposition of both SiO_2 and SiN_xH_y films by conventional PECVD. The saturation current was not affected by low pressure thermal CVD of GaO_xN_y films (without plasma excitation) or by RPECVD of SiN_xH_y films. These results indicate that the surface of a semiconductor undergoes much less degradation during a RPECVD processes compared with conventional PECVD methods. This considerable advantage of RPECVD is of particular importance for industrial applications.

RPECVD processes can also provide less interdependence of some process parameters, so these can be more readily controlled than with the conventional direct PECVD approach. This can be illustrated by considering the example of silicon nitride deposition. During conventional PECVD of silicon nitride films with the SiH_4–N_2 system, because of the dissociation of silane, and the

Table 12.1 Values of the saturation current of the gateless GaAs structures before and after deposition of various dielectric films by different deposition methods.[8]

Dielectric materials and deposition techniques	*Saturation current (mA)*	
	Before	*After*
SiO_2, one-step conventional PECVD in $C_6H_{18}OSi_2$– O_2 system $T_{substr} = 200\,°C$,	140	80–85
$W_{RF} = 1\,kW$	120	60–70
SiO_2, two-step conventional PECVD in $C_6H_{18}OSi_2$–O_2 system $T_{substr} = 200\,°C$,	120	90
$\tau_1 = 4\,min$, $W_{RF} = 0.3\,kW$, $\tau_2 = 8\,min$, $W_{RF} = 1\,kW$	140	115
SiN_xH_y conventional PECVD in SiH_4–NH_3–N_2 system $T_{substr} = 200\,°C$,	130	85
$W_{RF} = 1.5\,kW$	120	75
	120	80
SiN_xH_y RPECVD in SiH_4–N_2 system $T_{substr} = 250\,°C$, $W_{RF} = 0.5\,kW$	130	130
	125	120
	125	125
GaO_xN_y, low pressure thermal CVD (without plasma) in $GaCl_3.NH_3$–H_2O	120	120
system $T_{substr} = 400\,°C$	125	125

formation of various types of radicals and atoms such as SiH_3, SiH_2, SiH, Si, and H,[10] even small variations of silane flow rate can cause dramatic changes in the plasma phase composition. These plasma composition changes obviously occur in the discharge generation region, and this leads to variations of the level of RF power dissipated in the plasma. For a remote plasma process, when silane is introduced downstream from the discharge region (*cf.* Figures 12.1 and 12.2) and with gas flow conditions chosen to minimize or completely prevent back streaming of silane in the direction of the discharge generation zone, the variation of SiH_4 flow rate does not cause variation of RF power absorbed in the plasma.

Another example of the reduced interdependence of the process parameters during RPECVD is provided by a consideration of the processes occurring on the surface of the substrate and the growing film. The real temperature of the reaction surface is determined not only by the heat flux from the susceptor, but also by the flux of energy delivered to the reaction surface by energetic particles from the plasma.[3] Variations in the level of RF power absorbed in the plasma (such as those discussed above) must cause changes of reaction surface temperature for conventional PECVD. Clearly, this interdependence would be expected to be much less in the case of RPECVD. However, although RPECVD can minimize parameter interdependence, some interdependence will still occur. For example, variations in the total pressure in both remote and conventional PECVD lead to dramatic changes in RF power absorbed in the plasma, which affects the plasma phase composition for both techniques. Nevertheless, RPECVD undoubtedly provides less interdependence for some process parameters, and so processes can be more readily controlled.

A further feature of conventional PECVD mentioned above is the formation of a wide variety of reacting species in the plasma phase. Complex multi-pathway reactions can arise as a result of the simultaneous production of a range of reacting species in the plasma phase from inelastic collisions with accelerated electrons of various energies. The electron energy distribution function for this type of RF glow discharge is described satisfactorily by a Maxwellian distribution over the energy range 0.1–15 eV.[11] Typical mean values of electron energies are in the region 2–4 eV, but the fastest electrons can reach energies up to 30 eV.[12] Inelastic collisions of these electrons with precursor molecules directly introduced into the discharge generation region can lead to the simultaneous production of numerous different reacting species, and this is very dependent on process parameters. For example, more than 100 different species have been found in a silane based plasma.[13]

As just mentioned, this variety of species in the plasma can lead to many parallel reaction pathways, resulting in lack of control of layer composition. The number of reactive species can be significantly reduced and better control over layer properties attained if the penetration of charged particles downstream from the discharge generation region in the direction of the precursor inlet region is curbed.

Of course, in RPECVD formation of the active species required for film growth can occur both in the discharge generation region and by collisions of the precursor molecules introduced downstream from the plasma region with excited particles extracted from the discharge area. However, the number of particles available for reaction downstream from the plasma region can be drastically reduced if charged particles can be separated from neutral particles by using electric or magnetic fields (Figure 12.4).[8] Thus, the discharge generation region is essentially used as a source of only neutral particles. These neutral particles emerging from the discharge region will, of course, have various energies over a wide range, corresponding to their electronic excited states. Even these excited particles can undergo separation through their different life times, with the most energetic particles having relatively low life times and recombining first, and the longer-lived particles moving downstream to the precursor inlet region. In an ideal case, an appropriate location or the point of mixing of the precursor molecules with the neutral particles can be chosen. If two different precursors are to be used for film deposition and the neutral particles with an energy necessary for the excitation of one of the precursor molecules cause undesirable dissociation of the molecules of the other precursor, a reaction chamber with two separate zones of plasma excitation could be used (Figure 12.4b).

Notably, even though charged particles reaching the precursor inlet region can be greatly reduced by charge separation devices, there is still the possibility of generating electrons as a result of recombination of excited neutral particles, *via* a Penning type interaction. If this does occur,

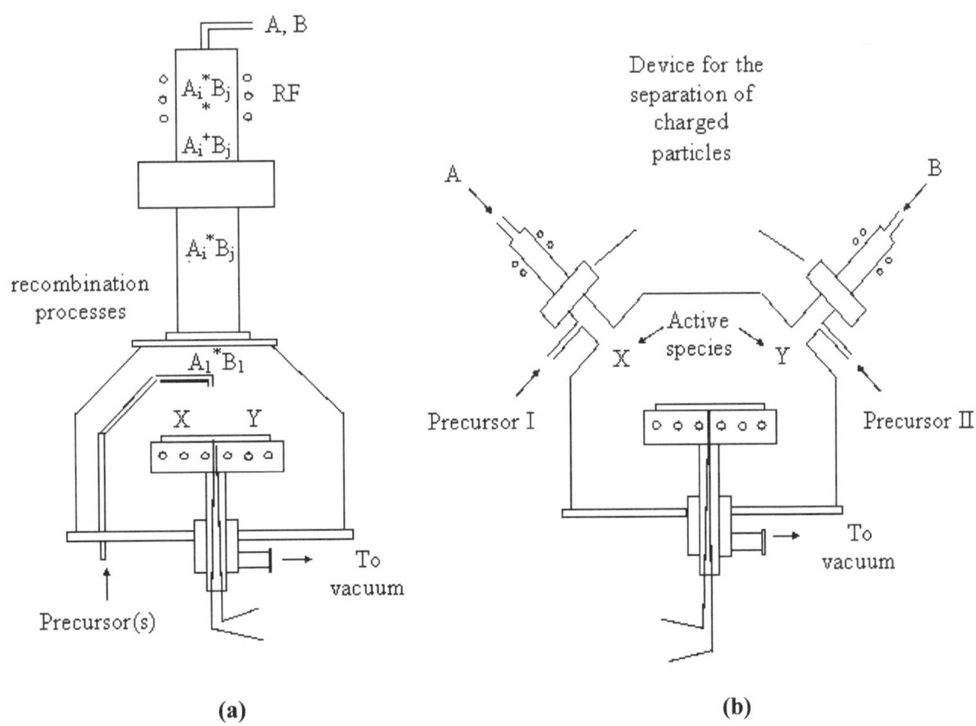

Figure 12.4 RPECVD systems with charge separation devices. (Reproduced with permission from ref. 8.)

though, the electrons formed by this mechanism usually have low energies, and in the absence of electric and magnetic fields their energies cannot be increased. Hence they can only enhance processes of excitation, dissociation or ionization with relatively low threshold energies, such as reactions involving silane which are characterized by energies in the range 2–5 eV[10] In many other cases, however, the energy of these electrons will not be enough to participate in the deposition processes.

Another issue is the nature of particles that can be used for excitation of precursor molecules. These exciting particles should have considerable life times, should be chemically inert with respect to the other reagents and reactor materials, and must have energies that are compatible with the energies required for the selective production of reacting species. Atoms of noble gases such as He and Ar, as well as stable molecules of other gases (for instance, nitrogen), in metastable electronic states, can have considerable radiative life times and are likely to be successful for transferring energy to precursor molecules. Results presented below for the deposition of silicon nitride films illustrate this.

In conclusion, the discussion and comparative analysis given above shows that RPECVD is a promising approach for plasma enhanced CVD since it overcomes some of the drawbacks of conventional direct PECVD techniques. One disadvantage of RPECVD compared with conventional PECVD is the much lower growth rates of the former. For example, a typical growth rate for RPECVD of silicon nitride in a SiH_4–NH_3 system is about 1–6 nm min^{-1}, which is significantly lower than the 10–25 nm min^{-1} achieved in a direct PECVD process.

12.2.3 RPECVD of Silicon Nitride Films

12.2.3.1 Precursor Systems and Precursor Chemistry for RPECVD of Silicon Nitride Films

Clearly, a precursor system for any form of the PECVD of SiN_xH_y films requires either separate nitrogen- and silicon-containing volatile substances or a so-called single source precursor – a gaseous compound involving Si and N atoms.

Ammonia (NH_3) is the most widely used nitrogen source for the PECVD of silicon nitride films. One reason for this is that it is a rather reactive gas and generally yields higher growth rates than the other most commonly used precursor source of nitrogen, molecular nitrogen (N_2).[14] Hydrazine (N_2H_4), dimethylhydrazine $(MeNH)_2$ and phenylhydrazine ($C_6H_5NHNH_2$) are all characterized by weaker N–H bonds that can lead to higher growth rates and lower deposition temperatures. However, the use of any of these nitrogen-containing precursors can cause uncontrolled doping of silicon nitride with hydrogen, in the case of NH_3 or N_2H_4, or carbon and hydrogen in the case of ($(MeNH)_2$) and ($C_6H_5NHNH_2$). From this point of view, molecular nitrogen is clearly the best precursor to try and achieve the deposition of silicon nitride with a near stoichiometric composition.

Another disadvantage of precursors with N–H bonds is that they can lead to the formation in the plasma phase of several active nitrogen-containing species. For example, in the case of PECVD with ammonia as the nitrogen-containing precursor, active species such as NH_2^+, NH_2, NH^+, NH, N, N^+, H, *etc.* can be formed in the plasma. Such a wide range of N–H containing species can provide several reaction pathways resulting in the formation of various intermediate compounds, such as silicon imides or amides, as well as the required silicon nitride.

A disadvantage of molecular nitrogen, though, is the extremely high strength of N–N bonds. Therefore, if nitrogen is to be used as a precursor for PECVD processes it is essential that the plasma conditions are capable of dissociating N_2 so that very active atomic nitrogen is formed. The results of studies[15,16] of the PECVD of silicon nitride films with molecular nitrogen as the nitrogen containing precursor show that a high concentration of atomic nitrogen in the deposition area is a necessary condition for the formation of near stoichiometric silicon nitride films. Understandably, therefore, in the particular case of RPECVD of silicon nitride where molecular nitrogen is used it is always this gas that is passed through the discharge generation region.[9,17,18]

A good way to enhance atomic nitrogen formation is to use metastable atoms of Ar for selective excitation of nitrogen molecules.[19] The nitrogen molecules are excited to specific vibrational levels of the $C_3\Pi_u$ electronic state, from which pre-dissociation of molecular nitrogen occurs (Figure 12.5). This causes an increase in the concentration of atomic nitrogen, which, in turn, leads to an increase in the nitrogen content in the deposited films. The exact influence of nitrogen dilution with argon depends on two factors: the deposition conditions used and the method of coupling the RF power into the reaction chamber.[20]

For the silicon-containing precursor for PECVD of silicon nitride films, silane (SiH_4), is the most widely used material.[7,9] However, problems with silane are that it is highly reactive, explosive and it induces a lot of hydrogen incorporation into the deposited film; also the step coverage of patterned surfaces is reported to be of poor quality.[21] For these reasons there has been a lot of research to try and find silicon-containing precursors that are better suited for silicon nitride PECVD. One example is the RPECVD of silicon oxynitride films with compositions ranging from SiO_2 to near stoichiometric silicon nitride, Si_3N_4, which were deposited at 250 °C using $SiCl_4$, O_2 and NH_3 as precursors.[22] Unfortunately, the reported results do not allow one to evaluate the benefits of this particular precursor mixture, but one potential disadvantage that one could anticipate from our own experience is that deposition of powdery intermediates of amidochlorides and imidochlorides of silicon[23] could be formed on the cold walls of the reactor downstream from the $SiCl_4$ inlet.

Two other precursor that have been used for RPECVD of silicon nitride are hexamethylsilane (Me_3Si-$SiMe_3$)[24] and hexamethyldisilazane [$(Me_3Si)_2NH$)].[25] However, since Si–C bonds are strong, unsurprisingly, the deposited films contain a significant amount of carbon. Therefore, the use of monomeric precursors that have no such bonds is preferable to minimize carbon

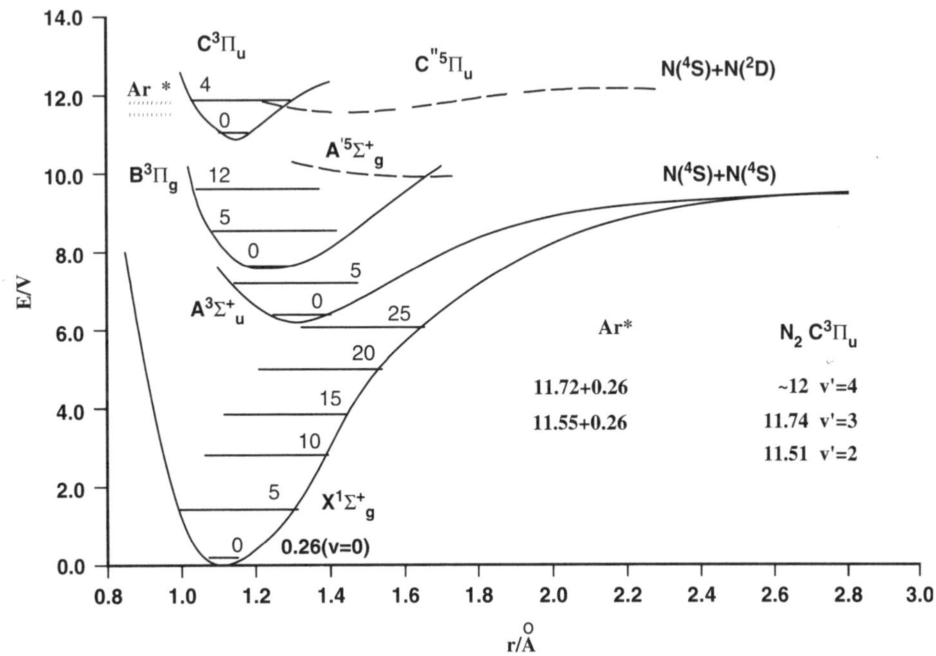

Figure 12.5 Potential energy curves of molecular nitrogen. (Reproduced with permission from ref. 19.)

contamination.[26] Organic compounds with weaker Si–N and Si–H bonds such as tetrakis(dimethylamino)silicon, $[Si(NMe_2)_4]$,[27,28] and tris(dimethylamino)silane, $[(Me_2N)_3SiH]$,[26] have both been used as precursors for the RPECVD of SiN_xH_y. Even so, remarkable amounts of carbon contamination (30–40 at.%) have been found with these precursors in films deposited at low substrate temperatures (150–200 °C). Although this contamination level decreased to 6–8 at.% when the growth temperature was above 400 °C[29] the density of the films was low, as indicated by very low values (1.62) of the refractive index. The published results suggest, therefore, that when organosilicon precursors are used for RPECVD of silicon nitride films, deposition of high quality layers should not be expected. So, despite other problems with SiH_4–NH_3 and SiH_4–N_2 mixtures they are still the preferred precursor systems for RPECVD of good quality silicon nitride films.

The reaction mechanisms of RPECVD processes of silicon nitride films from SiH_4–NH_3 and SiH_4–N_2 systems remain unclear despite several studies.[7,19,20] Optical emission spectroscopy (OES) of the plasma phase for a pure ammonia precursor has indicated the presence of NH and H species in the plasma, and FTIR spectroscopy has shown that deposited films contain Si–NH bonds.[30–32] The authors concluded that plasma-generated NH species are the precursors for the Si–NH bonds in the deposited films and that plasma-generated N species are the precursors for the Si–N–Si bonding arrangement. By contrast, mass spectrometry of the plasma with ammonia in conventional PECVD[33] suggests that aminosilanes – triaminosilane, $SiH(NH_2)_3$, and tetraminosilane, $Si(NH_2)_4$ – are formed homogeneously during the deposition process.

Mass spectrometry of the plasma phase formed in a SiH_4–N_2 RPECVD system with capacitive coupling[34] shows that silane introduced downstream from the discharge generation region is cracked by collisions with the excited species coming from the discharge generation region with the formation of SiH_x ($x = 1$–3) radicals. In that particular study, about 80% of the original silane introduced was dissociated. Further, the mass spectrometry showed no indication of Si-N moieties, such as aminosilanes. This suggests that RPECVD silicon nitride films are formed by heterogeneous reactions between excited nitrogen species (atomic nitrogen[19,20]) and SiH_x ($x = 1$–3) radicals. In addition, since the average rate of silicon nitride deposition of 3 nm min^{-1} for RPECVD in the SiH_4–N_2 system with capacitive coupling of the RF power is significantly higher than the 0.3 nm min^{-1} for RPECVD with inductive coupling this indicates more efficient power absorption with the former method than with the latter.

In conclusion, RPECVD with the SiH_4–N_2 precursor system is the most promising for deposition of high quality silicon nitride layers and if the highest high growth rate is desired then RPECVD with capacitive coupling is more attractive. The quality of silicon nitride layers deposited by RPECVD is considered in more detail in the next section.

12.2.3.2 Composition of Silicon Nitride Films Deposited using RPECVD

Having considered the chemistry of silicon nitride films deposited by RPECVD with SiH_4–N_2 and SiH_4–NH_3 systems, we now turn to two important characteristics of deposited films that reflect possible deviations from stoichiometry: the Si : N ratio and hydrogen content.

The results of several studies[14,15,19,20] indicate that atomic nitrogen is probably the main active nitrogen species interacting with gaseous silane radicals to form silicon nitride, and that the homogeneous $SiH_4 : N_{at}$ ratio determines the Si : N ratio in deposited films, and consequently their main physical properties.[20] So, for example, increasing the silane flow leads to a significant increase in the refractive index (Figure 12.6), reflecting an increase in the relative silicon content of the films.[9] At the same time, increased silane flow leads to a steady increase in the deposition rate, with this being proportional to the silane flow rate (Figure 12.6).

To obtain near-stoichiometric silicon nitride,[9] published results show that it is necessary to have a $N_2 : SiH_4$ ratio greater than about 250; such a remarkable excess of molecular nitrogen is required

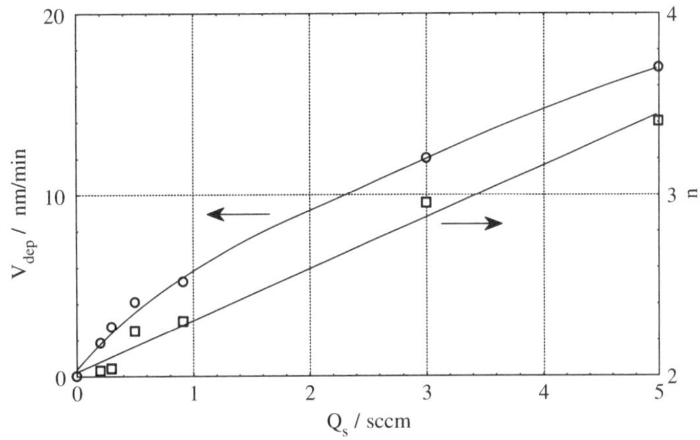

Figure 12.6 Dependence of growth rate and refractive index (n) of a silicon nitride film on silane flow rate. (Reproduced with permission from ref. 9.)

to provide a high enough concentration of atomic nitrogen in the deposition. However, the extremely high strength of N–N bonds in molecular nitrogen can cause some difficulties in providing such a high concentration of atomic nitrogen in the deposition zone and this often leads to the excess of silicon over stoichiometry in deposited films.[9,14] This is particularly the case for RPECVD, where formation of atomic nitrogen occurs mainly in the plasma generation region through which the nitrogen-containing precursor is passed. The concentration of atomic nitrogen near the substrate will depend on the initial concentration in the discharge generation region and on the rate of recombination. Thus, in RPECVD with both capacitive and inductive RF coupling[19,20] an increase of the RF power and of the flow rate of nitrogen passed through the discharge generation region, and a decrease of the total pressure, all cause an increase in the N : Si ratio in the film.[20,35,36]

Stoichiometry can also be affected by dilution with argon of nitrogen passed through the discharge generation region. Figure 12.7[20] shows the dependence of the growth rate and refractive index of silicon nitride films on the flow rate ratio $Q_{N2}/(Q_{N2}+Q_{Ar})$ through the discharge generation region in an inductively coupled system. A decrease in the refractive index of the deposited films corresponds to an increase in the nitrogen content in the films as the amount of argon in the gas phase is increased. This is understandable in terms of findings from results of emission spectroscopy.[19] Despite a decrease in the relative amount of nitrogen introduced into the reaction system, the deposited films become more "nitrogen-rich". This is because of the enhancement of atomic nitrogen formation by the selective excitation of molecular nitrogen by Ar atoms in metastable electronic states. From the point of view of the production of SiN films without Si excess using SiH_4–N_2 mixtures, this enhancement has an important bearing on the remote PECVD technique. Replacement of molecular nitrogen with ammonia does not significantly change the sensitivity of the Si : N ratio in the deposited films on process parameters. Nevertheless, this replacement does lead to a decrease in the NH_3 : SiH_4 ratio required, compared with when N_2 is used, for deposition of films with good stoichiometry.[7]

Hydrogen in SiN films in a chemically bonded form, such as Si–H_i ($i=1$–3) and N–H_j ($j=1, 2$), is another possible cause of stoichiometry deviations and it can dramatically influence the physical and chemical properties of the films. The use of silane with ammonia leads to the incorporation of significant amounts of up to 30 at.% of bonded hydrogen into deposited films,[37,38] which are more properly described as SiN_xH_y.

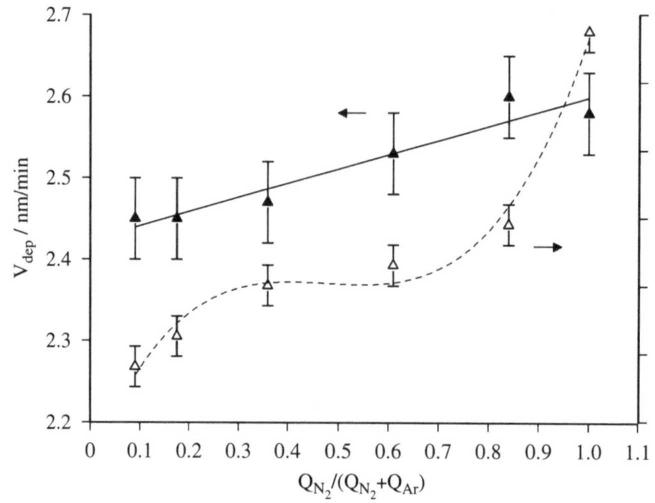

Figure 12.7 Dependence of growth rate and refractive index (n) of silicon nitride films on $Q_{N_2}/(Q_{N_2}+Q_{Ar})$ ratio. (Reproduced with permission from ref. 20.)

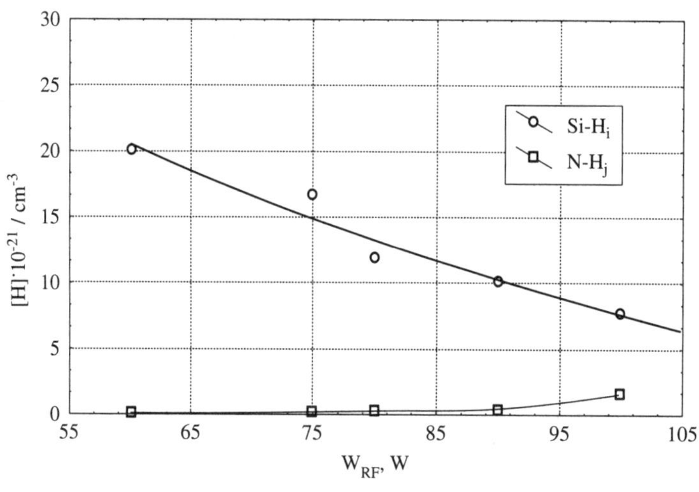

Figure 12.8 Dependence of hydrogen content in SiN_xH_y on RF power for capacitively coupled RPECVD. (Reproduced with permission ref. 39.)

The influence of process parameters on the concentration of bonded hydrogen in silicon nitride films deposited by RPECVD with capacitive and inductive RF coupling has been investigated.[20,35,39] For example, it was found that the concentration and configuration of bonded hydrogen in SiN_xH_y films are determined by gas phase interactions as well as by heterogeneous reactions on the substrate surface. The nature and concentration of the SiH_x species formed in the gas phase from silane largely determine the concentration and configuration of hydrogen bonded with silicon in the deposited films, while surface processes, dependent on the plasma energy delivered to the substrate by plasma particles, control the configuration and content of hydrogen bonded nitrogen in the films. An increase in the plasma power density causes an increase in the N–H_j bond concentration (Figure 12.8).[39] In general, for RPECVD, with both capacitive and inductive coupling, the concentration of bonded hydrogen in SiN_xH_y films decreases with an increase in deposition temperature and plasma power density. A decrease in the relative concentration of silane in the gas phase[20,36,39] leads to the same effect.

It is interesting to compare the growth rate and bonded hydrogen concentration in SiN_xH_y films deposited under similar conditions by inductively and capacitively coupled RPECVD. Films deposited by inductively coupled RPECVD have shown a growth rate of 0.18 nm min^{-1} and contained 3×10^{21} cm^{-3} bonded hydrogen[40]while films deposited by capacitively coupled RPECVD had a growth rate of 3.5 nm min^{-1} and a mean value of total hydrogen concentration of 9×10^{21} cm^{-3}.[39] These results indicate that capacitively coupled RPECVD provides reasonably low concentrations of bonded hydrogen in films, but at a much higher growth rate than for films deposited by inductively coupled RPECVD.

As already mentioned, the total concentration of bonded hydrogen is an important factor in determining the physical and chemical properties of silicon nitride films deposited by RPECVD.[41–43] However, the effect on the properties is different for Si–H_i and N–H_j bonds. For example, the formation of N–H_j bonds causes a greater relaxation of the amorphous structure of SiN_xH_y films than the formation of Si–H_i bonds. This is because blocking by hydrogen of any one of the three bonds of nitrogen leads to the formation of a more open silicon nitride structure than the blocking of any of the four silicon bonds. Thus, as the concentration of N–H_j bonds is increased, one can expect a more dramatic decrease of film density and corresponding changes in dielectric properties than for an increase of Si–H_i bonds. This is confirmed by experiment,[44] which has shown that even though the total concentration of bonded hydrogen decreases the values of the refractive index, density and dielectric breakdown strength of RPECVD SiN_xH_y films still decrease because of an increased amount of N–H_j bonding.

Clearly, to try and achieve deposits with characteristics close to those of stoichiometric Si_3N_4 the hydrogen content should be minimized. In addition to controlling the process conditions as discussed above, there are several other ways to lower the amount of hydrogen incorporated into the layers. One method is by introducing a hydrogen scavenger into the reaction gas mixture. Fluorine-containing reactants such as gaseous silicon fluorides ($SiH_{4-x}F_x$, $x = 0$–4)[45] or NF_3[46] have been used to remove hydrogen in the form of HF, although, at the same time, F atoms are introduced into the lattice. An illustration of what happens is provided by investigations[47,48] of the effects of dilution of nitrogen with NF_3 on the composition and properties of silicon nitride films deposited by RPECVD with both capacitive and inductive coupling. The main trends in the variation of amounts of bonded hydrogen and fluorine as well as in the film properties are very similar in both cases. As can be seen from Figure 12.9,[47] the introduction of small flows of NF_3 causes a considerable decrease in the content of Si–H_i bonds and a steady increase in the content of N–H_j groups. However, the rate of decrease of the Si–H_i bond content with increasing NF_3 flow rate in the range 0–5 sccm is more rapid than the rate of increase in the N–H_j content. So there is an overall decrease in the total content of bonded hydrogen.

The presence of even small amounts (<0.2%) of NF_3 in the gas phase results in the incorporation of about 2 at.% fluorine in the deposited films, but after this initial sharp increase in fluorine content any further increase is approximately linear with NF_3 input. The increase in fluorine concentration in the deposited films with increasing NF_3 flow rate in the range 0 5 sccm (less than 1% of the total flow) leads to a steady change in the Si-N-F matrix composition with corresponding variations of physical and chemical properties of the layers. For example, the refractive index of films decreases steadily from about 2.05 to about 1.6 as the F content increases.[47] At the same time, breakdown strength and the optical energy gap increase from 6 to 12×10^6 V cm^{-1} (ref. 47) and from 2.2 to 5.5 eV,[48] respectively, with increasing NF_3 flow rate. All these effects occur with the dilution of nitrogen by relatively small amounts (<1%) of NF_3 and this could lead to useful possibilities for industrial applications.

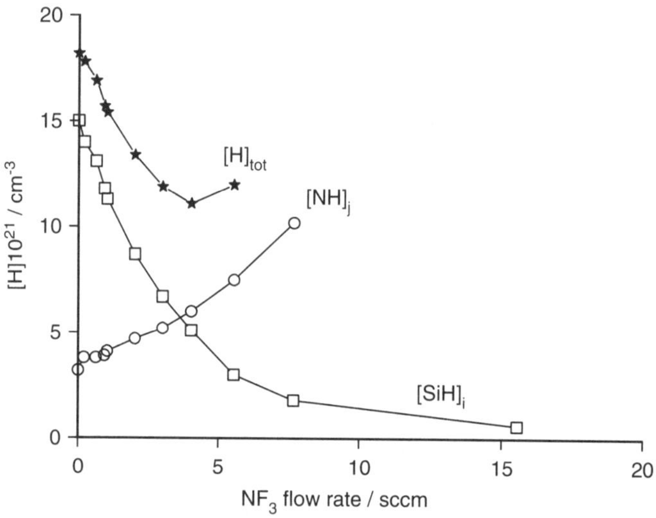

Figure 12.9 Dependence of hydrogen content of silicon nitride films on NF_3 flow rate for a capacitively coupled system. (Reproduced with permission from ref. 47.)

12.2.3.3 Effect of Noble Gas Dilution

Atoms of noble gases such as He and Ar with excited species having considerable radiative lifetimes have often been used in RPECVD to enhance the process of transferring energy from the discharge generation region.[7,44,49] It has been found, for example, that silicon nitride films can be deposited even when nitrogen is introduced downstream from the discharge generation region if argon is the only gas excited in the discharge. As in the case with a N_2 remote plasma, the formation of nitride with near stoichiometric composition requires very high N_2 flow rates to provide a high gas phase N : Si ratio (typically above 100).[8,30–32]

Although several papers have been published on the use of noble gases as diluents of nitrogen-containing precursors in the discharge generation region, the role of such gases during RPECVD is still not completely clear. Having said that, the influence of noble gas dilution on the deposition process is probably related to factors such as values of the plasma parameters, variation of mass transport conditions for the active species, and formation of atoms in metastable electronic states with considerable radiative lifetimes, which can contribute to the energy transfer from the discharge region to the surface of the growing film.

Studies of the influence of helium dilution on RPECVD with inductive[49] and capacitive[44] coupling of RF power in the SiH_4-N_2 system show that the effect of He depends on the process parameters under which the deposition is carried out and does not depend on the method of coupling the RF power. For example, if transport of excited particles from the discharge generation region to the substrate controls the deposition process, the introduction of helium together with nitrogen into the discharge region promotes an increase in growth rate since the enhancement of energy transfer by He metastables is itself a transport controlled process. This mostly corresponds to the RPECVD process with inductive coupling of the RF power[49] because the reactor design leads to transport control dominating. In contrast, for deposition with the SiH_4–N_2 system carried out under conditions promoting transport of active species and energy to the deposition zone, the influence of the helium dilution on the growth rate will be negligible. Again, because of reactor design, this corresponds to an RPECVD process with capacitive coupling.[44]

For both types of RF coupling, He dilution has a very strong influence on the composition of the deposited films. This is because of the promotion of atomic nitrogen formation in the gas phase and the assistance it provides in delivering additional energy to the surface of the growing film. This leads to changes in the relative amounts of Si–H$_i$ and N–H$_j$ bonds and in the Si : N ratio in the deposited films.

If the role of helium in RPECVD processes based SiH_4–N_2 is not completely understood, processes bases on the use of NH_3 are even less clear. From results of optical emission spectroscopy (OES) of the plasma phase it has been shown only[7] that:

- For intermediate and high dilutions of NH_3 with He, up to about 95 % He, OES shows the same NH and H species; also the films contain decreasing amount of Si–NH bonds.
- For still higher dilution, >95% He, OES indicates a narrow window with a significant decrease in NH emission and an increase in N emission; this corresponds to a region where stoichiometric films can be grown.
- At the highest He dilutions, >98% He, He emission dominates, and the films display significant stoichiometric deviations, with both Si–Si and Si–H bonding configurations.

The influence of Ar dilution on remote PECVD of SiN films in a SiH_4–N_2–Ar system has been carefully studied using OES.[19] As for the case of He dilution, with SiH_4–N_2–Ar a high concentration of atomic nitrogen is produced as a result of long-lived noble gas atoms in metastable electronic states. This leads to changes in the relative amounts of Si–H$_i$ and N–H$_j$ bonds (Figure 12.10)[20] and in the Si : N ratio in the deposited films. Such variations of film composition and related properties with dilution by argon of nitrogen passed through the discharge region are found for RPECVD with both inductive and capacitive RF coupling.[20]

For inductively coupled RPECVD of silicon nitride films using NH_3 as the nitrogen-containing precursor, increasing dilution of NH_3 by argon results in a decrease in the film deposition rate and in an increase in the concentration of Si–H$_i$ bonds in the deposited films.[30,31] The authors suggest that an increase in the degree of ammonia dilution "opens up the a-Si–H deposition channel".[7]

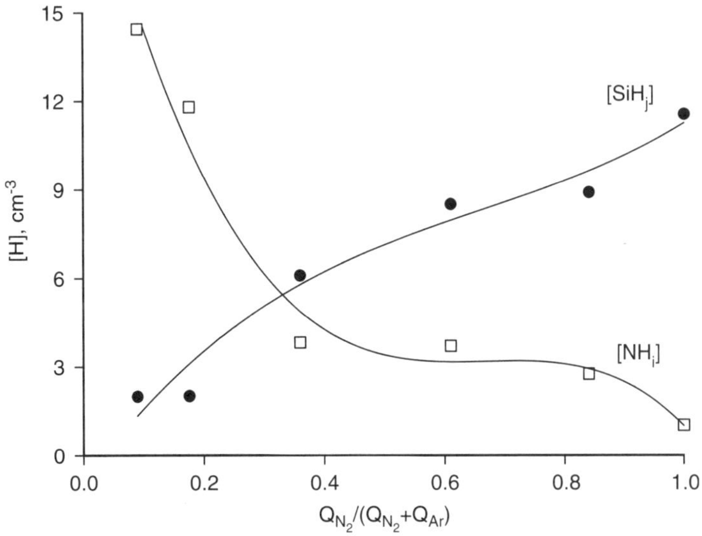

Figure 12.10 Content of Si–H$_i$ and N–H$_j$ bonds in SiN$_x$H$_y$ films as a function of $Q_{N_2}/(Q_{N_2}+Q_{Ar})$ ratio. (Reproduced with permission from ref. 20.)

12.2.3.4 Properties and Applications of Silicon Nitride Films Formed by RPECVD

Not unexpectedly, the main physical and chemical properties of silicon nitride films are determined by their composition.[9,36,50] A systematic study of etch rate in NH_4F buffered HF solutions and the electrical properties of nominally stoichiometric, Si-rich and hydrogenated silicon nitride films deposited by electron cyclotron resonance RPECVD[50] has confirmed conclusions reached about the properties and composition by many other papers. For example, silicon nitride films with near stoichiometric composition and minimal content of bonded hydrogen are characterized by properties similar to those of silicon nitride films deposited by high temperature CVD.

Property studies have also shown that, for reasons discussed above, silicon nitride films deposited by RPECVD have better characteristics than those obtained by the direct PECVD approach. Because of the advantages of RPECVD already mentioned, it is possible to deposit high quality dielectric silicon nitride films without any damage to semiconductor substrates, and the possibility of making a perfect interface to the semiconductor substrate is clearly an important benefit of RPECVD. However, the relatively low growth rates obtained with RPECVD, also mentioned above, could limit possible technological applications in microelectronics.

Because of the good interface with the substrate that can be achieved with RPECVD silicon nitride films, such films have been shown to provide excellent electronic passivation of silicon surfaces.[51] This property, together with high values of refractive index, make silicon nitride films deposited by RPECVD promising surface-passivating, antireflection coatings on silicon solar cells. However, because of an increase in the extinction coefficient with increasing refractive index, optimization is required for silicon nitride antireflection coatings based on growth by RPECVD.

In this context the dispersion of refractive indices and of extinction coefficients of RPECVD silicon nitride films, porous SiO_2 and several other relevant materials (*e.g.* MgF_2, TiO_x, ZnS, B270 crown glass, soda lime glass, ethylene vinyl acetate and resin as used in commercial photovoltaic modules) has been studied.[52] For example, an investigation of the thermal stability of electronic surface passivation by silicon nitride films deposited by RPECVD[53] found that such films provide an excellent, thermally stable surface passivation if they have a refractive index in the range 2.0–2.2. After a short annealing at temperatures above 850 °C the surface recombination velocity on Si (1.5 Ω cm, p-type, float-zone) stayed at a very low level of about 20 cm s^{-1}. The first bifacial silicon solar cells with screen-printed rear contacts on silicon substrates yielded a very promising rear efficiency of 13.4%. It has also been found[54] that hydrogen radical (H$^\bullet$) annealing of c-Si cells with RPECVD silicon nitride films improved the efficiency of the cells. A cell efficiency of 21.8% was obtained by applying a SiN_xH_y/SiO_2 double-layer structure on the emitter of a PERL-type solar cell. The authors demonstrated that H$^\bullet$ annealing caused two effects: reduction of surface recombination velocity and degradation of the bulk-lifetime in p-type Si crystals.

A rather different example of interfacial features is provided by results from a study of the interfacial properties of RPECVD silicon nitride films deposited on NMOSFETs and PMOSFETs.[55] Here, in contrast to the comments made above, in both cases interfacial properties were severely degraded, with the PMOS interfaces undergoing more significant degradation than NMOS interfaces. The authors suggested that this indicated a relatively high density of interface traps located below the Si mid-gap that inhibited hole channel formation. These traps originate from the intrinsic nature of the Si-"RPECVD silicon nitride" interface. However, results from a systematic study of the oxygen and hydrogen content in the silicon nitride films and their effect on electrical properties[52] allowed the authors to conclude that the presence of oxygen, either as a monolayer at the interface or within the silicon nitride film, can still produce high quality interfaces suitable for aggressively scaled CMOS devices.

In another study ultrathin (1.9 nm) nitride/oxide dual layer gate dielectrics have been prepared by RPECVD of silicon nitride onto oxides.[56] It was shown that compared to PMOSFET's with heavily doped p+-poly-Si gates and oxide dielectrics, devices incorporating RPECVD stacked nitrides display reduced tunnelling current, effectively no boron penetration from the poly-Si gate to the substrate, and improved interface characteristics. By preventing boron penetration into the bulk oxide and channel region, gate dielectric reliability and short channel effects were significantly improved. Also, compared with the oxide layer alone, the tunnelling current is lower by about an order of magnitude for nitride-oxide films structures with an approximately 1.9 nm equivalent oxide thickness. It was suggested that the suppression of boron transport could be explained by a percolation model in which B diffusion is blocked in a nitride that is sufficiently thick, with the blocking being proportional to the oxide fraction in oxynitride solutions.[56]

The effect of process parameters, such as deposition temperature and especially the NH_3 : SiH_4 ratio, on the properties in general of silicon nitride films deposited by RPECVD, and particularly on the Si-SiN_xH_y interface, has been investigated.[57] It was found that the interface trap density is related to the amount of silicon intermediate species and Si–NH bonds at the Si-SiN_xH_y interface, and the trap density can be minimized by reducing the intermediate Si species and Si–NH bonding states. The films showed a mid-gap interface trap density in the region 2×10^{11}–$2 \times 10^{12}\,cm^{-2}$, compared to a trap density at a good Si–SiO_2 interface of the order of $10^{10}\,cm^{-2}$.

An evaluation has been made of possible applications of RPECVD silicon nitride films for III-V semiconductor based metal insulator semiconductor (MIS) structures.[50] To determine the practical applicability of RPECVD silicon nitride films, various transfer characteristics of GaAs-based depletion-mode MISFETs were measured. From stringent conductance measurements, interface state densities in the high $10^{10}\,eV^{-1}\,cm^{-2}$ were obtained, and the performance of GaAs-based field effect transistors in switching and power applications was enhanced substantially by employing MIS structures with RPECVD silicon nitride films. Taking into account the numerous unsuccessful attempts in the development of GaAs-based MISFET technology, these results can be considered as very promising.[50]

12.2.3.5 Summary of RPECVD

From this section it can be concluded that RPECVD processes provide several advantages over direct PECVD processes. First, there is the protection of the substrate and growing film surfaces from degradation due to bombardment by energetic particles during deposition; this is particularly important for semiconductor applications. Then there is less interdependence of some process parameters and so they can be more readily independently controlled. This means, for example, that with RPECVD of SiN_xH_y films it is possible to achieve layers closer to stoichiometric composition than with direct PECVD, even though RPECVD gives a lower growth rate. For this particular case, RPECVD can also deposit layers with low concentrations of incorporated bonded hydrogen and excellent physical and chemical properties, comparable to those of Si_3N_4 prepared by high temperature CVD. A final benefit of RPECVD processes is that they provide an approach to controllable plasma deposition processes through the use of only neutral species from the discharge generation region for selective excitation of precursor molecules in the deposition zone.

The lower growth rates of RPECVD compared with direct PECVD is, of course, related to the transfer of energy from the plasma region to the growth surface. Use of atoms of various gases, particularly noble gases, with metastable long-lived electronic states, is one way of enhancing energy transfer from the plasma region. The example considered of the use of argon atoms in metastable states for the selective excitation of nitrogen molecules, with formation of atomic nitrogen, for enhancing the deposition of different nitride films shows the possibility of improving RPECVD deposition rates.

12.3 Atmospheric Pressure PECVD (AP-PECVD) using Non-thermal Plasmas

12.3.1 Introduction

Section 12.2, on RPECVD, shows that the technique can be used for different applications. It can also be used for a range of materials, both inorganic and organic, often with novel properties. However, the use of low pressures for PECVD processes brings limitations for various industrial applications. For many industrial products it is not practicable to use vacuum technology with load locks for large-scale processes, and even if it were practicable the high capital and running costs of vacuum equipment can become prohibitive. As a result of these limitations, there has been considerable interest in recent years in the development of atmospheric pressure, non-thermal plasma sources suitable for use with CVD technologies.[6] In this section we review some of the results achieved with atmospheric pressure PECVD (AP-PECVD) processes.

12.3.2 Sources of Atmospheric Pressure, Non-thermal Plasmas

The various types of electrical discharge that can be used to generate non-thermal plasmas at atmospheric pressure have been reviewed.[58] These discharges occur in suitable gaseous atmospheres and are, typically, low frequency, dielectric barrier, glow or filamentary discharges, various types of corona discharge, radio-frequency (RF) discharges in narrow gaps (which may include a dielectric barrier) or microwave (MW) discharges. The main aspect of these discharges and their common features are that with gas molecules passing through the discharge generation region the electrical input power generates plasmas with highly energetic electrons and yet remain "cold". Inelastic collisions of the energetic electrons with gas molecules produce chemically active species such as free radicals, atoms and ions that can be involved in chemical reactions leading to layer deposition.

There are several approaches for maintaining a low temperature of neutral species in a gas passing through an electrical discharge at atmospheric pressure:[59]

- Limiting the input energy absorbed in the discharge by restricting the voltage applied or the discharge current.
- Providing efficient cooling of the plasma stream by increasing the rate of heat transfer through turbulent flow or cooled electrodes or sidewalls.
- Decreasing the residence time of the gas stream in the discharge generation region.

Dielectric barrier, surface and corona discharges usually occur under conditions with limited absorbed power with the equipment being designed to limit the applied voltage or discharge current; the design can also help prevent arcing. Because of this limitation on input energy the absorbed power density (*i.e.* the watts per unit volume of the discharge generation region) is generally rather low (typically, $<100\,\mathrm{W\,cm^{-3}}$), and heavy particle temperatures do not usually exceed 200 °C; the mechanism of energy absorption in the case of RF or MW discharges is, however, quite different. These types of discharge provide high efficiency energy absorption in the volume of the discharge and this feature, coupled with the low thermal conductivity of plasmas at atmospheric pressure, means that cooling of the plasma stream is needed to obtain non-thermal plasma characteristics. Such plasmas generated by RF and MW gas discharges are usually sustained in narrow gaps between cooled electrodes or in narrow capillaries. In addition, even though direct current (dc) discharges have a different ignition mechanism to ac discharges, non-thermal plasmas can also be obtained by using intensive cooling of the sidewalls of the plasma source. For all types of discharge, high flow rates of gas streams through the discharge can assist in keeping the gas temperature low.

Even though there are various types of gaseous discharge at atmospheric pressure that can be used to generate non-thermal plasmas, the discharges are characterized by rather similar properties – Table 12.2.[2,60] For example, as can be seen, the electron temperature (T_e) for the various types of non-thermal plasma lies in a relatively narrow range of 1–10 eV, while for all cases the temperature of heavy particles (*i.e.* the gas temperature T_g) is relatively low and does not exceed 1000 K. Also, on the whole, the parameters of many types of non-thermal plasmas generated at atmospheric pressure are similar to those for low pressure PECVD processes.

Therefore, one can expect some similarities between many of the characteristics of low pressure (LP) and AP-PECVD processes. At the same time, though, the higher pressure associated with AP-PECVD processes would be expected to lead to some specific differences from LP processes. So, for example, for AP-PECVD:

- Partial pressures of precursors are about three orders of magnitude higher than in LP systems and this can lead to a significant increase in the rate of homogeneous reactions; in many cases these reactions are responsible for the formation of powders deposited at the same time as a layer grows heterogeneously.
- The significantly higher precursor partial pressures can lead to higher film growth rates compared with LP processes, which in turn can cause depletion of the precursor (or of intermediate active species) in the gas stream over the substrate, resulting in poor thickness uniformity.
- Mass transport limitations become much more significant at higher pressures and this can cause poor thickness uniformity as well.
- For remote AP-PECVD processes, where the plasma stream exiting the discharge generation region is mixed with a gas stream containing the precursor, the fast homogeneous reactions of the active plasma species necessary for layer growth means that the good intermixing required for uniform layer distribution over the substrate surface can be difficult to achieve.

These particular features of AP-PECVD processes need to be taken into account when developing equipment and choosing precursors. However, despite the potential problems arising from the use of high reaction pressures, AP-PECVD techniques present considerable opportunities for simple, cost-effective depositions. In particular, the absence of vacuum equipment means there are the possibilities of cost savings and significant simplifications for continuous sample treatment. Some particular examples of the promising benefits of AP-PECVD processes are discussed in the following sections.

12.3.3 Applications of Electrical Discharges for AP-PECVD

12.3.3.1 Dielectric Barrier Discharges (DBD)

Dielectric barrier discharges (DBDs) are one of the most promising sources for the generation of non-equilibrium plasmas used in AP-PECVD processes. They are characterized by the presence of one or more insulating layers in the current path through the discharge space between two metal electrodes;[61] planar or cylindrical configurations have been described.[59,62,63] A closely related technique involves surface discharge configurations in which a discharge is initiated at a dielectric surface as a result of strong electrical fields generated by embedded metal electrodes;[64] the presence of the dielectric prevents a dc discharge.

A typical range of operating frequencies lies between 50 Hz and 500 kHz, with gap spacings typically of a few mm, and ac driving voltages of 5–10 kV. For a DBD so-generated, breakdown is initiated in most gases by a large number of independent current filaments of microdischarges (streamers) with diameters of about 0.1 mm. At the dielectric surface the microdischarges spread

Table 12.2 Overview of parameters for various atmospheric pressure discharges.[2,60]

	Operating parameter				*Plasma characteristic*		
Discharge type	*Discharge frequency (MHz)*	*Power input (W)*	*Power density ($W\,cm^{-3}$)*	*Discharge gas flow rate (slm)*	*Electron temperature (T_e) (eV)*	*Electron density (n_e) (cm^{-3})*	*Gas temperature (T_g) (K)*
RF non-thermal	1–40	20–500	3–30	0.5–50	1–4	10^{11}–7×10^{12}	300–700
Glow discharge between two electrodes with dielectric	0.00005–0.05	10–2000	0.5–1	1–10	3–5	10^{11}–5×10^{11}	300–500
DBD	0.00005–0.05	20–1000	1–50	0.2–5	1–10	10^{14}–10^{15}	300–400
Corona	0.001–1		1–50	0.5–4	3–7	10^{9}–10^{13}	300–400

Table 12.3 Characteristics of DBD microdischarges.[58]

Operating parameter		*Discharge characteristic*	
Current density	100–1000 A cm^{-2}	Filament duration	1–10 ns
Peak current	0.1 A	Filament radius	0.1 mm
Total charge	0.1–1 nC	Electron density	10^{14}–10^{15} cm^{-3}
		Electron energy	1–10 eV
		Gas temperature	300–400 K

out into a surface discharge, covering the whole region. Table 12.3 summarizes typical micro-discharge properties for a 1 mm air gap at 1 bar.

Under certain conditions pulsed DBD discharges can show a homogeneous normal glow, or an atmospheric pressure glow discharge (APGD).[65] This can be influenced by frequency selection, power of excitation and gas type; the presence of He atoms[61] in metastable states promotes the sustaining of a uniform glow plasma. The electrode geometry can also have an influence with, for example, thin mesh between the insulators and metal electrodes allowing an APGD at even 50 Hz.[66] It has been suggested that perfect matching of the power supply and discharge assembly is the main condition leading to the appearance of a glow discharge,[67] but the mechanisms of DBD with APGD are still unclear.

The most widely used arrangement for DBD AP plasma processes is with the substrate having its surface orthogonal to the filamentary discharge. However, with this substrate orientation, undesirable surface damage or surface heating can occur. To overcome these drawbacks, planar type DBD systems with the substrate and discharge in a parallel orientation have been proposed.[64] This type of DBD system has been realized with coplanar, surface and insulated surface barrier discharges[64] and is promising for extended area AP-PECVD processes since these configurations can allow for the design of large, PECVD reactors with uniform plasma densities. Another approach to minimizing surface damage is to employ remote AP-PECVD, as discussed below.

Many papers on the application of DBD discharges for AP PECVD processes have been published, including several reviews.[58,59,61,62] Some of the earlier papers described the use of a DBD discharge for polymerizations, with the first attempt for the AP-PECVD of polyethylene coatings made in 1979.[68] A uniform glow discharge was sustained by a 60 Hz power supply in ethylene–He gas mixtures introduced between two flat electrodes covered with insulator layers. Uniform films were deposited on glass substrates oriented parallel to the surface of the electrodes. Later on, a filamentary type of discharge was successfully used for deposition of polymeric thin films, again on glass substrates,[69] and by substituting acetylene for ethylene deposition rates of up to 40 μm min^{-1} could be achieved. Polymerizations from hydrocarbons in a DBD at atmospheric pressure show, in general, fast kinetics,[70,71] and therefore are very promising for industrial processes for growing polymer-like materials for protection, lubrication, *etc.* Figure 12.11 shows a typical system for the study of plasma polymerizations at atmospheric pressure.[72,73]

The discharge arrangement consisted of two disk-shaped Al electrodes, each 150 mm diameter and both covered with Al_2O_3 plates 2 mm thick. The gap width between the un-cooled electrodes was typically limited to a few millimetres to ensure stable plasma operation. In some studies[72,73] the reaction chamber has been initially evacuated and then filled with the process gases to prevent the influence of water vapour, CO and CO_2.

With this type of equipment, polymerization of ethylene has been studied.[72,73] The flow rate was varied between 1 and 20 L min^{-1} and the DBD was produced by means of a 20 kV/200 mA power supply with a frequency between 1 and 4 kHz. With a plasma formed from He mixed with ethylene in the ratio of 10 : 0.5 the ethylene polymerized to give a low density, sticky, opaque polymer on the substrates of silicon, glass or stainless steel placed on the lower powered electrode. If Ar was used

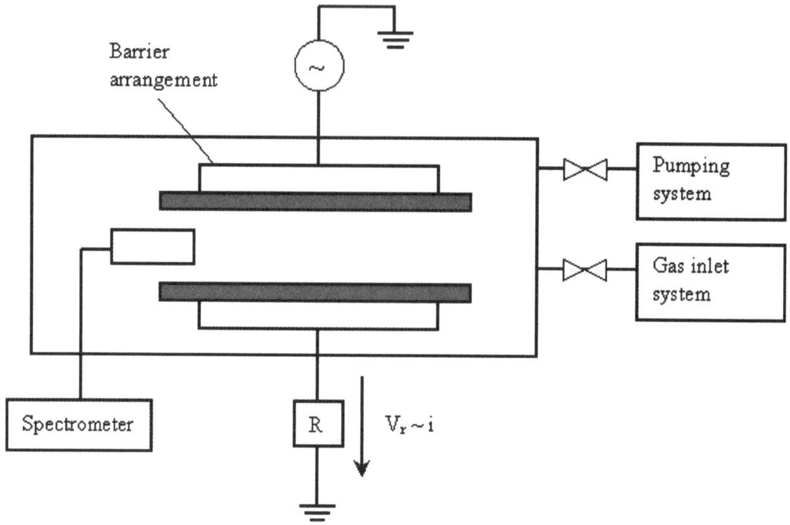

Figure 12.11 Schematic of an experimental arrangement for plasma polymerization by DBD AP-PECVD. (Reproduced with permission from ref. 72.)

instead of He then a more solid polymer with good adhesion to all substrates was obtained. The polymer from the He plasma dissolved readily in chloroform, in contrast to the one obtained from the Ar plasma, which was practically insoluble. Differences in physical appearance and solubility clearly indicate differences in molecular structure of the two types of material.

An interesting design of reactor (Figure 12.12) has been described for AP-PECVD based on the use of surface planar DBD discharge[64] for the formation of polymer films on the inner surfaces of PVC tubes.[74] This reactor had special electrodes consisting of two electrically conductive parallel structures – two copper plates, each <3 mm wide – helically wound around a cylindrical glass tube. This was embedded in a plastic jacket and the space between the electrode and jacket was filled with insulating silicone oil. The plastic jacket was sealed at both ends with the copper leads protruding at each end, and the high-voltage and ground leads of the power supply were connected to the electrode leads. The sample PVC tube was inserted into the glass tube of the electrode. The carrier gas, which was He (600 sccm), and the monomer – tetrafluoroethylene (TFE) or hexafluoropropylene (HFP) (2.5 sccm) – flowed directly into the PVC tube at atmospheric pressure; optimum gas flows were determined in a separate study.[75] When a voltage of 5 kV – with an alternating current of 20 kHz – was applied an APGD was generated that was limited to the inner surface of the PVC tube. A PTFE-like polymer was formed from both fluoromonomers with good uniformity and a thickness of about 1 mm. The fluoro-polymer layers deposited using AP-PECVD process showed similar properties to those obtained by more conventional low pressure PECVD.

By combining APGD technology with a unique precursor delivery system Dow Corning Plasma has developed a new coatings approach – Atmospheric Pressure Liquid Deposition.[76,77] Operating at atmospheric pressure and ambient temperature this process allows the use of a wide range of liquid precursors for achieving high coating rates onto flexible substrates. For example, polypropylene film with a line speed of 1 m min^{-1} was coated with highly hydrophilic polyacrylic acid by introducing acrylic acid liquid precursor into the chamber at 30 slm with a He APGD at a power 0.4 W cm^{-2}. Introduction of heptadecafluorodecene liquid precursor at 100 slm in He into the chamber, with a polypropylene film substrate line speed of 1 m min^{-1} and an APGD at a power of 0.6 W cm^{-2}, produced an oleophobic, fuel resistant surface.

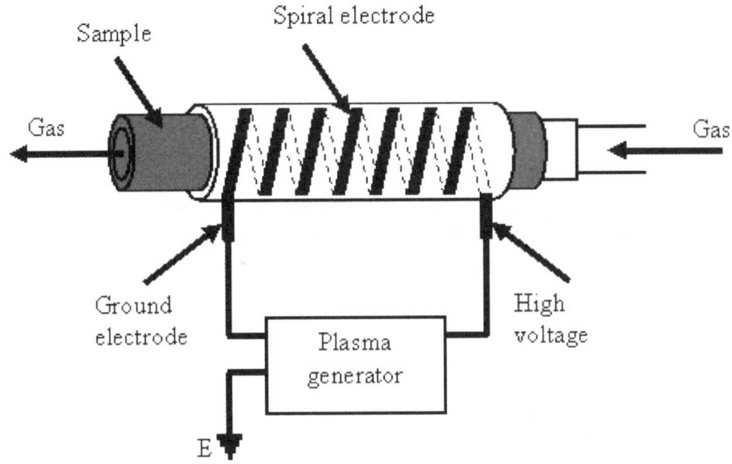

Figure 12.12 Experimental set-up for DBD AP-PECVD deposition of polymer films on the inner surfaces of tubes. (Reproduced with permission from ref. 74.)

Recently, barrier coatings have been obtained from hybrid organic–inorganic precursors as well as from purely organic compounds.[78,79] The DBD was produced between two parallel stainless steel electrodes, both covered with an insulating glass plate 3 mm thick. Plasma discharges were generated by a 20 kV/200 mA power supply with a frequency of 9 kHz. The output power was maintained at 150, 300 or 450 W, resulting, respectively, in a power density of approximately 0.5, 1 or 1.5 W cm^{-2} in the plasma zone. Nitrogen was used as a carrier gas as well as for generating a precursor aerosol. The total N_2 flow rate was 20 slm. A hybrid precursor, 1-hexene and vinyltriethoxysilane (VTES), was atomized and mixed with the carrier gas flow before entering the plasma zone where plasma polymerization took place. The deposition rate of polymeric films from 1-hexene was in the range 12–30 nm min^{-1}, while the growth rate from VTES was 180–320 nm min^{-1}. In both cases the growth rate increased linearly with power density. Atmospheric pressure plasma polymerization of hybrid polymer precursors leads to coatings with excellent barrier properties because of unique synergistic effects of the organic and the inorganic network structures that are formed in the plasma.

Layers of inorganic materials can also be successfully deposited by AP-PECVD based on the use of DBD discharges.[80–83] Silicon dioxide (or, more correctly, silicon dioxide-like) layers are the most widely studied layers deposited by AP-PECVD. These depositions have been based on the use of organosilicon precursors such as hexamethyldisiloxane (HMDSO), hexamethyldisilazane (HMDSN), diacetoxy-di-*t*-butoxysilane (DADBS) or tetraethoxysilane (TEOS), with TEOS being the most frequently used.[80,84–86]

With HMDSO, the reaction systems HMDSO–Ar, HMDSO–N_2, HMDSO–O_2–Ar and HMDSO–N_2O–Ar have been investigated.[80,84] Figure 12.13 shows a typical reactor for production of thin silicon oxide films with a barrier discharge. Here, an external high-voltage electrode (70 mm in diameter) was separated from the discharge space with a dielectric barrier of a quartz-glass Plate 1 mm thick. The heated grounded electrode was not shielded and the substrates were placed directly on its surface. The discharge gap was 1–2 mm and efficient discharges of the filamentary type could be obtained at moderate frequencies of 5 kHz.

A comparison of films obtained from mixtures of HMDSO and argon alone with those involving oxidants (*e.g.* oxygen, nitrous oxide)[80,84] has shown similar results for the two cases. The molar ratio of the oxidant to HMDSO has been varied over a wide range (*e.g.* O_2/HMDSO, 0–150; N_2O/HMDSO, 0–9.5). In all cases the films contained some organic fragments detectable by FTIR,

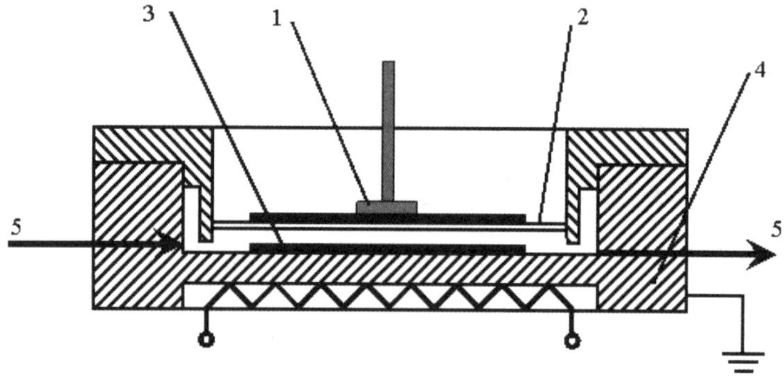

Figure 12.13 Typical reactor with an external high-voltage electrode used for deposition of silicon oxide films: (1) high-voltage electrode, (2) dielectric glass barrier, (3) substrate, (4) grounded, heated electrode and (5) gas flow. (Reproduced with permission from ref. 82.)

mainly as absorption bands of Si–CH_n groups; other workers have observed similar results under comparable conditions, but using a narrower range of O_2/HMDSO molar ratio.[87] However, not unsurprisingly, the films obtained in the presence of O_2 or N_2O contained less carbon than those without an oxidizer.[80,84,87]

Deposition rates from mixtures involving HMDSO (typically, greater than 10 nm min^{-1}) are generally higher than those observed when TEOS is used (not higher than about 5.5 nm min^{-1}). Similar trends are found for deposition rates based with APGD, *e.g.* 5–22 and 25–35 nm min^{-1} from TEOS–O_2 and HMDSO–O_2 mixtures, respectively.[88] Much higher growth rates (up to 120 nm min^{-1}) for silicon oxide-like films have been achieved with a homogeneous atmospheric pressure DBD in helium and small admixtures ($<10^{-3}$) of HMDSO.[89] In this case the DBD discharge was maintained between two plane electrodes with the powered one covered by a glass insulator layer. Typical electrical parameters were voltages of 1.4–3.2 kV at 100 kHz, discharge current densities up to 20 mA cm^{-2}, maximum gap voltages of 700 V and powers of 1–4 W. In the study an attempt was made to study gas phase composition using mass-spectrometry. A decrease of the HMDSO concentration after discharge ignition to about 20% of the value without discharge was observed. The authors suggested that the increase in m/z of 15, 16, 25, 26 and of peaks in the range of m/z 35–47 indicated the formation of hydrocarbons such as CH_4, C_2H_2 and C_3H_x. An enhanced signal of m/z 73 $[Si(CH_3)_3{}^+]$ could be attributed to the formation of tetramethylsilane.

A reactor used previously for the deposition of silicon oxide layers[80] has been used to study the possibility of depositing silicon nitride with the gas mixture HMDSN–N_2–NH_3.[84] Absorption peaks corresponding to Si–N and Si–NH–Si groups were observed in FTIR spectra of the deposit, but peaks related to incorporated organic and other fragments (*e.g.* Si–CH_n, Si–H) were also observed. Furthermore, Si–O–Si absorption bands were often seen as well. XPS analysis confirmed the presence of carbon and oxygen in some films with typical compositions beings Si: 51–54, O: 15–17, N: 20–28 and C: 4–6 at.%. In addition, the refractive index of the films varied between 1.23 and 1.68, showing their very non-stoichiometric nature. Clearly, careful regulation of the reactor environment is required if uncontrollable doping of deposited films with oxygen and other elements is to be prevented.

As indicated above, TEOS is widely used as a precursor for AP-PECVD of silicon oxide like layers.[85,86] Thin films containing Si, C and O have been deposited on polycarbonate from TEOS by pulsed DBD.[85] The influence on the deposition rate of plasma parameters, such as the energy of a single pulse and the position of polycarbonate sheets on the electrodes, has been investigated. Results showed that, not unexpectedly, the deposition rate increased with increasing energy of a single pulse; for example, deposition rate increased from 4.6 to 37 nm min^{-1} when the single pulse energy increased from 6.2 to 9.1 mJ. This plasma technology was demonstrated to successfully deposit transparent, smooth and homogeneous films.

Siloxane precursors such as tetramethylcyclotetrasiloxane and octamethylcyclotetrasiloxane can also be used to prepare SiO_x, particularly for hydrophilic coatings.[76] AP-PECVD with DBD has been used as well to improve the deep colouring effect of polyethylene terephthalate (PET) fabrics by the deposition of anti-reflective coatings using two different organo-silicon precursors such as hexamethyldisilane (HMDS) and tris(trimethylsilyloxy)vinylsilane (TTMSVS).[90] With both precursors, helium was used as the carrier gas and deposition was at room temperature with a He/O_2/HMDS or TTMSVS ratio of 8/2/0.04. Growth rates and properties of the deposited films were not reported. However, it was noted that while the morphology of the PET surface was not changed, the optical properties of the PET surface did vary. Minimum reflectance was observed when the deposit was 150–200 nm thick. FTIR analysis of the deposited films revealed the presence of a Si-O-Si peak in the spectra. It was suggested that the Si–O bonding process was enhanced by addition of O_2. As a result, the films obtained from TTMSVS (a precursor with intrinsic Si–O bonds) were characterized by a higher intensity of the Si-O-Si peak at 1050 cm^{-1} than films deposited from HMDS. It was demonstrated that PET with layers deposited from TTMSVS had better anti-reflection properties than those obtained from HMDS, and this was probably because the greater Si–O bonding structure of the former leads to a more silica-like structure with an inherent lower reflection.

As well as silicon dioxide-like layers, titanium oxide films have also been deposited by AP-PECVD with DBD.[91] A reactor that has been used is based on a laminar flow design where the reaction gases were introduced between parallel plate electrodes inside a silica tube (Figure 12.14). The electrodes consisted of rectangular metal plates, each with a surface area of 186 cm^2. The powered electrode at the top was run at 33 kHz and had a peak potential of 10 kV; it was separated from the reaction space by a silica plate.

The main working gas was helium, typically delivered at 3 slm, and a small amount (5 sccm) of oxygen was added. In addition, a helium flow of 0.5 slm was passed through titanium tetrachloride in the bubbler, which was kept at 40 °C. Amorphous titanium oxide films were grown for 0.5–3 min

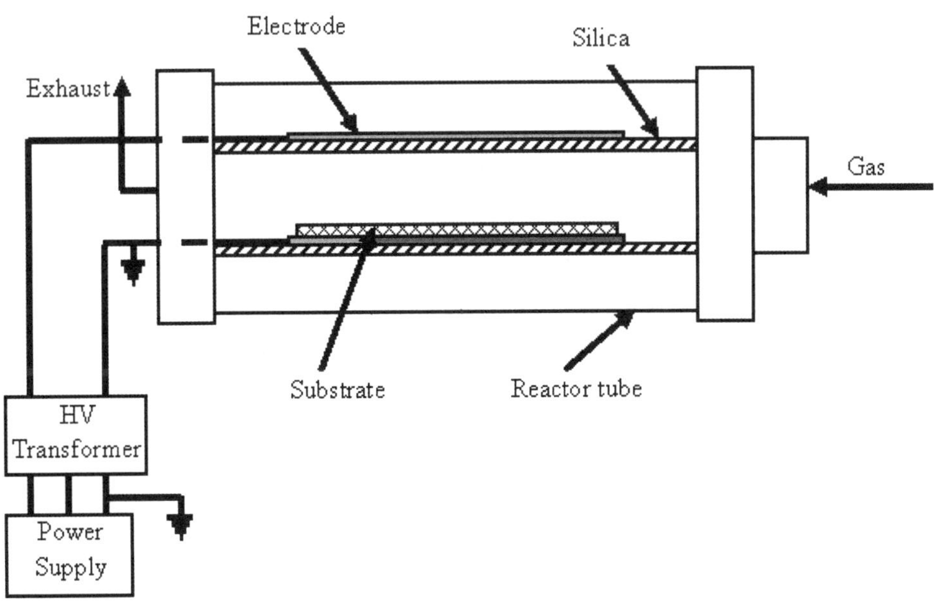

Figure 12.14 Experimental DBD laminar flow reactor design used for deposition of titania layers. (Reproduced with permission from ref. 91.)

onto soda-lime glass plate substrates or onto cellulose acetate films attached to the glass plates with adhesive tape. The growth rate was about 60–70 nm min^{-1}.

Another very simple design of a DBD reactor consisting of two parallel plate electrodes with a glass dielectric barrier, 1.5 mm thick, has been successfully employed for deposition of diamond like carbon (DLC) films using CH_4 and C_2H_2 as precursors.[83] The authors reported successful deposition of hydrogenated diamond-like films (α-C:H) with extremely high growth rates (tens of μm h^{-1}). The coatings with the best properties were deposited in the system 5% C_2H_2/95% H_2.

As discussed earlier, the conventional approach of PECVD where substrates are placed on one of the electrodes has drawbacks because of the possible generation of defects and pinholes in the deposited films, particularly when the discharge is filamentary. As a result of these problems, remote AP-PECVD processes have been investigated, where the plasma stream exiting the discharge generation region is mixed with a gas stream containing the precursor.

A systematic study of remote AP-PECVD with DBD for deposition of silicon oxide films has been carried out.[92] Two experimental AP-PECVD arrangements, one using a horizontal reactor and one with a vertical reactor, were developed, but most of the studies have been with the vertical reactor. The vertical glass reaction chamber consisted of two zones. The top zone of the reactor was used for the generation of the barrier discharge between two coaxial insulated copper electrodes. The distance between plasma generation region and susceptor was varied over 30–100 mm. Typical electrical characteristics were 3.5–15 kV at 50 Hz and 40–90 W of power. Electronic grade argon and oxygen were introduced into the top of the barrier discharge gap and passed through it in the direction of the substrate at a flow rate of about 1 slm. Three organosilicon compounds mentioned already, namely, hexamethyldisiloxane (HMDSO), hexamethyldisilazane (HMDSN) and diacetoxy-di-*t*-butoxysilane (DADBS) were studied as precursors for the AP-PECVD of silicon dioxide films. High quality silica films could be deposited with all three precursors, but HMDS had the potential for achieving high growth rates (up to 25–30 nm min^{-1}) while DADBS, although having a substantially lower growth rate ($<$3 nm min^{-1}), allowed deposition at the lowest temperatures – less than 300 °C.

A rather different type of remote AP-PECVD DBD system has been used for the deposition of polymeric films from two monomers, hexafluoropropene (C_3F_6) or trifluoroethylene (CF_2CFH).[93] A spray-type remote reactor (Figure 12.15) was used that consisted of electrodes covered with Pyrex glass tubes.

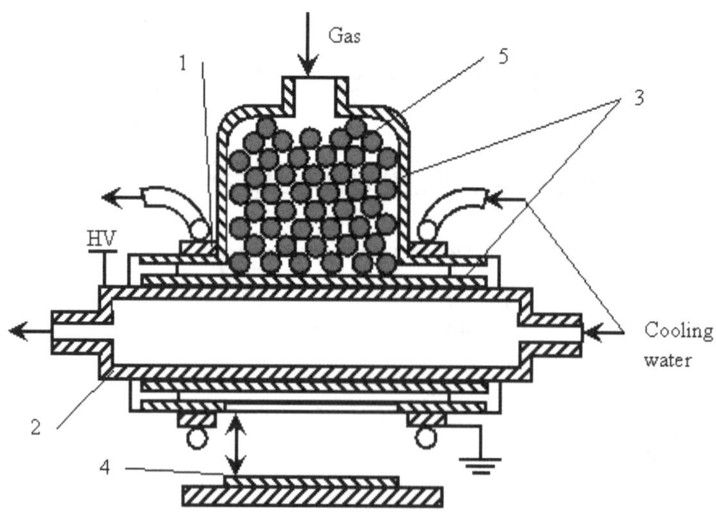

Figure 12.15 Schematic diagram of a spray-type remote AP-PECVD reactor: (1) outer electrode, (2) inner electrode, (3) Pyrex glass, (4) sample and (5) glass beads. (Reproduced with permission from ref. 93.)

The discharge part of the reactor had a double circular cylinder structure, and the rectangular nozzle (0.5 × 50 mm) was set in the sidewall of the outer glass tube. Deposition was carried out by introducing: a monomer diluted with He (or Ar), from the upper side of the reactor into the discharge area and then over the sample. Typical deposition conditions were a power supply frequency of 100 kHz, power 100 W, monomer flow rate 1–60 sccm and He flow rate 10 slm. The distance from nozzle to sample surface, as shown with a two-headed arrow in Figure 12.15, was 1.5 mm. Polymer films were deposited with growth rates of up to 60 nm min^{-1} with C_3F_6, and up to 200 nm min^{-1} for the CF_2CFH/He system.

From this brief survey of both direct and remote AP-PECVD with DBD it can be seen that layers of several polymeric and inorganic materials can be deposited with various material characteristics and over a range of growth rates. The suitability of the techniques for producing commercially interesting layers will need to be assessed for each particular application. However, their particular simplicity and the low power requirements make them an attractive technology for commercial applications, with remote AP-PECVD having the advantage of low damage to the substrate and the depositing layer.

12.3.3.2 Corona Discharges

AP-PECVD using a corona discharge has produced similar types of results to those obtained with DBD. One of the first elementary applications of corona discharge to AP-PECVD was in 1988 for the deposition of carbon films.[94] Figure 12.16 shows the simple design of the reactor used. The brush-style upper electrode consisted of 25 fine wires of stainless steel or tungsten. A continuous stable discharge was created by using 3 kHz or an RF voltage. Carbon thin films were deposited on

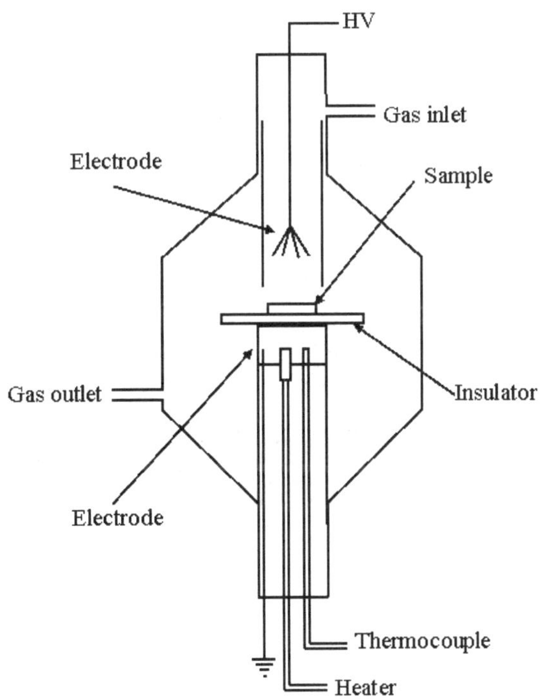

Figure 12.16 Schematic of an AP-PE-CVD corona discharge reactor. (Reproduced with permission from ref. 94.)

Pyrex or quartz substrates by decomposition of methane in H_2/He mixtures. Carbon films with the best properties were deposited at about 400 °C.

Another design based on AP-PECVD with corona discharge is shown in Figure 12.17. This has also been used for the deposition of organic and inorganic thin films at low temperatures.[95]

Here, the experimental setup consisted of four metal rods used as high voltage electrodes, with each covered by an alumina tube (diameter 12 mm and length 160 mm) as a dielectric barrier. The high-voltage electrodes were placed over a grounded and PTFE-covered aluminium plate, leaving a gas gap of about 2 mm. Process gases were introduced by two gas showers, each placed between two high voltage electrodes. To obtain homogeneous deposition the grounded substrate was mechanically moved back and forth. The discharge was driven by 10–20 kV at frequencies between 20 and 50 kHz.

With this equipment silicon oxide films have been deposited from tetramethylsilane (TMS) or TEOS. Increasing the discharge power and precursor concentration led to an increase of growth rate greater than 100 nm min^{-1}, but at the same time produced deposits of soft, polymer-like films. Also it was found that without movement of the substrate very rough and powdery deposits were obtained directly below the alumina tubes. For good quality films with a high scratch resistance the results of electron probe micro analysis (EPMA) and secondary ion mass spectrometry (SIMS) showed that the oxygen : silicon ratio was in the range 2.2–2.5, with, not unexpectedly, an increase in content of oxygen in the gas phase leading to deposition of films with lower carbon content. When oxygen was not used polymeric hydrogenated silicon-carbon films were deposited with TMS-Ar or TMS-N_2, as shown by the appearance of absorption bands corresponding to vibrations of Si-CH_3– bands in FTIR spectra of the films. These films also contained substantial amounts of oxygen, up to 27 at.%, possibly from the surrounding atmosphere.

With a corona discharge in C_2F_4, soft but smooth fluorocarbon, PTFE like coatings have been deposited with rather high growth rates of about 100–200 nm min^{-1}. The IR spectrum of the deposited films demonstrated the intense C–F absorption at 1200 cm^{-1} and the C : F ratio on the film surface was 5 : 8 (quite close that of PTFE).

A novel method, which combines a non-equilibrium plasma reaction with template-controlled growth technology, has been developed for synthesizing aligned carbon nanotubes at atmospheric

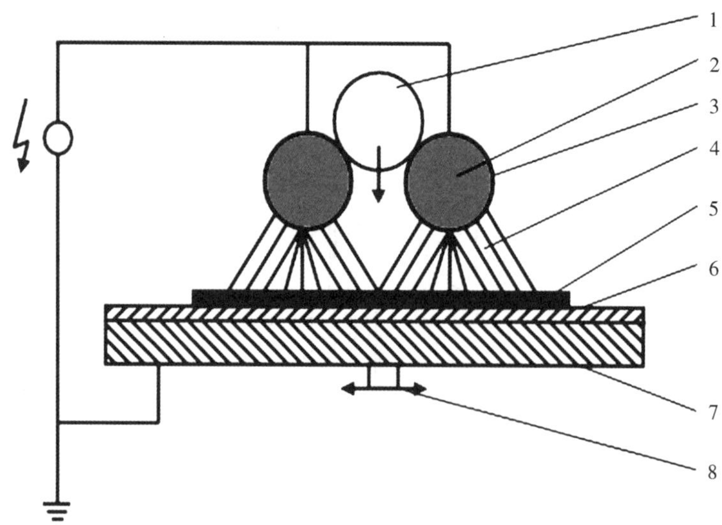

Figure 12.17 Another design for AP-PECVD by corona discharge:[95] (1) gas shower, (2) high-voltage electrodes, (3) alumina tube, (4) gas gap with microdischarges, (5) substrates, (6) PTFE, (7) grounded electrode and (8) substrate movement. (Reproduced with permission from ref. 97.)

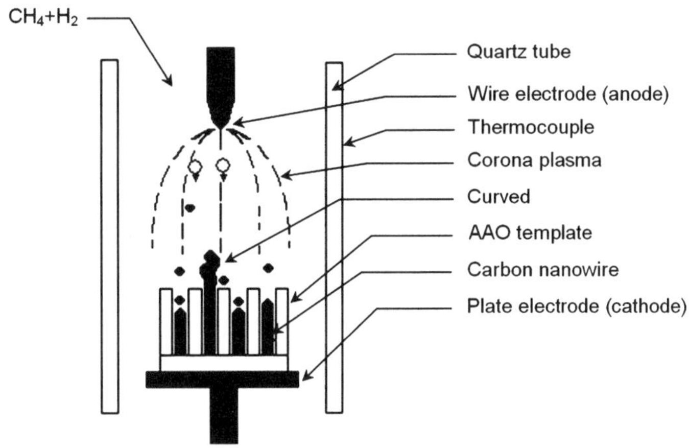

Figure 12.18 Schematic diagram of a template-directed synthesis of carbon nanotubes using atmospheric pressure corona discharge. (Reproduced with permission from ref. 96.)

pressure and low temperature.[96,97] Figure 12.18 shows a schematic presentation of the reactor. An anodized aluminium template with an area of 7×7 mm was placed on the plate electrode and the wire electrode was positioned with its tip 5 mm above the template surface. The gas reactant, a mixture of methane and hydrogen in a 1 : 10 ratio, was fed into the reactor at a total feed rate of 22 sccm. An ac high voltage generator operating at 8 kV and 25 kHz with power of 40 W was used to initiate corona discharge. The average power density of the corona discharge plasma was approximately 82 W cm^{-2} of template.

Utilizing this set-up, multiwall aligned carbon nanotubes with diameters of approximately 40 nm could be synthesized only in the channels of the anodic aluminium oxide template at a temperature below 200 °C. This method allowed carbon nanotubes to be fabricated at ambient conditions and the low synthesis temperature offers the possibility of practical applications as well as opening up a prospect for studying the growth mechanisms of carbon nanotubes.

Interestingly, the typical main parameters of plasmas generated by DBD and corona discharges are quite similar (*cf.* Table 12.2). Consequently, the similarities in the results of AP-PECVD processes based on the two types of discharge are not too surprising. However, because a DBD discharge requires a lower voltage than corona discharge and is more flexible from the point of view of reaction chamber design, the use of AP-PECVD with DBD has been more widespread than applications of corona discharges.

12.3.3.3 RF Discharges

The use of atmospheric pressure RF capacitive discharge sources for PECVD of thin films various materials has been reviewed recently.[98] In recent years a lot progress has been made in the development of RF atmospheric pressure plasma sources with planar (Figure 12.19)[99] and coaxial (Figure 12.20) electrodes.[98,100] As mentioned earlier, non-thermal plasmas under these conditions can be generated in narrow gaps by limiting the RF power or cooling the electrodes. In the planar electrode arrangement, He (50–300 sccm) can be flowed down over the alumina-coated electrodes, and a capacitively coupled plasma generated by 20–50 W of power.

Because of their simplicity of operation, a jet or torch are the main types of RF plasma sources that have been tested for AP PECVD processes. One of the first attempts to use an RF cold plasma torch was for deposition of silicon oxide films.[101] Figure 12.21 shows a schematic diagram of the

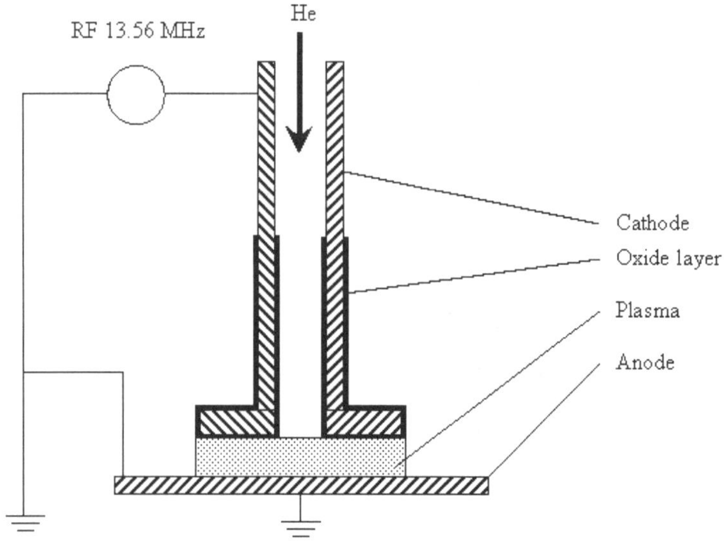

Figure 12.19 Schematic diagram of a planar RF AP-PECVD cold plasma reactor. (Reproduced with permission from ref. 99.)

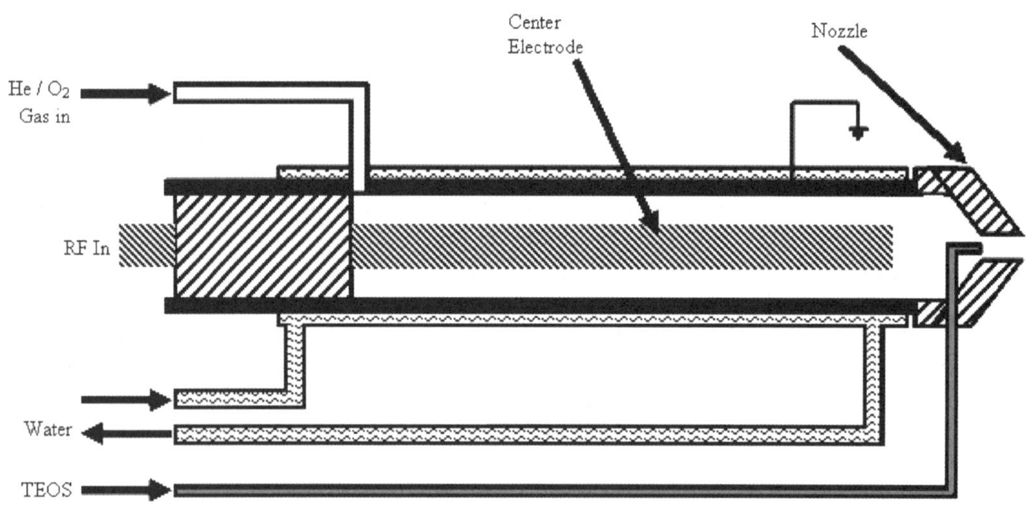

Figure 12.20 Coaxial RF plasma torch. (Reproduced with permission from ref. 100.)

experimental setup. The cathode was a tungsten needle connected to an RF (13.56 MHz) generator. The grounded anode was a stainless steel cylinder of 5 mm inner diameter and was separated from the cathode by alumina tubing. A plasma gas of He and oxygen was passed through the space between electrodes with a typical flow rate of 300 sccm. A reactive cold plasma was generated by mixing with tetramethoxysilane TMOS (0.2–1 sccm), an Ar carrier (5–25 sccm) and hydrogen (0–5 sccm). The applied RF power was 80–100 W. It was demonstrated that SiO_2 films could be deposited on various substrates exposed to air at a rate exceeding 600 nm min^{-1}. The

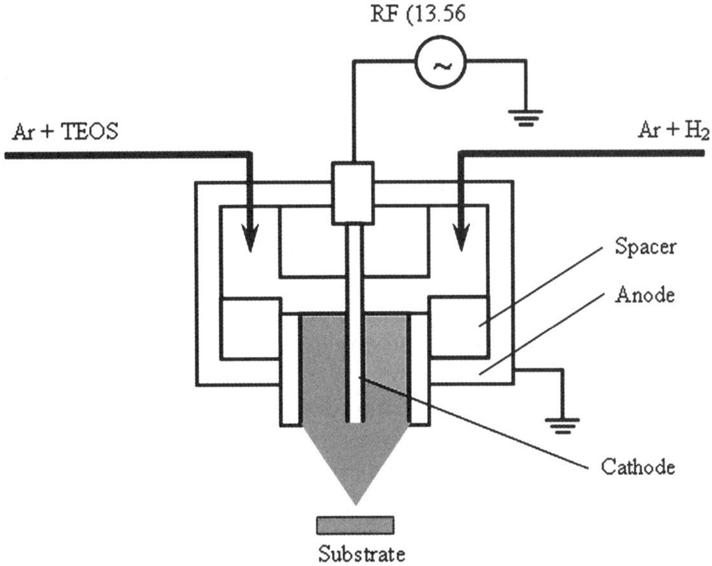

Figure 12.21 Schematic diagram of an RF cold plasma torch for deposition of SiO_2. (Reproduced with permission from ref. 101.)

characterization of the deposited layers by IR, XPS and SEM revealed that the mixing of hydrogen was quite effective in reducing the carbon contamination and improving the film quality.

Later, the same type of cold plasma torches was used for deposition of TiO_2 films on substrates exposed to air.[102] $Ti(OEt)_4$ or $Ti(O\text{-}^iPr)_4$ vapour was introduced into the plasma at the nozzle. A homogeneous plasma with an electron temperature of about 1.8 eV and gas temperature of about 200 °C resulted in decomposition of the precursor and active products of decomposition were directed onto the substrate surface. The film growth rate for both precursors was very high (up to 1 μm min^{-1}). The deposits were stoichiometric and amorphous TiO_2, but they contained some short-range crystallinity. The TiO_2 deposited from $Ti(OEt)_4$–H_2 mixtures had better dielectric properties.

Substantial contributions to the development of AP-PECVD processes based on the use of RF plasma jets have been made by Hicks *et al.*[98,100,103,104] A new plasma source based on the use of a capacitive RF discharge stabilized with inert gas has been developed and has been studied from the point of view discharge physics and chemistry; it has been tested for AP PECVD of hydrogenated silicon, silicon oxide and nitride layers. The plasma source consists of two closely spaced, perforated metal electrodes, the upper electrode being coupled to the RF (13.56 MHz) power, and the lower electrode being grounded.[98] Plasma forming gases (He, Ar) with small additions (<3 vol.%) of O_2, N_2 or H_2 are passed through the discharge generation region. An inlet distributor used for introduction of precursors is incorporated into the grounded electrode so that the precursors are introduced downstream from the discharge generation region. The precursors react with atoms and radicals coming from the discharge generation region, leading to the formation of excited active chemical species that participate in processes on the substrate surface and resulting in film formation. On the basis of the results obtained,[98,100,103,104] RF AP-PECVD clearly yields much higher growth rates than processes based on the use DBD or corona discharges, with film quality comparable to that obtained by traditional, low pressure PECVD. Because of the high values of growth rates the technology is very attractive for applications when the object to be covered spends a limited time in the deposition zone, as in reel-to-reel webs. Another advantage of RF AP-PECVD is that it can be used to coat thermally-sensitive materials, such as plastics and polymers.[98]

RF AP-PECVD has also been tested for deposition of thin layers on pigment particles.[105] Figure 12.22 shows a schematic presentation of the experimental setup developed. The high voltage electrode was an inner aluminium tube (diameter 6.5 mm) and the ground was a stainless steel mesh, 255 mm long, wrapped over an outer tube (diameter 16 mm). The discharge area was cooled by water to keep its temperature under 100 °C during discharge. The pigment powder was blown upwards through the inner tube and interacted with the plasma as it fell down between the inner and the outer tubes. Silica coatings have been deposited from TEOS using the technique on Fe_3O_4 black, FeOOH yellow and Lithol Rubine BCA RED powders.

A similar set-up has been described for the deposition of zirconia coatings on amorphous $Co_{70.3}Fe_{4.7}Si_{10}B_{15}$ powder with $Zr(OC_4H_9)_4$ as a precursor.[106] Figure 12.23 shows a schematic of the experimental reactor. An ultrasonic horn was used to separate and disperse the aggregated powder particles and also to assist in blowing the powder up into the plasma. In this case the high voltage electrode was a stainless steel rod (15 mm diameter and 260 mm long) with many minute grooves cut on its surface like a screw. In this way, it became possible to keep a stable APGD since the electric charge was evenly focused on the minute grooves. Inside the electrode was a water-cooled honeycomb structure that kept the discharge temperature below 100°C. The ground electrode was a copper plate (80–82 mm) wrapped around a Pyrex tube (diameter 25 mm). The treated

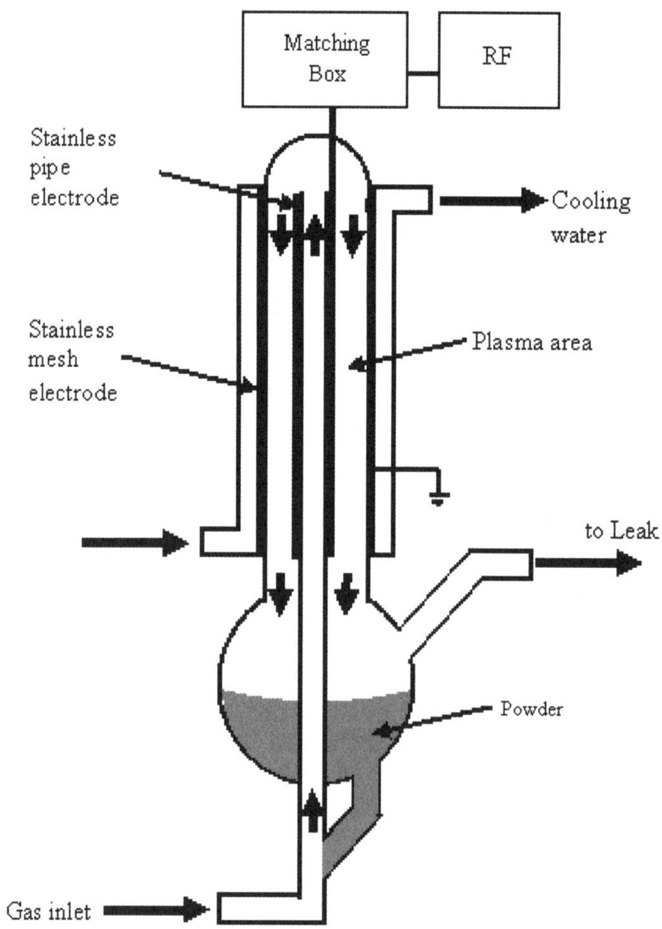

Figure 12.22 Experimental setup for RF AP-PECVD on pigments. (Reproduced with permission from ref. 105.)

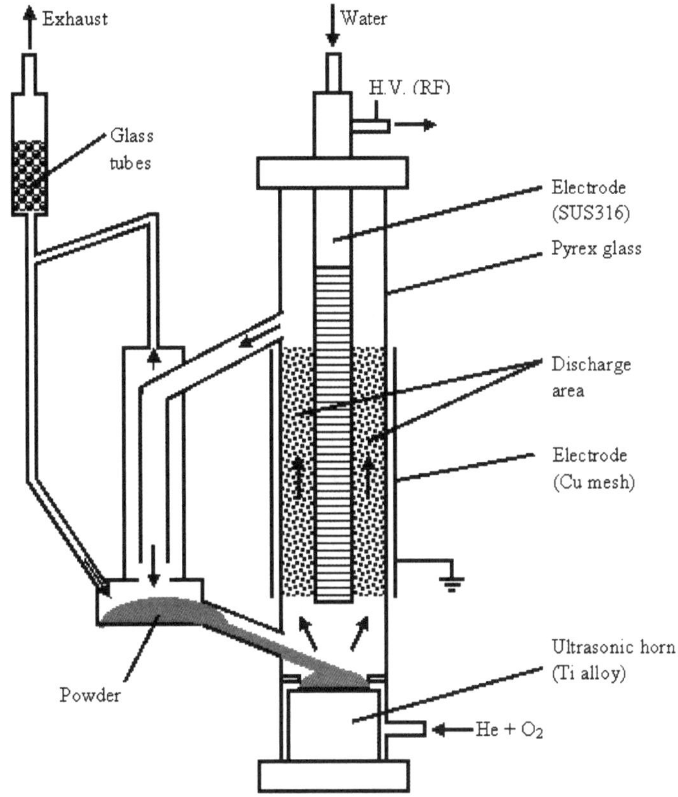

Figure 12.23 RF AP-PECVD reactor with an ultrasonic horn. (Reproduced with permission from ref. 106.)

powder was circulated through the reactor by a He/O_2 mixture and the process was repeated as often as required. The general discharge conditions were an RF frequency of 13.56 MHz, RF power of 400 W, He flow rate 6500 sccm and O_2 flow rate of 15 sccm. The treatment time was 20 min. The substrate powder was converted by the surface deposition of ZrO_2 from a conductor into a good electrical insulator. In addition, the treated powder showed a greater resistance to oxidation.

One problem associated with RF torches is overheating of the nozzle, which can lead to an arc discharge with hot electrodes. To prevent this undesirable phenomenon a dielectric layer can be used to cover the inner walls of the metallic nozzle[107,108] (Figure 12.24). The RF barrier-torch discharge with plasma jet channel can be considered as a modification of the well-known RF torch discharge with a metallic needle nozzle.[109]

An atmospheric pressure RF torch-barrier plasma jet system has been used for the deposition of CeO_2 thin films on polished aluminium substrates.[107] Helium was used as a working gas for generation of the RF torch discharge and a cerium-containing precursor was carried by an air flow to the plasma jet by the second nozzle. The precursor was prepared as an aerosol of a water solution of cerium acetate ($(CH_3COOH)_3Ce$). The He flow through the quartz nozzle was varied in the range 300–1890 sccm, while the flow rate of air with the precursor was in the range 500–900 sccm. Initially, the RF power was varied from 120 W to greater than 360 W; but results of preliminary experiments showed that the creation of CeO_x films was only possible when the RF power applied to the plasma jet was higher than 360 W. The plasma in the jet channel was strongly non-isothermal, with the temperature of the He being below 800 K, but with the vibrational temperature of the nitrogen molecules in the air being around 3600 K. However, the average temperature of the

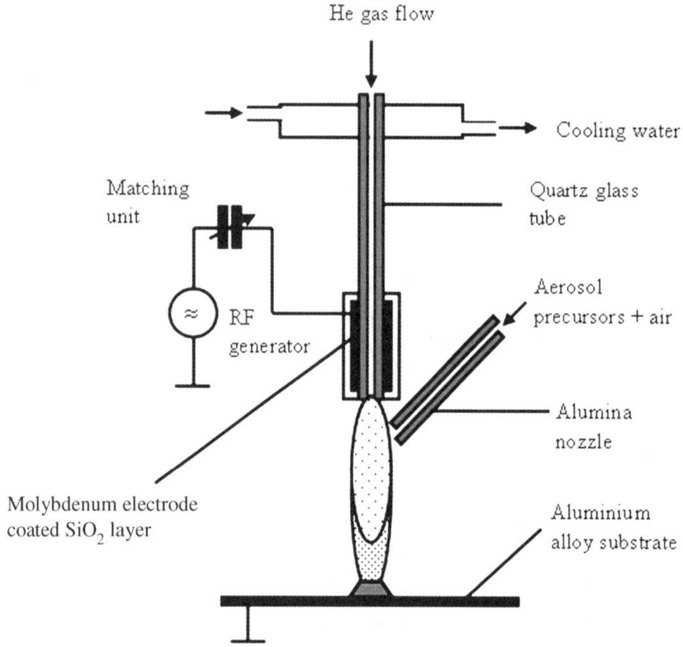

Figure 12.24 RF torch-barrier plasma jet set-up. (Reproduced with permission from ref. 107.)

substrate did not exceed 250 °C during the deposition. With these typical deposition conditions the deposition rate was approximately 500 nm h^{-1}. The stoichiometry of all the CeO_x films exhibited an excess of oxygen, with some contamination by carbon atoms. *e.g.* the thickest sample grown (220 nm) contained 10 at.% of carbon throughout its total thickness.

Barrier torch discharge has also been used[108] for low temperature deposition of In_xO_y and SnO_x thin films on polymer substrates. Vapours of Sn- and In-acetylacetonate were used as growth precursors for the deposition process. Transparent films with a conductivity of about 10 S cm^{-1} for SnO_x and 10^2 S cm^{-1} for In_xO_y were deposited under conditions when the atmospheric plasma jet directly interacted with the polymer substrate. SnO_x and In_xO_y films of thicknesses in the range 200–300 nm were deposited with typical deposition rates in the range 0.4–0.6 μm h^{-1}. Electron probe analysis did not find any detectable contamination by carbon or other measurable elements in In_xO_y films but, usually, carbon at the 2 at.% level was found in the SnO_x films. Analysis by electron microprobe showed that the films had chemical compositions close to SnO_2 and In_2O_3, and XRD showed no crystalline phases in any of the deposits.

Very exciting results have been obtained for the application of RF AP-PECVD for high-rate growth of epitaxial silicon at relatively low temperatures of 530–690 °C.[110] The authors developed an original experimental setup (Figure 12.25) with a cylindrical rotary electrode 300 mm in diameter and 200 mm long. The substrate was vacuum-chucked onto the SiC coated graphite susceptor, which could be heated up to 1000 °C. The plasma was generated in the gap (700 μm) between the rotary electrode and the substrate by applying 150 MHz VHF power through an impedance matching unit. Silicon films were deposited on (001) Si wafers from gas mixtures containing He, H_2 and SiH_4. High quality Si films with excellent crystallinity and surface flatness similar to or better than those of commercial Si wafers were grown in the area where the deposition gap between the substrate and rotary electrode was small. Examination by TEM of epitaxial Si films grown at 610 °C with an input plasma power of 2000 W showed no lattice defects. The maximum growth rate was approximately 6.6 μm min^{-1} at 690 °C with 1500 W,

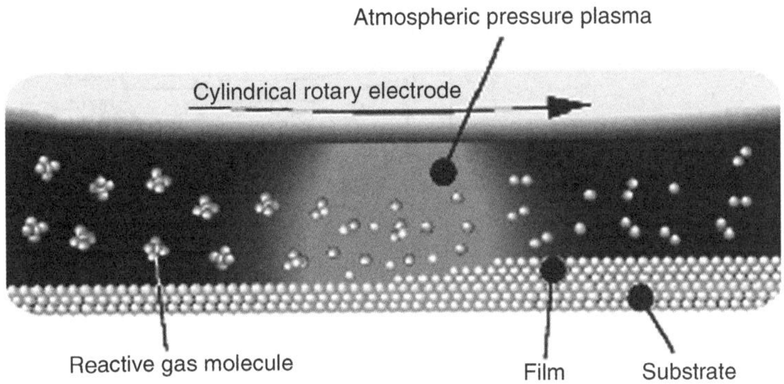

Figure 12.25 Schematic illustration of AP-PECVD with a rotary electrode. (Reproduced with permission from ref. 110.)

and 1.2 μm min^{-1} at 610 °C with 2000 W; these rates are, respectively, approximately 20–30 and 4–6 times faster than those obtained by thermal CVD at about 1100 °C. By scanning the substrate, a uniform Si film could be formed in the area having the dimensions of the plasma and the substrate scanning distance.

Finally, in this section on RF AP-PECVD mention should be made of an interesting system that uses an inductively coupled plasma (ICP) generated in micrometre scale quartz capillaries by applying a UHF voltage.[111] Deposition of carbon from CH_4–Ar mixtures on the substrate without additional heating in ambient air yielded spherical graphite on micron-sized regions at a growth rate of about 2 μm min^{-1}.

12.3.3.4 MW Discharges

The operating frequency of microwaves (MW) is very much higher than the sources for the other discharges we have considered so far, typically being about 2.54 GHz. In addition, electron densities can be as low as 10^7 cm^{-3} and gas temperatures as high as 1000 K, or even higher. The use of MW sources for generating non-equilibrium, non-thermal plasma for AP PECVD processes has been much less than for the other techniques discussed above. This is most probably connected with the difficulties of designing simple and convenient equipment for use in deposition processing and also with the difficulties in sustaining MW discharges at low powers.

One case where MW has been used is for continuously depositing silica coatings onto low modulus carbon fibres, in a tow form with a low cost, novel atmospheric pressure microwave plasma technique.[112] The objective was to improve the interfacial properties of aluminium/carbon fibre composites produced by liquid metal infiltration techniques. Amorphous silica coatings, approximately 50–400 nm thick, were successfully produced and a schematic diagram of the experimental setup is shown in Figure 12.26.

A plasma was initiated and maintained using a 2 kW microwave energy source at atmospheric pressure. The carbon fibre tow was spread out and passed from reel to reel through a quartz T-piece outside the microwave cavity (Figure 12.26). A controlled nitrogen flow between 1 and 10 slm was used to force the plasma into the T-piece, where different coatings could be produced using a suitable volatile precursor, such as hexamethyldisiloxane (HMDS), which was introduced into the plasma path, where it dissociated to form a silica coating on the carbon fibres.

A more complex MW AP-PECVD has been described[113] and results have been obtained for both direct and remote plasma deposition of SiO_x films. The plasma operated at a frequency of 2.45 GHz

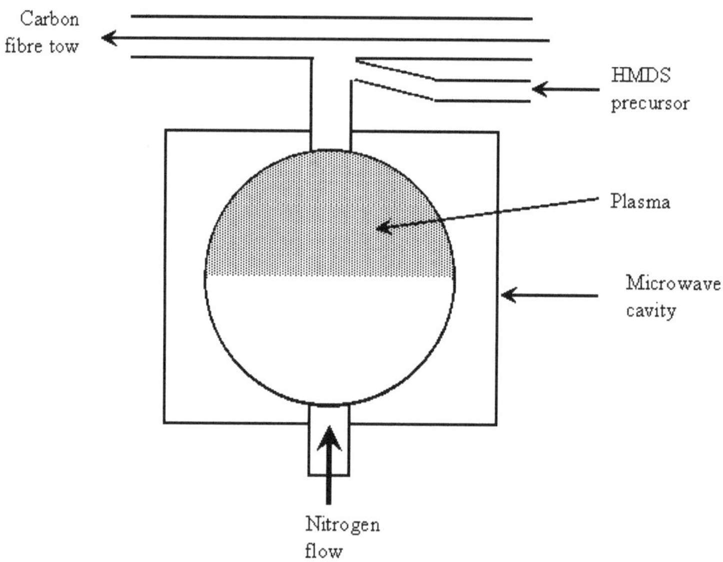

Figure 12.26 Schematic diagram of an AP-PECVD microwave reactor. (Reproduced with permission from ref. 112.)

and was a slot antenna type CYRANNUS-I source that was able to maintain a homogeneous plasma at atmospheric pressure. Figure 12.27 shows schematically the experimental arrangement. The microwave radiation was provided by a 6 kW generator and was channelled through a waveguide, a circulator and an E-H-Tuner to the annular resonator of the CYRANNUS-I source. Inside the annular resonator, slot antennas coupled the microwave radiation into a quartz chamber, with an inner diameter of about 150 mm, which was located in the centre of the source. For ignition of the plasma it was necessary to evacuate the chamber to low pressure.

The carrier and the monomer gases for the silica deposition were an argon–oxygen mixture (or air) and hexamethyldisiloxane (HMDSO), respectively. The precursor gas was fed through a nozzle arrangement that consisted of two separate inlets. One of those was for the HMDSO, at room temperature, with its argon carrier, and the other was for an additional argon flow supplied through a bypass; an argon–oxygen (70 : 30%) mixture as well as dried air was used for this line.

The CVD process in the direct mode proved to be unsuitable for achieving good quality film deposition because of polymerization of the HMDSO, but for SiO_x layer growth in the remote mode a suitable deposition regime was developed. In this case, activation of the additional argon flow took place within the plasma source while plasma interaction with the HMDSO was carried out in the afterglow area outside the source. This is illustrated in Figure 12.27, where the main argon flow is introduced into the plasma source from the top while the precursor gas stream is fed to the substrate through a nozzle. Short distances from the source exit to the substrate were found to be preferable to minimize polymerization effects. With this configuration, by choosing the right distance it was possible to deposit SiO_x thin films with sharp interfaces and refractive indexes in the region 1.27–2.17, depending on the distance from the injection nozzle. Table 12.4 summarizes typical deposition parameters for microwave induced atmospheric pressure plasma CVD in a remote mode.

An interesting phenomenon occurs with a new type of MW (2.45 GHz) source, providing generation of non-thermal plasmas[114] when perovskite-type oxide powders were placed in a quartz tube held orthogonal to the exit of an atmospheric pressure wave guide of the MW source and through which argon flowed. It was found that a plasma was generated around the powders with optical emissions and that the phenomenon was specific to electrically conductive and heat-resistive

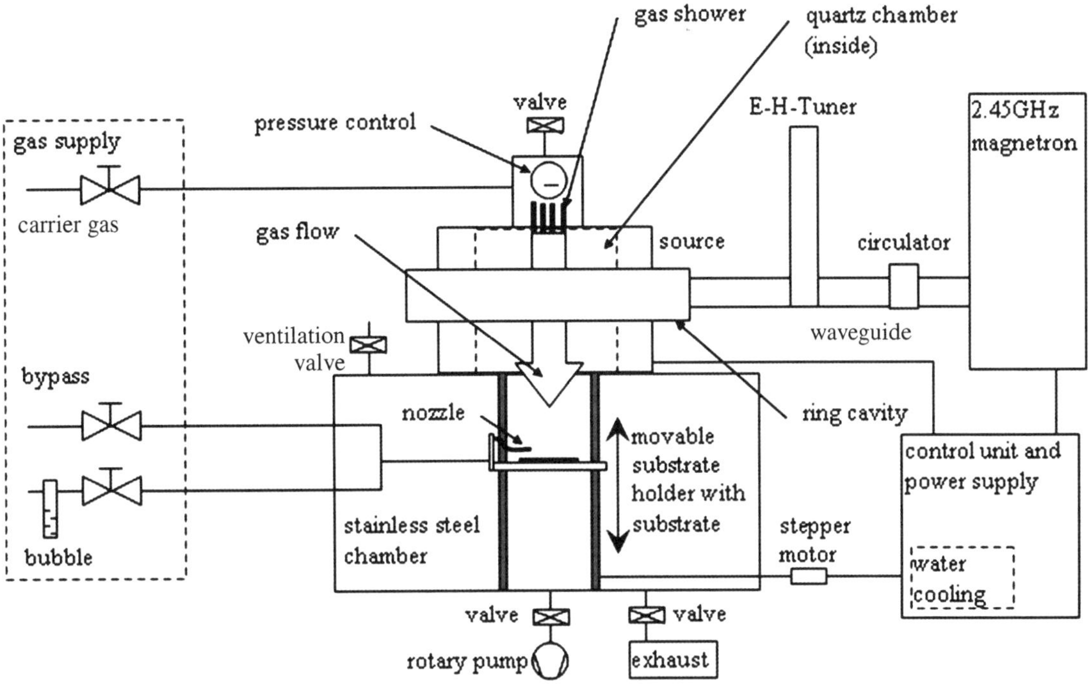

Figure 12.27 Schematic of the experimental arrangement for a MW AP-PECVD system. (Reproduced with permission from ref. 113.)

Table 12.4 Experimental conditions for MW AP-PECVD of SiO_x.[113]

Experimental parameter	*Experimental value*
Distance between the sample and bottom of the plasma source	20 mm
Position of the precursor nozzle	Nearly at an angle of 20° to the surface, directly above the substrate surface
Ar/O_2 carrier gas flow through the source	360 L h^{-1}
Ar/HMDSO precursor flow	4 L h^{-1}
Additional Ar flow	100 L h^{-1}
Anode current of the magnetron	0.5 A
Deposition time	Typically 10 min
Film growth rate	150 nm min^{-1}

materials such as $La_{0.8}Sr_{0.2}CoO_3$, graphite and TiN, LaB_6. For example, the quartz tube and the wave guide (aluminium, 110 × 55 mm) were crossed at right angles and about 1 g of $La_{0.8}Sr_{0.2}CoO_3$ powder, which acted as a the plasma-generating material, was put at the crossing point. After the $La_{0.8}Sr_{0.2}CoO_3$ powder sample was placed in the reactor, the argon flow was started, and the microwave power was turned on. The inner reactor tube space exhibited a plasma state and the plasma expanded in the same direction as the argon gas flow. With a MW power of 30 120 W and an argon flow rate of 120 mL min^{-1}, the temperature of the quartz tube walls varied from 370 to 400 K. The mechanism of the phenomenon is not clear, but it has been used successfully for surface modification of polymers,[114] surface treatment of steel[115] and decomposition of inorganic gases.[116]

It could also be promising for activation of CVD processes, but it has not been tested for that purpose as yet.

12.4 Conclusions

PECVD is a well established technology that has ever increasing commercial applicability. However, the traditional approach to PECVD of having the substrate on one of the electrodes in the discharge generation region has several drawbacks. Some of these disadvantages can be overcome by the technique of remote PECVD and the first part of this chapter has discussed aspects of the chemistry of RPECVD processes and has illustrated some of the benefits as well. Certainly, RPECVD is a promising approach to plasma assisted deposition and one would expect to see a greater use of the technique as thin film requirements and coatings demand higher functionality and performance.

In addition to improved layer quality and better operational characteristics for commercial materials, there is the ever increasing push to lower capital and operating costs. The possibility of achieving high quality layers without expensive vacuum systems is a very attractive one. Atmospheric pressure PECVD is a relatively new technology that offers exciting possibilities for achieving that goal. The second part of this chapter has concentrated on recent developments with AP-PECVD and while it may still be in its infancy the discussion has shown that it does have the potential for wide ranging and varied applications. One would expect to see it become an important industrial technology.

References

1. J. R. Roth, *Industrial Plasma Engineering, Vol. 1*, IOP, Bristol, 2001.
2. J. R. Roth, *Industrial Plasma Engineering, Vol. 2*, IOP, Bristol, 2001.
3. D. W. Hess and D. B. Graves, in *Chemical Vapor Deposition*, ed. M. L. Hitchman, K. F. Jensen, Academic Press, New York, 1989, Ch.7.
4. R. Reif and W. Kern, in *Thin Film Process II*, ed. J. L. Vossen, W. Kern, Academic Press, London, 1991, pp. 525.
5. M. A. Lieberman and A. J. Lichtenberg, *Principles of Plasma Discharges and Materials Processing*, Wiley Interscience, New York, 1994.
6. *Chem. Vap. Deposition*, Special Issue on Low Pressure Plasma Enhanced CVD (LP-PECVD), 6-7, 2007, ed. F. J. Gordillo-Vázquez.
7. G. Lucovsky, D. V. Tsu, R. A. Rudder and R. J. Markunas, in *Thin Film Process II*, ed. J. L. Vossen, W. Kern, Academic Press, London, 1991, pp. 565.
8. S. E. Alexandrov, *J Physique IV*, 1995, **C5**, 567.
9. S. E. Alexandrov, M. L. Hitchman and S. Shamlian, *Adv. Mat. Opt. Electron.*, 1993, **2**, 301.
10. F. J. Kampas, in *Semiconductors and Semimetals 21A*, ed. J. I. Pankove, Academic Press, Orlando, 1984, Ch. 8.
11. D. L. Smith, in *Plasma Technology in Manufacturing of VLSI*, ed. D.Braun, N. Ainsprooc, Mir, Moscow, 1987, pp. 245.
12. J. R. Hollahan and A. T. Bell, *Techniques and Applications of Plasma Chemistry*, John Wiley & Sons, New York, 1974.
13. H. A. Weakleim, in *Semiconductors and Semimetals 21A*, ed. J. I. Pankove, Academic Press, Orlando, 1984, Ch. 10.
14. D. L. Smith, A. S. Alimonda and F. J. Preissig, *J. Vac. Sci. Technol. B*, 1990, **8**, 551.
15. L. Bardos, J. Musil and P. Taras, *Thin Solid Films*, 1983, **102**, 107.
16. H. V. Boenig, *Adv. Low. Temp. Plasma Chem., Technol. Appl.*, 1984, **1**, 195.

17. S. E. Alexandrov and A. Y. Kovalgin, *J. Physique III*, 1992, **2**, 1723.
18. G. Lucovsky, D. V. Richard, D. V. Tsu, S. Y. Lin and R. J. Markunas, *J. Vac. Sci. Technol. A*, 1986, **4**, 681.
19. S. E. Alexandrov and A. Yu. Kovalgin, *Adv. Mat. Opt. Electron.*, 1998, **8**, 13.
20. S. E. Alexandrov, M. L. Hitchman and A. Yu. Kovalgin, *Adv. Mat. Opt. Electron.*, 1998, **8**, 23.
21. A. C. Adams, F. B. Alexander, C. D. Capio and T. E. Smith, *J. Electrochem. Soc.*, 1981, **128**, 1545.
22. O. Sanchez, M. A. Aguilar, C. Falcony, J. M. Martinez-Duart and M. Hernandez Velez, *Thin Solid Films*, 1998, **317**, 149.
23. S. S. Lin, *J. Electrochem. Soc.*, 1978, **125**, 1877.
24. S. Wickranamayaka, Y. Hatanaka, Y. Nakanishi and A. M. Wrobel, *J. Electrochem. Soc.*, 1994, **141**, 2910.
25. Y. Hatanaka, K. Kitamura, S. Wickramanayaka, Y. Nakanishi, J. Tyczkowski, *Proc. Int. Conf. Silicon Carbide and Related, Mat.*, 1995, Kyoto, p 1055.
26. M. Boudreau, M. Boumaerzoug, P. Mascher and P. E. Jessop, *Appl. Phys. Lett.*, 1993, **63**, 3014.
27. D. M. Hoffman, S. P. Rangarajan, S. D. Athavale, S. C. Desmukh, D. J. Liu, Z. Zheng and W. Chu, *J. Mater. Res.*, 1994, **9**, 3019.
28. D. M. Hoffman, S. P. Rangarajan, S. D. Athavale, D. J. Economou, D. J. Liu, Z. Zheng and W. Chu, *J. Vac. Sci. Technol. A.*, 1995, **13**, 820.
29. T. Aoki, T. Ogishima, A. M. Wrobel, Y. Nakanishi and Y. Hatanaka, *Vacuum*, 1998, **51**, 747.
30. D. V. Tsu, G. Lucovsky and M. Mantini, *Phys. Rev.*, 1985, **B33**, 7069.
31. D. V. Tsu and G. Lucovsky, *J. Vac. Sci. Technol A.*, 1986, **4**, 480.
32. G. Lucovsky and D. V. Tsu, *J. Cryst. Growth*, 1988, **86**, 804.
33. D. L. Smith, A. S. Alimonda, C. C. Chen, S. E. Ready and B. Wacker, *J. Electrochem. Soc.*, 1990, **137**, 614.
34. S. E. Alexandrov, M. L. Hitchman and S. Shamlian, *J. de Physique IV*, 1993, **C3**, 233.
35. S. Lee and A. Gopinath, *J. Vac. Sci. Technol. B*, 1990, **3**, 402.
36. S.-S. Han, B.-H. Jun, K. No and B.-S. Bae, *J. Electrochem. Soc.*, 1998, **145**, 652.
37. W. A. Lanford and M. J. Rand, *J. Appl. Phys.*, 1978, **49**, 2473.
38. W. A. P. Claassen, W. G. J. N. Valkenburg, M. F. C. Willemsen and W. M. Wijgert, *J Electrochem. Soc.*, 1985, **132**, 893.
39. S. E. Alexandrov, M. L. Hitchman and S. Shamlian, *J. Mater. Chem.*, 1994, **4**, 1843.
40. S. V. Hattagady, G. G. Fountain, R. A. Rudder and R. G. Markunas, *J. Vac. Sci. Technol.*, 1989, **7**, 570.
41. M. Gupta, V. K. Rathi, R. Thangaraj, O. P. Agnihotri and K. S. Chari, *Thin Solid Films*, 1991, **204**, 77.
42. D. Schalch, A. Scharmann and R. Wolfart, *Thin Solid Films*, 1985, **124**, 301.
43. A. Piccirillo and A. L. Gobbi, *J. Electrochem. Soc.*, 1990, **137**, 3910.
44. S. E. Alexandrov, M. L. Hitchman and S. Shamlian, *J. Mater. Chem.*, 1995, **5**, 457.
45. S. Fujita and A. Sasaki, *J Electrochem. Soc.*, 1988, **135**, 2566.
46. R. E. Livengood, M. A. Petrich, D. W. Hess and J. A. Reimer, *J. Appl. Phys.*, 1988, **63**, 2651.
47. S. E. Alexandrov and M. L. Hitchman, *Chem. Vap. Deposition*, 1997, **3**, 111.
48. B.-H. Jun, J. S. Lee, D. W. Kim, T.-H. Sung, B.-S. Bae and K. No, *J. Mater. Res.*, 1999, **14**, 995.
49. D. V. Tsu, G. N. Parsons and G. Lucovsky, *J. Vac. Sci. Technol. A*, 1988, **6**, 1849.
50. D. G. Park, M. Tao, A. E. Botchkarev, Z. Fan, Z. Wang, S. N. Mohammad, A. Rockett, J. R. Abeison and H. Morkos, *J. Vac. Sci. Technol. B*, 1996, **14**, 2674.
51. A. G. Aberle and R. Hezel, *Prog. in Photovoltaics: Res. & Appl.*, 1997, **5**, 29.

52. H. Nagel, A. G. Aberle and R. Hezel, *Prog. in Photovoltaics: Res. & Appl.*, 1999, **7**, 245.
53. B. Lenkeit, S. Steckemetz, F. Artuso and R. Hezel, *Sol. Energy Mat. and Sol. Cells*, 2001, **65**, 317.
54. S. Muramatsu, T. Uematsu, H. Ohtsuka, Y. Yazawa, T. Warabisako, H. Nagayoshi and K. Kamisako, *Sol. Energy Mat. and Sol. Cells*, 2001, **65**, 599.
55. V. Misra, H. Lazar, M. Kulkarni, Z. Wang, G. Lucovsky and J. R. Hauser, *Mat. Res. Soc. Symp. – Proc.*, 1999, **567**, 89.
56. Y. Wu, G. Lucovsky and Y.-M. Lee, *IEEE Trans. Electr. Dev.*, 2000, **47**, 1361.
57. Y.-B. Park and S.-W. Rhee, *J. Mat. Sci.: Mat. Electr.*, 2001, **12**, 515.
58. S. E. Alexandrov and M. L. Hitchman, *Chem. Vap. Deposition*, 2005, **11/12**, 457.
59. A. P. Napartovich, *Plasma and Polymers*, 2001, **6**, 1.
60. A. Schutze, J. Y. Jeong, S. E. Babayan, J. Park, G. S. Selwyn and R. F. Hicks, *IEEE Trans. Plasm. Sci.*, 1998, **26**, 1685.
61. J. Salge, *J. Phys. IV*, 1995, **5**, C5–583.
62. J. Salge, *Surf. Coat. Techn.*, 1996, **80**, 1.
63. U. Kogelschats, B. Eliasson and W. Egli, *J. Phys. IV*, 1997, **C4**, 47.
64. J. Engemann and D. Korzec, *Thin Solid Films*, 2003, **442**, 36.
65. J. R. Roth, J. Rahel, X. Dai and D. M. Sherman, *J. Phys. D: Appl. Phys.*, 2005, **38**, 555.
66. S. Okazaki, M. Kogoma, M. Uehara and Y. Kimura, *J. Phys. D: Appl. Phys.*, 1993, **26**, 889.
67. Z. Chen, *IEEE Trans. Plasm. Sci.*, 2002, **30**, 1922.
68. K. G. Donohoe, T. Wydeven, *Proc. 4th Int. Symp. on Plasma Chemistry*, eds, S. Veprek and J. Hertz, University of Zurich, Zurich, 1979, p. 765.
69. U. Reitz, J. G. H. Salge and R. Schwarz, *Surf.Coat. Technol.*, 1993, **59**, 144.
70. S. P. Bugaev, A. D. Korotaev, K. V. Oskomov and N. S. Sochugov, *Surf. Coat. Technol.*, 1997, **96**, 123.
71. T. Yokoyama, M. Kogoma, S. Kanazawa, T. Moriwaki and S. Okazaki, *J. Phys. D: Appl. Phys.*, 1990, **23**, 374.
72. O. Goossens, E. Dekempeneer, D. Vangeneugden, R. Van de Leest and C. Leys, *Surface and Coatings Technology*, 2001, **142–144**, 474.
73. O. Goossens, D. Vangeneugden, S. Paulussen, E. Dekempeneer, in *Proc. of Int. Symp. on High Pressure, Low Temperature Plasma Chemistry HAKONE VIII*, ed. A. Haljaste and T. Plank, *Univ. of Tartu*, Estonia, 2002, p 385.
74. R. Prat, Y. J. Koh, Y. Babukutty, M. Kogoma, S. Okazaki and M. Kodama, *Polymer*, 2000, **41**, 7355.
75. Y. Babukutty, R. Prat, K. Endo, M. Kogoma, S. Okazaki and M. Kodama, *Langmuir*, 1999, **15**, 7055.
76. A. Goodwin, T. Herbert, S. Leadley, F. Swallow, in *Proc. of Int. Symp. on High Pressure, Low Temperature Plasma Chemistry HAKONE VIII*, ed. A. Haljaste and T. Plank, *Univ. of Tartu*, Estonia, 2002, p. 441.
77. L. O'Neill, L.-A. O'Hare, S. R. Leadley and A. J. Goodwin, *Chem. Vap. Deposition*, 2005, **11**, 477.
78. S. Paulussen, R. Rego, O. Goossens, D. Vangeneugden and K. Rose, *J. Phys. D: Appl. Phys.*, 2005, **38**, 568.
79. D. Vangeneugden, S. Paulussen, O. Goossens, R. Rego and K. Rose, *Chem. Vap. Deposition*, 2005, **11**, 491.
80. K. Schmidt-Szalowski, Z. Ṙzanek-Boroch, J. Sentek, Z. Rymuza, Z. Kusznierewicz and M. Misiak, *Plasmas and Polymers*, 2000, **5**, 173.
81. R. Foest, F. Adler, F. Sigeneger and M. Schmidt, *Surface and Coatings Technology*, 2003, **163**, 323.

82. Z. Rymuza, M. Misiak, Z. Rzanek-Boroch, K. Schmidt-Szalowski and J. Janowska, *Thin Solid Films*, 2004, **466**, 158.
83. S. P. Bugaev, A. D. Korotaev, K. V. Oskomov and N. S. Sochugov, *J. Techn. Phys.*, 1997, **67**, 100.
84. Z. Rżanek-Boroch, K. Schmidt-Szałowski, J. Janowska, K. Dudziñski, A. Szymañska and M. Misiak, *in Proc. of Int. Symp. on High Pressure, Low Temperature Plasma Chemistry HAKONE VIII*, ed. A. Haljaste and T. Plank, Univ. of Tartu, Estonia, 2002, pp. 415.
85. T. Opaliñska, B. Ulejczyk, L. Karpiñski and K. Schmidt-Szaiowski, *in Proc. of Intern. Symp. on High Pressure, Low Temperature Plasma Chemistry HAKONE VIII*, ed. A. Haljaste and T. Plank, Univ. of Tartu, Estonia, 2002, p. 420.
86. J. F. Behnke, H. Steffen, A. Sonnenfeld, R. Foest, V. Lebedev and R. Hippler, *in Proc. of Int. Symp. on High Pressure, Low Temperature Plasma Chemistry HAKONE VIII*, ed. A. Haljaste and T. Plank, Univ. of Tartu, Estonia, 2002, p. 410.
87. A. Sonnenfeld, T. M. Tun, L. Zajickova, H.-E. Wagner, J. F. Behnke and R. Hippler, in *Proc. of Int. Symp. on High Pressure, Low Temperature Plasma Chemistry HAKONE VII*, ed. H.E. Wagner, Ernst-Moritz-Arndt Univ., Greifswald, Germany, 2000, p. 445.
88. Y. Sawada, S. Ogawa and M. Kogoma, *J. Phys. D: Appl. Phys.*, 1995, **28**, 1661.
89. R. Foest, F. Adler, F. Sigeneger and M. Schmidt, *Surface and Coatings Technology*, 2003, **163–164**, 323.
90. H.-R. Lee, D. Kim and K.-H. Lee, *Surface and Coatings Technology*, 2001, **142–144**, 468.
91. M. J. Davis, J. Lewis, M. E. Pemble, D W. Sheel, M. Tsanos, in *Proc. Int. CVD XVI & EUROCVD 14*, ed. M. D. Allendorf, F. Maury, F. Teyssandier, Electrochemical Society, Pennington, NJ, 2003-08, p. 668.
92. S. E. Alexandrov, M. L. Hitchman, N. McSporran, in *Proc. Int. CVD XVI & EUROCVD 14*, ed. M. D. Allendorf, F. Maury, F. Teyssandier, Electrochemical Society, Pennington, NJ, 2003-08, p. 630.
93. K. Tanaka, T. Inomata and M. Kogoma, in *Proc. of Int. Symp. on High Pressure, Low Temperature Plasma Chemistry HAKONE VII*, ed. H.E. Wagner, Ernst-Moritz-Arndt Univ, Greifswald, Germany, 2000, p.
94. S. Kanazawa, M. Kogoma, T. Moriwaki and S. Okazaki, *J. Phys. D.: Appl. Phys.*, 1988, **21**, 838.
95. R. Thyen, A. Weber and C.-P. Klages, *Surf. Coat.Techn.*, 1997, **97**, 426.
96. M. Li, Z. Hu, X. Wang, Q. Wu and Y. Chen, *Thin Solid Films*, 2003, **435**, 116.
97. M. Li, Z. Hu, X. Wang, Q. Wu, Y. Chen and Y. Tian, *Diamond and Related Materials*, 2004, **13**, 111.
98. M. Moravej and R. F. Hicks, *Chem. Vap. Deposition*, 2005, **11/12**, 469.
99. M. Koidea, T. Horiuchi, T. Inushima, B. J. Lee, M. Tobayama and H. Koinuma, *Thin Solid Films*, 1998, **316**, 65.
100. S. E. Babayan, J. Y. Jeongy, V. J. Tuy, J. Parkz, G. S. Selwynz and R. F. Hicks, *Plasma Sources Sci. Technol*, 1998, **7**, 286.
101. K. Inomata, H. Ha, K. A. Chaudhary and H. Koinuma, *Appl. Phys.Lett.*, 1994, **64**, 46.
102. H. Ha, B. K. Moon, T. Horiuchi, T. Inushima, H. Ishiwara and H. Koinuma, *Mater. Sci. Engineer.*, 1996, **B41**, 143.
103. G. R. Nowling, S. E. Babayan, V. Jankovic and R. F. Hicks, *Plasma Sources Sci. Technol.*, 2002, **11**, 97.
104. M. Moravej, S. E. Babayan, G. R. Nowling, X. Yang and R. F. Hicks, *Plasma Sources Sci. Technol.*, 2004, **13**, 8.
105. T. Mori, K. Tanaka, T. Inomata, A. Takeda and M. Kogoma, *Thin Solid Films*, 1998, **316**, 89.

106. S. Ogawa, A. Takeda, M. Oguchi, K. Tanaka, T. Inomata and M. Kogoma, *Thin Solid Films*, 2001, **386**, 213.
107. L. Soukup, Z. Hubicka, A. Churpita, M. Cada, P. Pokorny, J. Zemek, K. Jurek and L. Jastrabik, *Surface and Coatings Technology*, 2003, **169–170**, 571.
108. A. Churpita, Z. Hubicka, M. Cada, D. Chvostova, L. Soukup, L. Jastrabik and P. Ptacek, *Surface and Coatings Technology*, 2003, **174–175**, 1059.
109. V. Kapicka, M. Sicha, M. Klima, Z. Hubicka, J. Tous, A. Brablec, P. Slavicek, J. F. Behnke, M. Tichy and R. Vaculik, *Plasma Source Sci. Technol.*, 1999, **8**, 15.
110. Y. Mori, K. Yoshii, K. Yasutake, H. Kakiuchi, H. Ohmi and K. Wada, *Thin Solid Films*, 2003, **444**, 138.
111. Y. Shimizu, T. Sasaki, T. Ito, K. Terashima and N. Koshizaki, *J. Phys. D: Appl. Phys.*, 2003, **36**, 2940.
112. M. Brown, P. Hayes and P. Prangnell, *Composites: Part A*, 2002, **A33**, 1403.
113. A. Pfuch and R. Cihar, *Surface and Coatings Technology*, 2004, **183**, 134.
114. K. Sugiyama, K. Kiyokawa, H. Matsuoka, A. Itou, K. Hasegawa and K. Tsutsumi, *Thin Solid Films*, 1998, **316**, 117.
115. K. Kiyokawa, A. Itou, H. Matsuoka, M. Tomimatsu and K. Sugiyama, *Thin Solid Films*, 1999, **345**, 119.
116. K. Kiyokawa, H. Matsuoka, A. Itou, K. Hasegawa and K. Sugiyama, *Surface and Coatings Technology*, 1999, **112**, 25.

CHAPTER 13

Commercial Aspects of CVD

ALBERT BARRY LEESE[a] AND ALAN RODNEY MILLS[b]

[a] SAFC Hitech, Power Road, Bromborough, Wirral, CH62 3QF, UK; [b] Independent Market, Business & Technology Consultant, 980 Golden Way, Los Altos CA 94024, USA

13.1 CVD Industries Introduction

The chemical vapor deposition (CVD) process is a very versatile method for coating surfaces with uniform thin films. In its many forms it supports several worldwide industries and has been widely used in volume commercial production for many years. CVD can accurately provide critical materials depositions that are very difficult to achieve by other methods. Uniform thin crystalline films or coatings can be deposited over a wide range of materials and pressure regimes and can even be deposited with specific surface areas masked off. Hence, it is suitable for a wide range of industrial applications and for the deposition of a wide range of elements and compounds including semiconductors. Hundreds of precursor chemicals are already produced on the industrial scale to deposit commercially valuable films ranging from protective and anti-reflective coatings on glass (titanium and silicon oxides), protective coatings on metals (oxides and refractory metals), active and dielectric layers for semiconductor devices (silicon oxides and silicon), silicon nitride for masking and passivation, quantum dots for optoelectronic applications, multi-layer structures for UV, visible and IR LEDs and lasers, specific performance coatings on tools providing wear and endurance and economic benefits (molybdenum, tungsten, boron nitride). Molecular beam epitaxy (MBE) should also be included under the CVD umbrella, but as an ultrahigh vacuum modification used to manufacture some commercial products, including LEDs and electronic circuits. However, it is a chemical vapor process, but not typical CVD since the precursors, when used, are usually thermally cracked before being introduced into the vacuum deposition chamber to form the desired molecular beams. Additionally, the precursors used are not usually specifically made for MBE applications. Other specific variations, mostly used for semiconductor manufacture, are atomic layer deposition (ALD), organometallic vapor phase epitaxy (OMVPE) and hydride vapor phase epitaxy (HVPE).

Values in this commercial CVD chapter are provided in US dollars and at the time of going to press, US$1.00 was approximately equivalent to 0.49 UK pounds and 0.7 euros.

Chemical Vapour Deposition: Precursors, Processes and Applications
Edited by Anthony C. Jones and Michael L. Hitchman

Published by the Royal Society of Chemistry, www.rsc.org

Six of the largest volume and multi-billion dollar industries worldwide are listed below, most with sales of $3 billion per year, or higher:

1. The glass coatings industry where billions of square feet of glass are made every year, with much of this being CVD-coated with UV, IR and scratch resistant coatings.
2. The silicon semiconductor industry where critical CVD processes such as epitaxy (the CVD deposition of a growing film in the same crystal orientation as the substrate or wafer), silicon nitride passivation and oxide insulator deposition are now an integral part of the IC fabrication process for more than 95% of the trillions of silicon chips produced annually.
3. The compound semiconductor industry where billions of LEDs and optoelectronic devices are made, many of these being totally dependent on CVD, OMVPE, HVPE or MBE processes for their desirable properties.
4. The III-V high frequency device industry, which provides high-speed electronic devices (5 to 300+ GHz) from CVD, MOCVD and MBE processes, mostly from Group III-V compounds such as GaAs, InP, GaN and their alloys. Their operating frequency range continues to expand above the 300–400 GHz range for R&D devices concurrent with improving efficiencies and power outputs as device designs and CVD processes continue to be refined.
5. The diode laser industry, where CVD plus MOCVD and MBE processes presently produce devices that cover the 300–8000 nm wavelength range and their unit volumes make up over 99% of all lasers produced today. As research continues, this wavelength range is constantly being extended, using CVD layer deposition methods.
6. The solar cell industry, where most photovoltaic (PV) cells are CVD produced from either silicon or compound semiconductor materials. Some less expensive, but lower efficiency PV materials such as the diselenides $CuInSe_2$ and $CuInGaSe_2$, are approaching the high volume commercial product status.

There are other developing technology groups with products in the nano-dimension range that depend heavily on CVD and related (*e.g.* HVPE, MOCVD, MBE) processes and they are expected to become spectacular growth industries in the future. They include quantum dots, carbon nanotube and nano-wire product groups, all of which can exhibit quantum level physical properties and have at least one product dimension in the sub-20 nm range. Other nano-sized materials such as coatings, phosphors and powders are already manufactured, but most of these are produced by physical or precipitation processes.

13.1.1 Glass Coatings

Today, sheet glass coatings are now a very important part of the environmental control for all large buildings to admit or reflect heat and or sunlight. Consequently, billions of square feet of architectural glass are coated annually with various protective or active CVD films such as silicon oxide, carbon doped tin oxide and self-cleaning titanium oxide. These thin CVD coatings are deposited by efficient processes on the float glass production line and do not consume large quantities of precursor chemicals per unit area covered and therefore are effective and inexpensive. Compared to other CVD processes, there are only a few precursors used for the large area coatings and are therefore not as interesting from a CVD process technology or precursor synthesis point of view. The main chemicals used are silane, stannic chloride, ethylene and titanium tetrachloride. However, because of the large areas of glass consumed, the estimated worldwide total value of finished products is about $13 billion. CVD processes on glass are also used for transition metal oxide depositions on spectacles to make photochromic coatings that darken when exposed to sunlight and electrochromic coatings that change color or darken with an applied voltage. Only relatively small volumes of precursors are used for the latter processes, when compared with sheet glass or silicon process use.

13.1.2 Silicon Devices

Silicon CVD layers are part of the technical basis for the $250 billion silicon IC industry, without which there would be no modern electronics industry, and many of our wireless services and electronic appliances could not exist in their present form. The outstanding and continuing growth during the last 60 years has spawned a healthy CVD and specialty chemical business together with a supporting $45 billion materials business that show steady growth with the constantly expanding silicon IC markets. The volume of chemicals used is significant, but historically the variety of chemical precursors used has not been large when compared with compound semiconductor requirements. However, the continuing demand to increase performance and reduce component cost demands dielectric materials (mixed oxides), passivation layers and diffusion barriers (metal nitrides). New precursors are being added to support these requirements. Today's focus is on the synthesis of families of precursors that will provide the process engineer with a range of different deposition temperatures for any given layer, including the latest mixed oxide depositions and the ALD processes. Typical silicon device technology CVD precursors now include chlorosilanes, silane, ammonia, tungsten hexafluoride, tetraethyl orthosilicate and tetrakis(dimethylamino)titanium.

Originally, most semiconductor junctions were made by diffusion of the required active impurity into the semiconductor wafer (a slow high temperature process) and electrical isolation was provided by *in situ* growth of silicon dioxide from the underlying silicon in air or oxygen ambients. Today, some silicon junctions are grown by CVD epitaxy from (silane and chlorosilanes) and many oxide layers are rapidly grown by CVD from the same source chemicals (silane, chlorosilanes) and also from the relatively volatile liquid, silicon tetraethyl orthosilicate. Most CVD oxide depositions are followed by high temperature annealing and compaction processes. As circuit complexity has increased during the last 50 years, the application of CVD technology has expanded and semiconductor epitaxy has become an ever more important CVD process that requires a changing and expanding range of precursor chemicals. Each of these has ever higher quality demands, minimum vapor pressure requirements for vapor transport plus an industry base to supply the demand and chemical need in commercial quantities. In the latest silicon integrated circuit processes CVD diffusion barriers such as titanium or tantalum nitrides are often used and mixed oxide layers (silicon, aluminum, hafnium, titanium oxides and Group II titanates [barium, strontium]) are being investigated for more critical isolation requirements such as higher-k dielectric constant mixed-oxide barriers for ever thinner dielectric layers (where k is the dielectric constant of the material). Each of these requires a volatile, high purity, metal-organic precursor compound to be developed for materials transport to the layer deposition equipment or reactor.

Early CVD precursors were already available in commerce, since they were used for the purification and production of bulk materials such as polycrystalline silicon to form the melt for single crystal silicon ingot growth (and for arsenic, gallium and phosphorus *etc.* for bulk compound semiconductor ingot manufacture). These single-crystal ingots are then sliced and polished to form the wafers used for semiconductor chip and IC production. They also use a range of CVD precursors and processes in the subsequent IC manufacturing steps. Historically, one of the first CVD precursors for silicon epitaxy was silicon tetrachloride, but the precursor preferences for silicon deposition have changed to include di- and tri-chlorosilanes and more recently large volumes of silane are part of the precursor mix. During this period annual CVD demand and the related specialty chemical (or precursor sales) exhibited relatively high growth rates with worldwide sales running at an estimated value of $27 billion.

A key CVD process, recently adopted for mixed oxide deposition by the silicon IC industry, is "ALD" or atomic layer deposition. It is closely related to atomic layer epitaxy (ALE) used for compound semiconductors (where each precursor is also introduced separately into the growth reactor, *e.g.* in one to ten second bursts). ALE is already used for some high-performance IC production. Several other process names are also used, but they all provide alternating monolayer-type growth and high quality, well controlled, nanometre-dimension layers with the desired electrical characteristics. Prior ALD for silicon was started using thermal processes such as sputtering or laser ablation, but these are directional processes that do not provide conformal coatings for deep channels, trenches or vias whereas most chemical (CVD and MOCVD) processes do. These conformal coatings or layers are essential for the high performance of many semiconductor devices such as high capacity memory and multi-million-transistor microprocessor chips. Today's successful ALD process variations mimic the compound semiconductor processes to provide alternating monolayers from each chemical to develop thin, uniform and conformal atomic layer coatings of high quality purity. ALD offers higher levels of composition and thickness control, but at lower growth rates. They differ from standard CVD or MOCVD processes where all reactants are mixed above the substrate to deposit the layers at higher rates of growth. The ALD processes have one distinct advantage in that there can be no pre-reaction of the precursors before deposition on the wafer surface.

The early chemical ALD processes for silicon chips were developed to deposit thin layers of intermediate-*k* dielectrics by first applying a monolayer of water, or ozone, followed by a monolayer of aluminum from trimethylaluminium (TMA) or other Al-precursor and repeating these alternating atomic layer depositions until the required thicknesses were achieved. These processes were developed to modify the dielectric values of the old standby, silicon dioxide isolation, and to uniformly and conformally coat high aspect ratio trenches and vias now used in the manufacture of latest IC chips. Currently, ALD has started to use a wide range of custom-designed precursors for such applications as the deposition of hafnium and zirconium oxides for high-*k* dielectric materials, for titanium and/or tantalum nitrides for barrier and conductivity layers and ruthenium for conductive layers. These CVD processes are driving the chemistry for unique and process sensitive precursors that offer a range of deposition temperatures and will improve the performance of the "next generation" memory, microprocessor and other ICs.

13.1.3 Compound Semiconductors

Compound semiconductors are separated into three Periodic Group element pairs, II-VI, III-V and IV-IV, with the III-V compounds having by far the largest sales volumes. The most frequently used elements are the Group II or IIA metals, cadmium, magnesium, mercury and zinc (Cd, Mg, Hg, Zn) as their alkyls with Group VI, oxygen, sulfur, selenium and tellurium (O, S, Se, and Te) as hydride or alkyl derivatives, but their combined sales volume is small when compared with the III-Vs. Bulk zinc oxide is being proposed a substrate for II-VI and some III-V commercial device production and may require some custom zinc precursors. Other Group II-VI materials such as CdS, CdSe, ZnS, ZnSe and HgCdTe are used to fabricate devices, but only in relatively small volumes. The common metals used for II-VI compounds are magnesium, cadmium and zinc as alkyls and the Group VI elements are oxygen, selenium, sulfur and tellurium. HgCdTe has a long history as a wideband IR detector and it still is unique in this field, providing some of the best wide band infrared detectors when cooled to liquid nitrogen temperatures. Military, night vision and satellite uses are its main applications, but are not a large market on a volume basis.

Initial commercial applications of II-V materials were military or space driven. Gallium aluminum arsenide photo cathode windows were developed for image intensifiers for night vision. Gallium arsenide solar cells were widely used for both military and commercial satellite applications, gallium arsenide cells being preferred over silicon because of their higher efficiency and improved hardness to radiation.

The III-V compounds have the largest market volumes and values and the largest range of available precursors. Their LED, laser and other optoelectronic applications categories also make up the largest range of device types. For IV-IV semiconductors, the main elements used are carbon, silicon and germanium, with silicon carbide for high power electrical and some electronic high frequency devices. Thin layers of silicon-germanium on silicon ICs are providing ever higher frequencies (*e.g.* 300 GHz). However, SiGe devices usually consume more power than one based on GaAs and the ever-thinner active layers mean lower operating voltages (0.5 V and lower), which can lead to punch through of the device. At this time, the commercial volumes of the IV-IV based devices are not large.

The control of the deposition processes for compound semiconductor circuits is more critical and more demanding than for silicon devices, because their active device characteristics and performance are determined by up to hundreds of thin epitaxial CVD layers rather than up to 30 layers (or bulk properties) for silicon. For compound semiconductors, the atomic composition and the same Group element atomic ratios of the layers are used to define the device performance. Thus three or four elements may be used in any layer with differing compositions or Group ratios within each layer. Combine this requirement with multiple 2–10 nm layers for the active quantum well regions plus 50 or more layers for Bragg mirrors and 20 nm and thicker layers for blocking, confinement, contact improvement and other device functions and the manufacturing complexities of 100 layers or more become apparent.

The commercial precursor volumes for compound devices are currently smaller than for silicon technology, but the present range of precursor chemical types used for II-VI, III-V and IV-IV semiconductor production is much wider than for silicon CVD. In most cases more than one element in each of the compound Periodic groups is used for ternary or quaternary alloys for device manufacture. Today, most colored and white LEDs are Group III-V devices, although for many years ZnSe and ZnS were the favoured material systems.

13.1.3.1 III-V Compound Semiconductors

For III-V compound semiconductor processing, the most widely used elements are aluminum, gallium and indium from Group III in the form of metal alkyl compounds (or "metalorganics") in combination with nitrogen, phosphorus, arsenic or antimony from Group V (as alkyls or hydrides). High-speed indium antimonide devices are being developed, but no significant markets have developed to date. For the larger volume commercial devices, the active device layers may be binary, ternary or quaternary combinations of the III-V elements and their deposition process is commonly referred to as metal organic vapor phase epitaxy (MOVPE), a more specific form of CVD. With the need for multiple atom depositions, III-V or compound semiconductor epitaxial reactors are more complex than the silicon versions and are often designed for accurate control of four elemental sources and perhaps two dopants and up to 100 or more layers with three or four elements. This should be compared with silicon reactors with one large wafer that requires the deposition of one element, silicon, and two dopants and perhaps up to 25 layers of one element in total.

The value of the III-V high brightness LED (HBLED), white and colored, chip market segment (CVD-produced), already exceeds US$5 billion and could triple by 2020. Additionally, CVD processing accounts for $2 billion in low output LEDs, $3 billion for lasers and over $3 billion (mostly GaAs based) for the III-V electronic device market segment. The estimated total 2007 value for all CVD processed III-V device markets is almost $16 billion (Figure 13.1). Although there are many niche markets for compound semiconductor devices, the III-V HBLED chip markets are the largest compound semiconductor market group in both unit volumes (close to 100 billion) and dollar value. The HBLEDs have also demonstrated one of the highest CVD product based growth

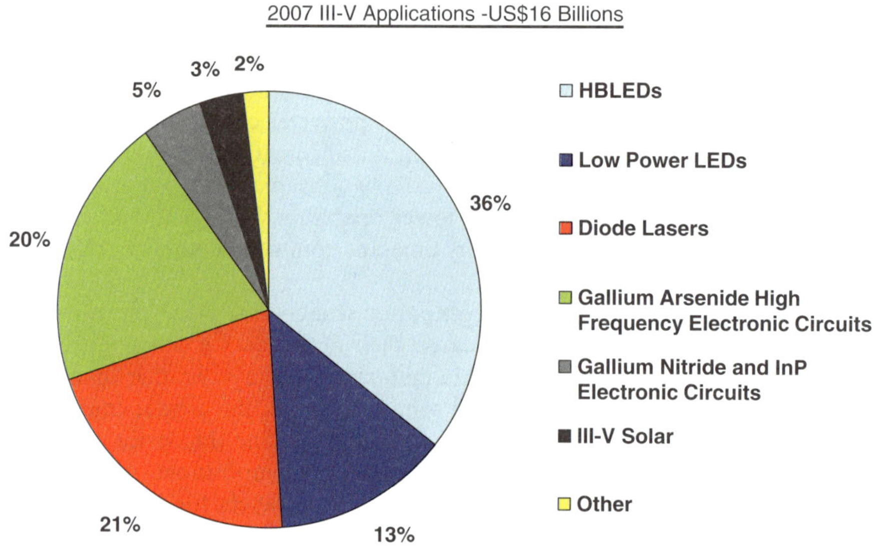

Figure 13.1 Estimated total 2007 value for all CVD processed III-V device markets.

rates during the last 10 years (over 20×) despite negative growth in 2001. HBLED sales rose from about \$200 million in 1996 to \$4.4 billion in 2006 (Figure 13.2).

Annual HBLED compound semiconductor CVD equipment sales are in the \$600 to \$700 million range although, notably, compound semiconductors produce higher device values per annual industry dollar spent on equipment. In comparison, the high perennial demand and high growth rates for silicon semiconductor ICs and chips during the past 50 years have also created semiconductor processes equipment demand and it has been a multi-billion dollar, silicon driven, market for decades and the performance, dollar value and the unit volumes continue to improve. In 2007 silicon chip manufacturing equipment sales are estimated by Gartner, Inc, to be almost \$47 billion.

13.1.3.2 II-VI Compound Semiconductors

II-VI Compound semiconductor materials usually include Group II or IIA metals, cadmium, magnesium, mercury and zinc (Cd, Mg, Hg, Zn) combined with Group VI, oxygen, sulfur, selenium and tellurium (O, S, Se and Te), again in binary, ternary and quaternary combinations. MgZnSSe elemental variations have been CVD grown and tested as blue and green emitters with ZnSe being the preferred blue emitter material for over 10 years, but successes with GaN blue LEDs convinced most laboratories to cease work on ZnSe. CVD grown ZnO/ZnMgO variations are being developed as a blue emitter and also as a substrate for gallium nitride device growth. Although there is some potential benefit from an alternate substrate for heterogeneous III-nitride growth, none of these have yet developed into volume CVD processes. Cadmium selenide and sulfide are widely used for light detection devices and zinc sulfides and selenides are used for transparent nose cones, usually with the harder and more expensive selenide as an exterior protective coating, but not all of these products use CVD deposition processes. A key II-VI compound semiconductor for infrared detection is mercury cadmium telluride (HgCdTe) and it has been made by several processes, including liquid phase epitaxy and CVD. For compound semiconductor device development and for transparent conductive windows, often with a dual function as contacts, new Zn precursors are

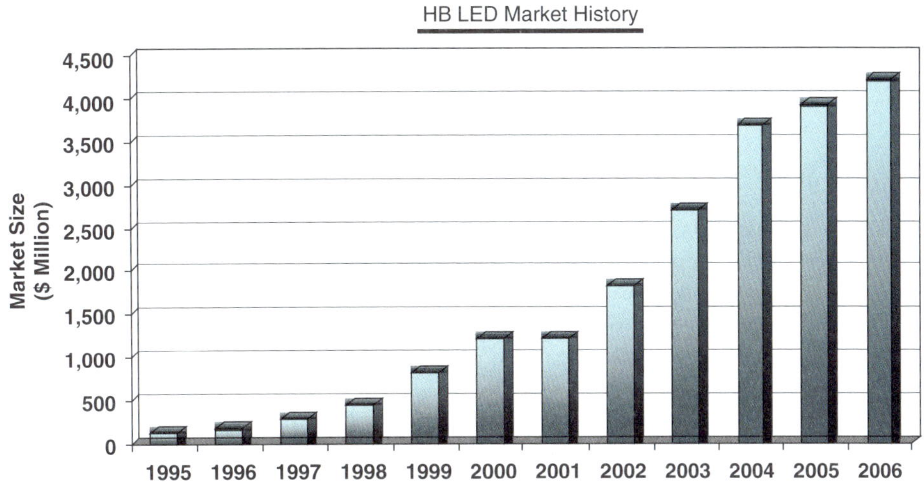

Figure 13.2 Market history of HBLED. (Reproduced courtesy of R. Steele at Strategies Unlimited.)

of interest for zinc oxide bulk crystal and layer growth, for modified indium-tin oxide (ITO) and for the deposition of other transparent conducting coatings. However, not all of these are CVD processes and II-VI compounds have not been a noteworthy growth market.

13.2 Industry Structure

13.2.1 The Sheet Glass Industry

Sheet glass production has been a high tonnage product for many years for windows and other flat glass and window products and it became a lower cost commodity with the introduction of the float glass process invented by the Pilkington Glass Company. CVD coated float glass production is a more recent growth segment and it meshes readily with the float process, because the CVD precursors for the desired coating can usually be introduced for deposition before the continuous glass sheet cools down. It is a \$13 billion market with only a few large trans national and vertically integrated companies such as Asahi Glass, Nippon Sheet Glass, Saint Gobain and Visteon operating worldwide. At the distribution end, the industry is a mix of large and small companies that supply a large and small user base. From a CVD point of view, the precursor chemical use is quite large and a growing market, but the precursors themselves usually had been produced for another process or application and the number of compounds in use for this application is not large.

13.2.2 The Silicon Industry

Since the discovery of the silicon transistor in the late 1940s, silicon device production has shown phenomenal growth over 60 years to become the \$250 billion industry of today and it relies heavily on CVD processes. It has spawned a string of startup companies over the years in both chip production and in silicon ingot, wafer, chemical, wafer process equipment and infrastructure support, most of these being initially in the USA. Fairchild Semiconductor spawned so many of these start up companies that SEMI, the industry supply association, issued a start-up family tree. Despite this industry size, startup companies continue to successfully enter the semiconductor business, but on a wider global distribution. Today, after many corporate consolidations, large international companies make more than 80% of silicon chip production. The development and size of this industry and its wide range of application markets is well documented and its detailed CVD development and growth requires separate treatment outside this text.

13.2.2.1 Group IV-IV Compound Semiconductors

There are IV-IV compounds, silicon carbide and silicon-germanium, that are used for semiconductor production but in much smaller volumes. However, silicon-germanium is usually processed as an adduct to a silicon production line to make epitaxially produced higher frequency devices that are not attainable by silicon alone and as such they are usually classed as a silicon products. Some Si-Ge devices are made by CVD and MBE processes and others use physical deposition processes.

Silicon carbide is an ideal material for the production of high power devices that can operate at high temperatures and at moderately high frequencies and CVD processes are used for epitaxially grown device layers and for some bulk growth silicon carbide processes. Silicon carbide is a compound semiconductor and will be discussed in that section. Compared to silicon, both of these IV-IV device industries are relatively small, and fragmented as to company size and structure and include a mix of startup, specialty and larger companies. Their present sales volumes are not believed to exceed \$300 million, although there are growth prospects for both material systems, particularly where their device performances are of interest for several applications, mainly for power for SiC and for high frequency devices for SiGe.

13.2.3 The III-V LED Industry

The HBLED industry has become a high growth phenomenon (Figure 13.2) only within the last decade and it is very dependent on CVD processes. It already has over 70 companies active worldwide and is a mix of captive or integrated and regionally distributed merchant epitaxy (or CVD companies) with the fraction of merchant epitaxy performed varying with the geographic region. Only about 15% of these companies are vertically integrated. In 2005, the USA and Taiwan had about equal volumes of merchant wafer supply (~40%), with the rest of Asia supplying another 15%, but the US CVD segment has dropped to about 25% as Taiwan has increased to about 50%, the rest of Asia 15% and Europe less than 10%. LED assembly requires labour-intensive packaging with specialist companies concentrating on packaging services for the whole industry being located in low labor cost economies and they make up about 50% of the industry players. About half of these are located in China with most of the others being in Malaysia and Thailand.

With the exception of Taiwan, Asian LED manufacturers tend to be large, well-established companies or their subsidiaries such as Lucky Goldstar (LG), Nichia, Samsung and Toyoda Gosei. However, Taiwan has over 20 LED companies, most being established on a venture basis within the last 15 years. With a mix of captive and merchant epitaxy companies, Taiwan claims over 50% of the world market. The Taiwanese companies include both captive and merchant epitaxy companies and they manufacture most types of LEDs, including the high volume, high brightness AlInGaP and GaN/InGaN based devices. There may be too many players for the size of this industry, a factor that may be responsible for several recent mergers and acquisitions such as the recent mergers in Asia and the purchase of Lumileds, Color Kinetics and TIR systems by Philips Lighting.

In Europe, most LED and Laser manufacturers are associated with large well-established international companies such as Infineon, Philips and Osram, although there are some merchant suppliers of LED epi-wafers and venture LED and laser manufacturers. In North America, LED suppliers range from small venture to larger diversified companies such as Cree and Philips

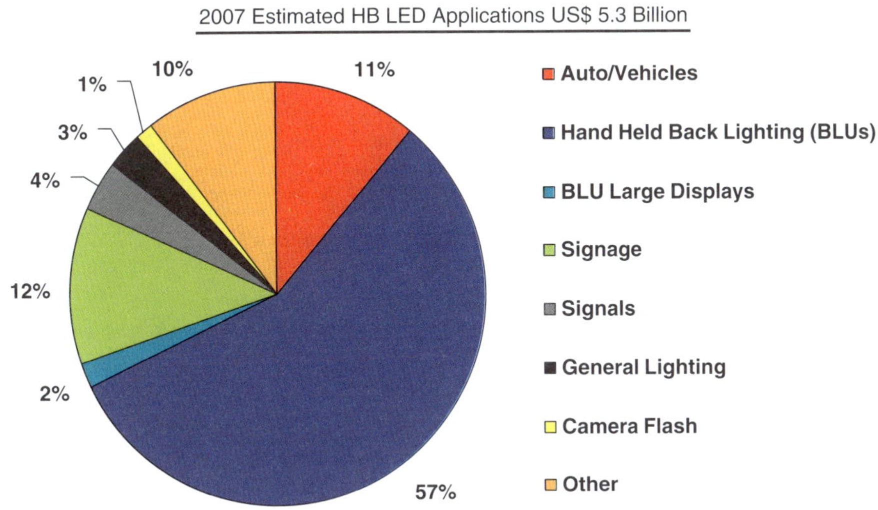

Figure 13.3 Estimated HBLED applications in 2007.

(after acquiring Hewlett Packard's interests in the Lumileds LED division). It is possible to have an LED business anywhere in the world without a wafer fabrication facility because of the many companies offering merchant CVD and LED wafer production services. Sales of HBLEDs in 2007 were estimated to be about \$5.3 billion and Figure 13.3 shows the major applications, with many of these uses having high growth rates for the foreseeable future. In addition to the LEDs, III-V compound semiconductor electronic devices, mostly GaAs and InP based, also have a market value exceeding \$3 billion. Because of the diverse substrate element use in III-V technology, the high number of layer requirements, the ongoing new precursor demands and the high market growth rates, these CVD-based products will receive extended coverage in the following applications sections of this chapter.

13.2.4 III-V Semiconductor Lasers

Group III-V semiconductor diode lasers are a subsection of the III-V LED growth phenomenon that are electrically or optically pumped LEDs. The development of the higher volume, CVD produced, devices started more than 20 years ago with the quaternary InGaAsP lasers for infrared fiber-optic telecommunications, followed by GaAs and InGaAs for red lasers for printers, solid state pumping, write and read optical storage compact discs, for bar code readers (supermarkets), pointers, plus multi-watt chips for welding and cutting, and finally to the latest two, front emitting VCSELs and VECSELs (vertical cavity surface emitting lasers and vertical external cavity surface emitting lasers) for chip to chip communications and InGaN blue lasers for high density DVD players. All have contributed or are contributing to the development of new CVD precursors, processes and markets and should provide continuing market growth and process demand well into the future. The 2007 market was estimated to be little changed from 2006 at about \$3 billion, with the only real value growth market being blue lasers for high capacity DVDs, which should reach about \$25 million in 2007. The two largest laser application categories in 2007, with about equal sales of 40%, are the lower power sub-700 nm and 950–1550 nm class lasers, with the 1–10 W and 10+ W laser groups each having about 10%. Of these, the lower power segment has the highest unit volumes, but with lower average selling prices. The unit volumes for most of these applications are increasing, but their price declines are eliminating most of the potential value growth in the laser diode market (excepting new diode laser markets). Although the diode lasers comprise more than 99% of all lasers produced, they represent less than 50% of the laser market value due to the low average unit price of the semiconductor diode lasers.

13.3 Precursor Selection

Precursors for CVD processes must have several desirable qualities:

- they should be volatile to allow transportation in the vapor form;
- stable in the supply container at the given storage and/or transportation temperatures;
- stable in the delivery system;
- breakdown cleanly on the substrate to give the desired material of the correct levels of purity;
- capable of being manufactured safely in high volume to the correct levels of purity and consistency;
- appropriate reductions in costs as volumes increase.

In general, as the atomic weight of the metal increases it becomes increasingly difficult to produce volatile materials, *e.g.*, Si, Ga, Ti and Ta:

- silane gas, vapor pressure 65 bar;
- trimethylgallium, vapor pressure at 20 °C is 182 mmHg;
- tetrakis(dimethylamino)titanium, vapor pressure at 20 °C is 0.09 mmHg;
- pentakis(dimethylamino)tantalum, vapor pressure at 20 °C is 0.003 mmHg.

Traditionally, the material has been introduced into the deposition equipment by "bubbling" a carrier gas through the material or by vapor drawn into a reduced pressure system. The less volatile precursors, which are often solids, will increasingly require new techniques to introduce the material into the vapor stream to be employed, such as direct liquid injection, whereby the neat precursor or the precursor dissolved in a suitable solvent is introduced by flash evaporation and purged with a suitable carrier gas.

This is particularly important in high volume applications such as the production of DRAM and Flash memory devices, where the economics require a high throughput of silicon device substrates through the CVD or ALD deposition equipment.

13.4 Commercial Considerations

13.4.1 Silicon

In the case of the silicon industry these requirements historically have been fairly readily achieved from silicon chlorides such as $SiCl_4$, $SiHCl_3$ and SiH_2Cl_2 and later from its hydride. Recently, however, the demands of the 65 nm node and smaller have introduced new atom requirements and process challenges for high-*k* dielectric oxide deposition.

Silane (SiH_4) has been one of the fundamental materials used to deposit both epitaxial and polycrystalline silicon but it is also widely used for silicon oxide or silicon nitride deposition by combining with N_2O or O_2, or NH_3. SiH_4 is a pyrophoric, poisonous compressed gas and whilst being an extremely hazardous material it is very convenient as a CVD precursor. It was initially manufactured in relatively small (10 tonne annum^{-1}) capacity plants based upon the reduction of chlorosilanes with lithium hydride. A number of these plants were installed globally during the

1970s and 1980s to service the growing requirements of the semiconductor industry. The process was an inherently costly one both in terms of labor and raw materials, with lithium hydride being expensive.

The big breakthrough in the pricing structure and availability of silane came with the development of large-scale plants to produce polysilicon, but which used silane as an intermediate to achieve the purity requirements. They both produced 1000+ tonnes per annum of silane and this made available large volumes of product at relatively low pricing.

13.4.1.1 Chlorosilanes

These materials, which have been fundamental to the silicon industry since its inception in producing both polysilicon and CVD growth, have been produced in large volumes since the 1940s as precursors for fumed silica, used in many industrial applications. Consequently, they have been readily available at low cost and in high volumes.

Many of the other precursors and dopants used traditionally in the silicon industry have followed a similar model, *e.g.*, tetraethyl orthosilicate (TEOS), and they are produced in large volumes for non-electronic applications and they can be relatively easily purified to the required semiconductor standards. Generally being silicon based (the second most abundant element on earth after oxygen at 25.7% weight) their inherent cost is low. TEOS is also used in electronic device production for CVD SiO_2 and in spin on dopant applications.

13.4.1.2 Future Precursors for Silicon

The drive to continue Moore's Law at 65 nm and beyond (with Intel having already started 45 nm production processes) is requiring new materials and processes to be used to grow the semiconductor structures high-*k* dielectric layers. Hafnium and zirconium amides grown by ALD are some of the more exotic materials being introduced at the 45 nm node. There is no other market for these materials and whilst the quantities required are significantly declining per chip, the cost of production and the inherent cost of the new elements increase, because of their rarity. (Hf for instance accounts for 0.00058% weight abundance of the earths crust.) In addition to the high level of advanced R&D required to arrive at the right molecules, the traditional low cost model is no longer sustainable going forward.

13.4.2 Compound Semiconductors

The development of precursor sources for this technology has by necessity followed a more complex path. The fundamental precursors have changed little since the inception of the MOVPE process outlined by H. M. Manasevit and consist of: trimethylgallium, triethylgallium, trimethylindium, trimethylaluminium, arsine, and phosphine with dimethylzinc, diethylzinc and cyclopentadienyl magnesium being used as dopants. More recently significant volumes of high purity ammonia are also consumed for the growth of gallium nitride based III-V materials.

Of the list of precursors, four are produced in significant volumes for other applications but at lower purities than demanded in these applications. Trimethylaluminium and diethylzinc are both produced in tonnage quantities as components of Zeigler–Natta co-catalysts used in the production of propylene. The purity requirement for this application is typically 99%. Considerable expertise is required to achieve the 99.9999% purity levels necessary for semiconductor chip manufacture. Trimethylaluminium can also be utilized as a raw material for the production of trimethylgallium and trimethylindium.

Phosphine as a high pressure compressed gas is produced in large volumes as a precursor for a whole range of phosphorous based chemicals used in many applications, including flame retardants and fumigants. Again the purity has to be greatly improved to meet semiconductor applications and phosphorus is often used as an n-dopant for silicon from either the oxychloride or chloride source.

Ammonia is a readily available commodity chemical and precursor, which has found a growing application in gallium nitride growth. Moisture or oxygen sources are a major detriment with this product, as produced by including oxygen in the growing nitride film, and great improvements have been made in reducing NH_3 moisture levels to extremely low levels.

Other critical precursors are only manufactured for their specific applications in much smaller quantities. Trimethylgallium, a pyrophoric liquid is a key precursor for most compound semiconductor materials. It is produced in multi-tonnage plants and its price is ultimately controlled by the price of gallium metal, a by-product of the production of aluminum.

Trimethylindium (TMI) is a pyrophoric solid that also has a nasty habit of free radical decomposition at temperatures above 150 °C. TMI has been one of the most difficult materials for which to achieve the required levels of purity and consistency. It is a key component of several III-V compound alloys and one of the additional difficulties has been obtaining a consistent pick up into the reactor gas-stream from this solid source.

Arsine, a highly poisonous liquefied compressed gas, is only produced for semiconductor application; it is used in small volumes as a dopant in silicon but by for the largest application in terms of volume is as a component of compound semiconductor materials.

Currently there are only two small plants producing this precursor from zinc arsenide.

13.4.3 Architectural Glass Coating

Precursors for these applications need to be available in large volumes and low prices. Much lower purity materials are also acceptable for these applications. Precursor materials used in the production of low emissivity glass on the float line include silane, tin chloride and ethylene, all now available in large volumes and at acceptable prices that can compete with off-line physical vapor deposition (PVD) processes. Self-cleaning glass utilizes titanium tetrachloride for titanium oxide coatings.

13.4.4 Thin Film Transistor (TFT) and Solar Applications

CVD processes are increasingly desirable in these technologies and purity requirements usually fall between the extreme of 99.9999% for semiconductor applications and 99.5% for large area architectural glass applications. Again, silane is a key precursor with diethylzinc increasingly used to produce a transparent conductive coating of zinc oxide.

13.5 Health, Safety and Analytical

13.5.1 Health and Safety

Many of the precursors used are sensitive to moisture, highly reactive and pyrophoric (silane and the Group III metal alkyls such as trimethylaluminium or trimethylgallium) or very toxic (chlorosilanes and Group V hydrides or alkyls such as arsine, phosphine). Despite these potential hazards, these very high purity chemicals are shipped and used around the world. Therefore, the handling, use and shipping of laboratory and commercial quantities must adhere to strict hazardous handling requirements under all circumstances. Not withstanding these properties there have been few reported problems for several decades. All of these factors are of even greater importance in the manufacturing of starting materials and precursors, because the quantities and production volumes are larger and because other reactants may be present. Consequently, government health and safety authorities are actively involved in these materials handling and manufacturing processes for all the industries. Extensive use of HAZOP and FEMA risk analysis to ensure production plants and containers are designed and operated to ensure safety standards of the highest levels are maintained.

Attempts have been made to introduce alternative precursors that are less toxic but ultimately they have not been widely accepted by the market since inevitably they are more costly. An example of this is the attempt to introduce *tert*-butylarsine and *tert*-butylphosphine as safer alternatives to arsine and phosphine. They have only found limited acceptance for niche applications due in major part to them being at least 15× more expensive.

13.5.2 Analytical Requirements

Purity of the precursors for semiconductor and optoelectronic applications is critical. Ultimately it is the electrical and/or optical quality of the grown layers that is the important factor. It is therefore important to have analytical sensitivity for impurities in the precursors that can be correlated to the electrical and optical qualities of the grown layers. This has led to a unique interaction between chemists, material scientists, electrical engineers and physicists taking place and it is a key to the successful development of the new precursors at the correct levels of purity and performance.

The very nature of the precursors means that even sample preparation becomes a major challenge. The measurement of trace oxygen species (down to 1 ppm) in a pyrophoric air sensitive material becomes a major challenge. It is also the case that the metal precursors often originate from the same ore, which means that element cross contamination is often present. Since they often have similar chemical properties, purification and the subsequent analysis become major challenges, an example would be analyzing for zirconium and hafnium. This becomes particularly difficult when the impurities act as dopants in the active material. Examples of this are zinc (a p-dopant) contamination of gallium and silicon contamination of indium.

In addition to the purity aspect, much analysis is carried out in the initial characterization of a new molecule for its temperature stability and suitability as a CVD precursor molecule.

Typical analytical tools are listed below.

- Molecular characterization:
  - micro analysis – carbon, hydrogen, nitrogen, oxygen
  - infrared
  - X-ray diffraction
  - thermogravimetric analysis
  - mass spectrometry
  - Fourier-transform nuclear magnetic resonance spectroscopy.
- Impurity analysis:
  - inductively coupled plasma mass spectroscopy (ICP-MS)
  - inductively coupled plasma optical emission spectroscopy (ICP-OES)
  - atomic adsorption spectroscopy (AA)
  - gas chromatography atomic emission spectroscopy (GC-AED)
  - Fourier-transform nuclear magnetic resonance spectroscopy (FT-NMR)
  - chloride analysis
  - Karl Fisher moisture analysis.
- Stability analysis:
- accelerated rate calorimetry
  - automated lab reactor – thermodynamics of manufacture of precursors

- Usability as CVD precursor:
  vapor pressure measurement
  Epison, Lorex – vapor transport measurement

13.6 Typical Precursors for the Silicon Semiconductor Industry

Table 13.1 lists precursors already in production, those that might be used in future and also typical precursors used for the compound semiconductor industry.

13.7 III-V CVD HB Light Emitting Diode Applications

13.7.1 Historical LED Production

The first evidence of electroluminescence was reported in 1907, when a British scientist, Henry Joseph Round, reported yellow, green and blue emissions when voltages in the 10–110 V range were applied to a selection of silicon carbide crystals. Unfortunately, these results were overlooked for many years with the first commercial LEDs only being offered about 40 years ago. Commercial LEDs started as feeble red milli-lumen emitters and were used for low wattage applications, such as indicator lights and eight segment alphanumeric chips. These types of LEDs are still available and are a separate LED market segment with annual sales in the $2 billion range. Other lower output LEDs in today's market include "off specification" HBLEDs that are sold by the kilo for toy and other low power uses.

The LED wavelength range has now expanded to offer emitters in the 230–15 000 nm range with a five to six orders of magnitude increased power output for the HBLEDs. For the UV and far-infrared regions they provide milliwatt outputs per chip, but for visible light some HBLEDs exceed 70 lumen per watt and 200 lumen per chip. In the near-infrared region, multiwatt LEDs are offered for high power applications such as welding and steel cutting. Progress in UV and blue LED manufacture is also driving new phosphor development for RGB (red, green and blue) and white LEDs from phosphors. In the electrically stimulated mode they also include lasers in the 300–5000 nm wavelength range, with the IR-lasers emitting at the 1.3 and 1.6 nm moisture transparent wavelengths providing the technical backbone of the high data rate fiber-optic communications industry. Up to the mid-1990s, LEDs provided only infrared (IR), red, orange, yellow and limited green wavelengths without high lumen blue or deep green colors. IR LEDs (with silicon receivers) have long been used as coupler-isolators in PC and computer data lines to protect from electrical cross talk and high voltage pulses and recently for such applications as medical heat therapy and night vision illumination.

Over a 20 year period, research in Japan by I. Akasaki, H. Amano and S. Nakamura eventually led to the introduction by Nichia of commercial blue LEDs in 1993 and green in 1995. White light from the RGB LED combination became feasible a couple of years later and therefore full color LED displays became a commercial reality as blue and green LEDs with acceptable light outputs were offered on the market. As white LEDs from phosphor converted blue LEDs were added to the wavelength mix by Nichia, most applications benefited from the new HBLED product range, which has grown rapidly during the last decade due to several large application categories. It now totals about $5.3 billion (Figure 13.3), with about 40 billion-units being produced in 2007. New market segments are being added as new applications discover the benefits of the increased power of the LED and HBLED unit volumes will continue to increase, and hence sales volumes are estimated to grow to $8.3 billion in 2011 (Figure 13.4). The largest HBLED segment is LCD backlighting units (BLUs) for hand held displays, with close to a 57% market share and 2007 sales of about $3 billion (Figure 13.3). InGaN LEDs comprise about 75% of the market value,

Table 13.1 Summary of precursors.

Precursors in production	*Potential future precursors*			*Typical precursors used for the compound semiconductor industry*	
Trimethylaluminium	Dimethylethylamine alane	Bis(2,4-dimethylpentadienyl)ruthenium	Tungsten hexafluoride	Trimethylaluminium	Solution Bis(h5-cyclopentadienyl)magnesium
Germane 10% in hydrogen	Dimethylaluminium hydride	Bis(isopropyl-h5-cyclopentadienyl)ruthenium	Tungsten hexacarbonyl	Dimethylethylamine-alane	Bis(methyl-h5-cyclopentadienyl)magnesium
Germanium tetrachloride	Diethylaluminium ethoxide	Ruthenium carbonyl	Tris(2,2,6,6-tetramethylheptane-3,5-dionato)yttrium	Diethylaluminium ethoxide	Bis(methyl-h5-cyclopentadienyl)manganese
Tetrakis(diethylamino)hafnium	Bis(2,2,6,6-tetramethylheptane-3,5-dionato)barium	Tris(2,2,6,6-tetramethylheptane-3,5-dionato)scandium	Tris(1-methoxy-2-methyl-2-propoxy)yttrium	Tris(dimethylamino)antimony	*tert*-Butylamine
Tetrakis(dimethylamino)hafnium	Tris(1-methoxy-2-methyl-2-propoxy)bismuth	Tris(dimethylamino)silane	Tetrakis(dimethylamino)-zirconium	Triethylantimony	1,1-Dimethylhydrazine
Silane	Tris(2,2,6,6-tetramethylheptane-3,5-dionato)bismuth	Tetrakis(dimethylamino)silane	Tetrakis(2,2,6,6-tetra-methylheptane-3,5-dionato)zirconium	Trimethylantimony	*tert*-Butylhydrazine
Dichlorosilane	Bis(2,2,6,6-tetramethylheptane-3,5-dionato)copper	Tetrakis(ethylmethylamino)silane	Tetrakis(*tert*-butoxy)zirconium	Tris(dimethylamino)arsine	Phosphine
Trichlorosilane	Tetraethoxygermanium	Trisilane (Silcore®)	Tris(isopropoxy)mono(2,2,6,6-tetramethylheptane-3,5-dionato)zirconium	Arsine	Tris(dimethylamino)phosphine
Tetraethylorthosilicate	Tetramethoxygermanium	Bis[pentakis(ethoxy)-dimethylaminoethoxy)-tantalum]strontium	Tetrakis-(ethylmethylamino)-zirconium	Trimethylbismuth	Bis(2,4-dimethylpentadienyl)ruthenium
Silicon tetrachloride	Tetrakis-(dimethylamino)-germanium	Bis(2,2,6,6-tetramethylheptane-3,5-dionato)strontium	Tetrakis(diethylamino)zirconium	Triethylboron	Bis(isopropyl-h5-cyclopentadienyl)ruthenium
Tetrakis(dimethylamino)titanium	Tetrakis(*tert*-butoxy)hafnium	Pentakis-(dimethylamino)-tantalum	Bis(*tert*-butoxy)bis(1-methoxy-2-methyl-2-propoxy)zirconium	Dimethylcadmium	Dimethylselenide
	Tetrakis-(ethylmethylamino)-hafnium	Pentakis(ethoxy)-tantalum	Bis(methyl-h5-cyclopentadienyl)-methoxymethylzirconium	Carbon tetrabromide	Di-*tert*-butylselenide
	Tetrakis(1-methoxy-2-methyl-2-propoxy)hafnium	Pentakis(butoxy)tantalum		Carbon tetrachloride	Di-*tert*-butylsulphide
	Bis(*tert*-butoxy)bis(1-methoxy-2-methyl-2-propoxy)hafnium	Tetraethoxy-(dimethylaminoethoxy)-tantalum		Carbon trichlorobromide	Dimethyltelluride
	Hafnium tetrachloride	Tris(diethylamino)(*tert*-butylimido)tantalum		Solution Carbon tetrabromide	Diethyltelluride
	Bis(methyl-h5-cyclopentadienyl)-dimethylhafnium	Tantalum pentachloride		*tert*-Butylchloride	Diisopropyltelluride
	Bis(methyl-h5-cyclopentadienyl)-methoxymethylhafnium	Titanium tetrachloride		Trimethylgallium	Tetramethyltin
	Tris(2,2,6,6-tetramethylheptane-3,5-dionato)lanthanum	Tetrakis(diethylamino)titanium		Triethylgallium	Tetraethyltin
	Solution Tris(1-methoxy-2-methyl-2-propoxy)lanthanum tetraglyme adduct	Bis(isopropoxy)bis(2,2,6,6-tetra-methylheptane-3,5-dionato)titanium		Trimethylindium	Diethylzinc
	Bis(2,2,6,6-tetramethylheptane-3,5-dionato)lead	Tetrakis(*tert*-butoxy)titanium		Solution TMI™	Dimethylzinc
	Pentakis(ethoxy)niobium	Tetrakis(1-methoxy-2-methyl-2-propoxy)titanium		Bis(h5-cyclopentadienyl)iron	Dimethylzinc triethylamine
	Pentakis(butoxy)niobium	Bis(isopropoxy)bis(1-methoxy-2-methyl-2-propoxy)titanium		Bis(h5-cyclopentadienyl)magnesium	

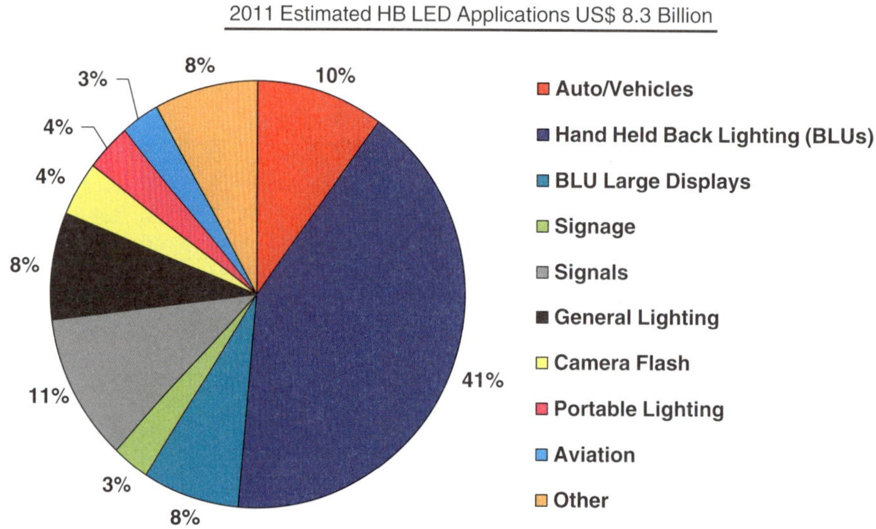

Figure 13.4 Projected HBLED applications for 2011.

with about two-thirds of these being white. The remaining market share is mostly AlInGaP-produced green, orange and red, plus a few RGB LEDs. All HBLED applications categories are currently growth markets, with the hand held segment having only unit volume growth plus medium-term possibilities of competition from organic light emitting diodes (OLEDs).

A small percentage of LEDs are produced by other processes, such as liquid phase epitaxy, hydride vapor phase epitaxy (HVPE) and molecular beam epitaxy (MBE), but the vast majority of the HBLED chips are MOVPE engineered and grown. Both MBE and HVPE processes should be regarded as variations of CVD technology and are used for the growth of some AlGaN and GaN devices. It is perhaps significant that some compound semiconductor bulk substrates are also CVD grown at high temperatures (>1000 °C), including HVPE grown AlN and GaN plus some silicon carbide ingots. SiC device wafers also use high temperature CVD processes for epitaxy with silane and alkene or alkane precursors.

13.7.2 Overview of CVD Processes

CVD silicon layer (epitaxial and polycrystalline) and CVD oxide depositions have been the backbone of the silicon IC industry for many years, but compound semiconductor production and some of the latest silicon IC processes use more technically sophisticated CVD process versions to support more rapid market growth, namely MOCVD, MBE and HVPE (*vide infra*). Because CVD epitaxial processes are so critical to the outstanding worldwide growth of today's silicon and compound semiconductor businesses, a brief process review is included prior to discussing the range of applications that offer so many benefits to society:

(1) *MOVPE*: The MOVPE process, metal organic vapor phase epitaxy (also known as OMVPE and MOCVD), is used to make most of today's diode lasers, HBLEDs, compound semiconductor devices and high-*k* mixed oxide insulators. For III-V compounds, volatile Group III vapor phase metal alkyl compounds (*e.g.*, trimethyl-aluminum or -gallium) and Group V elemental hydrides (*e.g.*, phosphine or arsine) plus the relevant dopants would be transported by carrier gases to a reaction chamber, where they mix above the wafers to deposit device layers on the substrate

wafers. For other materials depositions such as transition metal oxides, liquid or solution delivery methods may be required because of the lower volatility of the precursors. Reaction chamber capacities vary from 1 to 50 wafers, with typical single wafers being 8 and 12 inches in diameter for silicon processes and 1 to 6 inches for compound semiconductors. These wafer sizes vary with product demand, process maturity and deposition technology – hence the large sizes for silicon IC production. MOVPE is the preferred manufacturing process for HBLED and for compound semiconductor electronic devices, but recently its ALD (atomic layer deposition) version has been introduced for high-*k* mixed oxide insulators for silicon memory and microprocessor ICs. For MOVPE and for silicon depositions, most process temperatures are in the 500–1100 °C range, but for oxide and mixed oxide processes the temperatures may be as low as 250 °C.

(2) *MBE*: Molecular beam epitaxy (MBE) is also used for the growth of multiple thin layer structures. The main process difference between MBE and HVPE and MOVPE is that the MBE deposition of the required atoms occurs in an ultrahigh vacuum ambient ($\sim 10^{-13}$ Torr), atomic layer by atomic layer from atomic or molecular beam sources (with no pre-reaction). Although MBE is also used for the manufacture of commercial electronic circuits, it is also ideal for the growth of many device types on small substrates on a one of a kind basis and therefore is often used for device research. MBE deposition temperatures usually range from 400 to 700 °C and provide uniform deposition thicknesses. MBE has the lowest growth rates of the three CVD processes.

(3) *HVPE:* Hydride vapor phase epitaxy (HVPE) is an established, atmospheric pressure, device-growth process, where hydrogen chloride, transported by hydrogen or nitrogen carrier gases, is passed over a metal reservoir or source (*e.g.* gallium) to form a volatile metal chloride. This gas stream is then combined with a carrier transported Group V element chloride or hydride, such as arsenic trichloride or arsine, to the wafer or device substrate surface. Other gas streams transport the dopants for the growth of active device layers. This example would produce GaAs layers. Reaction temperatures for HVPE layer deposition are usually in the range 600–1100 °C, depending on the semiconductor composition. The HVPE process has low equipment costs and one of the highest deposition rates of any epitaxy processes at up to 600 micron h^{-1}, but it has not been widely used for device fabrication.

Recently, however, it is being employed with promising results for some III-nitride device fabrication processes and for materials bulk growth, especially for GaN and aluminum nitride (AlN). The HVPE process has the highest useful growth rates among the semiconductor CVD methods at up to 600 micron h^{-1}, followed by MOVPE and MBE with their respective 1–50 and 0.05–2 micron h^{-1} rates. To improve uniformity, the wafers or the platens are usually rotated at rates up to 1500 rpm; however, for MOVPE, the swept volume above the wafers is probably more important for layer deposition uniformity than fast rotation. The swept reactor volume typically decreases in the order from bell jar to planetary to showerhead reactor design, with deposition control improving in the same order. Hence, the consensus appears to be that the close coupling to the wafer surface and the multiple feeds of the showerhead design will provide the best layer thickness and composition control scenario of the leading reactor designs. Substrate temperature uniformities of better than ± 1 °C are needed for good control of layer composition and thickness uniformity for all these CVD processes.

Plasma assisted deposition is also used where precursor or chemical reactivities are not adequate either for the desired deposition temperature or for lower deposition temperatures. Examples of this are where low temperature silicon nitride passivation deposition is required below the normal silane-ammonia process temperature, where the deposition rates or the layer structure would be unacceptable. In this case a plasma assist will produce acceptable surface and layer structure plus suitable growth rates. It is also used for some III-nitride deposition processes.

13.7.3 Hand Held Devices and Display Backlighting

The hand held device segment, with typical display sizes in the 2.0–3.5 inch range, is the largest CVD based HBLED market and supports the huge LED backlighting unit (BLU) market segment. It has also been the volume driver and financial support for the most of the performance improvements of the white LED. The estimated sales for this segment in 2007 will be about \$3 billion and this, together with monitor and TV BLUs, will remain the largest LED use category (Figure 13.3) at least until the use of HBLEDs in the general lighting market or large BLU displays become well established. In 2006 there were over 2.5 billion hand held units in the field (mobile telephones, cameras and personal digital assistants [PDAs]) and with cellphone sales alone expected to add about 1.2 billion units in 2007 the total operating hand held devices in the field was expected to be over 4 billion units by the end of 2007. Larger screens (5–7 inch) for global positioning, car screens and games are also rapidly adding BLU market strength and forming a new growth segment. An encouraging note is the offer (in late 2007) of white side emitting LEDs in quantity in the \$0.20 to \$0.25 price range and this could open up new display markets for HBLED BLUs, although it could reduce the overall dollar growth of the market.

Present competition for the LED BLU market is small CCFL tubes (cold cathode fluorescent lamps) with a less than 10% market share, but they need diffusers and filters for full color compatibility and consume more space and power than the LED BLUs, which should gradually replace the CFLs as new and larger versions (*e.g.*, 3 and 3.5 in diagonal) are introduced. Full color LED BLUs are available from either RGB LEDs or blue LEDs with phosphors and produce good color rendition. The benefits of LED backlighting over compact fluorescent systems are higher definition, often no color filter required, thinner and smaller volume units possible, lower power use and improved color rendition. However, most of these applications are not obvious to the consumer, because the LEDs are internal to the equipment used. The use of camera HBLED flash units is also showing rapid growth, although this is only a hand held use, not a BLU application.

13.7.4 Large Display Backlighting

LED BLU use for the larger LCD displays (monitors, laptops and televisions) is a new growth market and, requiring many LEDs per display, it will drive CVD demand. With diagonal measurements in the 20 to 80 inch range, the required LED numbers are high but they will decline as more powerful CVD produced LEDs are being introduced and BLU designs are being improved. Large display BLUs use either the RGB LED combination or blue LEDs with phosphors and produce some of the best definition and color rendition available to the consumer, with the added advantage of providing up to more than 140% of the NTSC color specifications. Some projection TV BLU competition comes from Luminus Devices, where their large chip RGB-HBLEDs are used by Samsung and LG for DLP projection TV models.

For large LED BLUs, side-by-side comparisons with CCFLs and other technologies show obvious improvement in definition and color rendition and really support customer appeal for these models. Side emitting LEDs are also being used to improve display uniformity and form factors. Potential benefits from LED backlighting over the compact fluorescent systems are: no color filter required, thinner and smaller volume BLUs, lower power use, better picture definition quality and improved color rendition, partially justifying the higher cost in today's TV market.

For Osram RGB BLUs, "Golden Dragon" LED chips are used in groups of four for each pixel, one red (625 nm), one blue (458 nm) and two green at 527 nm. Several manufacturers, including Sony and Osram, have demonstrated an 80″ TV, the latter using Osram's proprietary LEDs. Samsung's second generation 40″ LED BLU TV exhibited excellent definition and color rendition and it received the "Display of the Year" award from the Society for Information Display in May of 2007. Sony already has 50–65 inch LED models on the market and has announced an 80 inch

LED prototype only 4 cm thick that uses a total of 1120 RGB Golden Dragon LEDs. In an extension of display dimensions, Osram has demonstrated an 82″ TV prototype using its own proprietary LEDs.

As these larger displays continue to increase in popularity and size and as the LED prices decline from increased production volumes, lower manufactured costs and increased competition, only picture quality and not price will determine consumer demand. Consequently, Sony 50″ prices are now much lower than the original $9000 price range. Current TV shipments are about 30 000 for the large LED based displays and were expected to grow to 300 000 units by 2008. LED BLUs are also rapidly gaining market share in the laptop display segment above 17 inches for power saving, battery life and picture quality reasons and for high-end monitors to improve definition and color rendition performance. As unit volumes increase and TV prices decline, LED BLU TVs may achieve as much as 60% of the large display BLU market share by 2011 (Figure 13.4).

13.7.5 Portable Lighting

Portable lighting covers a wide range of applications from flashlights to camera flash, to strobe lighting, to small colored sources and should also include portable traffic signs, amber warning lights and other mobile niche-type light sources. For these applications the low power demand from LEDs allows solar power to be considered for many uses. In the case of amber safety lights, their replacement with CVD produced HBLEDs greatly extends the useful lamp and battery life, improves flash brightness and potentially provides more user-friendly lighting that can be solar powered. In the small portable lamp category, *e.g.*, the flashlight and the new stick-on lamps, the LED has revolutionized the industry as battery and lamp life can be extended to years for some small lights instead of months. At $1.50 each, LED stick on lamps are now the lowest priced LED product and can be used in almost any location. Also popular are the various sizes of flashlights with one to four LEDs, front and rear multi-LED cycle lamps that can also serve as safety lights to protect people taking late evening strolls. Miniature versions such as key ring lights have become popular with the use of small LED chips and miniature batteries with existing battery lives being extended by factors of five to ten. Prior to the HBLED, portable lamps needed either heavy battery power or for large units, petrol powered generators plus fueling and related services. Camping and emergency lighting uses could soon be replacing fluorescent versions, as increased LED lumen outputs are available, again benefiting from extended lamp life. In such applications, the CVD process is enabling the consumer with many easy to use LED lighting solutions.

A significant key to the development of a new application segment is the availability of useful products for the consumer to purchase. This goal has been achieved, as a range of LED items can be selected from and purchased at recreational and department stores, with many being in the $1.5 to $50 price range and with some "stick on" LED lights filling the low-end price slot. There is competition for small LEDs from mini-krypton bulbs, which although less expensive, draw more power from the battery and have bulb replacement expenses. New LED flashlights have power level options not feasible with the filament lamps, including low level lighting (2.5 lumen with a standard LED) or 1–3-watt LED sources with corresponding 30 and 60 lumen beams reaching out to 30 m. Most of these early high power models were powered by Philips' one or three watt Luxeon® LEDs.

What is more significant for future LED market penetration is the recent availability of plug compatible units such as work lights, night light bulbs, MR-16 miniature flood replacements and (in the USA) a 120 V drop-light, complete with an extension cord and two flexible methods of attachment above the work surface. In this unit, 20 LEDs emit 1.5 W of white LED light at a price under $35, a value that is midway between two- and four-tube fluorescent models. Other portable lighting aids are work caps with up to four front facing LEDs (at a lower price than some caps without LEDs), safety or work goggles (with two LEDs and a battery per side) for work space illumination and magnetic "clip-on LEDs" for tools to illuminate work areas. All of these new applications are dependent on novel designs and the efficiency of the CVD manufacturing processes.

Other novel applications for "portable" LED lighting are becoming worldwide phenomena such as from Amish country in Pennsylvania, where mains electricity and fuel-powered generators are against the sect beliefs, although solar energy and wind power are acceptable. The Amish had been using 12 V batteries to serve filament bulbs for side lighting and high intensity discharge (HID) bulbs for the headlamps for their horse and buggy transportation. Under these circumstances, only 12–15 hour life was anticipated from typical car batteries and could cause lighting failure and potential safety problems on long journeys without fully charged batteries at the outset. However, about three years ago, LED vehicle headlamps (made locally in Lancaster, PA) were introduced by SunLion Solar and with six to eight LEDs and they extended battery use from 10–15 hours to about five days, with elimination of the former safety problem. The forward illumination was reported to be better than that of the HID lamps and SunLion became the first company to provide LED vehicle headlamps! Unfortunately, slow moving but well-lit horse-drawn vehicles, out of the line of sight, still provide a collision hazard to some cars and have caused multiple injury accidents. Now, the local authorities are using 4×4 foot square LED-illuminated road signs to warn of the potential hazard from slow moving horse-powered traffic.

Also from the same Amish counties, HBLEDs have satisfied a state lighting requirement for cattle barns with a four D-cell powered multi-LED portable lamp and the batteries last about one week, again made by SunLion Solar. In the social services field, Amish country midwifery was performed with very little outside assistance during night hours, other than flashlights or oil lamps, which can fail inconveniently. The same D-cell powered portable HBLED lamp was tested and found to be much more reliable than flashlights or oil and has became a welcome addition to their operating equipment.

Other commercial portable applications include LED camera strobe illuminators that are offered in 7- or 27-LED configurations by Integrated Design Tools Inc., in Florida. They provide relatively cool operating temperatures and high luminosity beams that can be pulsed in synchronization with high-speed cameras in both hand held and camera-mounted modes. Additionally, small multi-color light sources with heat-sinked packages provide nine different visible wavelength LEDs suitable as light sources for chemical and bio-analytical procedures. At longer wavelengths, infrared LEDs are being used for heat therapy, welding, fiber-optic communications and night vision illumination in both military and civilian applications. A perhaps surprising niche market is the provision of surface mounted HBLED lighting on the front of bulletproof vests for national and local security personnel use. The unit opens like batwings when being used and the chest surface mounted LEDs provide enough light to see potential hazards and assailants. This segment could generate sales of $310 million by 2011 (Figure 13.4).

Simple LED lamps, with solar, wind or water power charging of the batteries are also making positive changes to the lifestyles of third world regions, where mains power is both not available and not even on the horizon. Fires or smoky oil lamps normally provide light and cooking heat and heaters that often lead to burn and pollution hazards. The use of white HBLED lamps removes some of these hazards and improves health, safety and night lighting conditions. Self-contained LED lighting systems (*e.g.*, a 1 W multiple chip LED lamp plus battery with water, wind, bicycle or solar recharging) are being provided to residents in remote locations, often by a charitable institution. These simple systems totally change the way the recipients can live by allowing a family to do homework or read or play games and to enjoy nighttime living without the hazards of indoor oil lamps. Additionally, one outdoor nightlight, suitably placed, in a school or small village allows children to visit outside facilities in the dark. A welcome improvement in rural life and family benefits provided by simple HBLED lighting units!

13.7.5.1 Camera Flash Units

LED camera flash units are a recent addition to hand held applications markets and, by providing higher white LED outputs from pulsed power operation and reduced power consumption, they provide valuable advantages for any battery-powered flash system. HBLEDs were adequate for the first low-resolution phone-camera flash units, but as camera pixel counts increased with new model introduction (now in the 12 megapixel range) the demand for better subject illumination over a larger area requires more powerful LED flash units. Thus, market demand is driving the design of HBLED camera flash units for both professional cameras and phone models with a long list of critical requirements, including heat sinking, cycle times, lens types, overdrive currents and voltages, pulse lengths *versus* total power output, color temperature, chromaticity, chip life, unit cost, chip disguise and flash unit size reduction. In the long term this market could require special CVD processes to optimize the market needs of these LEDs. An assist to further LED market penetration may come from Seoul Semiconductor's late 2007 availability of a 2× brightness flash LED for a major camera manufacturer. There appears to be a ceramic phosphor plate that allows their white LED to match (and soon exceed) the output and lifetime of xenon camera flash bulbs. It is an example of improvements that are in the pipeline where hybrid micro-optics and high output HBLED chips are already improving unit volume, cost, light collection and beam uniformity for LED flash sources. With strong market demand from camera owners, their performance can only improve with market growth, HBLED outputs and field experience. The need for single and multiple-HBLED flash units is creating another hand held application growth market that should reach $350 million in 2011 (Figure 13.4).

13.7.6 Automotive and Vehicle Uses

Rear lighting for the red center high mounted stop lights (CHIMSELs) was the first external LED use on cars. They were initially introduced on higher priced models, but as costs were reduced new car models receive these LED CHIMSELs. Recently, all-LED rear lighting units have been introduced, again on high-end cars, but they are now a design goal for many standard models as they are updated or as new models are introduced. Notably, because red HBLEDs are now so bright, fewer LEDs are required and so optics and light diffusers are becoming a necessity in the rear LED lighting units in order not to dazzle a following motorist.

Internal LED lighting is also becoming a standard feature for many new models since it supplies glare-free lighting for the interiors and dashboards and HBLEDs are also being used as a brand recognition item for cars and commercial vehicles. Recent LED design advances allow the driver or passenger to select color changes in the passenger compartment and dashboard innovation includes LED lit instrument pointers manufactured by Denso using Nichia white LEDs. Continued high levels of LED use are assured because of the flexibility, reliability, reduced power consumption and pleasing design options offered to both the manufacturer and the consumer. These factors are leading to steady growth in the automotive and commercial vehicle HBLED applications, providing a wide range of LED products for internal and external vehicle lighting, which should contribute to sales of $860 million expected by 2011.

Leading car manufacturers continue to add interior and exterior LED lighting and segment growth continues in all geographic regions. The present average led count, when used, is about 200 LEDs per-car with over 70 production models now using them. Additionally, colored LED lighting and other after market decorative units are becoming popular add-ons. Europe has been the volume leader for HBLED automotive uses, followed by the Japanese and Asian manufacturers. Car manufacturers are very cost conscious and tend to use low cost LEDs, a factor that provides a steady rate of introduction, but low value growth rates. However, the high total unit demand and new model introductions are expected to provide at least 10% LED market growth in 2007, to a value of $590 million. Apart from low power requirements and low space demands LEDs provide the manufacturers with other potential benefits such as vibration resistance, front and rear design freedoms, brand identity, form lighting, age group appeal and new styling options. Even for emergency vehicles they allow the use of more compact and more efficient lighting units, with reduced power needs, reduced vibration sensitivity, and lower maintenance requirements and costs.

13.7.6.1 HBLED Front Auto Lighting

Front running lights have been gradually added to cars since 1995, when research by Rumar showed that daytime front lighting reduced accident rates by about 20%. As LED designs and CVD processes have increased LED output powers, the white HBLED has become an obvious candidate for front running lights and were one of the first white LED front uses, such as the LED illuminated fiber-optic front lights supplied by Osram for the 2007 Cadillac Escalade. A few models, including the Escalade and Lincoln, are using molded polycarbonate resin (Makrolon) for close-coupled optics for the LEDs including amber turn signals. Strong competition has existed between the auto-manufacturers as to which would be the first car with all front-end LED lighting or the first with HBLED headlamps. It now appears that both Lexus and the Lincoln MKR sports models were fitted with HBLED headlamps in 2007 with the Audi R8 expected to follow in the first quarter of 2008 using Lumileds' HBLED low and high beams. Also possible for late 2008 release will be LED headlamps for commercial and fleet vehicles. As HBLED module prices decrease and output powers improve by up to 50% from CVD process and packaging refinements, HBLED front lighting systems will gradually be introduced on lower priced model vehicles.

European automotive manufacturers have been leaders in the use of LED lighting, with Hella unveiling its prototype at the Frankfurt Auto show in 2005. The new LED headlamp will require advanced designs, LED modules with optics, integral with thin, flat or form-fitting packages and large heat sinks. At the same time, the advantages of space saving, styling, reduced maintenance, reliability and shock resistance will become apparent and customers and manufacturers alike will appreciate lower warranty costs. In 2007, the estimated number of HBLEDs needed for each front lighting function was 5, 10 and 20 for the respective fog, high and low beam functions. These numbers should decrease with increasing HBLED lumen outputs and Osram, as an example, has announced a higher brightness, 6000K color temperature, LED for headlamps with 620 lumen of output at a 700 mA forward current and that is timed for commercial release in late 2008.

Apart from custom designs specific to a single model, most previous models of auto headlamps were commodities made by large companies such as Hella, Osram-Sylvania and GE. One of the most anticipated automotive LED uses is front lighting, including running lights, turn signals and headlights. The biggest anticipated reward is the flexibility of car design when using HBLEDs for future generations of front lighting and headlights, with other benefits being design freedoms, reduced warranty liabilities and performance improvements, with such models as Audi, Cadillac, Ford, Lexus, Mitsubishi and Opel having exhibited concept models. Progress continues as amber-LED turn signals have been accepted and LED running lights are now close to adoption, with prototype headlamp units having been demonstrated by such companies as Hella in the USA and Germany, Koito in Japan and Visteon Corporation in the USA. The optimized new headlamp designs are expected to offer steady beam intensities, minimum glare and beam bending for cornering.

New designs will be the order of the day and the headlamp LED heat sink will probably be the largest and heaviest component of the new front lighting units. Even though LEDs run much cooler than HID headlamps and enable the use of plastic front lenses, active and passive cooling may be needed to meet the HBLED cooling loads. Market growth should increase steadily after 2010 as higher power LEDs, advanced designs and lower-cost LED array-packages will have been introduced. However, unlike commercial vehicles there is not an aftermarket for external LED lights on cars and it is possible that front LED lighting could upset the structure of the headlamp industry, since some packaged LED chip manufacturers are new to the car lighting business.

13.7.6.2 Commercial Vehicle and Fleet Use

HBLEDs are now the preferred exterior identification and lighting method for many commercial vehicle fleets because of their longer life, their multi-LED format that rarely fails catastrophically like bulbs, their reduced downtime costs, their lower power demand and improved safety and maintenance factors. The longer useful life of HBLEDs provide immediate savings over bulb replacement costs, the largest maintenance cost for fleet vehicles, easily justifying their initial higher cost. Because many commercial vehicle external lamps are standard sizes, there exists an aftermarket for these products, a situation that has also allowed an HBLED replacement market to be formed. Therefore, over 100 types of LED vehicle lamps (red, amber, tail, stop, marker lights and white LED lamps for interior, back-up and license plates, blue, purple and orange LED lights for decoration and detailing) are now available at reasonable costs ($5 to $20 range), as are blue and white strobe lights. An example would be Grote Industries Inc., which offers LED lamps for boat and vehicle transportation and it demonstrated a 6000-lumen LED forward lighting bar in the 2006 Baja 1000 off-road race. Grote also makes multi-operating-mode military lighting units that provide blue or white and/or night vision illumination. An LED based high lumen front lighting unit based on the Baja prototype is expected in 2008. New vehicle lighting products and aftermarket sales provide low cost of ownership products that support HBLED growth and penetration of the fleet vehicle market segment.

Transportation vehicles such as buses and trams are also in the fleet category and their LED lighting and information signs share the same solid-state lighting benefits, plus vibration resistance and easy passenger reading. LEDs external to vehicles and indicator equipment provide additional safety benefits because of their 0.2-second shorter rise time than filament lamps, giving additional reaction time to drivers and pedestrians alike.

Emergency vehicles also benefit greatly from LEDs and because of the extended range of colored HBLEDs available, they are becoming the lighting method of choice. Police vehicles with their multi-color requirements benefit from the high power efficiency of blue, red and orange colored LEDs and they are rapidly incorporating HBLED lighting into lighter, low profile roof racks or LED light bars. They reduce wind resistance at high speeds, have lower maintenance costs, can reduce the sizes of the batteries and alternators required and provide longer standby service, all reducing the costs of operation. LEDs have not yet taken over the rear lighting functions of polices cars because it is a car design option, but ambulances and fire engines are following the lead of the commercial vehicle industry. Industry growth is steady, but not enough to significantly affect vehicle LED market growth or CVD precursor demand.

13.7.7 Signage and Channel Letters

For about 100 years, many kinds of neon signs have denoted product brands, such as petrol, bread, beer, soft drinks, hot food locations and have displayed hours and types of business around the world. Many of the small open and closed signs were made by local entrepreneurial businesses, which made up over 80% of the neon sign market. The large signs were the domain of regional or national companies that could provide service over larger geographic areas and that could also install many of the large letter signs or channel letters. Channel letters are also part of the architecture for many national companies and they provide or add to the corporate image. Some of these were changed over to fluorescent tubes for more reliable operation, but letter outages are a too frequent occurrence. However, multi-color and flexible plug-together HBLED lighting strips have been devised that can be attached to the channel sidewalls or encased in plastic or glass tubes for even higher reliability channel letter lighting. In either case, neon kilovolt sources are not needed, but LEDs have an additional advantage over fluorescent lamps of not suffering from cold start problems, having lower power demand and longer lifetimes.

In the past, large indoor or outdoor video screens like the Jumbotron® used mini-CRTs. However, the signage industry is changing rapidly as large volumes of CVD produced LEDs are steadily penetrating these markets with their benefits of color range, safety (low voltages *versus* kilovolts), reliability, flexibility of design, long lifetimes and no glass breakage. Higher output HBLEDs from improved CVD processes have totally changed the large display industry (indoor and outdoor) as their RGB pixel flexibility, clarity and brightness levels have turned this market into one of the larger HBLED markets that provide full color video and fulfill many advertisers' dreams. The growth of LED video signage will continue, since research has shown that static signs cannot compete when video is close by and because it is easy for the sponsor to remotely change the message and message length as the passerby traffic changes with the time of day.

The worldwide signage market is quite large, in the region of \$10 billion, with the US market share being about 50% of this value, or \$5 billion, and the neon component has been an entrepreneurial market totaling about \$550 million with only about \$120 million in sales attributable to large suppliers with national or regional service. Companies with less than \$5 million turnover generate the major value segment, or about 80% of the market. This structure will probably change in the future as more video signs are installed and as larger resources are required to provide the video service. In 2007 the HBLED signage market was estimated to be about \$640 million, which with increasing HBLED penetration should grow to about \$950 million in 2011 (Figures 13.3 and 13.4).

13.7.7.1 Neon Sign Replacement

There was some resistance to LED use in this market from small manufacturers with little knowledge of the disruptive LED technology and customers who would accept the familiar neon lighting with its limitations, only one color per tube, kilovolt operating requirements, glass breakages, and unlit sign segments. There is little power advantage for the LED over neon signage, and thus little advantage is to be gained there, but as more information about the LEDs has become available (greater safety, reliability, flexibility of design, color selection and their ease of incorporation), increasing numbers of LED signs are being sold as the smaller neon sign manufacturers have "seen the LED light" and started to add them to their product lines.

Large sign users are often national or international companies with standardized advertising functions for their building and specific channel letter lighting requirements. These are often in inaccessible places and the 3000–15 000 V neon lighting installations are exposed to the elements and are not easy to service. The benefits of CVD grown LEDs are now better known and are driving signage market demand for replacements as reliable weatherproof drivers and controls and as flexible, plug together, HBLED lighting strips have become available. Typical users would include international oil companies, supermarkets, retailing and fast food chains and many of these are now considering HBLED replacements for signage installations.

13.7.7.2 Emergency and Safety Signs

Exit, fire and emergency signs are another small and growing segment of the LED sign market. Their long life, low power demand and government mandated power levels will make them a steady replacement in this market. Some state mandated 5 W LED signs are replacing 40 to 75 W bulbs or fluorescent tubes, which required larger battery capacities for power off operation. The HBLED will be a steady replacement in this market, since red, amber and green LED efficiencies provide adequate lighting at the 5 W power level. They require much less battery back up and in the USA they are government certified and NEMA approved. Amber LEDs are also widely used in the portable safety and caution sign markets, for road and construction work, where LED use can provide smaller units with longer battery life. In addition, the LEDs are brighter, the units are much more reliable in remote locations, have improved visibility and reliability, and their low power needs allow the use of solar operation. All of these factors provide benefits to the construction industry.

13.7.8 Signals – Traffic

HBLED traffic signals are now used in many parts of the world and have significant power savings (80–90% range). Initial installation and market growth was quite rapid although the growth rates are now lower with sales in the $200 million per year range worldwide. The main incentives for HBLED installation are the reduced signal maintenance costs that provided 1.5 to 3 year paybacks, a tempting prospect that municipalities and governments could not resist. In addition there is the large power saving factor, which in the case of California, was estimated to save the state about four megawatts per day after a large fraction of the installed traffic light base was converted into HBLED operation. The power savings and cost per signal head should improve with the latest models, since the early version LED signals required 100 to 200 LEDs per head, whereas the latest HBLED versions need only 10 to 60 chips (depending on the color) plus a diffusion screen to provide even illumination for the legal signal head diameter.

Solar powered LED traffic signals with battery backup are now a reality, and a good example is the city of Lancaster PA, USA, because of the diversity. The retrofits include some LED signals with solar power and battery back-up, using 12 V dc for new HBLED signal installations and 120 V ac driven HBLEDs for the existing installations. These were manufactured by a local company, SunLion Solar. The LED also has improved the clarity and visibility of pedestrian walk signs, where the white LED versions are popular with pedestrians and offer much reduced power drain and lower service requirements. Count down by the second walk signs and traffic lights are now widely used and HBLED illuminated street signs are a new application, from ACOL in Switzerland. All of these market segments will contribute to the growth of the HBLED market and CVD demand, with the total for the traffic signal segment being estimated at $200 million in 2007 and $260 million 2011 (Figure 13.4).

13.7.8.1 Signals – Rail

Red, orange and green lights are common to both road and rail transportation systems, but specifications differ for the two transport types. Rail signal requirements include longer distance visibility for high-speed tracks (often up to two miles) than do traffic signals. Fortunately for rail systems and the LED, tighter LED packing in a smaller head with good lens designs and narrow beam angles (*e.g.*, with 3° primary optics) will guarantee the long distance railway visibility requirement and offers the HBLED advantages that are also beneficial to railway operations, bright colors, long life, low maintenance and lower power demand plus the added benefit of no bulb type catastrophic failure. Single color HBLEDs will steadily take over the signal market with their superiority on a longevity and cost of ownership basis and solar power can also be considered for remote signal locations. Therefore, HBLED demand will steadily take over the traffic and rail signal markets.

13.7.9 Aviation Lighting

Penetration of the aviation lighting market segment by the HBLEDs has now passed the evaluation phase for several applications and their installation has become an ongoing process. Some LED application categories are large, with about nine uses for interior and exterior plane lighting plus another eight in the cockpit. These applications have taken longer for aviation lighting certification, because some technologies and equipment in this industry have to meet stringent testing standards before use and also because after market direct plug-ins were not available. In reality, companies such as Rockwell in the USA, already offer cockpit displays (using Nichia LEDs) and the potential LED benefits are offered for almost every aviation lighting application, including, plane interiors, exteriors, airfield runway, taxiway, aviation infrastructure lighting and safety lighting. Also important in today's economy are weight and power economies on the aircraft that lead to significant fuel and operating cost savings.

Early LED aviation uses were airport information screens and cabin emergency floor lighting, but cabin and customer lighting and exterior aircraft lighting are being installed in new aircraft and in upper class cabin retrofits and they include passenger controlled white HBLED reading lamps. Vibration resistant HBLEDs will eventually replace the rows of fluorescent tubes lining the cabin ceilings, facilitating the removal of their heavy ballast units and most of the new aircraft due to be delivered during the next five years, passenger and military, will use HBLEDs for cockpit, cabin and some exterior lighting. However, with an estimate of about 30 000 commercial and 20 000 business jets in service (plus almost the same total of military planes) the main HBLED market will be the replacement market segment of the existing \$500 million per year aircraft lighting business, and by 2011 the LED could achieve \$250 million in aviation sales – a 45% penetration.

HBLED vibration resistance and longer lamp lifetimes in all lighting categories will provide less weight (no ballasts needed), individual passenger controls and potential mood or ambiance lighting availability, improved safety from lower operating voltages, and fuel savings from weight reduction. The related weight reduction for a jumbo jet is estimated to result in about \$100 000 savings per year at today's high fuel cost. HBLEDs should soon replace external aircraft flashing lights, which have relatively short service lives (a few hundred hours), in part due to airframe vibration. These LED replacements should increase service lives to thousands of hours with much lower downtime and maintenance costs per aircraft, but their installation is a slower process due to lengthy testing procedures mandated by internationally recognized regulatory bodies such as the CAA and FAA.

Flashing lights also have mandatory safety requirements for night and day operation and aircraft may not take off without them being operational, a category in which the HBLED should excel. Additionally, the red-flashing infrastructure lighting for towers and buildings has relatively short service lives (3–6 months) and will benefit from future HBLED replacements. As the change over to HBLEDs occurs, it will become a worldwide growth market segment for both replacement and new equipment, with fluorescent and xenon strobe lights losing their present dominant positions. At this time, the aviation segment annual sales are not large and are estimated to be as high as \$25 million, but, as noted, rapid growth onward from 2008 could raise the 2011 value to \$250 million.

13.7.10 Marine Lighting

Historically, massive navigation buoys with tonnes of batteries and multiple self-replacing bulbs, often weighing as much as 15 tonnes, had been in place for general navigation for generations and underwater cabling was used for inshore applications. These usually required three to six month service visits from a large vessel plus cranes large enough to handle the weight to change batteries and bulbs. The maintenance of marine navigation lights has been a high expense business, because of the failsafe requirements for the safety of all shipping movements at sea, in rivers, docks or channels. Safe passage depends on buoys and fixed navigation lights, at sea, in the approaches and on the many sizes of vessels that use them. The red, white and green guidance lights must always be lit at night for both stationary and moving equipment. In the past, the operation of shipping channel navigation lights required on-line power or heavy battery loads and the heavy equipment to maintain them.

However, CVD fabricated LEDs have enabled sweeping changes in the field of marine lighting, where these HBLEDs have already allowed a two-orders of magnitude weight reduction and extended bulb lifetimes by a factor of 10×. The replacement buoys are now 100–300 lb versions (50–150 kilos) that can be solar powered and that can be adjusted or reprogrammed with hand held remotes from a small service boat. The HBLED replacements are expected to have at least a 5-year life, with little or no service being required. Solar panels are already in use to power remote HBLED-powered buoys, eliminating large battery requirements and placing fewer restrictions on navigational marker locations. Many advantages of the LED are useful to marine lighting, including longer life, lower battery drain, no ballasts needed, lower power demand per lumen output, lower maintenance costs, resistance to vibration, decreasing manufacturing costs per lumen, and higher outputs per watt, all now allowing HBLEDs to compete in most marine lighting applications. Thus, as more HBLEDs are used for marine lighting, the overall navigation safety factors improve and maintenance becomes much simpler, with operating cost reductions of 10× or more.

For small boat owners, such as yachts, sailing and motorboats, an owner or crewmember may have to climb the mast, often 6–9 m (or 20–30 feet) high, every three to six months to replace navigation lamps. Now, when using HBLEDs, such service should only be a once every five year task and because of their lower power demand battery charging can be reduced to every three to five days when in port instead of daily (with less exhaust pollution). On future mast lighting kits, the addition of an IC chip will allow the resetting or resequencing of the light timing and color changes from the deck or a dock with hand held remotes.

The incorporation of HBLEDS in marine lighting systems is well under way and companies from Australia to the USA and to the UK already offer white and colored HBLED marine lighting equipment. These include interior, navigation and underwater lights and many are now specified for after market replacements and new boats up to 50 m. Marine Lighting Ltd. in the UK has its bridge and chart lighting specified for the Tamar lifeboats and the US and European coast guards have recently certified marine LED lighting for large-scale use. Additionally, the underwater use of LEDs has been endorsed by the swimming pool industry, where low-voltage LED pool and Jacuzzi lighting is available off the shelf.

A similar situation is true for the US Navy where Energy Focus Inc.'s (EFOIs) LED luminaires are replacing fluorescent and incandescent fixtures fleet wide with 30–90% power savings and still meet −40 to 90 °C operating conditions. Lighting is one of the main power consuming services on board ship and the electric power saved by the LED installations on oil fueled ships can allow them to remain at sea longer between refueling stops. These improvements are occurring at the same time as the LED lamps provide less starting problems and electrical hazards yet still meet stringent specifications, including the provision of beam patterns required for reduced stray light, wavelength control for night vision requirements and low levels of external electronic ship signature. The new LED fixtures are reported to be very popular with the ship crews where they have been installed and are also the envy of visiting navy crews. LED lighting must become a basic requirement for all shipping in the future because of its lower power requirements, its improved safety from lower voltages and a zero mercury content, its fuel savings, its durability and resistance to vibration, and its lower service needs due to extended LED lifetimes. All these benefits and the availability of LED replacements will lead to a rapid growth in the marine LED lighting market, which was estimated by J. Rabello of iSupply to be $4 million in 2007 and with expected growth to about $10 million in 2011.

13.7.11 General Lighting and Illumination

The worldwide general lighting business was estimated to be $12 billion in 2007 and it is a long sought after goal for HBLED manufacturers. It is a historic market spanning about 150 years, progressing from carbon and refractory metal filaments to the higher efficiency, sodium vapor, compact fluorescent and HID lamps common in today's lighting stores. Most lamp types will feel competition from the LED within the next ten years as the output of the white LED continues to improve and as larger segments will be accessible to LED penetration. The major lighting companies are large international players such as General Electric, Osram, Philips and Ushio. In addition, most were not prepared for a disruptive product like the HBLED, which has significantly different sizes, lower operating voltages (dc *versus* ac), extended lifetimes, different light source dimensions and different regulatory compliance needs. However, these companies have either formed strategic partnerships with LED manufacturers for future LED lamps or have started their own competitive research on both LEDs and phosphors to be able to compete in this changing market place. Leading LED manufacturers with commercial product lines now include established companies such as Osram and Philips and LED suppliers such as Nichia LEDs (part of a phosphor chemical company), Toyoda Gosei and Cree, whose LED chip and Cotco packaging group sales exceed those of their original SiC ingot and wafer businesses.

The use of blue HBLEDs with phosphors has been the key to white LED growth, since they are less expensive than the RGB LED combinations. At present, most indium gallium nitride (InGaN) LEDs (blue and green) are made on sapphire wafers (heteroepitaxy) and the red yellow and orange HBLEDs are made with lattice matched AlInGaP active layers on gallium phosphide or gallium arsenide wafers. The only major supplier with a different CVD process for HBLEDs is Cree Inc., from Raleigh, NC, which manufactures high output III-nitride InGaN HBLEDs (blue green and white) on proprietary silicon carbide wafers.

The main competitors for HBLED-based general lighting are tungsten filament, high intensity discharge (HID), fluorescent tube and sodium vapor lamps. Of these, filament lamps have the lowest off the shelf cost, the highest heat output, the shortest lives and the lowest power efficiency with light outputs in the 10–15 lm W^{-1} range. Because of its lower price and well-established market base, the current incandescent lamp replacement market is large and makes up about one-third of the annual lighting market or \$4 billion. All the other lamp types have higher initial costs, but longer lifetimes and higher lumen efficiencies, most in the 30–90 lm W^{-1} range. Non-LED lamps also compete by having much higher light output per lighting unit, because existing HBLEDs may emit only 60–100 lumen compared with 500 to 5000 for most lamps. Higher output fluorescent tubes, cold cathode fluorescent lamps (CCFLs) and HID lamps are five to ten times more expensive, but with higher efficiencies and 10× longer lives. Some of the latest fluorescent lamp types have efficiencies in the 100 lm W^{-1} range, but they still use mercury (against the EU RoHS rules) and their costs are higher. However, developmental HBLED outputs have already passed these values with power outputs in the 120–150 lm W^{-1} range and these outputs should soon transfer to commercial products.

Early designs of multi-LED chip arrays with kilo-lumen outputs have already been used for street and building lighting projects and additional support for the HBLED is coming from small venture-type companies that have successfully entered the LED support businesses, drivers and controls, luminaire, packaging and phosphor manufacturing. Consequently, the structure of the lighting industry could be changing irreversibly. HBLED general lighting use was estimated to have a market value of about \$100 million in 2007 and could optimistically rise six-fold to at least \$650 million by 2011. LED competition for the incandescent segment can only become stronger, since all the key components for increased HBLED market penetration appear to be in place. The rate of the white HBLED acceptance in general lighting should increase annually as lumen outputs continue to rise. Support for this trend comes from the incandescent replacement market (~\$4 billion), which is showing signs of declining sales from CFL and LED competition. As an example, General Electric (USA) announced the closure of three incandescent lamp factories in 2007. Other bulb manufacturers are expected to follow suit as this market segment starts to decline at about 5% per year or more.

Specially designed 1000 lumen red, green and blue HBLED chips are already being used for some DLP rear projection TV sets and high output, single and multiple white-LED modules, with outputs in the 200 to 2000 lumen range also now available. The cost per installed LED lumen will continue to decrease for many years to come, based on increasing LED efficiencies, CVD process improvements and product volumes. The LED offers the greatest potential for power savings in our power hungry world and the LED general lighting market segment could reach over $7.5 billion annually by 2020 as billions of LEDs are installed for general lighting in the 2010 to 2020 time period.

The key to white high power LED growth has been the combination of blue HBLEDs with phosphors. In today's market, most of the white HBLEDs use a cerium doped YAG phosphor with excitation from the efficient InGaN blue LED chips. At the development level, their light output powers are already exceeding 150 lm W^{-1} and higher values are anticipated in the near future. RGB (red, green, blue) LED sourced white light packages have not been eliminated from general lighting contention, but the necessity of having three or four dissimilar chips per white output package (RGB or RGGB), with different intensity outputs and lifetimes, means that today's packaging and control circuit costs usually exceed those of phosphor coated blue HBLED lamps. This factor may delay large-scale use of the RGB white sources for general lighting for quite some time. Other white light source research is using AlGaN UV and near-UVLED excitation of different phosphors, but presently their efficiencies and lumen outputs are relatively low and not serious competition for the commercial white light market. Even so, there are specific niche-type markets already open to the available low power UV LEDs based on CVD AlGaN/AlInGaN materials systems, such as document validation, air and water sterilization, and DNA and missile detection. To obtain higher, commercial requirement output levels of UV optical power from the present UV chips, Opto Technology Inc. assembles arrays of 50 UVLEDs in TO-66 packages that produce a 250 mW UV beam at the 360 nm wavelength.

The higher efficiencies and output powers of the latest phosphor-converted-LEDs are key to the penetration of the general lighting market, a condition that should become easier as their power conversion efficiencies continue to improve. Because of the strong competitive situation and because of the relatively low lumen outputs per LED chip, the design of LED lamps or luminaires is extremely important at this time to support market growth. In order not to waste a single lumen of LED light output, micro-lenses have become an integral part of most luminaire designs. Lensed, high lumen, multi-LED packages are already in commercial use in the leading manufacturers packages and luminaires, such as one watt and higher power HBLED modules from suppliers such as ACOL, Cotco, Cree, Nichia, Philips, Osram and Toyoda Gosei.

For colored lighting, the HBLEDs are now the most efficient lighting sources and far exceed light from other multi-wavelength lamps with filters. As more efficient LED packages are offered, they are becoming the preferred colored lighting source. Efficient, reliable and less expensive drivers and controls have recently become available, and new plug-compatible luminaire designs have been introduced to the market, using either ac or dc power. LEDs have already been used for large and small lighting projects, where their cost of ownership, 50 000 hour lifetimes or the architectural benefits favor the LED and outweigh their initial higher costs. These include bridges, buildings, monuments and landmark lighting, such as the Thomas Jefferson Memorial in Washington, where LEDs were installed by Osram-Sylvania in the inaccessible locations. In the latter cases, ease of installation, low maintenance demand, low failure rates and long lifetimes (plus lower power costs) completely overturned the initial installed cost considerations. Additionally, with respect to the Jefferson Monument, about one yellow LED in 10 was added to provide a warmth effect to the interior monument lighting. Recent LED lighting installation projects of note include the use of Luxeon® LEDs by Philips Lighting for the Bosphorous Bridge in Istanbul, the Mega Bridge in Bangkok and Buckingham Palace in the UK, the use of Cree's XLamp LEDs on the Croatian Riviera (with Uitrusting Schreder of Belgium), for the City of Raleigh, NC and the lighting of

Shinjuku Alta by Mitsubishi using Nichia LEDs, and the Hunterian Museum at the University of Glasgow. The benefits of zero UV and low IR radiation provide desirable lighting ambients for art and historic displays and a non-degrading lighting growth market for food displays.

The designs for Hindu temples (mandirs) were made over 1000 years ago and in 2005 North Star Lighting, of Broadview Illinois, was contracted to design HBLED lighting for a new Swaminarayan Temple in Bartlett, Illinois. The typical building materials of limestone, sandstone and marble are usually carved in and imported from India, thus LED lighting would be a state of the art addition to highlight ancient temple designs. HBLEDs have already been used for many public lighting projects and now the lighting of temples, including in Toronto and a new temple opened in Lilburn, Atlanta in August 2007, should be added to the list. The North Star designs used about 2000 Lamina high output RGB "Titan" and "Atlas" luminaire light engines per system to provide internal lighting for walls and columns and indirect lighting for ceilings. These interior (and exterior) temple lighting systems have been well received by the congregations and their local communities and are leading to additional temple lighting projects that now include the first one for a Christian church. This rapid acceptance of HBLED lighting for temples and churches appears to be leading to a "religious building" HBLED growth market segment, again supporting future expansion of MOVPE process volumes.

HBLED benefits already accrue to the armed forces (army, marines and the navies), where higher voltage incandescent and fluorescent lighting is a hazard in moving platforms such as vehicles, tanks, submarines and ships. Breakage under rough ground, rough sea or battle conditions can leave live wires exposed in cramped spaces with the potential for serious hazards and consequences. For air forces, many types of CVD grown III-V and SiC-based devices are being considered for better performance and lower power demand in new equipment and for retrofits. Hence, some specifications already exist for 120 or 240 V lighting units to be replaced with solid-state lighting and at the same time these are safer, vibration resistant and require less power, with similar LED lights being used by the US Navy. Modifications of the EFOI LED lights designed for the Navy are also being used in commercial markets. Military equipment is a growing market for compound semiconductor HBLED lighting (and for efficient high frequency devices) and will provide increasing support for advanced MOVPE processes.

13.7.12 Group III-V High Frequency Devices

In addition to providing a wide range of HBLED products the Group III-V semiconductor materials, including those based on GaAs, AlGaAs, InGaAs, InP, InSb, AlN, GaN and the AlGaN/InGaN alloys, also support a range of high frequency and/or higher power devices. Although most commercial devices use standard CVD processes for their manufacture, a significant fraction of the commercial GaAs devices are made by MBE. All CVD grown III-V materials can normally operate at higher frequencies, temperatures and higher efficiencies than silicon, but III-nitride devices in particular can operate at above 350 °C. Most offer higher power levels than silicon, with the III-nitride devices able to perform electronic functions at power levels 5–10× those of GaAs and its ternary alloys and also higher than those of InP. Many commercial and military systems such as radars and satellite communications use devices based on GaAs, AlGaAs, InGaAs and InP and have been successfully introduced to the market during the last 15 years, with operating frequencies in the 5–300 GHz range. Military and commercial radars have been using GaAs based devices for more efficient and lighter weight radars and future use of GaN ICs for radar will further increase system power outputs and reduce system weights and air conditioning loads, all being ideal for military and commercial uses.

In the 2006/07 time frame, CVD based InSb devices such as infrared emitters and detectors were relatively new to the market and InSb high frequency ICs are only in the early stages of development, although InSb quantum well FETs provide unity gain cut off frequencies (f_t) exceeding 300 GHz at an operating voltage of only 0.5 V VDS. This is a 50% higher speed than silicon NMOS at <0.1 of the active dc power draw. Automobile radars may become a future growth market for GaAs ICs if they fulfill the noise and efficiency requirements for the 77 GHz vehicle detection systems, although SiGe circuits are possible alternatives. The high power and high temperature capabilities of III-nitride electronic devices are already being considered for commercial use by industry and the military. In addition to their ability to operate at high frequencies, most of these III-nitride based devices also provide high efficiency electrical characteristics, such as power densities and power added amplification and with the established III-V characteristics (higher temperature operation, higher efficiencies, higher power levels) will continue to open up new application areas and add growth to CVD device process volumes.

With very little contribution from the InSb and III-nitride devices and auto radars, the estimated 2007 market value for the CVD produced III-V electronic devices is over $3 billion, mostly GaAs circuits for cellphone and WiFi supported uses. They include HBT power amplifiers, PHEMT switches and microwave devices. Future growth is expected in most application areas and Strategy Analytics Ltd has forecast this GaAs device market to grow by ~75% during the next five years to $5 billion in 2011. Other III-V high frequency device markets should reach $1.5 billion in the same period for a total III-V electronic device market of at least $6.5 billion. However, even considering their future application growth forecasts, the III-V electronic device markets are not expected to match the III-V HBLED market size for many years to come, if at all. The $6.5 billion total represents a significant CVD process market, but the two basic reasons for the higher HBLED markets are the much larger and growing size of the total available world lighting market and the constant competition from ever improving silicon discrete and IC devices that eat away at the lower end of the III-V device high-frequency market window. However, it is possible that the higher efficiencies, the lower noise factors and the higher power factors required by ever more versatile hand held electronic appliances will keep III-Vs ahead of silicon in this competitive race.

The total Group III-V device applications market, including diode lasers and solar cells, is estimated to be about $16 billion in 2007 (Figure 13.1) and will grow to about $27 billion in 2011 (Figure 13.5).

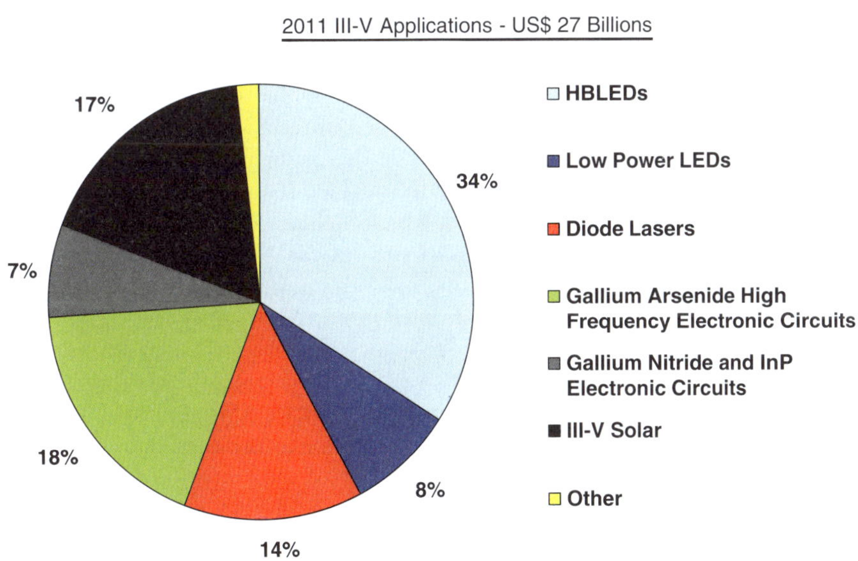

Figure 13.5 Projected III-V applications for 2011.

13.7.13 Group III-V Semiconductor Diode Lasers

Solid state lasers began with modified crystals such as neodymium doped YAG (yttrium aluminum garnet) in the 1980s and changed the whole performance and operating parameters of the industry. However, the development of semiconductor diode lasers (electronic solid state) took the laser industry a stage further by providing smaller, more efficient and more durable lasers that could be assembled in small packages and extended the whole range of potential applications. Lasers are basically an electrically pumped LED with mirrors and have been on the market for over 20 years, with the majority being produced from Periodic Group III-V elements. Most are CVD produced, although liquid phase epitaxy and MBE processing is also used for some laser manufacture. Today's diode lasers are usually more efficient than solid state lasers, are relatively inexpensive (some only a few cents) and consequently about one billion units will be produced in 2007, making up over 99% of all lasers sold. Initially, InP and GaAs compounds were used for the manufacture diode lasers, but the characteristics of other alloy compounds such as AlGaAs and quaternary InGaAsP and their lasers were found to be superior for many red and infrared applications, particularly for infrared fiber-optic telecommunications. Recently, other elements such as antimony and nitrogen have been added to the element mix for laser manufacture and GaN and its alloys have become the compounds of choice for green, blue and near-UV lasers that are already being used for high-capacity DVD recording systems. The latter blue/violet lasers have provided several novel CVD processes such as epitaxial lateral overgrowth (ELO) to make lower defect level substrates for higher yield device manufacture.

Diode laser device emission now extends over the 350–5000 nm wavelength range and supports a wide variety of applications with growing unit volumes. During the previous decade, CVD-based growth of the diode laser market achieved a \$6 billion value in 2000, but this value declined by about 50% in 2001 (to \$3 billion) due to the collapse of a bloated fiber-optic industry. Unit volumes have continued to increase, but little dollar growth has occurred since. According to R. Steele from Strategies Unlimited, the 2006 diode laser market remained in the \$3 billion range, with the two main segments responsible for over 80% of the market, optical storage accounting for about one-half and telecommunications about one-third of the value. Medical and surgery laser systems also use CVD produced diodes with a 50% segment penetration, but their total market value is quite small. Without the growth contribution from the Sony BlueRay® and the competing blue laser systems, the diode laser market may have shown a slight decline for 2006. The outlook for 2007 does not appear to be much better, since the year over year decline in the yen may be enough to offset the dollar value of the increased unit volume growth in both the red and blue laser markets and therefore the worldwide market is again expected to remain suspended in the \$3 billion range, with only the blue laser growth offsetting the continuing price declines in the major market segments.

The many III-V applications categories have in the past led to the development of new precursors and now support a large CVD production base, which includes GaAs and InGaAs red lasers (for printers), pumping of solid state lasers, read and write optical storage CDs, bar code readers (*e.g.*, supermarkets), plus the smaller market segments for laser pointers and levels and multi-watt, fluid cooled chips (welding, marking and cutting). For IR, InGaAsP covers telecommunication lasers, with other wavelengths for resin curing and frequency doubling (from 1.06 micron to 530 nm green). More recent developments are for short wavelength (blue) emitters for DVDs and visible and IR front emitting VCSELs and VECSELs (vertical cavity surface emitting lasers and vertical external cavity surface emitting lasers). The latter are easier to produce than edge emitting lasers and because the perpendicular surface emission is easier to feed into fiber-optic cable and other receivers they are gaining popularity and are supporting a growth market for chip-to-chip and other communications uses, again from CVD processes. IR lasers (2–5 kW) are used for cutting and welding, but the highest output diode laser power is also in the IR and currently in the 10 000 W per chip range, but these lasers can only be used in the pulsed mode with adequate cooling.

After intense research efforts covering more than a decade, blue-violet lasers were introduced in 2006 for high capacity DVDs (8–10× times the CDs), and the latter was probably the only laser growth market in 2006. With a current annual value of about \$25 million, it is expected to be the only significant value growth market in 2007. The lifetime and yield of these diode lasers critically depends on the material and substrate defect levels, such that the demand for blue LEDs has spawned new, lower defect level, GaN and AlN CVD processes. They provide a two-orders of magnitude or more defect level reduction (to 10^5 or better defects per cm^2) on the substrate wafer and on the subsequent growing layers, whether heteroepitaxy on sapphire or homoepitaxy on expensive GaN wafers is being used. These techniques may incorporate GaN or AlN buffer layers and are variations of ELO (epitaxial lateral overgrowth) processes, where narrow stripes of thin silicon dioxide or other mask material (*e.g.*, 5 micron wide with 10 micron spaces) are masked off on the GaN wafer, followed by epitaxial growth of GaN. Etched grooves into polished sapphire wafers serve the same purpose and the subsequent growth of GaN layers in either case cantilevers out over the mask or groove to join with the adjacent growing GaN layer from the opposite side of the spaces. When the laterally growing GaN regions meet above the masked area, they join and continue growth, forming a smooth continuous epitaxial surface, where the defect levels are much lower (*e.g.*, 10^{-2} or less per cm^2), and produce surface regions where the good (or best) lasers can be grown. These ELO CVD methods are used because low defect level gallium nitride or lattice matched substrates have not been available in quantity, although this GaN situation may be improving. Lower defect level aluminum nitride wafers are also being produced with a good lattice match, but both AlN and GaN substrates are believed to be priced at over \$1200 per 2″ diameter wafer. Owing to these high costs and device wafer quality problems, yields of stable, CVD-produced, GaN-based blue or violet laser diodes, emitting in the 400–450 nm wavelength range, have been low.

The two largest semiconductor laser application categories in 2007, each with about 40% market share, are the lower-power sub-700 nm wavelength and 950–1550 nm group lasers. The 1–10 W and 10+ watt laser groups each had about 10% market share. The unit volumes for these applications are increasing, but their price declines are eliminating potential value growth. Most of the shorter wavelength lasers are used for optical storage, with red lasers being used in CD optical storage systems. All these lasers have contributed or are contributing to the development of new CVD precursors, processes and markets and should continue to provide unit market growth and process demand well into the future. As noted earlier, in 2007, the laser market is again expected to be about \$3 billion, little changed from 2006 with the only real value growth markets being blue lasers for high capacity DVDs, which should grow to over \$25 million in 2007, plus the vertical cavity lasers. The market outlook for the diode laser is good from a unit growth point of view, but value growth will not be anything spectacular, with only about \$3.7 billion being expected for 2011. Most of the new growth is expected to come from the new market application categories including VCSELs and blue lasers. Research and development is on going on the shorter wavelength UV diode-lasers, and several niche-type applications await their success, even though their output powers will initially be quite low, in the milli- or sub milli-watt value range. As device design and CVD processes improve in the future and low power UV-lasers become available, multi-laser packages will be assembled to fulfill market demands for higher UV output powers.

13.7.14 Solar Cells

Silicon solar cells were first reported by researchers at Bell Laboratories in the 1950s, with commercial applications commencing about 10 years later. CVD-produced PV cells are made worldwide and are a growing industry with terrestrial products providing a wide range of efficiencies from a wide range of elemental and compound materials. The three largest markets worldwide are

Germany, Japan and California, with the latter having most of the US market share. The available cells include thick and thin film amorphous, polycrystalline and single crystal silicon, and thin film cadmium sulfide and telluride (CdS, CdTe), copper-indium and copper-indium-gallium diselenides ($CuInSe_2$, $CuInGaSe_2$), plus a range of higher efficiency gallium arsenide single and multiple junction III-V cells from Al, In, As, P, either on GaAs or germanium substrates. However, most terrestrial cells are silicon based and most satellite PV systems are now III-V-element multi-junction based.

For about 20 years, single crystal silicon cells provided the initial photovoltaic satellite power requirements plus the limited terrestrial installations with efficiencies in the 12–16% range. However, these PV efficiencies had increased to about 20% by 2007. Silicon cells are CVD produced and, even in the 1980s, 80% of satellite photovoltaic power systems were still using single crystal silicon despite higher efficiency GaAs cells being available. Polycrystalline silicon based cells have been available for over 20 years, but have lower efficiencies (8–10% range), and because of their lower cost are widely used in the terrestrial market, even though their peak watt costs are similar.

Many applications are suitable for PV power generation, with the consumer uses being residential, remote housing, water pumps and marine and recreational vehicles. The industrial use category is much broader and includes oil and gas recovery, traffic and railway signals, lighting, common grid supply, rural developments and solar powered hydrogen generators. As the reliability of CVD-grown GaAs solar cells was verified, their radiation hardness, their higher efficiencies of over 30% and their ability to operate at higher temperatures made them the material of choice for space, government and military satellite applications, but still too expensive for domestic use, even when used in concentrator systems. However, less expensive concentrator designs have been developed and these may open up the post-2008 terrestrial PV markets to the higher efficiency GaAs-based multi-junction PV cells and add to the III-V share of the terrestrial PV market.

The absorptivity of the PV cell materials can be economics and cost factors, since some materials like GaAs need only thin layers (*e.g.*, about 1 micron) for complete absorption, whereas silicon requires about 100 micron and a higher mass for complete absorption. These quantities alter the substrate cost equation – much thicker material leads to higher cost and for space use they also affect rocket payloads and fuel requirements, thus affecting launch costs.

Lower efficiency cells such as copper indium selenide and organic solar cells (4.5–5.5%) are being developed, but they have not been widely installed to date. Even though their installed peak costs are lower (~\$3.50 Wp), they require much larger solar area for an equivalent solar power generation. In 2007, Solarbuzz LLC, reported that there were three main countries rather than regional markets worldwide, Germany (55%), Japan (17%) and the USA [mostly California] (8%), with the remaining installations being shared between the ROW and the rest of Europe regions. By 2011, as additional large solar arrays are deployed, the German share is expected to fall to about 30% whilst the US share will increase to the 20–25% range together with growth in other markets such as Australia, Italy and Spain. The terrestrial PV cell market is growing at about 30% per year.

13.7.14.1 Group III-V Solar Cells

Owing to their physical properties and their efficiencies in the 30–35% range, the III-V multi junction cells were responsible for over 80% of space PV systems market by 2007.

They also hold the present record efficiency of 40.7% for all solar cells, multi-junction and concentrator solar cells, reported in late 2006 by the Boeing-Spectrolab consortium using III-V compound layers on a germanium substrate. Indium III-V compounds have also been used to make flexible PV cells. In 2006, the consolidated, worldwide installed PV cell base was estimated by the DOE to be 2.1 gigawatts and growing at an annual rate of 30%. During the last 30 years, the cost of installed terrestrial peak watt power has been high, but it is declining with improved manufacturing volumes and technology and for many applications it is now below \$3 per peak watt. CVD processes have been the main PV cell production process, but others such as spray and PLD have been used and are still being considered for the lower cost, low efficiency cell production.

13.7.14.2 Silicon Solar Cells

Single-crystal silicon PV cells, fabricated through CVD processes, were the mainstay of the satellite and terrestrial photovoltaic power generation systems for many years, with 12–17% efficiencies, but Sunpower Inc., of San Jose, reported increases to the 20–22% range in late 2007. Polycrystalline silicon solar cells have been produced for many years, but have lower efficiencies, in the 8–10% range. They have become popular in the terrestrial market because of their lower cost, although their peak watt costs, at about \$4.3 Wp, are similar to those of monocrystal silicon installations. European peak watt prices are currently about 40% higher than in North America, mostly because of recent exchange rate fluctuations.

As GaAs-based cells started to take over the satellite power generation more than 25 years ago, while silicon gradually became the terrestrial PV material of choice. Hence, most worldwide installed PV capacity is silicon based, including polycrystalline and amorphous silicon cells and in 2006 was estimated to be 5000 MW worldwide. Amorphous silicon is less efficient than the crystalline silicon cells, at about 5%, but it is less expensive to manufacture and can also be used to make flexible PV cells such as those using CVD processing onto thin stainless steel sheets.

13.7.15 Silicon Carbide Applications

Single crystal silicon carbide (SiC) has three main applications in which CVD processing is involved: (1) for the bulk growth of single crystal ingots as a source of sliced and polished wafers, (2) as a substrate for high power SiC circuits and (3) as substrate for HBLED manufacturing. The bulk growth processes use a high temperature (1000–1400 °C) CVD deposition of SiC onto a single crystal seed from a mixture of silane with an alkane (such as propane) or an alkene or chloromethane precursor. Typical ingot sizes in the 2007 market are 2, 3, or 4″ in diameter, which are then sliced into wafers, polished and used as substrates for electrical, optoelectronic or electronic device manufacture. The smaller wafer prices are in the \$1500–\$2000 range, with defect levels in the 10^4 per cm^2 range, but the Toyota Company reported laboratory results for 3″ diameter wafers with less than 100 defects per cm^2, and 4″ diameter "Zero-micropipe-free" grade wafers only became available from Cree Inc. in late 2007.

SiC has intrinsically higher electrical and thermal conductivities and a higher breakdown voltage than most semiconductors (*e.g.*, 10× that of silicon) and all of these properties are beneficial for high power electrical and electronic uses and also as substrates for high output III-V optoelectronic devices. High power, high voltage, SiC devices such as Schottky diodes, IGBTs, MESFETs and thyristors are commercially available with some being able to function at temperatures of up to 650 °C, but they have yet to achieve the status of high volume commercial products. Typical high performance operating conditions would be a 1200 V at 50 A DMOSFET and an 8.6 kV PiN diode from Cree or a 900 V MOSFET from Rohm. Devices with 10-kilovolt operation at 100 amp power levels are expected in the future.

The drawback to volume production has been that their high power performance is compromised by the presence of the micropipe and other defects; hence the requirement for improvements in the CVD production processes to routinely deliver zero micropipe defect ingots and wafers – a goal that has now been achieved for diameters up to 4 inches. Additionally, SiC circuits such as ultra-fast Schottky diodes switch faster than the equivalent silicon circuits, providing another reason why SiC devices are being used in new designs and upgrade circuits in the power transmission and power control industries. The SiC device market is estimated to be in the \$30 million range for 2007 and with good growth this value is expected to be about \$90 million in 2011, supporting CVD process diversity. By 2012 as many as 15 million hybrid cars could benefit from the use of SiC power control devices. Potential future competition for high power SiC devices may come from GaN/AlGaN based devices in the lower frequency range (*e.g.*, below 10 GHz), but in the higher frequency ranges GaN/AlGaN and InP based power devices will exhibit superior performance over SiC devices.

Subject Index

Page references to figures, tables and text boxes are shown in italics.